# Spin and Isospin
# in Nuclear Interactions

# Spin and Isospin
# in Nuclear Interactions

Edited by

## Scott W. Wissink, Charles D. Goodman,
## and George E. Walker

*Indiana University*
*Bloomington, Indiana*

Springer Science+Business Media, LLC

Library of Congress Cataloging-in-Publication Data

Spin and isospin in nuclear interactions / edited by Scott W. Wissink,
Charles D. Goodman, and George E. Walker.
        p.   cm.
    Proceedings of an international conference held March 11-15, 1991
in Telluride, Colo.
    Includes bibliographical references and index.
    ISBN 978-1-4613-6711-6     ISBN 978-1-4615-3834-9 (eBook)
    DOI 10.1007/978-1-4615-3834-9
    1. Nuclear reactions--Congresses.  2. Nuclear spin--Congresses.
3. Isobaric spin--Congresses.   I. Wissink, Scott W.  II. Goodman,
Charles D.  III. Walker, George E.
QC793.9.S64  1992
539.7'5--dc20                                            91-39670
                                                            CIP

Proceedings of an international conference on Spin and Isospin in
Nuclear Interactions, held March 11-15, 1991, in Telluride, Colorado

ISBN 978-1-4613-6711-6

© 1991 Springer Science+Business Media New York
Originally published by Plenum Press, New York in 1991
Softcover reprint of the hardcover 1st edition 1991

# PREFACE

This volume contains the proceedings of an International Conference on "Spin and Isospin in Nuclear Interactions", which was held in Telluride, Colorado USA, 11–15 March 1991. This was the fifth in a series of conferences held in Telluride every three years since 1979. In attendance at the conference were just under 100 participants, representing a total of 43 institutes from 12 different countries.

In keeping with previous Telluride conferences, the role of spin and isospin degrees of freedom in both nuclear structure and nuclear interactions remained an important theme. Topics covered included new results on the spin- and isospin-dependent terms in the free and effective nucleon-nucleon interaction, Gamow-Teller excitations, charge and spin exchange with hadronic probes, and spin measurements with leptonic probes. Recent progress in the development of polarized sources, polarized targets, and polarimetry was also discussed, as were applications to neutrino physics and astrophysics. Whereas earlier Telluride conferences had dealt primarily with nucleon–nucleus interactions, this meeting included extensive discussions on the role of spin and flavor in particle interactions, and on ways of "bridging the gap" between concepts usually associated with particle physics and the domain of more conventional nuclear physics.

The conference consisted of morning and evening scientific sessions, leaving the afternoons free for informal discussions, recreation, and enjoyment of the scenic beauty of the Telluride area. In addition to the invited talks, time was allotted for contributed talks on new results.

LOCAL ORGANIZING COMMITTEE:

Scott W. Wissink
Charles D. Goodman
George E. Walker

Indiana University
Cyclotron Facility
2401 Milo B. Sampson Lane
Bloomington, Indiana 47405
(812) 855-9365
FAX (812) 855-6645
e-mail username@IUCF

INTERNATIONAL ADVISING COMMITTEE:

A. Arima (Tokyo)
J. Arvieux (Saclay)
P. Barnes (Carnegie-Mellon)
V. Dmitriev (Novosibirsk)
C. Dover (Brookhaven)
B. Frois (Saclay)
G. Garvey (Los Alamos)
W. Haxton (Seattle)
K. Kilian (Jülich)
A. Miller (Alberta)
E. Redish (Maryland)
W. Weise (Regensburg)

The Advising Committee was instrumental in helping us to bring together at this conference some of the leaders in research from the international communities of both nuclear and particle physics.

We wish to acknowledge the sponsorship of the U.S. National Science Foundation and the Indiana University Cyclotron Facility. We are also grateful for the support of LeCroy Research Systems and Solon Technologies Inc. towards the conference reception.

The efforts of many people were crucial to the smooth operation of all phases of the conference administration. The Indiana University Conference Bureau was responsible for all pre-conference mailings and lodging agreements, as well as the local food and conference site arrangements. The great help of Heidi Hauan both before and especially during the conference is especially appreciated. Robert Dover deserves all the credit for the design and preparation of the conference poster, which also became the conference T-shirt logo. We also thank Luci Reeves for her help in dealing with many of the local crises that arose.

We would like to thank the session chairmen, many of whom were called to duty at the last minute: A. Bacher, M. Garçon, J. Londergan, P. Roos, W.G. Love, R. Lindgren, E. Henley, J. McClelland, and J. Sowinski. Their ability to balance lively and stimulating presentations and discussions within the time constraints of the conference schedule was much appreciated.

Finally, we would like to thank all of the speakers, to whom the success of the conference is due. The talks were not only very informative, but were always presented with enthusiasm and clarity. The manuscripts enclosed herein were carefully written and submitted in a timely fashion. We hope that the conference and these proceedings will serve to stimulate new developments in this field, which we anticipate will be reported at the next Telluride conference.

Scott W. Wissink
Charles D. Goodman
George E. Walker

# CONTENTS

(* Asterisk indicates author who presented the paper.)

# DEEP INELASTIC SCATTERING OF LEPTONS
# AND TESTS OF QUARK/PARTON MODELS

J. T. Londergan and S. Kumano

Dept. of Physics and Nuclear Theory Center
Indiana University
2401 Milo Sampson Lane, Bloomington, IN 47408-0768

## 1. Introduction

In this paper we will review the information which can be obtained from deep inelastic scattering (DIS) of leptons from hadrons. As is well known, it was through DIS that one was able to infer that nucleons are made up of apparently pointlike elementary constituents.[1] The latest Nobel Prize in Physics was awarded to Friedman, Kendall and Taylor for their leadership in these experiments. Since this time very detailed experiments have been carried out on both nucleon and nuclear targets, using either charged leptons (electrons or muons) or neutrinos.

In this paper we will begin by reviewing the basic properties of deep inelastic scattering. We will outline the quark/parton model [QPM] by which these experiments are described and review the predictions of quark/parton sum rules. These provide powerful tests of the QPM and can give considerable insight into the structure of the nucleon. There exist excellent reviews of deep inelastic scattering and the quark/parton model. In particular we refer the reader to the comprehensive paper of Mishra and Sciulli.[2] Our review will follow a format similar to that of Mishra and Sciulli. However, in the relatively short time following publication of that review there have been several recent experimental and theoretical developments. Our paper may be seen as an updating of that paper. Further, since the main topic of this conference is on spin and isospin in nuclear and particle physics, we will focus on those experiments which shed light on the spin and $SU(2)$ flavor dependence of nucleon structure functions. A detailed review of quark/parton structure functions and current issues is given in the Proceedings of the 1990 Fermilab Workshop on Structure Functions and Parton Distributions.[3] A detailed review of the QCD analysis of DIS is given in the article by Buras.[4]

In Section 2 we will first outline the kinematics of deep inelastic scattering. We will then describe the structure functions which can be extracted from DIS electromagnetic interactions probed with electron or muon beams, and also structure functions seen in charged–current neutrino interactions [we will not review neutral–current neutrino interactions, where we have far less experimental data]. We will then express each of these structure functions in the QPM. In Section 3, we will discuss the current status of five quark/parton sum rules: the Gross–Llewellyn Smith, Adler, Gottfried, Bjorken and Ellis–Jaffe sum rules. In each case we will derive the sum rule and show its implications for quark/parton models, and we will review the latest experimental evidence regarding each sum rule.

There have been recent measurements of several of the quark/parton sum rules. These have allowed us to make quantitative tests of these sum rules, where previous experiments provided only qualitative confirmation of the models. In some cases experiments appear to agree extremely well with the predictions of sum rules; in other cases the results either contradict our intuition or seem to require additional terms not present in most quark/parton models of the nucleon.

In Section 4, we will discuss QCD predictions for the behavior of structure functions. We will discuss recent experiments which test either QCD scaling violations in DIS, or which examine the $Q^2$ and $x$ dependence of the ratio $R = \sigma_L/\sigma_T$.

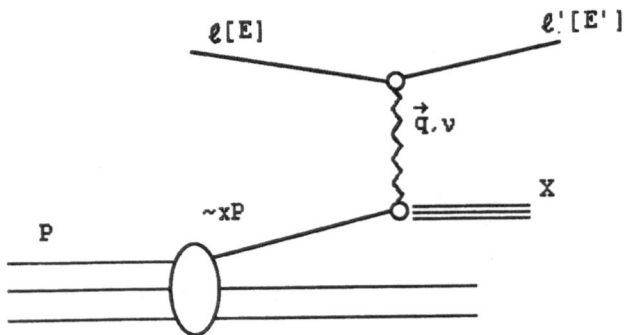

Fig. 1 Kinematics of deep inelastic scattering of leptons
from nucleons.

## 2. Structure Functions and DIS

### 2.1 DIS Kinematics

The basic kinematics for deep inelastic scattering are shown in Fig. 1. An incident lepton with energy $E$ emits a virtual particle [photon, $W$ or $Z$] with energy $\nu$ and three–momentum $\vec{q}$. The final lepton has energy $E' = E - \nu$, while the target hadron has four–momentum $P$. We can define three invariants $Q^2 \equiv -q^2$, $x \equiv Q^2/2P \cdot q$, and $y \equiv 2P \cdot q/s$, where $s$ is the square of the center of mass energy and $\nu$ is the energy loss of the lepton. In the target rest frame, $x = Q^2/2M\nu$, where $M$ is the target mass, and $y = \nu/E$ is the fractional energy loss of the lepton. In a frame where the hadron is travelling at extremely large momentum, the momentum carried by the struck constituent of the hadron is $xP$; therefore we traditionally speak of $x$ as the momentum fraction carried by the "parton" or struck constituent of the target hadron.

### 2.2 Structure Functions in $\mu/e$ DIS

We first consider the structure functions in deep inelastic scattering of charged leptons, of the form

$$e^- + N \rightarrow e^- + X$$
$$\mu^- + N \rightarrow \mu^- + X \qquad (2.1)$$

We assume that we are working at energies where the weak interactions can be neglected relative to the electromagnetic interaction. In this case the virtual photon strikes a constituent in the target nucleon, and the electron or muon–induced deep inelastic scattering for unpolarized particles can be described in terms of two structure functions

$$\frac{d^2\sigma^{eN}}{dx\,dy} = \frac{8\pi Em_N\alpha^2}{Q^4}[xy^2 F_1^{eN}(x,Q^2) + (1 - y - \frac{xym_N}{2E})F_2^{eN}(x,Q^2)] \qquad (2.2)$$

In Eq. (2.2), the DIS cross sections depend on the two structure functions $F_1$ and $F_2$. Bjorken[5] showed that in the limit of very large energies, as $Q^2 \rightarrow \infty$ and $\nu \rightarrow \infty$ [but $0 \leq x \leq 1$], the structure functions $F_1$ and $F_2$ become functions only of $x$.

In a series of seminal experiments at SLAC[1] involving DIS induced by electrons, it was seen that the extracted structure functions depended sensitively on $x$ but showed very little dependence on $Q^2$; furthermore, this "scaling" set in at remarkably small values of $Q^2$. This was interpreted as scattering from pointlike constituents ["partons"] in the nucleon.

One additional feature of DIS reactions was that the longitudinal to transverse cross section ratio $R = \sigma_L/\sigma_T$ was very small. For spin–zero elementary constituents, we would predict $R \rightarrow \infty$, whereas for spin–1/2 constituents, $R \rightarrow 0$. Therefore, the small experimental values of $R$ gave strong evidence that the elementary constituents had spin 1/2. The quantity $R$ is given by the approximate relation

$$R \equiv \frac{\sigma_L}{\sigma_T} \approx \frac{F_2(x,Q^2) - 2xF_1(x,Q^2)}{2xF_1(x,Q^2)} \leq 0.2 \qquad (2.3)$$

The very small experimental value of $R$ implies $F_2 \approx 2xF_1$, the Callan–Gross relation.[6] We will discuss the experimental value of $R$ and its $Q^2$ behavior in Section 4.2.

## 2.3 Structure Functions in $\nu$ CC DIS

The differential cross section for deep inelastic scattering induced by neutrinos [charged–current interactions] has the form

$$\frac{d^2\sigma^{\nu N}}{dx\,dy} = \frac{G^2 Em_N}{\pi}[xy^2 F_1^{\nu N}(x,Q^2) + (1 - y - \frac{xym_N}{2E})F_2^{\nu N}(x,Q^2) + (y - y^2/2)xF_3^{\nu N}(x,Q^2)] \qquad (2.4)$$

Neutrino–induced charged–current DIS depends on three structure functions. The third structure function $xF_3$ exists because the neutrino reactions violate parity. Antineutrino–induced reactions can be obtained from Eq. (2.4) by the replacement $F_{1,2}^{\nu} \rightarrow F_{1,2}^{\overline{\nu}}$; $F_3^{\nu} \rightarrow -F_3^{\overline{\nu}}$.

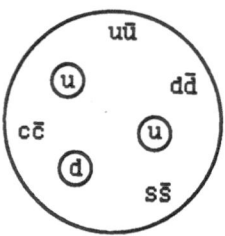

Fig. 2  Schematic picture of proton, composed of three
"valence" quarks plus a "sea" of $q - \bar{q}$ pairs.

## 2.4 Quark/Parton Distribution Functions

Our present picture of the structure of the nucleon is given schematically in Fig. 2. The nucleon is composed of three "valence" quarks and a "sea" of quark–antiquark pairs. The structure functions observed in deep inelastic scattering result from scattering of the virtual particle from the quark constituents; therefore the structure functions measure the momentum densities of quarks inside the nucleon. For example, the proton has two valence up quarks and a valence down quark, plus a sea of quark–antiquark pairs. Let us assume that there are four flavors of quark within the proton: up, down, strange and charmed quarks [we neglect the extremely heavy bottom and top quarks for the remainder of this paper; in addition, for many applications we will also neglect the charmed quark distribution in the nucleon]. We can then describe any structure function for the proton in terms of eight quark distribution functions: four for the quarks and four for the antiquarks.

For example, the proton structure function $F_2$ measured with electrons or muons is given by the form[8]

$$2xF_1^{\mu p}(x, Q^2) \cong F_2^{\mu p}(x, Q^2) = \sum_i e_i^2 \left[ q_i(x) + \bar{q}_i(x) \right]$$
$$= \frac{4}{9} \left[ u_p(x) + \bar{u}_p(x) + c_p(x) + \bar{c}_p(x) \right] + \frac{1}{9} \left[ d_p(x) + \bar{d}_p(x) + s_p(x) + \bar{s}_p(x) \right]$$
$$(2.5)$$

The electromagnetic structure functions have the following qualitative properties:

*i*) They are proportional to the squares of the quark charges.

*ii*) The quark and antiquark distribution functions enter symmetrically for each flavor. Thus electromagnetic DIS can determine only the sum of the quark and antiquark distribution functions for each flavor.

If we make no further assumptions, then we would have eight distribution functions for the proton, and an additional eight for the neutron. We can simplify this considerably by invoking sum rules based on additive quantum numbers, and by isospin symmetry. Since we assume the strong interactions are invariant under isospin, then neglecting Coulomb effects we can obtain the relations:

4

$$u_p(x) = d_n(x) \equiv u(x)$$
$$d_p(x) = u_n(x) \equiv d(x)$$
$$s_p(x) = s_n(x) \equiv s(x) \tag{2.6}$$
$$c_p(x) = c_n(x) \equiv c(x)$$

and analogous equations for the antiquark distributions. With this assumption we can then describe all distributions in terms of quark/parton distributions **in the proton.**[7]

We can obtain other restrictions on the quark distributions by using the fact that the proton has charge and baryon number one, and zero strangeness and charm. These absolute additive quantum numbers can be expressed in terms of integrals of the respective quark minus antiquark distributions

$$\int_0^1 dx[u(x) - \bar{u}(x)] = 2$$

$$\int_0^1 dx[d(x) - \bar{d}(x)] = 1 \tag{2.7}$$

$$\int_0^1 dx[s(x) - \bar{s}(x)] = \int_0^1 dx[c(x) - \bar{c}(x)] = 0$$

In Fig. 3 we show a compilation of the structure function $F_2(x, Q^2)$ as measured on hydrogen and deuterium. The structure functions are shown for various values of $x$ from 0.07 to 0.75, as a function of $Q^2$. The triangles are measurements from electron scattering at SLAC[9], solid circles are muon scattering measurements from the EMC group at CERN[10], and the open circles are muon measurements from the BCDMS group.[11] It is apparent from Fig. 3 that the structure functions have a slow, but consistent, variation with $Q^2$. In Section 4.1 we describe the scaling violations predicted by QCD and compare them with recent experimental data.

The structure functions measured in charged–current neutrino DIS have the following form in the quark/parton picture:

$$2xF_1^{\nu p}(x) \cong F_2^{\nu p}(x) = 2x\,[d(x) + s(x) + \bar{u}(x) + \bar{c}(x)]$$

$$xF_3^{\nu p}(x) = 2x\,[d(x) + s(x) - \bar{u}(x) - \bar{c}(x)] \tag{2.8}$$

Antineutrino structure functions are related to the neutrino structure functions of Eq. (2.8) under the replacements $q_j(x) \leftrightarrow \bar{q}_j(x)$ for $F_2^{\nu N} \to F_2^{\bar{\nu}N}$, and $q_j(x) \leftrightarrow -\bar{q}_j(x)$ for $F_3^{\nu N} \to F_3^{\bar{\nu}N}$.

The structure functions measured in neutrino scattering have the following qualitative properties:

 i) Different flavors enter with the same proportions [rather than being proportional to the squares of the quark charges as for electromagnetic structure functions];

*ii*) The structure functions are **asymmetric** in the quark and antiquark distributions. Therefore neutrino (or antineutrino) scattering can provide unique sensitivity to individual quark or antiquark distributions.

The quark distributions depend on the respective flavor considered. For the strange quarks, there is no valence contribution so the entire strange quark distribution is part of the "sea" distribution. The up and down quark distributions, however, contain one piece which comes from the valence distribution and another contribution from the sea quarks. For the purpose of this work, we define the valence distributions for up quarks as

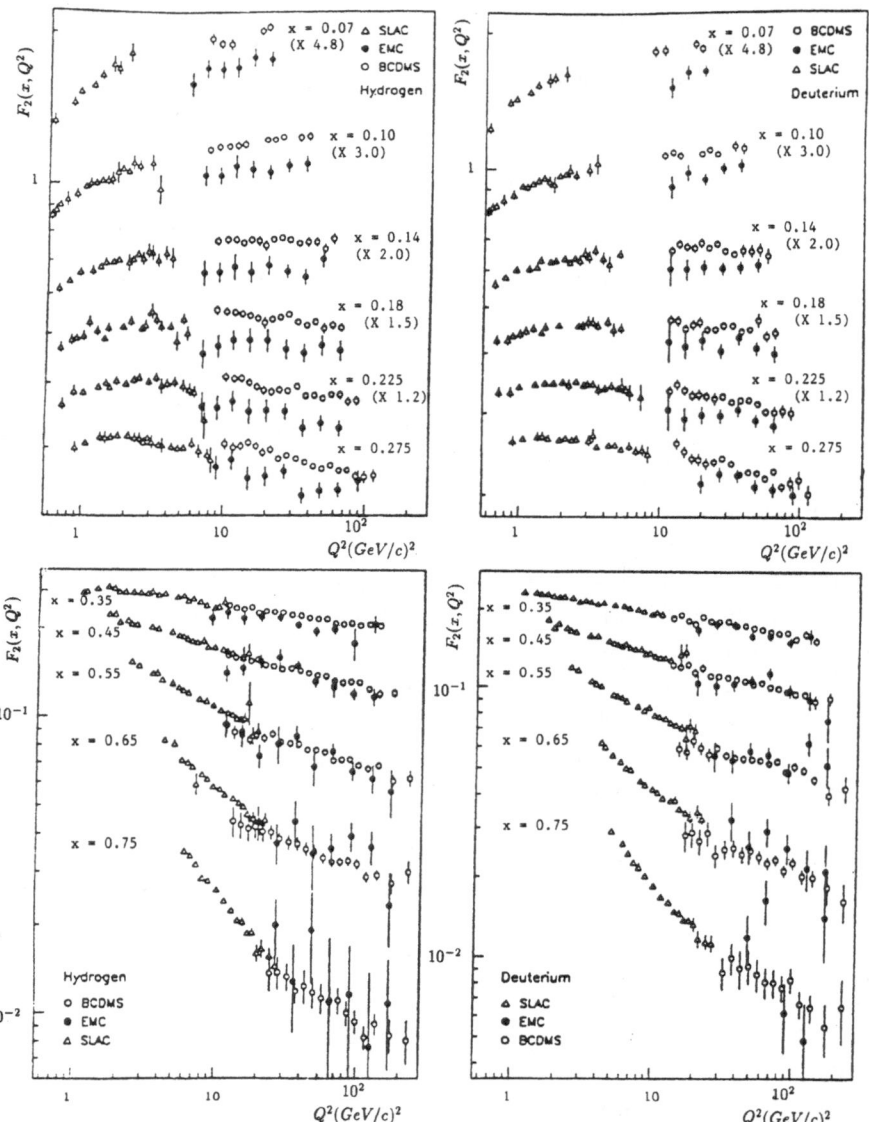

Fig. 3    Experimental measurements of $F_2(x, Q^2)$ from H [left] and D [right] by SLAC [triangles], BCDMS [open circles] and EMC [solid circles]. EMC data have been corrected for $R = R_{QCD}$ (from Ref. 12).

$$u_V(x) \equiv u(x) - \bar{u}(x) \quad , \tag{2.9}$$

with an identical equation for the down quarks.

In Fig. 4 we show valence up and down quark distributions which have been extracted from experiment. In Fig. 4a we show $xu_V(x)$ calculated at $Q^2 = 15\ GeV^2$. The data are obtained from the EMC muon experiment,[10] and from three different neutrino experiments.[13-15] Fig. 4b shows $xd_V(x)$ extracted at the same momentum transfer from the same experiments. It is difficult to obtain precise values for the up [down] quark distributions, since the relevant experiments measure linear combinations of quark and antiquark distributions [the experimental uncertainties are much smaller for the combinations of quark and antiquark densities measured by the experiments]. Also, from inspection of Fig. 4b, we see that the EMC results suggest quite a different down quark valence distribution from the neutrino experiments.

These valence quark distributions peak at values of $x \approx 1/3$. From our discussion of $xu_V$ as the momentum carried by the valence up quarks, we see that this result is consistent with our intuition that the valence quarks each carry roughly 1/3 of the nucleon's momentum. In Fig. 5 we show the antiquark density from an isoscalar target. The data, at $Q^2 = 5\ GeV^2$, are from the CDHSW[16] and CCFR[17] neutrino scattering measurements on iron targets [note that the ordinate scale is logarithmic]. Unlike the valence quark densities, the antiquark densities peak at $x \to 0$. In fact, since the antiquark densities of Fig. 5 are multiplied by $x$, the sea quark densities vary like $1/x$ at small $x$. The sea quarks thus represent a "soft" distribution; since the momentum density peaks as $x \to 0$, this represents a very large number of sea quarks which carry very little of the proton's momentum.

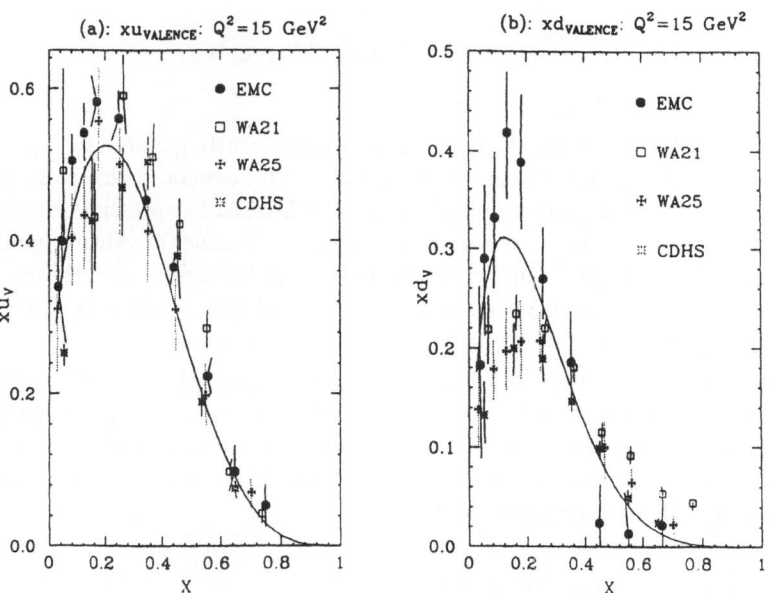

Fig. 4  Valence quark densities in the proton, vs. $x$. (a) $xu_V$; (b) $xd_V$. Data
are from EMC, [Ref. 10], WA21 [Ref. 13], WA25 [Ref. 14], and CDHS
[Ref. 15] (from Ref. 2).

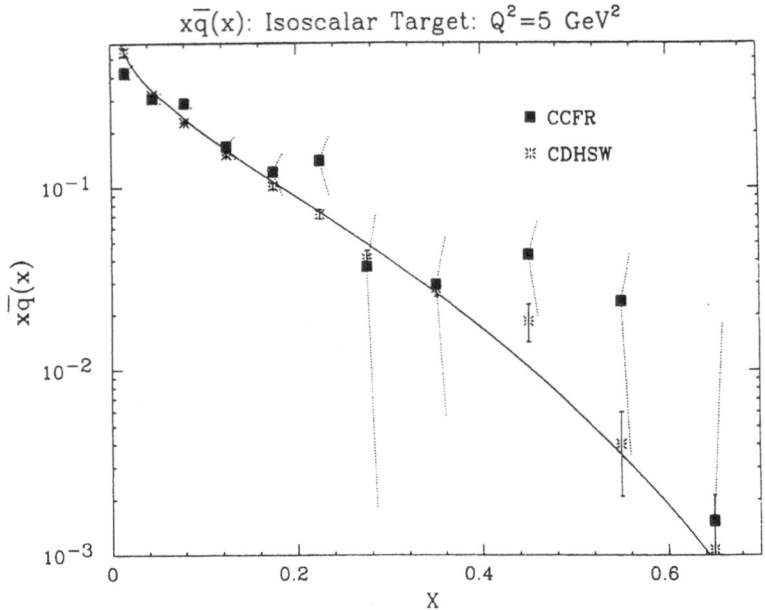

Fig. 5  The antiquark density from an isoscalar target. Data are from
CDHSW [Ref. 16] and CCFR [Ref. 17] collaborations (from Ref. 2).

## 3. Quark/Parton Sum Rules

We can obtain useful information regarding structure functions by considering the $N^{th}$ moment of a structure function $F$, e.g.

$$F_N(Q^2) \equiv \int_0^1 dx\, x^{N-1} F(x, Q^2)$$

The $N = 1$ moment is particularly useful as it is frequently possible to express this in terms of the matrix element of the commutator of two currents. If this is the case, then one can use current algebra to obtain a Lorentz-invariant expression for the matrix element. This then leads to expressions involving the number of valence quarks in the nucleon and their charges. Since this information can be derived directly from current algebra, many of the sum rules we review here were obtained before the quark/parton model was formulated.

However the resulting sum rules can all be expressed efficiently and easily in the quark/parton model. In the following sections we will give the sum rule in the quark/parton framework and discuss the most recent experimental evidence for each sum rule. We will focus on those sum rules which are related to spin and flavor dependence in the quark/parton model.

### 3.1 Gross–Llewellyn Smith Sum Rule

The Gross–Llewellyn Smith sum rule[18] measures the number of valence quarks in the nucleon. This sum rule is obtained by taking the average of the $xF_3$ structure functions for neutrinos and antineutrinos on a nucleon target. The Gross–Llewellyn Smith [GLS] sum rule has the form [using Eq. (2.8) for the $xF_3$ structure functions]

$$S_{GLS} = \int_0^1 x \left[ \frac{F_3^{\nu N}(x) + F_3^{\bar{\nu} N}(x)}{2} \right] \frac{dx}{x}$$

$$= \sum_i \int_0^1 dx \, [q_i(x) - \bar{q}_i(x)] = \int_0^1 dx \, [u(x) - \bar{u}(x) \qquad (3.1)$$

$$+ d(x) - \bar{d}(x) + s(x) - \bar{s}(x) + c(x) - \bar{c}(x)]$$

$$= 3 \left[ 1 - \frac{\alpha_S(Q^2)}{\pi} + \mathcal{O}\left( \frac{1}{Q^2} \right) \right]$$

In Eq. (3.1), the second term in square brackets is the leading–order QCD correction; here $\alpha_S$ is the running coupling constant of QCD, which has the leading order form

$$\alpha_S(Q^2) = \frac{4\pi}{[11 - 2n_f/3] \ln(Q^2/\Lambda^2)} \qquad (3.2)$$

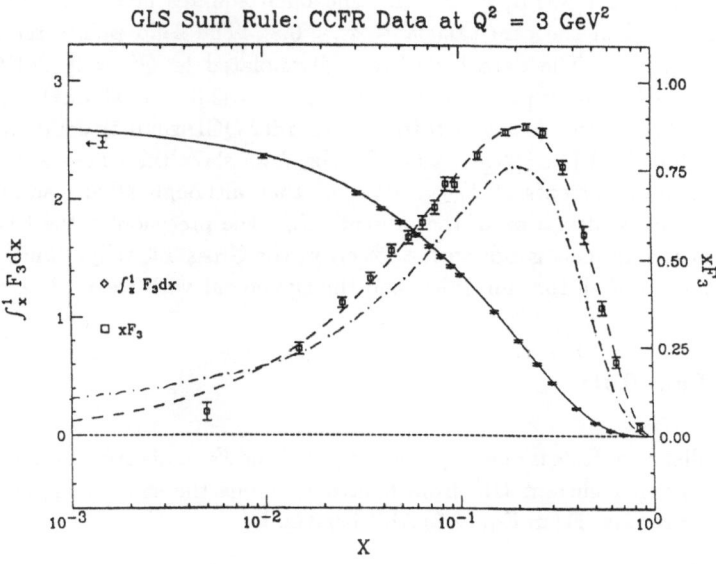

GLS Sum Rule: CCFR Data at $Q^2 = 3$ GeV$^2$

Fig. 6 The $xF_3$ function and the GLS sum rule, from the CCFR QTB neutrino beam, Ref. 17. $xF_3$: open squares and right scale; integral of $xF_3$: solid squares and left scale. Solid and dashed curves represent best fits to the data, see Ref. 17. Dot–dashed curve is prediction of the HMRS–B global fit of Ref. 23.

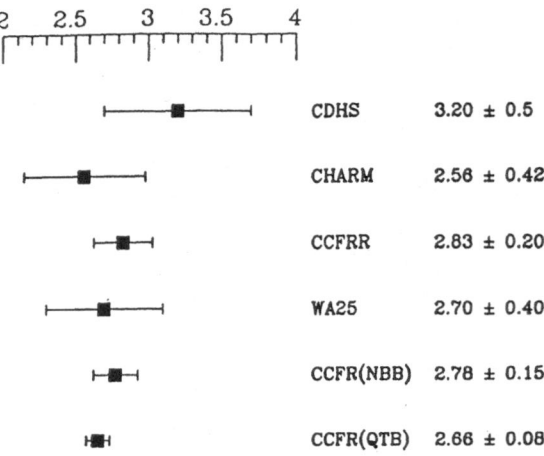

Fig. 7  Best values and accuracy for the GLS sum rule from
recent experiments [see discussion in Ref. 17]. Note
high precision of latest experiment on GLS sum rule.

In Eq. (3.2), $\Lambda_{QCD}$ is the QCD scale parameter; in recent studies of QCD, this has been found to be approximately in the range $200 - 250$ MeV.[19] $n_f$ represents the number of quark flavors. The third term in square brackets in Eq. (3.1) arises from higher–twist contributions. A numerical estimate of this term has been made by Jaffe and Iijima.[21,22]

The latest experimental data for the GLS sum rule are shown in Fig. 6. These were taken by the CCFR group from neutrino scattering on iron with the quadrupole triplet neutrino beam [QTB] at FNAL. The open squares give the data for $xF_3$, which is measured in the range $0.005 \leq x \leq 0.8$. The solid points represent the integral $\int_x^1 F_3(x)\,dx$. The data have been extrapolated to $Q^2 = 3 \ GeV^2$; at this momentum transfer, the experimental result is $S_{GLS} = 2.66 \pm 0.03(\text{stat}) \pm 0.08(\text{syst})$. The theoretical prediction [using next to leading order QCD running coupling constant with $\Lambda_{\overline{MS}} = 250 \ MeV$] is $S_{GLS}^{th} = 2.63$. In Fig. 7 we show the values and errors for experimental measurements of $S_{GLS}$. We see that although all measurements are within one standard deviation of the present value, the precision of the CCFR QTB measurement is extremely good: with a 4% error, the Gross–Llewellyn Smith sum rule is the best known of all the sum rules, and the agreement with theory is excellent.

## 3.2 Adler Sum Rule

The Adler Sum Rule is given by the integral of the $F_2$ structure function measured in neutrino charged–current DIS from neutrons, minus the same structure function measured on protons. From Eq. (2.8), this is given by

$$S_A = \frac{1}{2}\int_0^1 dx \, \frac{F_2^{\nu n}(x) - F_2^{\nu p}(x)}{x} = \int_0^1 dx \left[ u(x) - \bar{u}(x) - \left( d(x) - \bar{d}(x) \right) \right] = 1 \qquad (3.3)$$

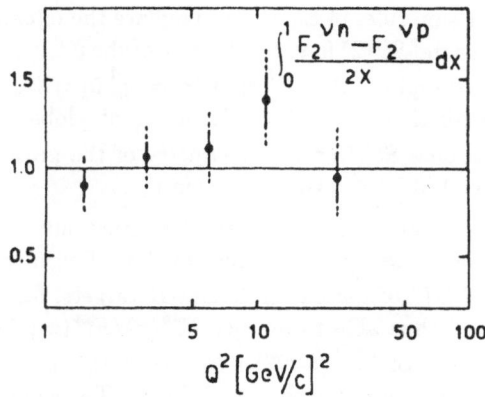

Fig. 8 Experimental results for the Adler sum rule,
from WA25 experiment [Ref. 14] at the
BEBC H and D bubble chambers, vs. $Q^2$.

Whereas the Gross–Llewellyn Smith sum rule has QCD corrections to all orders, the Adler sum rule [as a consequence of the algebra of SU(2) charges][22] has no QCD corrections. The latest experimental results are shown in Fig. 8. These come from the WA25 experiment at CERN[14], using the CERN–SPS wide band neutrino and antineutrino beams with the BEBC H and D bubble chambers. The experimental results are shown at several values of $Q^2$. The average value is $S_A = 1.01 \pm 0.20$. Within the rather large errors, the results are independent of $Q^2$. The Adler sum rule is the least precisely known of the sum rules which have been measured to date.

### 3.3 Gottfried Sum Rule

The Gottfried Sum Rule [$S_G$] is obtained by measuring the $F_2$ structure functions with electrons [or muons] on proton and deuteron targets; the Gottfried Sum Rule is then obtained by integrating the difference between proton and neutron $F_2$ structure functions, e.g.

$$S_G \equiv \int\limits_0^1 dx \, \frac{F_2^{\mu p}(x) - F_2^{\mu n}(x)}{x} \tag{3.4}$$

Inserting the quark/parton form for this structure function from Eq. (2.5) gives $S_G$ in terms of quark distribution functions

$$S_G = \int\limits_0^1 dx \sum_i e_i^2 \left[ q_i^p(x) + \bar{q}_i^p(x) - q_i^n(x) - \bar{q}_i^n(x) \right]$$

$$= \frac{1}{3} \int\limits_0^1 dx \left[ \underbrace{u(x) - \bar{u}(x)}_{2} - \underbrace{(d(x) - \bar{d}(x))}_{1} \right] + \frac{2}{3} \int\limits_0^1 dx \left[ \bar{d}(x) - \bar{u}(x) \right] \tag{3.5}$$

$$S_G = \frac{1}{3} + \frac{2}{3} \int\limits_0^1 dx \left[ \bar{d}(x) - \bar{u}(x) \right]$$

where we have used the sum rules of Eq. (2.7) to replace the integrals of the differences $u - \bar{u}$ and $d - \bar{d}$. If we assume $SU(2)$ flavor symmetry of the proton sea, i.e. $\bar{d}(x) = \bar{u}(x)$, or even the weaker assumption that $\int_0^1 \bar{d}(x)\,dx = \int_0^1 \bar{u}(x)\,dx$, then we obtain the Gottfried sum rule prediction $S_G = 1/3$. Most recent global fits to quark/parton distributions in fact assume $SU(2)$ flavor symmetry of the proton sea,[23–26] so they would naturally predict 1/3 for the Gottfried sum rule.

Like all of the sum rules we discuss, the Gottfried sum rule contains a factor $dx/x$ and hence is strongly weighted towards small $x$. Recently the NMC group at CERN[27] has measured DIS of muons on $H$ and $D$ targets, for muon energies of 90 and 280 GeV. They are thus able to extract $F_2^{\mu n}(x)/F_2^{\mu p}(x)$, for $0.004 \leq x \leq 0.8$. From a previous knowledge of $F_2^{\mu D} \approx F_2^{\mu n} + F_2^{\mu p}$, they can then obtain results for the Gottfried Sum Rule. The results are shown in Fig. 9. The solid circles and triangles represent the experimental data for $F_2^{\mu p}(x) - F_2^{\mu n}(x)$, while the open circles and triangles represent $\int_x^1 [F_2^{\mu p} - F_2^{\mu n}]\,dx'/x'$, interpolated to $Q^2 = 4\ GeV^2$. The circles and triangles arise from two independent methods for performing the interpolation, and agree quite closely.[27]

Using the measured data for $0.004 \leq x \leq 0.8$, the NMC group obtain

$$S_G(x = .004 \to 0.8) = 0.227 \pm 0.007\ \text{(stat)}\ \pm 0.014\ \text{(syst)} \qquad (3.6)$$

When they extrapolate their results from $x = 0.004 \to 0$ and $x = 0.8 \to 1$, the NMC group obtain $S_G = 0.240 \pm 0.016$ [the extrapolation to zero is by far the more important

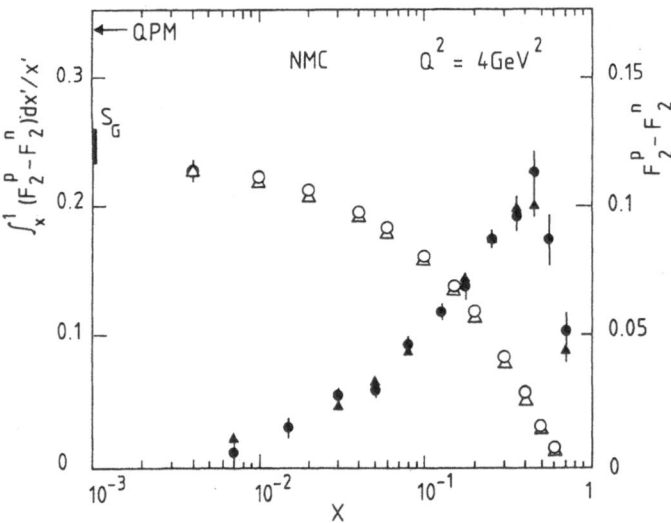

Fig. 9    Structure function differences $F_2^p - F_2^n$ vs. $x$, measured
by the NMC group, Ref. 27. $F_2^p - F_2^n$: solid points and
right scale; integral: open points and left scale. Circles
and triangles represent two different ways of performing
interpolation to $Q^2 = 4\ GeV^2$, as discussed in Ref. 27.

of the two]. Unlike the other sum rules we discuss, the Gottfried Sum Rule **has not** been derived from current algebra. It is a relation which holds in the quark/parton model. $S_G$ corresonds to the first moment of a flavor nonsinglet distribution and terms of order $\alpha_S$ should be absent[28]. Ross and Sachradja[29] have estimated the [very small] QCD corrections to $S_G$. The Gottfried Sum Rule is thus known to within about 6%. This is a considerable improvement over the previous best value $S_G = 0.24 \pm 0.11$.[2] It falls almost 30% short of the naive quark/parton model prediction of 1/3, and is more than five standard deviations away from that value. There have been several theoretical speculations as to the origin of this apparent discrepancy[30-33]

We will discuss two of these here.

Martin, Roberts and Stirling[30] claim that the Gottfried sum rule is obeyed. They suggest that the "missing strength" occurs in the unmeasured region $x < 0.004$. To support this they point to the HMRS global quark/parton fits[23], which are shown as the dashed curve in Fig. 10 vs. the NMC data. As the HMRS global fits assume $\bar{u}(x) = \bar{d}(x)$, they necessarily predict $S_G = 1/3$. Although the HMRS global fits do not give a particularly impressive quantitative fit to the NMC data, they do predict a value of 0.23 for the Gottfried sum rule integral over the measured region.

The fact that the HMRS fits predict that 30% of the Gottfried sum rule occurs for $x < 0.004$ appears to be a strong argument in favor of this model. However, the HMRS global fit **also** predicts that about 20% of the Gross–Llewellyn Smith sum rule is found for $x < 0.004$. The HMRS prediction for the GLS sum rule is shown as the dot–dashed curve in Fig. 6; it is in significant disagreement with experiment.

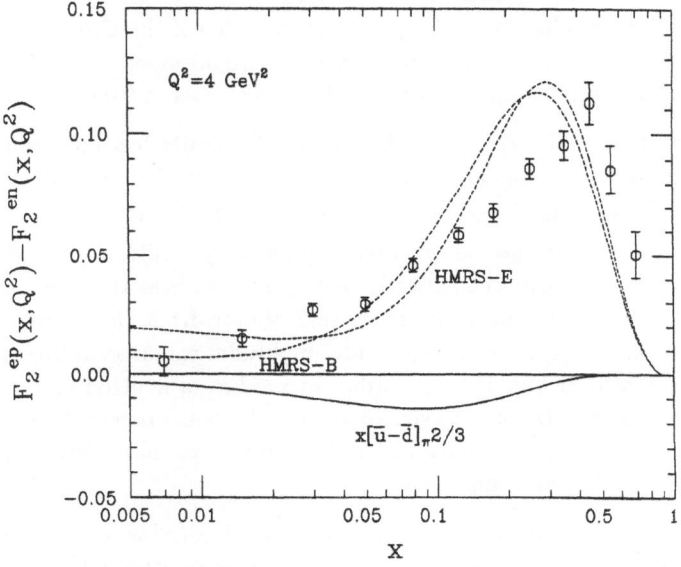

Fig. 10   NMC data of Fig. 9. Dashed curves: HMRS-B and HMRS-E
global predictions for $F_2^p - F_2^n$, Ref. 23. Solid curve:
"pionic" contribution to $F_2^p - F_2^n$ arising from $\bar{d} - \bar{u}$
as predicted by Kumano and Londergan, Ref. 33.

13

a) πNN process

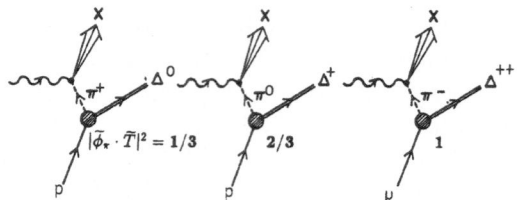

b) πNΔ process

Fig. 11 "Pionic" contributions to DIS from nucleons.
(a) $NN\pi$ process. (b) $\Delta N\pi$ process. Isospin
Clebsch–Gordan coefficients are listed for
each term.

Martin et al. suggest that the explanation for this lies in the fact that the CCFR neutrino experiment was performed on an iron target. They claim that nuclear shadowing on iron at very small $x$ will shift the measured $xF_3$ structure function to substantially larger values of $x$. They thus predict that if the Gross–Llewellyn Smith sum rule was measured on a deuterium target it would agree with their predictions, namely that 20% of the GLS sum rule would be seen at $x < 0.004$.

A second possible explanation for the NMC results has been put forward by Henley and Miller[31], Signal, Schreiber, and Thomas[32], and by Kumano and Londergan.[33] This model assumes that $\bar{u}(x) \neq \bar{d}(x)$ due to a nonperturbative mechanism, the **pionic** mechanism first suggested by Sullivan[35] twenty years ago. This mechanism is shown schematically in Fig. 11. It arises through the production of a virtual pion by the nucleon; the virtual photon from the muon then scatters inelastically from the pion. Several possible amplitudes can arise in this manner, some of which are shown in Fig. 11. A proton can produce a neutron and $\pi^+$, a $\Delta^0$ and $\pi^+$, or a $\Delta^{++}$ and $\pi^-$. Due to the spin–isospin Clebsch–Gordan coefficients, it is more probable to produce the $\pi^+$ [which in quark/parton terms has a large $u\bar{d}$ component] than the $\pi^-$ [with the $d\bar{u}$ component].

In Fig. 12 we show the pionic mechanism for $\bar{d}(x) - \bar{u}(x)$ as predicted by in Ref. 33. In this model one uses $NN\pi$ form factors extracted from other DIS experiments, and in addition pion structure functions measured in DIS. With this model one can account for half of the "missing" strength in the Gottfried sum rule. Hwang, Speth,

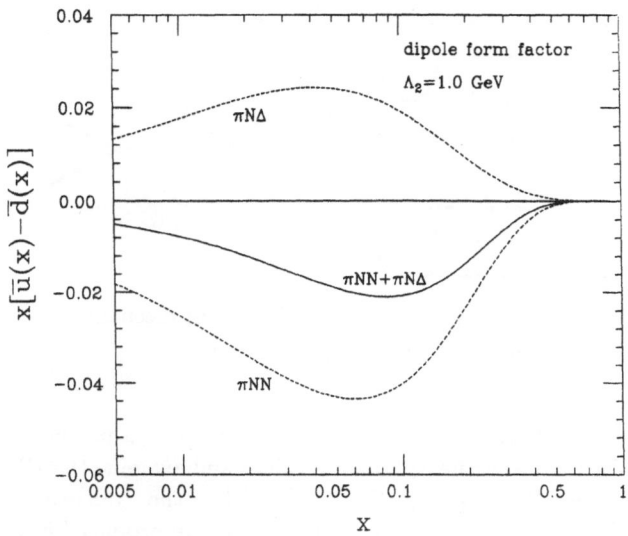

Fig. 12   Contributions to $\bar{u}(x) - \bar{d}(x)$ from pionic mechanism
as calculated in Ref. 33.

and Brown[34] have examined both the pionic term and a "kaonic" contribution which includes amplitudes arising from terms like $p \to K^+ + \Lambda$.

### 3.4 Experimental Methods to Measure $\bar{d}(x) - \bar{u}(x)$

Since the NMC experiments suggest that $\bar{d}(x) \neq \bar{u}(x)$, we review briefly two possible ways to isolate the quantity $\bar{d}(x) - \bar{u}(x)$. The first way is neutrino [antineutrino] DIS from protons and deuterium. If we extract the structure functions $F_2^\nu$ and $x F_3^\nu$ on both $p$ and $D$, then from Eq. (2.8) we see that the combination

$$\frac{1}{2}\left[ F_2^{\nu D}(x) - x F_3^{\nu D}(x) \right] - \left[ F_2^{\nu p}(x) - x F_3^{\nu p}(x) \right] = 2x \left[ \bar{d}(x) - \bar{u}(x) \right] \qquad (3.7)$$

is directly proportional to $\bar{d}(x) - \bar{u}(x)$. Exactly the opposite result holds if we replace $\nu \to \bar{\nu}$ everywhere in Eq. (3.7). From the pionic model predictions in Ref. 33 we predict a maximum value of approximately 0.04 for the right hand side of Eq. (3.7), for values $x \approx 0.1$ (see Fig. 12). The precision of present neutrino scattering measurements is not sufficient to determine a nonzero value for $\bar{d} - \bar{u}$.

A second possible way to measure differences $\bar{d} - \bar{u}$ is through Drell–Yan processes[36] induced by charged pions on light nuclei. Drell–Yan (DY) processes involve $\mu^+ - \mu^-$ pairs with large invariant mass produced in hadronic interactions. The basic processes are shown schematically in Fig. 13. A quark of flavor $i$ in the projectile annihilates an antiquark of the same flavor in the target [or vice versa], producing a virtual photon which then decays into a $\mu^+ - \mu^-$ pair. One feature of Drell–Yan processes is that one can separately vary the Bjorken variables $x_P$ for the projectile and $x_T$ for the target.

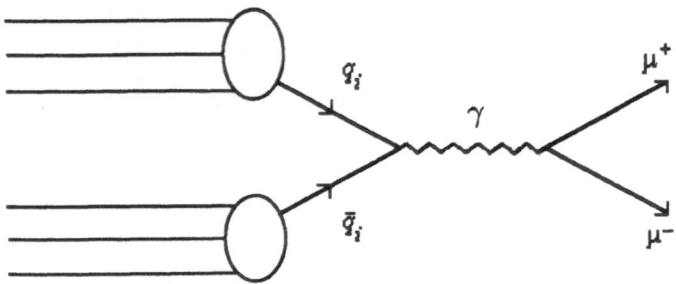

Fig. 13 Schematic picture of Drell–Yan mechanism for
production of $\mu^+ - \mu^-$ pairs.

For example, look at the Drell–Yan reaction for $\pi^+$ on protons, e.g. $\pi^+ + p \to \mu^+ + \mu^- + X$. Fix the projectile $x_\pi$ to large values and examine the DY cross sections as a function of $x_P$. For large $x_\pi$, the $\pi^+$ looks like the "valence" distribution for the pion, i.e. $u\bar{d}$. Therefore, the charged muon pairs are produced from $u_\pi \bar{u}_p$ or $\bar{d}_\pi d_p$ processes. Repeat the process for $\pi^-$ projectiles at large $x_\pi$. Assuming the $\pi^+$ and $\pi^-$ have identical valence distributions [at the same large values of $x_\pi$], we have

$$\tilde{\sigma}(\pi^+ p) \sim V_\pi(x_\pi)\left[d(x_p) + 4\bar{u}(x_p)\right]$$

$$\tilde{\sigma}(\pi^- p) \sim V_\pi(x_\pi)\left[4u(x_p) + \bar{d}(x_p)\right] \tag{3.8}$$

The factors of 4 occur because the relevant cross sections are proportional to the squares of the quark charges [as can be seen from Eq. (2.5)]. Repeating these same processes with deuterons gives

$$\frac{4\left[\tilde{\sigma}(\pi^+ p) - \tilde{\sigma}(\pi^+ D)\right] + \tilde{\sigma}(\pi^- p) - \tilde{\sigma}(\pi^- D)}{5\left[\tilde{\sigma}(\pi^+ D) - \tilde{\sigma}(\pi^- D)\right]} = \frac{\bar{d}(x) - \bar{u}(x)}{u_V(x) + d_V(x)} \tag{3.9}$$

The quantity $\tilde{\sigma}$ here refers to the Drell–Yan cross section per unit target hadron. The resulting ratio is proportional to $\bar{d} - \bar{u}$ divided by the sum of the valence quark distributions $u_V(x) + d_V(x)$. In deriving this equation we have assumed that the hard scattering "K–factor" in Drell–Yan processes[37] is the same for proton and deuteron-induced reactions. Implications for the production of hadrons from the asymmetric sea $\bar{u} \neq \bar{d}$ are investigated by Levelt, Mulders, and Schreiber.[38]

Although both processes we have outlined here are capable in principle of measuring the difference $\bar{d} - \bar{u}$, existing experimental measurements of these quantities are not sufficiently precise to extract this difference unambiguously. Determination of $\bar{d} - \bar{u}$ requires precise measurements of several quantities which must then be combined together. We believe that Drell–Yan processes are promising quantities for such precision measurements.

## 3.5 Spin Sum Rules

Spin structure functions can be obtained by scattering longitudinally polarized charged leptons from a polarized nucleon target. For a collinear beam and target polarization, if we take the differences between cross sections with parallel and antiparallel beam and target, we obtain

$$\frac{d^2 \left[ \sigma \uparrow\uparrow - \sigma \uparrow\downarrow \right]}{dx \, dy} = \frac{e^4}{2\pi Q^2} \left[ \left( 1 - \frac{y}{2} - \frac{y^2 \sqrt{Q^2}}{4\nu} \right) g_1(x, Q^2) - \frac{y\sqrt{Q^2}}{2\nu} g_2(x, Q^2) \right] \quad (3.10)$$

In Eq. (3.10), $\uparrow\uparrow$ represents the lepton beam polarization and $\uparrow$ represents the target polarization. We see that the cross section differences are sensitive to two polarized structure functions $g_1$ and $g_2$. In the Bjorken limit $[Q^2, \nu \to \infty]$, both $g_1$ and $g_2$ become functions only of $x$.

The present experimental situation is as follows. There exist two measurements of $g_1^p$ over a limited kinematic range for the proton: one a polarized electron experiment from SLAC[39] and the second a polarized muon experiment by the EMC group from CERN.[40] There is at present no experimental data for $g_1^n$. As we shall show, the current experimental situation makes it extremely important that the structure function $g_1^n$ be measured for the neutron, and such experiments are presently planned by at least two groups. We have essentially no information on the polarized structure function $g_2$.

In the quark/parton model, $g_1(x)$ has a simple interpretation. For a nucleon with a given helicity [and flavor "i"] define

$$q_i \uparrow (x, Q^2) \equiv \text{quark distribution with helicity parallel to that of the nucleon}$$
$$q_i \downarrow (x, Q^2) \equiv \text{quark distribution with helicity antiparallel to nucleon}$$

$$(3.11)$$

In this language the spin–averaged [spin–dependent] structure functions $F_1$ [$g_1$] are given by

$$F_1(x) = \frac{1}{2} \sum_i e_i^2 \left\{ q_i \uparrow (x) + q_i \downarrow (x) + \bar{q}_i \uparrow (x) + \bar{q}_i \downarrow (x) \right\} \equiv \frac{1}{2} \sum_i e_i^2 q_i(x)$$

$$g_1(x) = \frac{1}{2} \sum_i e_i^2 \left\{ q_i \uparrow (x) - q_i \downarrow (x) + \bar{q}_i \uparrow (x) - \bar{q}_i \downarrow (x) \right\} \equiv \frac{1}{2} \sum_i e_i^2 \Delta q_i(x)$$

$$(3.12)$$

and the measured asymmetry is given by $A(x) \equiv g_1(x)/F_1(x)$. In the proton we thus have

$$g_1^p(x) = \frac{1}{2} \left[ \frac{4}{9} \Delta u(x) + \frac{1}{9} \Delta d(x) + \frac{1}{9} \Delta s(x) \right] \quad (3.13)$$

and assuming isospin symmetry $g_1^p(x) \to g_1^n(x)$ under the interchange $\Delta u(x) \leftrightarrow \Delta d(x)$.

There is a new spin-dependent structure function $b_1(x)$, which was named by Hoodbhoy, Jaffe, and Manohar,[41] for spin-one hadrons. The sum rule for $b_1(x)$ was derived by Close and Kumano.[42] They obtain $\int dx \, b_1^D(x) = \lim_{t \to 0} -5t F_Q(t)/12 M_D^2 = 0$, where $F_Q(t \to 0)$ is the quadrupole moment of the deuteron, if the sea of quarks and antiquarks is unpolarized. This $b_1$ structure function will be measured at HERA.[48]

## 3.6 Bjorken Sum Rule

The Bjorken Sum Rule[43] gives the relationship between the spin structure functions $g_1^p$ and $g_1^n$ and the axial vector coupling constant measured in neutron $\beta$-decay

$$\int_0^1 \left[ g_1^p(x) - g_1^n(x) \right] dx = \frac{1}{6} \frac{g_A}{g_V} \left[ 1 - \frac{\alpha_S(Q^2)}{\pi} \right] \qquad (3.14)$$

Inserting the quark/parton model values for the spin structure functions from Eq. (3.13), we obtain

$$\int_0^1 \left[ \Delta u(x) - \Delta d(x) \right] dx \approx \frac{g_A}{g_V} \qquad (3.15)$$

Why should there be a direct relationship between the spin structure functions of proton and neutron, and the axial vector coupling constant? We follow here the arguments of Isgur.[44] We can write the axial vector coupling as the matrix element of the axial vector operator $A_+^3$ between polarized neutron and proton, i.e.

$$\frac{g_A}{g_V} = \langle p \uparrow | A_+^3 | n \uparrow \rangle = \langle p \uparrow | \left[ A_+^3, I_- \right] | p \uparrow \rangle \qquad (3.16)$$

The second term is derived using the isospin lowering operator to obtain a matrix element with protons in both initial and final states. As the axial operator $A^3$ is a vector in isospin, the commutator has the explicit form

$$\frac{g_A}{g_V} = \langle p \uparrow | \bar{u}\gamma^z\gamma^5 u - \bar{d}\gamma^z\gamma^5 d | p \uparrow \rangle \qquad (3.17)$$

This has a simple form in the nonrelativistic quark/parton model [QPM], where the operator $\gamma^z\gamma^5$ reduces to $\sigma_z$. In this limit the matrix element has the form

$$\frac{g_A}{g_V} \text{ [NR QPM]} \quad \simeq \langle p \uparrow | \sigma_z^u - \sigma_z^d | p \uparrow \rangle = \Delta u - \Delta d \qquad (3.18)$$

In the relativistic quark/parton model, Eq. (3.16) then becomes

$$\frac{g_A}{g_V} \text{ [REL QPM]} \quad \rightarrow \int_0^1 \left[ \Delta u(x) - \Delta d(x) \right] dx \qquad (3.19)$$

This establishes the desired relation between the axial vector coupling constant and the spin difference distribution measured in the proton.

As we mentioned previously, there is at present no experimental data on the neutron spin structure function. Given the EMC result on the spin structure of the proton [which we discuss in the following section], it is **very important** that a measurement be made of the neutron spin structure function. As the validity of the Bjorken Sum Rule depends only on the assumption of isospin, and as the axial vector coupling constant is known with good precision, any significant violation of the Bjorken Sum Rule would present severe problems for the quark/parton model.

## 3.7 Ellis–Jaffe Sum Rule

Ellis and Jaffe[45] calculated a sum rule which involved the proton spin structure function $g_1^p$ only. This necessarily involves additional information above that needed for the Bjorken sum rule. One can obtain part of this information from the strangeness–changing weak decay $\Sigma^- \to n$ [the SU(3) analog of neutron $\beta$ decay]:

$$G_A^{n\Sigma} = \langle n \uparrow | A^3_{\Delta S=1} | \Sigma^- \uparrow \rangle \qquad (3.20)$$

We can write this as a matrix element between polarized neutron states using the V–spin lowering operator, where V–spin is the SU(3) analog of isospin[46]

$$G_A^{n\Sigma} = \langle n \uparrow | [A^3_{\Delta S=1}, V_-] | n \uparrow \rangle \qquad (3.21)$$

With a suitable rotation in isospin space this can be transformed into an operator between polarized proton states

$$G_A^{n\Sigma} = \langle p \uparrow | \bar{d}\gamma^z\gamma^5 d - \bar{s}\gamma^z\gamma^5 s | p \uparrow \rangle \qquad (3.22)$$

The final results follow just as Eq. (3.19) from applying first the NR quark/parton model and then its relativistic version.

$$G_A^{n\Sigma} \xrightarrow[\substack{NR \\ QPM}]{} \Delta d - \Delta s \xrightarrow[\substack{REL \\ QPM}]{} \int_0^1 [\Delta d(x) - \Delta s(x)]dx \qquad (3.23)$$

We now have a relation between the axial vector coupling constant and the first moment of $\Delta u - \Delta d$ in the proton, and a corresponding relation between the axial vector coupling constant measured in $\Delta S = 1$ decays, and the first moment of $\Delta d - \Delta s$. This can give us a prediction for $I_1^p$, the integral of $g_1^p(x)$. We have

$$I_1^p \equiv \int_0^1 g_1^p(x)dx = \int_0^1 dx \left[ \frac{2}{9}\Delta u(x) + \frac{1}{18}\Delta d(x) + \frac{1}{18}\Delta s(x) \right]$$

$$= \int_0^1 dx \left[ \frac{2}{9}[\Delta u(x) - \Delta d(x)] + \frac{5}{18}[\Delta d(x) - \Delta s(x)] + \frac{1}{3}\Delta s(x) \right] \qquad (3.24)$$

$$I_1^p \simeq \frac{2}{9}\frac{g_A}{g_V} + \frac{5}{18}G_A^{n\Sigma} + \frac{1}{3}\int_0^1 \Delta s(x)dx$$

As a first approximation, **assume** that $\int_0^1 \Delta s(x)\,dx = 0$ [it is "natural" to assume that the polarized strange quark distribution in the proton is very small]. In this case we can see from Eq. (3.24) that

$$I_1^p \approx \frac{2}{9}\frac{g_A}{g_V} + \frac{5}{18}G_A^{n\Sigma} \tag{3.25}$$

Using the experimental values,

$$\begin{aligned} \frac{g_A}{g_V} &= 1.254 \pm 0.006 \\ G_A^{n\Sigma} &= -0.34 \pm 0.04 \end{aligned} \tag{3.26}$$

the best theoretical prediction for $I_1^p$ [including lowest order QCD corrections at $Q^2 = 10.7\ GeV^2$, and neglecting polarized strange quarks] is

$$I_1^p(th) = 0.189 \pm 0.005 \tag{3.27}$$

There are two experiments which have measured $g_1^p$: a SLAC experiment using longitudinally polarized electrons on a polarized butanol target,[39] and an EMC experiment using polarized muons on two frozen spin ammonia targets[40] [one polarized parallel to the muon helicity and the other polarized antiparallel to the muons]. The EMC experiment obtained data for $x \geq 0.01$. The combined result for the two experiments is

$$I_1^p(exp) = 0.126 \pm 0.010(\text{stat}) \pm 0.015(\text{syst}) \tag{3.28}$$

In Fig. 14 we show the EMC polarized muon result for $g_1^p(x)$. The squares denote the experimental values $xg_1^p(x)$ while the diamonds show $\int_x^1 g_1^p(y)dy$ as a function of $x$. We see that the experimental result is more than two standard deviations from the

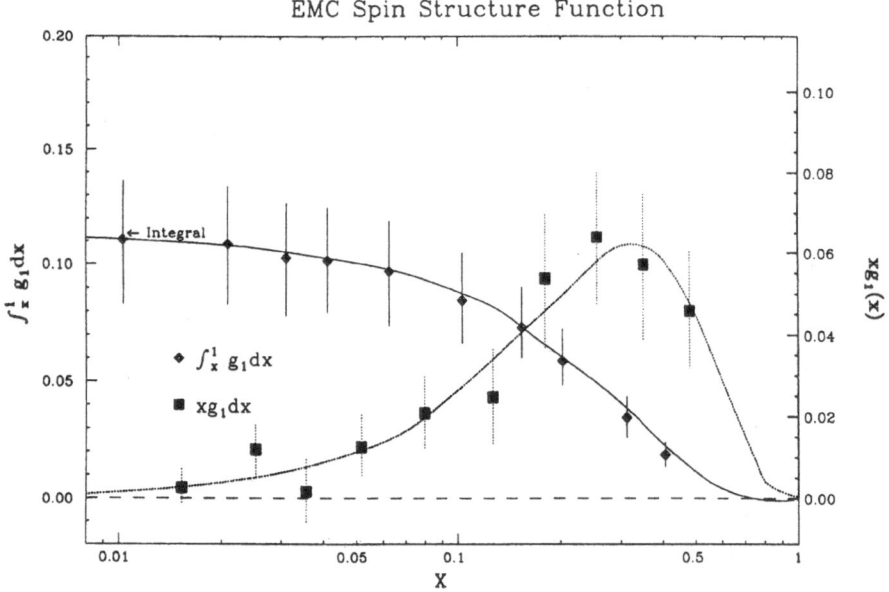

Fig. 14 The spin structure function of the proton measured by the EMC group, Ref. 40. $xg_1^p(x)$: solid squares and right scale; integral: diamonds and left scale, vs. $x$.

Ellis–Jaffe prediction. Assuming that the experiment is correct, and assuming the validity of the Bjorken sum rule, we can infer the integral of the neutron structure function $g_1^n$. The predicted value for the Bjorken sum rule is

$$I_1^p - I_1^n = 0.191 \pm 0.003 \qquad (3.29)$$

From the EMC result and the Bjorken sum rule, we can thus predict

$$I_1^n(pred) = -0.065 \pm 0.010 \pm 0.015$$

If the Bjorken and Ellis–Jaffe predictions were correct [Eqs. (3.27) and (3.29)], then we would predict $I_1^n \approx -0.002$. The EMC result suggests that $I_1^n$ is about thirty times larger than previously predicted; $I_1^n$ is also predicted to be predominantly negative. In Table I we have also included the result for the sum of these quantities, $\Sigma \equiv\ <\Delta u> + <\Delta d> + <\Delta s>$.

With the polarized EMC result we now have three independent pieces of data: $g_A/g_V, G_A^{n\Sigma}$ and $I_1^p$. These in turn depend linearly upon the averaged quantities $<\Delta u>$, $<\Delta d>$, and $<\Delta s>$ [where $<\Delta q_i> \equiv \int_0^1 \Delta q_i(x)\,dx$ for a given flavor "i"]. We can thus invert the experimental quantities to solve for $<\Delta u>$, $<\Delta d>$ and $<\Delta s>$. We obtain the values in Table I.

Table I shows the two surprises from the EMC polarized muon experiment.

1) $<\Delta s>$ is **much larger** than we would naively have assumed [remember that Ellis and Jaffe assumed $<\Delta s> = 0$ to make their prediction]. In Table I for comparison we have included the predictions of the naive QPM and relativistic QPM; note that the experimental prediction for $<\Delta s>$ is more than half as large as the prediction for $<\Delta d>$ in the Rel. QPM!

2) The experimental value for $\Sigma$ is **very small**. In the QPM this quantity is basically the "spin carried by the [valence] quarks." In the NR QPM $\Sigma$ should essentially be one, or twice the proton's spin. The best experimental value for this quantity is surprisingly close to zero!

Table I

|  | naive QPM | rel. QPM | Exp't. |
|---|---|---|---|
| $<\Delta u>$ | $4/3$ | $\sim 0.92$ | $0.78 \pm 0.05$ |
| $<\Delta d>$ | $-1/3$ | $\sim -0.32$ | $-0.47 \pm 0.05$ |
| $<\Delta s>$ | $0$ | $\sim 0$ | $-0.19 \pm 0.05$ |
| $\Sigma \equiv \Delta u + \Delta d + \Delta s$ | $1$ | $0.6$ | $0.120 \pm 0.094 \pm 0.138$ |

These two interpretations of the polarized EMC results have led to the characterization of this experiment as constituting the "proton spin crisis", or to the statement that "none of the spin of the proton is carried by the quarks." As the central theme of this Conference is spin and isospin in nuclear and particle physics, we shall **very briefly** list attempts to explain these results or interpret their significance.

### 3.8 Proposed Explanations of the EMC Spin Measurements

1) **The Experiment is Wrong**. The deviation between experiment and theory is just over two standard deviations. The experiment is a very difficult one, and error bars on the small–$x$ points are large. The SMC experimental group is planning to repeat the EMC spin measurement in the near future.[47] The new experiments will involve both polarized butanol and deuterated butanol targets, which will provide a determination of both $g_1^p$ and $g_1^n$, and hence test the Bjorken sum rule. In addition, there is a proposal by the HERMES collaboration to perform experiments using internal polarized targets in a polarized HERA electron beam.[48] The HERMES experiment, if performed, should provide extremely accurate results in a relatively short running time.

2) **The Ellis–Jaffe prediction is correct**. There is a large [unmeasured] contribution to $g_1^p$ from the region $x < 0.01$. This possibility was suggested by Close and Roberts.[49] It cannot be refuted by the available data.

3) **There is a large strange quark contribution to the proton spin**. Since the apparently large quantity $< \Delta s >$ arises from the difference between two positive probabilities, it thus stands to reason that the spin-**independent** strange quark probability $\int_0^1 s(x)\,dx$ must be even larger. Kaplan and Manohar[50] have suggested that this is the case, and that this large unpolarized strange quark distribution can be extracted from elastic neutrino–proton scattering.[51] Parity–violating elastic electron–proton scattering is also very sensitive to the strange quark distribution in the proton [through the interference term between electromagnetic and weak interactions which gives rise to parity violation]. An experiment to measure this will be carried out at the BATES electron accelerator.[52]

$$\gamma_\mu\,\gamma_5$$

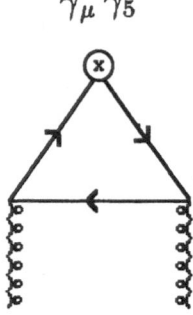

Fig. 15   Triangle diagram for axial current which gives rise to anomalous dimensions.

4) **Polarized Glue plays a significant rôle in the measured spin of the proton.** Several groups[53-55] have pointed out that, due to the Adler–Bell–Jackiw anomaly in the axial current, what is measured in polarized DIS might not be the averaged spin value $< \Delta q_i >$ for each flavor, but $< \Delta q_i > -\alpha_S \Delta\Gamma/2\pi$, where $\Delta\Gamma \equiv < \Delta G >$ is the averaged polarized gluon probability, which arises in polarized DIS through diagrams such as are shown in Fig. 15. If $\Delta\Gamma$ is sufficiently large, then one need not invoke any large polarized strange quarks to explain the EMC result. For example, if $\alpha_S\Delta\Gamma/2\pi = -0.19$ then from Table I we can see that $< \Delta s >= 0$. This dispenses with the mystery of why $< \Delta s >$ is so large, replacing it with the mystery of why $\Delta\Gamma$ is so large [for experimental values $Q^2 \approx 10.7$ GeV$^2$, this would require $\Delta\Gamma$ to be roughly 5].

There are several questions regarding the size of the polarized gluon contribution to the proton spin [or the reliability of calculations of this quantity].[56-58] Since the contribution from polarized glue is proportional to the running coupling constant $\alpha_S$ which goes to zero at sufficiently large values of $Q^2$, it would appear that this term would vanish at sufficiently large momentum transfer; however it was suggested[53-55] that the anomaly requires that $\alpha_S\Delta\Gamma$ becomes a **constant** in the limit $Q^2 \to \infty$, so that this term would give a constant correction even at extremely high $Q^2$.

5) **The quark spin content $\Sigma$ may vary rapidly with $Q^2$.** We may consider the spin carried by quarks $\Sigma = \Delta\mu + \Delta d + \Delta s$ as a function of $Q^2$. Glück and Reya[59] suggested that if $\Sigma$ varies extremely rapidly at small $Q^2$, this could produce a significant decrease in the measured value of $\Sigma$ and hence give agreement with experiment. The $Q^2$ evolution of $\Sigma$ is controlled by the axial anomaly shown in Fig. 15 which introduces an anomalous dimension determining scaling violations. This was originally determined by Kodaira.[60] Such rapid variation would be unexpected in that it is not observed for other known operators. Furthermore, it was shown by Schreiber, Thomas and Londergan[61] that with this assumption the shape of the polarized structure function $g_1^p$ is in qualitative disagreement with the EMC results.

6) **Quarks may carry only a small fraction of the spin of the nucleon,** with the bulk of the spin carried by the angular momentum of the constituents. For example, one could imagine a Cloudy Bag[62] or Little Bag[63] model of the nucleon with a core of quarks surrounded by a nonperturbative cloud of various mesons. If the bulk of the spin of the nucleon were carried by angular momentum of the mesons, this might conceivably produce a picture in accord with the EMC measurements. Meng et al.[64] have discussed possible experimental measurement of the fraction of nucleon spin carried by angular momentum.

7) It has been shown[65] that in the **canonical Skyrme model** constructed from $SU(3) \otimes SU(3) \otimes SU(N_c)$, where $N_c$ is the number of colors, the quantity $\Sigma \approx 1/N_c$. Hence in the limit $N_c \to \infty$ [necessary to justify truncation of the effective Lagrangian], one naturally obtains $\Sigma \to 0$ in rough agreement with experiment. However, in this same limit one obtains many observables which go to zero as $N_c \to \infty$; these include the ratio $2 < s\bar{s} > /[< u\bar{u} + d\bar{d} >]$, which has an experimental value close to 1/2, and $M_{\eta'}/M_N$ which is predicted to vanish as $1/N_c^2$.[22]

## 4. QCD Predictions for the Behavior of Structure Functions

As is well known, in QCD the running coupling constant $\alpha_S$ goes to zero at very high energies. In this limit perturbative QCD should give testable predictions for the behavior of structure functions, and we can check the behavior of observables with predictions of the theory. As the limits are reached logarithmically, data must be taken over a very wide range of energies in order to test these predictions. In this brief review we will cover only two aspects of QCD which have been tested in recent experiments. First, we will review the QCD predictions for the violations of scaling behavior in DIS structure functions. Second, we will review the $Q^2$ behavior of the R parameter [the ratio of longitudinal to transverse cross sections measured in DIS]. In each of these cases we have recent experimental data which allows us to make meaningful tests of QCD predictions.

### 4.1 QCD Predictions for Scaling Violations in DIS

Bjorken[5] showed that in the extreme high energy limit $[Q^2, \nu \to \infty]$, structure functions become functions only of $x$. The original experimental measurements at SLAC indeed showed structure functions which appeared to "scale", showing very little dependence on $Q^2$ for fixed $x$. In fact, these measurements were rather fortuitous in that the variation of structure functions with $Q^2$ is not necessarily small [see Fig. 3].

Quantum chromodynamics, along with the renormalization group equations satisfied by this gauge theory,[66] predicts that structure functions will exhibit a logarithmic dependence on $Q^2$ at fixed $x$. The QCD predictions can be expressed in terms of an integrodifferential equation first derived by Altarelli and Parisi:[67]

$$\frac{dq_{NS}(x, Q^2)}{d[\ln Q^2]} = \frac{\alpha_S(Q^2)}{2\pi} P \otimes q_{NS} \equiv \frac{\alpha_S(Q^2)}{2\pi} \int_x^1 \frac{dy}{y} P_{qq}(\frac{x}{y}) q_{NS}(y, Q^2) \qquad (4.1)$$

The Altarelli–Parisi equation for a quark distribution function expresses the dependence on $Q^2$ in the following way. A quark distribution function for momentum fraction $x$ and four–momentum transfer $Q^2$ could arise from a quark with a larger momentum fraction $y$ which radiates a gluon, as shown schematically in Fig. 16a. The "splitting function" $P_{qq}$ gives the probability for a quark to radiate a gluon. The entire process is proportional to the running coupling constant at that value of $Q^2$. The splitting functions can be calculated to a given order in perturbative QCD through the Wilson operator product expansion.[66] As the structure function is simply a linear combination of quark distribution functions, the structure function obeys the same Altarelli-Parisi equation as the quark distribution functions.

Eq. (4.1) is only true for the flavor–nonsinglet part of a quark distribution, or for a flavor–nonsinglet structure function [examples of flavor–nonsinglet structure functions are $xF_3^\nu$, or the difference $F_2^n - F_2^p$]. Flavor singlet quark distributions of structure functions couple to gluons. Therefore, instead of the uncoupled equations, the flavor–

Fig. 16  Mutual couplings of quarks and gluons to one another which give
rise to scaling violations in QCD. Each amplitude is shown with
its relevant splitting function.

singlet distributions satisfy **coupled** integrodifferential equations

$$
\begin{aligned}
\frac{dq_S(x, Q^2)}{d[\ln Q^2]} &= \frac{\alpha_S(Q^2)}{2\pi} \left[ P_{qq} \otimes q_S + P_{qG} \otimes G \right] \\
\frac{dG(x, Q^2)}{d[\ln Q^2]} &= \frac{\alpha_S(Q^2)}{2\pi} \left[ P_{Gq} \otimes q_S + P_{GG} \otimes G \right]
\end{aligned}
\tag{4.2}
$$

In Eq. (4.2), $G$ represents the gluon distributions. The additional terms in Eq. (4.2) correspond to the couplings shown schematically in Fig. 16b–16d. As the flavor singlet equations involve a knowledge of the gluon distributions they are more difficult to solve. It is thus more straightforward to predict the $Q^2$ dependence of flavor–nonsinglet structure functions.

Let us apply the QCD predictions for scaling violations to the nonsinglet structure function $x F_3^\nu$, for which we have recent precise measurements from the CCFR group.[17] From Eq. (4.1) we predict that the $Q^2$ dependence of the structure functions should obey the schematic equation

$$
\frac{d\, x F_3^\nu(x, Q^2)}{d \ln Q^2} = \frac{\alpha_S(Q^2)}{2\pi} P \otimes x F_3
\tag{4.3}
$$

The convolution on the right hand side of Eq. (4.3) involves an integral from $x$ to 1. The data is taken in various $x$ bins for many values of $Q^2$. Thus both left and right hand sides of the equation involve the same data [or interpolations of the data]. The splitting function $P$ can be calculated analytically, so the only unknown in Eq. (4.3) is the running coupling constant, which depends both on the expansion scheme and the QCD scale parameter $\Lambda_{QCD}$. For all of the curves shown here the running coupling constant has been calculated using the next to leading order $\overline{MS}$ scheme with four quark flavors.[20]

In Fig. 17 we show the $Q^2$ dependence of the logarithm of the $x F_3$ structure function measured by the CCFR group with the QTB neutrino beam on iron target at FermiLab.[17] The dashed curve is the QCD prediction with $\Lambda_{QCD} = 250$ MeV.

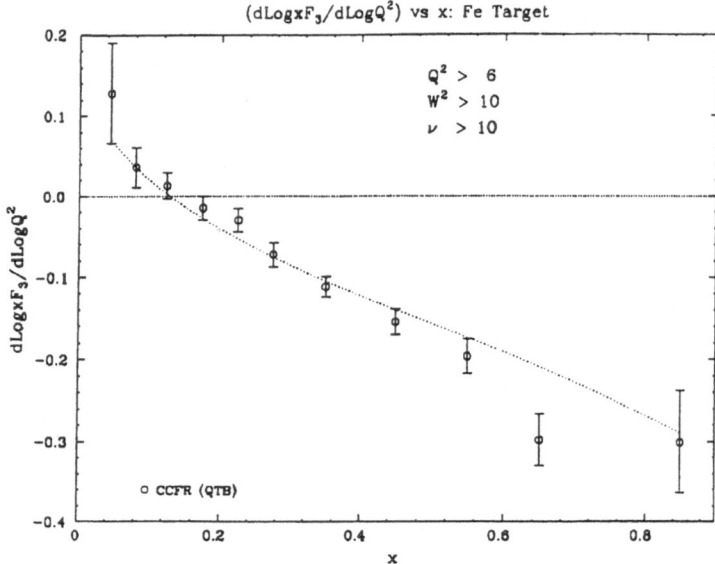

Fig. 17   Logarithmic slope of $xF_3$ with respect to $Q^2$ vs. $x$. Open
circles: CCFR QTB neutrino data of Ref. 17. Dotted
curve: QCD prediction for scaling violations.

The agreement with the theoretical prediction is quite extraordinary. Previous experimental neutrino data for this structure function from the CDHSW group[16] or from the narrow band CCFR data[2] showed rather poor agreement with QCD predictions [and also had much larger error bars]. The extremely large range of $Q^2$ values covered by the wide band CCFR experiment give them much smaller error bars on their data; also recent data from several experiments appears to be uniformly consistent with values of $\Lambda_{QCD}$ between 200–300 MeV. Thus the present data is not only more precise than previous data, but the agreement between theory and experiment appears excellent here and was uncompelling at best with older data.

Similar results occur with structure functions which contain a singlet component. In Fig. 18 we show BCDMS data for the $F_2$ structure function obtained from DIS of muons on H.[11] As we mentioned previously, this satisfies the considerably more complicated coupled integrodifferential equations of Eq. (4.2). In practice, one can start at large $x$ where these equations reduce to Eq. (4.3). This is because the gluon distribution, being soft, goes to zero rapidly at large $x$. Therefore at sufficiently large $x$ the coupled equations of Eq. (4.2) reduce to the flavor–nonsinglet Altarelli–Parisi equations. One can then move in to smaller values of $x$ where the gluon contribution becomes progressively more important.

In Fig. 18 the shaded area shows the effect of gluons. We see that gluons become less and less important as $x$ increases; for $x > 0.4$ there is no visible effect of gluons. The dotted curve in Fig. 18 shows the perturbative QCD predictions for this structure function. The solid curve shows the effect of adding target mass corrections [TMC],[68,69] and the dashed curve shows the additional effect of higher–twist corrections to the PQCD plus TMC terms. Again, the predictions are in excellent quantitative agreement with experimental data. Although we do not show them here, this excellent

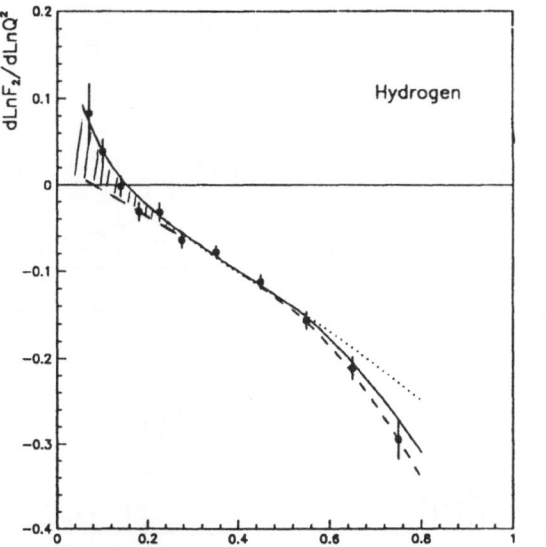

Fig. 18   Logarithmic derivative of $F_2$ structure function from
the BCDMS group on H, Ref. 11. Dotted line: perturbative QCD prediction; solid line, PQCD plus target
mass corrections [TMC]; dashed line: PQCD+TMC+
higher–twist corrections. Shaded area at small $x$:
effect of gluons on logarithmic derivatives.

agreement with QCD predictions **does not** hold for either the EMC muon results[10] or
the CDHSW neutrino results[16] for $F_2$ structure functions measured from iron targets.
In both cases the logarithmic derivatives with $Q^2$ of the experimental results are in
qualitative disagreement with the QCD predictions. This is discussed in detail in the
review article of Mishra and Sciulli.[2]

## 4.2   $Q^2$ Behavior of the Ratio $R = \sigma_L/\sigma_T$

To lowest order the longitudinal to transverse ratio $R$ is just given by the Callan–
Gross relation of Eq. (2.3),

$$R(x, Q^2) \equiv \sigma_L/\sigma_T = \frac{F_2(x, Q^2)\left[1 + Q^2/\nu^2\right] - 2xF_1(x, Q^2)}{2xF_1(x, Q^2)} \qquad (4.4)$$

However, QCD corrections to this equation produce an integral relation involving
radiative gluon corrections. The modified equation for $R$ has the form

$$R(x, Q^2) = \frac{\alpha_S(Q^2)x}{\pi} \int_x^1 \frac{dz}{z^3} \left[\frac{2}{3}F_2(z, Q^2) + f\left(1 - \frac{x}{z}\right)zG(z, Q^2)\right] / F_1(x, Q^2) \qquad (4.5)$$

In Eq. (4.5), $G$ is the gluon distribution, and the quantity $f$ is $n_f$ [the number of quark
flavors] for neutrino scattering and is the sum of squares of quark charges [$\sum e_i^2$] for
muon or electron scattering.

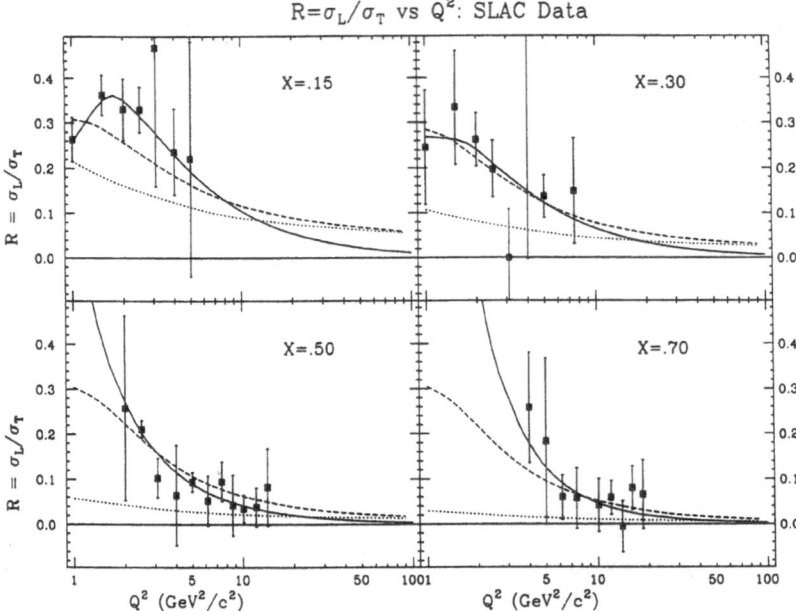

Fig. 19   Longitudinal to transverse ratio $R$ as a function of $Q^2$ for four $x$ bins. Data from SLAC, Ref. 71. Dotted curves: PQCD prediction. Dashed curves: PQCD plus higher twist contribution.

From Eq. (4.5) we should be able to predict the $Q^2$ dependence of $R$ for a given value of $x$. As was mentioned, the earliest experimental measurements of $R$[1] showed a value $R \leq 0.3$, which demonstrated clearly that DIS was occurring from spin-1/2 elementary constituents in the nucleon. However, later and more extensive measurements of $R$[70] did not confirm the $Q^2$ dependence expected from Eq. (4.5), e.g. $R$ should fall as a power of $1/Q^2$ for intermediate values of $Q^2$ and should go to zero logarithmically at very high $Q^2$. In fact, the experimental data appeared independent of $Q^2$ within the large experimental errors.

A recent experiment at SLAC by Dasu et al.[71] was carried out explicitly to examine the $Q^2$ behavior of $R$. We show the results in Fig. 19. The SLAC data is shown as a function of $Q^2$ in four different $x$ bins. The dotted curve in Fig. 19 is the perturbative QCD prediction for $R$. The dashed curves include both the PQCD prediction plus a higher-twist term [which varies like $1/Q^2$] with a coefficient fit to the data. The curves agree rather well with the data: at high $Q^2$, the data are consistent with the PQCD prediction, particularly at the higher values of $x$ where accurate data exists at large $Q^2$. Second, at somewhat lower values of $Q^2$ the experimental curves deviate from the PQCD prediction, roughly like $1/Q^2$.

However, Mishra and Sciulli,[2] pointed out that the data are also consistent with the hypothesis that **all** of the $Q^2$ dependence of $R$ is given by higher-twist terms. In any case, the new data show that $R$ falls off with increasing $Q^2$ as predicted by

QCD plus higher twist effects. This is an improvement over the previous experimental situation which appeared to show $R$ independent of $Q^2$.

## 5. Conclusions

In the two decades since the first DIS scattering measurements, much qualitative information has been gained from a variety of experiments with leptonic probes. The approximate scaling of DIS structure functions and the small value of $R$ were established early on. Since then a series of experiments have been carried out with electron, muon and neutrino beams. In many cases our knowledge of sum rules obtained from these experiments appeared to support the QCD predictions, but frequently with considerable errors. Furthermore, since the sum rules we have listed all contain factors $dx/x$, they emphasize the region of very small $x$; until recently experimental data in this low-$x$ region were not sufficiently precise to draw firm conclusions.

As we showed in this review, in the last few years there has been remarkable progress for several of these sum rules. For example, the Gross–Llewellyn Smith sum rule is now known to within 3% and appears to agree extremely well with the theoretical prediction. On the other hand the Gottfried sum rule is now known to about 6% accuracy and appears to **disagree** with the naive quark/parton model prediction. We discussed this case at some length as it suggests violation of $SU(2)$ flavor symmetry in the proton.

Recent measurements of the Ellis–Jaffe sum rule by the EMC group have also caused great interest, as the experimental value falls short of the Ellis–Jaffe prediction by more than two standard deviations. We discussed possible theoretical explanations for this result. We also emphasized the importance of testing the Bjorken sum rule at the earliest possible opportunity, as this may shed light on the origin of the violation of the Ellis–Jaffe sum rule.

We then reviewed QCD predictions for the behavior of structure functions. First we discussed QCD predictions for violations of scaling in the structure functions. Recent CCFR measurements of $xF_3$ from neutrino DIS on iron were shown to give excellent agreement with QCD predictions. Previous neutrino data were ambiguous and appeared to contradict the QCD predictions. Also, BCDMS measurements of $F_2$ on hydrogen give very good agreement with QCD predictions. The latter case was more difficult as the $F_2$ structure function contains a flavor–singlet piece which requires knowledge of the gluon distribution.

Finally we showed results of a recent SLAC experiment measuring the $Q^2$ dependence of $R = \sigma_L/\sigma_T$. The data showed a power–law decrease in $R$ vs. $Q^2$, in contradistinction to earlier measurements of this quantity. For the most part, the most recent and precise experiments show quantitative agreement with predictions of quark/parton sum rules or QCD. We are now able to make quantitative comparisons of predictions with data. In some experiments the results confirm the predictions of QCD: in others, we apparently see deviations from simple models of the nucleon. Newer and even more precise measurements of many quantities are anticipated. We expect them to give even more dramatic tests of our models of nucleon structure and QCD predictions.

## Acknowledgments

The authors would like to acknowledge useful conversations regarding the material in this review with D. Beck, Q.H.C. Ingram, R.L. Jaffe, J. Speth and A.W. Thomas. The authors have been supported in part by the U.S. National Science Foundation under contract NSF–PHY88–805640.

# References

1. J.I. Friedman and H. Kendall, Ann. Rev. Nucl. Part. Sci. **22**, 203 (1972).

2. S.R. Mishra and F. Sciulli, Ann. Rev. Nucl. Part. Sci. **39**, 259 (1989).

3. *Proceedings of the Workshop on Hadron Structure Functions and Parton Distributions*, eds. D.F. Geesaman, J. Morfin, C. Sazama, and W.K. Tung (World Scientific, Singapore, 1990).

4. A.J. Buras, Rev. Mod. Phys. **52**, 199 (1980).

5. J.D. Bjorken, Phys. Rev. **179**, 1547 (1969).

6. C. Callan and D. Gross, Phys. Rev. Lett. **22**, 156 (1969).

7. We neglect here the possibility that structure functions in nuclei are likely to differ from structure functions measured on free nucleons.

8. With leading-order QCD corrections the Callan-Gross relation $F_2(x, Q^2) = 2xF_1(x, Q^2)$ is maintained. However, higher order QCD corrections modify this relation. We can then **define** the quark densities as satisfying Eq. (2.5) for $F_2$; we then use the QCD-corrected formula Eq. (4.5) to express $F_1$ in terms of the experimentally measured $R$.

9. A. Bodek et al., Phys. Rev. **D 20**, 1471 (1979).

10. EM Collaboration, J.J. Aubert et al., Nucl. Phys. **B259**, 179 (1985).

11. BCDMS Collaboration, A.C. Benvenuti et al., Phys. Rev. Lett. **B195**, 97 (1987).

12. A. Milsztajn, A. Staude, K.M. Teichert, M. Virchaux and R. Voss, preprint CERN-PPE 190–135 (Sept. 1990).

13. WA–21 Collaboration, G.T. Jones et al., Preprint 87, Rec. Jul. (1987).

14. WA–25 Collaboration, D. Allasia et al., Phys. Lett. **B135**, 231 (1984); Z. Phys. **C28**, 321 (1985).

15. CDHS Collaboration, H. Abramowicz et al., Z. Phys. **C25**, 29 (1984).

16. CDHSW Collaboration, P. Berge, et al., A. Phys. **C35**, 443 (1987).

17. CCFR Collaboration, W.H. Smith et al., Proc. 14th Int. Conf. on Neutrino Physics and Astrophysics, CERN, Geneve, June 1990, to be published.

18. D.J. Gross and C.J. Llewellyn Smith, Nucl. Phys. **B14**, 337 (1969).

19. Most recent global fits to quark/parton distribution functions have been carried out using the next to leading order QCD coupling constant, in the minimal subtraction $[\overline{MS}]$ scheme, as described in Ref. 4 and 20.

20. W–K. Tung, in Ref. 3.

21. B.A. Iijima, MIT preprint CTP-993 (1983, unpublished).

22. R.L. Jaffe, personal communication.

23. P.N. Harriman, A.D. Martin, W.J. Stirling, and R.G. Roberts, Phys. Rev. **D42**, 798 (1990).

24. M. Diemoz, F. Ferroni, E. Longo and G. Martinelli, Z. Phys. **C39**, 21 (1988).

25. P. Aurenche, R. Baier, M. Fontannaz, J. F. Owens and M. Werlen, Phys. Rev. **D39**, 3275 (1989).

26. J. G. Morfin and W–K. Tung, Preprint Fermilab–Pub–90/24, IIT–90–11, and J. G. Morfin, in Ref. 3.

27. NM Collaboration, P. Amaudruz et al., CERN-PPR/91-05, submitted to Phys. Rev. Lett.

28. W.A. Bardeen, A.J. Buras, D.W. Duke, and T. Muta, Phys. Rev. **D18**, 3998 (1978).

29. D.A. Ross and C.T. Sachradja, Nucl. Phys. **B149**, 497 (1979).

30. A.D. Martin, R.G. Roberts, and W.J. Stirling, Phys. Lett. **B208**, 327 (1988); Phys. Rev. **D37**, 1161 (1988).

31. E.M. Henley and G.A. Miller, Phys. Lett. **B251**, 453 (1990).

32. A. Signal, A.W. Schreiber, and A.W. Thomas, Mod. Phys. Lett. **A6**, 271 (1991).

33. S. Kumano and J.T. Londergan, preprint IU/NTC 90–16, to be published, Phys. Rev. **D**; S. Kumano, Phys. Rev. **D43**, 59 (1991); to be published in Phys. Rev. **D43** (1991).

34. J.D. Sullivan, Phys. Rev. **D5**, 1732 (1972).

35. W–Y. P. Hwang, J. Speth and G.E. Brown, Jülich preprint (Z. Phys. **A**, to be published).

36. S.D. Drell and T.M. Yan, Ann. Phys. **66**, 595 (1991).

37. R.D. Field, *Applications of Perturbative QCD* (Addison–Wesley, 1989).

38. J. Levelt, P.J. Mulders, and A.W. Schreiber, preprint NIKHEF-91-P5 (1991).

39. M.J. Algurd et al., Phys. Rev. Lett. **37**, 1261 (1978); **41**, 70 (1978).

40. EM Collaboration, J. Ashman et al., Phys. Lett. **B206**, 364 (1988).

41. P. Hoodbhoy, R.L. Jaffe, and A. Manohar, Nucl. Phys. **B312**, 571 (1989).

42. F.E. Close and S. Kumano, Phys. Rev. **42**, 2377 (1990).

43. J.D. Bjorken, Phys. Rev. **148**, 1467 (1966).

44. N. Isgur, in *Physics with Polarized Beams on Polarized Targets*, eds. J. Sowinski and S.E. Vigdor, (World Scientific, 1990).

45. J. Ellis and R.L. Jaffe, Phys. Rev. **D9**, 1444 (1974).

46. D.C. Cheng and G.K. O'Neill, *Elementary Particle Physics*, (Addison-Wesley, 1979).

47. SMC Collaboration proposal to CERN.

48. HERMES Collaboration proposal at HERA: see R. McKeown talk at this conference.

49. F.E. Close and R.G. Roberts, Phys. Rev. Lett. **60**, 1471 (1988).

50. D.B. Kaplan and A. Manohar, Nucl. Phys. **B310**, 527 (1988).

51. L.H. Ahrens et al., Phys. Rev. **D35**, 785 (1987).

52. D.H. Beck, Phys. Rev. **D39**, 3248 (1989); SAMPLE Collaboration proposal to Bates.

53. A.V. Efremov and O.V. Teryaev, Dubna preprint E2-88-287 (1988).

54. G. Altarelli and G.G. Ross, Phys. Lett. **B212**, 391 (1988).

55. R.D. Carlitz et al., Phys. Lett. **B214**, 229 (1988).

56. G. Bodwin and J. Qiu, Phys. Rev. **D41**, 2755 (1990).

57. R.L. Jaffe and A.V. Manohar, Nucl. Phys. **B337**, 509 (1990).

58. A.V. Manohar, Phys. Rev. Lett. **66**, 289 (1991).

59. M. Glück and E. Reya, Z. Phys. **C43**, 678 (1989).

60. J. Kodaira et al., Phys. Rev. **D20**, 627 (1979); Nucl. Phys. **B159**, 99 (1979); J. Kodaira, Nucl. Phys. **B165**, 129 (1980).

61. A.W. Schreiber, A.W. Thomas and J.T. Londergan, Phys. Rev. **D42**, 2226 (1990).

62. G.A. Miller et al., Phys. Lett. **B91**, 192 (1980); A.W. Thomas, Adv. Nucl. Phys. **13**, 1 (1983) and references therein; see N.A. Thornqvist, Phys. Lett. **B221**, 701 (1989).

63. G.E. Brown and M. Rho, Phys. Lett. **B82**, 177 (1979).

64. Meng Ta-Chung et al., Phys. Rev. **D40**, 769 (1989).

65. S. Brodsky el al., Phys. Lett. **B206**, 309 (1988); J. Ellis and M. Karliner, Phys. Lett. **B213**, 731 (1988).

66. K. Wilson, Phys. Rev. **179**, 1499 (1969); see also Ref. 4.

67. G. Altarelli and G. Parisi, Nucl. Phys. **B126**, 298 (1977).

68. W.K. Tung, in Ref. 3.

69. H. Georgi and H.D. Politzer, Phys. Rev. **D14**, 1829 (1976); A. de Rujula et al., Phys. Rev. **D15**, 2495 (1977); R. Barbieri et al., Phys. Lett. **B64**, 171 (1976); see also Ref. 4.

70. M.D. Mestayer et al., Phys. Rev. **D27**, 285 (1983).

71. S. Dasu et al., Phys. Rev. Lett. **61**, 1061 (1988).

# MESON–EXCHANGE AND DEEP INELASTIC SCATTERING

W–Y. P. Hwang[1] and J. Speth

IKP (Theorie)
Forschungszentrum Jülich GmbH
5170 Jülich, Germany

[1]Department of Physics
National Taiwan University
Taipei, Taiwan, R.O.C.

## ABSTRACT

We present a dynamical model for meson–meson scattering and apply it to $\pi\pi$–scattering. The experimental phase shifts and the electric form factor of the pion are well reproduced using very short–range form factors for the strong interaction vertices. Within the same model we predict a strong energy dependence for the scalar form factor of the pion in the energy range between $E = 0$ and the Cheng–Deshen point. We also apply this model for understanding the sea quark distributions.

## INTRODUCTION

It is generally accepted that QCD is the fundamental theory of the strong interaction. At high momentum transfer and short distances, where perturbation theories are applicable (perturbative regime), the experimental data (with a few exceptions) can indeed be well explained within QCD, where quarks and gluons are the relevant degrees of freedom. At larger distances and lower momentum transfer, respectively, non–perturbative methods have to be applied to the corresponding strong interaction problems. Lattice gauge calculations represent the only way up to now to accomplish this. However, despite great efforts this method gives so far reliable answers only to a very limited class of problems. For that reason, models still have to be developed for describing experimental results in the non–perturbative regime. A large number of very different experimental facts show clearly that colorless objects, namely, baryons and mesons, are the relevant degrees of freedom at larger distances. The important question, however, is to find the range of applicability of baryons and mesons which is, as we shall point out in this contribution, connected with the size of the confining region of the valence quarks.

In low and medium energy physics the strong interaction between baryon–baryon [1,2] and baryon–meson [3] systems can be quantitatively reproduced within the meson–exchange picture in a surprisingly large energy and momentum range. Also the strong interaction between two mesons (meson–meson scattering) has recently been described in a very satisfactory way using a meson–exchange model [4], as I shall report in section 2.

Fig. 1 shows qualitatively the various meson—exchange contributions to the nucleon—nucleon (NN) potential. There is no discussion about the long—range part of the NN—interaction, which is provided by the pion exchange. There is also general agreement that the intermediate attraction is due to the exchange of two correlated pions in a relative s—state (there exists no obvious mechanism in the "QCD—inspir-ed" quark models). This very complicated process is in most of the models paramet-rized by an effective low—mass scalar meson. The controversy and also the question about the limitation of the meson—exchange model is connected with the short—

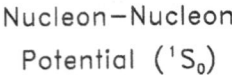

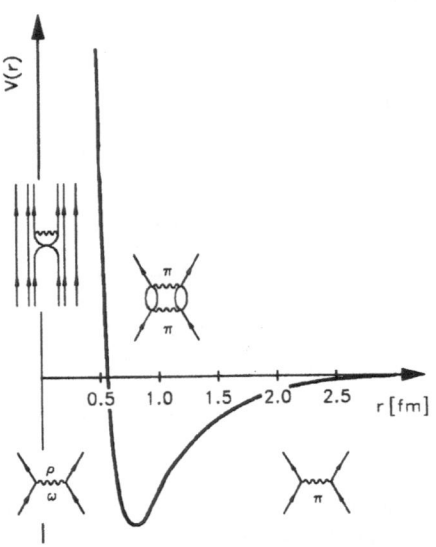

Fig. 1        Schematic representation of the nucleon—nucleon potential V(r) as a function of the relative distance r between the two nucleons.

range repulsion. In the meson—exchange picture the repulsion is provided by the exchange of the $\omega$— and $\rho$—vector mesons. On the other hand, at such short distance, one may expect a repulsive force which is connected with the substructure of the baryons (quark—gluon dynamics). At least in certain quark—models [5] and not too short distance the two mechanisms give similar results in the NN—case. A detailed analysis of the short—range behavior of the hadron—hadron interaction, which also includes the $K^+N$— and $K^-N$—scattering seems to indicate that there are deviations from the simple vector—meson exchange mechanism [6]. The range of applicability of

Fig. 2        "One–pion–exchange" in a nucleus. The spheres represent the electric charge radii of the proton $<r_p^2>^{\frac{1}{2}} = 0.86$ [fm] and the pion $<r_\pi^2>^{\frac{1}{2}} = 0.66$ [fm].

A mean distance of $\bar{d} = 1.8$ [fm] for the two nucleons in the nucleus has been assumed.

the meson–exchange model is directly connected with the size of the hadrons. If one considers the charge radii of the proton and the charged pions and takes into account the mean distance of the nucleons in nuclei (see Fig. 2), then indeed the meson exchange model is very questionable because the pion and the nucleons overlap. This argument, however, is misleading because the charge form factor of the pion is mainly given by the range of the correlated two–pion–exchange in the $I = 1$, $J^\pi = 1^-$–channel, as we shall discuss in section 2. The same is true also for the electromagnetic form factor of the proton (vector dominance). These facts are qualitatively shown in Fig. 3. In both cases, the strong interaction form factors, which give the relevant length–scale for the meson–exchange model, are of much shorter range than the electromagnetic ones.

The strong interaction form factors reveal the quark substructure of the hadrons which has been investigated experimentally in the framework of deep inelastic scattering as shown in Fig. 4. The differential cross section for inclusive electron scattering is given in the parton model by [7]:

$$\frac{d^2\sigma}{dxdy} = \frac{8\pi\alpha^2}{MEx^2y^2}\left[xy^2 G_1(x,\frac{q^2}{M}) + (1-y-\frac{M}{2E}xy)G_2(x,\frac{q^2}{M})\right] \qquad (1)$$

Here $Q^2 = -q^2 = \vec{q}^2 - v^2$ is the (negative) four momentum transfer, $x = \frac{Q^2}{2Mv}$ and $y = \frac{v}{E}$, where $v$ is the energy transfer and E the incident energy of the electrons. The structure functions $G_1$ and $G_2$ depend in general on x and $q^2$. In the Bjorken limit (eq. 2) they give the parton distribution functions $F_1(x)$ and $F_2(x)$ which depend on x only:

$$\lim_{\substack{Q^2\to\infty \\ x\ \text{fixed}}} G_i(x,\frac{q^2}{M}) \to F_i(x) . \qquad (2)$$

Within the quark parton model we obtain for the structure function of the proton in the case of electron scattering:

## Electro–Magnetic Form Factor
## of the Pion

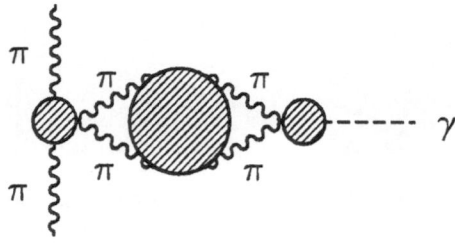

## Electro–Magnetic Form Factor
## of the Proton

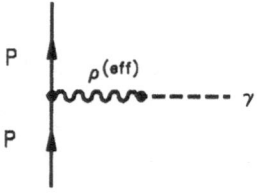

Fig. 3    Vector dominance model for the pion and the proton.
The effective $\rho$–meson in the lower part corresponds
to the correlated two pion exchange (in the $J^{\pi} = 1^-$
channel) which is shown in the upper part of the figure.

## Deep inelastic scattering in
## the parton model

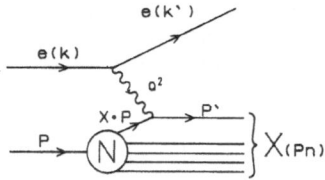

Fig. 4.    Electron–proton scattering in the parton model.

$$F_2^{ep}(x) = \sum_i e_i^2 \left[ q_i(x) + \bar{q}_i(x) \right], \tag{3}$$

where $q_i(x)$ and $\bar{q}_i(x)$ are the probabilities of finding a quark and antiquark, respectively, of flavor i with a momentum fraction between x and x+dx and $e_i$ is the corresponding charge. The momentum distribution function of the quarks is the sum of the valence distribution $q_v(x)$ and the sea distribution $q_s(x)$:

$$q(x) = q_v(x) + q_s(x). \tag{4}$$

In Fig. 5 the valence and sea–quark distribution of the proton at $Q^2 = 4$ GeV$^2$ is given [8].

Some years ago Thomas [9] connected the $\pi$NN–form factor with the SU(3) symmetry violating part of the sea–quark distribution using the deep inelastic scattering process from virtual pions, described originally by Sullivan [10]. He obtained a very soft form factor. This analysis has recently repeated [11], where the authors obtained even softer form factors. If one accepts the results of refs. [9,11] that the mesons contribute very little to the sea quark distribution, one has to begin with the QCD evolution at momentum transfer of the order of $Q \approx 1$ [fm$^{-1}$] [12]. The most recent measurement [13] of the tensor polarization of the deuteron up to $q \simeq 5$ [fm$^{-1}$], however, shows that the experimental result is in excellent agreement with the prediction of the meson–exchange picture, whereas the perturbative QCD prediction completely fails, as shown in Fig. 6. Therefore, there is no justification to use perturbative QCD at $Q \simeq 1$ [fm$^{-1}$], in fact, this contradicts completely not only the most recent experimental result but also earlier analyses of experimental data, which come to the conclusion that the meson–exchange model is valid well beyond $Q^2 = 1$ GeV$^2$. In section 3 it will be shown that the existing data from deep inelastic electron scattering also agree with much harder meson–baryon form factors if one considers also the strange mesons K and K*. In that case, the dominant part of the sea quarks up to moderate $Q^2$ are due to the contribution of mesons so that the QCD evolution has to contribute only at much higher $Q^2$ than up to now, where perturbative QCD is justified.

## 2. MESON–MESON SCATTERING AND PION FORM FACTORS

In the introduction we already pointed out that the meson exchange model of hadronic interaction can only be applicable in a limited energy range, which must be related to the size of the confining region of the quarks. The interesting question is: how large is this range, and at what point must one consider quark effects? To answer this question, it is necessary to investigate different hadronic systems within a consistent framework. Such a program has been started some time ago when the Bonn NN–potential [1] was extended to include the exchange of strange mesons [2], enabling the calculation of hyperon–nucleon and KN potentials. The further extension of the model to include also meson–meson scattering is also straightforward. The underlying Lagrangian is the nonlinear $\sigma$–model by Bando et al. [14], where the vector mesons are introduced into the theory as gauge bosons via the "hidden" symmetry. This Lagrangian is chiral invariant, the masses of the vector mesons are generated by a Higgs mechanism and it also has U(1)xSU(3) symmetry built in. This allows us to treat the complete pseudoscalar– and vector–meson nonet in a consistent way. The most important input into our calculations are the vertices $F_{PPV}$ between two pseudosclar and one vector meson.

Those vertices are functions of the momenta involved which reflects the quark substructure of the mesons. In principle, those vertex functions may be derived from quark models [15]. Here, however, we simply parametrize them in the form of a monopole form factor if the meson is exchanged in the t–channel:

## valence − quark

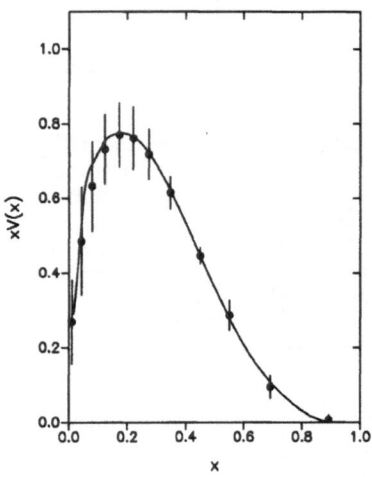

## sea − quark

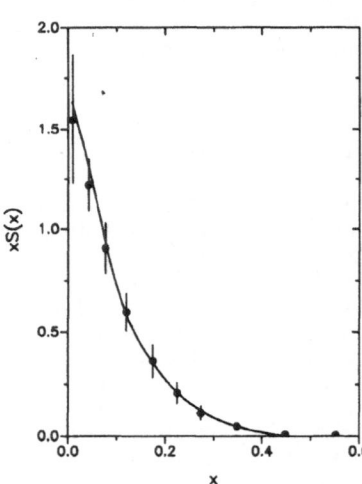

Fig. 5.  Valence− and sea−quark distribution of the proton
at $Q^2 = 4$ GeV$^2$ [8].

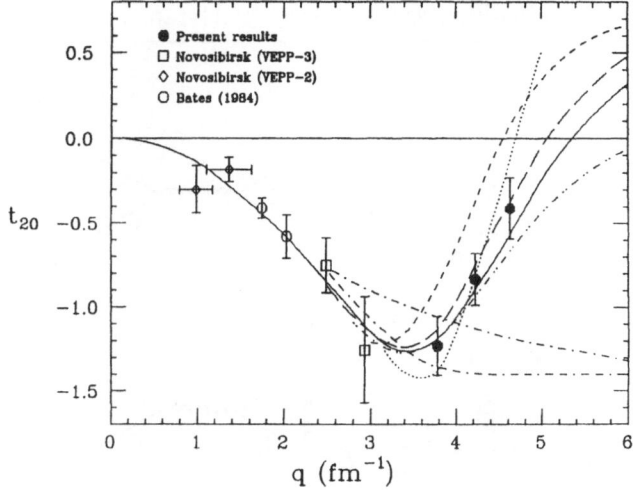

Fig. 6.    Experimentally determined tensor polarization $t_{20}(q)$ of the deuteron [13] compared with the meson–exchange prediction (long dashed line) [16], the IA, Argonne $V_{14}$ potential (full line) [17] and the perturbative QCD prediction (dashed–dotted line) [18].

$$F_\alpha^t(q^2) = f_\alpha \frac{\Lambda_\alpha^2 - m^2}{\Lambda_\alpha^2 + \vec{q}^2} \tag{5}$$

where m is the mass of the exchanged meson and $\vec{q}$ is the (three–) momentum transfer. For the s–channel exchange we chose

$$F_\alpha^s(q^2) = f_\alpha \frac{\Lambda_\alpha^4 + m^4}{\Lambda_\alpha^4 + E_q^2} . \tag{6}$$

Here $E_q$ is the total on–shell energy with CM momentum q. The various coupling constants $f_\alpha$ are connected through SU(3) symmetry relations, the cutoff parameters have to be adjusted to the empirical data.

In Fig. 7 the scattering process between two mesons is shown in a schematic picture. If the incoming and outgoing mesons are the same than we have a one channel problem; if the outgoing mesons can be of a different kind than the incoming ones we have to solve a coupled channel problem. We shall see in the following that channel coupling plays an important role. We can view the scattering process within the meson–picture as an initial state interaction and a final state interaction between the incoming and outgoing mesons and in between the two mesons can interact via an s–channel resonance, if there exists one in the considered channel. In the quark–picture (lower part) we have two quarks–two antiquarks (in the incoming channel) which can interact with each other, they can be transformed into a correlated quark–antiquark pair; and, finally, in the outgoing channel, we have again correlated two quarks– two antiquarks. From this point of view the resonances seen in meson–meson scattering which one usually interprets as mesons are of a more complicated structure: they are in general a superposition of a genuine (bare) meson (correlated $q\bar{q}$–pairs) and correlated two meson states.

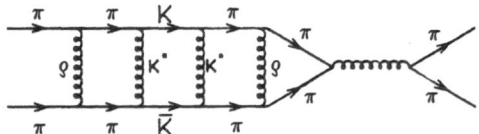

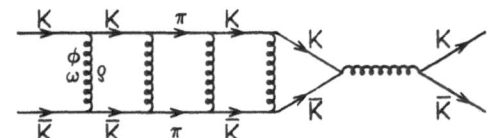

Fig. 7.　　The various contributions to the scattering
process of two pions and two kaons, respectively, which are included in the present
calculations. In the scalar and vector
channel, these two processes are coupled.
The mesons exchanged in the s–channel
depend on the quantum numbers of the
scattering channel considered.

In the actual calculation we determine the transition operator T(z) from an
integral equation of the Lippmann–Schwinger type:

$$T(z) = V(z) + V(z) \frac{1}{z - H_0 + i\epsilon} T(z) \qquad (7)$$

where z is the starting energy and V(z) the quasi–potential. In the t–channel we
consider the complete vector mesons nonet and in the s–channel the $\rho$– and $\epsilon$–mesons, respectively. The masses, coupling parameters and form factors may be found
in Ref. [4]. Here, it is important to note that the strong interaction cut–off parameters are between $\Lambda = 1.6$ GeV and 4.5 GeV. The corresponding phase shifts are
shown in Fig. 8. The $I = 1$, $J^\pi = 1^-$ ($\rho$–meson)–channel is dominated by the "bare"
$\rho^{(0)}$ in the s–channel, whereas the t–channel plays a less important role in the
energy regime considered. (Due to interference effects it nevertheless has an important influence on the width of the $\rho$–resonance in the $\pi\pi$–channel). The situation is
very different in the $I = 0$ and $J^\pi = 0^+$ case. First of all the t–channel interaction is
the dominant part up to 800 MeV. In addition, the coupling to the $K\overline{K}$–channel
produces the resonance structure at 980 MeV. The scalar meson, which has a mass
around 1.4 GeV due to the present analyses, influences the phase shifts mainly
beyond 1 GeV.

With the help of the transition operator T we can calculate the form factors
for scalar and vector fields, respectively [19]. In Fig. 9 we show graphically the
various contributions, which enter in such a calculation. The hadronic part (strong
form factors, coupling constants and $T_{\pi\pi}$–transition matrix) is fixed from the
previous calculation – what is still open is a possible electromagnetic form factor for
the coupling of the photon to the bare $\rho^{(0)}$. Preliminary results for the electromag-

netic form factor of the pion are shown in Fig. 10a. The dashed line represents the result of the vector dominance model (VDM), where the experimental $\rho$–meson data have been used. The full line is the present result without the $\rho^{(0)}$–electromagnetic form factor, the dotted line is the result which includes such a (phenomenological) factor of the form

$$F_{\gamma v}(q^2) = \frac{\Lambda^2}{\Lambda^2 - q^2} , \qquad (8)$$

where $\Lambda = 1.95$ GeV is adjusted in such a way that we reproduce quantitatively the experimental results if we include the $\rho$–$\omega$–mixing into our calculation, as shown in Fig. 10b.

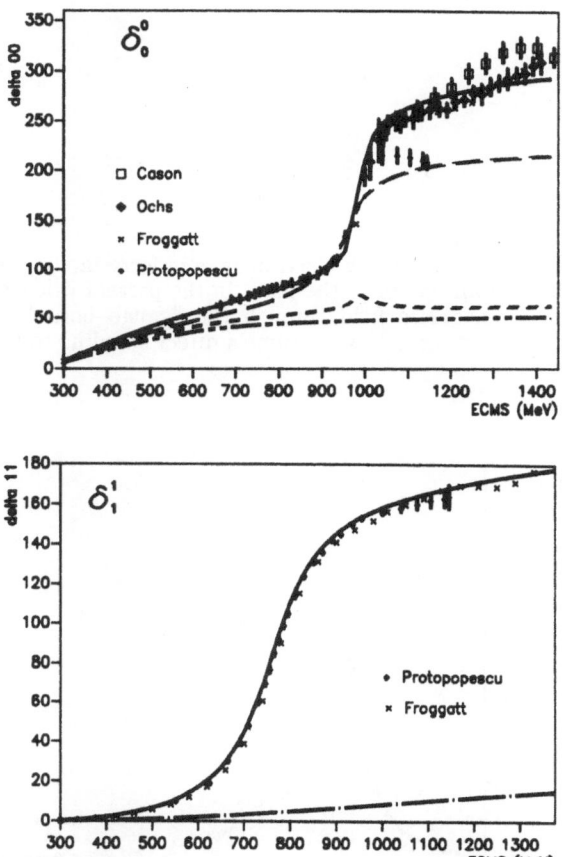

Fig. 8.        Comparison of experimental and (different) theo–
               retical $I = 0$, $J^\pi = 0^+$ and $I = 1$, $J^\pi = 1^-$ $\pi\pi$–phase
               shifts. Upper part: The dot–dashed curve is the re–
               sult for $\rho$–exchange only. The short–dashed curve in–
               cludes the effect of coupling to the $K\overline{K}$–channel
               with no diagonal $K\overline{K}$ interaction. The long–dashed
               curve includes strong t–channel exchange contribu–
               tions in the $K\overline{K}$–channel. The solid curve adds the $\epsilon$
               as an s–channel process. Lower part: The dot–dashed
               curve shows the results with only t–channel exchange.
               The solid curve shows the result when the $\rho^{(0)}$–meson
               is included as an s–channel contribution in the potential.

a) Electro Magnetic Form Factor

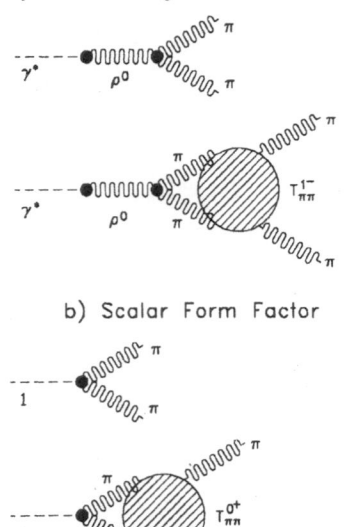

b) Scalar Form Factor

Fig. 9.     Contributions to the electromagnetic form factor (upper part) and the scalar form factor of the pion. In the present calculation we couple the electromagnetic field through the "gauge–boson" $\rho^{(0)}$ to the pions. For the scalar field we assume a direct coupling to the pions.

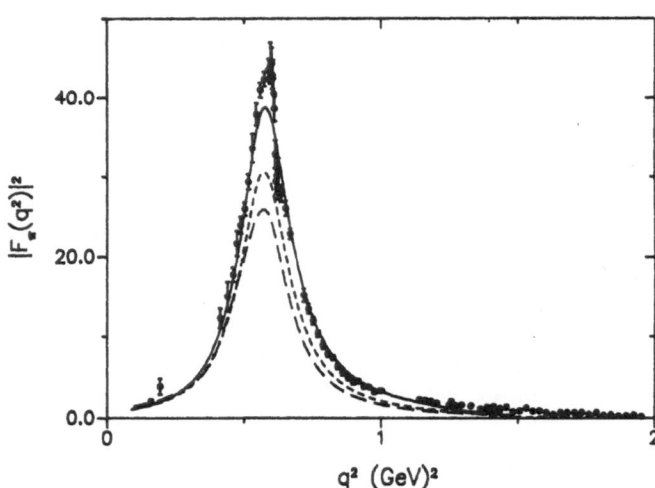

Fig. 10.    (a) Comparison of the experimental electromagnetic form factor of the pion with various theoretical models [19]: the dashed line is the phenomenological vector dominance model, the full line is the theoretical result using the model described in Fig. 9. The dot–dashed results include in addition the $\gamma\rho^{(0)}$ form factor as described in the text.

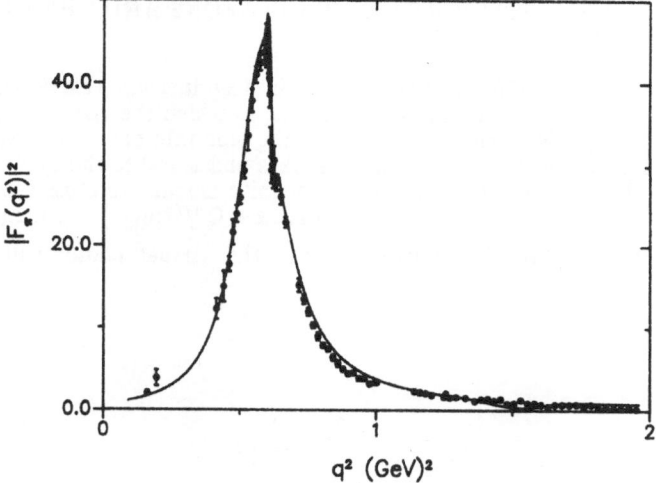

Fig. 10.    (b) The complete theoretical model which includes also the $\omega$–$\rho$ mixing.

The scalar form factor of the pion (full line in Fig. 11) shows a strong energy dependence below the $2\pi$ threshold and around 1 GeV. The latter one is connected with the $K\overline{K}$–molecule at 980 MeV, which also shows up in the scalar form factor of the K–meson (dashed line). The energy dependence at the lower end is of interest in connection with the $\sigma$–term of the nucleon. The famous discrepancy between its value extracted from the masses of the baryon octet and that determined from $\pi N$–scattering is to a large extent due to the difference of the scalar form factor at the Cheng–Dashen point at E = 0. (The scalar form factor of the pion and the nucleon has the same energy behavior). See also a similar discussion in Ref. [21].

To summarize this section: we have shown that the electromagnetic form factor of the pion (and correspondingly also of the nucleon) can be well reproduced with very short–ranged strong interaction form factors.

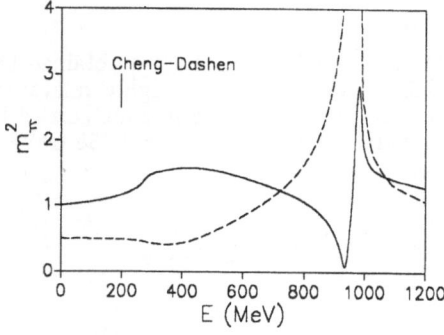

Fig. 11.    Theoretical predictions [20] for the isoscalar form factors of the pion (full line) and the kaon. The Cheng–Dashen point is also indicated in the figure.

## 3.  SEA–QUARK DISTRIBUTION OF NUCLEONS AND THE GOTTFRIED SUM RULE

In 1972, Sullivan [10] pointed out that, in deep inelastic scattering (DIS) of a nucleon by leptons, the process shown in Fig. 12, in which the virtual photon strikes the pion emitted by the nucleon and smashes the pion into debris, will scale like the original process, where the virtual photon strikes and smashes the nucleon itself. In other words, the process will contribute by a finite amount to cross sections in the Bjorken limit, $Q^2 \to \infty$ amd $\nu \equiv E_\ell - E'_\ell \to \infty$ with $x \equiv Q^2/(2m_N\nu)$ fixed. In 1983, Thomas [9] observed that in the Sullivan process the virtual photon will see most of

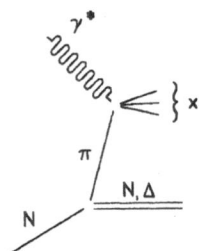

Fig. 12.     The original Sullivan process which scales in the Bjorken limit.

the time the valence distributions in the pion as the probability of the pion carrying the momentum fraction y of the nucleon, $f_\pi(y)$ to be given by Eq. (10 b) below, peaks at $y \approx 0.3$, a region where only valence quarks and antiquarks are relevant. By attributing to the Sullivan process the excess of the momentum fractions carried by $\bar{u}$ and $\bar{d}$ quarks as compared to that of the $\bar{s}$ quarks, Thomas was then able to set a limit on the momentum fraction carried by those pions which surround the nucleon. The limit is such that a chiral bag radius cannot be too small, say $R = 0.87 \pm 0.10$ fm. In 1989, the idea of Thomas was elaborated somewhat further by Frankfurt, Mankiewicz, and Strikman [11], who obtained, with both the $\pi NN$ and $\pi N\Lambda$ couplings taken into account, that, if the $\pi NN$ coupling is parametrized by the form factor in a monopole form, the cutoff mass must be less than 0.5 GeV, which is reminiscent of a large bag radius.

Very recently, there is new information [21] obtained by the CCFR Collaboration at Fermilab which, as we shall see, is highly relevant for the argument of Thomas. Denoting by $<x>_i$ the momentum fraction carried by partons of flavor i, the CCFR Collaboration obtained, with $<Q^2> = 16.85$ GeV$^2$,

$$\kappa \equiv \frac{2<x>_s}{<x>_{\bar{u}}+<x>_{\bar{d}}} = 0.44^{+0.09+0.07}_{-0.07-0.02} , \qquad (9.a)$$

$$\eta_s \equiv \frac{2<x>_s}{<x>_u+<x>_d} = 0.057^{+0.010+0.007}_{-0.008-0.002} , \qquad (9.b)$$

$$R_{\overline{Q}} \equiv \frac{<x>_{\bar{u}}+<x>_{\bar{d}}+<x>_{\bar{s}}}{<x>_u+<x>_d+<x>_s} = 0.153 \pm 0.034. \qquad (9.c)$$

Here Eq. (9.c) comes from an earlier measurement with errors determined from cross section results [22]. As one would expect $<x>_{\bar{s}} \approx <x>_s$ inside a nucleon, the value for $\kappa$ (together with the antiquark distribution of Field and Feynman for a free proton [23] constitutes the basis for the argument of Thomas [9]. Here the lower value reported for the ratio may loosen slightly the bounds obtained earlier [9, 11], but not by any significant amount.

Assuming the standard wisdom that gluons carry (45–50)% of the nucleon momentum, we obtain from Eq. (9) that the total momentum fraction carried by antiquarks of all kinds must be in the vicinity of (6–7)% while $<x>_s \approx <x>_{\bar{s}} \approx 1\%$.

These values will play an important role in the following calculation. The basic difference between the previous approaches [9, 11] and ours [24] is that we introduce kaons as is required [2] by the hyperon–nucleon scattering data or by the validity of approximate flavor SU(3) symmetry. In that case there will be more room for pions, thereby pushing the allowed cutoff to well beyond 1 GeV. In other words, we consider the strange sea quarks at moderate $Q^2$ as essentially nonperturbative. As we shall see, cutoffs in the range of near 1 GeV, when used in our meson–exchange model for parton distributions, yield results in good agreement with the CCFR data, thereby not only questioning the validity of the earlier claim [9, 11] but offering the validity of using the meson–exchange picture to generate the sea distributions of a hadron at moderate $Q^2$.

Perhaps we should clarify our statements a little further. As there is little reason to deny the necessity to include processes as suggested by Sullivan (which scale in the Bjorken limit), nor to deny the successes of using coupling constants given by approximate flavor SU(3) symmetry and relatively hard form factors to obtain excellent fits in a variety of nuclear physics phenomena (such as nucleon–nucleon or hyperon–nucleon scatterings), it is imperative to know how much of the sea content in a hadron can be associated with the meson cloud.

Thus, there is a very basic difference between what we are trying to do here and what Thomas [9] and Frankfurt et al. [11] have considered. They considered the original Sullivan process as the new source for making additional $\bar{u}$ and $\bar{d}$ sea quarks, which are then added to the existing sea distributions of the nucleon. On the other hand, we take out all the sea distributions in the baryons involved in the "generalized" Sullivan processes and then find that the sea distributions of the nucleon at moderate $Q^2$ can even be attributed entirely to the meson cloud.

The significance of our conjecture of attributing the sea distributions of a hadron at moderate $Q^2$ to its associated meson cloud as generated by strong interaction processes at the hadron level is that we are now able to determine the parton distributions of a hadron from the knowledge of the valence distributions of the various hadrons, which are calculable from quark models such as a bag model. The QCD evolution equations then take us from moderate $Q^2$ to very high $Q^2$. The previously very "fuzzy" gap between low $Q^2$ (nuclear) physics and large $Q^2$ (particle) physics is now linked nicely together, should our conjecture be substantiated by future experiments.

More specifically, we have so far considered the "generalized" Sullivan processes including those shown in Fig. 13 where the meson and baryon pair (M,B) includes $(\pi,N)$, $(\rho,N)$, $(\omega,N)$, $(\sigma,N)$, $(K,\Lambda)$, $(K,\Sigma)$, $(K^*,\Lambda)$, $(K^*,\Sigma)$, $(\pi,\Delta)$, and $(\rho,\Delta)$. It turns out that all f(y)'s are such that, as the virtual photon strikes the meson in the "cloud", it sees essentially only the valence partons. Such contributions just look like a "sea" to the nucleon which we thought the experiment would be probing. The value of $R_{\bar{Q}}$, which determines the total momentum fraction carried by antiquarks, thus place stringent limits on the form factors associated with the various meson–nucleon strong couplings.

The well–known formula [10] for the Sullivan process is given by

$$\delta F_{2N}^{\pi}(x, Q^2) = \int_x^1 dy f_{\pi}(y) F_{2\pi}(\tfrac{x}{y}, Q^2), \tag{10.a}$$

$$f_{\pi}(y) = \frac{3}{4\pi} \frac{1}{4\pi} (f_{\pi NN} \frac{2m}{\mu})^2 y \int_{-\infty}^{t^m} dt \frac{(-t)|F_{\pi}(t)|^2}{(-t+\mu^2)^2}, \tag{10.b}$$

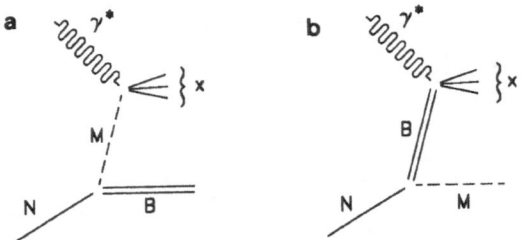

Fig. 13.  The generalized Sullivan processes: (a) the virtual photon strikes the cloud meson, and (b) the virtual photon stikes the recoiling baryons. Both scale in the Bjorken limit. The meson and baryon pair (M,B) includes $(\pi,N)$, $(\rho,N)$, $(\omega,N)$, $(\sigma,N)$, $(K,\Lambda)$, $(K,\Sigma)$, $(K^*,\Lambda)$, $(K^*,\Sigma)$, $(\pi,\Delta)$, and $(\rho,\Delta)$.

where $t^m = -m^2 y^2 / 1-y)$ with m the nucleon mass. $F_{2\pi}(x)$ is the pion structure function as would be measured in deep inelastic electron (or muon) scattering. $\delta F_{2N}(x)$ is the correction to the nucleon structure function due to the Sullivan process. $f_{\pi}(y)$ is the probability of finding a pion carrying the nucleon momentum fraction y. $\mu$ is the pion mass. $f_{\pi NN}$ is the $\pi NN$ coupling in the form of a pseudovector coupling (as dictated by chiral symmetry) with F(t) characterizing its t–dependence. As a specific attempt, we consider

$$F(t) = \left[ \frac{\Lambda_{\alpha}^2 - m_{\alpha}^2}{\Lambda_{\alpha}^2 - t} \right]^{n_{\alpha}}. \tag{11}$$

To simplify the situation, we use $n_{\alpha} = 2$ for all couplings and find that a universal cutoff mass of 1150 MeV in the $\Delta/N$ sector and 1400 MeV in the $\Lambda/\Sigma$ sector already yields very reasonable results. Nevertheless, we have also considered the cases where all couplings are characterized by exponential cutoffs. In addition, we have also adopted in Eq. (11) $n_{\alpha} = 1$ for pseudoscalar or scalar mesons and $n_{\alpha} = 2$ for vector mesons and for all the couplings in the $\Delta$ sector. Results are similar. In all cases, the quality of the fit to the experiments can be made better by adjusting the various cutoffs. Introduction of heavy mesons can be made by replacing $f_{\pi}(y)$ by suitable probability functions [24].

At first sight, there might appear a good number of parameters in our model.

This is in fact not true. All the coupling constants and masses are taken to be the same as those obtained or used in Ref. [2]. Although it is obvious [9, 11] that the predictions are extremely sensitive to the various cutoffs (or form factors), we decided after a few numerical trials to set all the t–channel cutoff masses $\Lambda_\alpha$'s in the $\Delta/N$ and $\Lambda/\Sigma$ sectors respectively to universal values of 1150 MeV and 14 MeV.

The structure functions for the pions, especially the valence distributions, which are the essential input information, are known experimentally [25]. Using flavor SU(3) rotations while taking spin–isospin average, we obtain structure functions for all the other mesons. We take into consideration the slight difference [26] between the kaon and pion valence distributions. For the structure functions of a nucleon, we have considered distributions obtained by Duke and Owens [27] and by Eichten et al. [28] EHLQ). At $Q^2 = 16.85$ GeV$^2$, we find that, while the sea content of the Duke–Owens distributions is too strong in comparison with the CCFR data (Eq. (9)), the EHLQ distributions are in good agreement with the CCFR data. As we shall need only the valence and glue distributions in our theory, however, both distributions give rise to very similar results. The distribution functions for $\Lambda$, $\Sigma$ and $\Delta$ are then obtained again by flavor SU(3) rotations. Small breaking of flavor SU(3) symmetry does not affect our major conclusions.

In addition to the sea quarks which can be generated in this manner, we also expect to have renormalizations over the valence distributions. For example, kaons provide u or d quarks in addition to $\bar{s}$ quarks. Thus we renormalize the original valence distributions before we add the "sea" to it, in order to make certain that the number sum rules are maintained. In the same spitit, we renormalize the original glue distributions before we add the new "sea glue" to it, so that the momentum sum rules are satisfied by ansatz. In principle, this renormalization will also reduce the sea distributions, which means that we may use even harder form factors. A consistent calculation where we introduce renormalization parameters $z$ ($z < 1$) from the beginning and determine them within the theoretical model is in progress.

The predictions of the meson–exchange model using dipole form factors are listed in eq. (12). The EHLQ valence and glue distributions are used but the results from using Dukes–Owens distributions differ y at most a few percent

$$\kappa \equiv \frac{2<x>_s}{<x>_{\bar{u}}+<x>_{\bar{d}}} = 0.40$$

$$\mu_s \equiv \frac{2<x)_s}{<x>_u+<x>_d} = 0.056 \tag{12}$$

$$R_{\bar{Q}} \equiv \frac{<x>_{\bar{u}}+<x>_{\bar{d}}+<x>_{\bar{s}}}{<x>_u+<x>_d+<x>_s} = 0.161$$

It is also remarkable to find that the shape of the various parton distributions obtained in this way are very similar to that in the EHLQ distributions at $Q^2 = 16.85$ GeV$^2$, lending support towards our conjecture that sea distributions of a hadron at moderate $Q^2$ come almost entirely from the meson cloud. In Fig. 14 we compare our result for the $\bar{s}$–quark distribution (solid curve) with the EHLQ–results (in dotted curve) and Duke–Owens (in dashed curve) at $Q^2 = 16.85$ GeV. In Fig. 15 the ratio $\frac{1}{2}\{\bar{u}(x)+\bar{d}(x)\}/\{u_v(x)+d_v(x)\}$ as a function of x for $Q^2 = 16.85$ GeV$^2$ which may be compared with, e.g., the sea–to–valence ratio extracted from the CDHS data (in triangles) [29] and the E615 experiment (in solid squares) [30]. This ratio depends also on the valence distributions used. If we take the most recent one [8], the discrepancy at very small x is greatly reduced.

We may use our model to determine possible deviation from the Gottfried sum rule. In the dipole scenario, we obtain

$$\int_{0.005}^{1} \frac{dx}{x}\{F_2^{ep}(x) - F_2^{en}(x)\} = 0.25 \ , \tag{13}$$

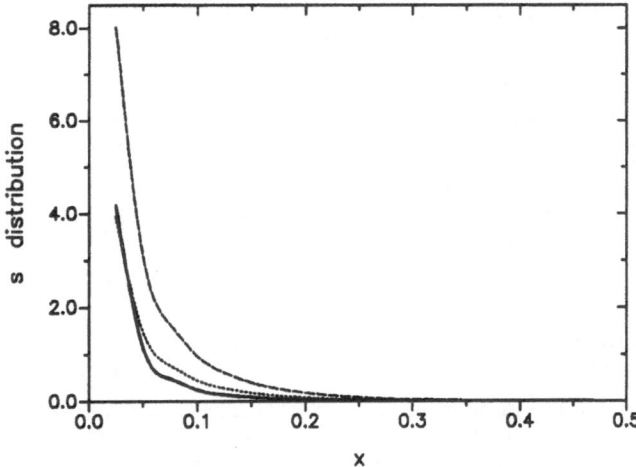

Fig. 14.     The calculated strange parton distribution (in solid curves), as explained in the text, and the corresponding ones from EHLQ (in dotted curves) and Duke—Owens (in dashed curves) at $Q^2 = 16.85$ GeV plotted as function of x.

which compares well with a preliminary value [31] reported by the NMC group:

$$\int_{0.04}^{1} \frac{dx}{x}\{F_2^{ep}(x) - F_2^{en}(x)\} = 0.218 \pm 0.008(\text{stat.}) \pm 0.021(\text{syst.}) \ . \tag{14}$$

In fact, a better agreement can easily be obtained — either by increasing the cutoff in the nucleon sector by about 50 MeV or by decreasing that in the $\Delta$ sector by a slightly larger amount — while maintaining the quality of the agreement between the model predictions and the data mentioned in the text.

We should emphasize that, despite the fact that the integrated value as listed in Eq. (14) may come close to the data, it is nontrivial to reproduce as well the shape of the experimental data as a function of x. The curves shown in Fig. 16 reflect directly the shape of the proposed valence distribution convoluted according to Sullivan processes. It is clear that the shape of the EHLQ valence distributions performs better than that of the HMSR ones. The QCD evolution softens the valence distributions slightly (from $Q^2 = 4$ GeV² to 10 GeV²) so that the results from the NC and CC neutrino data are more or less consistent with the EHLQ prediction. In any event, the general agreement may be taken as additional evidence toward our suggestion that the sea distributions of a hadron, at low and moderate $Q^2$ (at least up to a few GeV²), may be attributed primarily to generalized Sullivan processes.

An important aspect of nuclear physics is that quark distributions of nucleons in nuclei are expected to be different from quark distributions of nucleons in free space. Several phenomena are present in nuclei which can lead to modified quark distributions as measured by deep inelastic lepton–nucleus scattering [31] or lepton pair production [32] in hadron–nucleus collisions. Our meson–exchange model for nucleon parton distributions provides a framework for understanding how parton distributions of a nucleon are modified in a nucleus. These applications are in progress.

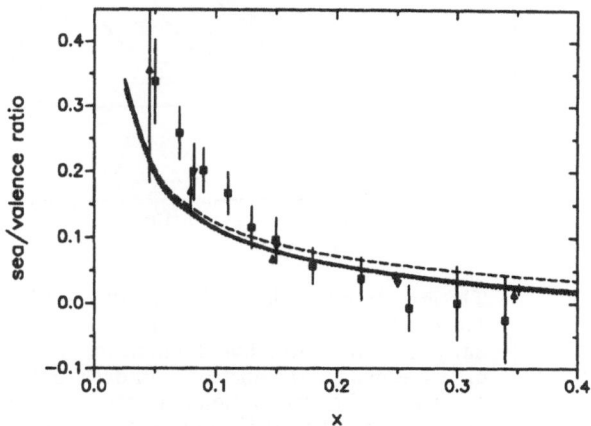

Fig. 15.  The ratio $\frac{1}{2}\{\bar{u}(x)+\bar{d}(x)\}/\{u_v(x)+d_v(x)\}$ shown as a function of x for $Q^2 = 16.85$ GeV$^2$, which may be compared with, e.g., the sea–to–valence ratio extracted from the CDHS data (in triangles) [29] and the E615 data (in solid squared) [30]. The solid curve is obtained by using dipole form factors for all couplings; the dotted curve is for exponential cutoffs; and the dashed curve is for the mixed monopole–dipole scenario.

4. CONCLUSIONS

In this contribution we have first shown that short–range vertices for the strong interaction do not contradict long–ranged electric form factors of the pion and the proton. The large cutoff parameters which were necessary to explain the experimental phase shifts reproduce the empirical electric form factor of the pion even better than the phenomenological vector dominance model. Within the same approach we also calculated the scalar form factor of the pion and predict a strong energy dependence in the range between E=0 and the Cheng–Dashen point. This may explain to a large extent the well–known discrepancy between the two methods used to determine the $\sigma(t)$–term of the nucleon.

In the second part we have applied the meson–exchange model for understanding the sea quark distributions of a nucleon at moderate $Q^2$, by taking into account effects of the various mesons including $\pi$, $\rho$, $\omega$, $\sigma$, K and K*, with the coupling constants fixed by the low–energy nucleon–nucleon and hyperon–nucleon scattering data. Contrary to earlier claims that the $\pi$NN and $\pi$N$\Delta$ form factors must be very soft, we used much harder form factors and obtained parton distributions which are very similar to the corresponding ones in existing parametrized parton distributions. The magnitude of the momentum transfer of $Q^2 = 16.85$ GeV$^2$ which we have chosen for convenience may be somewhat too large. But as shown in Fig. 17, the differences in the sea–quark distributions at $Q^2 = 20$ GeV$^2$ and $Q^2 = 4$ GeV$^2$

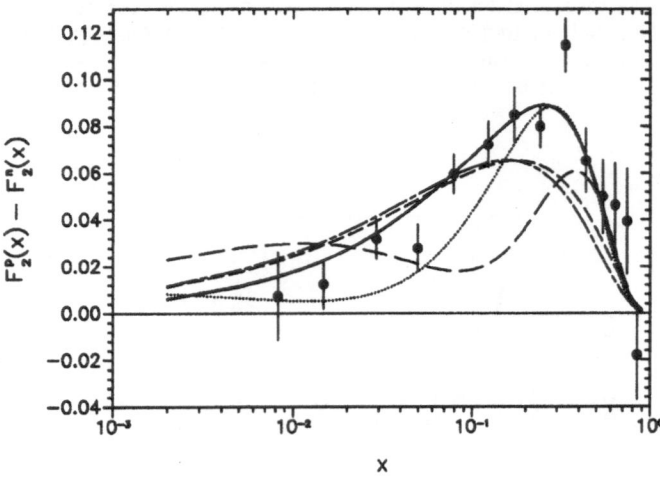

Fig. 16.　　The structure function difference $F_2^p(x)-F_2^n(x)$ shown as a function of x. The five curves are our predictions using five different input valence distributions for the nucleon — in dash–dotted curve from the distribution extracted from the neutral–current neutrino data [33], in dashed curve from the charge–current neutrino data [33], in dotted curve from the input valence distribution of Harriman et al. [8], in solid curve from the valence distributions of Eichten et al. [28], and in long dash curve from the valence distributions of Duke and Owens [27].

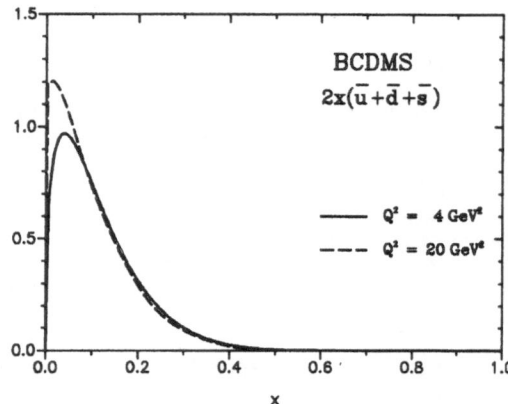

Fig. 17.　　Sea–quark distributions at $Q^2 = 4$ GeV and $Q^2 = 20$ GeV$^2$ obtained from the analysis by Ref. [8].

are small and in addition, the important ratio of the momentum fraction carried by sea quarks to that carried by valence quarks is in the most recent analysis [8] $R_{\overline{Q}} = 0.18$ at $Q^2 = 4$ GeV$^2$. This is nearly 20% larger than what we have used here at $Q^2 = 16.85$ MeV. Therefore, we expect that our present result will be confirmed by a corresponding analysis at $Q^2 = 4$ GeV$^2$, which is at present under investigation.

W–Y. P. Hwang wishes to acknowledge the Alexander von Humboldt Foundation for a fellowship to visit Jülich to conducting research.

## References

1. R. Machleidt, K. Holinde and C. Elster, Phys. Rep. 149 (1987) 1
2. R. Büttgen, K. Holinde, B. Holzenkamp and J. Speth, Nucl. Phys. A450 (1986) 403;
   B. Holzenkamp, K. Holinde and J. Speth, Nucl. Phys. A500 (1989) 485
3. R. Büttgen, K. Holinde and J. Speth, Phys. Lett. B163 (1985) 305;
   R. Büttgen, K. Holinde, A. Müller–Groeling, J. Speth and P. Wyborny, Nucl. Phys. A506 (1990) 586
4. D. Lohse, J.W. Durso, K. Holinde and J. Speth, Nucl. Phys. A516 (1990) 513
5. K. Holinde, Nucl. Phys. A415 (1984) 477
6. J. Speth, preprint KFA–IKP(TH)–1991–2, Nucl. Phys. A, in print
7. e.g. T.–P. Cheng and L.–F. Li, Gauge theory of elementary particle physics, Clarendon Press, Oxford
8. P.N. Harriman, A.D. Martin, W.J. Stirling and R.G. Roberts, Phys. Rev. D42 (1990) 798
9. A.W. Thomas, Phys. Lett. 126B (1983) 97
10. J.D. Sullivan, Phys. Rev. D5 (1972) 1732
11. L.L. Frankfurt, L. Mankiewicz, and M.I. Strikman, Z. Phys. A 334 (1989) 343
12. M. Glück, R.M. Godbole and E. Reya, Z. Phys. C41 (1989) 667
13. I. The et al., Phys. Rev. Lett., in print
14. M. Bando et al., Phys. Rev. Lett. 54 (1985) 1215
15. J. Speth and R. Tegen, Nucl. Phys. A511 (1990) 716
16. R. Dymarz and F.C. Khanna, Nucl. Phys. A507 (1990) 560; Phys. Rev. C41 (1990) 2438
17. R. Schiavilla and P.O. Riska, Phys. Rev. C43 (1991) 437
18. C.E. Carlson, Nucl. Phys. A508 (1990) 481c
19. R. Tegen, J. Nitschkowski, J.W. Durso and J. Speth, to be published
20. B.C. Pearce, private communication
21. C. Foudas et al., CCFR Collaboration, Phys. Rev. Lett. 64 (1990) 1207
22. D. MacFarlane et al., Z. Phys. C26 (1984) 1;
    E. Oltman, in The Storrs Meeting: Proc. of the Division of Particles and Fields of the American Physical Society, 1988, ed. by K. Hall et al. (World Scientific, Singapore, 1989)
23. R.D. Field and R.P. Feynman, Phys. Rev. D15 (1977) 2590
24. W–Y. P. Hwang, J. Speth and G.E. Brown, Z. Phys. A, in print
25. J. Badier et al., Z. Phys. 18 (1983) 281
26. J. Badier et al., Phys. Lett. 93B (1980) 354
27. D.W. Duke and J.F. Owens, Phys. Rev. D30 (1984) 49
28. E. Eichten, I. Hinchliffe, K. Lane, and C. Quigg, Rev. Mod. Phys. 56 (1984) 579; Erratum: 58 (1986) 1065
29. H. Abramowicz et al., CDHS Collaboration, Z. Phys. C17 (1983) 283
30. J.G. Heinrich et al., E615 Collaboration, Phys. Rev. Lett. 63 (1989) 356
31. J.J. Aubert et al., Phys. Lett. 123B (1983) 275;
    U. Landgraf, New Muon Collaboration, "Nuclear Effects in Deep Inelastic Scattering", Talk at PANIC XII, M.I.T., June 25–29, 1990
32. D.M. Alde et al., E772 Collaboration, Phys. Rev. Lett. 64 (1990) 2479
33. T.S. Mattison et al., Phys. Rev. D42 (1990) 1311

# ELECTRON SCATTERING WITH POLARIZED $^3$He

R. D. McKeown

Kellogg Radiation Laboratory
California Institute of Technology
Pasadena, CA 91125, USA

Inclusive electron scattering has been a tremendously useful tool for the study of hadronic and nuclear structure. Although there is substantial kinematic flexibility, one can only access two experimental quantities by using the technique of Rosenbluth separation: the longitudinal response function and the transverse response function. More detailed information on the electromagnetic response requires either detection of final state hadrons in coincidence with the scattered electron or utilization of polarization degrees of freedom (or both). Many CW electron accelerators are being built at present to exploit the possibility of coincidence measurements. In addition, recent developments in polarized beams and targets have opened many new opportunities for studies of spin-dependent electromagnetic response functions. I will concentrate here on the possiblities for using polarized $^3$He, primarily in inclusive electron scattering experiments.

$^3$He is a particularly interesting nucleus for polarization studies. The spin-dependent properties should be dominated by the neutron because the $^3$He wavefunction is predominantly a spatially symmetric S-state and antisymmetrization of the wavefunction requires that the protons be in a spin singlet state. If the $^3$He wavefunction were entirely S-state, the spin of the nucleus would be carried solely by the unpaired neutron. In this case measurement of the spin-dependent quantities in inclusive scattering of polarized electrons from polarized $^3$He would directly yield information on the neutron electromagnetic structure. There are small admixtures of other states in the $^3$He wavefunction which introduce a dependence on the proton electromagnetic form factors, but realistic calculations for the three-body system can give a reliable estimate of these contributions. In addition, as shown below, further coincidence measurements with polarized $^3$He can be used to test and constrain these small and interesting pieces of the three nucleon wave function.

## Polarized $^3$He Target Development

There are two techniques which have been recently developed for laser optically pumped polarized $^3$He targets. One has been developed at Caltech, with further work now taking place at MIT as well as Caltech. The other was developed at Princeton and later Harvard by T. Chupp and his co-workers.

The technique used by our group at Caltech[1] involves optical pumping of the $2^3S$ metastable state of the He atom. These states are created with a weak discharge at a concentration of about $10^{-6}$ and are polarized by pumping to the $2^3P$ states with 1083 nm polarized laser light. The $^3$He nuclei become polarized through the hyperfine interaction. The metastables then collide with ground state atoms, exchange atomic states, but leave the polarized nucleus in an atomic ground state with no electronic angular momentum. In this way, the nuclear polarization is transferred to the much larger population of atoms in the ground state. Substantial polarizations ($50 - 80\%$) are achieved at low pressures of at most a few Torr.

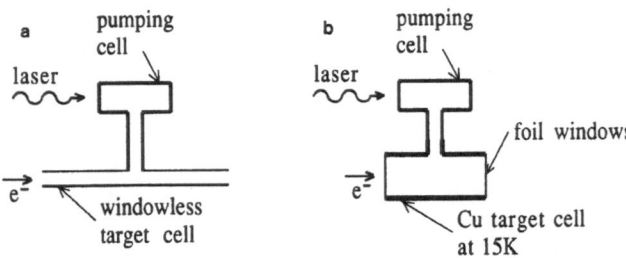

Fig. 1. Schematic diagrams showing the target configuration for (a) an internal target for a storage ring and (b) an target for an external beam.

Figure 1 shows the two configurations for using this method to polarize a gaseous target for an experiment with incident electron beam. In figure 1a the cell where the optical pumping takes place is connected via a capillary to a long tube that forms a windowless target in an electron storage ring. The polarized atoms continually flow from the pumping cell to the target cell and out into the ring where they are pumped away. The target density is thin, only $10^{14}$cm$^{-2}$, as it must be in order not to perturb the stored electrons. However, the stored electron beam current is very high: typically 10-100 mA. This high current then facilitates reasonable counting rates with the thin windowless target. In figure 1b, a target for a less intense external electron beam is shown. Here the pumping cell is connected to a copper target cell with foil windows. The cell is cooled to 15K in order to increase the number density in the path of the electron beam: we typically produce target thicknesses of order $10^{19}$cm$^{-2}$. Target polarizations of over 50% are routine with higher power lasers that have recently been developed.

Chupp and his co-workers[2] have developed a technique that involves optical pumping of Rb vapor in the $^3$He gas. The polarized Rb atoms transfer angular momentum to the $^3$He nuclei during collisions. A nitrogen buffer gas is necessary to reduce radiation trapping. Higher densities of $> 1$ atm are achievable with this technique. This group also employs a two-cell system to decouple the optical pumping region from the electron beam.

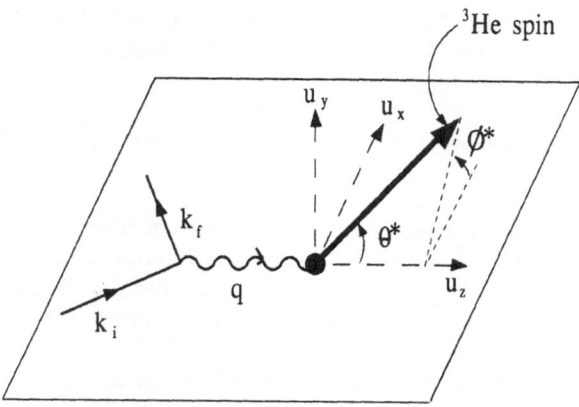

Fig. 2. Kinematic variable definitions for electron scattering from polarized targets. Here $\mathbf{u}_z$ is along the direction of momentum transfer $\mathbf{q}$. The vector $\mathbf{u}_y$ is normal to the electron scattering plane and $\mathbf{u}_x = \mathbf{u}_y \times \mathbf{u}_z$ lies in the scattering plane. The target polarization direction is specified by the angles $(\theta^*, \phi^*)$ in this coordinate system.

## Inclusive Quasielastic Scattering

According to the arguments given above, one would expect that at the kinematics corresponding to quasifree nucleon knockout (quasielastic scattering) the spin dependent cross section would contain information about the elastic electromagnetic form factors of the neutron. Calculations using a Fadeev wavefunction indicate that in the vicinity of the quasielastic peak the neutron properties do dominate the spin dependent scattering.[3] There is particular interest in studying the neutron electric form factor, as the presently available experimental information is based on a rather model-dependent analysis of elastic $e$-$d$ scattering.[4]

The general form for inclusive scattering of longitudinally polarized electrons from a polarized spin 1/2 target is given by[5]

$$\frac{d\sigma}{d\Omega d\omega} = \Sigma \pm \Delta(\theta^*, \phi^*), \tag{1}$$

where $\omega$ is the energy transfer and the angles $\theta^*$ and $\phi^*$ define the target spin direction as shown in figure 2. The plus(minus) sign correspond to positive (negative) helicity incident electrons. The spin-independent cross section $\Sigma$ is given by the usual Rosenbluth formula

$$\Sigma = 4\pi\sigma_{Mott}[v_L R_L(q,\omega) + v_T R_T(q,\omega)], \tag{2}$$

in which $q$ is the momentum tranfer, $v_L$ and $v_T$ are kinematic factors, $R_L$ is the longitudinal response function, and $R_T$ is the transverse response function. These response functions contain the nuclear electromagnetic structure information. The spin-dependent cross section $\Delta$ contains two new response functions:

$$\Delta = -4\pi\sigma_{Mott}[\cos\theta^* v_{T'} R_{T'}(q,\omega) + 2\sin\theta^* \cos\phi^* v_{TL'} R_{TL'}(q,\omega)]. \qquad (3)$$

In particular, the $R_{TL'}$ response function results from interference of longitudinal and transverse amplitudes. Thus it is of special interest in helping to determine longitudinal amplitudes when the transverse amplitudes are dominant. This is the case in quasielastic scattering in $^3$He where $R_{TL'}$ is very sensitive to the value of the neutron electric form factor.[3] On the other hand, the response function $R_{T'}$ in quasielastic scattering from $^3$He is predicted to be determined by known quantities and serves as a useful check on the whole procedure. As mentioned above, the values of these response functions are affected by the $^3$He wave function. The $S'$ and $D$ state components cause the neutron to be less than 100% polarized and also cause the protons to contribute via spin triplet configurations. Generally, these small components generate 10-50% changes in the spin dependent response functions.[6]

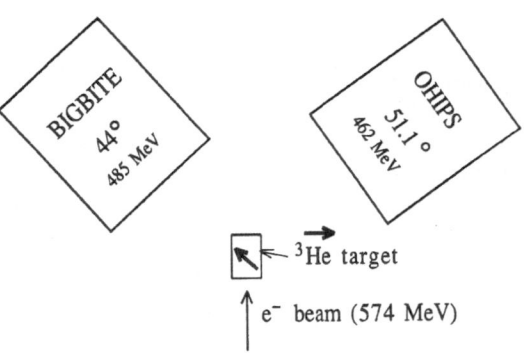

Fig. 3 Schematic diagram of the experimental setup at Bates. BIGBITE and OHIPS are magnetic spectrometers used to detect the scattered electrons.

We have performed the first measurement of spin dependent electron scattering from a polarized $^3$He target at Bates. The experimental geometry is schematically shown in figure 3. The OHIPS spectrometer was at a scattering angle $\theta = 51.1°$ so that $\theta^* \sim 0°$ selecting $R_{T'}$ at a squared 4-momentum transfer of $Q^2 = 0.2\ (\text{GeV}/c)^2$ in quasielastic kinematics. Experimentally one measures the asymmetry in the cross section under reversal of electron helicity:

$$A \equiv \frac{\Sigma}{\Delta}. \qquad (4)$$

The measured asymmetries are shown in table 1 for each of the three different settings for the target spin direction. It is important to note that reversing the target polarization causes the asymmetry to change sign. We reverse the sign of the $\theta^* = 172°$ data and combine it with the other two to give the result labeled "combined". Theoretical predictions of this quantity using the formalism of Blankleider and Woloshyn[3] and Friar, et al.[6] are indicated in the table also and are in good agreement with the experimental result. The poor statistics are due to insufficient beam delivered to the experiment because of many technical problems with the Bates accelerator. These results were reported in an earlier publication.[7]

Table 1. Results of OHIPS asymmetry measurements.

| Charge | $\theta^*$ | $\phi^*$ | A |
|---|---|---|---|
| $\mu$A·hours | degrees | degrees | % |
| 310 | 0.9 | 180 | $-3.5 \pm 2.4$ |
| 342 | 7.9 | 180 | $-2.1 \pm 2.7$ |
| 826 | 172.1 | 0 | $+4.2 \pm 1.7$ |
| 1478 (combined) | | | $-3.49 \pm 1.23 \pm 0.54$ |
| Theory (reference 3) | | | $-4.3 \pm 0.2$ |
| Theory (reference 6) | | | $-4.5 \pm 0.2$ |

The other spectrometer, BIGBITE, was employed near $\theta^* \sim 90°$ at $\theta = 44°$ where the $R_{TL'}$ response function is dominant. This spectrometer has a large momentum acceptance which allowed measurement over the entire quasielastic peak as can be seen in the experimental energy loss spectrum in figure 4. The calculated spectrum in a quasielastic model is also shown in the figure and is in reasonably good agreement with the experimental data. The average $Q^2$ for the asymmetry measurement (integrated over the quasielastic peak) is 0.16 $(GeV/c)^2$. The measured asymmetries are presented in table 2. By using the Blankleider and Woloshyn formalism, we computed the expected asymmetries for these conditions as a function of the form factor $G_E^n$. We then fit the predictions to the data to extract the best value of $G_E^n$ to be consistent with our measurements. The result is $G_E^n = +0.070 \pm 0.100 \pm 0.035$ where the first uncertainty is statistical and the second is systematic. These results have recently been submitted for publication.

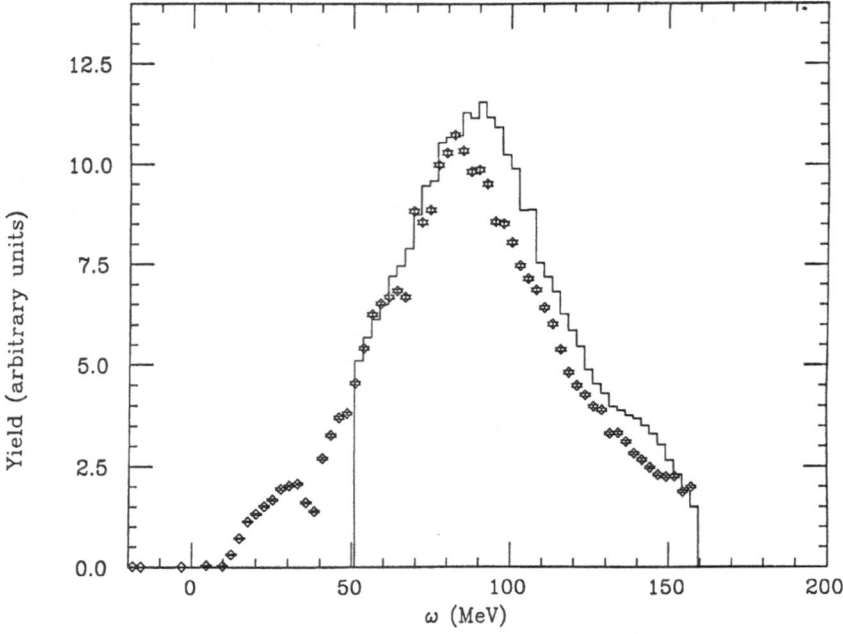

Fig. 4. The experimental energy loss spectrum measured with the BIGBITE spectrometer at $\theta = 44°$ for 574 MeV incident electrons. The histogram is a quasielastic model calculation with radiative effects included for comparison.

Table 2. Results of BIGBITE asymmetry measurements.

| Charge | $\theta^\star$ | $\phi^\star$ | A |
|---|---|---|---|
| $\mu$A·hours | degrees | degrees | % |
| 228 | 108.4 | 0 | $+3.2 \pm 2.7$ |
| 336 | 101.4 | 0 | $+2.8 \pm 2.6$ |
| 808 | 78.6 | 180 | $-1.9 \pm 1.7$ |
| 1372 (combined) | | | $+2.38 \pm 1.27 \pm 0.44$ |

Additional corrections to the spin dependent response functions are expected from final state interactions of the knocked-out nucleon as well as meson exhange currents. These should be relatively small corrections, but clearly one needs quantitative calculations. There is a large theoretical effort at present to apply Fadeev techniques to the three body continuum. In fact, first calculations of the quasielastic response in the $A = 3$ system by solving the Fadeev equations in the continuum have recently been published.[8] Therefore, one expects to have reliable predictions for the final state effects soon. The issue of meson exchange currents remains to be examined quantitatively.

## Coincidence Measurements with a Polarized $^3$He Target

The detection of final state hadrons in coincidence with the scattered electron from a polarized target offers a powerful technique for studying the multipole structure of these response functions.[9] Such experiments that use the spin degrees of freedom and correlations of final state particles in coincidence with the scattered electron will fully exploit the polarized, high-intensity, CW beams that will be available in the near future at CEBAF, Bates, and Mainz.

We note that the small amplitudes with the protons in spin $S = 1$ states referred to above can be studied in a rather direct fashion using the $(\vec{e}, e'p)$ reaction on a polarized $^3$He target. (In the plane-wave impulse approximation the quasielastic asymmetry would vanish if the protons are in spin $S = 0$ states only.) These amplitudes are of fundamental interest themselves and can be studied in detail using this technique. This type of study has been proposed at both CEBAF and Bates.[10]

A more exotic component of the $^3$He ground state wave-function consists of the presence of a $\Delta$ with the other two nucleons coupled to $L = 2$, $S = 0$ and $T = 1$. It was noted by Lipkin and Lee [11] that this component would cause a small anomaly in the ratio of $\pi^+$ to $\pi^-$ production. More recently, Milner and Donnelly [12] have shown that the ratio of *asymmetries* (again $\pi^+/\pi^-$) from polarized $^3$He is much more sensitive to this part of the wave-function: the presence of $\Delta$ components in the ground state with probability of 2% would cause changes in this ratio of order factor of 2. This can also be studied at Bates as well as CEBAF.[10]

## Deep Inelastic Scattering

The fundamental structure of hadrons in terms of quarks and gluons was first established in deep inelastic scattering of leptons from nucleon targets. The elucidation of this picture of the nucleon has continued with more detailed experimental study and further analysis grounded in the theory of strong interactions, Quantum Chromodynamics (QCD).

In spin-independent deep inelastic electron (or muon) scattering one measures the structure functions that are independent of $Q^2$ in the high $Q^2$ limit:

$$F_1(x) = \frac{1}{2} \sum_i q_i^2 \left[ f_i^\uparrow(x) + f_i^\downarrow(x) \right], \qquad (5)$$

where $x$ is the momentum fraction of the struck quark, the sum is over quark flavors with $q_i$ being the quark charge, and $f_i^{\uparrow(\downarrow)}(x)$ is the probability of the quark have momentum fraction $x$ and spin parallel (antiparallel) to the nucleon spin, and

$$F_2(x) = 2x F_1(x). \tag{6}$$

(Here and in the following I will only discuss the structure functions in the scaling limit; there are corrections for scaling violations and radiative corrections which I ignore for simplicity.) Using a polarized target and measuring the asymmetry for longitudinally polarized electrons (or muons) incident on a longitudinally polarized target, one can measure the spin-dependent structure function[13]:

$$g_1(x) = \frac{1}{2} \sum_i q_i^2 \left[ f_i^{\uparrow}(x) - f_i^{\downarrow}(x) \right]. \tag{7}$$

By measuring the spin-dependent structure functions of both the proton and neutron, one can test a fundamental prediction of QCD, the Bjorken sum rule[14]:

$$\int_0^1 [g_1^p(x) - g_1^n(x)] dx = \frac{1}{6} g_A, \tag{8}$$

where $g_A = 1.262$ is the axial vector coupling constant of the nucleon measured in neutron beta decay.

Recent $g_1^p$ data from EMC[15] have generated much interest in the spin structure of the proton. Those data (combined with earlier SLAC data[16]) give the result

$$\int_0^1 g_1^p(x)\, dx = 0.126 \pm 0.010 \pm 0.015, \tag{9}$$

where the first error is statistical and the second is systematic. When combined with the value of $g_A$ and information from hyperon beta decays one can use the result of eq. (9) to extract a value for $\Delta s$, the fraction of the proton spin carried by strange quark-antiquark pairs:[15]

$$\Delta s = -0.190 \pm 0.032 \pm 0.046. \tag{10}$$

This is non-zero and surprisingly large. In addition, one can solve for the total fraction of the proton's spin carried by quarks[15] as $(12 \pm 10 \pm 16)\%$, a result often referred to as the "proton spin crisis". Needless to say, these results are the subject of much discussion and the interpretations are somewhat controversial for various reasons. However, there is general agreement that more and better experimental data are desirable for the spin-dependent structure function of the proton. In addition, there is more motivation than ever to study the neutron as well to test the Bjorken sum rule. Various additional experimental efforts are in preparation.

The HERMES proposal is to use thin internal polarized targets in the HERA electron storage ring (35 GeV) at DESY to study these effects with high precision. The proposal is to use pure atomic H, D, and $^3$He targets to study $g_1^p$, $g_1^n$, and other spin dependent structure functions involving other combinations of polarization variables [e.g. $g_2(x)$]. The expected statistical precision for $g_1^p$ is shown in figure 5 along with the much publicized EMC data discussed above. Clearly this type of improvement will bring this subject into a new era of precision.

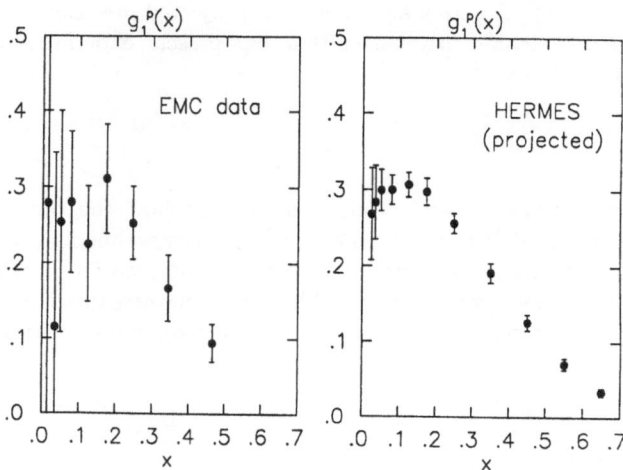

Fig. 5 Results for $g_1^p(x)$ from the EMC experiment[15] along with the projected precision expected for the HERMES experiment. Only statistical errors are shown.

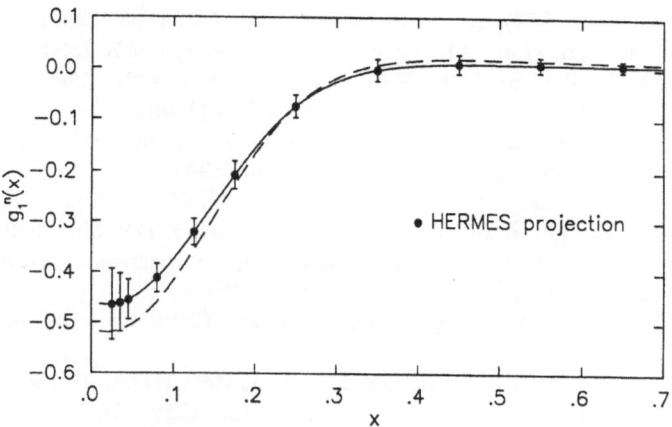

Fig. 6. Projected results for $g_1^n(x)$ from the HERMES experiment along with theoretical predictions of Woloshyn[17] for the neutron (dashed curve) and $^3$He (solid curve). Only the projected statistical errors are shown.

It should be noted that HERMES will study the neutron with both D and $^3$He, so that neutron effects can be separated from nuclear effects with greater confidence. Since one cannot polarize the neutron in D without also polarizing the proton, a large proton subtraction is necessary to extract the neutron structure function. However, as calculated by Woloshyn[17] and shown in figure 6, $g_1$ for $^3$He is a close approximation to the desired quantity $g_1^n$. Figure 6 also shows the projected precision of the HERMES experiment which is significantly better than our present experimental information (see fig. 5) on the proton!

## Conclusion

In this brief survey, I have attempted to show how the present and future applications of polarized $^3$He targets in electronuclear experiments will yield exciting information on a wide variety of physics topics. Clearly, we are just beginning and I expect that as the experiments discussed here are completed and as the techniques improve their will be a next generation of new topics to explore with these powerful tools.

## Acknowledgements

I would like to thank S. Wissink and C. Goodman for the opportunity to present this paper. This work is supported in part by the NSF grant number PHY88-17296.

## References

1. R. G. Milner, R. D. McKeown, and C. E. Woodward, Nucl. Instr. Meth. **A257**, 286 (1987); and Nucl. Instr. Meth. **A274**, 56 (1989).
2. T. E. Chupp, et al., AIP Conference Proceedings **187**, 1320 (1989).
3. B. Blankleider and R. M. Woloshyn, Phys. Rev. **C29**, 538(1984).
4. S. Platchkov, et al., Nucl. Phys. **A510**, 740 (1990).
5. T. W. Donnelly and A. S. Raskin, Annals of Physics **169**, 247(1986).
6. J. L.. Friar, et al., Phys. Rev. **C43**, 2310 (1990).
7. C. E. Woodward, et al., Phys. Rev. Lett. 65, 698 (1990).
8. E. Van Meijgaard and J. A. Tjon, Phys. Lett. **B228**, 307(1989).
9. T.W. Donnelly, Proceedings of Workshop in Electronuclear Physics with Internal Targets, SLAC January 1987, p. 28.
10. CEBAF proposal 89-007, R. D. McKeown, 1989; Bates proposal 89-12, R. G. Milner and J. van den Brand, 1989.
11. H. J. Lipkin and T. -S. H. Lee, Physics Letters **B183**, 22(1987).
12. R. G. Milner and T.W.Donnelly, Phys. Rev. **C37**, 870 (1988)
13. V. W. Hughes and J. Kuti, Ann. Rev. of Nuc. and Part. Sci. **33**, 611(1983).
14. J. D. Bjorken, Phys. Rev. **179**, 1547(1969).
15. J. Ashman, et al., Nucl. Phys. **B328**, 1(1989).
16. G. Baum, et al., Phys. Rev. Lett. **51**, 1135(1983).
17. R. Woloshyn, Nucl. Phys. **A496**, 749(1989).

# HADRONIC REACTIONS IN THE QUASI-ELASTIC PEAK REGION

A. De Pace

Istituto Nazionale di Fisica Nucleare
Sezione di Torino
I-10125 Turin, Italy

## INTRODUCTION

By now, quite a large amount of data has been collected with hadronic probes at intermediate energies. In the Quasi-Elastic Peak (QEP) region, experiments with the inclusive $(p,p')$ reaction[1] have soon been followed by extensive experimental researches by means of charge-exchange reactions: from the $(^3\mathrm{He},t)$[2] and $(d,2p)$[3] experiments at Saturne to the $(p,n)$[4] and $(\pi^\pm,\pi^0)$[5] ones at Los Alamos.

The typical projectile energy in these reactions is in the range from 500 MeV to 1 GeV and the nuclear response has been probed at momentum transfers of roughly $1-3$ fm$^{-1}$. In some of these experiments also the polarization observables have been measured: however, in the following we shall be mainly concerned with a specific feature of the problem, namely the position of the QEP in the double-differential inclusive cross-section $d^2\sigma/d\omega d\Omega$. The QEP position, in fact, had been suggested to be a good indicator of dynamical effects in the nuclear medium[6].

On the experimental side, the situation is rather puzzling. At intermediate energies, the charge-exchange reactions probe essentially the spin-isospin channels and, indeed, they had been proposed as a tool to explore pion correlations in the nuclear medium. However, the experimental outcome depends strongly on the specific probe: in the $(p,n)$ reaction, a hardening, i. e. a shift of the QEP at higher energy (with respect to the quasi-free position $q_\mu^2/2M$), is observed, which is practically independent of the momentum transfer; in the $(^3\mathrm{He},t)$ reaction, on the contrary, there is a smooth transition from a hardening to a softening of the response with increasing momentum transfers.

Since both reactions probe a combination of the isovector spin-longitudinal and spin-transverse channels (although, not at the same nuclear density), it is definitely not easy to ascribe these effects to nuclear correlations. On top of that, one has to cope with the $(p,p')$ data, where the QEP is found at the quasi-free position.

The theoretical treatment of hadronic reactions at intermediate energies is usually based on two ingredients: a model for the reaction mechanism, either Glauber theory[7-11] or Distorted Wave Born Approximation[12] (DWBA), and a framework for calculating the nuclear response, either bound state[7,8,10,11] or continuum[9,12] Random Phase Approximation (RPA). In the following, we present the results of a systematic analysis of the above mentioned reactions. Our model is based on RPA and Glauber theory: we have preferred the latter to DWBA, since it provides a well-defined framework for calculating multi-step processes, although, at the moment, only one-step contributions have been estimated.

*Spin and Isospin in Nuclear Interactions*
Edited by S.W. Wissink *et al.*, Plenum Press, New York, 1991

## NUCLEAR RESPONSES WITH HADRONIC PROBES

The nuclear response function to an external probe is obtained through the imaginary part of the polarization propagator

$$\Pi(\boldsymbol{q}, \boldsymbol{q}'; \omega) = \sum_{n \neq 0} <\psi_0|\hat{O}(\boldsymbol{q})|\psi_n><\psi_n|\hat{O}^\dagger(\boldsymbol{q}')|\psi_0>$$

$$\times \left[ \frac{1}{\hbar\omega - (E_n - E_0) + i\eta} - \frac{1}{\hbar\omega + (E_n - E_0) - i\eta} \right], \qquad (1)$$

where the sum is extended to the excited states of the nuclear Hamiltonian and $\hat{O}(\boldsymbol{q})$ is the vertex operator.

If, for definiteness, we consider the *longitudinal* $[O_L \propto \tau_a l(\boldsymbol{\sigma} \cdot \hat{\boldsymbol{q}})]$ and *transverse* $[O_T \propto \tau_a l(\boldsymbol{\sigma} \times \hat{\boldsymbol{q}})]$ spin-isospin operators, then we can define the following response functions

$$R_{L,T}(q, \omega) \propto \text{Im} \sum_J (2J+1) \Pi_{J(L,T)}(q, q; \omega). \qquad (2)$$

The $J$th multipoles $\Pi_{J(L)}$ and $\Pi_{J(T)}$ can be evaluated in RPA by solving a unique set of coupled integral equations[13]: in the calculation we use the standard particle-hole (ph) interaction "$g' + \pi + \rho$", which consists of pion and rho exchange plus the Landau-Migdal parameter $g'$.

The distortion of the impinging probe is treated, within the Glauber theory at the one-step level, by replacing the vertex operators $O_{L,T}$ with

$$O_{L,T}(\boldsymbol{q}, \boldsymbol{r}) \longrightarrow \frac{1}{(2\pi)^2 f_{L,T}(q)} \int d^{(2)}\boldsymbol{b}\, d^{(2)}\boldsymbol{\lambda}\, e^{i\chi_{opt}(b)}\, e^{i(\boldsymbol{q}-\boldsymbol{\lambda})\cdot\boldsymbol{b}} f_{L,T}(\lambda)\, O_{L,T}(\boldsymbol{\lambda}, \boldsymbol{r}). \qquad (3)$$

Formally, (3) corresponds to the spectral decomposition of the vertex operators in the bidimensional impact parameter $\boldsymbol{b}$ and momentum $\boldsymbol{\lambda}$ transferred in a single scattering process, weighted with the isovector spin-longitudinal and spin-transverse nucleon-nucleon (NN) scattering amplitudes $f_L(\lambda)$ and $f_T(\lambda)$. The quantity

$$\chi_{opt}(b) = \frac{2\pi}{k} A \tilde{f}(0) \int_{-\infty}^{+\infty} dz\, \rho(r = \sqrt{b^2 + z^2}) \qquad (4)$$

accounts for the distortion of the probe, $\rho(r)$ being the nuclear density, $k$ the projectile wave number pointing in the $z$ direction, $A$ the nuclear mass number and $\tilde{f}(0)$ the forward total projectile-nucleon scattering amplitude.

Inserting (3) in the general expression (1) and performing a multipole expansion, one finds that the surface nuclear responses $R_{L,T}^{surf}(q, \omega)$ can be obtained by replacing $\Pi_{J(L)}$ and $\Pi_{J(T)}$ in eq. (2) with

$$\Pi_{J(L)}^{surf}(q, q; \omega) = \Pi_{J(L)}(q, q; \omega)$$

$$+ \frac{1}{|f_L(q)|^2} \int_0^\infty d\lambda\, \lambda \int_0^\infty d\lambda'\, \lambda'\, \text{Re}[f_L^*(\lambda) f_L(\lambda')\, \tilde{G}_J(\lambda, \lambda'; q)] \Pi_{J(L)}(\lambda, \lambda'; \omega)$$

$$- 2\frac{1}{|f_L(q)|^2} \int_0^\infty d\lambda\, \lambda\, \text{Re}[f_L^*(q) f_L(\lambda)\, \tilde{H}_J(\lambda; q)] \Pi_{J(L)}(q, \lambda; \omega) \qquad (5a)$$

and

$$\Pi^{surf}_{J(T)}(q,q;\omega) = \Pi_{J(T)}(q,q;\omega)$$

$$+ \frac{1}{|f_T(q)|^2} \int_0^\infty d\lambda\,\lambda \int_0^\infty d\lambda'\,\lambda' \sum_{J'} \mathrm{Re}[f_T^*(\lambda)\,f_T(\lambda')\,\widetilde{G}_{J'}(\lambda,\lambda';q)]\Pi_{JJ'}(\lambda,\lambda';\omega)$$

$$- 2\frac{1}{|f_T(q)|^2} \int_0^\infty d\lambda\,\lambda \sum_{J'} \mathrm{Re}[f_T^*(q)\,f_T(\lambda)\,\widetilde{H}_{J'}(\lambda;q)]\Pi_{JJ'}(q,\lambda;\omega). \qquad (5b)$$

In eq.s (5) we have set

$$\widetilde{G}_J(\lambda,\lambda';q) = \sum_{lm} c_{Jlm}\,\widetilde{g}_m^*(\lambda,q)\,\widetilde{g}_m(\lambda',q), \qquad \widetilde{H}_J(\lambda;q) = \sum_{lm} c_{Jlm}\widetilde{g}_m(\lambda,q), \qquad (6)$$

where

$$\widetilde{g}_m(\lambda,q) = \int_0^\infty db\,b\,\{1 - \exp[i\chi_{opt}(b)]\}\,J_m(\lambda b)\,J_m(qb) \qquad (7)$$

and

$$c_{Jlm} = I_{l+m}\,a_{Jl}^2\,\frac{(l-m-1)!!(l+m-1)!!}{(l+m)!!(l-m)!!}, \quad \text{with } I_{l+m} = \begin{cases} 0, & l+m \quad \text{odd} \\ 1, & l+m \quad \text{even} \end{cases}$$

$$a_{Jl} = (-1)^l\sqrt{2l+1}\begin{pmatrix} l & 1 & J \\ 0 & 0 & 0 \end{pmatrix}. \qquad (8)$$

We refer the reader to ref. 10 for the definition of $\Pi_{JJ'}$ in (5b). As mentioned above, the distortion is taken into account by the phase shift function $\chi_{opt}$: at high energies it is purely imaginary and proportional to the total effective projectile-nucleon cross-section (which may be different from the free one because of Pauli blocking effects) $\chi_{opt} \propto i\widetilde{\sigma}_{tot}$.


## DISTORTION OF THE QUASI-ELASTIC PEAK

We have calculated[11] the response functions for momenta between 1 and 3 $fm^{-1}$ and for values of $\widetilde{\sigma}_{tot}$ of 30 mb (suitable for protons at $\sim$ 500 MeV), 55 mb (suitable for $^3$He at 2 GeV) and 90 mb. In Fig. 1, we show the calculated shift, $\Delta E_{MAX}$, of the QEP with respect to the non-relativistic free value $q^2/2M$, for the longitudinal and transverse spin-isospin responses. The results for the free and RPA (with $g' = 0.7$) cases are shown separately.

One notes a pattern that is rather independent of the spin channel, i. e. a positive shift in the low $q$ region and a negative shift at higher momenta, with quite a sudden jump from hardening to softening around $1.5 - 2$ $fm^{-1}$; this effect saturates with increasing $\widetilde{\sigma}_{tot}$. This pattern is not modified by RPA correlations, even in the transverse channel, where the strongly repulsive ph interaction might be expected to produce a sizable hardening. This happens because of the combined effects of low density (which implies small collectivity) and high momenta (which imply a decreasing ph interaction).

The behaviour of Fig. 1 is also independent of the specific channel: it is the same, for instance, in the scalar-isoscalar channel, the one that dominates $(p,p')$ scattering at high energies. It depends only on the cilindrical geometry introduced by $\chi_{opt}(b)$, which modifies the relative weight of the partial wave contributions to the response.

Thus, if one-step inelastic scattering and RPA were a good model for hadronic reactions, one should observe the above mentioned pattern in all experiments. As already remerked in the Introduction, this is not the case and in Fig. 2 we show a comparison of the experimental $\Delta E_{MAX}$[1,4] with our calculations for three different

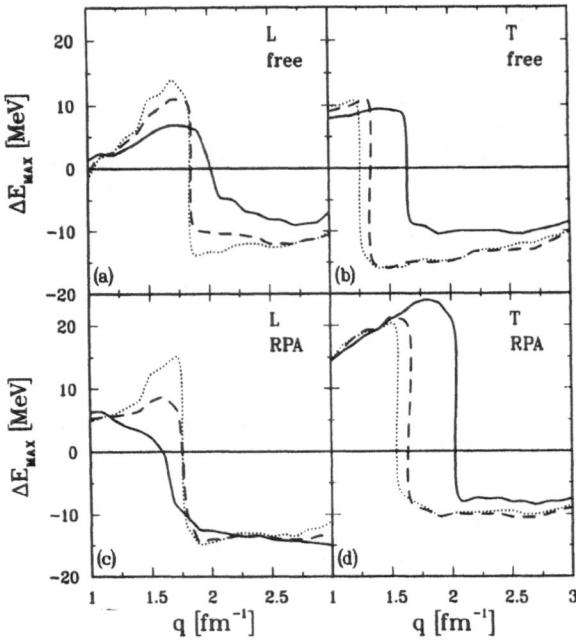

Fig. 1. $\Delta E_{MAX}$ as a function of the transferred momentum $q$ for the spin-longitudinal (L) and spin-transverse (T) isovector responses, both free and RPA; $\tilde{\sigma}_{tot}=30$ mb (solid line), $\tilde{\sigma}_{tot}=55$ mb (dashed line) and $\tilde{\sigma}_{tot}=90$ mb (dotted line).

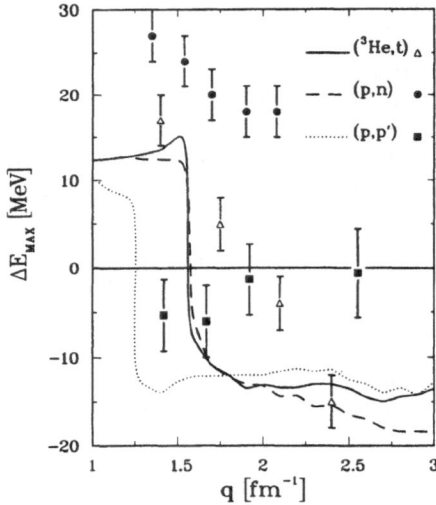

Fig. 2. $\Delta E_{MAX}$ as a function of the transferred momentum $q$ for the cross-sections of three hadronic reactions: $(^3He,t)$ at 2 GeV ($\tilde{\sigma}_{tot} = 55$ mb), $(p,n)$ at 500 MeV ($\tilde{\sigma}_{tot} = 30$ mb) and $(p,p')$ at 800 MeV ($\tilde{\sigma}_{tot} = 40$ mb).

reactions. The trend of the data is reproduced only for the ($^3$He,t) reaction, which is the most peripheral: in this case, collectivity should play a minor role and a quasi-free model, with the distortion properly accounted for, may be expected to work well. In the $(p,n)$ and $(p,p')$ reactions, RPA effects are bigger, but still not enough to change the quasi-free behaviour. Actually, the strong hardening observed in the $(p,n)$ experiment cannot be reached even with a more strongly repulsive ph interaction $(g' \sim 0.9)$[11].

A possible way out might be provided on the one hand by multi-step contributions, which may be expected to be important at high momenta; on the other hand, by 2p-2h ground state correlations, which have been shown to play a crucial role in electron scattering[14]. Both of these contributions harden the response and are strongly density-dependent: hence, they should affect sizably the $(p,n)$ reaction, but only mildly the ($^3$He,t) one.

Multi-step processes and 2p-2h ground state correlations are present also in $(p,p')$ scattering: here, however, their effect might be contrasted by the contribution of another class of 2p-2h correlations, which has been shown, for $(e,e')$ scattering, to be sizable and attractive in the isoscalar channel and much smaller in the isovector one[15]. If this were the case, we would have a unified picture of intermediate energy hadronic reactions, although much more complex than expected, and a powerful tool to explore the nuclear many-body problem.

# REFERENCES

1. R. E. Chrien et al., Proton spectra from 800 MeV protons on selected nuclides, *Phys. Rev. C* 21:1014 (1980).
2. I. Bergqvist *et al.*, The ($^3$He,t) reaction at intermediate energies, *Nucl. Phys. A* 469:648 (1987).
3. T. Sams, "The nuclear spin isospin response", thesis, Niels Bohr Institute, Copenhagen (1990).
4. C. Gaarde, Spin-isospin excitations. *Nucl. Phys. A* 507:79c (1990).
5. S. Høibraten et al., Nuclear isovector response from quasi-free pion single-charge exchange at 500 MeV, unpublished (1991).
6. W.M. Alberico, M. Ericson and A. Molinari, Quenching and hardening in the transverse quasi-elastic peak, *Nucl. Phys. A* 379:429 (1982).
7. Y. Okuhara. B. Castel, I.P. Johnstone and H. Toki, Nuclear spin response to inelastic proton scattering: finite geometry and absorption effects, *Phys. Lett. B* 186:113 (1987).
8. W.M. Alberico, A. De Pace, M. Ericson, M.B. Johnson and A. Molinari, Spin-isospin nuclear responses with hadronic probes, *Phys. Rev. C* 38:109 (1988).
9. T. Shigehara, K. Shimizu and A. Arima, The isovector spin response in a finite nucleus. *Nucl. Phys. A* 477:583 (1988).
10. A. De Pace and M. Viviani, Spin-isospin hadronic response in the Glauber theory, *Phys. Lett. B* 236:397 (1990).
11. A. De Pace and M. Viviani, Distortion of the quasi-elastic peak in hadron scattering, *Phys. Lett. B* 254:20 (1991).
12. M. Ichimura, K. Kawahigashi, T.S. Jørgensen and C. Gaarde, Excitation of spin-isospin modes in the quasi-free scattering region, *Phys. Rev.* C39:1446 (1989).
13. W.M. Alberico, A. De Pace and A. Molinari, Spin isovector responses in finite nuclei, *Phys. Rev.* C31:2007 (1985).
14. W.M. Alberico, M. Ericson and A. Molinari, The role of 2p-2h excitations in the spin-isospin nuclear response, *Ann. Phys.* (N.Y.) 154:356 (1984).
15. W.M. Alberico, R. Cenni, A. Molinari and P. Saracco, Dynamical pion propagation in the functional approach to the charge response, *Phys. Rev. Lett.* 65:1845 (1990).

# Δ EXCITATIONS IN NUCLEI

T. Udagawa

Department of Physics
University of Texas
Austin, Texas 78712


P. Oltmanns and F. Osterfeld

Institut für Kernphsik
Forschungszentrum
D-5170 Jülich, Germany

It has been known for sometime that the peak position of the $\Delta$ (1232) -resonance spectra observed in the (p,n)-[1], ($^3$He,t)-[2], and ($\vec{d}$,2p)-[3] charge-exchange reactions at intermediate energies is systematically shifted downward for a target with $A \geq 12$ as compared to the peak position for a proton target. In contrast to this, in the case of $\gamma$-absorption[4] and inelastic electron scattering experiments,[5] the $\Delta$-peak does not show such a pronounced displacement. The electromagnetic probes excite the $\Delta$ transversely, i.e., by the transition operator $\vec{S} \times \vec{q}\,\vec{T}$ ($\vec{S}$ and $\vec{T}$ are the spin and isospin transition operators, respectively.), while the hadronic probes measure both the transverse (TR) and the longitudinal (LO) spin-isospin response. It has thus been speculated that the shift of the $\Delta$-peak would be due to a nuclear medium effect in the isovector spin LO ($\vec{S} \cdot \vec{q}\,\vec{T}$) channel.[6,7] That is, if the delta particle-nucleon hole ($\Delta N^{-1}$) interaction becomes strongly attractive at large momentum transfers $|\vec{q}| \approx 1-2$ fm$^{-1}$ in this channel, then this attraction might lead to a lowering of the $\Delta$ mass produced in the target. Along this line of reasoning, no shift of the $\Delta$-peak position is to be observed with the electromagnetic probes. Recently, we have made some realistic calculations of the spectra of the $^{12}$C(p,n) and ($^3$He,t) reactions at the $\Delta$-excitation region, and succeeded to show that the LO response is indeed shifted downwards in energy and that this shift is caused by the energy-dependent $\pi$-exchange interaction.[8]

The calculations were based on the isobar model[9-11] and the distorted wave impulse approximation (DWIA), assuming a $\pi + \rho + g'$ model[12] for the $\Delta N^{-1}$ interaction and essentially a $\delta$-function type $t_{NNN\Delta}$-amplitude for the external force field applied to the target in the (p,n) and ($^3$He,t) reaction processes. A one body complex mean field for the excited $\Delta$ was also taken into account. The calculations thus properly included the nuclear medium effects on the $\Delta$ and the distortion effects on the projectile wave functions.

The aim of this report is to present some of the developments made after Ref. 8 was published. The first point concerns the spin- and momentum-dependence of the $t_{NN\Delta}$-amplitude, for which we made in Ref. 8 the following simple ansazt (in the momentum representation):

$$t_{NN,N\Delta} = g'_{N\Delta} J_{\pi N\Delta} \left( \frac{\Lambda'_{\pi 2} - m_{\pi 2}}{\Lambda'_{\pi 2} - t} \right)^2 \left[ (\vec{\sigma}_1 \cdot \hat{q})(\vec{S}_2^\dagger \cdot \hat{q}) + (\vec{\sigma}_1 \times \hat{q}) \cdot (\vec{S}_2^\dagger \times \hat{q}) \right] \vec{\tau}_1 \cdot \vec{T}_2^\dagger \quad (1)$$

with $J_{\pi N\Delta} = 4\pi\hbar c f_{\pi NN} f_{\pi N\Delta}/m_\pi^2 \approx 800$ MeV fm$^3$, $g'_{N\Delta} = 0.335$ and $\Lambda'_\pi = 650$ MeV. We made this choice since studies of the $t_{NN,N\Delta}$-interaction by means of the ($\vec{d}$,2p)-reaction indicated[13] that there is a strong TR component in $t_{NN,N\Delta}$ with the ratio TR/LO=2 and that the momentum-dependence is rather weak. The $t_{NN,N\Delta}$-amplitude assumed may be tested in a more quantative manner by comparing theoretical predictions of the tensor analyzing power with experimental data. We did such a test recently[14] for the $p(\vec{d},2p)\Delta^0$ reaction data with $E_{lab} = 2\,GeV$ taken at various angles. Fig. 1a and 1b

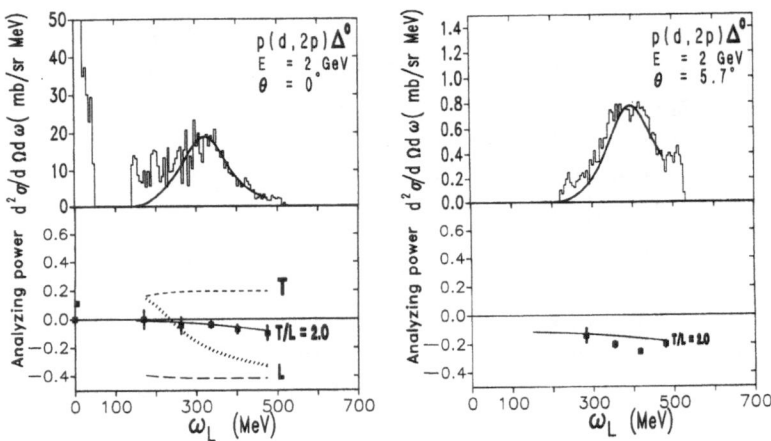

Fig. 1 Comparison of the calculated cross section and tensor analyzing power for the $p(\vec{d},2p)$ reaction at 2 GeV.

show the results for $\theta = 0^0$ and $5.7^0$, respectively. Fig. 1 includes the results for the cross section also. As seen, both data of the tensor analyzing power and the cross section are well reproduced by the calculations. It should be emphasized that if one assumes the $\pi + \rho + g'$ model for the $t_{NN,N\Delta}$-amplitude[15], the fit obtained to the tensor polarization data becomes poor as illustrated by the dotted line in Fig. 1a. The short (long broken) line shown there are theoretical predictions obtained when only the LO (TR) component of $t_{NN,N\Delta}$ is taken into account in the calculation.

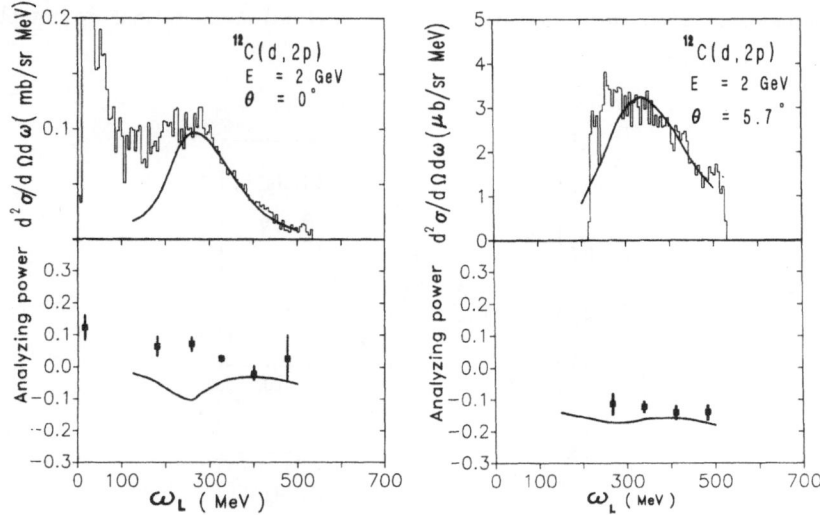

Fig. 2 Comparison of the calculated cross section and tensor analyzing power for the $^{12}C(\vec{d}, 2p)$ reaction at 2 GeV.

The calculations of the tensor analyzing power as well as the cross section have also been made for the nuclear target case, i.e., for the $^{12}C(\vec{d}, 2p)$ reaction at $E_{lab} = 2\ GeV$. We used the same $\Delta$-hole interaction as employed in Ref. 8. The results are presented in Fig. 2. The fit of the calculated results with experiment is good except at the lower excitation energies at $\theta = 0^0$. There the fit is not good in both cross section and tensor analyzing power. Similar failures were also observed in the results of the previous analyses of the $(^3He, t)$ and $(p, n)$ reaction cross section data[8].

We may ascribe these failures to "background" components coming from nucleon knockout processes that are not taken into account in the calculation. We may expect that a better fit would be obtained if one subtracts from the experimental data the background. We have attempted[16] such subtraction of the background to generate the background free data for the $^{12}C(^3He, t)$ reaction at $\theta = 0^0$ for three different incident energies of $E_{lab}$=1.5, 2.0 and 2.3 GeV. Fig. 3a shows experimental $0^o$-spectra at these energies. As seen, the cross sections in the $\Delta$-excitation region increase rather dramatically as $E_{lab}$ increases. In contrast, the cross sections below $\omega \approx 150$ MeV remain essentially unchanged. This indicates that the excitation mechanisms involved in the spectra below and above $\omega \approx 150$ MeV are different. Naturally, we may ascribe the mechanism involved in $\omega < 150$ MeV to the nucleon knockout processes. This is further supported by the fact that, while the magnitudes of the observed cross sections in the

Δ-excitation region are proportional[2] to $(3Z + N)$ (where $Z$ and $N$ are the proton and neutron number of the target), those below $\omega \approx 150$ MeV depend on the mass number A. The $(3Z+N)$-dependence is a characteristic feature of the Δ-excitation coming from the fact that the cross section of the $p + p \rightarrow n + \Delta^{++}$ process is three times stronger than the $p + n \rightarrow n + \Delta^{+}$ process. Such a characteristic feature is not seen in the cross sections below $\omega \approx 150$ MeV, indicating that the (virtual) Δ-excitations do not play a significant role there.

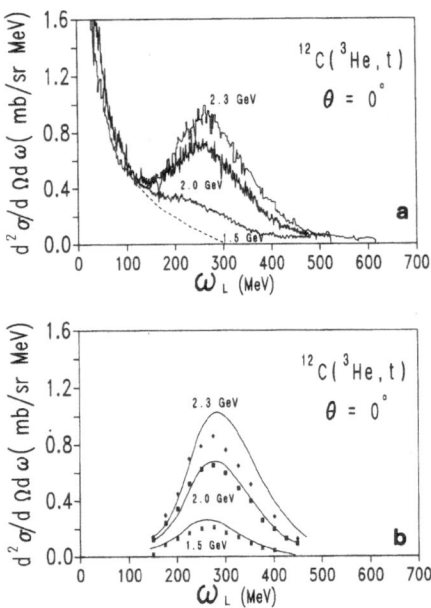

Fig. 3. Spectra at $0^{o}$ from the $(^{3}He, t)$ reaction on carbon at 1.5, 2.0 and 2.3 GeV incident energies; a) experimental data; b) background subtracted data in comparison with calculations.

The tail of the nucleon knockout (background) spectra may be present in the region higher than $\omega \approx 150$ MeV, i.e., into the Δ-excitation region. This means that even spectra for $\omega > 150$ MeV includes a certain amount of background, which we estimated by assuming that the background spectra are the same for all three incident energies, and further by requiring that the extracted Δ-resonance spectra at 1.5 GeV agree reasonably well with the calculated spectra. The background spectra thus estimated are shown in Fig. 3a by the broken line. The Δ-excitation spectra obtained by subtracting the above estimated background are presented and compared with the calculated spectra in Fig. 3b. The calculated spectra now fit the data rather well. The fit is much better than that obtained before in Ref. 8. At the same time, we are now able to reproduce the data at three different energies reasonably well.

In summary, we have extended the calculation of the Δ-excitation spectra of the $^{12}C(p, n)$ and $(^{3}He, t)$ reactions made earlier in Ref. 8 for only one angle and a fixed incident energy to other angles and energies, and also to the $(\vec{d}, 2p)$ reaction. The results show that the simple $t_{NN,N\Delta}$ amplitude used in Ref. 8 seems to be justified and reproduce the observed data of both cross section and tensor analyzing power fairly well except those at very small angle. It has been demonstrated then that the fit at very small angles can be improved if the background due to the nucleon knockout process is subtrated from the measured data.

ACKNOWLEDGMENT

The work is supported in part by the U.S. Department of Energy.

# References

[1] D. A. Lind, Can. J. Phys. **65**, 637 (1987).

[2] D. Contardo et al., Phys. Lett. **168B**, 331 (1986).

[3] C. Ellegaard et al., Phys. Lett. **154B**, 110 (1985).

[4] B. Mecking, Proc. Int. Conf. on Nucl. Phys. with Electromagnetic Interactions, Mainz 1979, Lecture Notes in Phys. **108** 382 (1979).

[5] P. Barreau et al., Nucl. Phys. **A402**, 515 (1983).

[6] J. S. O'Connell et al., Phys. Rev. Lett. **53**, 1627 (1984).

[7] G. Chanfray and M. Ericson, Phys. Lett. **141B**, 163 (1984).

[8] T. Udagawa, S.-W. Hong, and F. Osterfeld, Phys. Lett. **245B**, 1 (1990).

[9] M. Hirata, J. H. Koch, F. Lenz and E. J. Moniz, Phys. Lett. **70B**, 281 (1977); Ann. of Phys. **120**, 205 (1979).

[10] E. Oset, H. Toki and W. Weise, Phys. Rep. **83**, 281 (1982).

[11] J. H. Koch, E. J. Moniz and N. Ohtsuka, Ann. of Phys. **154**, 99 (1984).

[12] M. R. Anastasio and G. E. Brown, Nucl. Phys. **A285**, 516 (1977).

[13] C. Gaarde, Private communication.

[14] P. Oltmanns, F. Osterfeld, T. Udagawa and S.-W. Hong, to be published.

[15] S. Mundigl and W. Weise, Phys. Rev. **C39**, 710 (1989).

[16] T. Udagawa, F. Osterfeld, S.-W. Hong and P. Oltmanns, to be published.

# GLUONS, SPIN AND FLAVOUR IN THE *LEP*
# (LOW ENERGY PROTON)

F.E. Close

Rutherford Appleton Laboratory
Chilton, Didcot, Oxon, OX11 0QX, England

## ABSTRACT

This talk discusses the rôle of gluons in the proton at low energies. Spin dependent effects, magnetic moments and the possible excitation of hybrid (gluonic) baryons are considered. The relation between a low energy electron program (such as at CEBAF) and deep inelastic polarisation is examined.

## INTRODUCTION

The remarkable results from EMC on deep inelastic polarised leptoproduction have generated intense interest on the spin content of the proton. Londergan[1] here has reported how polarised gluons may be responsible for "hiding" the intrinsic quarks' spin polarisations. In my talk I shall consider the impact of these ideas on low energy observables and contemplate the rôle, or evidence, for gluons in the "low energy" proton.

There are three parts to this. First there is the proton of the naive constituent quark model (NCQM) - three quarks whose flavours and spins are correlated by the Pauli principle. This wavefunction describes the ratio of magnetic moments ($\mu_n/\mu_p$) well - how does this survive? Second, gluons implicitly play a rôle at $O(\alpha_s)$ in $pQCD$ through their exchange, mediating a $\vec{S}.\vec{S}$ "hyperfine" spin dependent energy shift. This splits the masses of $\Delta$ and $N$, may cause the transition form factor for $\Delta$ excitation to fall faster with $Q^2$ than does the elastic $G_M(Q^2)$ and, via Pauli, distorts the $x$ dependence of the parton distributions $u(x)/d(x)$ at $x \to 1$ as revealed in deep inelastic scattering.

The possibility that there is polarised glue in the (deep inelastic) proton, $\Delta G \neq 0$, raises a question as to whether the static properties, such as magnetic moments, may need reconsideration. And in turn, there is a question of how much of a dynamical rôle gluons play - are there "hybrid baryons" where both quark and gluonic degrees of freedom are excited?

## MAGNETIC MOMENTS

Define $\Delta q \equiv \int dx \Delta q(x)$ whose $\Delta q(x) \equiv q^\uparrow(x) - q^\downarrow(x)$ is the spin polarisation distribution of quarks in a polarised proton. Defining $\Delta G$ similarly (see Londergan's talk) the deep inelastic polarised experiment measures $\Delta q' \equiv \Delta q - \frac{\alpha}{2\pi}\Delta G$. The EMC results are

$$\Delta u' \simeq 0.78$$
$$\Delta d' \simeq -0.48$$
$$\Delta s' \simeq -0.18$$
$$\Delta q' = \sum_i \Delta'_{q_2} = 0.13 \pm 0.10 \pm 0.15$$

In an attempt to make contact with one's prejudices about the spin polarisation of constituent quarks (for which $\Delta S = 0$), suppose that $-\frac{\alpha}{2\pi}\Delta G \equiv \Delta S' = -0.18$. Thus one would deduce

$$\Delta u = 0.96 \pm 0.07$$
$$\Delta d = -0.30 \pm 0.07$$

These results are rather different than those of the NCQM but, nonetheless, are quite consistent with the ratio of magnetic moments as we shall now demonstrate.

The wavefunction for the VNP (very naive proton) is formed by noting that two identical quarks (the $uu$) will be in net $S = 1$ by Pauli (being color antisymmetric). Thus from the Clebsch's

$$p^\uparrow = \sqrt{\frac{2}{3}}(u^\uparrow u^\uparrow)d_\downarrow + \sqrt{\frac{1}{3}}(u^\uparrow u^\downarrow)d_\uparrow$$

and so the spin weighted probabilities are

$$\Delta u = \frac{4}{3}, \qquad \Delta d = -\frac{1}{3};$$

Note that $\Delta u + \Delta d = 1$, confirming the fact that for this wavefunction the quarks carry the entire $J_z$ of the proton, and $\Delta u - \Delta d (\equiv \frac{g_A}{g_V}) = \frac{5}{3}$. Notice also that $\Delta u = -4\Delta d$. If we ignore $L_z$ and consider only valence quarks, the ratio of magnetic moments may be written

$$\frac{\mu_p}{\mu_n} = \frac{\frac{2}{3}\Delta u - \frac{1}{3}\Delta d}{\frac{2}{3}\Delta d - \frac{1}{3}\Delta u} = -\frac{3}{2}$$

where the empirically satisfactory result has followed by inputting the $\Delta u, \Delta d$ above. However, we can turn this around. Given the result $-\frac{3}{2}$ for the *ratio*, all that is constrained is the *ratio* of $\frac{\Delta u}{\Delta d}$. The magnetic moments constrain

$$\Delta u = -4\Delta d$$

This is true for the VNP but is true for many other wavefunctions including some with constituent gluons. Notice also that the $\frac{\Delta u}{\Delta d}$ extracted from the EMC deep inelastic measurement are consistent with this ratio. There is a certain measure of consistency *if* we insist upon polarised gluons.

Among recent interest in including (polarised) gluons in static proton wavefunctions there is the work of Lipkin[2]. He considers three quarks with $S = \frac{1}{2}$ coupled with a gluon to form overall $J = \frac{1}{2}$.

Eight years ago Wagner[3] and also Barnes and Close[3] noted that a particular coherent superposition of states - $3q$ in $S = \frac{1}{2}$ *and* $S = \frac{3}{2}$ coupled to a gluon -

preserves $\mu_n/\mu_p(\Delta u/\Delta d = -4)$. As far as ratios of static properties are concerned this wavefunction is indistinguishable from the VNP (physically, if one radiates a single gluon additively from each quark in VNP, one obtains the above wavefunction). I have not investigated what is the most general $3q+$ glue configuration that preserves $\Delta u/\Delta d$.

These results may suggest that we understand the static wavefunction less than we thought - the VNP may well be an oversimplified description. This suggests the following question: what rules out (or "rules in") "dynamical glue" in the low energy proton, and are there dynamical gluonic hybrids (as suggested in the MIT bag model)? If so, can we hope to identify them in $\gamma N \to N^*g$ at low energies?

## GLUONIC BARYONS (very model dependent)

There is rather general expectation that the lightest glueballs should exist with masses in the 1-2 GeV region, and in turn this raises the question as to whether the gluonic degrees of freedom can be excited in a system of quarks but "independent" of these quarks. These states are known as hybrids or gluonic hadrons. Specific models predict their existence as low as 0.5 - 1 GeV above the mass of the lightest conventional quark - hadrons and so one might hope that the lightest gluonic baryons are below 2 GeV in mass, certainly accessible to excitation at low energy facilities such as CEBAF.

There are some tantalising theoretical questions about the interpretation and existence of these states. One suggestion has been that these states are dual to, or even misidentified as, the well known states that are traditionally interpreted as orbital and radial excitations in the constituent quark model. For example adding a gluon with $J^P = 1^-$ to these quarks generates the same set of $J^P$ quantum numbers as if a quark had been excited to $L = 1$. In the latter case the spatial symmetry forces the three quarks to be in $\underset{\sim}{70}$ of SU(6) spin-flavour; in the gluonic case the colour $\underset{\sim}{8}$ of the three quarks, with Pauli, generates the $\underset{\sim}{70}$ too.

Detailed study of the spectroscopy and excitations can distinguish between these two pictures. First, the two spectroscopies are not identical: the radial excitations of the NRQM do not precisely correlate with gluonic excitations. And where they do appear to overlap, the spin dependent mass shifts at $O(\alpha_s)$ in perturbative QCD have quite different patterns. For example, consider the lightest hybrids which, according to the bag model, could be an $J^P = \frac{1}{2}^+, I = \frac{1}{2}$ state (e.g. the Roper resonance $P_{11}(1470)$). If the Roper resonance is a radially excited state then it will be partnered by a $\Delta$ state analogous to the way that $\Delta(1232)$ partners the nucleon; there is a possible candidate for such a state but its existence is by no means established. Contrast this with the hybrid spectroscopy[4] where the $\frac{1}{2}^+(Ng)$ is alone, the next heaviest being another $\frac{1}{2}^+$ and $\frac{3}{2}^+(Ng)$ but with $\Delta g$ states pushed up to high masses by the $O(\alpha_s)$ mass shifts. Thus the presence or absence of the $\Delta(1630)$ would respectively eliminate or support $N(1470)$ as a hybrid.

Photo and electroproduction of the resonances also helps to probe the internal constitution of the hadrons. The $N(1470)$ and $\Delta(1630)$ will be excited by M1 radiation with characteristic strengths from proton and neutron targets if they are radial excitations of $N$ and $\Delta$. However, there is a selection rule that the lightest hybrid (the $\frac{1}{2}^+ Ng$) may be photoexcited from neutrons but not from protons[3]. This makes it seem unlikely that $N(1470)$ is a hybrid as it is excited from both $n$ and $p$ with relative amplitudes consistent with the ratio $-\frac{2}{3}$ as befits the ratio of magnetic moments.

There is a possibility that the "Roper" resonance actually consists of two nearly degenerate states in which case one could be the hybrid. Photoexcitations from protons and comparison with $\pi^- p \rightarrow N^* \rightarrow n\gamma$ may help to settle this issue.

If the $N(1470 \neq Ng$ then the next candidate would be the $N(1710)$. This has no clear $\Delta$ partner and also is consistent with zero photoproduction from protons. One possibility is that this zero (if indeed it is zero) is a result of an "accidental" cancellation between competing electric and magnetic multipoles (e.g. as in a NRQM assignment); if this is so, one would expect that the cancellation does not survive as one varies $Q^2$. Hence it will be important to study the photo and electroproduction of the $N(1710)$: a hybrid will have a vanishing excitation amplitude (from protons) at all $Q^2$ whereas a NRQM excitation will be expected to show a $Q^2$ dependent production[5].

## THE CONNECTION BETWEEN LOW ENERGY AND DEEP INELASTIC

The $O(\alpha_s)$ single gluon exchange gives spin dependent effects and, via the Pauli principle, flavour dependent effects. Its direct spin dependence elevates the mass of the $J = \frac{3}{2}\Delta(1232)$ relative to the $J = \frac{1}{2}$ nucleon. This also feeds more "energy" or, in the light cone momentum distributions a larger $< x >$, into quarks that are correlated with $S = 1$ relative to $S = 0$. The results is that as $x \rightarrow 1$ in deep inelastic polarised leptoproduction the polarised quark tends to spin parallel to the target and hence the polarisation asymmetry $A(x) \equiv g_1(x)/F_1(x) \rightarrow 1$ for both proton and neutron targets. Such predictions are consistent with the trend of the data for protons and we eagerly await the data for neutrons.

Combining the above with the Pauli principle implies that $< x >_u > < x >_d$ and hence $u(x)$ dominates over $d(x)$ as $x \rightarrow 1$. This is qualitatively seen in data though there is still an open question as to whether $F_1^n/F_1^p(x \rightarrow 1) \rightarrow \frac{3}{7}$ or $\frac{1}{4}$. In the bag model it has proven possible to correlate these two very different regions of physics: setting the $O(\alpha_s)$ strength by the $\Delta - N$ mass splitting, the $x$ dependence of deep inelastic scattering, both spin dependent and spin independent, is rather well described[6].

The $N - \Delta$ mass difference also leads to interesting flavour dependent effects in the sea, namely that $\bar{u}(x) \neq \bar{d}(x)$ in general.

At some level there is a pion cloud in the nucleon[7] and this will contribute $\bar{q}$ to the sea. Indeed, kinematically, it is at $x \simeq M_\pi/M_N \simeq 0.15$ that this begins to be felt. It gives a flavour dependent distortion because

$$p \rightarrow \pi^+(\bar{d})n; p \rightarrow \pi^-(\bar{u})\Delta^{++}$$

and $m_\Delta > M_N$ pushes the $\bar{u}$ to smaller values of $x(x \lesssim 0.05)$ than is the case for $\bar{d}$ (the $\pi^0 p$ cloud feeds both $\bar{u}$ and $\bar{d}$ and dilutes the effect but does not eliminate it). So there is a window centred on the range $0.05 \lesssim x \lesssim 0.15$ where $\bar{d}(x) > \bar{u}(x)$ (this will be spread over a larger $x$ range in practice due to $\pi NN$ form factors, see e.g. ref 8).

## QUARKS PRODUCING RESONANCES

The relation between these two apparently disparate regimes of deep inelastic and resonance production may be made by Bloom-Gilman duality[9]. For the unpolarised data this relates the shape of $F(x)$ and the elastic form factor. The key is that when $\gamma(Q^2)N \rightarrow W$, where $W$ is the mass of a final state be it inclusive or exclusive, then since $x' \simeq W^2/Q^2$ one can either measure:

Inclusive: $F(x')$ where $x' = W^2$ (varies)$/Q^2$ (fixed)

Exclusive: $F(x')$ where $x' = W^2$ (fixed)$/Q^2$ (varies).

In the latter case one is essentially measuring the squared elastic or transition form factors.

Thus $F(x' \to 1) \sim (1 - x')^3 \to G(Q^2 \to \infty) \sim Q^{-4}$ the well known counting rules.

These ideas appear to work in some detail. The flavour asymmetry ($u >> d$ as $x' \to 1$) appears to be related to the asymmetry in resonance excitation form factors ($\Delta$ excitation dies out faster with $Q^2$ than does $N^*, I = 1/2$). Stoler has shown[10] how for $Q^2 \gtrsim 3 GeV^2$ the bump (resonance plus background) in the $\Delta(1232)$ region dies faster than the elastic nucleon and $I = 1/2$ dominated region around 1500 MeV.

That this duality may be quantitative may be seen by the work of Close and Thomas[6] who, in the MIT bag model, have shown that gluon exchange induced hyperfine forces in QCD can generate these asymmetries.

Thus low energy excitation is complementary to deep inelastic in probing certain QCD induced phenomena.

The above gluon exchange affects flavours due to the Pauli statistics correlating the constituents' flavours and spins. Its most direct manifestation is in its coupling to the spins and hence to spin-dependent phenomena such as those manifested in deep inelastic or low energy polarised scattering or resonance excitation.

## HIGH-ENERGY ELECTRON PRODUCTION

There have been claims recently that a dramatic failure of the canonical quark model of nucleons has been manifested by deep inelastic scattering of polarized leptons from polarized protons[11]. The claim is that quarks may carry none of the spin of the polarized proton. If this is indeed true, then the quark model will be called into question and will at least have to be extended.

One interesting feature of these experiments is that the polarization asymmetry $A(x, Q^2)$

$$A \equiv \frac{\sigma_{1/2} - \sigma_{3/2}}{\sigma_{1/2} + \sigma_{3/2}}$$

(where $\sigma_{1/2,3/2}$ are the photoproduction cross sections for transversely polarized photon and the polarized target to have net $J_Z = 1/2, 3/2$) appears to maximize in the kinematic limit of Bjorken $x \to 1$, i.e. $A^p(x \to 1) \to 1$, at least for a proton target (the only target so far studied). In modern QCD we believe that the asymmetry tending to unity as $\to 1$ is due to chromomagnetic effects[6], in the wave function (one gluon exchange). If the (yet to be measured) neutron asymmetry also maximizes in this limit, then we will have important confirmation of these ideas. These chromomagnetic effects are also manifested in low-energy experiments. Chromomagnetic forces split the $\Delta(P_{33}(1232))$ and nucleon masses - the well-known hyperfine splitting of energy levels. This already shows us that QCD can make clean-cut predictions for low-energy experiments. Less well known, perhaps, is that QCD perturbations induce mixing in the quark wave functions which may be accessed most sharply by electromagnetic probes. Furthermore, if the $x \to 1$ behaviour of inclusive high-energy data relates to the low-energy exclusive behaviour, $Q^2 \to \infty$, then we may look for some characteristic differences in the $Q^2$ dependence of $\Delta/N^*$ form factors, $G_M^p/G_M^n$ - the elastic magnetic form factors, and other specific final states such as $ep \to e\Lambda K/e\Sigma K$.

Spin-dependent sum rules imply that nontrivial $Q^2$ dependence must occur, probably in the low-energy region accessible to a milli-TeV machine. To illustrate this, consider the Drell-Hearn-Gerasimov (DHG) sum rule for real photons[12]

$$\frac{-2\pi^2\alpha\kappa^2}{M^2} = \int \frac{d\nu}{\nu}(\sigma_{1/2} - \sigma_{3/2}).$$ (1)

($\kappa$ is the anomalous magnetic moment of the target mass $M$) and the Bjorken[13] sum rule (for $Q^2 \to \infty$).

$$\frac{1}{6}\left|\frac{g_A}{g_V}\right| = \int_0^1 dx(g_1^p - g_1^n)$$ (2)

We can write these in a similar form if we define

$$G(\nu) \equiv \frac{m^2(\sigma_{1/2} - \sigma_{3/2})}{8\pi\sigma^2}$$ (3)

Thus the DHG sum rule becomes for protons

$$-\frac{1}{4}\kappa_p^2 = \int \frac{d\nu}{\nu}G^p(\nu, Q^2 = 0)$$ (4)

and for the proton-neutron difference

$$\int \frac{d\nu}{\nu}(G^p - G^n)(Q^2 = 0) = -\frac{1}{4}[\kappa_p^2 - \kappa_n^2] = 0.112$$ (5)

The Bjorken sum rule (Eq. 2) becomes

$$\int_{Q^2/2m}^{\infty} \frac{d\nu}{\nu}(G^p - G^n)(Q^2) = \frac{1}{3}\frac{M^2}{Q^2}\frac{g_A}{g_V} \simeq \frac{0.37}{Q^2}$$ (6)

The EMC data for protons $\int_0^1 dx g_1^p(x, Q^2) = 0.114 \pm 0.029$ would suggest that

$$\int_{Q^2/2m}^{\infty} \frac{d\nu}{\nu}G^p(Q^2) \simeq \frac{0.2}{Q^2}$$ (7)

Thus, we can compare the proton-neutron difference in Fig. 1 as a function of $Q^2$. That the $Q^2$ dependence would be most interesting was noted long ago by Gilman Karliner and myself qualitatively,[14] and the above quantification follows that of Ioffe and collaborators[15].

Some comments about the significance of these figures is called for. Equation 2 applies in the scaling region. We know from unpolarized scattering that scaling is still rather good modulo logarithms at $Q^2 = 5 GeV^2$. The logarithmic breaking of scaling only changes the curves in Fig. 1 very slightly and cannot make them approach the value for real photons. Indeed the curves in Fig. 1 have already overshot the real photon value when $Q^2$ is as low as $4\ GeV^2$. Thus, nonperturbative effects must enter at low $Q^2$ in order to make correspondence with the real photon value. How is this achieved? As yet, we do not know. All of the above remarks have concentrated on the proton-neutron difference; but for proton targets alone, things promise to be even more interesting. As noted long ago[14], there is an interesting change of sign for the polarization asymmetry $A^p$. For real photons, this asymmetry will be negative over some range of energy, whereas for highly virtual photons, the asymmetry is positive over a considerable range of $x$ as shown by the recent data from EMC, confirming the old SLAC data [16]. While one might dispute errors in the EMC integral[17], a very

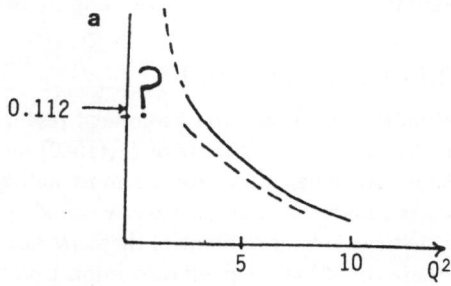

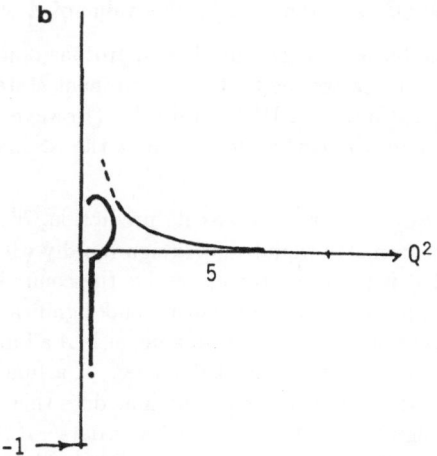

Figure 1. $Q^2$ dependence of the Bjorken integral (Eqs 6 and 7) for (a) $p-n$ difference and (b) proton target assuming the validity of EMC data (Ref 20). The values at $Q^2 = 0$ according to the DHG sum rule are indicated.

dramatic behaviour would have to ensure as $x \to 0$ if the integral would be negative! Indeed, if that was the resolution of the above "paradox", it would, by itself, be most interesting! If the integral is of the order of 0.1 when $Q^2$ is greater than 4 $GeV^2$, then Fig. 1 shows that something significant must occur as one proceeds into the nonperturbative regime of small $Q^2$ physics. This is not just a slight smoothing effect; here nonperturbative effects seem to have to change coalitions radically! So with hints that something "has to give", combined with the fact that the integrals are weighted towards small energies which hints that the resonance region may be relevant in the sum rule, let's see what we know so far about these spin-dependent effects and then delineate the questions that arise for a milli-TeV electronic machine.

## PHOTOPRODUCTION OF $N^*$: EMPIRICAL

Above pion production threshold are three prominent resonance bumps. The first resonance is the $P_{33}(1232)$; the second consists of $S_{11}(1550)$ and $D_{13}(1520)$, and the third is mainly $F_{15}(1690)$. Although these are the most noticeable states, a glance at the particle data tables shows that there are many other resonances within these bumps. The electromagnetic couplings of many of these are not well known. A better knowledge of some of these could have a significant impact on theory.

The spin dependence of the above prominent resonances is known, at least qualitatively. The $P_{33}$ is dominantly photoproduced by M1 radiation and so $A = -1/2$. If M1 dominates for all $Q^2$ as in the NRQM, this value of $A$ will be preserved.

The $D_{13}$ and $F_{15}$ are both photoproduced from protons dominantly in the helicity 3/2 mode, hence $A = -1$. So we see that the prominent states at $Q^2 = 0$ have the right sign of $A$ to help saturate the DHG sum rule. (However, the quantitative test of this sum rule, especially the contributions above the resonance region, is not yet done.)

One of the successes of the NRQM was its prediction,[14,18] confirmed by data,[19] that the polarization asymmetry would change sign rapidly with $Q^2$. Thus, these resonances seem to do what is required to help resolve the conundrum mentioned in the previous section. However, there are interesting and significant questions outstanding. Is the $Q^2$ dependence of $D_{13}$ and $F_{15}$ the same, or is $A$ a function of $Q^2/W^2$ where $W$ is the resonance mass, or some other behaviour? If a function of $Q^2/W^2(\simeq x')$, then at what value of $x'$ does $A$ change sign and how does this correlate with the possibility that $A$ change sign as $x \to 0$ in deep inelastic data? If $A$ changes sign at fixed $Q^2 \simeq 0.5 GeV^2$, say, does this correlate with a possible similar behaviour in the high $W$ inelastic region? Interesting structure seems likely for all $W$ and $Q^2$; we need to map it out to the best possible accuracy in order to understand the nonperturbative dynamics that are at work.

These spin dependences, and especially their $Q^2$ dependences, are among the sharpest probes of the nonperturbative dynamics. This is my challenge for experiment. Now I will address a parallel challenge for theory.

The above are all ideas, motivated by current excitement in the experimental community. The questions are potentially answerable at moderate $Q^2$, such as those accessible in Bonn and at CEBAF. Though not immediately requiring a higher energy facility their answers will stimulate yet deeper questions for that facility. There is a school that applies perturbative QCD to exclusive or resonance production as $Q^2 \to \infty$. These results imply significant changes in the helicity structure, even in the $\Delta(1232)$ region where the naive quark model does not immediately require new physics. The naive quark model must fail on rather general grounds probably within the region accessible at CEBAF/ELSA. The application of exclusive QCD probably will not be applicable at these energies. Thus there is likely a gap in the $Q^2 - W^2$ plane where no suitable theory framework yet exists. We must be led to that empirically and a high energy facility capable of resolving details of final state hadrons and resonances is required.

There is a possible role for nuclei as targets in the case of nucleon resonance excitation. Stimulated by the ideas on colour transparency Ralston has emphasised that nuclear targets may emphasise the "mini-hadron" (3 quark wavefunction) where perturbative QCD works better. Thus the predicted charge in helicity structure in the $\Delta(1232)$ region may be manifested more clearly, or at lower $Q^2$, when the $\Delta$ excitation takes place in a nuclear target.

Another possible use for nuclei comes from ideas stimulated by the first EMC effect which has led to suggestions that the nucleon is "larger" when bound in nuclei than when free. This could have observable consequences for the helicity structure of the excitation of the $D(1550)$.

This resonance can be excited by E1 and M2 multipoles and the quark model specifies their ratio (at $Q^2 = 0$ the helicity 1/2 amplitude off protons $\simeq 0$). One can weigh $\vec{q}$ against the size $R$ by varying either of them. When one varies $Q^2$ (hence $\vec{q}$) the quark model predicted successfully[7] that the magnetic multipole rapidly dominates, as manifested by the changing helicity structure. The possible change in $R$ for a nucleon in a nucleus could lead to an analogous change in the relative strength of the two multipoles. As this change is known to be so sensitive in $\Delta q^2$, so it may be in $\Delta R$.

There is also the possibility that a change in $R$ leads to a change in the excitation energies, and hence of resonance masses, in nuclei. Thus there may be rich phenomena to be accessed by resonance electroproduction in nuclei.

# References

[1] T. Londergan (these proceedings)

[2] H.J. Lipkin, Phys. Lett. B251, 613 (1990)

[3] F. Wagner, Proc. XVI Rencontre de Moriond (1981);
    T. Barnes and F.E. Close, Phys. Lett. 128B, 277 (1983)

[4] T. Barnes and F.E. Close, Phys. Lett. 123B, 89 (1983)

[5] F.E. Close and Z.P. Li, Phys. Rev. D42, 2194 (1990)
    ibid 2207 (1990)

[6] F.E. Close, and A.W. Thomas, Phys. Lett. B212, 227 (1988)

[7] J. Sullivan, Phys. Rev. D5, 1732 (1972)
    C. Llewellyn Smith, Phys. Lett. 128B, 107 (1983)
    M. Ericson and A.W. Thomas, Phys. Lett. 126B, 97 (1983)
    L. Frankfurt et al. Z. Phys. A334, 343 (1989)

[8] A. Signal et al. Adelaide Univ. ADP-90-145/T89

[9]  E. Bloom and F. Gilman, Phys. Rev. D4, 2901 (1971)

[10]  P. Stoler, Phys. Rev. Lett. 66, 1003 (1991)

[11]  See F. Close, Phys. Rev. Lett. 64, 361 (1990) and references therein

[12]  S. Drell and A. Hearn, Phys. Rev. Lett. 16, 908 (1966)
      S.B. Gerasimov, J. Nucl. Phys. (USSR) 2, 598 (1966)

[13]  J.D. Bjorken, Phys. Rev. 148, 1467 (1966); D1, 1376 (1970)

[14]  F. Close in Proc. of IX Rencontre de Moriond, Vol 2, ed J. Tran Thanh Van
      (CNRS France 1974) p285

[15]  B. Ioffe et al, Hard Processes, Vol I (North Holland, Amsterdam 1984)

[16]  V. Hughes and J. Kuti, Ann. Rev. Nuc. Part. Sci. 33, 64 (1983)

[17]  F. Close and R. Roberts, Phys. Rev. Lett. 60, 1471 (1988)

[18]  F. Close and F. Gilman, Phys. Lett. 33B, 541 (1972)

[19]  V. Burkert, Electroproduction of N* at CEBAF energies; CEBAF-R-86-011
      (1986)

[20]  J. Ashman et al (EMC) Phys. Lett. B206, 364 (1988)

# TENSOR POLARIZATION MEASUREMENTS OF THE RECOIL DEUTERONS
# IN ELASTIC ELECTRON-DEUTERON SCATTERING

I. The,[6] J. Arvieux,[5] D. H. Beck,[2,8] E. J. Beise,[6] A. Boudard,[3]
E. B. Cairns,[1] J. M. Cameron,[1,4] G. W. Dodson,[6] K. A. Dow,[6]
M. Farkhondeh,[6] H. W. Fielding,[1] J. B. Flanz,[6] M. Garçon,[3,6]
R. Goloskie,[9] S. Høibråten,[6] J. Jourdan,[2] S. Kowalski,[6]
C. Lapointe,[1] W. J. McDonald,[1] B. Ni,[4] L. D. Pham,[6]
R. P. Redwine,[6] N. L. Rodning,[1] G. Roy,[1] M. E. Schulze,[7]
P. A. Souder,[7] J. Soukup,[1] W. E. Turchinetz,[6] C. F. Williamson,[6]
K. E. Wilson,[6] S. A. Wood,[6,8] and W. Ziegler[1]

*(1) University of Alberta, Edmonton, Alberta, Canada T6G 2N5*
*(2) California Institute of Technology, Pasadena, California 91125*
*(3) DPhN/Saclay, 91191 Gif-sur-Yvette, France*
*(4) Indiana University Cyclotron Facility, Bloomington, Indiana 47405*
*(5) Laboratoire National Saturne, 91191 Gif-sur-Yvette, France*
*(6) Massachusetts Institute of Technology, Cambridge, Mass. 02139*
*(7) Syracuse University, Syracuse, New York 13210*
*(8) University of Illinois, Champaign, Illinois 61820*
*(9) Worcester Polytechnic Institute, Worcester, Massachusetts 01609*

## ABSTRACT

The tensor polarization $t_{20}$ of the recoil deuteron in elastic $e$–$d$ scattering has
been measured for three values of four-momentum transfer, $q = 3.78, 4.22$, and
$4.62$ fm$^{-1}$. The results have been used to separate the charge monopole and
charge quadrupole form factors of the deuteron and to locate the first node in
the charge monopole form factor at $q = 4.39 \pm 0.16$ fm$^{-1}$. The extracted $t_{20}$
values are compared with several theoretical predictions.

## INTRODUCTION

The short range behavior of the deuteron wave function is of particular interest
because of its expected sensitivity to non-nucleonic degrees of freedom, relativistic
effects, and meson-exchange currents (MEC). In the past several years, considerable
theoretical effort has been expended to explore the relevance of a wide range of models
to the deuteron electromagnetic structure. The more firmly based models are those
which use relativistic operators and wave functions and which incorporate MEC, while

some which include quark degrees of freedom, for example in a hybrid model, might be considered among the most speculative. Existing calculations vary widely in the predicted behavior of the charge monopole form factor and the tensor polarization $t_{20}$ in the four-momentum transfer range $3<q<6$ fm$^{-1}$. Measurements of the $t_{20}$ tensor polarization in this momentum transfer range can be used to provide strong constraints on the possible models of the deuteron electromagnetic structure.

The cross section for elastic electron-deuteron scattering in the one-photon-exchange approximation is given by

$$\frac{d\sigma}{d\Omega} = \frac{d\sigma}{d\Omega_{Mott}}\left[A(q) + B(q)\tan(\theta_e/2)\right], \tag{1}$$

where $q$ is the four-momentum transfer, $\theta_e$ is the electron scattering angle in the laboratory frame, and $d\sigma/d\Omega_{Mott}$ is the Mott cross section for scattering of electrons from point particles and is given by

$$\frac{d\sigma}{d\Omega_{Mott}} = \frac{\alpha^2\cos^2(\theta_e/2)}{4E^2\sin^4(\theta_e/2)[1 + 2(E/M_d)\sin^2(\theta_e/2)]} \tag{2}$$

where $\alpha$ is the fine structure constant, $E$ is the electron energy, and $M_d$ is the deuteron mass.

By imposing time-reversal and parity invariance, the $A(q)$ and $B(q)$ structure functions are given in terms of the individual electromagnetic form factors of the spin-one deuteron: charge monopole ($F_C$), charge quadrupole ($F_Q$), and magnetic dipole ($F_M$), and can be written as follows:

$$A(q) = F_C^2(q) + \tfrac{8}{9}\eta^2 F_Q^2(q) + \tfrac{2}{3}\eta F_M^2(q), \tag{3}$$

$$B(q) = \tfrac{4}{3}\eta(1 + \eta)F_M^2(q), \tag{4}$$

where $\eta = q^2/4M_d^2$.

The $A(q)$ and $B(q)$ structure functions have been measured[1] up to 10 fm$^{-1}$ and 8.4 fm$^{-1}$, respectively. By measuring the elastic cross section, the magnetic dipole form factor $F_M$ can be separated from the charge form factors $F_C$ and $F_Q$. However, it is not possible to separate further $F_C$ and $F_Q$ without measuring another observable which depends on the tensor polarization of the deuteron.

The $t_{20}$ tensor polarization of the deuteron is also a function of the form factors $F_C$, $F_Q$, and $F_M$ but with a different combination from those for the $A(q)$ and $B(q)$ structure functions. The tensor polarization of the deuteron is given by[2]

$$t_{20} = -\sqrt{2}\frac{[x(x + 2) + y/2]}{[1 + 2(x^2 + y)]}, \tag{5}$$

where $x = \tfrac{2}{3}\eta F_Q/F_C$, $y = \tfrac{2}{3}\eta f(\theta_e)F_M^2/F_C^2$, $f(\theta_e) = \tfrac{1}{2}+(1+\eta)\tan^2(\theta_e/2)$. The quantity $\tilde{t}_{20} \equiv t_{20}(y = 0)$, derived from Eq. (5) by neglecting the small magnetic contribution, is often used in the literature. However, it should be noted that the small magnetic contribution due to $F_M$ is always present in the extracted $t_{20}$ values from experiments. In the impulse approximation, the deuteron electromagnetic form factors are given as a product of the isoscalar nucleon form factor and the integral of S- and D-state wave functions. As shown in Eq. (5), $t_{20}$ is given in terms of the ratio of form factors so that the dependence on the isoscalar nucleon form factor cancels out. Hence, the tensor polarization $t_{20}$ is not sensitive to the parametrization of the nucleon form factors, in particular to that of the poorly known electric form factor of the neutron.

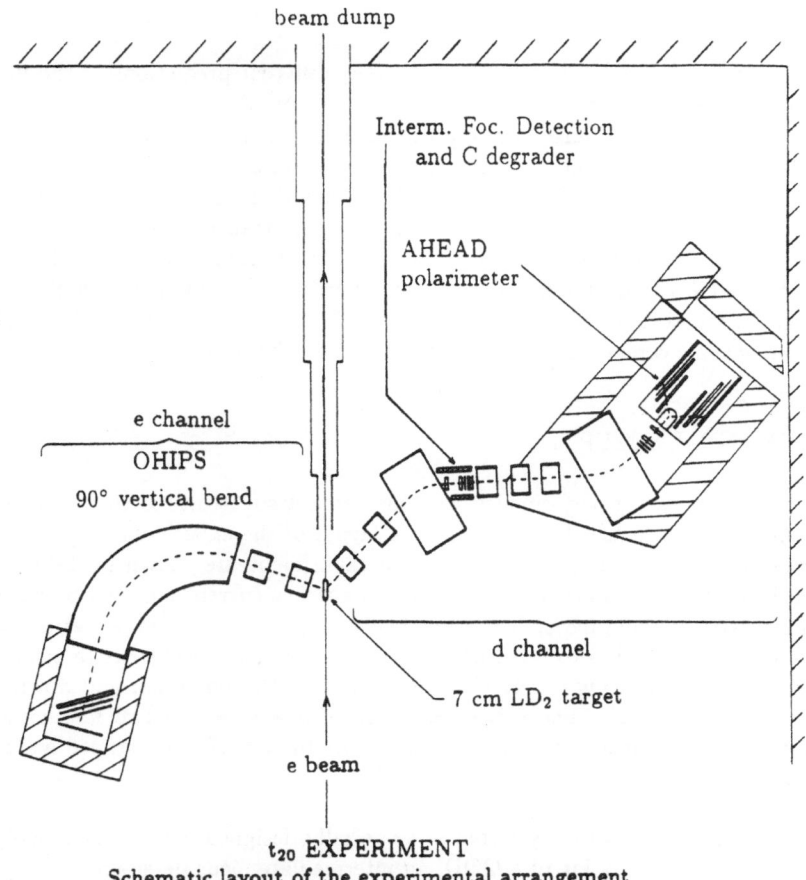

beam dump

Interm. Foc. Detection
and C degrader

AHEAD
polarimeter

e channel

OHIPS
90° vertical bend

d channel

7 cm $LD_2$ target

e beam

$t_{20}$ EXPERIMENT
Schematic layout of the experimental arrangement

FIG. 1 The electron channel: QQD, three plastic scintillators, and a drift chamber. The deuteron channel: QQD, IFD and carbon degrader, and QQQD.

In this paper, we present the results of measurements of the tensor polarization of the recoil deuterons in elastic electron-deuteron scattering in the four-momentum transfer range of $3.8 \leq q \leq 4.6$ fm$^{-1}$. Previous measurements[3-5] of $t_{20}$ have been reported at lower momentum transfers ($q < 3$ fm$^{-1}$). The measurements reported here cover a range of momentum transfer where the short distance structure of the deuteron and non-nucleonic degrees of freedom are expected to become important. Various models of the charge monopole form factor $F_C$ agree with each other in their predictions

Table I. Kinematics and some parameters of the experiment: electron beam energy, electron laboratory angle, deuteron laboratory angle, four-momentum transfer, luminosity, number of deuterons incident on polarimeter.

| E (MeV) | $\theta_e$ (deg.) | $\theta_d$ (deg.) | $q$ (fm$^{-1}$) | L (cm$^2$ s$^{-1}$) | $N_d^{inc}$ |
|---------|-------------------|-------------------|-----------------|---------------------|-------------|
| $653 \pm 3$ | 80.9 | 41.0 | $3.78 \pm 0.02$ | $7.5 \times 10^{36}$ | $3.6 \times 10^5$ |
| $755 \pm 4$ | 78.7 | 41.0 | $4.22 \pm 0.02$ | $3.0 \times 10^{37}$ | $7.6 \times 10^5$ |
| $853 \pm 5$ | 76.7 | 41.0 | $4.62 \pm 0.02$ | $2.2 \times 10^{37}$ | $4.9 \times 10^5$ |

at low momentum transfer, but they have widely different predictions in the momentum transfer range $3 < q < 6$ fm$^{-1}$. Most models predict a zero-crossing of $F_C$ in this momentum transfer range, while a model based on a quark description predicts a smooth fall-off in $F_C$ without passing through zero in the same momentum transfer range. The charge quadrupole form factor $F_Q$, which dominates over $F_C$ and $F_M$ and largely determines the cross section in this momentum transfer range, is less sensitive to the details of various models. Hence, it is very important to separate the charge form factors $F_C$ and $F_Q$ to obtain a complete characterization of the deuteron electromagnetic structure and to learn what corrections, such as relativistic effects, MEC effects, and non-nucleonic components, are needed to extend the impulse approximation.

## EXPERIMENTAL SETUP

The experiment was performed at the Massachusetts Institute of Technology–Bates Linear Accelerator Center. Figure 1 shows a layout of the experimental arrangement. The kinematics and some parameters of the experiment are shown in Table I. The electron beam from the linac was used with an average current between 5 and 30 $\mu$A and a duty cycle of about 0.8%. It was incident on a 7-cm long liquid deuterium target. The target was collimated to 5-cm useful length using tungsten blocks to stop events coming from the scattering of the electron beam and the upstream and downstream parts of the target wall. The target consisted of a loop where two fans circulated the liquid through a heat exchanger cooled to 19K by a 200W pressurized helium gas refrigerator.

The recoil deuterons were selected in a specially designed magnetic channel, fixed at an angle of 41°, consisting of a QQD, an intermediate focus detection (IFD), and a QQQD system. Each dipole in the deuteron channel bent the deuteron beam by 35° but in opposite directions. This dipole configuration minimized the precession effect on the tensor polarization of the deuteron beam. The bending of the deuteron beam in the second dipole cancelled most of the precession of the beam due to the bending in the first dipole. The scattered electrons were analyzed with a magnetic spectrometer (OHIPS), having a solid angle of 18 msr, and detected by three scintillators in coincidence. OHIPS was positioned at three angles, given in Table I, which were determined by the corresponding three values of incident electron energy and the fixed angle of the deuteron channel. The IFD was composed of 2 multiwire proportional chambers (MWPC) used for tuning only and a hodoscope of 10 scintillators (12.6 cm × 1.26 cm × 0.2 cm). The "Alberta High Efficiency Analyzer for Deuterons" (AHEAD) was installed in a shielded hut at the end of the deuteron channel. A schematic of the side view of the AHEAD polarimeter is shown in Fig. 2.

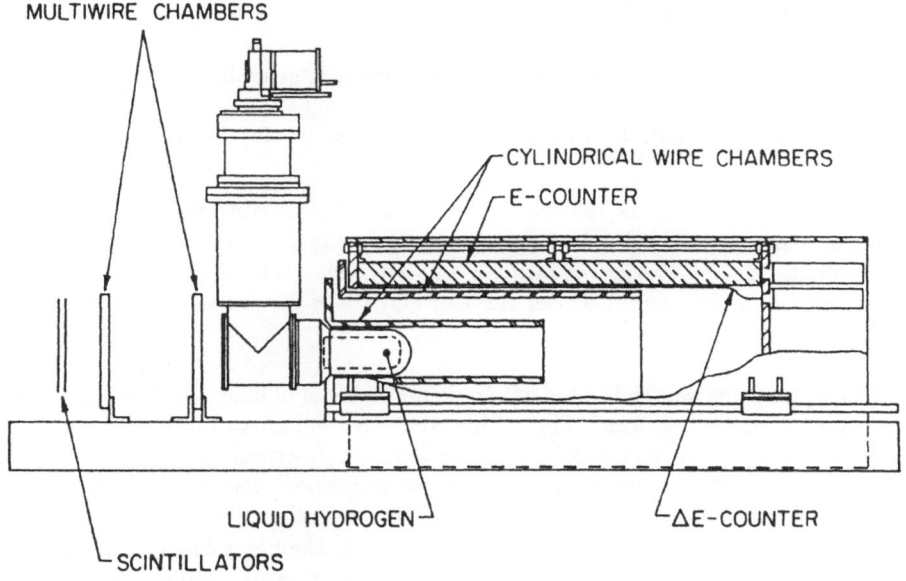

MULTIWIRE CHAMBERS

CYLINDRICAL WIRE CHAMBERS

E-COUNTER

LIQUID HYDROGEN

ΔE-COUNTER

SCINTILLATORS

FIG. 2. Schematic of side view of AHEAD polarimeter.

For the runs taken at the two higher electron energies, a carbon degrader was installed in front of the hodoscope. This degrader was used to reduce the flux of protons entering the polarimeter and to lower the deuteron energies to values within the maximum efficiency range of AHEAD. The coincidence between OHIPS, IFD, and AHEAD led to an unambiguous identification of $e$–$d$ elastic events.

The polarimeter was based on $d$–$p$ elastic scattering.[6] In the deuteron energy range of interest (60-170 MeV), the angular distribution of the analyzing power $T_{20}$ exhibits a characteristic minimum ($T_{20} \simeq -0.45$) at $\theta_{CM} \simeq 115°$ and a maximum ($T_{20} \simeq 0.3$) at $\theta_{CM} \simeq 150°$. The deuterons from $e$–$d$ elastic scattering events were incident upon a 27-cm long liquid hydrogen (LH$_2$) target with a 10-cm diameter. Their incident trajectories were determined by two MWPC's triggered by two scintillators in front of the MWPC's. Depending on the angles and energies of the scattered particles, a particle (a proton or a deuteron) or sometimes two particles in coincidence (two protons or a proton and a deuteron) were detected in an assembly of cylindrical detectors. Two cylindrical wire chambers (CWC), positioned concentrically around the LH$_2$ target, were used to measure the polar and azimuthal angles of the scattered particles. The CWC's were surrounded by an array of 6 ΔE (150 cm × 22 cm × 0.3 cm) and 18 E plastic scintillators (7.6 cm thick) to measure the energies of detected particles and identify them through their energy losses.

The recoil deuteron energies in the Bates experiment ($q^2/2M_d$) were degraded by various materials (target wall, spectrometer windows, air, IFD detectors, and a carbon degrader for the runs taken at the two higher $q$ points) between the liquid deuterium target and the second dipole of the deuteron channel. The central deuteron energies were 133 MeV for the lowest $q$ point and 160 MeV for the two higher $q$ points as measured by a magnetic-field probe installed in the second dipole of the deuteron channel.

## DATA ANALYSIS AND RESULTS

The yield for the scattering of polarized deuterons is given by[7]

$$N_C(\theta, \phi) = kN_0(\theta, \phi)[1 + t_{20}T_{20}(\theta) + 2t_{21}T_{21}(\theta)\cos(\phi) + 2t_{22}T_{22}(\theta)\cos(2\phi)], \quad (6)$$

where $N_0$ is the yield for the scattering of unpolarized deuterons; $t_{20}$, $t_{21}$, and $t_{22}$ are the tensor moments; $T_{20}$, $T_{21}$, and $T_{22}$ are the analyzing powers; $\theta$ is the angle in the center of mass frame between the incident and outgoing tracks of the deuterons in the polarimeter; $\phi$ is the angle between the $e$–$d$ and the $d$–$p$ scattering planes; $k$ is an irrelevant normalization factor.

The polarimeter was calibrated with a deuteron beam of known polarization at the Laboratoire National Saturne.[8] The calibration runs relevant to this experiment were performed at 120, 145, and 170 MeV. More detailed descriptions of the polarimeter and the analysis of the calibration data will be published[9] elsewhere. The yield for the scattering of unpolarized deuterons on protons and the analyzing powers for $\vec{d}$-$p$ scattering were measured in the Saturne experiment. In the Bates experiment, the yield for the scattering of tensor-polarized deuterons coming from the $e$–$d$ elastic scattering was measured.

The analyses for the calibration and the Bates data proceeded in the same way. The incident and outgoing tracks were reconstructed to determine the scattering vertices. The $\Delta E$ and E scintillators were gain matched to identify the particles detected in the polarimeter, and the angular distribution of detected protons was measured. A cut on the proton missing-energy spectrum ("E-diff cut") was applied to the data to eliminate highly inelastic protons coming from deuteron breakup reactions. The inclusion of some inelastic protons resulted in analyzing powers about 20% smaller than the values quoted above. In addition, because the deuteron energies incident on the polarimeter at Bates had large spreads, 14–23 MeV (full width at half max), and were not centered at any of the energies in the calibration experiment, the appropriate calibration was obtained by interpolation of the Saturne data using a Monte-Carlo technique to simulate the energy and vertex position dependence of geometrical and absorption effects in the polarimeter. The actual incident deuteron spatial distributions of the Bates data were used to generate the interpolated-calibration data, while the energy distributions, not being measured accurately, were simulated using Gaussian distributions. The final results were found insensitive to reasonable changes in the above distributions.

The tensor moments were obtained using a fitting procedure which adjusted the $t_{kq}$ in Eq. (6) such that the angular distribution of the interpolated-calibration data best reproduced the relative angular distribution of the Bates data. This procedure avoids the necessity for measuring the normalization factor $k$ in the Bates experiment. The measured yield was divided into 12 bins each for $\theta$ ($100° - 160°$) and $\phi$ ($0° - 360°$). Because a large number of these bins contained very few counts (5 counts or less), the use of Poisson statistics was necessary to obtain the best fit to the Bates data. The following expression was derived using Poisson statistics and the maximum likelihood method:[10]

$$\xi^2 = \sum_{\theta, \phi} 2\left[N_B(\theta, \phi)\ln\left(\frac{N_B(\theta, \phi)}{N_C(\theta, \phi)}\right) - N_B(\theta, \phi) + N_C(\theta, \phi)\right], \quad (7)$$

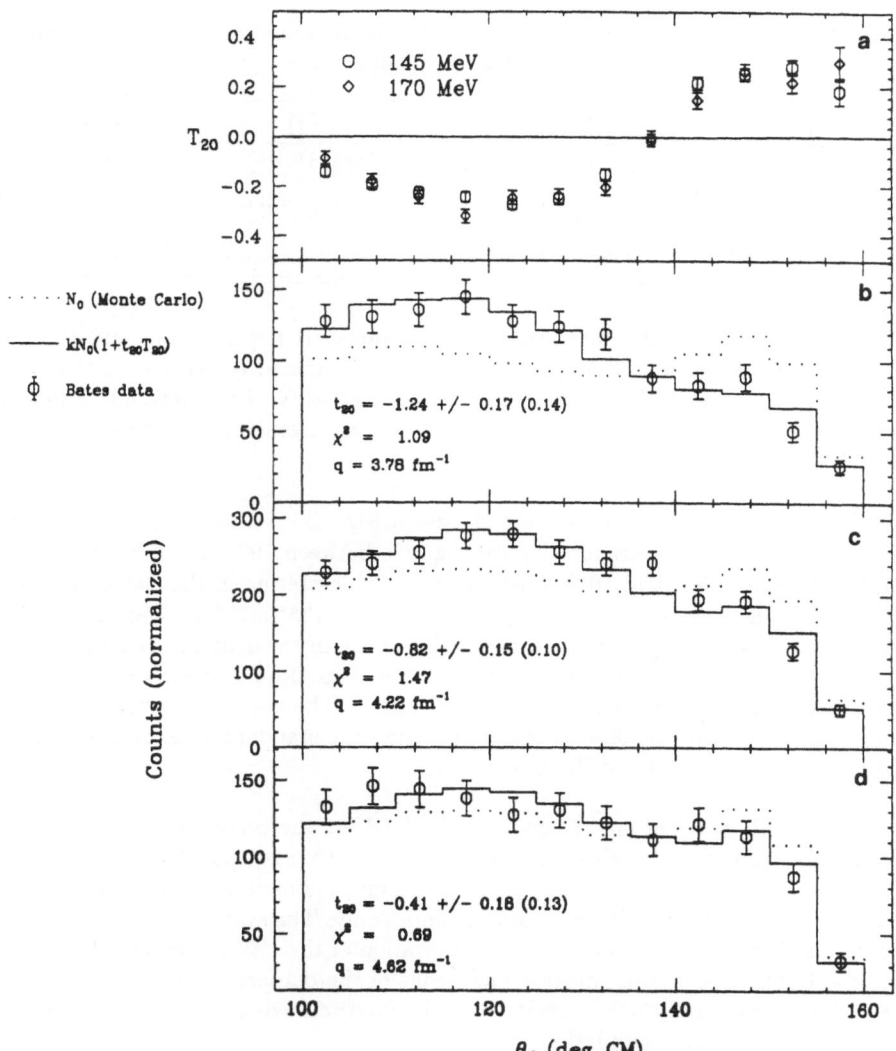

FIG. 3 a) Analyzing powers for $T_{20}$ measured at 145 and 170 MeV; Bates data and fit for b) $q = 3.78$ fm$^{-1}$, c) $q = 4.22$ fm$^{-1}$, d) $q = 4.62$ fm$^{-1}$. The dotted histograms are the yields for scattering of unpolarized deuterons on protons from Monte Carlo multiplied by the normalization factor $k$. The solid histograms are the results of the fit to the Bates data shown as open circles with error bars. Errors on $t_{20}$ are total errors from systematic and statistical errors added in quadrature while those in parentheses are statistical errors only. The $\chi^2$'s are calculated using the standard $\chi^2$ formula.

where $N_B$ are the Bates data and $N_C$ is given by Eq. (6). The $\xi^2$ in Eq. (7) was then minimized with respect to $t_{20}$, $t_{21}$, $t_{22}$, and $k$ using the minimization routine MINUIT.[11]

Figure 3a shows the analyzing powers for $T_{20}$ taken in the Saturne experiment at 145 MeV and 170 MeV. As can be seen in Fig. 3a, $T_{20}$ has very little dependence on the deuteron energy in this energy range. The analyzing power for $T_{20}$ measured at

Table II. Results of the $t_{20}$ measurements. Four-momentum transfer, $t_{20}$, charge monopole form factor, charge quadrupole form factor.

| $q$ (fm$^{-1}$) | $t_{20}(q, \theta_e)$ | $F_C(q)$ | $F_Q(q)$ |
|---|---|---|---|
| $3.78 \pm 0.02$ | $-1.24 \pm 0.17(0.10)$ | $0.0127^{+0.0047}_{-0.0056}$ | $0.482^{+0.077}_{-0.116}$ |
| $4.22 \pm 0.02$ | $-0.82 \pm 0.15(0.11)$ | $0.00166^{+0.00161}_{-0.00142}$ | $0.315^{+0.010}_{-0.011}$ |
| $4.62 \pm 0.02$ | $-0.41 \pm 0.18(0.13)$ | $-0.00147^{+0.00106}_{-0.00104}$ | $0.189^{+0.007}_{-0.008}$ |

120 MeV (not shown) also looks similar to $T_{20}$ measured at 145 MeV and 170 MeV. The analyzing power at 133 MeV, which was used in the fit to extract $t_{20}$ at $q = 3.78$ fm$^{-1}$, is an average of $T_{20}$ measured at 120 and 145 MeV. The analyzing power used in the fit at $q = 4.22$ fm$^{-1}$ and $q = 4.62$ fm$^{-1}$ is an average of $T_{20}$ measured at 145 MeV and 170 MeV.

The results of the fit for $t_{20}$ are shown in Figs. 3b-d after summing over the $\phi$ bins. The analyzing powers for $T_{20}$ are negative between 100° and 135° and becomes positive between 135° and 160°. The yields of $\vec{d}$-$p$ scattering in the Bates experiment are enhanced and reduced for the angular ranges 100°-135° and 135°-160°, respectively. These effects indicate that the $t_{20}$ tensor polarization of the deuteron beam is negative. One can see in the fits to the Bates data shown in Figs. 3b-d that the magnitude of $t_{20}$ decreases as the four-momentum transfer increases. The goodness-of-fit $\chi^2$ per degree of freedom shown in Fig. 3b-d is calculated using the standard $\chi^2$ expression between the fit (solid histogram) and the Bates data.

The fitted values of $t_{20}$ are given in Table II for the corresponding $q$ values and electron scattering angles given in Table I. The total uncertainties in $t_{20}$ given in the table combine statistical and systematic uncertainties in quadrature while the values in parentheses indicate the systematic uncertainties only. The systematic uncertainties are due to the E-diff cut, differences in the calibration of the CWC's between the Saturne and the Bates experiments, inexact knowledge of the deuteron energy spreads for the Bates kinematics, and the uncertainties in the interpolated-calibration data generated using the Monte Carlo simulation.

In Table II, we also give the values of $F_C$ and $F_Q$ extracted from our measurements of $t_{20}$ and from parametrizations[12] of the world data for the $A$ and $B$ structure functions. The separation of the form factors of the deuteron at each $q$ point was done using the fitting routine MINUIT[11] and the following expression:

$$\chi^2 = \left[ \frac{A_p - A(F_C, F_Q, F_M)}{\Delta A} \right]^2 + \left[ \frac{B_p - B(F_M)}{\Delta B} \right]^2 + \left[ \frac{t_{20}^m - t_{20}(F_C, F_Q, F_M)}{\Delta t_{20}} \right]^2 \quad (8)$$

where $A_p$ and $B_p$ are obtained from parametrizations of the world data for the $A$ and $B$ structure functions, $t_{20}^m$ is the measured $t_{20}$; $\Delta A$, $\Delta B$, and $\Delta t_{20}$ are the uncertainties in $A_p$, $B_p$, and $t_{20}^m$, respectively; $A(F_C, F_Q, F_M)$, $B(F_M)$, and $t_{20}(F_C, F_Q, F_M)$ are given in terms of the usual form factors of the deuteron defined in Eqs. 3–5. The normalizations of $F_C$ and $F_Q$ at $q = 0$ fm$^{-1}$ are defined as follows: $F_C(0) = 1$ and $F_Q(0) = M_d^2 Q$ where $Q$ is the quadrupole moment of the deuteron. The uncertainties in $F_C$ and $F_Q$ correspond to a $\chi^2$ change from $\chi^2_{min}$ to $\chi^2_{min} + 1$. A change of one in $\chi^2$ from its minimum value represents a one-standard deviation uncertainty.

The present results and previous $t_{20}$ measurements are plotted in Fig. 4. They have been adjusted using $B(F_M)$, the extracted $F_C$ and $F_Q$, and Eq. (5) to give the values for $\theta_e = 70°$, the angle at which some of the theoretical curves shown in Fig. 4 were calculated. Those calculated in the Skyrme, six-quark, and perturbative quantum chromodynamics (PQCD) models are, however, for $\tilde{t}_{20}$. Setting $y = 0$ in Eq. (5) lowers the value of $t_{20}$ around the minimum by about 0.2.

An impulse-approximation (IA) prediction calculated using conventional non-relativistic nuclear dynamics and a representative modern potential (Argonne $V_{14}$ ($AV_{14}$) potential)[13] for the NN interaction is shown by the solid curve in Fig. 4. The addition of meson-exchange currents (MEC) (short-dash curve) results in a prediction which clearly lies above the data for $q > 3.5$ fm$^{-1}$. In this IA model, the operators for the model-independent part of the MEC are constructed to be consistent with the potential. Other IA calculations using various realistic NN potentials (Reid soft core, Bonn, Paris)[2] show similar features to those using the $AV_{14}$ potential. On the other hand, the $t_{20}$ prediction (double-dot-dash curve) calculated using the new Bonn potential,[14] which included an energy dependence in the potential and pion-pair MEC contributions, is in good agreement with the data except at the highest measured $q$ where it falls slightly below the data.

A recent relativistic one-boson-exchange (OBE) prediction[15] of $t_{20}$ gives reasonable agreement with the data and is similar to the non-relativistic IA predictions. In this case, MEC contributions from $\rho\pi\gamma$ and $\omega\epsilon\gamma$, evaluated relativistically to be consistent with the NN dynamics, were found to cancel each other partially. The $\epsilon$ meson used in this model is actually the fictitious $\sigma$ meson. The agreement of the $t_{20}$ prediction with the data depends on the choice of the coupling constants for the $\rho\pi\gamma$ and the $\omega\epsilon\gamma$ contributions. The coupling constant for the $\rho\pi\gamma$ has a large uncertainty while that for the $\omega\epsilon\gamma$ is unknown experimentally. Other relativistic calculations using the formalism of light-cone quantum mechanics[16] also give predictions for $t_{20}$ similar to those of the relativistic OBE model. However, these models do not include MEC contributions.

Isobar degrees of freedom have been included in two coupled-channel (CC) models. These CC models include different amount of $\Delta\Delta$ components in the deuteron wave function. In a model with very small $\Delta\Delta$ admixtures (0.36%),[17] the effect of the addition of $\Delta\Delta$ components on $t_{20}$ for $q < 5$ fm$^{-1}$ is very small, and the predicted $t_{20}$ (long-dash curve) is in good accord with the data. The $t_{20}$ prediction of another CC model having large $\Delta\Delta$ admixtures (1.8–7.2%) and using a boundary condition radius[18] of $r_0 = 0.74$ fm lies above the data (not shown in Fig. 4) and is almost indistinguishable from the prediction of the IA+MEC calculation using the $AV_{14}$ potential (short-dash curve). By imposing a larger radius ($r_0 = 1.05$ fm), the predicted $t_{20}$ values are even more positive for $q > 3.5$ fm$^{-1}$ than those using the smaller radius. Both CC models include MEC contributions in the $t_{20}$ calculations, but it is noted that the $t_{20}$ prediction for the CC model with $r_0 = 0.74$ fm agrees with the data if the MEC contributions are removed.

The predictions of several quark-cluster models,[19] calculated using the resonating group method and a quark-bag radius of about 0.5 fm, are all in qualitative agreement with the data. One quark description excluded by the present data is a simple six-quark prediction[20] (dot-double-dash curve), in which a six-quark component was added

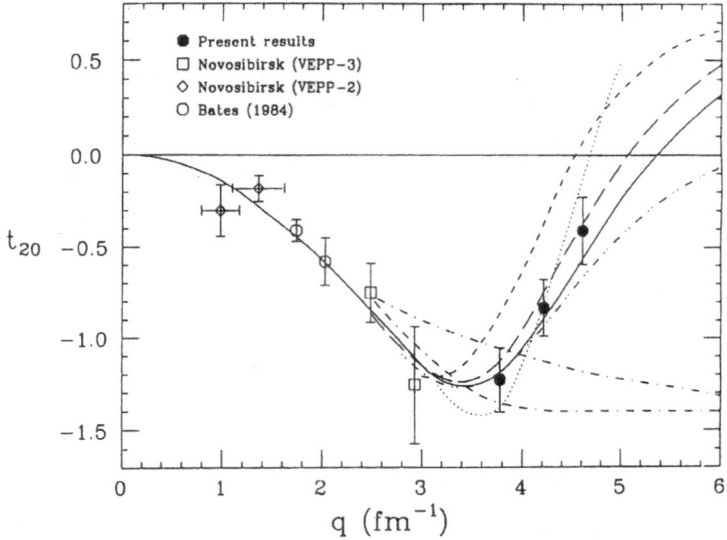

FIG. 4 Data and theoretical predictions of $t_{20}$ as a function of four-momentum transfer.
Data points: open circles (Ref. 3), diamonds (Ref. 4), squares (Ref. 5), and
solid circles (present results). All data points have been adjusted to give values
for $\theta_e = 70°$ (see text). The curves are predictions of different models (see text):
a) $\theta_e = 70°$ in Eq. (1): IA+MEC AV$_{14}$: short-dash (Ref. 13), CC: long-dash
(Ref. 17), IA AV$_{14}$: solid (Ref. 13), new Bonn: double-dot-dash (Ref. 14); b)
$y = 0$ in Eq. (1): Skyrme: dot (Ref. 21), PQCD: dot-dash (Ref. 22), six-quark:
dot-double-dash (Ref. 20).

directly to the deuteron wave function. In this model, the node in $F_C$ is missing so
that the predicted $t_{20}$ remains negative for $q > 3$ fm$^{-1}$.

Non-nucleonic effects are included in a totally different approach in the Skyrme
model[21] where nucleons and nuclei are formed as topological solitons of pion and scalar
meson fields. The model determines the forms of the isoscalar charge and current
operators. In this model, the wave functions of the deuteron are generated using the
Paris potential. The value of $t_{20}$ predicted by this model (dot curve) is in reasonable
agreement with the data. The sharp rise in the $t_{20}$ curve is due to smaller predicted $F_Q$
compared to $F_Q$ calculated in other models. Finally, the prediction of PQCD model[22]
(dot-dash curve) clearly disagrees with the data. This failure of the model shows that
the $q$ range of the present measurements is still far from the asymptotic region where
PQCD model is applicable.

The present data unambiguously show a sharp rise of $t_{20}$ from a minimum toward
less negative values. This behavior is due to the node in $F_C$ which is determined using
a polynomial fit to be at $q = 4.39 \pm 0.16$ fm$^{-1}$. The extracted values of $F_C$ and $F_Q$ are
plotted in Figs. 5 and 6 along with predictions for $F_C$ and $F_Q$, corresponding to the

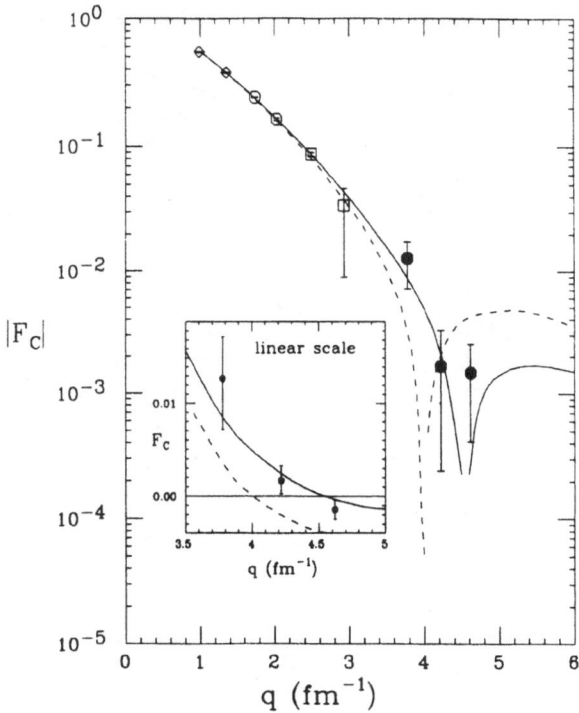

FIG. 5 Charge monopole form factor of the deuteron as a function of four-momentum transfer. Same notations as in Fig. 4. The solid and dashed curves correspond to the solid and dashed curves in Fig. 4.

solid and short-dash curves in Fig. 4. The data for $F_Q$ are described well by a non-relativistic IA model[13] using the $AV_{14}$ potential with and without MEC. The addition of MEC has small effects in the predicted $F_Q$. On the other hand, the contributions of MEC in the calculated $F_C$ are quite large. As can be seen in Fig. 5, the non-relativistic IA prediction[13] for $F_C$ of the deuteron without MEC contributions is in good agreement with the data. However, it is noted that this same non-relativistic IA model[23] requires MEC contributions to give good agreement with the data for the form factors of the three- and four-body nuclei. The failure of this non-relativistic IA model to describe consistently the electromagnetic structure of the few-body nuclei may indicate the range of validity of the impulse approximation.

CONCLUSIONS

The $t_{20}$ tensor polarization of the recoil deuterons in elastic $e$-$d$ scattering has been measured at three values of four-momentum transfer: 3.78, 4.22, and 4.62 fm$^{-1}$. Conventional nucleon-meson dynamics based on non-relativistic and relativistic models

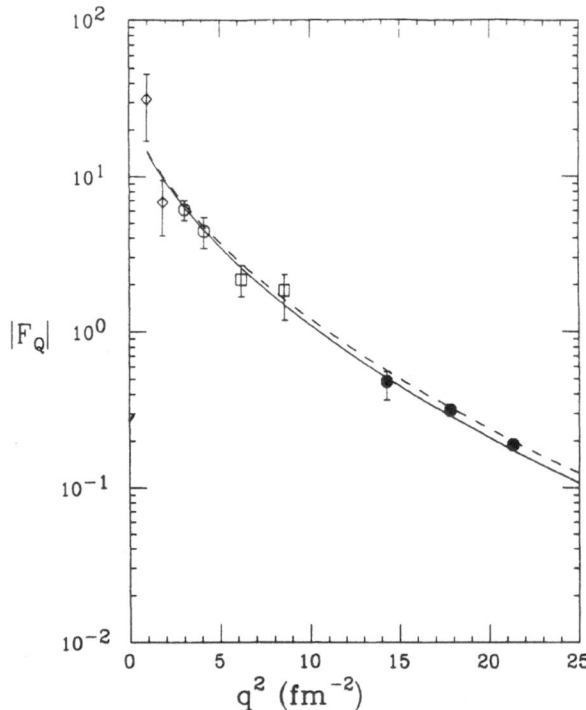

FIG. 6 Charge quadrupole form factor of the deuteron as a function of four-momentum transfer. Same notations as in Fig. 5.

provides good descriptions of the data. Further constraints on these models require either a reduction of the size of the error bars ($\Delta t_{20} < 0.1$) or new measurements at higher momentum transfer. Even though the predictions of some quark models are in good agreement with the data, there is no "smoking gun" signature to show that the inclusion of quark degress of freedom is absolutely required to describe the data. The perturbative QCD model fails to describe the data in this still relatively low momentum transfer region. The charge monopole and charge quadrupole form factors of the deuteron have been separated, and a node in the charge monopole form factor is seen at $4.39 \pm 0.16$ fm$^{-1}$. The position of this node provides a strong constraint on the possible models for the electromagnetic structure of the deuteron.

ACKNOWLEDGMENT

We gratefully acknowledge assistance from L. Antonuk, J. Pasos, G. van der Steenhoven, and A. J. Wagner. We also thank the Bates staff for their help in mounting and running this experiment. This work was supported by the U. S. Department of Energy under Contract No. DE-AC02-76ER03069, the National Science Foundation, the Natural Sciences and Engineering Research Council of Canada, and NATO Collaborative Research Grant No. 860669.

# REFERENCES

[1] R. G. Arnold *et al.*, Phys. Rev. Lett. **35**, 776 (1975); S. Auffret *et al.*, *ibid.* **54**, 649 (1985); P. E. Bosted *et al.*, Phys. Rev. C **42**, 38 (1990); S. Platchkov *et al.*, Nucl. Phys. **A510**, 740 (1990).

[2] M. I. Haftel, L. Mathelitsch, and H. F. K. Zingl, Phys. Rev. C **22**, 1285 (1980).

[3] M. E. Schulze *et al.*, Phys. Rev. Lett. **52**, 597 (1984).

[4] V. F. Dmitriev *et al.*, Phys. Lett. **157B**, 143 (1985); B.B. Voitsekhovskii *et al.*, Pis'ma Zh. Eksp. Teor. Fiz. **43**, 567 (1986) [JETP Lett. **43**, 733 (1986)].

[5] R. Gilman *et al.*, Phys. Rev. Lett. **65**, 1733 (1990).

[6] M. Garçon *et al.*, Nucl. Phys. **A458**, 287 (1986); E. J. Stephenson *et al.*, IUCF Scient. and Techn. Report, 58 (1983).

[7] R. G. Arnold, C. E. Carlson, and F. Gross, Phys. Rev. C **23**, 363 (1981).

[8] J. Arvieux *et al.*, Nucl. Instrum. Methods **A273**, 48 (1988).

[9] J. M. Cameron *et al.* (to be published).

[10] C. E. Hyde-Wright, Ph.D. thesis, Massachusetts Institute of Technology, 1984; Particle Data Group, Phys. Lett. **B239**, III.33 (1990).

[11] F. James and M. Roos, Comput. Phys. Commun. **10**, 343 (1975).

[12] S. Platchkov (private communication).

[13] R. Schiavilla and D. O. Riska, Phys. Rev. C **43**, 437 (1991).

[14] J. Pauschenwein, L. Mathelitsch, and W. Plessas (to be published).

[15] E. Hummel and J. A. Tjon, Phys. Rev. Lett. **63**, 1788 (1989); Phys. Rev. C **42**, 423 (1990).

[16] P. L. Chung, F. Coester, B. D. Keister, and W. N. Polyzou, Phys. Rev. C **37**, 2000 (1988); L. L. Frankfurt, I. L. Grach, L. A. Kondratyuk, and M. I. Strikman, Phys. Rev. Lett. **62**, 387 (1989).

[17] R. Dymarz and F. C. Khanna, Nucl. Phys. **A507**, 560 (1990); Phys. Rev. C **41**, 2438 (1990).

[18] W. P. Sitarski, P. G. Blunden, and E. L. Lomon, Phys. Rev. C **36**, 2479 (1987); P. G. Blunden, W. R. Greenberg, and E. L. Lomon, *ibid.* **40**, 1541 (1989).

[19] Y. Yamauchi and M. Wakamatsu, Nucl. Phys. **A457**, 621 (1986); H. Ito and A. Faessler, *ibid.* **A470**, 626 (1987); H. Ito and L. S. Kisslinger, Phys. Rev. C **40**, 887 (1989).

[20] V. V. Burov and V. N. Dostovalov, Z. Phys. A – Atomic Nuclei **326**, 245 (1987).

[21] E. M. Nyman and D. O. Riska, Nucl. Phys. **A468**, 473 (1987).

[22] C. E. Carlson and F. Gross, Phys. Rev. Lett. **53**, 127 (1984); C. E. Carlson, Nucl. Phys. **A508**, 481c (1990).

[23] R. Schiavilla, V. R. Pandharipande, and D. O. Riska, Phys. Rev. C **40**, 2294 (1989); *ibid.* **41**, 309 (1990).

# ELECTRON SCATTERING FROM TENSOR-POLARIZED

# DEUTERONS IN THE VEPP-3 ELECTRON STORAGE RING

David H. Potterveld

Argonne National Laboratory
9700 S. Cass Ave.
Argonne, Il. 60439

## ELASTIC SCATTERING

In the plane wave impulse approximation, the differential cross section for unpolarized elastic electron-deuteron scattering may be written in the familiar Rosenbluth form:

$$\frac{d\sigma}{d\Omega} = \sigma_m \left[ A(Q^2) + B(Q^2) \tan^2(\theta/2) \right] \equiv \sigma_0 \qquad (1)$$

where $\sigma_m$ is the Mott cross section, and $A$ and $B$ are given in terms of the deuteron charge ($G_c$), quadrupole ($G_q$) and magnetic ($G_m$) form factors:

$$A(Q^2) = G_c{}^2 + (8\,\tau^2/9)\,G_q{}^2 + (2\tau/3)\,G_m{}^2$$
$$B(Q^2) = 4\tau(1+\tau)\,G_m{}^2/3 \qquad (2)$$

with $Q^2$ the square of the 4-momentum transfer and $\tau = Q^2/4m_d{}^2$. Thus, a Rosenbluth separation may be used to extract $A$ and $B$ (and hence $G_m$) from scattering data, but $G_c$ and $G_q$ are not separated. To isolate these form factors requires the use of polarization techniques. Following the Madison convention[1], the scattering of unpolarized electrons from a tensor polarized deuteron is described by the cross section:[2]

$$\frac{d\sigma}{d\Omega} = \sigma_0 \left[ 1 + T_{20}\, t_{20} + 2\,T_{21}\, \mathrm{Re}(t_{21}) + 2\,T_{22}\, \mathrm{Re}(t_{22}) \right] \qquad (3)$$

in which $T_{2i}$ and $t_{2i}$ are, respectively, the components of the analyzing power and polarization tensors in a spherical basis. For moderate momentum transfers and suitably chosen polarization directions, the terms involving $T_{21}$ and $T_{22}$ are small, and may be ignored for the moment. The tensor analyzing power, $T_{20}$, is given by:

$$T_{20} = -\sqrt{2}\left[ X(X+2) + Y/2 \right] / \left[ 2(X^2+1) + 1 \right]$$
$$X = \frac{2}{3}\tau \left( \frac{G_q}{G_c} \right)$$
$$Y = \frac{1}{3}\tau \left( \frac{G_m}{G_c} \right)^2 \left[ 1 + 2(1+\tau) \tan^2(\theta/2) \right] \qquad (4)$$

while the tensor polarization, $t_{20}$, is:

$$t_{20} = p_{zz}\, P_2(\hat{n} \cdot \hat{q})/\sqrt{2} \qquad (5)$$

in which $P_2$ is the second Legendre polynomial, $\hat{n}$ is the polarization direction, $\hat{q}$ is the momentum transfer direction, and $p_{zz}$, the polarization in a cartesian basis, is $1 - 3\,n_0$, with $n_0$ being the fraction of deuterons with zero spin projection.

The effect of $T_{20}$ (and also $T_{21}$ and $T_{22}$) on the cross section is manifested as an asymmetry observed when either the target polarization, $p_{zz}$, the polarization direction, $\hat{n}$, or the azimuthal scattering angle is changed. At moderate momentum transfers and scattering angles, the terms involving $X$ are dominant, and $T_{20}$ becomes very sensitive to a predicted zero crossing of $G_c$, near $Q^2 = 0.6$ GeV$^2$. The location of this zero depends on details of the deuteron wave function and meson exchange effects. Thus, a measurement of $T_{20}$ can be combined with existing data on $A$ and $B$ to extract all three deuteron form factors, and determine the deuteron wave function. Additionally, $T_{20}$ is insensitive to nucleonic electromagnetic form factors, such as the poorly known $G_{en}$ which cancels out of the expression for $X$. A measurement of $T_{20}$ therefore probes the deuteron's structure directly.

These ideas are illustrated in figures 1 and 2, that show selected world data for $A$ and $B$[3-7], and the data for $T_{20}$ published prior to this experiment[8,9], together with a representative sample of theoretical curves[10-14]. The curves represent "realistic" calculations involving different nucleon potentials, mesonic degrees of freedom, relativistic effects, and nucleon form factors. In addition, we show the perturbative QCD prediction of Carlson, et al.[15] for $T_{20}$. The theoretical calculations will be discussed in more detail

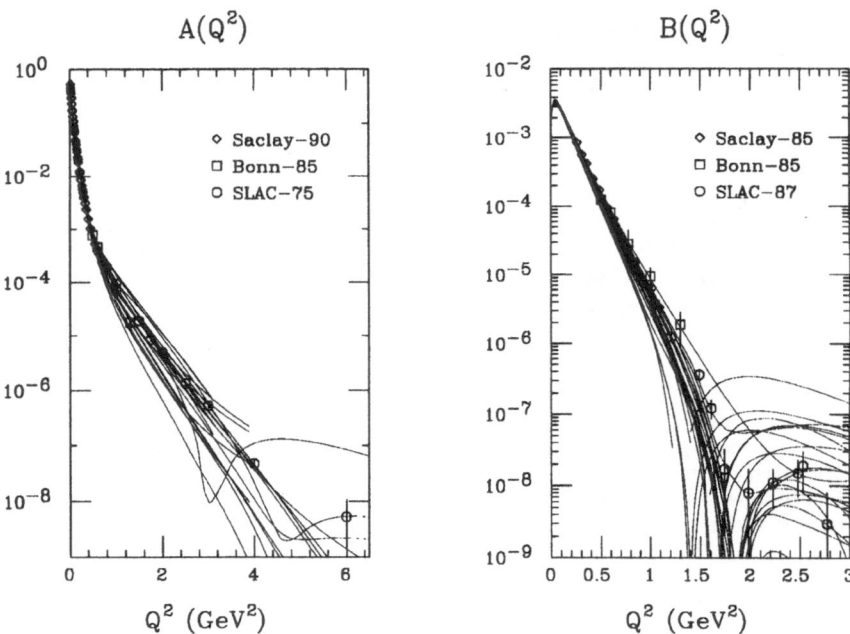

Fig. 1. Selected world data [3-7] and theoretical calculations [10 -14] for $A$ and $B$.

below, but two points may now be observed. First, the curves for $A$ and $B$ spread over a large range of values. As we shall see later, there are relatively few calculations that simultaneously fit the data for $A$ and $B$. Second, the data for $T_{20}$ do not extend to a high enough $Q^2$ to distinguish between any of the theories. Above a $Q^2$ of $\sim 0.6$ GeV$^2$, however, the curves for $T_{20}$ diverge quite widely, reflecting the sensitivity to $G_c$, and meson exchange currents.

In order to resolve the situation, accurate data at these momentum transfers has been desired for many years. We have undertaken a series of measurements at the VEPP-3 accelerator in Novosibirsk to measure $T_{20}$ out to $Q^2 \sim 1$ GeV$^2$. The experiment has been divided into three phases, which will provide results at increasingly higher momentum transfers. The first phase of this experiment is now complete[16], with data for $Q^2 < 0.35$ GeV$^2$, the second phase is underway, and the third is expected to run in the next two years. In addition to the physics goals, this work is notable in that it represents the first use of polarized atoms in a storage tube as an internal target in an electron storage ring.

## EXPERIMENT

This experiment was performed by the ANL/INP collaboration at the VEPP-3 electron storage ring at the Institute of Nuclear Physics in Novosibirsk, USSR. Electrons of 2.0 GeV were circulated around the ring at a frequency of 4 Mhz in a 1.0 ns bunch to form a current of up to 0.2 amps. In a straight section of the ring, the electrons passed

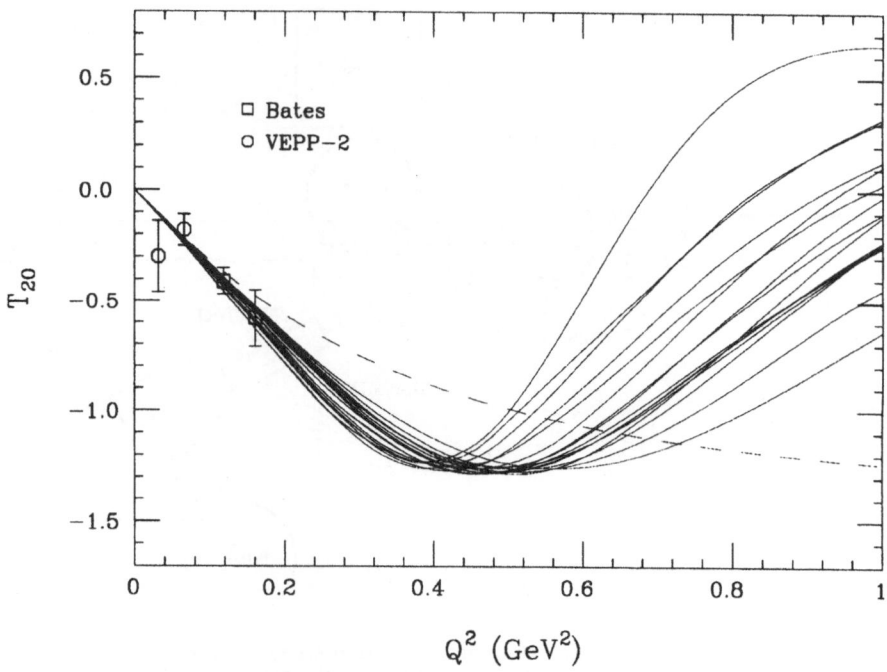

Fig. 2. World data (prior to this work) and theoretical calculations for $T_{20}$. The curves are from the same calculations as fig. 1. The dashed line is from [15].

through a fixed, windowless, Al storage tube, as shown in figure 3. The tube was 940 mm long, with an elliptical cross section of 24 mm × 46 mm, slightly larger than the cross section of the electron beam at injection. (The beam cools to a profile of 0.8 × 2.4 mm after injection.) Polarized deuterium atoms with $|p_{zz}| \sim 1$ from a conventional atomic beam source[17] were injected into the tube through an inlet pipe at the side. A small portion of this beam passed through an exit slit in the storage tube and continued to a polarization monitor consisting of a Rabi magnet followed by a beam profile monitor. Most of the atoms entering the storage tube randomly bounced from wall to wall until exiting out either end of the tube. The mean storage time was 7.5 ms, during which an atom made 400 bounces, on average. To inhibit depolarization during collisions with the wall, the storage tube was coated with drifilm[18]. Based on measurements of target polarization, the depolarization probability per bounce was estimated to be less than $1 \times 10^{-3}$.

The atomic beam source produced a flux of $1 \times 10^{16}$ atoms/s with polarization $|p_{zz}| = 1$. This tensor polarization was achieved by passing the beam through a sextupole magnet followed by an RF unit to induce transitions between the hyperfine states. The sign of $p_{zz}$ could be chosen by inducing different sets of transitions within the RF section. 50% of the atoms in the beam were injected into the storage cell, forming a gas target of total thickness $3 \times 10^{12}$ atoms/cm$^2$. This is 15 times thicker than was previously possible with the atomic beam-jet alone. However, only the central 6 cm were visible to the detectors, so that the useful target thickness was $6 \times 10^{11}$ atoms/cm$^2$.

The average polarization of atoms in the cell, $\bar{p}_{zz}$, was determined to be $0.57 \pm .05$, as discussed below. This value is somewhat higher than the wall bounce depolarization

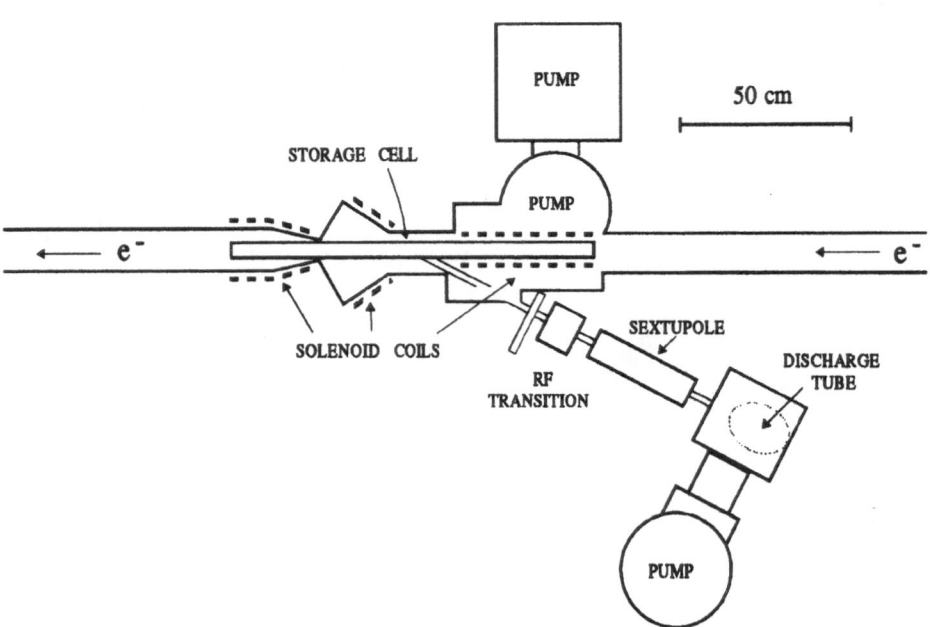

Fig. 3. Storage cell, scattering chamber, and atomic beam source. Not shown is the polarization monitoring equipment, the transverse field magnet surrounding the center of the cell, or the detectors for the scattered particles.

would predict. It is speculated that the longitudinal holding field of $\sim 300$ gauss created by solenoidal coils wound around the cell suppressed the wall bounce depolarization even further. The average polarization was monitored throughout the experiment and was stable throughout the 6 months of running, during which 1 MC of electrons passed through the cell. This clearly shows that drifilm coated storage cells can survive the harsh radiation environment of an electron ring.

Scattered particles left the storage cell through 80 $\mu$m windows etched in the sides, passed through 100 $\mu$m Ti vacuum windows, and entered four nearly identical detector systems[19] shown in figure 4. Each system consisted of separate arms to detect the scattered electrons and deuterons. Both arms consisted of six planes of drift chambers (DC) to determine the particle trajectory, followed by scintillation counters to identify the particle. In the electron arm, the counter was a 1 cm thick plastic scintillator (PS), viewed by a phototube, and preceded by a 5 radiation length thick Pb plate. The deuteron counters were composed of three layers of plastic scintillator of thickness 0.4, 1.0, and 1.0 cm, followed by a segmented counter of either 20 cm of plastic scintillator, or 16 cm of NaI, all viewed by phototubes. The detectors accepted electrons scattered between 10° and 22°, in a 40° range of $\varphi$, and deuterons in the corresponding range $68° \leq \theta_d \leq 80°$.

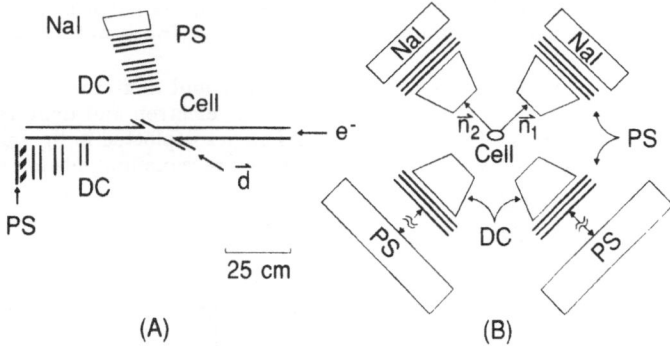

Fig. 4. Storage cell and detector systems. (A) Side view. Only one detector system is shown. (B) Axial view. Only the deuteron arms are shown. Not shown: central magnet to provide transverse field along $\vec{n}_1$ or $\vec{n}_2$.

In the portion of the cell visible to the detectors, the polarization direction of the deuterons was aligned in a transverse direction ($\vec{n}_1$ or $\vec{n}_2$) by applying a transverse magnetic field of 700 gauss. Thus, for a given system, the polarization direction and momentum transfer direction are nearly parallel or nearly perpendicular, so that the $t_{20}$ term of eq. (3) induces an azimuthal asymmetry in the scattering rates of the four systems. This asymmetry may be rotated by 90 degrees by switching the magnetic field between $\vec{n}_1$ and $\vec{n}_2$, or reversed by changing the sign of $p_{zz}$. With four systems, two signs of $p_{zz}$, and two polarization directions, there are 16 scattering rates that may be observed. By combining all 16 rates into a global asymmetry, a large reduction in systematic uncertainties is achieved. Systematic uncertainties were further reduced by reversing the sign of $p_{zz}$ every two minutes during the 1 to 2 hr runs following beam injection into the ring, and by switching the magnetic field between each run.

## DATA ANALYSIS

The trigger for this experiment consisted of a coincidence between the scintillators in the electron and deuteron arms, as well as a minimum number of hits in the drift chambers. (A tagged photon trigger also existed to study photodisintegration events, but this work will not be considered here.) This trigger effectively suppressed background due to stray beam particles, but did not distinguish between elastic electron-deuteron scattering and $(e, e'p)$ events. In fact, the elastic events comprised less than 3% of the data, so that a principal task of the data analysis was to separate deuterons from the much larger proton background.

This separation relied on kinematical correlations between the scattered electron and deuteron, and on the differences in specific ionization between the proton and deuteron. Figure 5a is a scatterplot of the electron scattering angle vs. the amplitude of the first thin scintillation counter in a deuteron arm ($A_1$), corrected for position variations. The events cluster into two bands corresponding to deuterons and protons. Fig. 5b plots the amplitude of the second vs. the first thin counter, for events that penetrate further than the first counter. Particles of the lowest energy deposit negligible energy in the second counter, and a maximum in the first. Particles of increasingly higher energy deposit less energy in the first layer and more in the second, so that the events cluster into bands. The deuteron and proton bands are distinct because of their different specific ionizations. Particles with enough energy to reach the third counter deposit less energy in both the first and second, so that the bands change slope. Although the high energy proton and deuteron bands overlap here, the particles can be identified by a similar technique using the amplitudes from the last two counters penetrated.

Figure 6 shows the correlation between electron and deuteron scattering angles. After all other cuts have been applied, the sum of the electron and deuteron scattering angles is plotted in fig. 6a and the difference in phi angles is plotted in fig. 6b. The elastic events cluster in a large peak, with a small background of remaining proton events. After

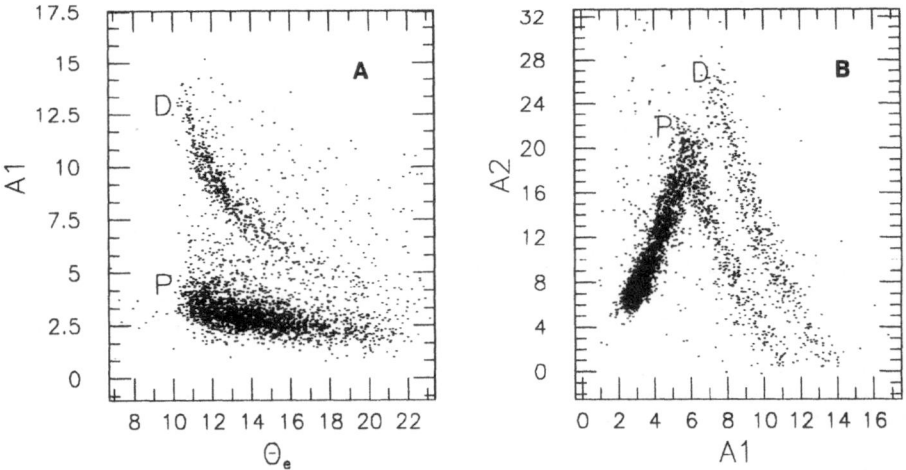

Fig. 5. Event distributions used in particle identification. (A) Correlation between electron scattering angle and amplitude from the first thin counter. (B) Correlation between amplitudes from the first and second thin counters.

a final cut on these correlations, the background is sufficiently suppressed that it may be ignored.

The remaining elastic events were then grouped into three momentum transfer bins, depending on the counter in which the particle stopped, ($A_2$, $A_3$, or the thick counter), and further grouped according to the magnetic field direction and sign of $p_{zz}$. The number of events in each bin was normalized by the collected charge, and the resulting scattering yields were reduced to an experimental asymmetry defined as:

$$a_{\text{exp}} = \left[ (S^2_{1+} - S^2_{1-} - S^2_{2+} + S^2_{2-}) - (S^1_{1+} - S^1_{1-} - S^1_{2+} + S^1_{2-}) \right] / \sum S^i_{jk} \qquad (6)$$

in which $S^i_{jk}$ is the sum of the counting rates in the detector systems facing the directions $\pm n_i$, with the magnetic guide field pointing in direction $n_j$, and the sign of $p_{zz}$ given by $k$. The values of $a_{\text{exp}}$ so obtained are listed in table 1.

To extract $T_{20}$ from these asymmetries it is necessary to know the average target polarization $\bar{p}_{zz}$. This quantity was not measured directly, but instead was determined by normalizing $a_{\text{exp}}$ at the lowest $Q^2$ to the prediction of theory. For this purpose, we used a PWIA calculation based on the non-relativistic wavefunctions of the Paris potential, including such effects as detector acceptances and non-zero values for $t_{21}$ and $t_{22}$. At this momentum transfer, the theoretical ambiguity is small, with a systematic uncertainty of $< 5\%$. We thus determined $\bar{p}_{zz} = 0.57 \pm .05$. This was used to compute $a'_{\text{exp}} = a_{\text{exp}}/\bar{p}_{zz}$, the asymmetry we would have observed in the case of perfect target polarization. $T_{20}$ is then given by:

$$T_{20} = a'_{\text{exp}} / \bar{t}_{20} + [\text{corrections for } T_{21}, T_{22} \text{ terms}]. \qquad (7)$$

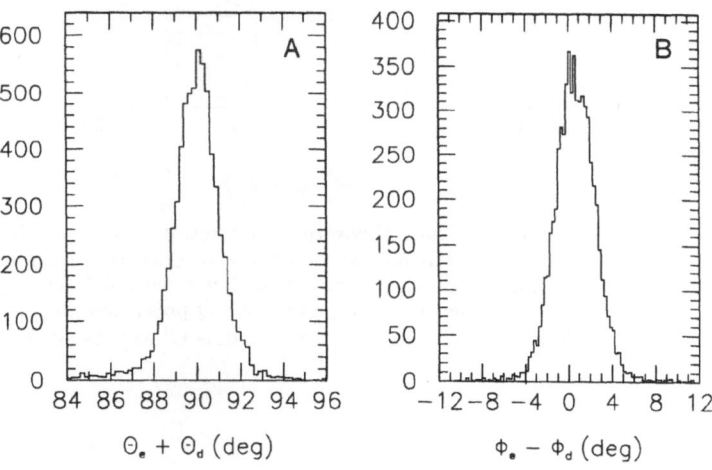

Fig. 6. Correlation between the electron and deuteron scattering angles, after application of all other cuts. (A) The sum of $\theta_e$ and $\theta_d$. (B) The difference $\varphi_e - \varphi_d$.

Table 1. Experimental results. The results at the lowest $Q$ for $a'_{exp}$ and $T_{20}$ are normalized to predictions from the Paris potential.

| $\bar{Q}^2$ (GeV$^2$) | $a_{exp}$ | $a'_{exp}$ | $T_{20}$ |
|---|---|---|---|
| 0.150 | $0.140 \pm .013$ | 0.245 | $-0.538$ |
| 0.242 | $0.189 \pm .038$ | $0.330 \pm .073$ | $-0.77 \pm .16 \pm .07$ |
| 0.335 | $0.309 \pm .077$ | $0.539 \pm .14$ | $-1.32 \pm .32 \pm .11$ |

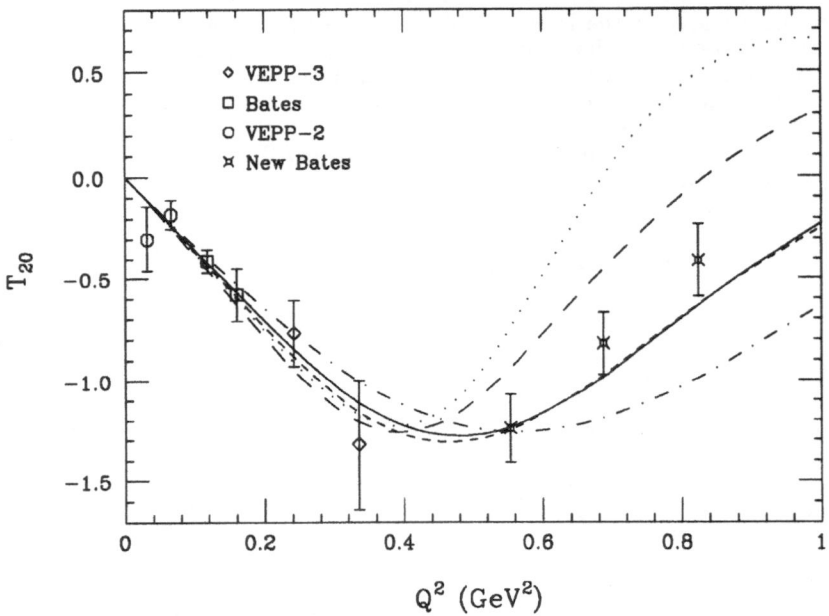

Fig. 7. Experimental results and theoretical predictions for $T_{20}$. The error bars represent statistical and systematic uncertainties added in quadrature. The solid, short-dashed, and dot-dashed lines are from [14], and represent the Paris, $V_{14}$, and Bonn Q potentials, respectively. The dotted and dashed lines are from models $C'$ and $D'$ of [11].

The effective tensor polarization components, $\bar{t}_{20}$, $\bar{t}_{21}$, and $\bar{t}_{22}$ were obtained by averaging the values for a fully polarized deuteron over the acceptance of the apparatus. The latter two were combined with predictions for $T_{21}$ and $T_{22}$ from the non-relativistic Paris potential to compute the correction terms in eq. 7. These respectively amount to +10% and −1.7% of the total asymmetry for $Q = 2.93$ fm$^{-1}$, the worst case. The resulting values for $T_{20}$ are also listed in table 1, and plotted in figure 7. Also shown are the results at lower $Q$ from previous experiments at Novosibirsk[9] and Bates[8], and the results at higher $Q$ from the Bates/Ahead collaboration[20], discussed elsewhere in these proceedings.

Several theoretical predictions for $T_{20}$, based on realistic nucleon potentials, as well as a prediction based on perturbative QCD, overlay the data in figure 7. Above a $Q^2$ of 0.4 GeV$^2$, the potential-based curves are strongly divergent, largely due to differences in $G_c$. The PQCD result is even more divergent, falling towards an asymptotic value of $-\sqrt{2}$ for large $Q^2$. The data from our experiment do not extend to a high enough $Q^2$ to distinguish between the curves, whereas the new data from Bates clearly rule out the PQCD curve and are also inconsistent with the $C'$ and $D'$ models of Blunden et al. and the Bonn-Q result of Chung et al. Our results are currently limited to $Q^2 < .35$ GeV$^2$ by the luminosity available at VEPP-3 with this target. However, improvements are underway to use a higher density storage cell, new detectors, and an optically pumped source of polarized deuterons to increase the luminosity by a factor of $\sim 50$. This will allow us to measure $T_{20}$ for $Q^2 \sim 1$ GeV$^2$ with smaller error bars than the present data from Bates.

## COMPARISON OF THEORY WITH DATA

With data for $T_{20}$ now extending to 0.82 GeV$^2$, it is quite interesting to simultaneously compare theoretical predictions with data for all three form factors. At issue are such questions as the form of the nucleon potential, the importance of relativistic effects and meson exchange currents, and the poorly known nucleon form factors, in particular $G_{en}$. We shall consider five representative sets of models which address these effects: the work of Blunden et al.[11], Dymarz et al.[12], Schiavilla-Riska[10], Chung et al.[14], and Hummel-Tjon[13].

The first three are non-relativistic calculations. Blunden et al. present a coupled channel model (NN, N$\Delta$, $\Delta\Delta$, NN*) explicitly including isoscalar meson exchange currents. We consider the $C'$ and $D'$ models with Hoehler (H) and Gari-Krumpelmann (GK) form factors. Dymarz et al. also use the coupled channel formalism in which virtual $\Delta$ isobars are explicitly included. The work is based on a OBE model, also including MEC. Schiavilla and Riska use the Argonne $V_{14}$ potential to construct an exchange-current consistent with current conservation, with MEC explicitly included.

The other two calculations are formulated in a relativistic framework. Hummel and Tjon use covariant wavefunctions based on the one-boson-exchange model, using either H or GK form factors, with MEC explicitly included. Chung et al. use Hamiltonian light-front dynamics to derive wavefunctions based on non-relativistic NN potentials, but explicit isoscalar MEC are not included. They investigated a variety of commonly used NN potentials, with both H and GK form factors.

The comparison of the predictions for $A$, $B$, and $T_{20}$ with the data is presented in table 2. Each calculation was judged by eye to be either in good agreement (+), clear disagreement (-), or questionable agreement (?). Although subjective, this ranking illustrates several points. First, no model is able to obtain an excellent fit (all pluses) to all the data. Second, only two models have no clear disagreement with the data (no minuses). Interestingly, these are two relativistic models which use Gari-Krumpelmann form factors, although in other details they are quite different. Lastly, neither of the coupled channel calculations is able to simultaneously fit $A$ and $B$.

# Table 2.

Table 2. Comparison of theoretical models with experiment. The quality of agreement between models and data for $T_{20}$, $A$, and $B$ was judged by eye. A '+' signifies good agreement, a '-' signifies clear disagreement, and a '?' signifies questionable agreement. $A$ and $B$ are judged separately for $Q^2$ below and above 1.5 Gev$^2$. Models without clear disagreement are flagged by a star at the left.

| Theory | A Hi $Q^2$ | A Low $Q^2$ | B Hi $Q^2$ | B Low $Q^2$ | $T_{20}$ |
|---|---|---|---|---|---|
| Schiavilla-Riska | + | + | + | + | - |
| Blunden et al. | | | | | |
| $\quad C'$ (H) | ? | ? | - | ? | - |
| $\quad C'$ (GK) | - | - | + | + | - |
| $\quad D'$ (H) | + | ? | - | - | - |
| $\quad D'$ (GK) | - | + | + | ? | - |
| Dymarz et al. (W1) | | | | | |
| $\quad$ NN | ? | + | - | - | ? |
| $\quad$ NN+$\Delta\Delta$ | ? | + | - | - | ? |
| $\quad$ NN+$\Delta\Delta + \pi\pi$ | - | - | + | + | ? |
| $\quad$ NN+$\Delta\Delta + \pi\pi + \pi\rho\gamma$ | ? | - | - | - | + |
| Hummel-Tjon | | | | | |
| $\quad$ IA (H) | - | - | - | - | + |
| $\quad \pi\rho\gamma$ (H) | ? | - | - | - | + |
| * $\quad \pi\rho\gamma + \omega\epsilon\gamma$ (H) | - | - | + | ? | ? |
| * $\quad \pi\rho\gamma + \omega\epsilon\gamma$ (GK) | + | + | + | ? | ? |
| Chung et al. | | | | | |
| $\quad$ Paris (H) | - | + | ? | - | ? |
| $\quad$ Paris (GK) | + | ? | ? | - | ? |
| $\quad$ AV14 (H) | - | + | + | ? | ? |
| * $\quad$ AV14 (GK) | + | ? | + | ? | ? |
| $\quad$ Bonn-E (H) | - | - | ? | - | - |
| $\quad$ Bonn-E (GK) | - | ? | + | ? | - |
| $\quad$ RSC (GK) | + | ? | - | - | + |

## SEPARATION INTO $G_c$ AND $G_q$

With data in hand for $A$, $B$, and $T_{20}$, it is possible to extract values for $G_c$ and $G_q$ with which to compare these models. The world data for $A$ and $B$ was fit with smooth functions to remove statistical fluctuations, and the resulting values were combined with the $T_{20}$ data to extract $|G_c|$, $|G_q|$, and $G_c \cdot G_q$. These values are plotted in figure 8. The error bars are given by:

$$\Delta G^2 = (\partial G/\partial T_{20})^2 \, \Delta T_{20}^2 + (\partial G/\partial A)^2 \, \Delta A^2 + (\partial G/\partial B)^2 \, \Delta B^2 \qquad (8)$$

except for the points at $Q^2 = 0.34$ GeV$^2$ and 0.55 GeV$^2$, for which $T_{20}$ is very close to the minimum value physically allowed by $A$ and $B$. In these cases the likelihood function does not have a gaussian shape, and error bars are, strictly speaking, not appropriate. Moreover, the application of eq. 8 for $Q^2 = 0.55$ GeV$^2$ is not possible because the data point lies just outside the physically allowed region! (The data point is consistent with the physically allowed region, within the statistical uncertainty.) In these cases we have mapped the upper $T_{20}$ error bar to an error bar for $G_c$ and $G_q$, and reflected it symmetrically about the $G_c$ or $G_q$ data point, which we take to be the border of the physically allowed region for $Q^2 = 0.55$ GeV$^2$.

From a change in the sign of $G_c \cdot G_q$, we infer a change in the sign of $G_c$ for the highest $Q^2$ of the new Bates data, that indicates a zero-crossing of $G_c$. The three curves in figure 8 correspond to the three models which can simultaneously fit $A$ and $B$. Interestingly, the $G_c$ predictions of Chung et al. and Hummel-Tjon lie almost directly on top of each other! They also predict a zero-crossing that agrees well with the data. On the other hand, the curve for Schiavilla-Riska places the zero-crossing at too low a momentum transfer. Schiavilla-Riska has an excellent fit to $A$ and $B$, but the misplacement of the $G_c$ zero-crossing evidently disturbs the agreement with $T_{20}$. In general, we find that models with a $G_c$ zero-crossing at too low a value of $Q^2$ result in $T_{20}$ predictions that lie above the new data, and models with a zero-crossing at too high a value of $Q^2$ lie below the $T_{20}$ data. Thus, the $T_{20}$ data strongly constrains the location of this zero.

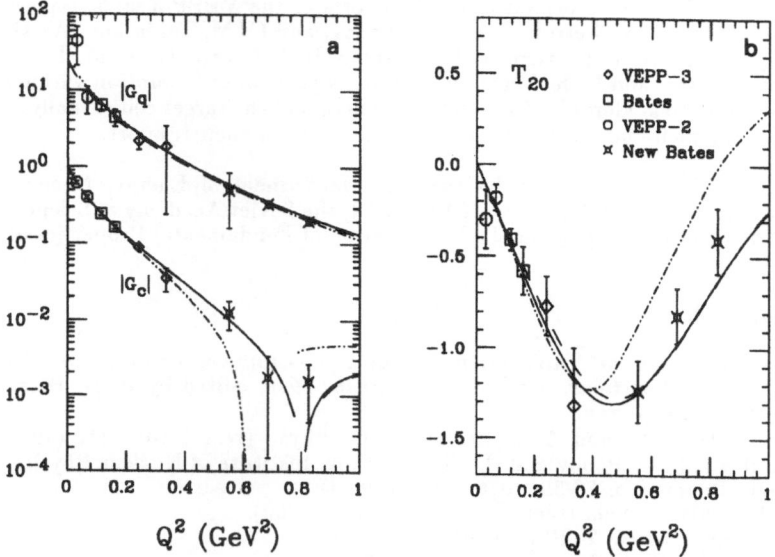

Fig. 8. Values for $G_c$ and $G_q$ extracted from the data for $A$, $B$, and $T_{20}$ are plotted in (a). $T_{20}$ is plotted in (b). The solid curve is the prediction of Chung et al. (Argonne $V_{14}$, GK), the dashed curve is the prediction of Hummel-Tjon $(\pi\rho\gamma + \omega\epsilon\gamma,$ GK), and the dash-dotted curve is the prediction of Schiavilla-Riska.

## CONCLUSIONS

In conclusion, we have measured the asymmetry in electron-deuteron elastic scattering to determine $T_{20}$ for momentum transfers $Q^2 < .34$ GeV$^2$. This work represents the first use of a polarized atoms in a storage tube as an internal target in an electron storage ring. Work is in progress to extend these measurements to $Q^2 \sim 1$ GeV$^2$ through the use of a high density target cell and a high-flux, optically pumped source of polarized deuterons. Although the present results cannot distinguish between theoretical models, they may be combined with recent results from Bates to show a clear minimum in $T_{20}$

for $Q^2 \sim 0.45$ GeV$^2$, in contrast to the PQCD predictions of ref. 15. A comparison between these data and the predictions of several theoretical models shows that it is difficult to simultaneously fit the data for $A$, $B$, and $T_{20}$. Separation of the data into $G_c$ and $G_q$ reveals that $G_c$ passes through zero near $Q^2 = 0.75$ GeV$^2$. The theoretical predictions for $T_{20}$ are very sensitive to the placement of this zero, and some models are in clear disagreement with the new results. The situation remains ambiguous, however, and further experimental and theoretical results are eagerly awaited.

## ACKNOWLEGEMENTS

This work was performed by the the Argonne/Novosibirsk phase-1 collaboration: R. Gilman, R. J. Holt, E. R. Kinney, R. S. Kowalczyk, S. I. Mishnev, J. Napolitano, D. M. Nikolenko, S. G. Popov, D. H. Potterveld, I. A. Rachek, A. B. Temnykh, D. K. Toporkov, E. P. Tsentalovich, B. B. Wojtsekhowski, and L. Young. We wish to thank Drs. F. Coester, V. F. Dmitriev, and V. G. Zelevinsky for many useful discussions. It is a pleasure to acknowledge the efforts of the VEPP-3 staff, especially Drs. Yu. I. Eidelman, V. M. Petrov, J. Y. Protopopov, and G. M. Tumajkin. We also thank Drs. P. I. Baturin, A. V. Evstigneev, L. G. Isaeva, B. A. Lazarenko, E. M. Trakhtenberg, Yu. G. Ukraintsev, and D. K. Vesnovsky for many years of collaboration with our group. We are grateful to J. Goral for the mechanical design of the target cell. Finally, we thank Dr. D. S. Gemmell and Academician A. N. Skrinsky for their support.

This work was supported by the U.S. Department of Energy, Nuclear Physics Division, under contract No. W-31-109-ENG-38, the Soviet Academy of Science, and the Soviet-American Committee for the Investigation of Fundamental Properties of Matter.

## REFERENCES

1. *Proceedings of the Third International Symposium on Polarization Phenomena in Nuclear Reactions, Madison, Wisconsin, 1970*, edited by H. H. Barschall and W. Haeberli, p. xxv.
2. T. W. Donnelly and A. S. Raskin, Ann. Phys. (N.Y.) **169**, 247 (1986); V. F. Dmitriev, S. G. Popov and, D. K. Toporkov, Institute of Nuclear Physics, Novosibirsk, Report No. 76-85, 1976 (unpublished).
3. S. Platchkov, *et al.*, Nucl. Phys. **A510**, 740 (1990).
4. R. Cramer, *et al.*, Z. Phys. C **29**, 513 (1985).
5. R. Arnold, *et al.*, Phys. Rev. Lett. **35**, 776 (1975).
6. S. Auffret, *et al.*, Phys. Rev. Lett. **54**, 649 (1985).
7. R. G. Arnold, *et al.*, Phys. Rev. Lett. **58**, 1723 (1987).
8. M. E. Schulze, *et al.*, Phys. Rev. Lett. **52**, 597 (1984).
9. V. F. Dmitriev, *et al.*, Phys. Lett. **157B**, 143 (1985); B. B. Wojtsekhowski, *et al.*, Pis'ma Zh. Eksp. Teo. Fiz. **43**, 567 (1986) [JETP Lett. **43**, 733 (1986)].
10. R. Schiavilla and D. O. Riska, Phys. Rev. C **43**, 437 (1991).
11. P. G. Blunden, *et al.*, Phys. Rev. C **40**, 1541 (1989).
12. R. Dymarz and F. C. Khanna, Nucl. Phys. **A507**, 560 (1990).
13. E. Hummel and J. A. Tjon, Phys. Rev. Lett. **63**, 1788 (1989).
14. P. L. Chung itet al., Phys. Rev. C **37**, 2000 (1988).
15. C. Carlson, Nucl. Phys. **A508**, 481c (1990).
16. R. Gilman *et al.*, Phys. Rev. Lett. **65**, 1733 (1991).
17. A. V. Evstigneev, S. G. Popov and D. K. Toporkov, Nucl. Instrum. Methods Phys. Res., Sect. **A 238**, 12 (1985).
18. L. Young *et al.*, Nucl. Instrum. Methods Phys. Res., Sect. **B 24/25**, 963 (1987); D. R. Swenson and L. W. Anderson, Nucl. Instrum. Methods Phys. Res., Sect. **29**, 627 (1988).
19. B. B. Wojtsekhowski *et al.*, Institute of Nuclear Physics, Novosibirsk, Report No. 88-120 (to be published).
20. I. The, *et al.*, Phys. Rev. Lett. **67** 173 (1991).

# STUDY OF DECAY AND ABSORPTION OF Δ RESONANCE

# IN NUCLEI WITH A 4π DETECTOR

B.Ramstein[a], D.Bachelier[a], H.G.Bohlen[b], J.L.Boyard[a,b], C.Ellegaard[d],
C.Gaarde[d], J.Gosset[c], T.Hennino[a], J.C.Jourdain[a], J.S.Larsen[d],
M.C.Lemaire[b], D.L'Hote[c], H.P.Morsch[b], M.Osterlund[e], J.Poitou[c],
P.Radvanyi[b], M.Roy-Stephan[a], T.Sams[b], O.Valette[c], P.Zupranski[a,b]

[a] Institut de Physique Nucléaire 91406 Orsay Cedex, France
[b] LN Saturne CEN-Saclay, 91191 Gif sur Yvette Cedex, France
[c] DPHN/SEPN,CEN-Saclay 91191 Gif sur Yvette Cedex, France
[d] Niels Bohr Institute, Blegdamsvej 17, DK-2100 Copenhagen, Denmark
[e] University of Lund, S-22362 Lund, Sweden

Charge exchange reactions have been investigated at Saturne with various projectiles ($^3$He, $\vec{d}$, $^{12}$C, $^{16}$O, $^{20}$Ne, $^{40}$Ar) and incident energies between .5 and 1.1 GeV/nucleon[1]. These experiments have given clear indication of the selectivity of charge exchange reactions for spin isospin excitations. In particular, a strong excitation of the Δ resonance shows up for energy transfers higher than the pion mass. The most striking common feature of all these charge exchange reactions is a shift of about 70 MeV between the positions of the Δ bump on nuclei and on hydrogen targets (fig.1). This result has been interpreted by theoreticians as a collective effect in the spin longitudinal channel[2] and has been quantitatively reproduced by recent calculations[3]. An alternative qualitative explanation is the projectile excitation[4].

In this context, the experimental

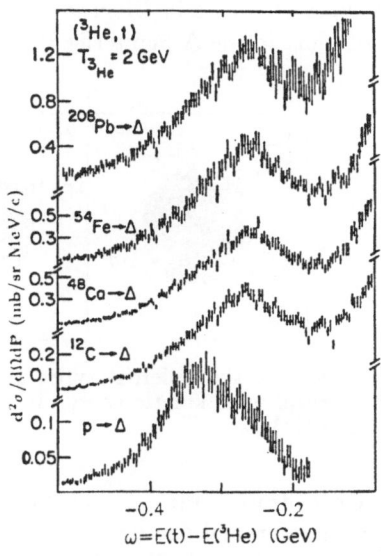

Fig.1 Spectra of the energy transfer in ($^3$He,t) reaction at 2 GeV on hydrogen and on nuclei

study of absorption and decay channels of the $\Delta$ resonance in nuclei is of great interest because it can teach us if one particular decay mode is responsible for the shift observed in the inclusive experiments and help to evaluate the possible contribution of projectile excitation.

For that purpose, we performed at Saturne an "exclusive" ($^3$He,t) experiment at 2 GeV where, in coincidence with the forward emitted triton, we detected the charged pions and protons produced by the decay of the target nucleus in the large acceptance Diogène detector consisting of 10 trapeze-shaped drift chambers in a 1 T magnetic field[5] (fig.2). The triton energy and angle were measured in drift chambers in the range 1.4 to 2 GeV and $0^0$ to $4^0$, while Diogène allowed particle identification, as well as momentum and angles $(\theta, \phi)$ measurements. Energy cuts were about 15 MeV for pions and 35 MeV for protons. The $\theta$ acceptance ranged from $20^0$ to $132^0$. The experiment has been performed on liquid hydrogen, liquid deuterium and carbon targets.

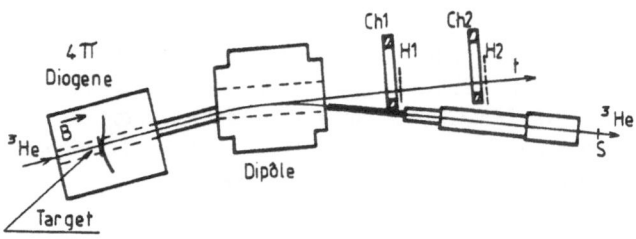

Fig.2 Experimental set-up

The simplest result we can extract from our data is the amount of the different types of events in Diogène in coincidence with a triton with kinetic energy corresponding to the $\Delta$ bump (fig.3).

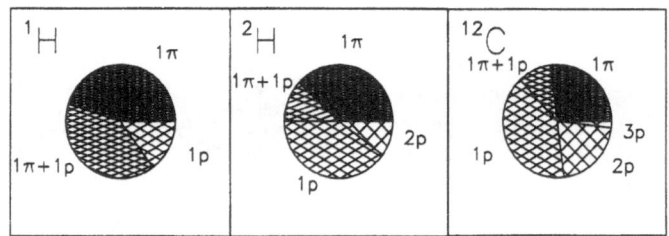

Fig.3 Yields of the different types of events detected in Diogène in coincidence with a triton with kinetic energy in the region of excitation of the $\Delta$ resonance ($\omega \leq -.140$ GeV)

On hydrogen, the rates of events with only one pion ($1\pi$), only one proton ($1p$) or one pion and one proton ($1\pi+1p$) are only due to the acceptance cuts and ray-tracing efficiency in the Diogène detector. The only possible process is indeed $\Delta^{++} \rightarrow \pi^+ + p$. A larger efficiency will be achieved in later stages of the analysis.

On deuterium target, the number of events with both one pion and one proton detected is much reduced, while two protons events ($2p$) appear. On the carbon target, we have more 2p events and even a significant fraction of events with three protons (3p) detected. This trend is qualitatively in agreement with our expectations of the different reaction processes on nuclei. On deuterium, the quasi-free process leads to excitation either of a $\Delta^{++}$ on the proton, the neutron being spectator, or of the $\Delta^+$ on the neutron, the proton being spectator. As the $\Delta^+$ deexcite into $\pi^0 + p$ or $\pi^+ + n$, this leads to more 1p and $1\pi$ events than on the hydrogen target, as observed experimentally. A fraction of the $1\pi$ events may also be due to the coherent process, where the deuterium is left in its ground state and a pion is emitted. Eventually, 2p events are expected to be due to the absorption process ($N\Delta \to pp$). On carbon, the absorption and coherent processes are likely to be more important than on the deuterium target. Moreover, the absorption process can involve several nucleons, which could explain the 3p events.

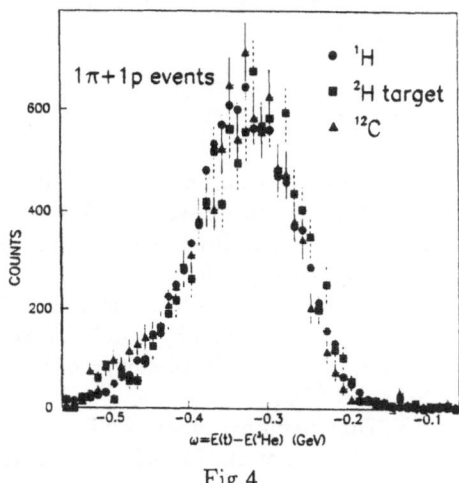

Fig.4

If we now look at a very specific type of events, namely the $1\pi+1p$ events, the energy transfer ($\omega$) spectrum is found to have the same position and width on the three targets within 15 MeV (fig.4). This result is very important because it means that these $1\pi + 1p$ events really select a quasi-free process where the $\Delta$ is excited with the same excitation energy and width than on the free nucleon.

The $\omega$ spectra for tritons in coincidence with 2p events are shown as dashed line on fig.5 for the deuterium and carbon targets. On deuterium, no $\Delta$ bump shows up, so it seems that very few of these 2p events are due to $\Delta$ absorption process. On carbon, on the contrary, a very clear structure can be seen. A structure is also present in the spectrum obtained in coincidence with 3 protons (dotted line) and is located at higher energy transfers, as expected due to the energy threshold for one more proton to be detected in Diogène. For both types of events, the missing mass spectra show that no pion is emitted. This suggests that these events come from absorption of the $\Delta$ resonance in $^{12}$C. The position of the bump associated with 2 protons in Diogène is shifted by 100 MeV in comparison with that obtained in coincidence with $1\pi+1p$ events. This big shift

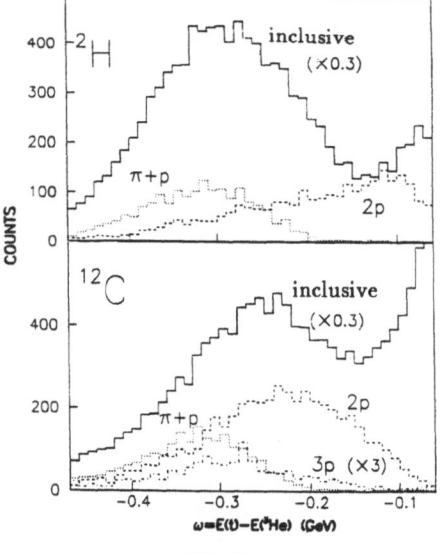

Fig.5

could be a very interesting indication of a collective effect in the absorption channel. It is however important to evaluate the tail due to the non resonant process (as seen on the deuterium).

As a conclusion, preliminary results on $\Delta$ decay modes in $(^3\text{He},t)$ reaction at 2 GeV on $^1\text{H}, ^2\text{H}$ and $^{12}\text{C}$ targets show two important features:
- . Events where one pion and one proton are detected in coincidence with the forward emitted triton select a quasi-free excitation of the $\Delta$ resonance with the same excitation energy and width as on the free nucleon.
- . A 100 MeV shift towards lower energy transfers is obtained when selecting events with 2 protons emitted, which could be an indication of a very strong collective effect in the absorption process of the $\Delta$ resonance.

Both results might give very interesting clues for the understanding of medium effects on pionic modes.

## REFERENCES

1) D.Contardo et al.,Phys.Lett. 168B (1986) 331.
   C.Ellegaard,Phys.Lett. 59 (1987) 974.
   M.Roy-Stephan Nucl. Phys. A488 (1988) 187c .
   C.Gaarde,this conference.
2) G.Chanfray et al., Phys. Lett. 141B (1984) 163.
   V. Dmitriev,Nucl. Phys. A459 (1986) 503.
3) J.Delorme et al.,Proceedings of the 10th Session d'etudes d'Aussois 1989, LYCEN Report 8902.
   V. Dmitriev,Phys. Lett. 226B (1989) 219.
   T.Udagawa et al., Phys.Lett. 245B (1990) 1.
   T.Udagawa, this conference.
4) E.Oset et al., Phys. Rev. Lett. 224B (1989) 249.
5) J.P.Alard et al., Nucl. Instr. Meth. A261 (1987) 379.

# FIRST MEASUREMENTS OF SPIN OBSERVABLES IN NN̄

# SCATTERING EXPERIMENTS AT LEAR

F. Bradamante

*Dipartimento di Fisica dell'Università, Trieste, Italy*
*Sezione di Trieste dell'INFN, Trieste, Italy*

## 1. Introduction

This talk summarizes results on spin observables in the p̄p elastic and charge exchange reactions as measured at the Low Energy Antiproton Ring (LEAR) at CERN. In particular I will concentrate on the charge-exchange reaction, where the analyzing power has been measured for the first time. This study is part of a general investigation of the NN̄ interaction through NN̄ scattering, which in the past was pursued in parallel also in Brookhaven and KEK. Before the commissioning of LEAR, most of the available data were cross-section data, and analyzing power data essentially did not exist. The quality of the LEAR beem is so much superior that presently experiments in this field are performed only at CERN, and analyzing power data of good quality have been measured both in the elastic channel and (for the first time) in the charge-exchange channel.

Several reviews exist by now on the LEAR results[1,2]. Also, the Proceedings of the dedicated LEAR Workshops[3,4] contain detailed descriptions of the experiments and of the new results, and specifically NN̄ scattering results have been reviewed recently[5]. Still, for completeness, before talking about the charge-exchange data, I will briefly mention also the scattering data and the polarization data in the elastic channel.

## 2. Motivation for NN̄ physics

The study of the NN̄ interaction at low energy is complementary to the study of the NN interaction, and one cannot say that one understands the latter without being able to explain the former. Although the ultimate goal is the understanding of hadron dynamics at the microscopic level, the comparison between data and theory is usually done by using potential models.

The basic ingredients of the NN̄ interaction are
— meson exchanges at large distances (r ≳ 0.8 fm). These are assumed to be the same

as determined in the NN interaction in the framework of the one-boson-exchange (OBE) models, and the sign of each contribution is obtained from the $G$-parity rule[6]:

$$V_{N\overline{N}} = \sum_{\pi\rho\omega\sigma...} (-1)^G V_{NN}(OBE);$$

- annihilation at short range, usually parametrized as an optical potential, eventually state and energy dependent.

It is well-known that the $G$-parity rule changes dramatically the overall $N\overline{N}$ meson-exchange potential as compared with the NN case. As an example, I will just recall the special role of $\omega$-exchange, which is believed to give rise to a repulsive force in NN interactions, and to a strong attraction in the $N\overline{N}$ case. In this sense the study of the long-range part of the $N\overline{N}$ interaction is complementary to that of the NN channels, and could provide new constraints on the OBE potential models. In addition to this, $N\overline{N}$ physics has sone unique features, i.e.

- the quarks' degrees of freedom play a much more prominent role than in NN (annihilation in $N\overline{N}$ as compared with core repulsion in the latter case);
- annihilation itself is a challenge for any theory of hadron costituents;
- both $N\overline{N}$ scattering and annihilation are expected to give privileged access to exotic states in the s-channel (Fig. 1).

The first point is particularly attractive: the onset of annihilation should allow to determine experimentally the distance at which the quark degrees of freedom take over the nucleonic degrees of freedom.

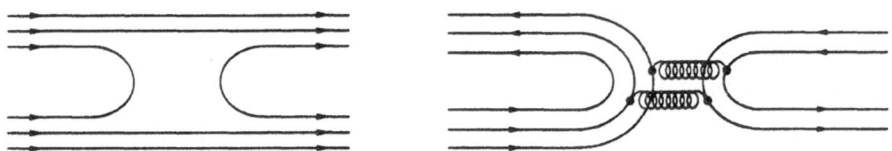

Fig. 1   Quark diagrams for exotic states in $N\overline{N}$ interactions (baryonium and gluonium).

## 3. Status of LEAR experiments

Table 1 summarizes the contribution of LEAR to $N\overline{N}$ scattering. Most of the results are published by now, and all of the experiments are over. The only experiment which is still on the floor is PS 199, the experiment I will talk about, which had its last run in December 1990. A proposed extension of the experiment[23] has not been approved by the SPSLC Committee, so very likely the study of this field at LEAR has come to an end.

Although I will not describe the cross-section data, which were already available at the time of the last Telluride conference, let me just remind briefly their main features:

- the $\bar{p}p$ total cross-section and annihilation cross-section data are very smooth with energy, and no narrow peaks are seen, thus confirming the KEK and BNL results demonstrating the non-existence of a narrow "baryonium" peak at 500 MeV/c, the S(1936), previously seen in several experiments.
- the annihilation cross-section at low momenta is about twice the elastic cross-section, indicating that annihilation is not just absorption by a black disc. This is due to the fact that the meson exchange potential is so strong and attractive in the $N\bar{N}$ system, that it pulls the $N\bar{N}$ wave function into the annihilation region, and effectively increases the absorption radius[24].
- the differential cross-sections have been measured with good precision down to low momenta (180 MeV/c) both in the elastic and in the charge-exchange channels. The main feature of the elastic channel is the strong P-wave enhancement (again due to the strong $N\bar{N}$ nuclear potential), which manifests itself as the forward peak present even at the lowest energies.

## Table 1

|  | Experiment | Measured Momenta (MeV/c) | Reference |
|---|---|---|---|
| **Cross Sections:** |  |  |  |
| $\sigma_{tot}$ ($\bar{p}p$) | PS 172 | 220 ⊢——⊣ 600, 74 momenta | 7, 8 |
| $\sigma_{ann}$ ($\bar{p}p$) | PS 173 | 180 ⊢——⊣ 600, 53 momenta | 9 |
| $\sigma_{tot}$ ($\bar{n}p$) | PS 178 | 100 ⊢——⊣ 350 MeV/c |  |
| $\bar{p}p \rightarrow \bar{p}p$ **Elastic scattering:** |  |  |  |
| $\rho$ | PS 172 | 233, 272, 550, 800, 1100 | 10, 11 |
|  | PS 173 | 181, 219, 239, 261, 287, 505, 590 | 12, 13 |
| $d\sigma/d\Omega$ | PS 173 | 181, 287, 505 | 14 |
|  | PS 172 | 529 ⊢——⊣ 1550, 15 momenta | 15, 16 |
|  | PS 198 | 439, 544, 697 | 17, 18 |
| $A_{0n}$ | PS 172 | 529 ⊢——⊣ 1550, 15 momenta | 15, 16 |
|  | PS 198 | 439, 544, 697 | 17, 18 |
| $D_{0n0n}$ | PS 172 | 988, 1089, 1291, 1359 | 19 |
|  | PS 198 | 697 |  |
| $\bar{p}p \rightarrow \bar{n}n$ **Charge exchange:** |  |  |  |
| $d\sigma/d\Omega$ | PS 173 | 183, 287, 505, 590 | 20 |
|  | PS 199 | 600, ⊢——⊣ 1300, 8 momenta | 21, 22 |
| $A_{0n}$ | PS 199 | 600, ⊢——⊣ 1300, 8 momenta | 21, 22 |
| $D_{0n0n}$ | PS 199 | 600, 900 |  |

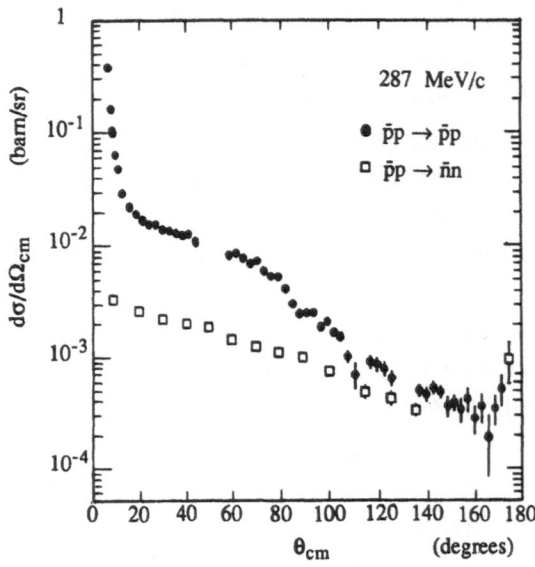

**Fig. 2**  Differential cross-section for $\bar{p}p \to \bar{p}p$ elastic[14] (full points) and $\bar{p}p \to \bar{n}n$ charge-exchange[20] (open squares) as measured from Experiment PS173 at 287 MeV/c.

Fig. 2 shows typical differential cross-sections for the elastic and the charge-exchange channel at a low momentum.

## 4. Comparison with theory

All the cross-sections data in Table 1 can be described surprisingly well by a variety of potential models: apart from the original Bryan and Phillips model, popular names are the Bonn model[25], the Dover-Richard[26] model (in its two versions), the Dalkarov-Myhrer[27] model, the Kohno-Weise[28] model, the Nijmegen model[29], and the Paris model[30]. All these models differ either in the parametrization of the boson exchange part or in the treatment of annihilation (or in both ingredients), but the predictions for the cross-sections are rather stable, which is an indication that the $N\bar{N}$ process is dominated by absorption.

In particular, only $\pi$-exchange and a suitably parametrized annihilation potential, with an annihilation radius of $\sim 0.8$ fm, already give a satisfactory description of the data. For instance, a Wood-Saxon type annihilation potential, with

$$\mathrm{Re}\, V_{\mathrm{ann}} = \frac{-c_0}{1 + \exp[(r-r_0)/a_0]} \;, \qquad \mathrm{Im}\, V_{\mathrm{ann}} = \frac{-c_1}{1 + \exp[(r-r_1)/a_1]} \;,$$

and no energy, spin, or isospin dependence, already gives very good fits with $c_0 = c_1 = 500$ MeV, $r_0 = r_1 = 0.74$ fm, and $a_0 = a_1 = 0.2$ fm[31].

The conclusion is that cross-sections data are only sensitive to the gross features of the interaction, namely to the strength and to the range of the potential describing the annihilation. To learn about the details of the interaction more sophisticated data are needed, i.e. data on spin observables. I was told that R. Bertini had promised them at the previous Telluride Conference, and here they come.

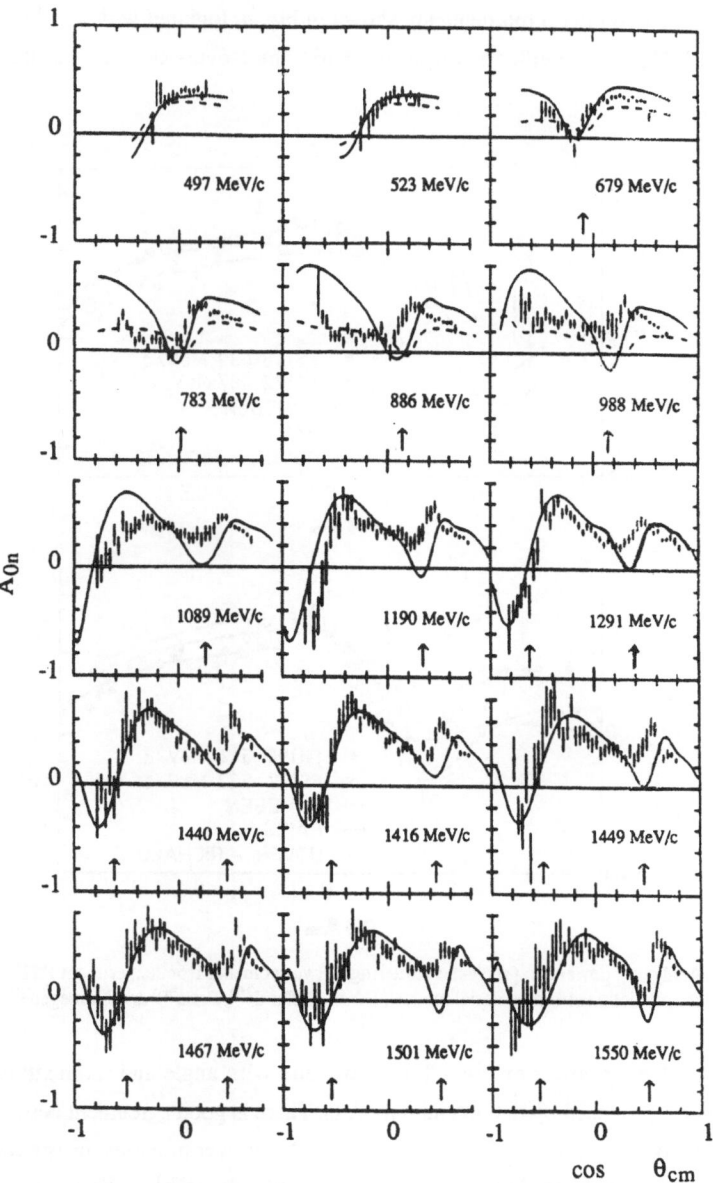

**Fig. 3**   Analyzing power for $\bar{p}p$ elastic scattering at fifteen momenta from experiment PS172. The dashed and full curves are predictions from the model of Ref. 26 and 30, respectively.

## 5. The elastic p̄p channel

The analyzing power $A_{0n}$ has been measured over most of the angular range by experiment PS172[16] at fifteen momenta, ranging from 530 to 1550 MeV/c. The measurements are shown in Fig. 3 and compared with the predictions of the Dover-Richard model and of the Paris model. Experiment PS198 has measured $A_{0n}$ over the entire angular range at 497, 523 and 697 MeV/c. The data at 697 MeV/c are already published[17]: new results[18] at the two lowest momenta are shown in Fig. 4, together with previous data from experiment PS172 and predictions from the Paris, the Dover-Richard, and the Nijmegen models.

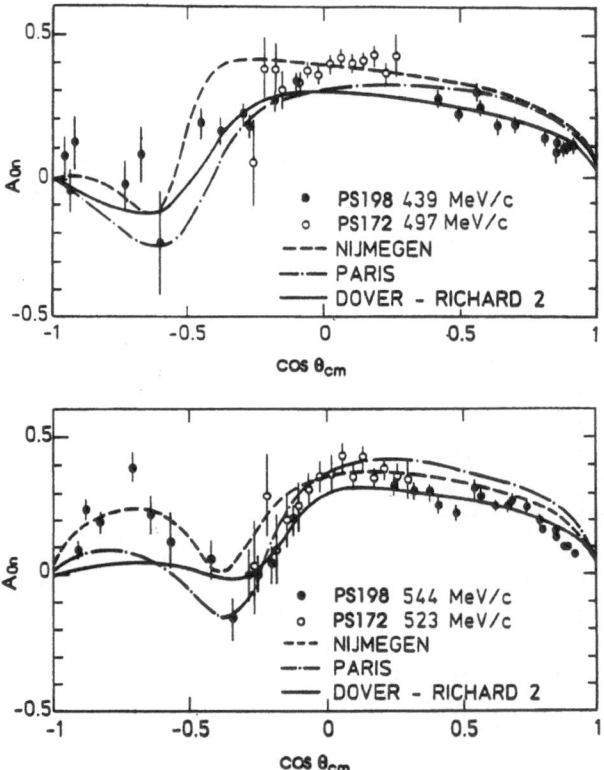

**Fig. 4**    Analyzing power for p̄p elastic scattering at two momenta from experiment PS173. The curves are predictions from potential models, Dover-Richard[26], Paris[30] and Nijmegen[29].

The analyzing power exibits a lot of structure with angle and momentum, which is rather poorly reproduced by the potential models. There is good agreement between the new data, which anyway represent either a vast improvement over previous measurements[32], or cover a region where data did not exist (momenta smaller than 910 MeV/c).

By analyzing the polarization of the scattered proton with a carbon polarimeter, experiment PS172 could obtain some $D_{0n0n}$ data in the backward hemisphere and in the higher momentum range (from 1000 to 1550 MeV/c). Some of this data[19] are shown in Fig. 5: although the error bars are large, the result is interesting because it suggests either zero or negative values for $D_{0n0n}$ (except at 1291 MeV/c), while the potential models would like this parameter to be close to 1. Data on $D_{0n0n}$ have been collected also by experiment PS198 at 700 MeV/c, but the analysis is still in progress.

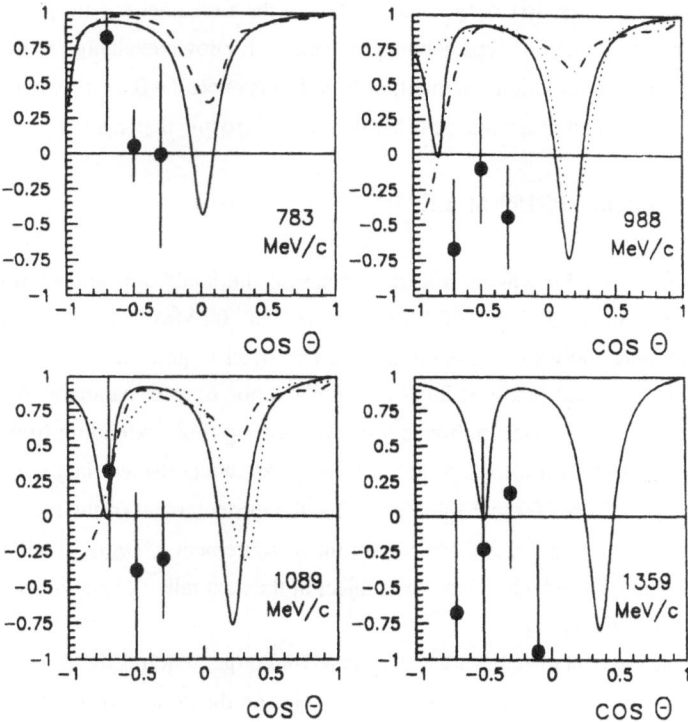

Fig. 5    $D_{0n0n}$ as a functionof $\cos\theta_{cm}$ for four momenta. The curves are the predictions of the relativistic Dover-Richard I model (solid line)[26], the Bonn model (dashed line)[25], the Nijgemen model (dashed-dotted line)[29] and the Paris model (dotted line)[30].

## 6. The charge exchange channel

At variance with the elastic channel, where several measurements (even of polarization!) existed even before LEAR, the charge exchange channel was known very poorly. In particular, no measurements existed of the Analyzing Power. This situation was a consequence of the fact that this channel is more difficult to measure experimentally, due to the presence of two neutral particles in the final state and to the necessity of distinguishing the neutron from the antineutron.

The knowledge of the charge exchange channel is essential to resolve the isospin structure of the $N\overline{N}$ interaction. In contrast to the pp case, the $\overline{p}p$ system is not a pure isospin state. The amplitude for p̄p elastic is given by the sum of the I=0 and the I=1 amplitudes, while the p̄p → n̄n charge-exchange amplitude is given by the difference of the two. Clearly a complete analysis of the system requires the measurements of the two channels.

The charge-exchange reaction is expected to be a particularly sensitive channel to probe the nucleon-antinucleon force. The long range part of the interaction should be dominated by pion exchange, a well known "classical" term in any boson exchange potential model, and the fact that I=1 only is exchanged in the t-channel should protect from the complications of "Pomeron" type exchanges. This well known exchange term can thus be used to probe the annihilation potential, at least to distances (~ 0.8 fm) where absorption becomes dominant, and determine possibly its spin and isospin dependence.

## 7. The experiment  PS199 at LEAR

Experiment PS199 was approved to measure the analyzing power in the charge-exchange reaction from 500 to 1500 MeV/c, in steps of 100 MeV/c, and the spin parameter $D_{0n0n}$ at 500 and 1000 MeV/c, using a solid polarized target; the measurements of the differential cross-section (and possibly of the polarization transfer parameter $K_{n00n}$) can be obtained from the same data. In three periods of running time, extending from September 1989 to August 1990, for a total of about five weeks, we collected data at eight p̄ beam momenta, ranging from 600 to 1300 MeV/c, for the measurement of the analyzing power, and at two momenta, 600 and 900 MeV/c, for the measurement of $D_{0n0n}$. I will not mention the latter measurement which will be the subject of the next talk[33], and will concentrate on the measurement of $A_{0n}$.

The measurement of $A_{0n}$ with a solid polarized target requires the detection of both n and n̄ produced in the reaction, to select the events on the polarized free hydrogen of the target using time of flight, coplanarity and angular correlation information. Also, a complete n / n̄ separation is needed.

The layout of the experiment is shown in Fig. 6. $NC_1$, $NC_2$ and $NC_3$ are the neutron detectors, and $ANC_1$ and $ANC_2$ are the n̄ detectors. The dashed line indicates the p̄ beam direction. The target sits in the nose of 1 m long cryostat (shown in the figure, together with the polarized target magnet (PTM)) and to reach the target the beam travels along its axis.

The pentanol polarized target (PT), 12 cm long, is operated in the frozen-spin mode and a very high polarization (~90%) can be achieved in 2-3 hours; with the holding magnetic field of 3.785 kGauss, the polarization decays of about 2% per hour. During data taking, the target polarization was ~80%, and, to reduce systematic effects, the spin orientation was reversed every four one-hour spills. For background evaluation the PT was replaced by a dummy target (DT) having the same mass and density as the PT but all H replaced by F. For calibration purposes and consistency checks a liquid hydrogen target (LHT), 12 cm long, has also been used.

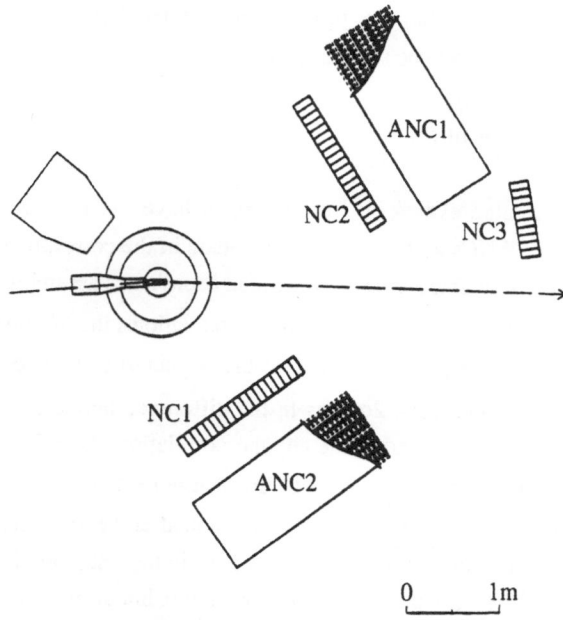

**Fig. 6**    Schematic layout of experiment PS199: NC labels the neutron counter hodoscope, ANC the antineutron detectors.

The neutron counters (NC), are made up of a total of 53 vertical scintillator counters. Each scintillator bar is 8 cm large, 20 cm thick, and its length varies from 40 to 130 cm; it is viewed from each end by a photomultiplier and a coincidence between the top and bottom PM is required for a bar hit. In the off-line analysis the neutron candidates are selected by asking no more than two adjacent bars hit in any of the NCs.

The $\bar{n}$ counters (ANC)[34] have been designed to detect the $\bar{n}$'s by looking for charged particles produced in their annihilation. This method has been already used in other experiments (see, for instance, Ref. 35). The new feature of our detector is that the material in which the $\bar{n}$'s annihilate has been concentrated as much as possible and surrounded by two telescopes of limited streamer tubes (LST) planes to reconstruct the tracks of the charged products of the annihilation; the thickness of the iron absorber (3 cm) has been chosen to maximize the number of outgoing charged particles, taking into account the $\pi^{\pm}$ absorption and the conversion probability of $\gamma$'s. With this geometry, the $\bar{n}$ gives a characteristic "star" pattern, so that a very good $n / \bar{n}$ separation can be obtained without loosing too much in efficiency. To increase the efficiency this modular structure has been repeated four times. The modules are made up of four planes of LST and one hodoscope of scintillator counters, with a sensible surface of $166 \cdot 200 \ cm^2$.

In the off-line analysis (see Ref. 22 for details), the $\bar{n}$ candidates are identified by requiring at least three tracks in the horizontal (or vertical) projection of the two LST modules sandwiching any of the four iron slabs, crossing the slab itself in a common point and at least two crossing tracks in the other projection. This selection allows to identify the n̄'s with an efficiency of ~20%; the probability of misindentifying a neutron as an antineutron has been measured to be smaller than $10^{-4}$.

## 8. Data analysis and results

The charge-exchange events on free hydrogen have been selected requiring a n̄ candidate in one of the ANCs and a neutron candidate in the corresponding NC counter. The neutron angle $\theta_n$ is obtained from the measured coordinates assuming that the reaction took place in the center of the target. Using kinematic relations, the n and n̄ time of flight and the expected n̄ angle $\theta_{n̄}^c$ are calculated; the coplanarity $\varepsilon$ is obtained from the measured n and n̄ coordinates. $2\sigma$ cuts in the difference between the measured and calculated TOFs, in coplanarity and in the angular correlation $\Delta\theta = \theta_{n̄}^c - \theta_{n̄}^m$ eliminate completely the $\gamma$'s from the target and most of the charge-exchange events on the nuclear content of the target. Fig. 7 shows the angular correlation after the TOF and coplanarity cuts for the PT and the DT data: the residual background left in the final sample after the last cut in angular correlation ($|\Delta\theta| < 5°$) increases with momentum, but even at the largest measured momentum is rather small.

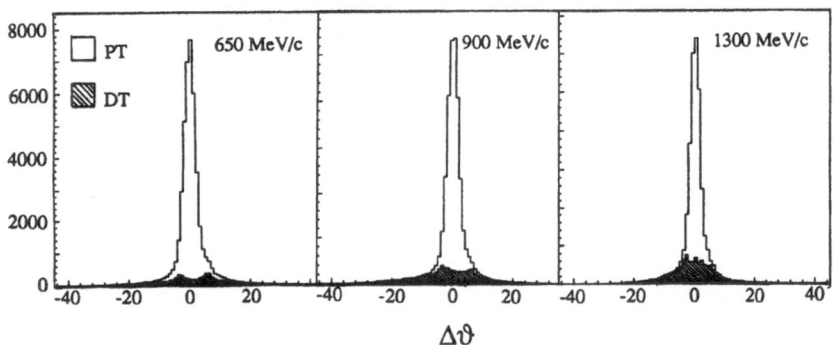

**Fig. 7** Angular correlations $\Delta\vartheta$ for samples of PT events and DT events at three different momenta.

Fig. 8 shows a measurement of the differential cross-section, obtained from a calibration with a liquid hydrogen target. The comparison with a previous measurement done at KEK[36] at the same momentum is satisfactory.

Fig. 9 shows the measured analysing power at 656 MeV/c, the mean p̄ momentum in the PT when the beam was 700 MeV/c. The results are already published[21], and exibit a remarkable angular dependence, reaching quite large positive values both in the forward and in the backward regions. Preliminary results at a nearby beam momentum (600 MeV/c) confirm this trend, but with better statistics. Also shown in Fig. 8 are the predictions of some potential models: apart from the model of Ref. 27, the agreement between data and models predictins is not too good. In particular, it is somewhat of a surprise that the various

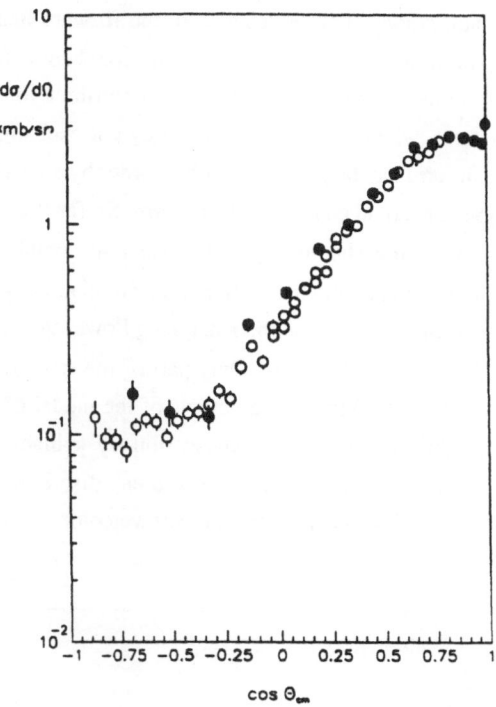

**Fig. 8** $\bar{p}p \to \bar{n}n$ differential cross-section data from KEK (full points) and from PS199 at LEAR (open points at 690 MeV/c.

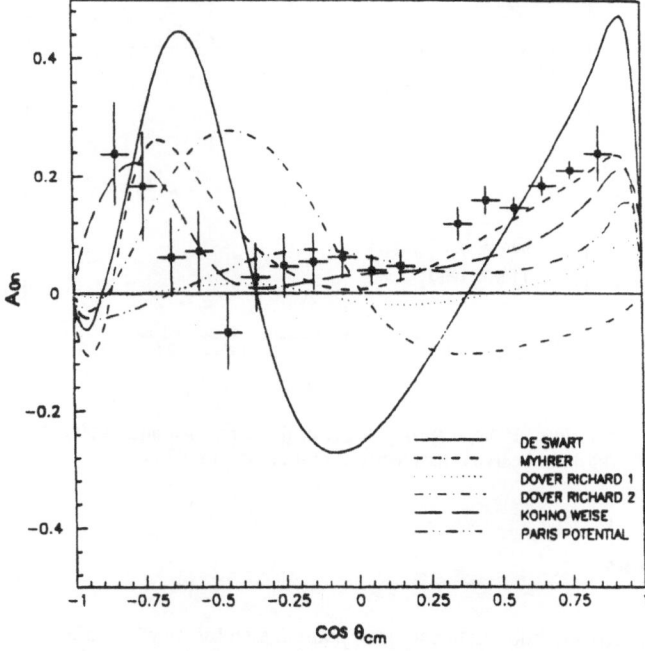

**Fig. 9** Analyzing power for $\bar{p}p \to \bar{n}n$ at 656 MeV/c, as measured by experiment PS199, compared with potential model calculations.

predictions differ so much among themselves also in the forward hemisphere, where $\pi$-exchange is expected to dominate. A possible explanation could come from the fact that, to first order in Born approximation, $\pi$-exchange does not contribute to $A_{0n}$. Polarization is contributed by an $\vec{L} \cdot \vec{S}$ term, which could be generated by a $\pi$-exchange only by iteration. Alternatively, a spin-orbit term can be generated to first order by a vector exchange (f.i. $\rho$). Also, the annihilation potential could contain an $\vec{L} \cdot \vec{S}$ term. So far it does not seem easy to draw a definite conclusion, but this should be possible when the complete set of $A_{0n}$ data, at all momenta, is available. A hint on which direction to go could already come from the data shown in Fig. 10, the preliminary results on the Analyzing Power $A_{0n}$ coming from the run for $D_{0n0n}$ at 900 MeV/c we had last August. Only part of the statistics is shown, still the result is significant when compared to the predictions of the model of Ref. 27. While that model (which treats annihilation simply as absorption by a black disc) was in good agreement with the data at 700 MeV/c, there is a clear discrepancy with the higher momentum data, and more work on the models is clearly welcome.

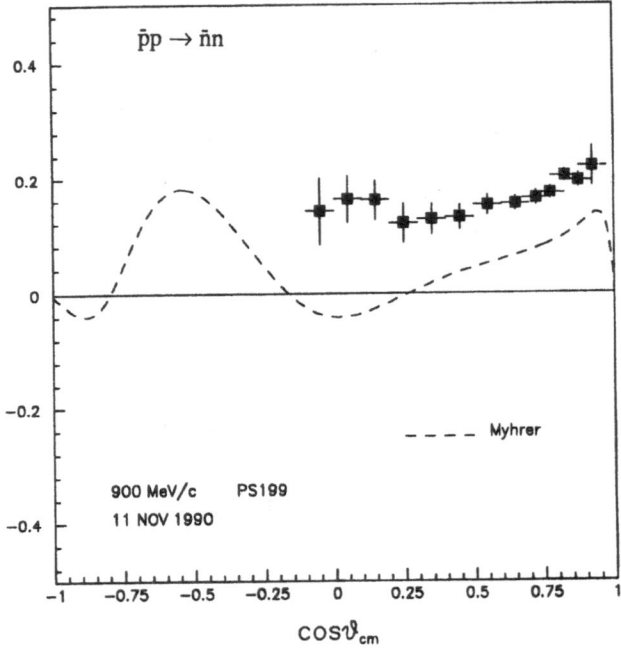

**Fig. 10**  Preliminary results for the analyzing power of $\bar{p}p \rightarrow \bar{n}n$, measured at 900 MeV/c by experiment PS199. The dashed curve is the prediction of the model from Ref. 27.

## 9. Conclusions

While the cross-section data can be explained also by very "simple" models, at present the set of $A_{0n}$ data for $\bar{p}p \rightarrow \bar{p}p$ and the new $A_{0n}$ data for $\bar{p}p \rightarrow \bar{n}n$ are not reproduced by any of the existing models. Even though some model agrees reasonably well with the data for one reaction at some energy, no model reproduces all the existing data for the two isospin combinations at once. In particular, it does not seem possible to reproduce any of the $A_{0n}$ data with an OBE potential model in which only a $\pi$ is exchanged[37].

The availability of the new LEAR data has stimulated theoretical work on the models, where, so far, several parameters were only loosely constrained. This is the case, for instance, for the Paris model[38], where a slight readjustment of the core parameters of both the real and the imaginary parts of the potential could improve the calculation of both charge-exchange cross-section and the analysing power of the elastic channel. This is true also in the case of the Dover-Richard model. In order to enhance the spin effects, namely the difference in the strength of the singlet and the triplet part of the meson-exchange potential, they reduced[37] somewhat $r_{cut}$, the radius at which they regularize the Yukawa potentials. Changing $r_{cut}$ from 0.8 to 0.74 fm, they could obtain a substantial improvement in their calculation of for $A_{0n}$ charge exchange and at the same time a good agreement for the elastic channel. Also, impressive fits are obtained on some of the data using the Bonn potential[39], or the Nijmegen coupled-channel approach[40].

Improving the models by taking into account the new data is surely a very important and necessary programme. Still, to my mind the main motivation for studying $N\overline{N}$ scattering is the assessment of the potential model approach, based on the meson exchange picture. Once that is established, one can

      – extract the global properties of annihilation;

      – to learn about the s-channel effects.

In this respect, the amount of theoretical work to be done is still considerable, and a crucial step is a test against the data of the basic ingredients of the theory, i.e. the $G$-parity rule. The comparison of NN data and the $N\overline{N}$ data should allow to test the $G$-parity rule, and, for instance, the very special role of $\omega$-exchange in the description of the nuclear force. I believe that clarifying in an independent way the role of the vector mesons in the nucleon-nucleon interaction is of such an importance to fully justify the experimental and theoretical work which is going on in the $N\overline{N}$ system.

## 10. Acknowledgements

The experiment PS199, measurement of spin effects in the $\overline{p}p \rightarrow \overline{n}n$ charge-exchange reaction at LEAR, is the effort of a Cagliari - Geneva - Saclay - Trieste - Turin Collaboration, and I would like to thank all the members, M.Agnello, A.Ahmidouch, J.Arvieux, R.Bertini, R.Birsa, T.Bressani, H.Catz, E.Chiavassa, S.Dalla Torre-Colautti, N.De Marco, J.C.Faivre, M.Giorgi, E.Heer, R.Hess, F.Iazzi, R.A.Kunne, M.Lamanna, C.Lechanoine-Leluc, P.Maciotta, A.Martin, C.Mascarini, A.Masoni, B.Minetti, A.Musso, A.Penzo, F.Perrot-Kunne, A.Piccotti, G.Puddu, D.Rapin, P.Schiavon, S.Serci, F.Tessarotto, for permission to present results prior to their publication.

# References

1. Th. Walcher,, Ann. Rev. Nucl. Part. Sci. **38** (1988) 67.
2. C. Amsler and F. Myhrer, Low Energy Antiproton Physics, CERN-PPE/91-29, 11 Feb. 1991, to be published in Ann. Rev. Nucl. Part. Sci. **41** (1991).
3. Physics at LEAR with Low-Energy Antiprotons, Proc. LEAR Workshop, Villars-sur-Ollon, 1987 (to be referred to here as Villars 87), eds. C. Amsler et al., (Harwood Academic Publishers).
4. Low Energy Antiproton Physics, Proc. First Biennial Conference, Stockholm, 2-6 July 1990 (to be referred to here as LEAP 90), eds. P. Carlson et al. (World Scientific Publiscers).
5. F. Bradamante, Proceedings of the 7th Int. Conf. on Polarization Phenomena in Nuclear Physics (Paris 90), Paris 9-13 July 1990, eds. A.Boudard and Y.Terrien (Les Edition de Physique), p. 66-299.
6. R.A. Bryan and R.J.N. Phillips, Nucl. Phys. **B5** (1968) 201.
7. A. Clough et al., Phys. Lett. **146B** (1984) 299.
8. D.V. Bugg et al., Phys. Lett. **194B** (1987) 563.
9. W. Brückner et al., Z. Phys. **A335** (1990) 217.
10. L.Linssen et al., Nucl. Phys. **A469** (1987) 726.
11. P. Schiavon et al., Nucl. Phys. **A505** (1989) 595.
12. W. Brückner et al., Phys. Lett. **B158** (1985) 180.
13. W. Brückner et al., in Villars 87, p.277.
14. W. Brückner et al., Phys. Lett. **B197** (1987) 463 and **B199** (1987) 596.
15. R. Kunne et al., Phys. Lett. **B206** (1988) 557.
16. R. Kunne et al., Nucl. Phys. **B323** (1989) 1.
17. R. Bertini et al., Phys. Lett. **B228** (1989) 531.
18. F.Perrot-Kunne et al., in LEAP 90, p. 251.
19. R. Kunne et al., *"First measurement of $D_{0n0n}$ in $\bar{p}p$ elastic scattering"*, Int. Rep. LNS/Ph/91-05, submitted for pubblication to Phys. Letters B.
20. W. Brückner, et al., Phys. Lett. **B169** (1986) 302.
21. R. Birsa et al., Phys. Lett. **B246** (1990) 267.
22. A. Martin et al., in LEAP 90, p. 257.
23. M.P. Macciotta et al., *"Extension of experiment PS199: further study of the spin structure of $\bar{p}N$ scattering at LEAR"*, proposal CERN-PSCC/90-16, PSCC/P93 Add. 2, 4 July 1990.
24. F. Myhrer, Nucl. Phys. **A508** (1990) 513c.
25. T. Hippchen et al., Nucl. Phys. (Proc. Suppl.) **B8** (1989) 116.
26. C. Dover and J.M. Richard, Phys. Rev. **C21** (1980) 1466.
27. O.D. Dalkarov and F. Myhrer, Nuovo Cimento **A40** (1977) 152.
28. M. Kohno and W. Weise, Nucl. Phys. **A454** (1986) 429.

29. P.H. Timmers et al., Phys. Rev. **D29** (1984) 1928.

30. J. Cote et al., Phys. Rev. Lett. **48** (1982) 1319.

31. T. Shibata, Phys. Lett. **B189** (1987) 232.

32. M.G. Albrow et al., Nucl. Phys. **B37** (1972) 349.

33. M. Lamanna, contributed talk to this Conference.

34. R. Birsa et al., Nucl. Instr. and Methods **A300** (1991) 43.

35. T. Bressani et al., Nucl. Instr. and Methods **A292** (1990) 563.

36. K. Nakamura et al., Phys. Rev. Lett. **53** (1984) 885.

37. Private communication from G. Ihle and J.M. Richard.

38. M. Pignone et al., *"Recent antiproton-proton data and the Paris N$\bar{\text{N}}$ potential"*, in LEAP 90, p. 90.

39. K. Holinde et al., *"Annihilation of the N$\bar{\text{N}}$ System into tw Mesons"*, in LEAP 90, p. 92.

40. R.G. Timmermaus et al., *"Lower Energy Coupled-channels Antinucleon-Nucleon Potential"*, n LEAP 90, p. 84.

# FLAVOR PRODUCTION AT LOW ENERGIES

Robert A. Eisenstein

Nuclear Physics Laboratory
University of Illinois
Champaign, IL 61821

A report on work done by the PS185 Collaboration [1] at LEAR/CERN, and also with Frank Tabakin and Yang Lu at the University of Pittsburgh.

## ABSTRACT

A brief examination of the physics we hope to learn from flavor production processes at low energies is presented. Results from the PS185 hyperon (strangeness) production experiment [2] at LEAR are discussed in a framework suitable for reactions taking place near threshold. Possible directions for future work in this field are outlined at the end.

## INTRODUCTION

"Flavor production" is the term applied to reactions generating final states containing one or more flavor-antiflavor quark pairs (e.g. $s\bar{s}$, $c\bar{c}$) that were not present in the initial state. As an example, one can cite the "associated production" reaction $\pi^+ n \rightarrow K^+ \Lambda$ (Fig. 1), where the initial strangeness-zero state consists of two $S=0$ particles, but the final state consists of particles with $S = \pm 1$. In such a reaction, one or more quark pairs in the initial state must be annihilated and one or more new pairs must be created in the final state (see Fig. 1). The emphasis is thus on the creation of these $q\bar{q}$ pairs, and on the associated reaction dynamics. In certain special cases (e.g. $\bar{p}p \rightarrow \overline{\Lambda}\Lambda$) we can hope to use the measurement of spin observables to learn about the underlying spin structure of the final state particles (here the $\Lambda$) and of the $q\bar{q}$ production process. We expect that such investigations will be all the more fruitful if they are carried out in the threshold region where only a few partial waves contribute, thus making the analysis easier.

Production processes such as these are to be distinguished from those involving purely a rearrangement of the quarks in the initial state. Such processes are themselves quite interesting, but are not the subject of this talk.

As indicated in Fig. 1, flavor production can equivalently be represented by $t$-channel meson/baryon exchange diagrams. In this description, one of the exchanged particles must carry the new quantum number. Calculations using this formalism, while not involving the quark degrees of freedom explicitly, are on somewhat firmer (albeit partly phenomenological) ground at the present time than are quark-based descriptions.

The high quality and precision of the PS185 data [2] for the $\bar{p}p \rightarrow \overline{\Lambda}\Lambda$ reaction has stimulated a great deal of activity in the theoretical arena. (Reasonably complete reference lists

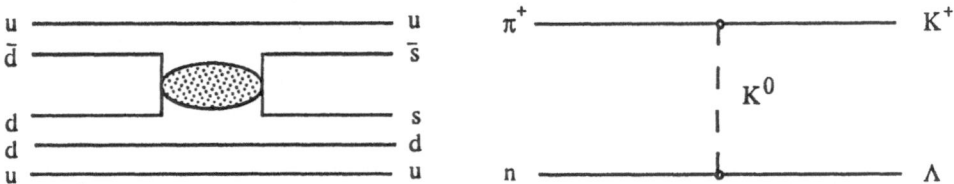

Fig. 1. Quark-line and meson exchange diagrams for the process $\pi^+ n \rightarrow K^+ \Lambda$.

can be found in Refs. [3-5]). The reasons for this are contained in the intrinsic interest of flavor production physics, and the quite complete set of spin observables and cross sections measured at several energies near the reaction threshold. The possibility of measuring these observables with such extraordinary precision is due to the self-analyzing feature [6] of the reaction and the remarkable properties of the LEAR/CERN complex [7].

## FLAVOR PRODUCTION PROCESSES

The simplest basic mechanisms for flavor production are shown in Fig. 2. We focus on processes in which the initial state baryon number is zero. The left panel shows that at very high momenta the process is simple due to asymptotic freedom, involving the exchange of only one qluon (the $^3S_1$ model). At low energies (right panel) the mechanism is complicated due to the confinement mechanism. In that case, the multiple gluon exchange is represented by a single exchange with $^3P_{0^+}$ quantum numbers. In either case the quantum numbers correspond to the simplest possible quark exchange mechanisms; however, it should be kept in mind that all amplitudes consistent with general conservation laws are allowed [5].

Fig. 3 shows how the basic physics of Fig. 2 appears in several reactions of varying degrees of complexity. Panels (a) and (b) both involve single $s\bar{s}$ creation, but with different quarks being annihilated in the initial state. That gives rise to different charge state possibilities in the final state, as shown. Panels (c) and (d) illustrate processes involving the annihilation and creation of more than one $q\bar{q}$ pair. Figs. 3(a)-(c) involve the production of final-state particles with "bare flavor" (the hyperons), while process 3(d) involves the production of "hidden flavor" states (the $\phi$ particles). To the right of each quark diagram is the corresponding diagram in the meson-nucleon sector.

In Fig. 3(a), the observation of the $\Lambda$ in the final state forces the spectator diquark (ud) pairs to have $I = 0$ and $\vec{S} = 0$. This because of the color Pauli principle. In the simplest constituent quark model, this means in turn that the spin properties of the $\Lambda$ are all due to the strange quark. Thus, we have the possibility that measurement of the spin observables for this reaction will lead to knowledge of the spin degrees of freedom for the (s) quark in the $\Lambda$. We return to this topic below.

If instead we measure outgoing $\Sigma^0$ particles, which have the same quark content as the $\Lambda$, the spectator diquark pairs in Fig. 3(a) are forced to be in I=1, $\vec{S} = 1$ states. In this way we can test two different configurations of the (uds) quark wavefunctions. Fig. 3(b) illustrates $\Sigma^+(uus)$ production in the final state via the annihilation and creation of a single $q\bar{q}$ pair. but the quark pair annihilated in the initial state is $d\bar{d}$ rather than $u\bar{u}$. Fig. 3(c) shows that to produce the $\Sigma^-\bar{\Sigma}^-$ final state requires final state production of a $d\bar{d}$ and an $s\bar{s}$ pair. In addition, both $u\bar{u}$ pairs in the initial state must be annihilated.

Lastly, Fig. 3(d) shows a process in which all quarks from the initial state are annihilated, while two $s\bar{s}$ pairs are created for the final state. In the simplest quark models, the zero overlap of the final and initial states means that the intermediate state is "pure glue". The hope is that this reaction will therefore be a rich hunting ground for specifically gluonic

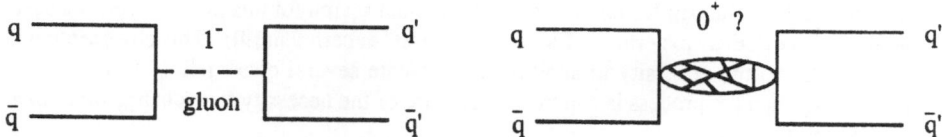

Fig. 2. The basic mechanism for gluon-mediated $q\bar{q}$ annihilation and creation, the essential
physics of Figs. 1 and 3 (see text), expressed in the quark-gluon sector. The diagram
on the left is characteristic of asymptotically free (high energy) processes, while the one
on the right is characteristic of the low-energy domain. The solid lines in the bubble
indicate multiple interactions between gluons, and interactions of a gluon with itself.

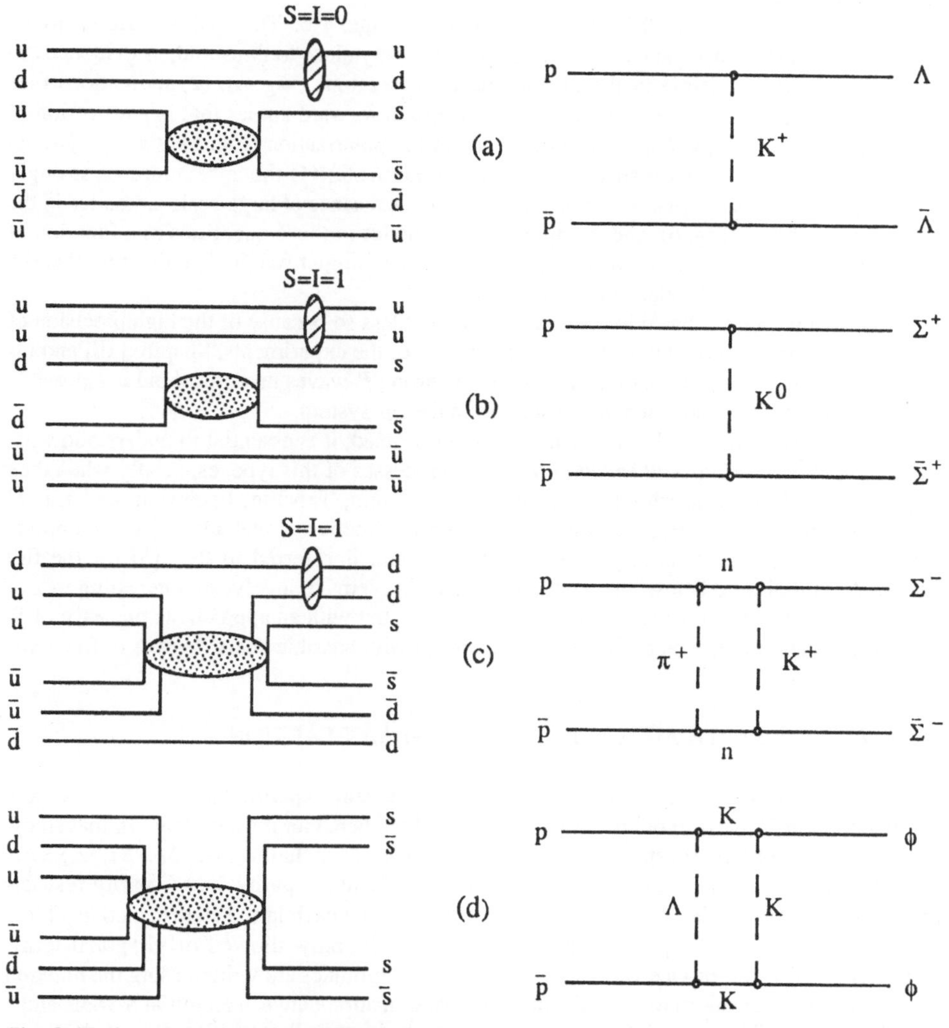

Fig. 3. Various hyperon production diagrams [(a)--(c)] that are being studied in PS185, and a
diagram of the $\bar{p}p \to \phi\phi$ reaction (d) being studied in PS202 (Jetset). Each quark-
gluon diagram is paired with the equivalent picture in the meson-nucleon sector.

phenomena such as glueball formation. An experimental search for this process, $p\bar{p} \rightarrow \phi\phi$, is the goal of LEAR/CERN experiment PS202 (the "Jetset" experiment [9]). This cross section is quite small due to the necessity to annihilate and create several quark pairs. In the meson exchange language, the process is suppressed because of the necessity to exchange more than one meson or baryon.

## THE $\bar{p}p \rightarrow \bar{\Lambda}\Lambda$ CASE

The PS185 experiment at LEAR has collected data on this reaction at several momenta near the threshold at 1435.28 Mev/c. The center-of-mass kinetic energy $\varepsilon = \sqrt{s} - 2m_\Lambda$ ranges from 0.236 to 91.7 MeV in the data published [2] so far. The data have been discussed in several places (see Refs.[2-5] and the work cited there), and several theoretical explanations of the results have been offered.

The data up to $\varepsilon = 39.0$ MeV are shown in Figs. 4-6. The major features to be described are: (1) the appearance of $\ell > 0$ waves even very close to threshold, as evidenced in the forward peaking visible in the differential cross sections (Fig. 4); (2) indications (see Ref. [3]) in the differential cross sections of behavior at forward angles that is a reflection of strong absorption ; (3) the appearance of a node in the polarization (Fig. 5) at a nearly fixed value of $t' \approx 0.2$ (GeV/c)$^2$ irrespective of incident momenta; (4) the appearance of strongly negative polarizations backward from this node; (5) the existence of large-scale structures in the spin correlation data (Fig. 6), and the appearance of nodes in those quantities as a function of $\cos\theta_{\bar{\Lambda}}^*$ and (6) the virtually complete suppression of the "singlet fraction", indicating that the outgoing $\Lambda$'s are always produced in a triplet state.

The observations listed here are intriguing, not less so because of the high precision of the data. Among the more interesting interpretations of the experiments, Shapiro [10] and his co-workers have suggested that the appearance of strong $P$-waves near threshold is a possible indication of sub-threshold resonance structures in the $\bar{p}p$ system.

However, before such ideas can be firmly endorsed, it is essential to understand what is to be expected *from normal reaction theory* in processes of this type, especially when they take place so close to threshold. With this idea in mind, Tabakin, Eisenstein and Lu [5] undertook a threshold analysis of the data based on "effective range" ideas. The complete discussion of these ideas is in progress, and the reader is referred to Ref. [5] for the full details. As is indicated below, the reaction mechanism seems to involve an interesting mix of ideas from several domains: strong absorption physics, but with an appreciable presence of $S$-wave amplitudes, and meson exchange notions mixed with possible quark degrees of freedom.

## SCATTERING LENGTH ANALYSIS OF THE $\bar{p}p \rightarrow \bar{\Lambda}\Lambda$ REACTION

In the analysis given in Ref. [5], all of the relevant experimental observables were expressed in the $LS$ basis in order to expose the threshold behavior most clearly. In that study, a restricted basis of only seven ($\bar{p}p$ or $\bar{\Lambda}\Lambda$) states was used, including the $^1S_0, {}^3S_1, {}^1P_1, {}^3P_0,$ $^3P_1, {}^3P_2,$ and $^3D_1$ waves. Conservation of angular momentum, parity, and $C$-parity restricts the transitions between initial and final states; it is clear that each initial state can connect to a final state with the same quantum numbers, but that the only allowed off-diagonal terms connect $S$- to $D$-states, or vice versa. Thus, transition amplitudes are written using the notation $T_{3S1}$ (a diagonal transition) or $T_{DS}$ (the off-diagonal transition between an initial $S$-wave and a final $D$-wave). The complete expressions for all of the observables (cross sections, polarization, and spin correlation coefficients) can be found in Ref. [5].

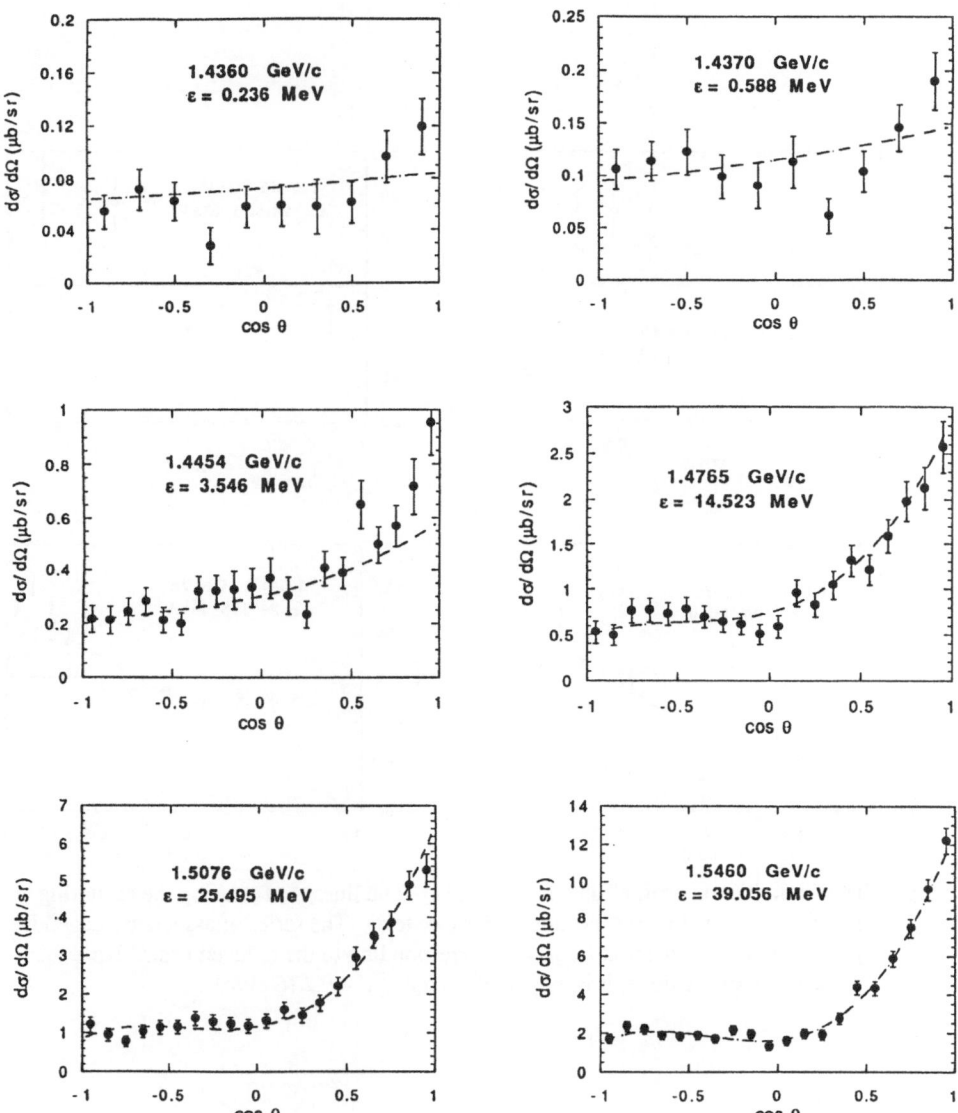

Fig. 4. The differential cross section data from Ref. [2]. The lines are fits to the data generated from a scattering length approximation to the amplitudes (see text). The Legendre series terminates at $L_{max} = 3$, corresponding to the restricted basis.

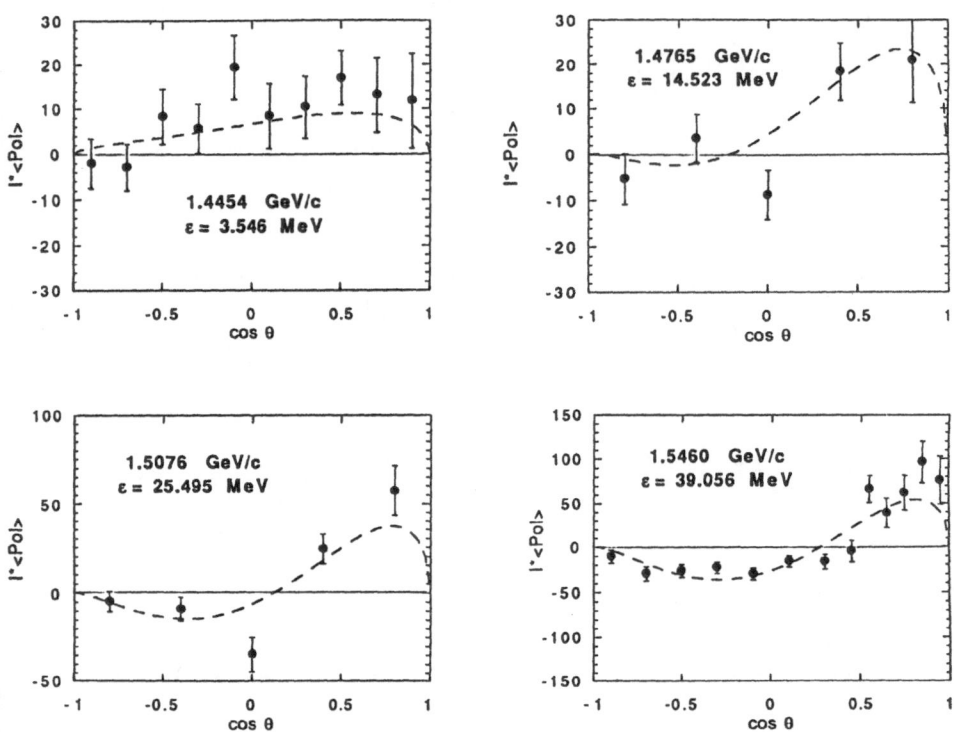

Fig. 5. The "polarization profile" data from Ref. [2]. The lines are fits using the scattering length approximation for the amplitudes (see text). The series of associated Legendre polynomials terminates at $L_{max} = 3$ corresponding to the reduced basis. Note the appearance of the node; it is essentially fixed at $t' \approx 0.2 \ (GeV/c)^2$.

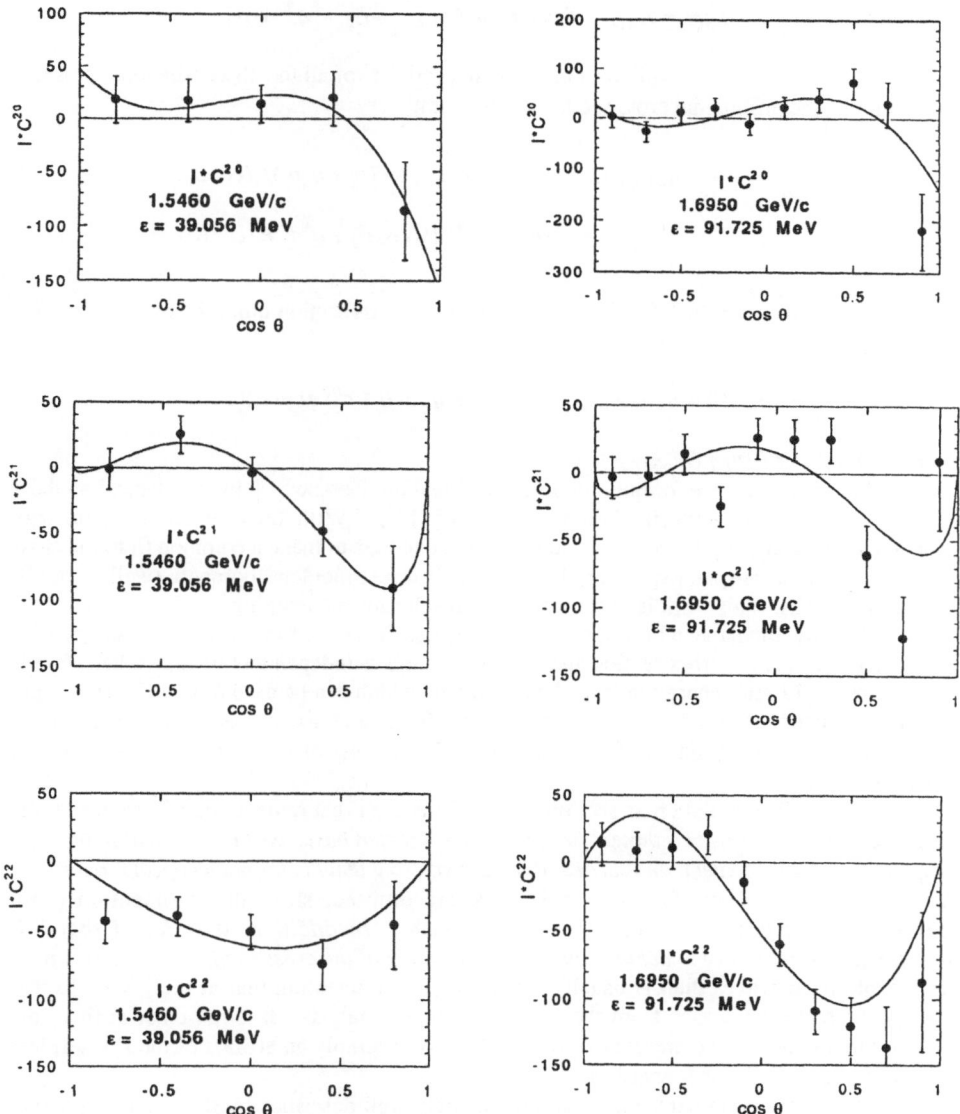

Fig. 6. The "correlation profile" data from Ref. [1],. Here the lines are simple fits to the appropriate Legendre series for each profile treated independently.

It is important to note that each amplitude will exhibit, near threshold, a strong dependence on the final state COM momentum, $q$, due to the centrifugal barrier. This leads naturally to the introduction of scattering length ideas [11], which means replacing each amplitude with a constant (complex) "scattering length" times $q^L$, *viz.*

$$T_{SD} = A_{SD}, \quad T_{3P1} = q\, A_{3P1}, \quad T_{DS} = q^2\, A_{DS},$$

and so on. In this way, the explicit momentum dependence of all the observables can be seen. For example, the differential cross section has the form:

$$\frac{d\sigma}{d\Omega} = \frac{q}{2ps}(m_p m_\Lambda)^2 [(a_0^T + a_2^T q^2 + a_4^T q^4) + q(b_1 + q^2 b_3) P_1(\cos\theta)$$
$$+ q^2(c_2 + c_4 q^2)\, P_2(\cos\theta) + q^4 d_3\, P_3(\cos\theta)]$$

and the "polarization profile" (basically the differential cross section times the polarization) for the outgoing $\overline{\Lambda}$ is:

$$\overline{P}_y(\theta) I(\theta) = q\sin\theta[(a_0^P + a_2^P q^2) + b_1^P q\cos\theta + c_2^P (q\cos\theta)^2]$$

Here $I(\theta) = 2ps\, \sigma(\theta)/[q\, (m_p m_\Lambda)^2]$.

It is seen that these formulas depend on 12 real numbers, which in turn depend on the 9 underlying complex scattering lengths (see Ref. [5]). With those 9 complex (energy independent!) scattering lengths (18 real numbers), one tries to make a common fit to the cross sections, polarizations, and spin correlations at 6 different incident momenta (well over 150 data points in all). This work is in progress; first results are encouraging.

On a less fundamental level, one can simply use the above formulas as they are, treating the coefficients for the cross section and the polarization as independent of each other. In this way, there are 12 real (energy independent) numbers which can be used to fit 133 data. Eight of the parameters are used for the cross section (100 data at six momenta) and four for the polarization profile (33 data at four momenta). The results of this approach are shown in Figs. 4 and 5.

For the differential cross section (Fig. 4), it is seen that quite a good fit to the data is obtained. *This confirms two things: first, that the truncated basis used is sufficient in terms of angular momentum content, and second, that the scattering length approach appears to be valid for $\varepsilon > 30$ MeV.* (The data at 1.695 GeV/c are omitted, since that momentum (with $\varepsilon = 91.7$ MeV) is just too high for this approach.) *In addition, it is found that S-P interference is necessary to achieve the forward peaking of the cross section.* This confirms a larger role for $S$-waves than is usually assumed. It is interesting that at the lowest excess energy there are deviations from the scattering length analysis. It is not clear at this time whether this indicates the presence of new physics or is simply an artifact of the sparser, less precise data sample at those energies.

The polarization data (Fig. 5) are moderately well described by the scattering length analysis. However, the poorer fit probably indicates that the polarization is a more sensitive test of small, higher angular momentum, components in the wavefunctions. This is not unexpected. As indicated in the introduction, there is a node in each data set at roughly constant value of $t' \approx 0.2$ (GeV/c)$^2$, and the polarization is sizeable at backward angles. *It is shown in Ref. [5] that the motion of this node with angle is firm evidence that there must be sizeable P-wave splitting among the P-wave amplitudes used to describe the process, and that the S-wave amplitudes play an essential role in the physics.* Since large $P$-wave splitting is an indication that strong tensor and/or $LS$ forces are present, this could be an indication that there is indeed a coherent amplification of the tensor component in the $\overline{N}N$ force, as originally conjectured by Buck, Dover and Richard [12]. It could also be an indication of the presence of vector ($K*(890)$) and/or tensor ($K_2^*(1430)$) meson exhange in this high momentum transfer process. In addition, it is shown in Ref. [5] that based on this analysis the (pure) Lund $^3P_0$ and $^3S_1$ exchange model is ruled out as a description of the $\overline{p}p \to \overline{\Lambda}\Lambda$ reaction at these energies.

The polarization node at fixed momentum transfer is reminiscent of similar phenomena observed in strong-interaction nuclear physics many years ago [13] by Austern, Blair, de Shalit, Hufner, Gubkin, and others. It was shown in those works that the angular distributions in elastic scattering would behave as $|J_1(kr)/kr|^2$, especially at forward angles. In addition, the "polarization profile" behaves as the derivative of the cross section, that is as $J_1(kr) J_0(kr)$. Here $(kr)$ is the momentum transfer times the strong interaction radius. These forms indicate exponential fall-off of the differential cross section at forward angles, and predict a zero in the polarization at a fixed momentum transfer. Both of these effects are seen in the PS185 data (Figs. 4 and 5).

Fig. 6 shows the spin correlation profiles (expressed in a spherical tensor basis [5]) fit to simple Legendre expansions. Due to the high momentum of the 1.695 GeV/c data, it was not used in the scattering length analysis described above. Nonetheless, one sees in the nodal structure of these quantities all of the physics mentioned above, as is described in Ref. [5]. In particular, the position and evolution of the nodal structure depends sensitively on the $P$-wave splitting and the $S$-$P$ interference.

Finally, we mention the singlet fraction, which is measured at several energies [2, 3] and found to be essentially zero. This indicates that the outgoing $\Lambda$ particles are always produced in a triplet state, irrespective of the fact that the incident particles are unpolarized. Such behavior is consistent once again with the notion of a strongly coherent tensor force in the $\overline{N}N$ interaction [14]. It will be recalled that a pure tensor force has no overlap between singlet states.

The above remarks show that the studies of the $\overline{p}p \to \overline{\Lambda}\Lambda$ reaction at threshold have borne fruit. The highly precise experimental work, coupled with careful theoretical analyses, is beginning to teach us the dynamical details of the transition from the $\overline{p}p$ state to the $\overline{\Lambda}\Lambda$ state. In particular, we have learned that both the $S$- and the $P$-waves play a crucial role in the threshold region, and that there is appreciable $P$-wave splitting due to tensor and/or $LS$ forces. We have learned also that very simple models (such as the Lund model [8]) of the transition are insufficient to describe the $\overline{p}p \to \overline{\Lambda}\Lambda$ reaction at these energies. In a short time we hope to have the scattering amplitudes (in the scattering length limit) for the first few partial waves available for comparison with dynamical models.

FUTURE DIRECTIONS

There are several obvious future directions in flavor production physics. Fig. 7 shows the $SU(4)$ multiplets for the $J = 1/2$ baryons, and Fig. 8 indicates roughly the thresholds for various channels of interest. In the $C = 0$, $S = 1$ sector, we have made substantial progress in the threshold region (in experiments of the kind discussed here), but only for the lowest-lying $\Lambda$ and $\Sigma$ states. Completing our study of the hyperons seems a very worthwhile goal. Unfortunately, the LEAR ring is not energetic enough to allow investigation of the $\Xi$ or the $\Omega^-$ ($J = 3/2$) regions.

In the $C = 1$ sector, the nearest reaction of interest is the $D\overline{D}$ channel. These reactions could be studied using an appropriate detector at the Fermilab antiproton complex. Rather little is known about $C = 1$ particles, so such studies could have a major impact. Other higher-lying channels of interest are indicated on the diagram. The thresholds are all above

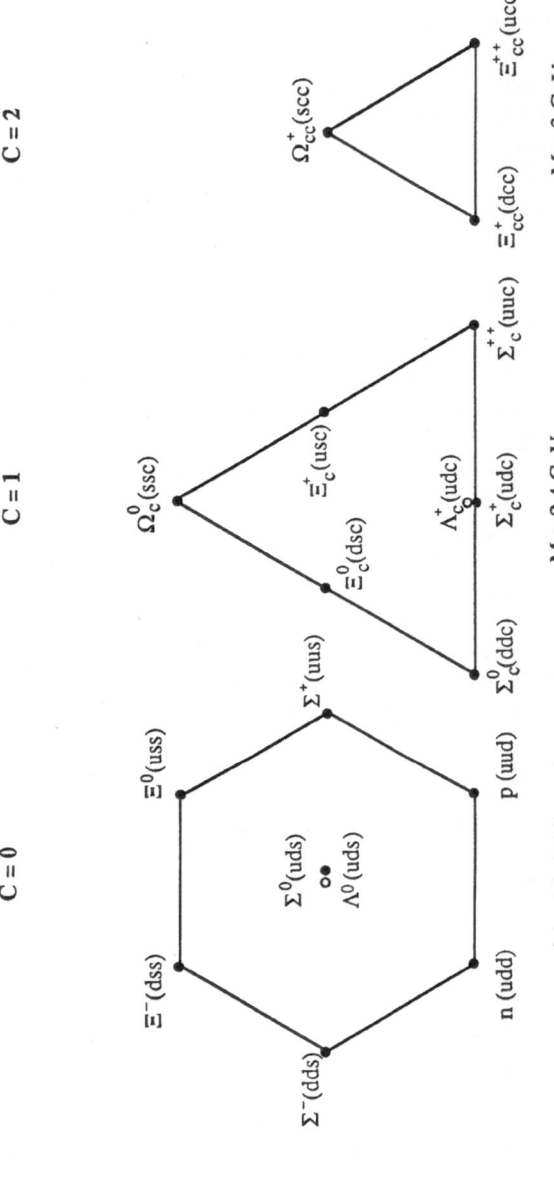

Fig. 7. The *SU(4)* *J* = *1/2* baryons. Typical masses are: about *1 GeV* for *C = 0* baryons; about *2.4 GeV* for *C = 1* baryons, and above *3 GeV* for *C = 2* baryons.

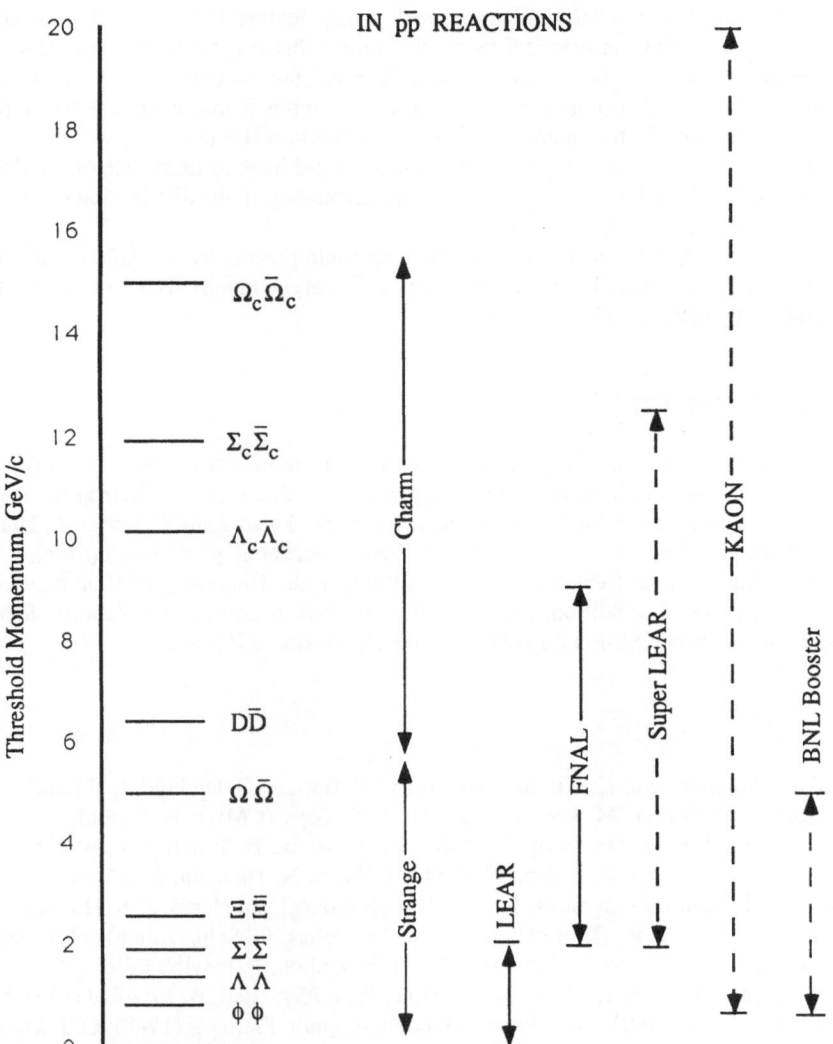

FLAVOR PRODUCTION THRESHOLDS

IN p̄p REACTIONS

Fig. 8. The leftmost column shows several interesting physics thresholds as a function of lab momentum. The other columns show roughly the momentum ranges available at various existing (solid lines) and possible future (dashed lines) facilities.

10 GeV/c and would require the beams of one of the possible new facilities being discussed for CERN (SuperLEAR), TRIUMF (KAON), or BNL.

The study of such reactions will not be easy, however. The production of heavy quarks (*e.g.* charmed quarks here) is strongly penalized by large phase space weighting factors that depend on the ratio of the quark masses involved to a high power. In addition, the measurement of the spin observables will be possible only for those particles whose lifetime is long enough to utilize the self-analyzing weak decay feature that made the $\bar{p}p \to \bar{\Lambda}\Lambda$ experiment so successful. In practical terms, this means that particle flight paths of several centimeters are required. Without such flight paths between the strong interaction and the weak decays it difficult to determine the reaction plane; without it one cannot determine the polarization directions. Unfortunately, particles in the charm sector possess quite short values of $ct$. Thus, for example, to analyze the decay one would have to make use of the double weak-decay chain $\Lambda_c^+ \to \Lambda \pi^+ \to p\pi^-\pi^+$, using any asymmetry in the distribution of the final decay products to measure the $\Lambda_c$ polarization.

Finally, it is noted that reactions of this type could potentially be explored in *proton* beams from future facilities like "LISS", although a net baryon number B = 2 makes the final states significantly more complicated.

ACKNOWLEDGEMENTS

I acknowledge with pleasure the many contributions of my colleagues on PS185, and at the University of Pittsburgh, to this work. In particular, I thank David Hertzog for several useful suggestions and a careful reading of the manuscript. I also thank C. Dover, E. Henley, and J. Speth for helpful comments. During the 1990-91 academic year I was the recipient of an Associateship from the Center for Advanced Study at the University of Illinois, which I gratefully recognize. In addition, this work is supported in part by the National Science Foundation under grant NSF-PHY-89-21146 to the University of Illinois.

REFERENCES

1. PS185 Collaboration, K. Kilian, spokesman:  P. Barnes, G. Diebold, G. Franklin, C. Maher, B. Quinn, M. Rozen, J. Seydoux, V. Zeps (CMU); R. Besold, W. Eyrich, R. v. Frankenberg, A. Hofmann, D. Malz, F. Stinzing, P. Woldt (Erlangen-Nürnberg); P. Birien, W. Dutty, J. Franz, N. Hamann, E. Rössle, H. Schledermann, H. Schmitt, H.-J. Urban (Freiburg); D. Hertzog, S. Hughes, P. Reimer, R. Tayloe, (Illinois); K. Kilian, W. Oelert, G. Sehl (Jülich); G. Ericsson, S. Ohlsson,  T. Johansson (Uppsala); W. H. Bruenlich, P. Pawlek (Vienna).
2. P. D. Barnes *et al.* Phys. Lett. B189: 249 (1987); Phys. Lett. B229: 432 (1989); Nuc. Phys. A526: 575 (1991); W. Dutty, Dissertation, Univ. Freiburg (1988); C. J. Maher, Dissertation, Carnegie-Mellon Univ. (1986); S. Ohlsson, Dissertation, Univ. Uppsala, (1990); G. Sehl, Dissertation, Univ. Bonn, Jül-Spez-535 (1989); R. von Frankenberg, Dissertation, Univ. Erlangen--Nürnberg (1987).
3. T. Johansson, in "Proceedings of the First Biennial Conference on Low Energy Antiproton Physics", P. Carlson, A. Kerek, and S. Szilagyi, eds., World Scientific Publishers, New Jersey (1991).
4. N. Hamann, work in progress, to be submitted to the University of Freiburg.
5. F. Tabakin, R. A. Eisenstein, and Y. Lu, "Spin Observables at Threshold for the $\bar{p}p \to \bar{\Lambda}\Lambda$ Reaction ", submitted to Physical Review C, May 1991.
6. L. Durand and J. Sandweiss, Phys. Rev. 135: B540 (1964).

7. P. Lefevre, in "Physics at LEAR with Low Energy Antiprotons", C. Amsler, G. Backenstoss, R. Klapisch, C. Leluc, D. Simon, and L. Tauscher, eds., Harwood Academic Publishers, Chur, Switzerland (1988).

8. B. Andersson, et al., Phys. Lett. 85B: 417 (1979); and Univ. Lund preprint LUTP 82-6 (1982).

9. "The Jetset Experiment at LEAR" (PS202), CERN/PSCC 86-23.

10. I. S. Shapiro, in "Antiproton--Nucleon and Antiproton--Nucleus Interactions", F. Bradamante, J.-M. Richard and R. Klapisch, eds., Plenum Press, New York (1990); O. Dalkarov, K. Protasov, and I. Shapiro, Int. J. Mod. Phys. A5: 2155 (1990); I. S. Shapiro, Nucl. Phys. A478: 665c (1988).

11. J. M. Blatt and V. F. Weisskopf, "Theoretical Nuclear Physics", John Wiley and Sons, New York (1952).

12. W. Buck, C. Dover, and J.-M. Richard, Ann. Phys. 121: 47 (1979).

13. N. Austern and J. Blair, Ann. Phys. 33:15 (1965); J. Hüfner and A. de Shalit, Phys. Lett. 15: 52 (1965); and A. Gubkin, Nucl. Phys. A111: 605 (1968).

14. C. Dover and E. Henley, private communications.

# CHIRAL-ODD PARTON DISTRIBUTIONS AND POLARIZED DRELL-YAN*

R. L. Jaffe

Center for Theoretical Physics
Laboratory for Nuclear Science and Department of Physics
Massachusetts Institute of Technology
Cambridge, Massachusetts 02139

The work I will present was carried out in collaboration with Xiangdong Ji. A compressed version will be published in *Physical Review Letters*.[1]

The nucleon's parton distributions characterize its properties in hard scattering processes. Measurement of these distributions, which has been undertaken for over 20 years, provides us with considerable insight into the quark-gluon substructure of the nucleon. The spin-independent quark and gluon distributions have been measured in a variety of experiments and with high accuracy, and their evolution as functions of the renormalization point is a decisive test for perturbative Quantum Chromodynamics (QCD). The longitudinal quark-spin distribution, $g_1(x)$, has been measured at both SLAC and CERN, and the data has prompted much theoretical work on the spin structure of the nucleon. The aim of this talk is to characterize a new class of nucleon structure functions, the chiral-odd spin-structure functions $h_1(x)$ and $h_2(x)$, to study their properties and to describe how they can be measured in lepton pair production with polarized beam and target ("polarized Drell–Yan").

We begin by systematically reviewing the bi-linear quark correlation function of the nucleon on the light cone. We concentrate on the two spin-dependent chiral-odd densities, $h_1(x)$ and $h_2(x)$. We identify $h_1(x)$ as the polarized-quark *transversity* distribution. We explain the distinction between $h_1(x)$ and the distributions $g_1(x)$ and $g_2(x)$ which are related to the quarks' spin. We analyze its Regge behavior and derive sum rules for $h_1(x)$ in terms of local operators. We compare its first moment with that of $g_1(x)$ and estimate it in quark models. For $h_2(x)$, we derive a relation which connects $h_2(x)$ with $h_1(x)$ and the matrix elements of a tower of twist-3 local operators. To obtain a qualitative estimate of $h_1(x)$ and $h_2(x)$, we calculate them in the bag model. Finally, we present the polarized Drell–Yan cross-section up to order $1/Q$. The result involves five of the twist-2 and 3 structure functions.

The structure function $h_1(x)$ appears to have been defined first by Ralston and Soper[2] in their systematic study of polarized Drell–Yan, where it is called $h^T(x)$. [Kodaria *et al.* and Bukhvostov *et al.*,[3] noticed that $h_1(x)$ contributes to $g_2(x)$, but did not appear to have recognized its general role in hard processes.] More recently, Artru and Mekhfi[4] apparently rediscovered $h_1(x)$ — called $\Delta_1 q(x)$ by them — calculated its QCD evolution and mentioned its place in polarized Drell–Yan. Collins[5]

---

* This work is supported in part by funds provided by the U.S. Department of Energy (D.O.E.) under contract #DE-AC02-76ER03069.

has summarized and extended the work of Refs. [2] and [4]. Some of our discussion of $h_1(x)$ overlaps the work of Refs. [2-5]. We cannot find any mention of the structure function $h_2(x)$ in the literature, nor can we find any systematic exploration of sum rules, Regge behavior, model dependence, *etc.*, for either $h_1(x)$ or $h_2(x)$. Finally, similarities and differences between the *chiral–even* spin-dependent structure functions $g_1(x)$ and $g_2(x)$ on the one hand, and the *chiral–odd* spin-dependent structure functions $h_1(x)$ and $h_2(x)$ on the other have never been carefully drawn and has caused considerable confusion surrounding (especially transverse) spin effects in hard processes. We hope to clarify the matter in this talk.

The parton distributions in QCD are defined by the light-cone Fourier transformation of field operator products between nucleon states. The simplest quark-parton distributions are related to the matrix elements of bi-linear quark operators,

$$\int \frac{d\lambda}{2\pi} e^{i\lambda x} \langle PS|\bar{\psi}(0)\Gamma\psi(\lambda n)|PS\rangle \tag{1}$$

where $n$ is a null vector of dimension $[\text{mass}]^{-1}$ ($n^2 = 0, n^+ = 0$) and $\Gamma$ is a Dirac matrix. $P$ and $S$ are the nucleon momentum and spin vectors ($P^2 = M^2$, $S^2 = -M^2$, $P \cdot S = 0$), respectively. For simplicity, we choose a light-cone gauge for gluon fields ($n \cdot A \sim A^+ = 0$) so that (1) is manifestly gauge invariant. We suppress the renormalization scale label, $\mu^2$, necessary to render (1) well-defined in perturbative QCD.

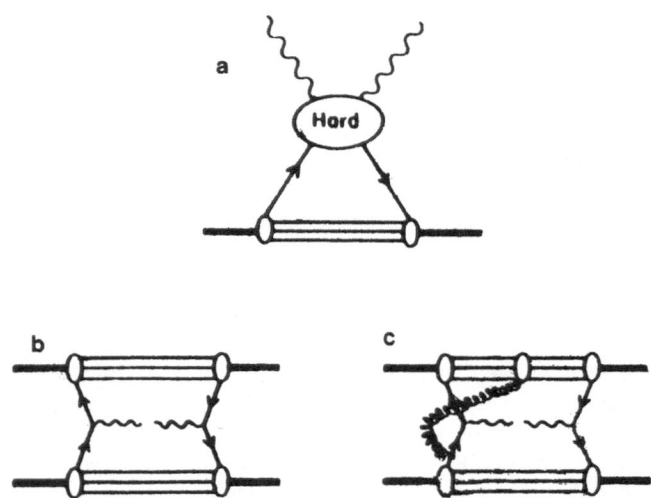

*Fig. 1.  Parton model diagram a) for deep inelastic scattering; b) for Drell–Yan at twist-2; c) one of four gluon correction diagrams which enter Drell–Yan at twist-3.*

Only chiral–even distributions (defined with a chiral-even $\Gamma$ in (1)) contribute to deep inelastic scattering when small quark mass effects are ignored. This is obvious from the hard part of Fig. 1a where weak or electromagnetic and strong currents preserve the chirality of the single active quark. These are obtained by setting $\Gamma = \gamma_\mu$ or $\gamma_\mu\gamma_5$ in (1). For $\Gamma = \gamma_\mu$

$$\int \frac{d\lambda}{2\pi} e^{i\lambda x} \langle P\hat{S}|\bar{\psi}(0)\gamma_\mu\psi(\lambda n)|PS\rangle = 2[f_1(x)p_\mu + M^2 f_4(x)n_\mu] \tag{2}$$

where $p$ is another null vector of dimension [mass]$^1$ ($p^2 = 0, p^- = 0, p \cdot n = 1$, and $P = p + \frac{M^2}{2}n$). [Note: for a target moving in the $\hat{e}_z$-direction: $p = \frac{1}{\sqrt{2}}(\mathcal{P}, 0, 0, \mathcal{P})$, $n = \frac{1}{\sqrt{2}}(1/\mathcal{P}, 0, 0, -1/\mathcal{P})$; the normalization for $f_1(x)$ is chosen such that $F_1^{eN} = \frac{1}{2}\sum_a e_a^2 f^a(x)$ and $f^a(x) = -f^a(-x)$]. From dimensional analysis, we know that $f_1(x)$ and $f_4(x)$ will contribute to a hard process at the order of $f_1(x)$ and $f_4(x)/Q^2$, and are hence twist-2 and twist-4, respectively. The twist-2 distribution, $f_1(x)$, has a simple parton-model interpretation: it measures the distribution of quarks with momentum $k^+ = xp^+$ and any $\vec{k}_\perp$. For $\Gamma = \gamma_\mu \gamma_5$, we have,

$$\int \frac{d\lambda}{2\pi} e^{i\lambda x} \langle PS | \bar{\psi}(0) \gamma_\mu \gamma_5 \psi(\lambda n) | PS \rangle$$
$$= 2[g_1(x)p_\mu(S \cdot n) + (g_1(x) + g_2(x))S_{\perp\mu} + M^2 g_3(x)n \cdot Sn_\mu] \tag{3}$$

where we write $S_\mu = S \cdot n p_\mu + S \cdot p n_\mu + S_{\perp\mu}$. These distribution functions, $g_1(x)$, $g_2(x)$, and $g_3(x)$, contribute to a hard process at order $g_1(x), g_2(x)/Q, g_3(x)/Q^2$, and hence have twist-2, 3, and 4, respectively. In the parton model, $g_1(x)$ measures the quark helicity distribution in a longitudinally polarized nucleon. Such an interpretation can be made because the quark helicity operator commutes with the free-quark Hamiltonian ($H = \alpha_z p_z$). In a transversely-polarized nucleon, however, the quark-spin operator projected along the nucleon spin, $\hat{\Sigma}_\perp = \gamma_0 \gamma_5 \mathcal{S}_\perp$, does not commute with the Hamiltonian, i.e. there exists no energy eigen-spinor $U(p_z)$ such that $\hat{\Sigma}_\perp U(p_z) = \lambda_\perp U(p_z)$. In the light-cone formalism, the transverse spin operator is a "bad" operator and depends on dynamics. Nevertheless, a transverse-spin average can still be defined in the nucleon state and according to (3), it is just $g_T(x) = g_1(x) + g_2(x)$. $g_T(x)$ is twist-3 and is sensitive to the quark-gluon interactions, a clear sign that no simple parton interpretation can be made for it.[6] The Burkhardt–Cottingham sum rule, $\int g_2(x)dx = 0$, guarantees that the quark-spin contribution to the nucleon spin be the same for any polarization.[6]

In certain physical processes such as Drell–Yan production of muon pairs, a new class of chiral-odd structure functions can be measured and studied. It is clear from the standard Drell–Yan diagram (Fig. 1b) that the chiralities of the quark lines originating in a single hadron are uncorrelated. The simplest example arises when $\Gamma = 1$ in (1),

$$\int \frac{d\lambda}{2\pi} e^{i\lambda x} \langle PS | \bar{\psi}(0) \psi(\lambda n) | PS \rangle = 2e(x) \tag{4}$$

where $e(x)$ is a twist-3 spin-independent structure function. More importantly, when $\Gamma = \sigma_{\mu\nu} i \gamma_5$,

$$\int \frac{d\lambda}{2\pi} e^{i\lambda x} \langle PS | \bar{\psi}(0) \sigma_{\mu\nu} i \gamma_5 \psi(\lambda n) | PS \rangle = 2[h_1(x)(S_{\perp\mu} p_\nu - S_{\perp\nu} p_\mu)/M$$
$$+ (h_2(x) + h_1(x)/2)M(p_\mu n_\nu - p_\nu n_\mu)(S \cdot n) + h_3(x)M(S_{\perp\mu} n_\nu - S_{\perp\nu} n_\mu] \tag{5}$$

where $h_1(x)$, $h_2(x)$, and $h_3(x)$ are twist-2, 3 and 4, respectively. Recently, Qiu and Sterman[7] have argued that the unpolarized Drell–Yan cross section for muon pair production through twist-4 can be expressed entirely in terms of parton distributions measurable in lepton scattering. In fact, $e(x)$ provides a counter example to Qiu and Sterman's argument since it enters *unpolarized* Drell–Yan like $\sim e(x)\bar{e}(y)/Q^2$.[8] The appearance of $h_1(x)$ and $h_2(x)$ in *polarized* Drell–Yan at $\mathcal{O}(1)$ and $\mathcal{O}(1/Q)$ makes it clear that no such statement can be made in this case.

According to (2)-(5), the complete specification of the light-cone quark correlation function requires nine distribution functions: three twist-2 ($f_1, g_1, h_1$), three twist-3 $(e, g_2, h_2)$, and three twist-4 $(f_4, g_3, h_3)$. At each twist there is one spin-average distribution $(f_1, e, f_4)$, one chiral-even spin distribution $(g_1, g_2, g_3)$, and one chiral-odd spin-dependent distribution $(h_1, h_2, h_3)$. This simple pattern can be understood in the light-cone formalism of Kogut and Soper,[9] in which the quark field $\psi$ is decomposed into "good" and "bad" components, $\psi_+$ and $\psi_-$ ($\psi_\pm = P_\pm \psi, P_\pm = \frac{1}{2}\gamma^\mp \gamma^\pm$). Correlators of the form $\psi_+^\dagger \psi_+$ are twist-2, $\psi_+^\dagger \psi_- \pm \psi_-^\dagger \psi_+$ are twist-3, and $\psi_-^\dagger \psi_-$ are twist-4. For each light-cone projection, the counting of the helicity amplitudes, $A_{h_1 H_1, h_2 H_2}$, for forward quark-nucleon scattering is the same. [$h$ and $H$ are helicities of quark and nucleon, respectively.] Using parity and time-reversal invariance, we find that there are three independent helicity amplitudes, $A_{\uparrow\uparrow,\uparrow\uparrow}$, $A_{\uparrow\downarrow,\uparrow\downarrow}$, and $A_{\uparrow\downarrow,\downarrow\uparrow}$. For the appropriate light-cone components, $(f_1, e, f_4) \sim A_{\uparrow\uparrow,\uparrow\uparrow} + A_{\uparrow\downarrow,\uparrow\downarrow}$, $(g_1, h_2, g_3) \sim A_{\uparrow\uparrow,\uparrow\uparrow} - A_{\uparrow\downarrow,\uparrow\downarrow}$, and $(h_1, g_2, h_3) \sim A_{\uparrow\downarrow,\downarrow\uparrow}$. Further discussion can be found in Ref. [8].

Thus, a complete quark parton model of the nucleon at leading twist requires *three* quark distributions: $f_1$, $g_1$, and $h_1$. The physical meaning of them can be made clear by introducing projection operators for the "good" component of the Dirac spinor. The usual choice is

$$P_{\substack{L\\R}} \equiv \frac{1}{2}\left(1 \mp \gamma^5\right) \ . \tag{6}$$

which project on chirality. This approach was taken in Ref. [2]. After performing a momentum decomposition of the Dirac field and a projection with $P_{L,R}$,[9] we obtain,

$$f_1(x) = \frac{1}{x}\left\langle P|R^\dagger(xP)R(xP) + L^\dagger(xP)L(xP)|P\right\rangle$$

$$g_1(x) = \frac{1}{x}\left\langle P\hat{e}_z|R^\dagger(xP)R(xP) - L^\dagger(xP)L(xP)|P\hat{e}_z\right\rangle \tag{7}$$

$$h_1(x) = \frac{2}{x}\mathrm{Re}\left\langle P\hat{e}_\perp|L^\dagger(xP)R(xP)|P\hat{e}_\perp\right\rangle$$

for $x > 0$. Here $R_+(xP)(L_+(xP))$ annihilates a right- (left-) handed quark with $k^+ = xP^+$ and any $k_\perp$. For $x < 0$ we find $f_1(-x) = -\bar{f}_1(x), g_1(-x) = +\bar{g}_1(x)$, and $h_1(-x) = -\bar{h}_1(x)$, where the overbar denotes the replacement of $R$ and $L$ by $\bar{R}$ and $\bar{L}$ which annihilate right- and left-handed *antiquarks*, respectively. The parton interpretation of $f_1(x)$ and $g_1(x)$ is self-evident from (7): they directly count right- and left-handed quarks.

The interpretation of $h_1(x)$ is obscure in a chiral basis. It is revealed by using the projection operators

$$Q_\pm = \frac{1}{2}\left(1 \mp \gamma^5 \gamma^\perp\right) = \frac{1}{2}\left(1 \pm \gamma^5 \rlap{/}S_\perp\right) \tag{8}$$

instead of $P_L$, $P_R$. In terms of the $Q_\pm$ basis:

$$f_1(x) = \frac{1}{x}\left\langle P|\alpha^\dagger(xP)\alpha(xP) + \beta^\dagger(xP)\beta(xP)|P\right\rangle$$
$$h_1(x) = \frac{1}{x}\left\langle P\hat{e}_\perp|\alpha^\dagger(xP)\alpha(xP) - \beta^\dagger(xP)\beta(xP)|P\hat{e}_\perp\right\rangle \tag{9}$$

and $g_1(x)$ is off-diagonal. Here $\alpha$ ($\beta$) annihilates a quark with $Q_+\alpha = \alpha$ ($Q_-\beta = \beta$). From (7) and (9) we obtain the inequalities

$$|g_1(x)| \le f_1(x); \qquad |h_1(x)| \le f_1(x) \tag{10}$$

which hold for each quark flavor.

Apparently, $h_1(x)$ measures the probability to find a quark in an eigenstate of the transversely-projected Pauli-Lubanski operator, $\mathcal{J}_\perp\,\gamma_5$, in a nucleon likewise polarized. Unlike the transverse spin operator $\hat{\Sigma}_\perp$, $\mathcal{J}_\perp\,\gamma_5$ commutes with the free-quark Hamiltonian and is a light-cone "good" operator. For that reason, a simple parton model can be made to interpret the distribution $h_1(x)$. This basis has been used for years in analysis of polarization in hadron-hadron scattering, where it is known as the "transversity" basis.[10] Hence we name $h_1(x)$ the quark *transversity distribution*.

To derive sum rules for $h_1(x)$, we introduce a set of twist-2 operators,

$$O_n \equiv \mathcal{S}_n \bar{\psi}\sigma^{\sigma\mu_1} i\gamma_5 iD^{\mu_2}\ldots iD^{\mu_n}\psi - \text{trace} \;, \qquad \text{for } n = 1, 2, \ldots \qquad (11)$$

and their matrix elements,

$$\langle PS | \mathcal{O}_n | PS \rangle \equiv 2a_n \mathcal{S}_n \left( S^\sigma P^{\mu_1} - S^{\mu_1} P^\sigma \right) P^{\mu_2} \ldots P^{\mu_n}/M - \text{trace} \;, \qquad (12)$$

where $\mathcal{S}_n$ symmetrizes the indices $\mu_1, \mu_2 \ldots \mu_n$. Then, from (5) and using $h_1(x) = 0$ for $|x| \geq 1$, we derive

$$\int_{-\infty}^{\infty} dx\, x^{n-1}\, h_1(x) = \int_0^1 dx\, x^{n-1}\left( h_1(x) - (-1)^{n-1}\bar{h}_1(x) \right) = a_n \qquad (13)$$

if above integral is convergent. In fact, a simple Regge analysis shows that as $x \to 0$, $h_1(x) \to x^{-\alpha}$, where $\alpha$ is the relevant Regge intercept. Like $g_1$, the Pomeron does not contribute to $h_1(x)$, so we expect the moments of $h_1(x)$ to be convergent even for $n = 1$.

Let us consider $n = 1$ sum rule in some detail. To clarify the flavor content, we rewrite the relevant matrix element $a_1$, hereafter called the tensor charge, as $\delta q$, where $q = u, d, s$, etc. To understand its physical meaning, we make a close comparison between $\delta q$ with $\Delta q$, the quark helicity contribution to the nucleon spin. In the rest frame of the nucleon, $P_\mu = (M, 0, 0, 0)$ and $S_\mu = (0, \mathbf{S})$, so (12) becomes

$$\langle PS | \bar{q}\Sigma_i q | PS \rangle = 2\delta q S_i \;. \qquad (14)$$

From (9) and (13), we find,

$$\int_0^1 dx [h_1(x) - \bar{h}_1(x)] = \delta q$$
$$= \int_0^\infty \frac{dk^+}{k^+} \langle P\hat{e}_\perp | \alpha^\dagger(k)\alpha(k) - \bar{\alpha}^\dagger(k)\bar{\alpha}(k) - \beta^\dagger(k)\beta(k) + \bar{\beta}^\dagger(k)\bar{\beta}(k) | P\hat{e}_\perp \rangle \qquad (15)$$

Therefore, $\delta q$ counts valence quarks (quarks *minus* antiquarks) of opposite transversity. The sea quarks do not contribute because the operator $\mathcal{O}^{\mu\nu} \equiv \bar{\psi}i\sigma^{\mu\nu}\gamma_5\psi$ is odd under charge conjugation. In contrast, the quark spin operator, $A^\mu = \bar{\psi}\gamma^\mu\gamma^5\psi$, is even under charge conjugation. The corresponding equations for $\Delta q$ are,

$$\langle PS | q^\dagger\Sigma_i q | PS \rangle = 2\Delta q S_i \;. \qquad (16)$$

and

$$\int_0^1 dx\, [g_1(x) + \bar{g}_1(x)] = \Delta q$$
$$= \int_0^\infty \frac{dk^+}{k^+} \langle P\hat{e}_z | R^\dagger(k)R(k) + \bar{R}^\dagger(k)\bar{R}(k) - L^\dagger(k)L(k) - \bar{L}^\dagger(k)\bar{L}(k) | P\hat{e}_z \rangle \qquad (17)$$

Obviously, $\Delta q$ includes the helicity of the sea. The opposite charge conjugation and chiral properties of $A^\mu$ and $O^{\mu\nu}$ make it clear that $h_1(x)$ is unrelated to quark spin. [Formally, the spin tensor density, $M_{\text{spin}}^{\mu\nu\lambda}$, is just $\frac{1}{2}\epsilon^{\mu\nu\lambda\sigma}A_\sigma$.] The name transversity distribution for $h_1(x)$ avoids possible confusion with the transverse spin distribution $g_T$.

Returning to the sum rule, if one writes $\Delta q = \Delta q^v + \Delta q^s$, the sum of valence and sea contributions, one might speculate that $\delta q = \Delta q^v$. This is indeed the case in the non-relativistic quark model, which predicts $\delta q = \Delta q^v = \Delta q^{\text{N.R.}}$, where $\Delta u^{\text{N.R.}} = 4/3, \Delta d^{\text{N.R.}} = -1/3$ and $\Delta s^{\text{N.R.}} = 0$. This can also be seen by comparing (14) with (16) and setting $\gamma^0 = 1$. However, in a relativistic valence quark model such as the MIT bag model, the speculation is false. Instead,

$$\Delta q^v = c \int \left(f^2 - \frac{1}{3}g^2\right) r^2 dr \; ; \qquad \delta q = c \int \left(f^2 + \frac{1}{3}g^2\right) r^2 dr \qquad (18)$$

where $c$ is a constant and $f$ and $g$ are upper and lower components of the quark wave function (in a $\gamma^0$-diagonal basis). Thus, relativity introduces a deviation of $\Delta q$ and $\delta q^v$ from $\Delta q^{\text{N.R.}}$ and a splitting between $\Delta q$ and $\delta q$: $\delta q - \Delta q^v = c\frac{2}{3}\int g^2 r^2 dr$.

Of course, these estimates about $\delta q$ are made in the context of rather naive models of the nucleon. However, it does appear that the tensor charge $\delta q$, free from sea quarks, provides a middle ground between the experimental data on $\Delta q$ and the non-relativistic valence quark model estimate $\Delta q^{\text{N.R.}}$, and a measurement of the former may provide us some knowledge about the relevant importance of the relativistic and sea-quark effects on $\Delta q$. To illustrate the relativistic effect, we have calculated $h_1(x)$ for a single quark in a nucleon in the MIT bag model. The result is shown in Fig. 2a where it is compared to $g_1(x)$.

High-twist structure functions cannot be interpreted as simple quark distributions. Precisely for this reason, they are potentially useful to understand the quark-gluon dynamics of confinement in QCD. The twist-4 parton distributions have been studied extensively in the literature.[11] The study of twist-3 structure functions has been limited only to $g_2(x)$.[6] There are two important advantages in studying twist-3 structure functions compared with twist-4 or beyond. Firstly, $g_2(x)$ and $h_2(x)$ contribute to certain spin asymmetries at leading order in $1/Q$ and therefore, they can be extracted straightforwardly from data. Secondly, although twist-3 structure functions couple to complicated quark-gluon correlation functions under QCD evolution, they have no explicit dependence on gluon fields. This second feature renders it possible to calculate them in valence quark models without dynamical gluons.

As it is defined in (5), $h_2(x)$ is not completely determined by matrix elements of twist-3 operators. Instead, $h_2(x)$ receives a contribution from the same operators as $h_1(x)$. The same phenomenon was recognized in the case of $g_2(x)$ by Wandzura and Wilczek.[12]. We find,[8]

$$h_2(x) = -h_1(x)/2 + 2x \left[\theta(x) \int_x^1 \frac{h_1(y)}{y^2}dy - \theta(-x) \int_{-1}^x \frac{h_1(y)}{y^2}dy\right] + \bar{h}_2(x) \qquad (19)$$

where $\bar{h}_2(x)$ is defined as follows. Define

$$\mathcal{O}_{n,\ell} = -\frac{1}{2}\mathcal{S}_{n-1}\bar{\psi}\sigma^{\alpha\mu_1}i\gamma_5 iD^{\mu_2}\ldots(igF_\alpha^{\mu_\ell})\ldots iD^{\mu_{n-1}}\psi$$
$$R_{n,\ell} = \mathcal{O}_\ell - \mathcal{O}_{n-\ell} \qquad (\ell = 2,\ldots,n-1) \qquad (20)$$

and its matrix elements,

$$\langle PS|R_{n,\ell}|PS\rangle = b_{n,\ell}\mathcal{S}_{n-1}S^{\mu_1}P^{\mu_2}\ldots P^{\mu_{n-1}} \qquad (21)$$

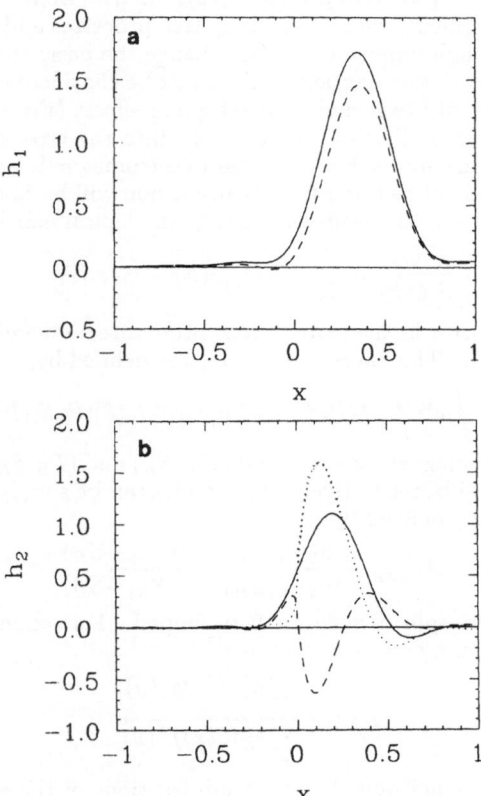

Fig. 2. *Bag model estimates for, a) $h_1(x)$ (solid), compared to $g_1(x)$ (dashes) and; b) $h_2(x)$(solid) and $\bar{h}_2(x)$ (dashes). The dotted line represents the difference.*

then, the sum rules for $\bar{h}_2$ are

$$\int \bar{x}^n h_2(x)dx = \sum_{\ell=2}^{n/2-1} (n - 2\ell + 1)b_{n,\ell} \tag{22}$$

The appearance of the combinations of $\mathcal{O}_\ell$ in $R_\ell$ ensures that each sum rule has definite charge conjugation. The first two terms in (19) give a trivial or kinematic contribution to $h_2(x)$. The last term, $\bar{h}_2(x)$, contains the truly new and dynamical information in $h_2(x)$. Like $g_2(x)$, $h_2(x)$ is hard to model without insight into the quark gluon correlations which determine $\bar{h}_2(x)$. In Fig. 2b we show $h_2(x)$ and $\bar{h}_2(x)$ in a bag model where boundary effects replace explicit gluon degrees of freedom.[6]

Now, we quote the lepton pair production symmetries from the collision of polarized nucleons. For simplicity, we consider only tree processes and ignore all radiative corrections which, though important, do not change the basic structure of the cross-section. Through twist-3, the diagrams which must be considered are those of Fig. 1b and 1c. Note the necessity to include explicit gluon effects (Fig 1c) at twist-3. They do not introduce any new distribution functions into the cross-section. Their only role is to render the first-order tree-diagram (electromagnetic and color) gauge invariant. This technical subject and further discussion will be found in Ref. [8]. The differential cross-section in the center-of-mass of the lepton pair is,

$$\frac{d\sigma}{d^4Qd\Omega} = \frac{\alpha^2}{2(2\pi^4)^2sQ^2}(\delta_{ij} - \hat{\ell}_i\hat{\ell}_j)W_{ij} \tag{23}$$

where $\hat{\ell}_i$ is the unit vector in the lepton-momentum direction and $Q^2$ is the squared-mass of the lepton pair. The hadron tensor $W_{\mu\nu}$ is defined by,

$$W_{\mu\nu} = \int e^{i\xi\cdot Q}d^4\xi\langle P_A S_A P_B S_B|J_\mu(0)J_\nu(\xi)|P_A S_A P_B S_B\rangle \tag{24}$$

where $J_\mu$ is the electromagnetic current and $(P_A, S_A)$ and $(P_B, S_B)$ are momenta and spins of nucleons A and B, respectively. The results can be simply expressed in terms of the spin asymmetries defined by

$$A_{S_A S_B} = \frac{\sigma(S_A, S_B) - \sigma(S_A, -S_B)}{\sigma(S_A, S_B) + \sigma(S_A, -S_B)} \tag{25}$$

where $-S_B$ means the spin of nucleon B is flipped. For longitudinal-longitudinal collisions, the spin-asymmetry is well-known[13]

$$A_{LL} = \frac{\sum_a e_a^2 g_1^a(x)g_1^{\bar{a}}(y)}{\sum_a e_a^2 f_1^a(x)f_1^{\bar{a}}(y)} \tag{26}$$

Here $x$ and $y$ are the longitudinal momentum fractions of the annihilating quarks. The sum over "$a$" covers all quark and antiquark flavors. We suppress the beam and target labels $A$ and $B$: by convention the structure function with argument $x(y)$ refers to hadron $A(B)$. For transverse-transverse collision,

$$A_{TT} = \frac{\sin\theta\cos 2\phi}{1 + \cos^2\theta} \frac{\sum_a e_a^2 h_1^a(x)h_1^{\bar{a}}(y)}{\sum_a e_a^2 f_1^a(x)f_1^{\bar{a}}(y)} \tag{27}$$

which was first obtained by Ralston and Soper.[2] And for longitudinal- transverse collision,

$$A_{LT} = \frac{2\sin 2\theta\cos\phi}{1 + \cos^2\theta}\frac{M}{Q}\frac{\sum_a e_a^2[g_1^a(x)yg_T^{\bar{a}}(y) - xh_L^a(x)h_1^{\bar{a}}(y)]}{\sum_a e_a^2 f_1^a(x)f_1^{\bar{a}}(y)} \tag{28}$$

which is our new result. Here, $h_L(x) = h_1(x)/2 + h_2(x)$. The asymmetry $A_{LT}$ depends on both twist-3 structure functions $g_T(x)$ and $h_L(x)$ and is down by a factor of $M/Q$.

ACKNOWLEDGEMENTS

I would like to thank Xiangdong Ji for his collaboration on this project. We, in turn, would like to thank John Collins, Jianwei Qiu and John Ralston for stimulating and helpful conversations and correspondence.

REFERENCES

1. R. L. Jaffe and Z. Ji, MIT preprint CTP#1952 (March 1991), to be published in *Phys. Rev. Lett.*

2. J. Ralston and D. E. Soper, *Nucl. Phys.* **B152** (1979) 109.

3. I. Kodaira, S. Matruda, K. Sasaki, and T. Uenatsu, *Nucl. Phys.* **B159** (1979) 99; A. P. Bukhvostov, E. A. Kuraev, and L. N. Lipatov, *Sov. Phys. JETP* **60** (1984) 22.

4. X. Artru and M. Mekhfi, *Z. Physik* **C45** (1990) 669.

5. J. Collins, Penn State Preprint (1990) and private communication.

6. R. L. Jaffe and X. Ji, *Phys. Rev.* **D43** (1991) 724; R. L. Jaffe, *Comm. in Nucl. Part. Phys.* **14** (1990) 239.

7. J. Qiu and G. Sterman, SUNY at Stony Brook preprint, ITP-SB-90-49, 1990.

8. R. L. Jaffe and X. Ji, to be published.

9. J. Kogut and D. E. Soper, *Phys. Rev.* **D1** (1970) 2901.

10. G. R. Goldstein and M. J. Moravcsik, *Ann. Phys. (N.Y.)* **98** (1976) 128; **142** (1982) 219; **195** (1989) 213.

11. R. L. Jaffe and M. Soldate, *Phys. Lett.* **B105** (1981) 467; A. Vainshteyn and E. Shuryak, *Nucl. Phys.* **B201** (1982) (1982) 141; R. K. Ellis, W. Furmanski, and R. Petronzio, *Nucl. Phys.* **B212** (1983) 29; J. Qiu, *Phys. Rev.* **D42** (1990) 30.

12. W. Wandzura and F. Wilczek, *Phys. Lett* **B172** (1977) 195.

13. F. E. Close and D. Sivers, *Phys. Rev. Lett.* **39** (1977) 1116; J. C. Collins and D. E. Soper, *Phys. Rev.* **D16** (1977) 2219.

# MEASUREMENT OF SPIN TRANSFER PARAMETERS IN THE

# p̄p → n̄n CHARGE-EXCHANGE REACTION AT LEAR

M. Agnello,[5] A. Ahmidouch,[2] J. Arvieux,[3] R. Bertini,[3] R. Birsa,[4] F. Bradamante,[4] T. Bressani,[6] H. Catz,[3] E. Chiavassa,[6] S. Dalla Torre-Colautti,[4] N. De Marco,[6] J.C. Faivre,[3] M. Giorgi,[4] E. Heer,[2] R. Hess,[2] F. Iazzi,[5] R.A. Kunne,[3] M. Lamanna,[4] C. Lechanoine-Le Luc,[2] M.P. Macciotta,[1] A. Martin,[4] C. Mascarini,[2] A. Masoni,[1] B. Minetti,[5] A. Musso,[6] A. Penzo,[4] F. Perrot-Kunne,[3] A. Piccotti,[6] G. Puddu,[1] D. Rapin,[2] P. Schiavon,[4] S. Serci,[1] and F. Tessarotto [4]

[1]INFN Cagliari and University of Cagliari, Cagliari, Italy
[2]DPNC, University of Geneva, Geneva, Switzerland
[3]CEN DPhN and LNS, Saclay, Gif-sur-Yvette, France
[4]INFN Trieste and University of Trieste, Trieste, Italy
[5]INFN Turin and Turin Polytechnic, Turin, Italy
[6]INFN Turin and University of Turin, Turin, Italy

**ABSTRACT**

The experiment PS199 at Lear investigates the spin structure of the charge-exchange reaction p̄p→n̄n. Data for the measurement of spin transfer parameters for both neutron and antineutron have been collected during 1990 and preliminary results are presented.

**INTRODUCTION**

The program of the collaboration PS199 is an important attempt to get new insight in the $\overline{N}N$ reactions using the $\overline{p}p \rightarrow \overline{n}n$ charge-exchange channel. A possible approach to study the detailed structure of the $\overline{N}N$ reactions is to measure the 5 independent complex amplitudes determining the scattering matrix (Ref.1 and 2).

A possible parametrization, for each isospin part, is listed in Ref.1 ($\sigma^j$ are the projections of the Pauli matrices on the fundamental frame of reference directions defined by the incoming and outcoming momenta in the centre of mass):

*Spin and Isospin in Nuclear Interactions*
Edited by S.W. Wissink *et al.*, Plenum Press, New York, 1991

$$M(s,t) = (a + b) + (a - b)\, \sigma^n_1 \sigma^n_2 + (c + d)\, \sigma^m_1 \sigma^m_2 +$$

$$+ (c - d)\, \sigma^l_1 \sigma^l_2 + e\, (\sigma^n_1 + \sigma^n_2)$$

The observables can be written down in terms of the amplitudes:

$$\frac{d\sigma}{d\Omega} = I = \frac{1}{2}\left\{ |a|^2 + |b|^2 + |c|^2 + |d|^2 + |e|^2 \right\} \qquad A_{ooon} \cdot I = \mathrm{Re}\left( a^* \cdot e \right)$$

$$( K_{noon} - 1 ) \cdot I = \frac{1}{2}\left\{ |b|^2 + |d|^2 \right\} \qquad\qquad ( D_{onon} - 1 ) \cdot I = \frac{1}{2}\left\{ |c|^2 + |d|^2 \right\}$$

The need of many measurements to disentangle the different components of the scattering matrix is evident; PS199 is the first experiment measuring even spin-dependent observables in the charge-exchange channel in this energy range (Ref.4).

Dramatic spin effects are predicted for the polarization transfer from the proton to the neutron and antineutron $D_{onon}$ and $K_{noon}$ by theoretical models (see Fig.1; recall that $D_{onon}$ and $K_{noon}$ are equal to +1 when there is no spin dependence in the scattering matrix). The measurement of these observables is a way to access the two-spin part of the interaction and to test the operators depending on two Pauli matrices as the tensor part of the $\pi$-exchange contribution.

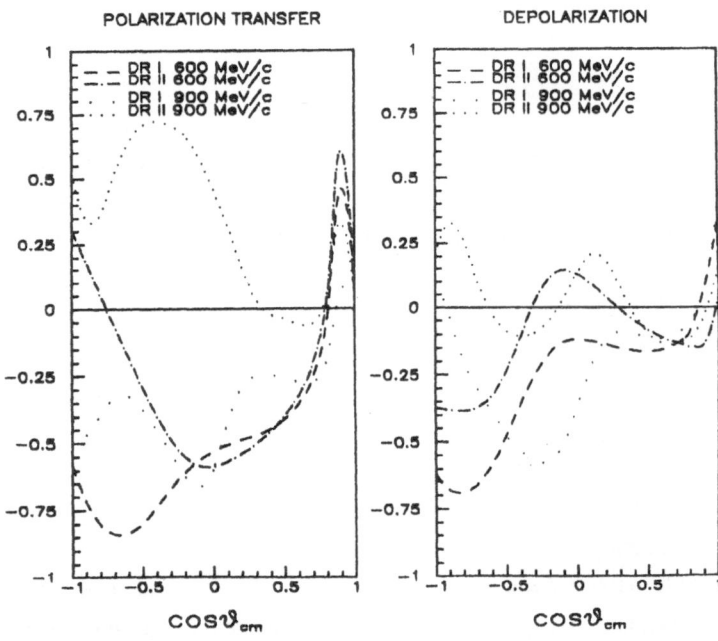

**Figure 1.** Predictions for D and K. DR I = Dover-Richard Model I. DR II = Dover-Richard Model II (Ref. 3).

# EXPERIMENTAL SET-UP

The quantities $D_{onon}$ and $K_{noon}$ can be sorted out by measuring the transfer of polarization from a proton target to the neutron and antineutron in the final state, respectively; the polarization of these particles has to be measured using polarimetry techniques (Ref. 5).

The neutron and the antineutron, produced by the $\bar{p}p$ interaction in a polarized target by the antiproton beam extracted from Lear, can interact first in the neutron counters. The analyzing power of these scattering reactions (very well known only for np elastic scattering while precise data for $\bar{n}p$ are lacking) is used to evaluate the polarization of the incoming particles.

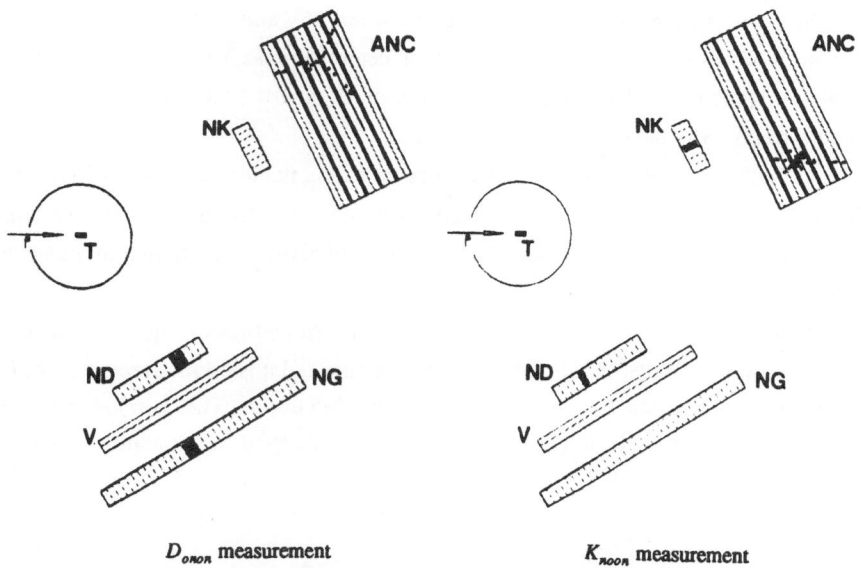

$D_{onon}$ measurement                    $K_{noon}$ measurement

**Figure 2.** PS199: Online display (900 MeV/c, Aug. 90)—Candidates for double-scattering events. ANC = antineutron detector. NK, ND, and NG = neutron counters. T = polarized target and veto box. V = charged veto for the neutron polarimeter (Ref. 4).

This technique, already developed in nuclear physics for neutron measurements, is for the first time employed even for antineutron polarimetry. The polarized target helps also to keep under control the systematic effects producing false asymmetries and to factorize out the absolute value of the efficiencies of the counters. Double scattering measurements require high statistics data samples, therefore we decided to measure only at two values of momentum of the incoming $\bar{p}$ (500 MeV/c and 900 MeV/c) with a dedicated set-up (Fig.2).

## PRELIMINARY RESULTS

The analysis of the samples collected with the dedicated set-up is just starting. To verify the feasibility of these kind of measurements, a well known data sample has been chosen, namely the one used to measure $A_{on}$ at 654 MeV/c (Ref.6).

The trigger used to perform the asymmetry measurement selected $\bar{n}$ double scattering events as well, which were discarded to measure the analyzing power of the reaction, but which can be used to extract preliminary values for $K_{noon}$ (the polarization transfer parameter from the proton target to the outcoming antineutron). The available statistics with this sample is only one tenth of the one collected for each of the two energies with the dedicated set-up.

The pattern of fired counters needed to define a charge-exchange reaction with a double scattering of the antineutron is given by a neutron candidate on each side and an annihilation star in the ANC (Fig.2).

Charge-exchange events in the range between 26° and 35° of the angle of the antineutron (from 0.6 to 0.3 in the cosine of centre of mass) are selected using the information (geometrical parameters and time of flight for both the neutron and the antineutron) given by the neutron counters (Fig.2).

The hypothesis of a double scattering event in the neutron counter in front of the antineutron counter where the $\bar{n}$ has been finally detected can be tested computing the time of flight belonging to this final point with the assumption of elastic $\bar{n}p$ scattering in the neutron counter (Fig.3a).

The difference between the computed and the measured time of flight gives a further test of the procedure to tag the charge-exchange reaction and it is useful to reject part of the background for the $\bar{n}p$ reaction, namely the $\bar{n}C$ scattering events. The correlation between the double scattering angle $\vartheta_{ds}$ and the recoil energy detected in the neutron counter is shown in figure 3b.

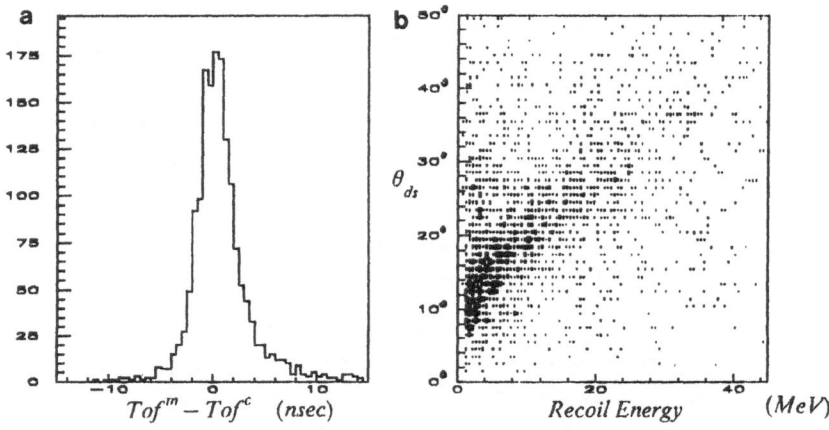

**Figure 3.** Difference in TOF for the $\bar{n}$ signal (left). Double scattering angle vs. recoil energy (right).

The distribution of events along the azimuthal angle $\varphi_{ds}$, depending on the polarization (*up* and *down* ) of the target, can be written down as:

$$N^{\pm} \propto \left\{ 1 \pm A_1 \cdot P_t^{\pm} + A_2 \cdot \left[ A_1 \pm K_{noon} \cdot P_t^{\pm} \right] \cdot \cos \varphi_{ds} \right\}$$

where $P_t^{\pm}$ is the absolute value of the target polarization, $A_1$ and $A_2$ are the analyzing power of the $\bar{p}p$ charge-exchange and the $\bar{p}n$ elastic reactions. The measured asymmetry is:

$$\frac{N^+ - N^-}{N^+ + N^-} = A_1 \cdot P_t \cdot \left[ 1 + \frac{A_2 \cdot K_{noon}}{A_1} \cdot \cos \varphi_{ds} \right]$$

The factor $A_1 \cdot P_t$ is known (Ref.6) and its evaluation provides an useful counter check, while $A_2 \cdot K_{onno}$ is the new quantity to be extracted from the data. In figure 4 the experimental data are shown; the fitting function $\alpha [ 1 + \varepsilon \cos \varphi_{ds} ]$ and its mean value are superimposed on the data. The results are, for the mean value: $\alpha = 0.080 \pm 0.036$, compatible with the measured $A_1 \cdot P_t = 0.10$, and the factor measuring $(A_2 \cdot K_{onno})/A_1$ is: $\varepsilon = 1.129 \pm 0.36$.

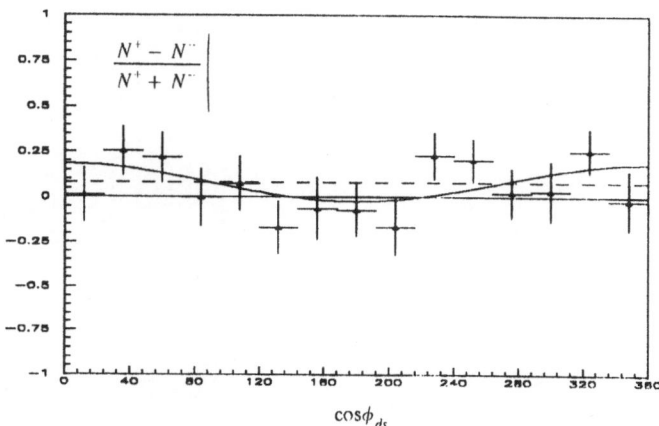

**Figure 4.** PS199: K measurement—preliminary results.

## CONCLUSIONS

Although this result is very preliminary and can be regarded as an useful exercise for the final measurements of both $D_{onon}$ and $K_{noon}$, the procedure to identify double scattering events and to evaluate the spin transfer parameters has been tested successfully on an experimental data sample.

Are the spin dependent observables the missing key for a deeper understanding of the $\bar{N}N$ reactions? We hope that the data provided by the PS199 collaboration constitute a new relevant contribution in this field.

## REFERENCES

1.  J.Bystricky et al., Journ.de Phys. (Paris) **39** (1978) 1.
2.  L.Puzikov et al., Nucl.Phys. **3** (1957) 436.
3.  C. Dover and J.M. Richard, Phys. Rev. **C21** (1980) 1466.
4.  F.Bradamante, invited talk to this conference, these proceedings.
5.  T.Taddeucci et al., NIM **A241** (1985) 448.
6.  R. Birsa et al., Phys. Lett. **B246** (1990) 267.

# STUDY OF THE np→ppπ⁻ REACTION

## WITH THE VERTEX DETECTOR ARCOLE

Y. TERRIEN, G. BRUGE, P. COUVERT, B. FABBRO, J.-C. FAIVRE, C. KERBOUL, M. ROUGER, F. WELLERS, R. BEURTEY* and J. SAUDINOS*

DPhN/SEPN ( *: Lab. Nat. Saturne)
CEN-Saclay
91191 Gif-sur Yvette, France

## INTRODUCTION

The study of inelastic channels induced by the nucleon-nucleon (NN) system has a two-fold interest: first, this elementary process has to be understood in itself, for example through the isobar model mechanism, second, its knowledge is necessary to interpret more complicated reactions induced by a nucleon probe with nuclei; for example, any microscopic calculation of the (n,p) charge exchange reaction in the $\Delta$ region requires the knowledge of np and nn→NNπ⁺ in this region of the phase space. Thus, the study of the one-pion production channel in NN interaction must be undertaken as carefully as the study of the elastic channel to be able to treat all channels on the same footing in a global description of these elementary hadronic processes. In the energy region ranging up to the GeV, a large body of data already exists. Extensive studies of proton-proton and neutron-proton elastic scattering involving complete determination of amplitudes from the measurement of spin observables by using polarized beams and targets, have been made. Many data have been obtained also for the pp→dπ⁺ and the pp→pnπ⁺ inelastic reactions, both proceeding through a mechanism dominated by the existence of an intermediate N$\Delta$ state, as can be seen from the success of the various approaches using such an isobar model[1].

## CASE OF THE np→ppπ⁻ REACTION

The np→ppπ⁻ reaction presents a particular interest in this respect. The total isospin of the entrance channel can be either I=0 or I=1. For I=0, the reaction cannot proceed through a $\Delta$N intermediate state at all. For

I=1, the possible Δ⁰p intermediate state is weakened by antisymmetrization. Thus, one can expect that some other mechanisms like existence of NN* intermediate states or non-resonant contributions will be seen, contrarily to the case of pp induced reactions where they are hidden by the strong Δ⁺⁺n intermediate state contribution.

There are only a few data for this np→ppπ⁻ reaction. Some of them come from bubble-chamber experiments with very low statistics, others were taken with detectors covering very small parts of the phase space, like in the recent experiment[2] on the quasi-2-body reaction np→π⁻(pp)¹S₀. This experiment was presented by the authors as an evidence for pion absoption by a 6-quarks cluster, but, in a recent brief report, M. Bachman et al[3] showed that the results could be reasonably well explained in the frame of a unitary NN→NNπ model containing only the one-pion exchange driving force. However, these interpretations deal with a very restricted part of the phase space open for the np→ppπ⁻ channel, since the two protons are in a state of a zero value for the relative momentum.

## DESCRIPTION OF THE EXPERIMENT

We present here the first results of an experiment undertaken at the Laboratoire National Saturne (LNS-Saclay) to study this np→ppπ⁻ channel in the full phase space, using a free polarized neutron beam of energies ranging from 400 MeV to 1150 MeV, and a 4π-solid angle detector to caracterize the outgoing channel completely. We shall give here a short description of the free neutron beam and of the principle and the apparatus used to do the experiment. Then we shall present some experimental results.

Polarized neutron beam facility has been used at Saturne for some time already[4]. Neutrons are produced by break-up of polarized deuterons on a Be target followed by a magnetic sweeping of the charged particles and by a strong collimation (1.5 cm-diameter on 8 meters long) to obtain a beam of usable and well defined size. Intensity of more than $10^6$ neutrons/s can be reached. It has been shown that the neutron beams obtained in such a way are reasonably monochromatic[5] (20 to 50 MeV FWHM, according to energy) and that the polarization percentage of the neutrons is the same as the one of the deuteron before break-up[6] ($P_{max}$=0.66 due to the polarized source structure, in practice P~0.60). To get absolute normalization of the cross-sections measured using such a beam, it is of course necessary to measure its absolute intensity. This is done with a neutron monitor[7] simply made of plastic scintillators telescopes looking at a 10-cm thick graphite target and which was calibrated using a dedicated method[8].

The experiment that we have set up to study the np→ppπ⁻ reaction is based on simple kinematics. For a given event of an identified reaction, to define completely the 3-body final-state, one needs to know 9 independant kinematical variables. Given the 4 equations of momentum and energy conservation, the measurement of 5 quantities only is

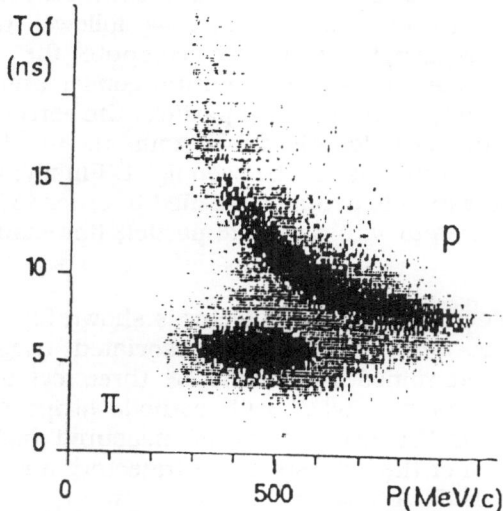

Fig. 1  Proton-pion separation by the TOF-momentum correlation.

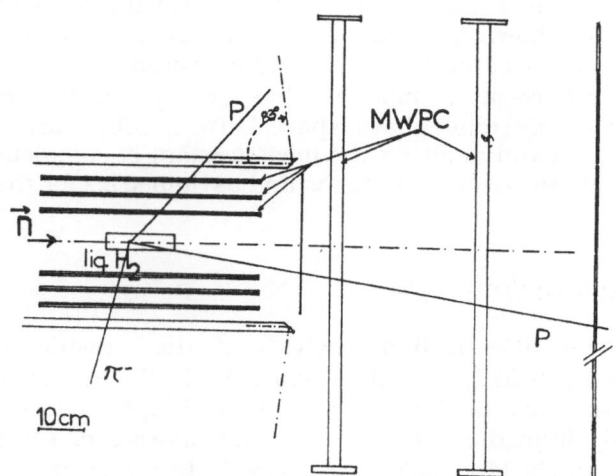

Fig. 2  The vertex detector ARCOLE: general layout.

required. Our set-up allows us to measure the 3×2=6 angles of the trajectories of the 3-particle final state, which permits the complete determination of kinematics and even provides an over-determination which is helpful to identify the reaction. This identification is also done using additional information obtained from measurements of energy loss and time-of-flight. In fact, we proceed as follows: using the angles measured for a 3-particle event, we compute the 3 corresponding momenta from the equations of momentum conservation, where masses do not enter explicitly; then, for each particle, the correlation between the momentum and the TOF (or dE/dx) permits us to discriminate easily between pion and protron, as can be seen fig. 1. Finally, we check that the total energy conservation is actually satisfied in order to eliminate random events or physical events with a neutral particle (for example, np→npπ+π- or np→ppπ-π+π0).

The principle scheme of our detector is shown fig. 2. It is essentially a quasi-4π vertex detector with an associated trigger system. The trajectories are determined by means of three cylindrical Multi-Wire Proportionnal Chambers (MWPC) with cathode strips read-out and two forward plane MWPC. The accuracy of the measured angles varies slightly with the orientation of the corresponding trajectory and is of the order of less than 1°, FWHM. A 2×2 m$^2$ plane of twenty 2-cm thick plastic scintillators set at forward angles and a barrel of eight 1-cm thick plastic scintillators surrounding the cylindrical MWPC constitute the trigger and are used to get the dE/dx and TOF information. Electronics is designed to read positions, dE/dx and TOF information out of MWPC and plastic scintillators for up to 4 particles detected. This set-up, ARCOLE, is almost a 4π detector as shown by Monte-Carlo simulations that we did for the reaction np→ppπ- at several energies and for various arrangments of the detector elements. As an example, the best geometry of the set-up gives a 98% coverage of the total phase space between 600 MeV and 1 GeV. Therefore, this apparatus allows the measurement of complete differential angular distributions of cross-sections and asymmetries for the reaction of interest.


FIRST PRELIMINARY RESULTS

We present here a first analysis of the  results obtained in experiments done at 572, 784, 1012 and 1134 MeV. Let's discuss here, as an example, the case of the 784 MeV data. Simple observables can be derived directly from the data, like invariant masses of the pp and pπ- pairs. The threshold $M_{pp}=2×M_p$ corresponds to the situation where the two protons are emitted together with the same momentum (no relative momentum). In this case, the np→(pp)π- reaction is the isospin-analog of the pp→dπ+ reaction, which has been very extensively studied. To get the asymmetry at this limit, we have cut the $M_{pp}$ histogram in bins of $M_{pp}$ and, for each bin, extracted the asymmetry of the data. Fig. 3 presents the angular distribution of the measured asymmetries for the 3 first bins. Then, a phenomenological fit of the asymmetry $A=f(\theta_\pi, M_{pp})$ has been

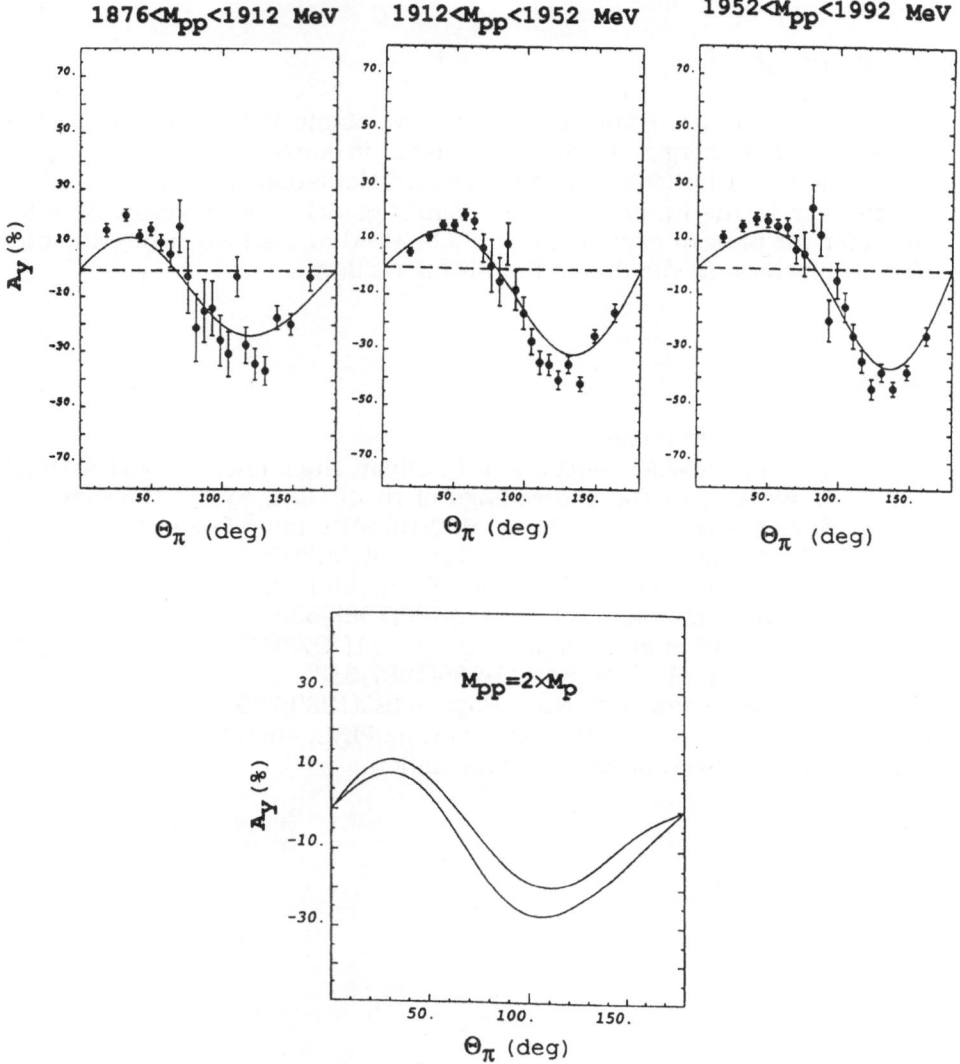

Fig. 3    Asymmetries measured

at 784 MeV (see the text) for the

reaction np→ppπ⁻.

done to extrapolate the asymmetry at the $M_{pp}=2{\times}M_p$ limit. Resulting experimental asymmetries are also presented in fig. 3, under a form representing the corridor of errors. The same procedure has also been applied to the data obtained at the 3 other energies.

## CONCLUSION

To summarize, I will say that we have obtained data for the reaction $np{\rightarrow}pp\pi^-$ at 4 energies in full solid angle, in such a way that complete differential distributions can be obtained. Calculations by J.-M. Laget[9] are under way in the framework of an isobar model. The $4\pi$ vertex detector used for the present experiment could be used in the near future to study the strangeness production in the NN interaction.

## REFERENCES

1 - See, for example:
J. Dubach, W.M. Kloet and R.R. Silbar, Nucl. Phys. A466(1987)573
F. Wellers, in the Proceedings of th 4th Int. Symp. "Mesons and Nuclei", Czec. Jour. of Phys. B39(1989)72 and references therein.
2 - C. Ponting et al, Phys. Rev. Lett. 63(1989)1792.
3 - M. Bachman et al, Phys. Rev. C42(1990)1751.
4 - Y. Terrien et al, Nucl. Phys. A478(1988)533c.
5 - G. Bizard et al, Nucl. Inst. Meth. 111(1973)451.
6 - J. Ball et al, Nucl. Phys. B286(1987)635.
7 - B. Silverman et al, Nucl. Phys. A499(1989)763.
8 - Y. Terrien and F. Wellers, Jour. de Phys. 46(1985)1873.
9 - J.-M. Laget, private communication.

# THREE DECADES OF MISSING GAMOW-TELLER STRENGTH

Charles D. Goodman

Indiana University, Bloomington, IN 47408

## ABSTRACT

Experimentally measured Gamow-Teller (GT) strength is generally less than that expected from models thought to be appropriate. This observation applies to transitions between single, defined energy levels as well as to entire GT strength functions. Over the past thirty years the scope of measurements and the nature of the interest in the problem have changed. However, the discrepancy persists and it signifies a fundamental difficulty in nuclear models. Various corrections have been proposed for the shell model such as high order configuration mixing and intrinsic excitations of the nucleons. A fully adequate model of GT transitions does not yet exist.

## PAUCITY OF BETA DECAY STRENGTH IN HEAVY NUCLEI

In the early 60s the focus of the missing strength problem concerned the measured ft values for allowed beta decay transitions. All the known transitions for medium-weight and heavy nuclei proceeded at rates that were orders of magnitude slower than what one would expect if a significant fraction of the strength available from the neutron excess were observed. (See fig. 1) The discovery of isobaric analog states provided an explanation of why Fermi strength was not observed in heavy nuclei, and Fujita, Fujii and Ikeda proposed a similar explanation for the lack of GT strength.[1] They suggested that if the GT strength were contained in a giant resonance, the coulomb displacement energy would move the center of the resonance out of the energy window of beta decay. As long as the spreading width were not too large, very little GT strength would appear within the beta decay window. (See fig. 2)

These authors went further to suggest that one could search outside the narrow energy window of beta decay with the (p,n) reaction. In fact, their suggestion was followed and the giant GT resonance was show to be a general feature of nuclear structure. However, as we shall see, the giant resonance does not contain all of the strength.

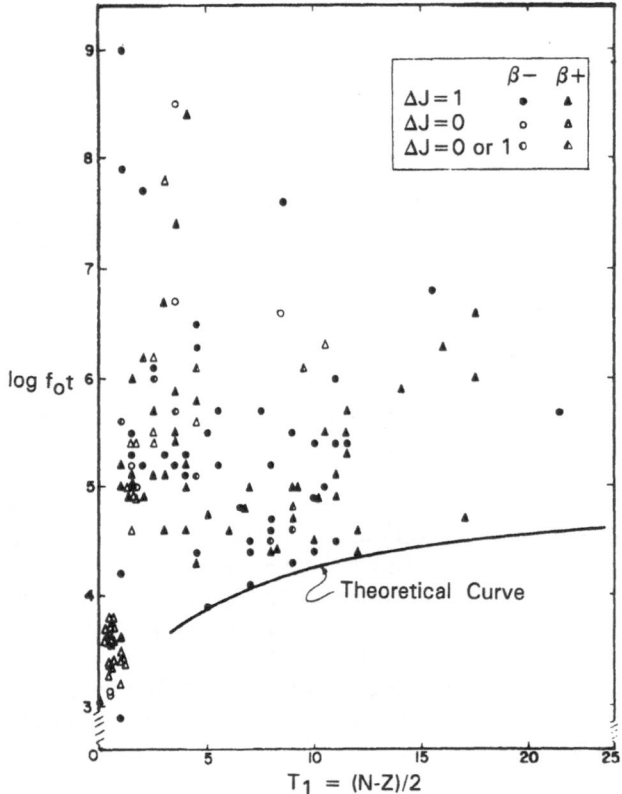

Fig. 1. Distribution of log ft values for allowed $\beta$ decay transitions between discrete states.

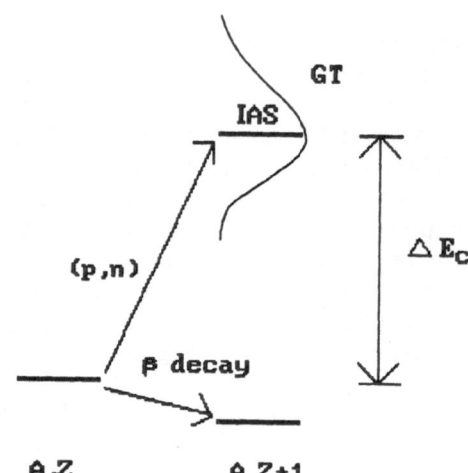

Fig. 2. Schematic representation of the giant Gamow-Teller resonance showing how it is inaccesssible to $\beta$ decay from neutron rich nuclei.

# IS $g_A$ RENORMALIZED IN NUCLEI?

In the meantime the question of the appropriate description of the nature of the axial vector interaction came under intense theoretical scrutiny, and the burning issue of the 70s was whether nucleons in the nucleus beta decay as though they were free nucleons. The problem is well expressed by Wilkinson. "Even if we were in possession of perfect nuclear (nucleonic) wave functions we should not expect the rate of Gamow-Teller $\beta$ decay to be given by the conventional expression incorporating the free-nucleon axial coupling constant $g_A$."[2]

At the time evidence concerning the possible difference between decay rates of bound and free nucleons depended on comparing measured rates for individual transitions with rates calculated with the shell model. "... the average decay amplitude found in practice is $(7.3 \pm 3.8)\%$ below that predicted by the best explicit wave-functions that are confined to the nominal major shell."[3]

Even the best available wave functions might be faulty. The complication of configuration mixing within the major shell can be avoided by examining transitions of the doubly closed LS shell $\pm$ 1 nuclei. Table 1 shows a summary of the measured GT strengths for these transitions and the calculated values assuming a single particle or a single hole. It is apparent that strength is really missing.

Table 1. B(GT) for L-S closed shells $\pm$ 1

| Transition | B(GT) sm | B(GT) exp[*] | B(GT) exp/ B(GT) sm |
|---|---|---|---|
| $^{15}O \to {}^{15}N$ | 1/3 | 0.27 | 0.81 + .01 |
| $^{17}F \to {}^{17}O$ | 7/5 | 1.09 | 0.78 + .01 |
| $^{39}Ca \to {}^{39}K$ | 3/5 | 0.28 | 0.47 + .01 |
| $^{41}Sc \to {}^{41}Ca$ | 9/7 | 0.74 | 0.58 + .01 |

[*] Based on ft values from ref. 4 and eq. (5).

These numbers are based on measured decay rates. For example, consider $^{41}Sc(\beta^+)^{41}Ca$. ft = 2869. Take B(F) = 1. Then B(GT) = 0.74. $0.74/(9/7) = 0.58$. Strongly quenched! This certainly suggests a renormalization of $g_A$.

This evidence was strong enough to induce a number of investigators to take the missing strength problem seriously and to propose ways to understand it. Magda Ericson and her collaborators viewed the problem in analogy with electromagnetism. "Our aim is to gain a physical insight into the physical origin of the renormalization for $\beta$ decay transitions. We want in particular to draw attention to the existence of a classical electromagnetic analogy: the strength of an electric dipole embedded in a dielectric medium."[5]

Rho and his collaborators emphasized coupling of nucleon states to Δ states in the nucleus.[6] Bohr and Mottelson cast the problem quite lucidly in terms of a constituent quark model.[7] Towner and Khanna turned their attention to doing the best calculations they could on corrections to the shell model for the LS closed shell ± 1 nuclei including core polarization and meson currents.[8] All of these approaches admit to the coupling to subnucleonic degrees of freedom, and pions in the nucleus are implicated through PCAC.[9]

Arima and his collaborators held to retaining ordinary, unexcited nucleons as the nuclear constituents. They attempted to account for the missing strength by implicating correlations brought about through the tensor force.[10]

## EXPLORATION OF THE "ENTIRE" GAMOW-TELLER STRENGTH FUNCTION

By the late 70s we had begun to explore the GT strength strength function over a broad energy range with the (p,n) reaction.

The incident that piqued my own interest in the missing strength problem happened in 1979 at a Giant Resonance conference in Oak Ridge. I showed a transparency of a spectrum from $^{90}$Zr(p,n) that illustrated the giant Gamow-Teller resonance. I remarked that the total strength in the giant resonance was only about 40% of the strength one expected from ten neutrons in the g-9/2 shell.[11] After the talk, Carl Gaarde, whom I had not met before, cornered me and told me that I should be much more worried than I seemed to be about finding only 40% of the strength in the giant resonance. After all, from the appearance of the spectrum, I wasn't going to find much strength outside of the giant resonance. He then pointed out something of immense importance.

Carl showed me that no matter how one manipulated the shell model configurations the total strength would never be *less* than that which I had calculated for ten neutrons in the g-9/2 shell. He showed me what we now call the Gamow-Teller sum rule.[12]

Carl carried the discussion yet further suggesting that, since all the strength must be there, my normalization of the spectrum must somehow be wrong and the total apparent strength in the spectrum was as good as anything to normalize to.

## GT VECTORS AND THE SUM RULE

It is convenient to discuss the sum rule in terms of state vectors and operators. We write | P> for the state vector of the parent nucleus. If we operate on this with the isospin lowering operator, or Fermi beta decay operator, we get a new vector that we may call the "collective Fermi vector."

$$| CF> = \Sigma t_i | P> \qquad (1)$$

Since the operation makes sense only when summed over all equivalent nucleons, we shall assume summation implicitly and drop the subscript, i, in the notation. The product of a spin and an isospin operator, $st^-$ or $st^+$, will, of course, also create a new vector which is our main interest here. We name it the "collective Gamow-Teller vector."

$$| CGT> = st^- | P>  \text{ or } st^+ | P> \tag{2}$$

This vector has three components, the three projections of the spin.

It is useful also to define the quantities $B(F)$ and $B(GT)$, the reduced F and GT transition probabilities. These are the scalar products of the collective F or GT vectors with the daughter states. To the jth daughter state:

$$B_j(F) = <D_j | CF> = <F>^2 \tag{3}$$

$$B_j(GT) = <D_j | CGT>^2 = <GT>^2 \tag{4}$$

These are the usual Fermi and Gamow-Teller matrix elements. In the case of the GT matrix element, summation over the spin substates is assumed.

It follows that the scalar products $<CF | CF>$ and $<CGT | CGT>$ are the total Fermi and the total Gamow-Teller strengths, respectively.

$$<CF | CF> = \Sigma<D_j | CF>^2 = \Sigma B_j(F) \tag{5}$$

$$<CGT | CGT> = \Sigma<D_j | CGT>^2 = \Sigma B_j(GT) \tag{6}$$

It should be noted that the total F and GT strengths are properties of $| P>$ alone. If one can calculate $| CF>$ and $| CGT>$ one need not calculate the vectors $<D_j |$ to find the total strength.

The sum rule is a statement about the lengths of $| CF>$ and $| CGT>$ and limits how *short* they can be.

$$<CF^- | CF^-> - <CF^+ | CF^+> = N - Z \tag{7}$$

$$<CGT^- | CGT^-> - <CGT^+ | CGT^+> = 3(N - Z) \tag{8}$$

It follows that:

$$<CGT^- | CGT^-> \geq 3(N - Z) \tag{9}$$

Shell model codes can be used to calculate $| CGT>$ and its projections on the eigenstates of the daughter nucleus. Legitimate questions to ask are how big are $<CGT^- | CGT^->$ and $<CGT^+ | CGT^+>$, and how is the strength distributed over the daughter states, that is, what does the strength function look like. It is important to keep in mind that the *total* strength found by projecting $| CGT>$ on a complete set of daughter states will be $>3(N-Z)$ as required by the sum rule. In general, if the open proton shell is not already filled for neutrons, $<CGT^+ \mid CGT^+> \; > 0$ and $<CGT^- \mid CGT^-> \; > 3(N-Z)$.

The sum rule is closely connected to the question of Pauli blocking of GT transitions. It tells us that given N neutrons and Z protons, we may block all of the neutron to proton transitions for Z of the neutrons, but no matter how we rearrange the protons, we cannot block all N transitions, if N > Z.

## THE "SMOKING GUN" DECADE

For the second Telluride conference in 1982 Stew Bloom and I collaborated on a paper summarizing the magnitude of the missing strength.[13] We found that shell model calculations using the full basis of states within a major shell could produce GT strength distributions that had a good resemblance in shape to the distributions measured with the (p,n) reaction. The calculations told us, however, that we should find all of the strength in the first 20 or so MeV of excitation, but, for all nuclei, only about 55% of the required minimum GT strength had ever been found. This indicated that something was going on beyond the shell model as we were using it.

Missing GT strength was a major topic in the 1982 Telluride conference, and the discussion took on an air of a shootout. The search was on for the "smoking gun" that would implicate or exonerate subnucleonic degrees of freedom.

"The importance of tensor correlations is pointed out to explain the quenching of spin magnetic moments and Gamow-Teller transitions. The delta-hole contribution is shown to be small at least in nuclei lighter than 41." --Arima[14]

"I would like to present here my argument again why the delta *must be* the principal agent for shifting a large amount of strength from the low-energy regime to the delta region." --Rho[15]

"...It is shown that most of the background subtracted in the experimental analysis of the $^{48}$Ca(p,n) -Gamow-Teller resonance is actually Gamow-Teller strength." --Osterfeld[16]

Meanwhile, Brown and Wildenthal, unconcerned about the smoking gun, made a systematic comparison of $\beta$-decay ft values with shell model calculations for 64 parent states in the sd shell.[17] They concluded that the data were consistent with the renormalization of $g_A$ by the factor $0.76 \pm 0.03$. This factor is also consistent with the overall quenching factor deduced from (p,n) data.

As we have remarked, $\beta$ decay explores only a narrow band of excitation energy. The band has been widened in a few cases by using very proton-rich nuclei.[18-20] In the case reported at this conference[20] the authors report a summed B(GT) of 2 up to 8 MeV of excitation for $^{37}$Ca($\beta^+$). The remaining 7 units required by the sum rule must appear at higher excitation beyond the region that can be explored by $\beta$ decay.

Most of the information we have about GT strength functions comes from 0 degree (p,n) spectra. For a light nucleus where we can resolve the individual levels, we see a pattern that we can interpret as the distribution of GT strength among the resolved levels. In the case of $^{18}O(p,n)^{18}F$ a peak corresponding to each of the known $1^+$ levels is seen in the spectrum (fig. 3). There is little else in the spectrum. This spectrum was obtained with a thick ice target with $^{18}O$ enriched (97.6%) water. No background subtraction is required. The peak width is due to target thickness. In this case ft values can be found in the literature for the inverse transition,[21] $^{18}F(\beta^+)^{18}O$ and for the isobaric mirror transition,[22] $^{18}Ne(\beta^+)^{18}F$. These are, respectively, ft = 3581, and ft = 1247 ± 11. The value for the inverse transition must be divided by 3 for the spin weighting. We use here ft = 1200 and find that B(GT) = 3.23 for the ground state. We normalize the spectrum to this value. The summed counts in the total spectrum less the counts in the IAS peak give us a summed B(GT) of 3.94, or 66% of 3(N-Z).

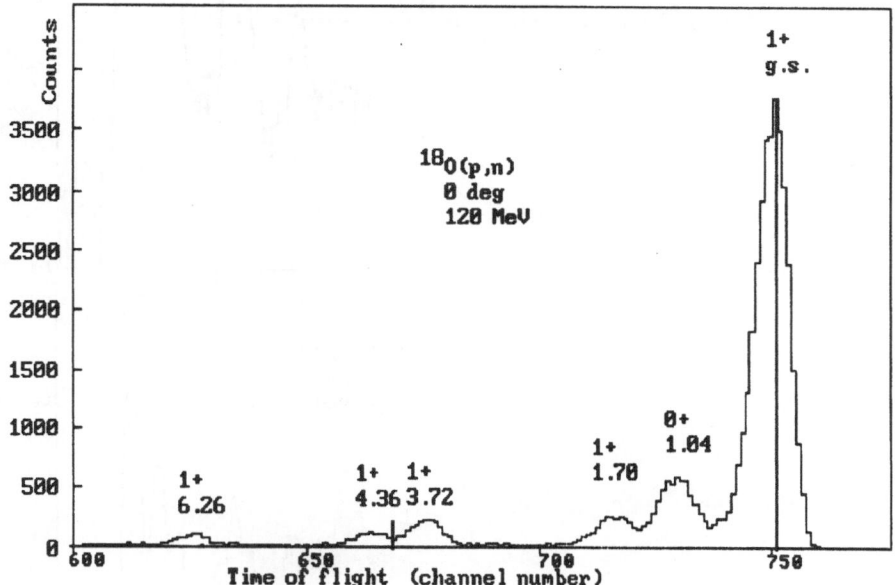

Fig. 3. Energy spectrum for $^{18}O(p,n)^{18}F$. All of the known $1^+$ states are indicated. The peak width is due to target thickness. (see text)

The solid bars in fig. 3 show the results of a shell model calculation using the Brown-Wildenthal interaction.[17] The model misses the $1^+$ state at 1.70 MeV altogether and shows only a single state in the 4 MeV region. A small amount of additional strength is calculated for T=1 states around 11 MeV. Other $^{18}O(p,n)$ data and the shell model calculation are also given by Anderson et al.[23] That experiment and analysis also shows 66% of 3(N-Z) strength found.

There is some isotropic background at large energy loss in the spectra. This can best be seen in logarithmic plots (fig. 4). This "background" may contain various multipole resonances and well as neutrons from multiparticle emission, and it is difficult to identify GT strength in this excitation region, if it exists.

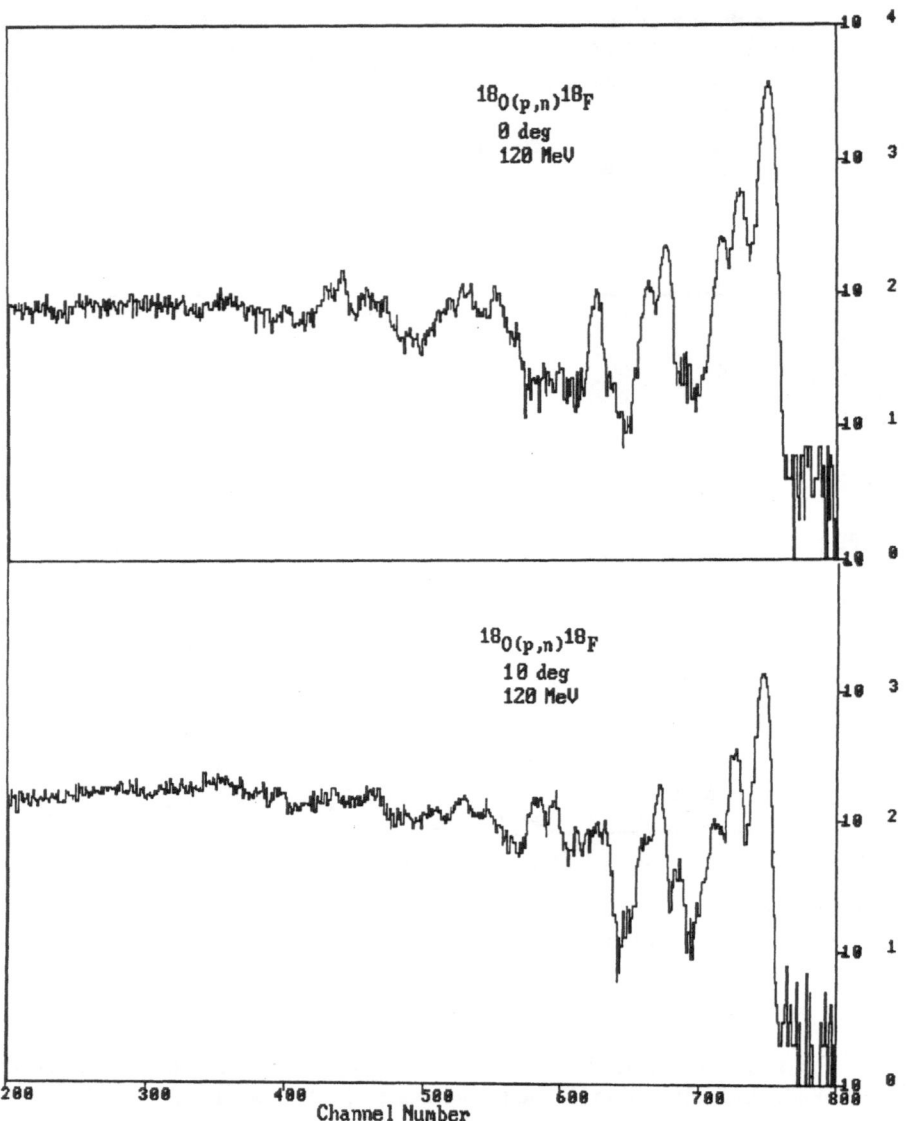

Fig. 4. Semi-log plots of energy spectra for $^{18}$O(p,n)$^{18}$F at 0 deg. and 10 deg. Note in particular that the cross section for the $1^+$ states has decreased with angle while the cross section for parts of the continuum has increased. These spectra are not normalized to integrated beam current.

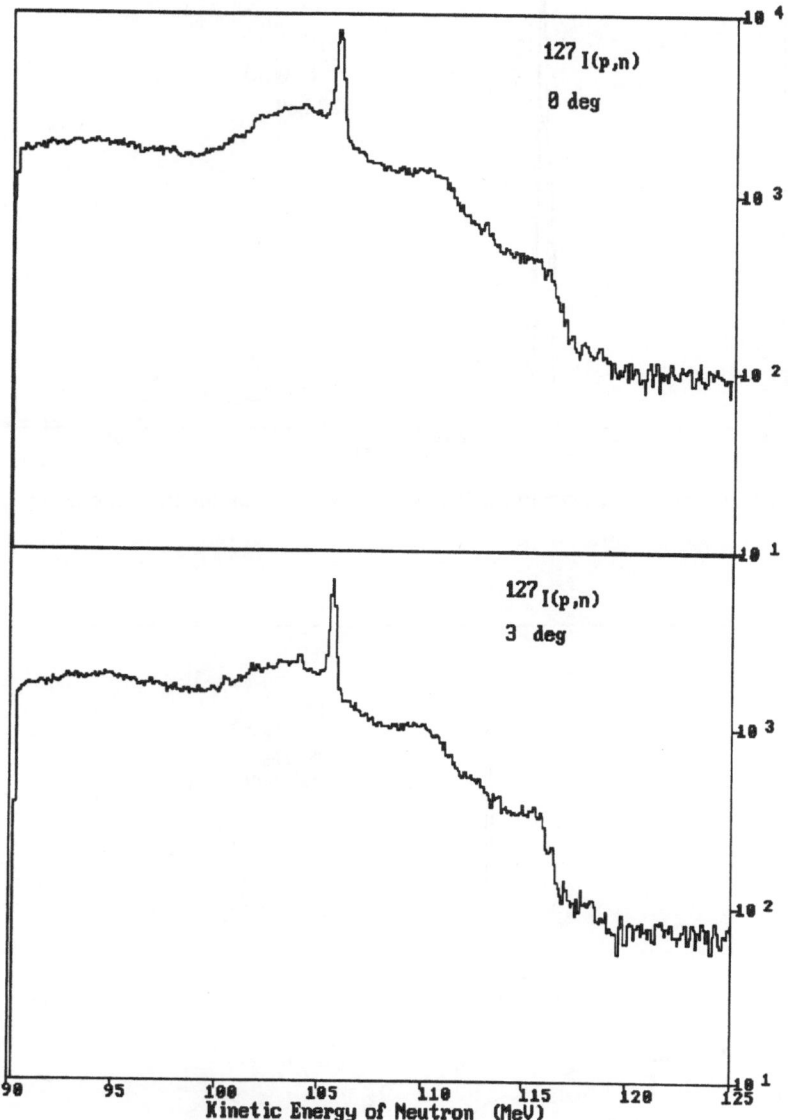

Fig. 5.　Neutron energy spectra for ¹²⁷I(p,n) at 0 degrees and three degrees. Copy these spectra. Overlay them and match the IAS peaks to see that the region below about 98 MeV is not forward peaked like the GT resonance and the IAS peak. The drop at 90 MeV is the end of the range of the time-of-flight spectrum.

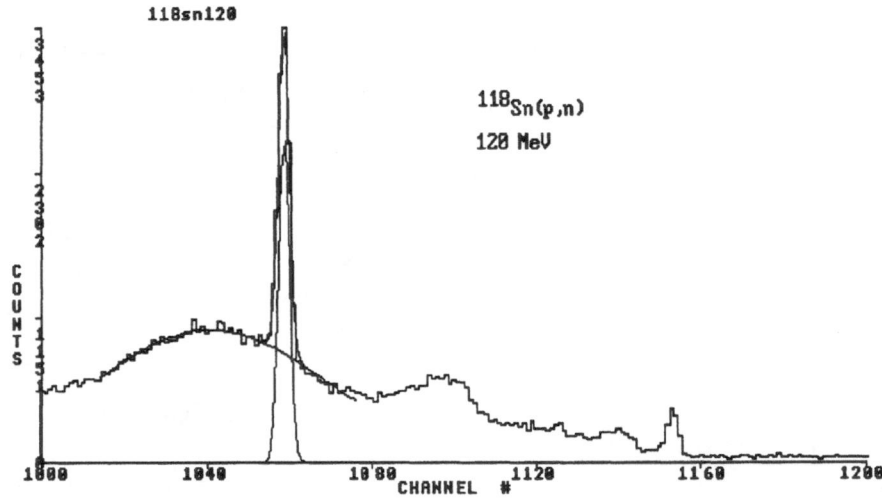

Fig. 6. Energy spectrum for $^{118}$Sn(p,n)$^{118}$Sb showing the fit to the IAS peak.

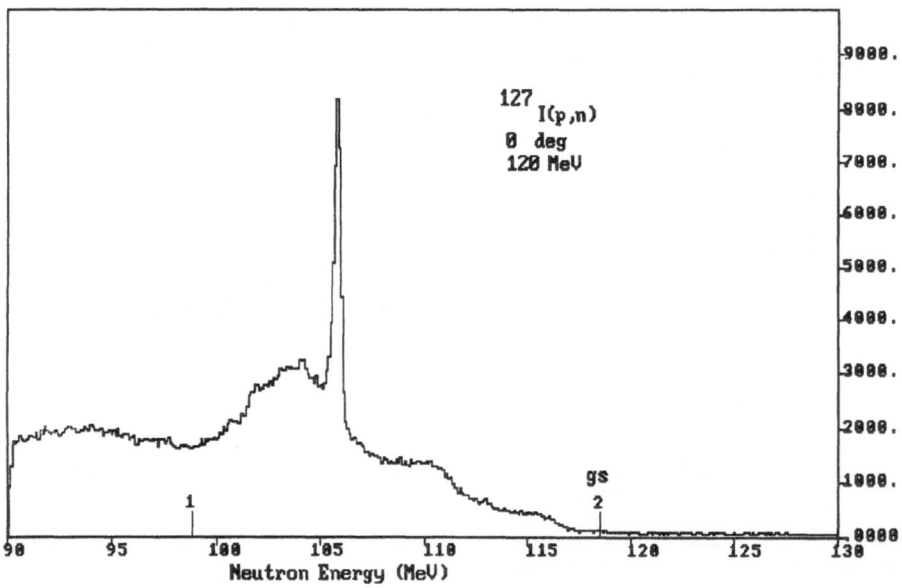

Fig. 7. Energy spectrum for $^{127}$I(p,n)$^{127}$Xe. The markers show the region in which the giant GT resonance appears to lie.

In heavy nuclei the spectrum looks essentially continuous. The most prominent feature is the giant GT resonance. Above the GT resonance (lower neutron energy) there is considerable strength in the spectrum. That component is not forward peaked in angle as can be seen by comparing zero and three degree spectra (see fig. 5). The only distinct feature that can be used to normalize the spectrum is the IAS peak. This contains all of the Fermi strength. What we do to extract the GT strength is fit a Gaussian shape to the IAS peak, sum the spectrum, and subtract the area of the IAS peak. This is shown, for example, in fig. 6. We must know the relative GT to F coupling strengths in the reaction to convert the GT area to B(GT). The empirical relation

$$\frac{\sigma(GT)/ B(GT)}{\sigma(F)/ B(F)} = (E_p / E_0)^2 \qquad (10)$$

$$E_0 = 55.1 \pm 1.4\, MeV$$

works very well for even mass nuclei. The normalization of (p,n) spectra to $\beta$-decay strength is discussed in detail by Taddeucci et al.[24] There is new evidence that $E_0 \approx 45$ provides a better fit for odd mass nuclei, although we do not have a real model to explain that. Fig. 7 shows a spectrum for an odd mass nucleus for which we have no information from $\beta$ decay for normalization. With $E_0 = 55$ MeV the total strength between the markers is 80% of 3(N-Z) while with $E_0 = 45$, the total strength is 55% of 3(N-Z).

The obvious feature that distinguishes odd-mass nuclei is that the target spin is greater than zero. The presence of an interaction of the form $(t_p \cdot T)( s_p \cdot I)$, where $t_p$ is the isospin of the proton, T is the isospin of the target, $s_p$ is the projectile spin and I is the target spin, would have the qualitative characteristic of altering the cross section and the spin transfer for the IAS peak in odd-mass nuclei. This interaction couples only to the IAS since it cannot alter the internal wave function of the nucleus. However, there may be conceptual difficulties in reconciling the inclusion of such an interaction in addition to the $(t_p \cdot t_i)(s_p \cdot s_i)$ Gamow-Teller interaction, since the sum of the nucleon spins forms part of the total nuclear spin. In the usual impulse approximation approach to the GT interaction the proton sees the nucleus as a group of individual nucleons. On the other hand, the wavelength of a 100 MeV proton is nearly 3 fermis. Yet if the proton cannot see the individual nucleons, our usual formulation of the GT interaction cannot be valid.

Two conclusions that we can safely draw are that 1) the shell model is not adequate to tell us where all the strength lies, and 2) a lot of strength has evaded detection. Perhaps it is at higher excitation than where we looked, or perhaps it is so thinly dispersed that we don't recognize it, or perhaps both effects play roles.

The constituent quark model can give us a simple, qualitative picture of how nucleon structure can change the GT strength function and shift strength out of the region where we have looked for it. We can retain much of the shell model picture. We don't have to take the drastic step of converting the A-body problem to a 3A-body problem with a nucleus composed of quasi-free quarks. Rather, let us retain the picture of the nucleus as made of nucleons, but say that each nucleon is made of three quarks. We note that there are only two color singlet states of three quarks in a relative s-state. These have the spin-isospin quantum numbers 1/2-1/2 and 3/2-3/2. We associate the first with protons and neutrons and the second with deltas. If we assume that the GT operator acts at the quark level, the operator couples nucleons to deltas. This puts a new light on the GT problem. The sum rule becomes

$$<CGT \mid CGT>^- - <CGT \mid CGT>^+ = 3(D - U) = 3(N - Z) \quad (11)$$

where D and U are the numbers of down and up quarks, but the complete GT strength function extends into the delta region above 300 MeV. The $\mid$CGT$>$ for the neutron is no longer identical to the proton state vector but contains a delta component in addition to the proton component. Also, the $\mid$CGT$>^-$ for the proton is not zero but is the state vector of the $\Delta^{++}$. Bohr and Mottelson[7] point out that the nucleon to delta transitions are not Pauli blocked, so all A nucleons participate in creating GT strength.[12] There is much more total strength, but almost all of it is at very high excitation where the present experiments would not detect it. How much strength would remain at low excitation has not yet been calculated in a realistic way.

I think what we need at this stage is a complete model that allows us to take into account nucleon excitations and can still allow us to calculate a good state vector for the ground state of the nucleus.

We will not find the "smoking gun" because the missing strength is not a consequence of a single omission that can be isolated in the shell model. There is a legitimate question of whether we can ignore nucleon structure to make our calculations easier. Since arguments have already been made to the effect that nucleon structure is relevant to the GT strength function, we cannot dismiss it without definitive evidence that its effect is not important. Unfortunately, although (p,n) spectra up to several hundreds of MeV excitation might provide some useful information, such spectra are not likely to settle the question of the involvement of delta coupling. A problem is that the high excitation region can be observed only at high momentum transfer, and it will be difficult to find the GT component in the midst of other processes that produce neutrons at high energy loss. We are dealing with a very complex many-body problem, and I think we haven't yet got a good model.

## ACKNOWLEDGEMENT

I wish to thank Stewart Bloom for helpful discussions and for calculating the GT strength function for $^{18}O$(p,n) with the Brown-Wildenthal interaction.

## REFERENCES

1.  J. I. Fujita, S. Fujii, and K. Iear core polarization effect on beta decay," Phys. Rev. **133**, BB549-B555 (1964)
2.  D. H. Wilkinson, "Renormalization of the axial-vector coupling constant in nuclear beta decay," Phys. Rev. C **7**, 930-936 (1973)
3.  D. H. Wilkinson, "Renormalization of the axial-vector coupling constant in nuclear beta decay (III)," Nucl. Phys. **A225**, 365-381 (1974)
4.  S. Raman, C. A. Houser, T. A. Walkiewicz, and I. S. Towner, "Mixed Fermi and Gamow-Teller beta transitions and isoscalar magnetic moments," Atomic Data and Nuclear Data Tables **21** (1978) 567-620
5.  M. Ericson, A. Figureau, and C. Thévenet, "Pionic field renormalization of the axial coupling constant in nuclei," Phys. Lett. **45B**, 19-22 (1973)
6.  Eulogio Oset and Mannque Rho, "Axial currents in nuclei: the Gamow-Teller matrix element," Phys Rev. Lett. **42**, 47-50 (1979); Mannque Rho, "Quenching of axial-vector coupling constant in beta-decay and pion-nucleus optical potential," Nucl. Phys. **A231**, 493-503 (1974)

178

7.  Aage Bohr and Ben R. Mottelson "On the role of the delta resonance in the effective spin-dependent moments of nuclei," Phys. Lett. **100B**, 10-12 (1981)

8.  I. S. Towner, and F. C. Khanna, "Corrections to the single-particle M1 and Gamow-Teller matrix elements," Nucl. Phys. **A399**, 334-364 (1983)

9.  R. J. Blin-Stoyle, and Myo Tint, "Theory of partially conserved axial-vector current and mesonic exchange effects in nuclear beta decay," Phys. Rev., **160**, 803-808 (1967); R. J. Blin-Stoyle, "Renormalization of the axial-vector coupling constant in beta decay," Nucl. Phys. **A254**, 353-369 (1975)

10. K. Shimizu, M. Ichimura, and A. Arima, "Magnetic moments and GT-type beta-decay matrix elements in nuclei with a LS doubly closed shell plus or minus one nucleon," Nucl.Phys. **A226**, 282-318 (1974); Akito Arima, "Nuclear magnetic moments and Gamow-Teller matrix elements - configuration mixing vs mesonic effects," Prog. in Part. Nucl. Phys. **1**, 41-65 (1978); K. Takayanagi, K. Shimizu and A. Arima, "Gamow-Teller strength function and the missing strength," Nucl. Phys. **A444**, 436-444 (1985)

11. C. D. Goodman, "Excitation of giant resonances via charge exchange reactions," in *Giant multipole resonances*, F. E. Bertrand, ed., Harwood Academic Publishers, Chur, (1980) 419-432

12. C. Gaarde, J. S. Larsen, M. N. Harakeh, S. Y. Van de Werf, M. Igarashi, and A. Muller-Arnke, "The $^{48}$Ca($^3$He,t)$^{48}$Sc reaction at 66 and 70 MeV, reaction mechanisms and Gamow-Teller strength," Nucl. Phys. **A334**, 248-268 (1980)

13. C. D. Goodman, and S. D. Bloom, in *Spin Excitations in Nuclei*, F. Petrovich et al. eds., Plenum, New York, (1983), 143-160

14. A. Arima, in *Spin Excitations in Nuclei*, F. Petrovich et al., eds., Plenum, New York, (1983)

15. M. Rho, in *Spin Excitations in Nuclei*, F. Petrovich et al., eds., Plenum, New York, (1983)

16. F. Osterfeld, in *Spin Excitations in Nuclei*, F. Petrovich et al., eds., Plenum, New York, (1983)

17. B. A. Brown and B. H. Wildenthal, "Experimental and theoretical Gamow-Teller beta-decay observables for the sd-shell nuclei," Atomic Data and Nuclear Data Tables **33**, (1985) 347-404

18. M. J. G. Borge, P. Dessagne, G. T. Ewan, P. G. Hansen, A. Huck, B. Jonson, G. Klotz, A. Knipper, S. Mattsson, G. Nyman, C. Richard-Serre, K. Rissager, G. Walter and the ISOLDE Collaboration, Study of the giant Gamow-Teller Resonance in nuclear beta decay: the case of $^{33}$Ar," Physica Scripta **36** (1987) 218-223

19. M.J.G. Borge, P. G. Hansen, B. Jonson, S. Mattsson, G. Nyman, A. Richter, K. Riisager, and the ISOLDE collaboration, "The axial-vector strength in the proton-rich argon isotopes," Z. Phys. **A332** 413 (1989)

20. Garcia et al. These proceedings

21. F. Ajzenberg, "Energy levels of light nuclei A = 18-20," Nucl. Phys. **A475**, p. 39 (1987)

22. E. G. Adelberger, M. M. Hindi, C. D. Hoyle, H. E. Swanson, R. D. Von Lintig, and W. C. Haxton, "Beta decays of $^{18}$F and $^{19}$F," Phys. Rev. C **27** (1983) 2833-2856

23. B. D. Anderson, A. Fazely, R. J. McCarthy, P. C. Tandy, J. W. Watson,R. Madey, W. Bertozzi, T. N. Buti, J. M. Finn, J. Kelly, M. A. Kovash, B. Pugh, B. H. Wildenthal, and C. C. Foster, "Gamow-Teller strength in the $^{18}$O(p,n)$^{18}$F reaction at 135 MeV," Phys. Rev. C **27** (1983) 1387-1393

24. T. N. Taddeucci, C. A. Goulding, T. A. Carey, R. C. Byrd, C. D. Goodman, C. Gaarde, J. Larsen, D. Horen, J. Rapaport, and E. Sugarbaker, "The (p,n) reaction as a probe of beta decay strength," Nucl. Phys. **A469** (1987) 125-172

25. Weidong Huang, thesis, Indiana Univeristy, 1991, (unpublished)

# CHIRAL SYMMETRY AND AXIAL CHARGE SUM RULES

M. Kirchbach

Institut für Kernphysik
Technische Hochschule Darmstadt
D-6100 Darmstadt, Germany

## 1 Introduction

Strong interactions are known to be characterized by two important features. From analyzing $(p,p)$ and $(p,n)$ as well as $(\pi^-, n)$ and $(\pi^+, p)$ scattering data electric charge independence follows ($SU_V(2)$ isospin symmetry). The second property is the intrinsic-parity independence of strong interactions indicated, for example, through the equality between the coupling constants of vector ($\rho$), and axial vector ($A_1$) mesons to the nucleon (axial $SU_A(2)$ isospin invariance). The combined $SU_V(2) \times SU_A(2)$ invariance is named handedness or chiral invariance. The electric charge independence applies to fermions and mesons separately and is also common to their composite systems. On the contrary, the intrinsic parity invariance applies only to systems of mesons and nucleons interacting in a proper way as there is no duplication of the nucleon with respect to parity. The observed splitting of the fermion mass spectrum into parity doublets (Gell–Mann mode of axial isospin symmetry [1]) occurs in the mass–region around 1500 MeV. For this reason, chiral symmetry is common in the low mass region only to interacting meson–nucleon systems. The notions *parity* and *handedness* (or equivalently *chirality*) characterize the object–image overlap. Consider, for concreteness, the case of reflecting the $(x, y)$–plane with respect to the $y$–axis (Fig. 1). By the reflection each point of the plane is carried into a new one, placed symmetrically on the normal to the $y$–axis. An object is characterized by positive parity if it can be brought to overlap with its mirror image via parallel translation like the black–white segment $AB$. If the object–image overlap is reached via inversion and a subsequent parallel translation as for the case of the congruent segment $CD$ one speaks of negative parity. Two congruent objects of opposite parities are called *parity partners*.

Obviously, there is no rotation in the plane considered by which a combination of two parity partners like the configuration $ACBD$ could be brought to coincide with its mirror image (compare $ADBC$). The only possibility would be the "forbidden" 180°–rotation orthogonal to the plane within the three–dimensional space. The reflection within a finite dimensional space therefore imitates a rotation in the space of the next higher dimension. Combinations of parity partners like the figures $ACBD$ and $ADBC$ possess the property of left and right hands being transformed into each other by a mirror reflection and are characterized by the respective left and right handedness (chiralities). After the arbitrary choice for what is meant by left- and right-handedness one can define left- and right-handed coordinate systems. Two configurations with handedness which have the property of being transformed into each other by a mirror reflection are called *enantiomorphic* or *chiral partners* This simple example illustrates the *duality* between *parity* and *handedness partners*. A further example for chiral partners are the circular components of a linearly polarized electromagnetic wave.

The notion *chirality* for handedness has been introduced by Lord Kelvin 1884 in his Baltimore lectures in connection with the demarcation line between living and dead matter found

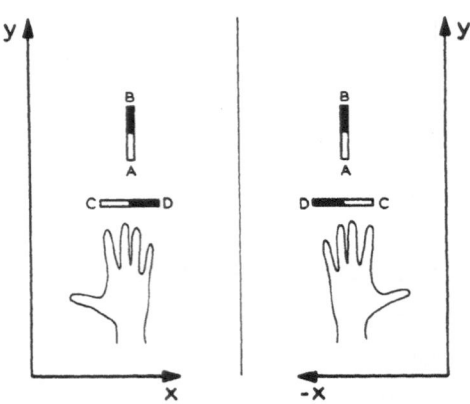

Figure 1. Parity and handedness partners.

1848 by Pasteur. In the terminology of Lord Kelvin the living matter which exists in only one of two possible mirror isomeric forms is chirally pure. On the contrary, the dead matter is chirally symmetric as it appears in nature in both possible mirror isomeric forms.

In general, a function in coordinate space has per definition negative or positive parity if it responds to the inversion of the coordinates by changing or preserving the sign, respectively. The inversion of the coordinates in three dimensional space is obtained by a reflection at the $(x, z)$–plane and a subsequent $180°$–rotation around the $y$–axis.

Chiral symmetry of strong interaction is one of the basic concepts of modern field theory and acquired great importance for nuclear physics in describing effects of mesonic exchange currents (MEC). As it is well known the observed partial half-lifes of $\beta$–transitions between states of equal angular momenta and opposite parities in nuclei near closed shells appear several times up to one order of magnitude shorter as compared to the respective predictions of common theories in which the $\beta$–decay of such nuclei is ascribed to the decay of a single–nucleon [2]. The impression appears, that the probability for $\beta$–decay of a valence neutron undergoing a transition between equal spin orbitals of opposite parities should be much larger then the $\beta$–decay probability of a free neutron. One is tempted to assume that a bound nucleon couples stronger to the axial charge of the external lepton field then the free one. In reality, however, the probability for emitting an electron-antineutrino pair by a single neutron embedded within the nuclear matter is only slightly modified by the medium density as compared to the corresponding decay probability of a free space neutron [3]. The total nuclear many–body weak axial current is, however, not identical to the single current of the valence nucleon, as the external lepton field can in addition couple to each core nucleon via two–body correlations. The coupling strength between the valence nucleon– and the external leptonic weak axial charges is only seemingly enhanced as it incorporates the lepton–core couplings and reflects the many-body aspects of the nuclear $\beta$–decay. It has been an important simplification to predict the size of this "effective" coupling in terms of a transparent algebraic relation–the axial charge sum rule for complex nuclei [4].

The aim of this talk is to review this impressive evidence for the predictive power of chiral symmetry for nuclear processes. The presentation is organized as follows. In section 2 the chiral components of an s-state nucleon, a case important for nuclear as well as atomic physics, are interpreted in terms of the spin–helix. Section 3 contains the concept of chiral symmetry as applied to quantum field theory together with the idea of its spontaneous breaking. In section 4 the consistency condition connecting the weak axial couplings of the manifest and the hidden modes of chiral symmetry is presented. The last section in which the axial sum rule for nuclear targets is given is followed by a short summary and conclusions.

## 2   The chirality of the nucleon

The nucleon, the basic constituent of atomic nuclei, is known to possess an intrinsic angular momentum (spin) determining the orientation of its dipole moments. The magnetic dipole moment $(\vec{\mu})$ is given as

$$\vec{\mu} = \mu_N \vec{\sigma}. \tag{2.1}$$

The energy contribution due to the interaction of this dipole with an external magnetic field $\vec{H}$ then reads

$$W_{\mu H} = -\vec{\mu} \cdot \vec{H}. \tag{2.2}$$

The possible existence of an electric dipole moment

$$\vec{\mathcal{D}}_e = d_e \vec{\sigma}, \tag{2.3}$$

with the respective energy contribution of

$$W_{\mathcal{D}_e E} = -\vec{\mathcal{D}}_e \cdot \vec{E}, \tag{2.4}$$

due to the interaction with an external electric field $\vec{E}$ has already been discussed in 1956 in the classical paper of Lee and Yang [5] in connection with the search for a parity violating electromagnetic interaction. The interaction of the electric dipole moment with the electric field responds to both mirror reflection $(P)$ and time reversal $(T)$ by a sign change (socalled parity– and time–odd observable),

$$P d_e \vec{\sigma} \cdot \vec{E} = d_e \vec{\sigma} \cdot (-\vec{E}), \tag{2.5}$$

and

$$T d_e \vec{\sigma} \cdot \vec{E} = (-d_e \vec{\sigma}) \cdot \vec{E}. \tag{2.6}$$

Since the laws of nature are in general believed to be invariant under time reversal (with the only known exception of the rare T–odd neutral–kaon decays) the electric dipole moment of the nucleon (if it exists at all) is expected to play a minor role for parity violating electromagnetic processes. Due to a rather original observation made by Zel'dovich in 1957 [6] a further P–odd but T–invariant interaction with the electromagnetic field should be possible for the nucleon. This interaction should be determined by the coupling of a third nucleon moment (termed by Zel'dovich as the "anapole" $\vec{a} = a\vec{\sigma}$) with an external electromagnetic current density. The relevant energy contribution is given by

$$W_{aj} = -a\vec{\sigma} \cdot \vec{j}^{\,ext}. \tag{2.7}$$

One should not be surprised by the statement that a new family of multipole moments of a current distribution can exist in addition to the well known electric and magnetic moments. Indeed, it is known that one is allowed to add an arbitrary divergenceless current configuration of vanishing electric and magnetic multipole moments of any order to a given current without changing the electromagnetic fields. A simple example for such a current configuration is given by a toroidal solenoid. Indeed, such a configuration has no electric and magnetic moments and therefore no long range electromagnetic interactions. The magnetic field is completely confined inside the toroid and can be detected only by a contact interaction. Consider, for example, the case in which the solenoid is immersed in an electrolyte in a way enabling the electrolytic current to pass the interior of the toroid. In this case a mechanical moment due to the Lorentz force will act on the solenoid. Obviously, the toroidal solenoid is a further example for a chiral configuration. As we shall see in the following such configurations will be rather helpful for giving an interpretation of the chiral components of the nucleon.

From the considerations presented above it follows that the nucleon is fully characterized by its energy–momentum vector $(E, \vec{p})$ and the spin–momentum angular correlation

$$\omega_{\sigma p} = \vec{\sigma} \cdot \vec{p}. \tag{2.8}$$

Information on both observables can be obtained from two spinor–functions (denoted by $\Phi^\sigma(\vec{r}, t) = f(\vec{r}, t)\chi^\sigma_{\pm 1/2}$ and $\Theta^\sigma(\vec{r}, t) = g(\vec{r}, t)\chi^\sigma_{\pm 1/2}$, respectively). Here $\chi^\sigma_{\pm 1/2} = \binom{1}{0}/\binom{0}{1}$ denotes the conventional two component Pauli–spinor determining the polarization state (parallel $(+)$ or antiparallel $(-)$ to the direction of the external magnetic field). Pictorially, the first function (the large component) describes the nucleon "trajectory" whereas the second one (the small component) gives the orientation of the intrinsic angular momentum (the spin $\vec{\sigma}$) with respect to the nucleon velocity (the tangent to the trajectory),

$$\Theta^\sigma(\vec{r}, t) = \frac{c\vec{\sigma} \cdot \vec{p}}{E + mc^2}\Phi^\sigma(\vec{r}, t). \tag{2.9}$$

As $\omega_{\vec{\sigma}, \vec{p}}$ changes sign under coordinate inversion (it is a pseudoscalar) one immediately notes that $\Theta^\sigma$ and $\Phi^\sigma$ have opposite responses to coordinate inversion and therefore they appear as parity partners. In order to ensure energy conservation $(E^2 = m^2c^4 + p^2c^2)$ the spinor $\Phi^\sigma$ has to be related to $\Theta^\sigma$ by

$$\Phi^\sigma(\vec{r}, t) = \frac{c\vec{\sigma} \cdot \vec{p}}{E - mc^2}\Theta^\sigma(\vec{r}, t). \tag{2.10}$$

One remarkable property of the equations of motion of a fermion is that for massless particles they appear symmetric to the parity partners $\Phi^\sigma \leftrightarrow \Theta^\sigma$. In this case the large and the small components are simply given by plane waves which enter the coupled equations (2.9) and (2.10) on equal footing. This symmetry allows one to decouple eqs.(2.9) and (2.10) in replacing them by two independent eigenvalue problems for the helicity operator with parity mixed eigenstates $\Psi^{R/L} = \Phi^\sigma \pm \Theta^\sigma$ according to

$$\frac{c\vec{\sigma} \cdot \vec{p}}{E}(\Phi^\sigma \pm \Theta^\sigma) = h^{R/L}(\Phi^\sigma \pm \Theta^\sigma). \tag{2.11}$$

Here $\Psi^R$, and $\Psi^L$ describe in turn particles with spin parallel $(h^R = 1)$, or antiparallel $(h^L = -1)$ to the boost direction. These states of right and left helicities build a chiral doublet. Note that a chiral state is obtained by the action of the helicity projection operator $\mathcal{P}^{R/L} = \frac{1}{2}(1 \pm \vec{\sigma} \cdot \frac{\vec{v}}{c})$ on $\Phi^\sigma$ as $\Psi^{R/L} = 2\mathcal{P}^{R/L}\Phi^\sigma$. The degeneracy of the energy spectra of massless fermions with respect to helicity is known as manifest realization of chiral symmetry. This socalled *Wigner mode* corresponds to a racemic system, i.e., a system consisting of equal parts of mirror conjugate objects.

Chiral symmetry requires the symmetry of the equations of motion of the nucleon with respect to the large and small components acting as parity partners. Subsequently we shall see how to extend the Dirac equations for massive nucleons in order to ensure this symmetry. Before that let us illustrate the chirality of a massive nucleon for the simple but for nuclear and atomic physics very interesting case of a pure radial wave "trajectory" ( $1s$–nucleon bound in a spherically symmetric potential $U(r)$):

$$\Phi_{1s_{1/2}}(\vec{r}) = b\, e^{-\frac{r^2}{2}}\chi^\sigma_{\pm 1/2}, \tag{2.12}$$

with $b$ being a constant. In the following the standard notation $Y^l_m$ is used for the spherical harmonics describing states of given orbital angular momentum. The parity of such states is defined as

$$P\Phi_{nljm}(\vec{r}) = (-1)^l\Phi_{nljm}(\vec{r}). \tag{2.13}$$

Therefore, the 1s–state (zero orbital angular momentum) has positive parity. Here $n$ stays for the radial quantum number, $j$ denotes the total angular momentum, and $m$ is the magnetic quantum number. The chiral projection operator for the case of radial waves becomes

$$\mathcal{P}^{R/L} = \frac{1}{2}(1 \mp i\hbar c \frac{\vec{\sigma} \cdot \hat{r}}{D(r)} \partial_r), \tag{2.14}$$

as the action of the gradient operator on an arbitrary radial function $f(r)$ results in

$$\vec{\nabla} f(r) = \hat{r} \, \partial_r f(r). \tag{2.15}$$

Here $D(r) \equiv E - U(r) - mc^2$, and the spinor function $\Theta$ (negative parity) reads

$$\Theta_{1p_{1/2}}(r, \theta, \phi) = i\hbar c \frac{b\,r}{D(r)} \vec{\sigma} \cdot \hat{r} \, e^{-\frac{r^2}{2}} \chi_m^\sigma, \tag{2.16}$$

where $\Theta$ describes an 1p–state as the unit radius vector is simply related to the rank–1 spherical harmonic via $Y_\mu^1(\hat{r}) = \sqrt{\frac{4\pi}{3}} \hat{r}_\mu$. Now, the chiral spinors $\Psi^{R/L}$ for an 1s state nucleon introduced via

$$\Psi^{R/L} = 2\mathcal{P}^{R/L} \Phi_{1s_{1/2}} = (\Phi_{1s_{1/2}} \pm \Theta_{1p_{1/2}}) \approx [1 \pm i\eta(r)\frac{\vec{\sigma}}{2} \cdot \hat{r}] \chi_{\pm 1/2}^\sigma \Phi_{1s_{1/2}}, \tag{2.17}$$

have the property of the right and left hands of being transformed into each other by a mirror reflection,

$$\Psi^L \overset{P}{\leftrightarrow} \Psi^R. \tag{2.18}$$

According to (2.17) a chiral component contains either the right or left helix–spinor

$$\chi_{\pm 1/2}^{R/L} = \frac{1}{2}[1 \mp i\eta(r)\frac{\vec{\sigma} \cdot \hat{r}}{2}] \chi_{\pm 1/2}^\sigma, \tag{2.19}$$

describing a spin rotation around $\hat{r}$ through an angle $\eta(r)$ common for all points on the sphere of radius $r$. Within the $(x, y)$–plane a chiral component is simply characterized by a circular spin distribution since through the rotation of spin by the angle $\eta(r)$ at all points placed on the circle of given radius $r$ the spin gets a component along the corresponding tangent. The socalled *spin–helix* appears (Fig. 2).

Such a circular distribution of magnetic dipoles can be found in ferromagnetic media as well as inside the toroidal solenoid considered above. The quantitative characteristics of the spin helix is a non–vanishing integral of the type

$$\vec{T} = \oint_{r=R} \vec{\sigma} \times \vec{r} d\vec{r} \neq 0. \tag{2.20}$$

The vector $\vec{T} \neq 0$ can be viewed as a signature for chirality.

Chiral nuclear states containing the helix–spinor have been known under the name "parity–mixed" states for a long time [7]. One of the most popular examples for a parity mixed state is the $\frac{1}{2}$–ground state of $^{19}F$. This state is indeed a chiral $2s_{1/2}$–hole in the $^{20}Ne$–core,

$$|\,^{19}F(1/2, gs) > = [1 - i\alpha(r)\frac{\vec{\sigma}}{2} \cdot \hat{r}] a_{2s_{1/2}}^p \,|\,^{20}Ne > \tag{2.21}$$

Due to the spin helix contained in its ground state, the $^{19}F$–nucleus acquires a non–vanishing toroidal (anapole) moment.

Now, according to $V - A$–theory of $\beta$–decay developed by Feynman and Gell–Mann the chirally pure neutrino interacts with the chirally symmetric nucleon by coupling only to its left chiral component [8]. In a similar way a left quartz crystal favours the left circular component of linearly polarized light. The chiral components of an $s$–state nucleon which contain the spin helix and therefore circular spin distributions appear analogous to the circular components of a linearly polarized electromagnetic wave (see Fig.3), and the $V - A$–theory resembles optical activity.

For the sake of compactness it is of common use in describing the nucleon to replace the two coupled equations (2.9)-(2.10) by an eigenvalue problem (the Dirac–equation) formulated in some abstract bi–spinor space after introducing the straightforward four–component spinor $\Psi = \begin{pmatrix} \Phi^\sigma \\ \Theta^\sigma \end{pmatrix}$,

$$(c\vec{\gamma} \cdot \vec{p} + mc^2)\begin{pmatrix} \Phi^\sigma \\ \Theta^\sigma \end{pmatrix} = E\begin{pmatrix} \Phi^\sigma \\ \Theta^\sigma \end{pmatrix}. \tag{2.22}$$

In terms of bi–spinors chiral symmetry means that the Dirac–lagrangian should remain unchanged after replacing the straightforward spinor $\Psi = \begin{pmatrix} \Phi^\sigma \\ \Theta^\sigma \end{pmatrix}$ by the inverted spinor $\Psi^P = \begin{pmatrix} \Theta^\sigma \\ \Phi^\sigma \end{pmatrix}$. Obviously, the Dirac–equation appears chirally invariant only in the special case of massless fermions as for massive nucleons $\Psi$ and $\Psi^P$ satisfy equations differing by the sign of the mass term. Here the well-known Dirac matrices

$$\vec{\gamma} = \begin{pmatrix} 0 & \vec{\sigma} \\ -\vec{\sigma} & 0 \end{pmatrix}, \tag{2.23}$$

and

$$\gamma_4 = \begin{pmatrix} 1 & 0 \\ 0 & -1 \end{pmatrix}, \qquad \gamma_5 = \begin{pmatrix} 0 & -1 \\ -1 & 0 \end{pmatrix}, \tag{2.24}$$

have been introduced. The $\gamma$–matrices have different properties within the abstract space of the Dirac spinors. The $\gamma_4$–matrix defines the spinor $P\Psi$ which satisfies the Dirac equation after inversion of the usual coordinate space has been performed, $P\Psi(\vec{r}, t) = \gamma_4\Psi(-\vec{r}, t) = \eta^\pi\Psi(\vec{r}, t)$, with $\eta^\pi$ being the parity of the nucleon. Therefore, the straightforward Dirac spinor $\Psi(\vec{r}, t)$ is of positive parity whereas negative parity has to be ascribed to the inverted spinor $\Psi^P$. Indeed, from eqs. (2.22) and (2.24) follows that

$$\gamma_4\Psi^P(-\vec{r}, t) = \begin{pmatrix} 1 & 0 \\ 0 & -1 \end{pmatrix}\begin{pmatrix} -\Theta^\sigma \\ \Phi^\sigma \end{pmatrix} = -\Psi^P(\vec{r}, t). \tag{2.25}$$

Therefore, it is natural to consider the states

$$\Psi \overset{parity\ partners}{\longleftrightarrow} \Psi^P, \tag{2.26}$$

as *parity partners*. For massless fermions these states appear degenerate in energy and build therefore a *parity doublet*. In terms of the Dirac matrices the following covariant (coordinate independent) representation of the chiral components of the nucleon is obtained:

$$\Psi^L = \frac{1}{2}(\Psi - \Psi^P), \qquad \Psi^R = \frac{1}{2}(\Psi + \Psi^P). \tag{2.27}$$

The left and right chiral combinationss $\Psi^L$ and $\Psi^R$ are then the corresponding *chiral partners*,

$$\Psi^L \overset{chiral\ partners}{\longleftrightarrow} \Psi^R. \tag{2.28}$$

## 3 The dynamical realization of chiral symmetry and the intrinsic parity partners to the nucleon

The way of realizing chiral symmetry is to achieve hidden multiplicity of the one–nucleon state with respect to intrinsic parity (instead of its explicit duplication) by producing virtually an infinite number one–nucleon–many–pions states degenerated in mass. This new possibility (the socalled *Goldstone* mode) has been suggested by the $\beta$ decay of the lightest particle of negative intrinsic parity, the pion

$$| \pi^+ > \rightarrow | 0 > + \mu^+ + \nu_\mu. \tag{3.1}$$

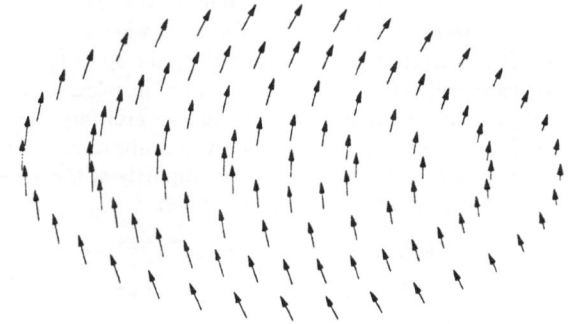

Figure 2. The spin helix.

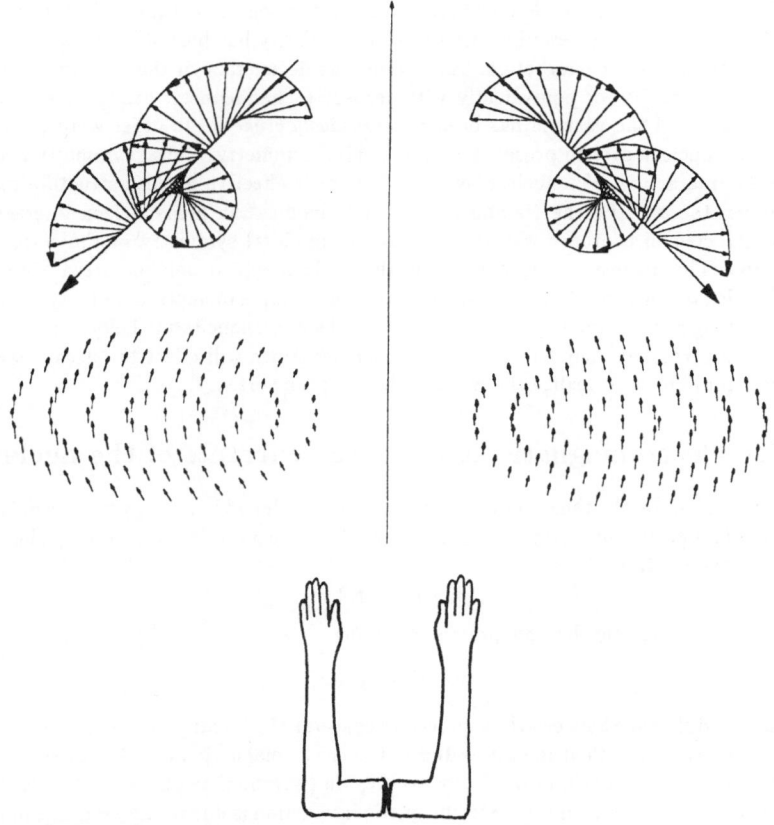

Figure 3. The chiral components of a linearly polarized electromagnetic wave (above) and an s–state nucleon (in the middle) together with the egyptian hieroglyph "ka" (below).

The notation $\mid 0 >$ stays for the hadronic vacuum. On the hadron mass scale the energy gap of $\sim 0.1\,GeV$ between the vacuum and the pion state is negligible. The splitting between the one–pion state and the hadronic vacuum resembles to the neutron–proton mass splitting which accounts for the $\beta$–decay of the neutron. However, whereas the neutron-proton mass difference is possibly due to electromagnetic interaction, the splitting between the one–pion and the vacuum state lacks still explanation. In the limit of massless pions of vanishing energy and momentum (*soft pion*–limit) one is allowed to add an arbitrary number of pions to the vacuum and so to produce its degeneracy with respect to intrinsic parity. In creating nucleons on these degenerate vacua one is led to an infinite multiplicity of the one–nucleon state with respect to intrinsic parity due to different soft pion content,

$$\cdots \leftrightarrow p\pi^+\pi^- \leftrightarrow p \leftrightarrow n\pi^+ \leftrightarrow n\pi^+\pi^-\pi^+ \leftrightarrow \cdots \tag{3.2}$$

$$\cdots \leftrightarrow n\pi^+\pi^- \leftrightarrow n \leftrightarrow p\pi^- \leftrightarrow p\pi^-\pi^+\pi^- \leftrightarrow \cdots \tag{3.3}$$

In a similar way the multiplicity of mesonic states appears,

$$\rho \leftrightarrow \rho\pi \leftrightarrow A_1, \ \pi \leftrightarrow \pi\pi \leftrightarrow \sigma. \tag{3.4}$$

From the duality between parity and handedness partners the multiplicity of the one–nucleon state with respect to handedness and therefore chiral symmetry follows. This way to ensure chiral symmetry is also termed *hidden (dynamical)* realization. The real physical vacuum which contains no pions and therefore has positive intrinsic parity has been arbitrarily favoured due to some small external perturbation. As a result the degeneracy of the vacuum with respect to the intrinsic parity, or equivalently with respect to handedness, has been removed (the chiral symmetry of the vacuum has been *spontaneously broken*). In other words, neither the pion nor the nucleon system possess separately chiral symmetry. Chiral symmetry is common only to the interacting pion–nucleon system. In various effective nucleon structure models as the Skyrme, the $\sigma$–, and the Nambu–Jona-Lasinio models the pion–nucleon interaction has been constructed in a proper way in order to ensure chiral symmetry of the corresponding lagrangians. The intrinsic parity partners of the nucleon appear only as virtual intermediate states (hidden intrinsic parity partners), as visible for example in neutron $\beta$–decay. This means that a decaying neutron can exist some time as a nucleon resonance state before the final proton state has been reached. Obviously, such intermediate states will affect the decay–parameters of the neutron, i.e., renormalize the weak nucleon–lepton vertex.

## 4  The electromagnetic and $\beta$–decay currents of the nucleon

As mentioned before, the theory of $\beta$–decay has been developed in analogy to electrodynamics. This is valid for parity conserving as well as for parity non–conserving transitions. The (inverse) $\beta$–decay of the neutron

$$n + \nu_e \to p + e^- \tag{4.1}$$

appears similar to elastic electron–proton scattering

$$p + e^- \to p + e^-. \tag{4.2}$$

The essential difference between these processes concerns the identity of the particles. Whereas particles do not change their identities during the electromagnetic process (flavour conserving electromagnetic interaction), in weak process they do (flavour changing weak decays) (Fig. 4). The long-ranged character of the electromagnetic interaction is due to the exchange of massless photons, whereas the almost contact character of the weak interaction is due to the exchange of heavy, charged $W^\pm$–bosons. The scattering of electrons by protons is naturally explained in electrodynamics as the interaction between the corresponding electromagnetic convection (vector) currents between states of good parities

$$\vec{j}^V = e\left((\Theta^\sigma)^\dagger \vec{\sigma} \Phi^\sigma + (\Phi^\sigma)^\dagger \vec{\sigma} \Theta^\sigma\right) \tag{4.3}$$

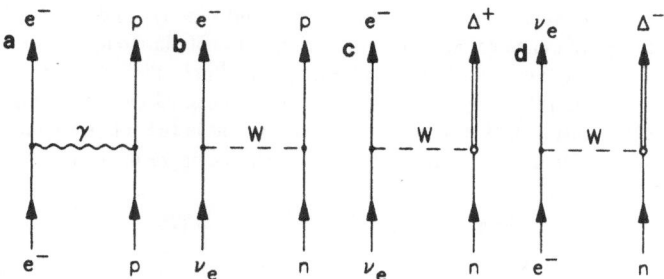

Figure 4. Flavour conserving electromagnetic (a) and flavour changing weak (b–d) lepton-nucleon scattering.

with the electric charge density being

$$j_4^V = e\Big((\Phi^\sigma)^\dagger \Phi^\sigma + (\Theta^\sigma)^\dagger \Theta^\sigma\Big) = e(\Phi^\sigma)^\dagger \Big(1 + \frac{\vec{p}^2}{4m^2 c^2}\Big)\Phi^\sigma. \tag{4.4}$$

Similarly, $\beta$–decay is explained by the interaction between two socalled weak flavour changing currents. The weak $\beta$–decay current $J_\omega$ of the nucleon is a convection current between states of left chiralities. Equivalenly, it can be considered as a left–chiral current defined by the difference between the weak convection $\{\vec{V}, V_4\}$ and the weak circular currents $\{\vec{A}, A_4\}$ connecting states of good parities. The spatial part of the weak convection (vector) current is obtained from eq.(4.3) by simply replacing $e$ by $G_V$, whereas the weak circular (axial vector) current is determined as

$$\vec{A} = G_A\Big((\Phi^\sigma)^\dagger \vec{\sigma}\Phi^\sigma + (\Theta^\sigma)^\dagger \vec{\sigma}\Theta^\sigma\Big), \tag{4.5}$$

with the corresponding axial charge density being the pseudoscalar

$$A_4 = G_A\Big(\big(\Phi^\sigma\big)^\dagger \Theta^\sigma + (\Theta^\sigma)^\dagger \Phi^\sigma\Big) \xrightarrow{E \approx mc^2} G_A(\Phi^\sigma)^\dagger \frac{\vec{\sigma}\cdot \vec{v}}{c}\Phi^\sigma. \tag{4.6}$$

Note, that the weak axial charge coupling constant $G_A$ appears in analogy to the spin–current coupling (compare eq.(2.7)) known as anapole. It is commonly accepted to normalize the value of the weak vector coupling at zero–momentum transfer as $G_V(\vec{q}^2 = 0) = 1$. For the axial charge coupling constant the value $G_A(\vec{q}^2 = 0) = -1.25$ has been obtained in analyzing a variety of $\beta$–decay data. The deviation of the axial charge coupling from unity is not a "chiral defect" but appears as a consequence of the hidden character of chiral symmetry. Indeed, the square of the matrix element of an arbitrary operator T between some initial ($| i >$) and final ($| f >$) states can be decomposed using closure into the expectation value of the commutator between T and its hermitean conjugate $T^\dagger$, and the difference between the strengths for all possible $T^\dagger$–, and T–excitations of the initial state leading to the states $| \tilde{X}_j >$ and $| X_l >$, respectively. From the latter the final state $| f >$ has to be excluded,

$$|< f | T | i >|^2 = < i | [T^\dagger, T] | i > + \sum_{\tilde{X}_j} |< \tilde{X}_j | T^\dagger | i >|^2 - \sum_{X_l \neq f} |< X_l | T | i >|^2. \tag{4.7}$$

For T being a generator of a symmetry group the commutator is model independently determined by the group algebra and the intermediate $X(\tilde{X})$– states exist only if the symmetry is realized in the hidden mode (the one–nucleon state is virtually infinetely multiplicated). In this way sum rules are generated

$$S_T - S_{T^\dagger} = < i | [T^\dagger, T] | i >. \tag{4.8}$$

The possibly velocity frame dependent strength–contributions $S_{T\uparrow}$ and $S_T$ rearrange in each frame so that their difference equals the expectation value of the commutator on the rhs in eq.(4.8). It is clear that there is a frame containing a minimal number of intermediate states contributing to the sum rule. For the matrix element of the axial charge raising operator $A_4^+$ between the single neutron and proton states such a frame is the infinite momentum frame. There, the single particle matrix element of the axial charge operator is maximal

$$|< p \mid A_4^+ \mid n >|^2 = G_A^2 \frac{v^2}{c^2}|_{\frac{v}{c}\to 1} \to G_A^2 \qquad (4.9)$$

and equals the value of the axial charge coupling constant at zero momentum transfer ($\vec{q} \to 0$). The expectation value of the commutator is identical to the square of the nucleon's helicity which is the weak axial charge coupling $G_A^{bare}$ of the nucleon corresponding to the manifest mode of chiral symmetry,

$$< n(\vec{p}) \mid [A_4^-, A_4^+] \mid n(\vec{p}) >= h^2 =\mid G_A^{bare} \mid^2 . \qquad (4.10)$$

On the other hand, chiral algebra connects the axial charge raising and lowering operators with the third component of the operator of isospin via

$$< n(\vec{p}) \mid [A_4^-, A_4^+] \mid n(\vec{p}) >= -2G_V < n(\vec{p}) \mid t_3 \mid n(\vec{p}) >= G_V = h^2. \qquad (4.11)$$

This equation shows that the charge couplings corresponding to the manifest modes of $SU_A(2)$ and $SU_V(2)$ isospin symmetries can simultaneously be normalized to unity. From eq.(4.7) then follows that the deviation of the axial charge coupling constant $G_A$ from unity is completely due to the contribution of the many–particle intermediate states appearing in the hidden mode of axial isospin symmetry. This deviation is most readily calculated in the infinite momentum frame where the axial charge is known to be a socalled good operator. This means that the axial charge $L_4$ of the external leptonic current couples neither to pions nor to antinucleons but only contact $L_4 N\pi$–terms survive. Pictorially, the axial charge on the light cone is carried only by free one–nucleon states differing by their soft pion content. In other words, the intermediate states $X$ and $\tilde{X}$ in eq. (4.7) are the nucleon resonances treated as degenerate in mass to the nucleon in accordance to the multiplicity of the one–nucleon state in the Goldstone mode of chiral symmetry. The contribution of the many–particle states can be evaluated either via a dispersion integral over the difference between the total elastic scattering cross sections for negatively and positively charged pions (treated as massless particles) on the neutron (Adler–Weisberger relation) [9, 10], or be ascribed to the contributions of the resonances observed between threshold and approximately 3 GeV. In the SU(6)–limit only the weak axial couplings $G_A^j$ of the two lowest nucleon excitations, the $\Delta(\frac{3}{2}^+; \frac{3}{2})$ and the Roper $N^*(\frac{1}{2}^+; \frac{1}{2})$-resonances to the external leptonic current contribute to give [11]

$$G_A^2 - \frac{(G_A^\Delta)^2}{(3/\sqrt{2})^2} + \frac{(G_A^R)^2}{2^2} = h^2. \qquad (4.12)$$

Chiral symmetry constrains the axial couplings of the nucleon and the weak axial $N \to \Delta(1232)$ and $N \to N^*(1440)$ transition couplings to a three dimensional hyperboloid (on the *helicity shell*). These results are valid in the limit of massless pions and therefore in the case of conserved axial vector currents (CVC). The non–zero pion mass leads to a spreading of the contributions of the intermediate many–particle states in eq.(4.7) over 12 resonances with the resulting $G_A$ being only about 5% smaller as compared to the CVC–limit (see [11] for details). Consequently, eq.(4.12) can be used to control chiral symmetry preservation in nucleon structure models by checking the consistency between the weak axial couplings corresponding to the manifest and hidden modes. Effective models of nucleon structure pretending to take care for chiral symmetry of strong NN–interaction should satisfy eq.(4.12). However, as it has been shown in [11] most of these models are not able to reproduce the axial coupling of the nucleon in the manifest mode as they cannot predict the correct expectation value of the commutator between the axial charges for the nucleon state. For example, in the case of the Skyrme model with a rigid rotor quantization the nucleon's helicity vanishes and the weak axial couplings of the nucleon, the $\Delta$ and the Roper resonances lie on a cone instead on the three–hyperboloid. The deviation of the weak axial charge coupling of the nucleon from unity is one of the most spectacular evidences for the hidden character of chiral symmetry of nucleon systems.

## 5   The axial charge sum rule for complex nuclei

Let us assume that the true nuclear axial charges describing nucleonic and mesonic degrees of freedom (subsequently denoted by $Q_5^{\pm}$) satisfy the chiral algebra commutation relation

$$[Q_5^-, Q_5^+] = -2T_3, \tag{5.1}$$

with $T_3$ being the third component of nuclear isospin. Let us consider for concreteness the $\beta$–decay of the ground state of $^{15}C_{gs}(\frac{1}{2}^+)$ to the ground state of $^{15}N_{gs}(\frac{1}{2}^-)$. The transition

$$^{15}C_{gs}(\frac{1}{2}^+) \rightarrow ^{15}N_{gs}(\frac{1}{2}^-) + e^- + \bar{\nu}_e \tag{5.2}$$

is mainly due to the operator of the axial charge. In terms of one–body operators (impulse approximation) this decay is exclusively ascribed to the $\beta$-transition of the $2s_{1/2}$–valence neutron in $^{15}C_{gs}$ to the $1p_{1/2}$–shell of the $^{14}C(0^+; 1, -1)$-core. It turned out that in order to reproduce the data the coupling constant of the axial charge of the $2s_{1/2}$–valence neutron to the leptonic axial charge should be enhanced by about 50% as compared to the free neutron value of $\mid G_A \mid = 1.25$. This in order to account for the fact that also each nucleon from the $^{14}C_{gs}(0^+; 1)$-core can undergo a $\beta$–transition via some two–particle mechanism. Thus the enhanced effective axial charge coupling of the $2s_{1/2}$–valence neutron incorporates the effect of the mesonic degrees of freedom of the $^{15}C$-nucleus. The nuclear weak axial current is given via matrix elements of the corresponding operator between nuclear states. It is therefore a non–relativistic object. In the impulse approximation it is of common use to express the nuclear matrix elements as linear combinations of single–particle matrix elements of transition operators acting on nucleons only (socalled single–particle $\beta$–moments) with the coefficients being the one–body transition–density matrices $(\rho_{\alpha,\beta}^{i \to f})$ containing the spectroscopic information,

$$< f \mid \{\vec{A}_{GT}^+, Q_5^+\} \mid i > = \{G_A^{GT} \rho_{\alpha,\beta}^{i \to f} < \beta \mid \vec{\sigma} \tau_+ \mid \alpha >, G_A^c \rho_{\delta,\eta}^{i \to f} < \eta \mid \vec{\sigma} \cdot \frac{\vec{v}}{c} \mid \delta >\}. \tag{5.3}$$

Here summation over equal indexes is assumed. It is clear that such a current density is not Lorentz invariant and there is no need for common effective couplings scaling the axial single–particle $\beta$–moments to the true many–body ones. Indeed, whereas the effective axial charge coupling $G_A^c$ measured in first–forbidden $\beta$–decays appears enhanced, the effective axial vector coupling $G_A^{GT}$ measured in Gamow-Teller $\beta$–decays has to be quenched. The enhanced $G_A^c$–value has been well understood in terms of meson–exchange corrections to the single particle $\beta$–moments. A lot of cumbersome detailed analyses of first–forbidden $(FF)$ $\beta$–decays have been performed by several groups during the last decade [2, 12]. The most transparent way for explaining this enhancement, however, has been obtained in [4] from the sum–rule based on eq.(5.1) obtained as

$$S_{\beta-}^{FF} - S_{\beta+}^{FF} = N - Z. \tag{5.4}$$

For the case of $^{15}C$ one is led to

$$S_{\beta-}^{FF}(^{15}C) - S_{\beta+}^{FF}(^{15}C) = 3, \tag{5.5}$$

with

$$S_{\beta+}^{FF}(^{15}C) = \sum_{\tilde{X}_i} \mid < \tilde{X}_i \mid Q_5^- \mid ^{15}C >\mid^2, \tag{5.6}$$

and

$$S_{\beta-}^{FF}(^{15}C) = \sum_{X_j} |< X_j \, | \, Q_5^+ \, |^{15}C >|^2 \,. \tag{5.7}$$

The $\beta^-$–strength contains the transition to the $^{15}N$–ground state considered as a proton–hole within the $1p_{1/2}$–shell in the $^{16}O$–core. It is the only observed $\frac{1}{2}^-$–state in the $^{15}N$–nucleus below the ground state of $^{15}C$ and the corresponding transition–matrix element reads

$$|<^{15}N_{gs} \, | \, \sum_i G_A \vec{\sigma}_i \cdot \frac{\vec{v}_i}{c} \tau_+ \, |^{15}C_{gs} >|^2 = |\, G_A^c(^{15}C) \,|^2 |\, \rho_{2s\to1p}^{^{15}C\to{^{15}}N} \,|^2 \, \frac{\vec{v}^2}{c^2} \,. \tag{5.8}$$

Further, the $\beta^-$–strength contains the excitations of the $^{15}C\pi^+$–continuum. Due to a vanishing overlap between the $^{15}C$– ground state and some possible $\frac{1}{2}^-$–excitations of the (experimentally non–observed) $^{15}B$–nucleus the $\beta^+$–strength should contain only the excitations of the $^{15}C\pi^-$–continuum. The excitation of the pion–nucleus continuum is most readily calculated in the infinite momentum frame in terms of an integral over the total elastic scattering cross sections of the pions treated as massless particles on the nuclear target. Within this frame $(\bar{v}^2/c^2 \to 1)$ one is led to the following relation for the effective axial charge coupling $G_A^c$:

$$|\, \rho_{2s\to1p}^{^{15}C\to{^{15}}N} \,|^2 |\, G_A^c(^{15}C) \,|^2 = 3 + I(^{15}C) \,. \tag{5.9}$$

Here $I(A(Z,N))$ denotes the Adler–Weisberger integral for nuclear targets

$$I(A(Z,N)) = \int_{M_A m_\pi}^{\infty} (\sigma_{A(Z,N),\pi^-}^{tot}(\nu, m_\pi^2 \to 0) - (\pi^- \leftrightarrow \pi^+)) \frac{d\nu}{\nu} \,. \tag{5.10}$$

Such integrals are rather small as compared to the neutron excess and can in practice be neglected. From eq.(5.9) then follows

$$|\, G_A^c(^{15}C) \,| \simeq \sqrt{3} \,. \tag{5.11}$$

Similarly, for the $\beta$–decay of the $^{16}C$–ground state to the first excited $0^-$–state in $^{16}N$ one obtains

$$|\, \rho_{2s\to1p}^{^{16}C\to{^{16}}N} \,|^2 |\, G_a^c(^{16}C) \,|^2 = 4 + I(^{16}C) \,, \qquad |\, G_A^c(^{16}C) \,| \simeq 2 \,. \tag{5.12}$$

The interpretation of these results is rather transparent. In the hypothetical limit of the infinite momentum frame where no pions and no antinucleons are "seen" by the axial charge of the external weak lepton current the left–bound nuclei of the isobaric multiplets appear similar to asymmetric nuclear matter as they look approximately like $N - Z$–free neutrons decoupled from the isoscalar boson–like core. The event space describing the probability for the nuclear $\beta$–decay contains therefore $N - Z$–independent elementary events. The size of the true event–space for the nuclear first–forbidden $\beta$–decays is determined by the relevant neutron excess. In the impulse approximation the true event space is simulated by a restricted one containing only the decays of the valence neutrons of the most outer shell coupled to the isovector (isotensor etc) boson–like core. In order to account for the decay of the excess neutrons of the core one has to introduce an enhanced probability impacted by the enhanced axial–charge coupling of the valence neutrons. This effective axial charge will change within a given isobar multiplet as it depends on the neutron excess of the decaying nucleus. For example, in evaluating the sum rule for the $^{16}N(0_1^-)$–state one obtains the following relation:

$$|\, \rho_{2s\to1p}^{^{16}C\to{^{16}}N} \,|^2 |\, G_A^c(^{16}C) \,|^2 - |\, \rho_{2s\to1p}^{^{16}N\to{^{16}}O} \,|^2 |\, G_A^c(^{16}N) \,|^2 \simeq 2 \,, \qquad |\, G_A^c(^{16}N) \,| \simeq \sqrt{6} \,, \tag{5.13}$$

which is by about 20% larger as compared to the effective axial charge coupling in the neighbouring $^{16}C$–nucleus. In general, the values for the effective single–particle axial charges obtained from the sum rule overestimate the corresponding results of the microscopic MEC–analyses by about 20% to 40%.

# 6 Summary and conclusions

In this talk we emphasized the deviation of the weak axial charge coupling constant of the nucleon from unity to be the most spectacular evidence for the hidden character of chiral symmetry. Chiral symmetry has been shown [11] to constrain the weak axial couplings of the nucleon and its two lowest resonances to lie on a hyperboloid (the helicity–shell). In assuming validity of chiral algebra commutation relations for nuclear charges we explained the need for enhancing the effective axial charge single–particle $\beta$–moments through simple algebraic relations following from the axial charge sum rule for complex nuclei. The quenching of the effective Gamow-Teller single–particle $\beta$–moments can intuitively be understood by considering axial charge–axial current commutators $[Q_5^+, \vec{A}_{GT}^-]$ for nuclear targets. Such commutators would lead to less model independent constraints because of the model dependent Schwinger–terms contained there. Nevertheless, they would lead to constraints on the product $G_A^{GT} G_A^c$ and one could possibly see that the approximately 30%– quenching of $G_A^{GT}$ is naturally related to the inverse 50%–enhancement of $G_A^c$. For this reason also the quenching of $G_A^{GT}$ would depend on the neutron excess (deficiency) in accordance with the regularity found for example in eq. (5.13).

The separation of the spatial and and temporal components of the pion decay constant within the medium due to the absence of Lorentz invariance has first been obtained by Akhmedov et al. [3] in studying the PCAC condition in finite Fermi systems as

$$\tilde{F}_\pi = \{\tilde{f}_\pi^0(\omega, \vec{q})\,\omega, \tilde{f}_\pi(\omega, \vec{q})\vec{q}\}. \tag{6.1}$$

The space part of the pion decay constant within the nuclear medium appears quenched by about 70% − 90% according to

$$\tilde{f}_\pi = f_\pi[1 + \frac{\Pi^{(P)}(\omega, \vec{q})}{\vec{q}^2}], \qquad \Pi^{(P)}(\omega, \vec{q}) \approx -0.27\frac{\rho}{\rho_0}\vec{q}^2, \tag{6.2}$$

whereas the time part is by about 20% enhanced as seen from

$$\tilde{f}_\pi^0 = f_\pi[1 - \frac{\Pi^{(S)}(\omega) - \Pi^{(S)}(0)}{\omega^2}], \qquad \Pi^{(S)}(\omega) = 0.35 m_\pi^2 \frac{\rho}{\rho_0} - 0.2\omega^2 \frac{\rho}{\rho_0}. \tag{6.3}$$

Here $\Pi^{(S)}(\omega)$ and $\Pi^{(P)}(\omega, \vec{q})$ denote in turn the S– and P– wave parts of the pion polarization operator in the nuclear medium. The numerical results obtained correspond to a normal medium density of $\rho = 0.17 fm^{-3}$. From the Goldberger-Treiman relation $G_A M = f_\pi g_{\pi N}$ the separation between the space and time parts $G_A$ and $G_A^0$, respectively, of the fundamental lepton–nucleon weak axial vertex (not to be confused with the effective couplings $G_A^{GT}$ and $G_A^c$ scaling the axial single–particle $\beta$–moments to the many–body ones) within the medium follows. The results on the effective axial couplings $G_A^c$ and $G_A^{GT}$ following from the nuclear axial charge sum rule are therefore in complete agreement with the results on the medium renormalized weak axial vertex constants $G_A^0$ and $G_A$ obtained in finite Fermi systems.

### Acknowledgement

I would like to acknowledge the collaboration with Dan Olof Riska and for part of this work with Rostislav Mach.

# References

[1] M. Gell-Mann, Physics **1** (1964) 63

[2] M. Kirchbach, Proc.Int.Symp. "Modern Developments of Nuclear Physics", Novosibirsk, 1987 ed. O.P. Sushkov (World Scientific, Singapore–New Jersey–Hong Kong, 1988) p.145

[3] E.H. Akhmedov, Yu.V. Gaponov and I.N. Mishustin, Phys. Lett. **92B** (1980) 261;
    E.H. Akhmedov, Sov. Phys. Pis'ma JETP **34** (1981) 141

[4] M. Kirchbach, R. Mach and D.O. Riska, Nucl. Phys. **A511** (1990) 592

[5] T.D. Lee and C.N. Yang, Phys. Rev. **104** (1956) 254

[6] J.B. Zel'dovich, Sov. Phys. JETP **33** (1957) 1184

[7] I.V. Khriplovich, Parity Non–Conservation in Atomic Phenomena, (Nauka, Moscow, 1981)
    (in Russian)

[8] R.P. Feynman, The Character of Physical Law (Cox and Wyman, London, 1965)

[9] S.L. Adler, Phys. Rev. **140** (1965) B736

[10] W.I. Weisberger, Phys. Rev. **143** (1966) 1302

[11] M. Kirchbach and D.O. Riska, preprint University of Helsinki, HU–TU–90–82

[12] E.K. Warburton, J.A. Becker, B.A. Brown and D.J. Millener, Ann. Phys. (NY) **187** (1988)
     471

# SPONTANEOUS-SYMMETRY BREAKING AND

# GAMOW-TELLER STATES

F. C. Khanna*, H.X. He[†] and H. Umezawa[§]

[§]Theoretical Physics Institute
Department of Physics
University of Alberta
Edmonton, Alberta, Canada, T6G 2J1

and

TRIUMF – 4004 Wesbrook Mall
Vancouver, British Columbia, Canada, V6T 2A3

## ABSTRACT

Spin-isospin excitations are common to all nuclei. The charge exchange reactions, (n,p) or (p,n), have been dominant in isolating these excitations. The quenching of the strength may be attributed to the excitation of $\Delta$-isobar particle states. This is established by using two-body Ward-Takahashi relations that are a consequence of PCAC to derive consistency conditions in a many-body environment. Here we conjecture that the universality of the spin-isospin excitations suggests a deeper dynamical reason for their origin. A spontaneous breakdown of SU(4) symmetry would lead to 6 Nambu-Goldstone bosons that have a close correspondence to the dominant spin and isospin excitations in nuclei. Nature and structure of these excitations may be established by an analysis of the response function and studying in detail the decay of the excitations.

## 1. INTRODUCTION

Spin-isospin excitations in nuclei[1] are known through the periodic table with an excitation energy that is not very large. Such states appear very strongly in charge-exchange reactions (p,n) or (n,p). Total Gamow-Teller (G-T) strength $S^-$ and $S^+$ in the reactions (p,n) and (n,p) respectively satisfies the sum rule[2]

$$S^- - S^+ = 3(N - Z) \qquad (1)$$

with

$$S^{\mp} = \Sigma \ B^{\mp} \ ,$$

---

* Talk given by F.C. Khanna

[†] Permanent address: Institute of Atomic Energy, P.O. Box 257, Beijing, P.R. China.

*Spin and Isospin in Nuclear Interactions*
Edited by S.W. Wissink *et al.*, Plenum Press, New York, 1991

where

$$B^{\mp} = \frac{1}{(2J_i + 1)} \left| < f \, || \, \sigma \, \tau^{\pm} \, || \, i > \right|^2 .$$

At small momentum transfers i.e. $q^2 \rightarrow 0$, the excitation strength is normalised to the Gamow-Teller $\beta$-transitions. In all cases the observed strength in charge-exchange reactions is quenched.

Model calculations[3] at $q^2 = 0$ point, i.e. for G-T $\beta$-transitions, indicate that contributions due to one-body operator using many-body functions are not sufficient to understand the decay rates. Two-body contributions[4], meson-exchange currents (MEC) and excitation of $\Delta$-particle nucleon-hole states, play an essential role in understanding the quenching of G-T strength. The soft-pion theorems help to define MEC[5]. PCAC helps to constrain the two-body contributions by defining Ward-Takahashi (W-T) relations[6] between the wavefunction (W) and vertex ($\Lambda$) functions for the weak interaction and similar functions for $\pi$-production. With a definition for PCAC given as $\partial_\mu J_5^\mu = \mu^2 f_\pi \pi(x)$, where $J_5^\mu$ is the axial current, $\mu$ and $f_\pi$ are mass and decay constant for the pion respectively and $\pi(x)$ is the pion field. W-T relations define mathematical relations between amplitudes.

One body W-T relation follows from the PCAC and is given as

$$q_\mu \, \Gamma_{5\mu\alpha} (k; k+q; q) = i \, \mu^2 \, f_\pi \, \Gamma_{\pi\beta} (k; k+q; q) \, \Delta_{\beta a}(q)$$

$$-i \, [S^{-1}(k) \, \gamma_5 \tau_\alpha + \gamma_5 \tau_\alpha \, S^{-1}(k+q)] \ . \tag{2}$$

This is the analog of the well-known W-T relation in QED[7]

$$q_\mu \, \Gamma_{\mu\alpha}(k; k+q; q) = -i[S^{-1}(k) \, \tau_\alpha - \tau_\alpha \, S^{-1}(k+q)] \tag{3}$$

where $S(k)$ is the one-particle Green's function for a fermion, $\Delta_{\beta\alpha}$ is the pion propagator, $f_\pi$ and $\mu$ are the pion decay constant and pion mass and $\tau$ refers to the isospin of the fermion. In the $q \rightarrow 0$ limit, the axial W-T relations yield a consistency condition

$$\mu^2 \, f_\pi \, \Gamma_{\pi\beta} \, (k;k;0) \, \Delta_{\beta\alpha}(0) = \{ S^{-1}(k), \, \gamma^5 \, \tau_\alpha \} \tag{4}$$

that was derived by Adler and Dothan[8]. Similar consistency conditions may be obtained for two-particle W-T relation. For example

$$\mu^2 \, f_\pi \, \Gamma_{\pi\beta} \, (k_1, k_2 \, ; k_1^{'}, k_2^{'} \, ; 0) = \sum_{i=1,2} \, \{ S^{-1}(k_1,k_2 \, ; k_1^{'}, k_2^{'}) \, , \, \gamma_5^i \, \tau_\alpha^i \} \ . \tag{5}$$

Similarly, terms linear in q give the first order W-T relation for the axial-vector amplitude. Such linear terms may be written for the vertex function $(\Lambda_{5\mu})$ as well as the wave-function renormalisation $(W_{5\mu})$ term. It should be stressed that W-T relations would yield only the longitudinal components of $\Lambda_{5\mu}$ and $W_{5\mu}$. As usual $\Lambda_{5\mu}$ and $W_{5\mu}$ obtained from the two-body W-T relations may be combined for the weak process to obtain a relationship between the weak decay axial-amplitude and the pion bremsstrahlung amplitude by

$$W_{5\mu\alpha}(k_1,k_2; k_1'k_2'; 0) + \Lambda_{5\mu\alpha}(k_1,k_2; k_1'k_2'; 0) = i\,\mu^2\,f_\pi\frac{\partial}{\partial q_\mu} \times$$

$$\left[\{W_{\pi\beta}(k_1,k_2; k_1'k_2'; q) + \Lambda_{\pi\beta}(k_1,k_2; k_1'k_2'; q)\}\,\Delta_{\beta\alpha}(q)\right]_{q\to 0} \tag{6}$$

The pion-production amplitude is dominated by the $\Delta$-resonance. Therefore the weak two-body amplitude relates to the derivative of $\pi$-production amplitude at $q\to 0$. It has been stated that the quenching of the G-T strength may be attributed directly to the role of $\Delta$-resonance in the excitations of nuclei. The relation given above suggests large cancellation among terms in a perturbative approach. The residual effects relate to the isobars. One to one correspondence to the non-relativistic perturbative calculations is not obvious. Certainly a qualitative correspondence may be established.

What have we learnt in the study of Gamow-Teller states? It may be summarised as:
(i)   G-T excitations are universal;
(ii)  excitation energy is small ($E_{GT} \sim 10$–$20$ MeV) but depends on A;
(iii) the strength is spread out and width depends on A; and
(iv)  the strength is quenched and the quenching depends weakly on A.

No doubt, the excitation energy, the spreading width of the resonance and the quenching are related to the model while the universality of the G-T excitations is rather intriguing. It is interesting to note that the excitation energy (10-20 MeV) is small compared to the total binding of a large nucleus i.e. for $^{90}$Zr with a total binding of $\sim 700$ MeV, $E_{GT}$ is $\sim 2\%$ of the total binding. The combination of universality and low excitation energy suggest that there may well be a deeper dynamical reason for the Gamow-Teller excitations. In this note we wish to speculate on dynamical aspects that lead to such consequences and wish to explore their impact on the dynamical response function of the system. The discussion will be pedagogical. Numerical consequences and the impact of exact W-T relations to constrain the perturbative approach will be discussed later on.

## 2. SPONTANEOUS-SYMMETRY BREAKING

The concept of spontaneous-symmetry breaking is foreign to quantum mechanics. The idea arises naturally in quantum field theory[9]. A simple example[10] to understand this is to consider the case of a simple phase transformation defined in terms of a generator $Q$ ($= \int \psi^\dagger\psi\,d^3x$). With a unitary operator $U = \exp(iQ\theta)$, the invariance of the Hamiltonian under phase transformation implies $[H,U] = 0$. A transformation of the field $\psi$ may be written as

$$\psi'(x) = U\,\psi(x)\,U^\dagger = e^{-i\theta}\,\psi . \tag{7}$$

Using the property $[Q,\psi] = -\psi$, $\psi$ being a fermion field, and taking vacuum expectation value we get

$$< 0 \mid \psi'(x) \mid 0 > = e^{-i\theta} < 0 \mid \psi(x) \mid 0 > = < 0 \mid U\,\psi\,U^\dagger \mid 0 >. \tag{8}$$

If we assume that the vacuum is invariant under the phase transformation i.e. $U^\dagger | 0 > = | 0 >$, then RHS becomes $< 0 | \psi | 0 >$. Obviously there is a contradiction. Therefore the ground state $| 0 >$ and $U^\dagger | 0 >$ are degenerate but are not the same. In fact, there are infinitely many degenerate ground states. We need a function $\alpha(x)$ to distinguish various states. Oscillations in the local field variable, $\alpha(x)$, may be classified as Goldstone bosons[11]. Overall this example provides a simple case where the Hamiltonian is invariant under a phase transformation but the ground state is not invariant under the same transformation. Superconductivity[11] provides a simple example that corresponds to the spontaneous-breaking of phase symmetry (U(1) symmetry). The BCS theory provides a basis for detailed quantitative calculations of the properties of superconductors. However exact consequences for the theory of superconductivity must follow general considerations of the spontaneous breakdown of symmetry.

Before consideration of spontaneous-breakdown of larger symmetry groups, a brief comment about Nambu-Goldstone[9] (N-G) bosons is essential. For a system with relativistic invariance, mass and helicity are Casimir operators of the Poincaré group. The Nambu-Goldstone bosons are massless and spinless. However for a non-relativistic system mass is a parameter that relates energy and momentum i.e. $\omega_k = k^2/2m$. Then the energy of N-G bosons vanishes in the limit of zero momentum i.e. $\omega_k \to 0$ as $k \to 0$. This provides a general definition of the N-G boson. In such a case boson energy is zero without being massless. This corresponds to the case of superfluidity. In the present case, for a non-relativistic system the generalised definition of the N-G bosons is to be assumed. Consequences of spontaneous breaking of symmetry in a system that is invariant under a larger symmetry group will be examined now.

## 3. INTERNAL SU(2) SYMMETRY

The three generators of the group are the Pauli spinors $\sigma_1$, $\sigma_2$, $\sigma_3$ that satisfy the algebra

$$[\sigma_i, \sigma_j] = 2i \, \varepsilon_{ijk} \, \sigma_k \; . \tag{9}$$

Assume that a system, ferromagnetic or paramagnetic, is described by a Lagrangian that is invariant under spin rotation, rotations in space and under translations. The spin-rotational symmetry may be spontaneously broken by adding a term like

$$h \, \psi^\dagger(x) \, \sigma_3 \, \psi(x) \equiv h \, \sigma_3(x) \tag{10}$$

with the parameter $h \to 0$. Physically this may correspond to an external magnetic field along z-axis. Then it has been shown[12] that there is one N-G mode that is obtained with the operator $\sigma_+$ or its conjugate $\sigma_-$. This N-G boson has the property that its energy $\omega_k \to 0$ as the momentum $k \to 0$. This N-G mode may be identified with spin waves in ferromagnets and with paramagnon mode in paramagnets.

To prove the existence of spin waves, consider the Green's function

$$< \sigma_+(x) \, \sigma_-(y) >_h \tag{11}$$

with its Fourier transform defined as

$$< \sigma_+(x) \, \sigma_-(y) >_h = \frac{1}{(2\pi)^4} \int d^4p \, e^{-ip \cdot (x-y)} \Delta_B(p) \tag{12}$$

with $\sigma_\pm = \frac{1}{2}(\sigma_1 \pm i\sigma_2)$ and $\sigma_\pm(x) = \psi^\dagger(x) \, \sigma_\pm \, \psi(x)$. Here and in the following $<A>$ implies the ground state expectation value of the time-ordered product i.e. $<A> = <0 | \, T[A] \, | 0>$. $\Delta_B(p)$ is defined by

$$\Delta_B(p) = \frac{<0 | \sigma_+(0) | B_-(\vec{p})> <B_-(\vec{p}) | \sigma_-(0) | 0>}{p_o - [\omega_B^-(p^2) + h] + i\varepsilon}$$

$$- \frac{<0 | \sigma_-(0) | B_+(\vec{p})> <B_+(\vec{p}) | \sigma_+(0) | 0>}{p_o + [\omega_B^+(p^2) + h] - i\varepsilon} + \Delta^c(p) \tag{13}$$

where $|B_\pm(p)>$ is the state with one spin wave quantum with $S_3 = \pm 1$ and momentum $\vec{p}$, $\Delta^c(p)$ is the part that contains cut singularity, and $\omega_B^\pm(p^2)$ is the energy of the spin-wave state. In the limit $p \to 0$ i.e.

$$\lim_{\substack{h \to 0 \\ p_o \to 0}} \lim_{\vec{p} \to 0} \, h \, \Delta_B(p) \to 0 \tag{14}$$

unless $\omega_B(p) \to 0$ as $\vec{p} \to 0$ when we find that

$$h \, \Delta_B(0) = | <B_-(0) | \sigma_-(0) | 0 > |^2 - | <B_+(0) | \sigma_+(0) | 0 > |^2 . \tag{15}$$

The order parameter is defined as $M = - \, h \, \Delta_B(0)$ and it is argued that with the choice $M > 0$, there appears a bound state, the spin-wave state, with energy $\omega_B(p)$ such that if $\vec{p} \to 0$ $\omega_B(p) \to 0$. This defines the N-G boson for the spontaneous breakdown of SU(2) symmetry of a large spin system. One obtains from W-T relations expressions of the type

$$S_-^{-1}(p) - S_+^{-1}(p) = M \, \Gamma_-^{(3)}(p,p; 0) \tag{16}$$

where

$$< \psi_\downarrow(x) \, \psi_\downarrow^\dagger(y) > = \frac{i}{(2\pi)^4} \int d^4p \, e^{-ip \cdot (x-y)} \, S_-(p)$$

$$< \psi_\uparrow(x) \, \psi_\downarrow^\dagger(y) > = \frac{i}{(2\pi)^4} \int d^4p \, e^{-ip \cdot (x-y)} \, S_+(p)$$

and $\Gamma_-^{(3)}(p,p; 0)$ is the three point vertex function. Relations of this type are very useful in writing perturbative scheme like RPA, and beyond, that are consistent with the underlying symmetry properties.

Such a theory has been used for the study of spin waves[12] in itinerant electron ferromagnetic system and in a paramagnetic system. The collective mode, the NG mode, is a bound state of two particles. The use of W-T relations helps to define a consistent theory to study the response at low momenta.

The nucleus may be considered as a paramagnetic system. In the case of nuclei, SU(2) symmetry is insufficient. The isoscalar spin waves in nuclei are not very prominent and their excitation in inelastic scattering is not an important feature of the nuclear excitation spectrum.

One may use the SU(2) isospin symmetry for large nuclei. This would lead to excitation modes $\tau_\pm |0>$ that may be associated with the isobaric analog states[13] that have been prominent in (p,n) reactions. It was a surprise to find them in heavy nuclei. SU(2) isospin symmetry is explicitly broken by the Coulomb interaction. It was anticipated that for nuclei with large value of Z, isospin would not be a good symmetry. However, rather sharp states that were related to the parent ground state by the operator $\tau_\pm |0>$ were found in heavy nuclei. The state is displaced by an energy that is very close to the expectation value of the Coulomb interaction. If the conjecture of spontaneous symmetry breaking is correct, in the absence of Coulomb interaction, the isobaric analog state would have a structure that may be described in terms of pair of nucleons. The pair of particles in the Coulomb interaction would acquire a finite mass. It would be interesting to find if the level spacing can be obtained using such an argument.

## 4. INTERNAL SU(4) SYMMETRY

For non-relativistic nuclear systems, the idea of SU(2) symmetry has to be extended to SU(4) symmetry, with spin ($\sigma$) and isospin ($\tau$) of the particles. In this case the generators of the group are

$$\vec{\sigma}, \vec{\tau} \text{ and } \vec{\sigma}\vec{\tau}$$

with both $\sigma$ and $\tau$ satisfying the algebra

$$[\sigma_i, \sigma_j] = 2i\, \varepsilon_{ijk}\, \sigma_k$$

and

$$[\tau_i, \tau_j] = 2i\, \varepsilon_{ijk}\, \tau_k . \tag{17}$$

Again the system may be described by a Lagrangian, $\mathcal{L}$, that is invariant under SU(4) symmetry, i.e. spin and isospin rotations as well as rotations and translations in space. The symmetry may be spontaneously broken in spin-isospin space by terms like

$$h\, \psi^\dagger(x)\, \sigma_3\, \psi(x) = h\, \sigma_3(x) ,$$

$$h\, \psi^\dagger(x)\, \tau_3\, \psi(x) = h\, \tau_3(x) \text{ and } h\, \psi^\dagger(x)\, \tau_3\, \sigma_3\, \psi(x) = h\, Q_{33}(x) . \tag{18}$$

At the end the limit $h \to 0$ is performed. Functional formalism may be used to obtain W-T relations[14]

$$< \sigma_3(x) >_h = 2ih \int d^4y < \sigma_+(y)\, \sigma_-(x) >_h \tag{19}$$

$$< \tau_3(x) >_h = 2ih \int d^4y < \tau_+(y)\, \tau_-(x) >_h \tag{20}$$

$$< Q_{33}(x) >_h = 2ih \int d^4y < Q_{+3}(y) \, Q_{-3}(x) >_h \qquad (21)$$

$$< Q_{33}(x) >_h = 2ih \int d^4y < Q_{3+}(y) \, Q_{3-}(x) >_h \qquad (22)$$

$$< Q_{33}(x) >_h = -4ih \int d^4y < Q_{++}(y) \, Q_{--}(x) >_h$$

$$+ 2ih \int d^4y < Q_{+3}(y) \, Q_{-3}(x) >_h$$

$$+ 2ih \int d^4y < Q_{3+}(y) \, Q_{3-}(x) >_h$$

$$+ 4ih^2 \int d^4y \, d^4z < Q_{+3}(y) \, Q_{3+}(z) \, Q_{--}(x) >_h \qquad (23)$$

and

$$< Q_{33}(y)_h > = -4ih \int d^4y < Q_{+-}(y) \, Q_{-+}(x) >_h$$

$$+ 2ih \int d^4y < Q_{+3}(y) \, Q_{-3}(x) >_h$$

$$+ 2ih^2 \int d^4y < Q_{3-}(y) \, Q_{3+}(x) >_h$$

$$+ 4ih \int d^4y \, d^4z < Q_{+3}(y) \, Q_{3-}(z) \, Q_{-+}(x) >_h \qquad (24)$$

where

$$Q_{ij}(x) = \psi^\dagger(x) \, \sigma_i \, \tau_j \, \psi(x) \qquad (25)$$

with i,j being spherical vector indices +, - and 3. The order parameters are defined as

$$M_{ST} = < Q_{33}(x) > , \quad M_S = < \sigma_3(x) > \quad \text{and} \quad M_T = < \tau_3(x) > \ .$$

In momentum space, the order parameters[14] may be related to the Fourier transform of the Green's functions given as

$$< \sigma_-(x) \, \sigma_+(y) > = \frac{i}{(2\pi)^4} \int d^4q \, e^{-iq \cdot (x-y)} \, \Delta_S(q) \qquad (26)$$

$$\langle \tau_-(x)\, \tau_+(y) \rangle = \frac{i}{(2\pi)^4} \int d^4q \; e^{-iq\cdot(x-y)} \, \Delta_T(q) \tag{27}$$

$$\langle Q_{-3}(x)\, Q_{+3}(y) \rangle = \frac{i}{(2\pi)^4} \int d^4q \; e^{-iq\cdot(x-y)} \, \Delta_{ST_3}(q)) \tag{28}$$

$$\langle Q_{3-}(x)\, Q_{3+}(y) \rangle = \frac{i}{(2\pi)^4} \int d^4q \; e^{-iq\cdot(x-y)} \, \Delta_{S_3T}(q) \tag{29}$$

$$\langle Q_{--}(x)\, Q_{++}(y) \rangle = \frac{i}{(2\pi)^4} \int d^4q \; e^{-iq\cdot(x-y)} \, \Delta'_{ST}(q) \tag{30}$$

$$\langle Q_{-+}(x)\, Q_{+-}(y) \rangle = \frac{i}{(2\pi)^4} \int d^4q \; e^{-iq\cdot(x-y)} \, \Delta_{ST}(q) \tag{31}$$

Now the order parameters may be obtained from the W-T relations by taking the limit $q \equiv (q_0, \vec{q})$ tends to zero. As established for the SU(2) case, by writing the spectral representation for each of the Green's function defined above, it may be shown that unless the energy of the bosonic excitations tends to zero as $q \to 0$, the Green's functions $\Delta(q)$ would be zero in each of the channel defined above. The bosonic excitations are bound states of the quasiparticles. Then it may be established that the order parameter for each of the N-G bosons are distinct and may be written as

$$M_S = -2h\,\Delta_S(0)\,, \quad M_T = -2h\,\Delta_T(0)\,, \quad M_{ST_3}(0) = -2h\,\Delta_{ST_3}(0)$$

$$M_{S_3T} = -2h\,\Delta_{ST_3}(0)\,, \quad M_{ST} = 4h\,\Delta_{ST}(0) - 2h\,\Delta_{S_3T}(0)$$

$$-2h\,\Delta_{ST_3}(0) \quad \text{and} \quad M'_{ST} = 4h\,\Delta'_{ST}(0) - 2h\,\Delta_{ST_3}(0) - 2h\,\Delta_{S_3T}(0)\,.$$

These six distinct order parameters may be associated with the six distinct ordered states in the nuclear system. Then using an analysis similar to the case of SU(2), six distinct N-G bosons or collective excitations may be classified as follows:

| | |
|---|---|
| $\sigma_\pm$ | Isoscalar-spin waves |
| $\tau_\pm$ | Isobaric analog states |
| $\sigma_3\,\tau_\pm$ | Spin-isobaric analog state |
| $\tau_3\,\sigma_\pm$ | Isovector-spin waves |
| $\sigma_\pm\,\tau_\pm$ | Gamow-Teller states. |

It is obvious some of these excitation modes are the strongest spin, isospin and spin-isospin excitations in the nuclei. Isobaric analog states have been a surprise in nuclei. Isovector spin waves and Gamow-Teller are some of the strongest excitations in inelastic scattering, (p,p'), (e,e') etc., and in charge-exchange reactions, (p,n), (n,p) etc. respectively. The remaining three excitation modes are not very strong and are not found strongly in any of the scattering reactions. Decay of these strong excitations is a good indication of the type of excitations these might be. We are suggesting that the structure of these states is based on the bound states of pairs of particles. These have a rather specific dispersion relation i.e. $\omega_k \to 0$ as $k \to 0$. No doubt all of these excitation modes are universal in nature and are present in all nuclei.

We have already indicated that strong isobaric analog states are observed in heavy nuclei. Inelastic scattering, e, p etc., would lead to excitation of isovector spin waves that have been identified in many nuclei. In general it is not easy to identify these states and to measure the transition strength.

Gamow-Teller states have been studied in detail for nuclei through the periodic table. Detailed model calculations[15] have been carried out to understand the nature and structure of these resonances. Measurements with polarised particles have provided even more detailed information.

The underlying $SU(4)$ symmetry of the Lagrangian would lead to model-independent relations among Green's functions i.e. Ward-Takahashi relations. Such relations put constraints on any perturbative calculation for the system. In particular these relations are useful in formulating a consistent approach to calculating a response function. One such relation may be written as

$$S^{-1}_{\downarrow\uparrow}(p) - S^{-1}_{\uparrow\uparrow}(p) = M\, \Gamma_S(p,p,0) \ ,$$

$$(32)$$

$$S^{-1}_{\uparrow\downarrow}(p) - S^{-1}_{\downarrow\downarrow}(p) = M\, \Gamma_S(p,p,0) \ ,$$

where, for the case of spin waves, $S_{\alpha\beta}(p)$ is the nucleon propagator with $\alpha$ and $\beta$ indicating directions of spin and isospin vectors respectively. Three point function is defined by $\Gamma_S$. Similar relations for many-point Green's functions may be written for each of the N-G mode or the collective mode in the system. The expressions for the Gamow-Teller excitation mode are most complicated since both the spin and the isospin of the system have to be changed. The spin-isospin mode in nuclei is like pionic excitation in the nuclei and corresponds to the excitation of unnatural parity states. For an even-even nucleus, with a $0^+$ ground state, the excitations would have spin and parity $1^+$. For G-T transitions such a relationship is between states of neighboring nuclei, while for isovector spin excitations are within the same nucleus. Within $SU(4)$ these two transitions are related by a rotation in isospin space. Axial-vector current in non-relativistic form corresponds to operators, $\sigma_-\tau_+$ and $\sigma_+\tau_-$. PCAC provides a relationship between the axial current and the pion. The spin-isospin mode manifests itself as the Gamow-Teller excitation and relates to pion emission if energy-momentum conservation permits it.

## 5. TWO QUESTIONS?

It is highly beneficial to obtain model-independent relations based on an assumed symmetry i.e. $SU(4)$ of the system. The spontaneous symmetry-breaking leads to a set of N-G bosons that appear as excitations of the system and have been identified with specific excitations of the nucleus. However there are two distinct specific questions that must be addressed.

(a) *Explicit symmetry breaking.* The two-body nucleon-nucleon interaction includes terms like spin-orbit, tensor and spin and isospin dependant that do not preserve $SU(4)$ symmetry. It is possible to calculate the self-energy of the N-G bosons and this would lead to a finite energy of these bosons. As stated earlier the energy of these excitations is small i.e. 10-20 MeV, which is small compared to the total binding energy (or total mass) of the nucleus. The magnitude of the excitation energy indicates that explicit symmetry breaking effects are small. It may well be that tensor and spin-orbit interactions are quenched in a nuclear medium leaving a central interaction to be dominant. A parametrisation of the effective interaction in terms of Landau parameters is a strong indication along this line.

An effective interaction that has no tensor or spin-orbit interaction is able to account for many of the static properties of nuclei through the periodic table.

This consideration is similar to the spontaneous breakdown of chiral symmetry that has a N-G boson that acquires mass due to the explicit breaking of chiral symmetry. This explains the case of the pion with a finite mass. It will be interesting to write low-energy theorems for the case of the 6 N-G bosons that are a consequence of the spontaneous symmetry breaking of SU(4). The derivation would be similar to the soft pion relations in the broken SU(2) x SU(2) chiral symmetry. In analogy it should be stressed that explicit symmetry breaking gives finite energy to the N-G bosons. The magnitude of the energy is consistent with the size of a large nucleus.

(b) *Finite size effect*. The spontaneous breakdown requires the system to have infinite degrees of freedom. This is really the reason that it works in quantum field theory rather than in quantum mechanics. It has been frequently said that the spontaneous breakdown of symmetries requires an infinite volume. It should be noted that the statement is based on the assumption that the vacuum is homogeneous in space. When a system creates a boundary surface which naturally encloses the system, the surface itself carries a variety of modes such as surface vibration, surface spin wave, and so on, and these modes are associated with infinite number of degrees of freedom. When the size of a nucleus is large, the homogeneous vacuum calculation should well-describe the situation inside the surface. However, the domain near the boundary surface deviates from the computational result. These surface effects have some merits. For example, occurrence of spontaneous breakdown of SU(4) changes the total spin and isospin of the system. However, these spin and isospin deviations are accumulated around the surface. The entire system including the boundary surface carries good quantum numbers of spin and isospin. Actual mechanism for the formation of boundary is an open problem that requires careful study. A possible explanation may follow from soliton-like extended objects.

The N-G bosons are to be interpreted as bound states of two quasiparticles. As in the case of superconductivity, use of a pair-state would lead to a non-conservation of the particle number. For an infinite system this problem may not be serious but for a system with a small number of particles this can have important consequences. There is no simple method to resolve it.

## 6. CONCLUSIONS

Many-body system in a field-theoretic basis may have spontaneous-breaking of an internal symmetry leading to a set of N-G bosons that depend on the basic symmetry group. For SU(4) symmetry, spin and isospin, there are 6 N-G bosons that may be identified to well-known low-energy spin-isospin collective excitations of the nucleus. Such collective states are excited strongly in charge-exchange, (p,n) and (n,p), reactions, inelastic proton and electron scattering and inelastic scattering of heavier projectiles. The decay mechanism for these states would help to determine the microscopic nature of these states. Experimental studies of the G-T state and the isovector magnetic dipole state have helped us to learn about the structure of these states. Classification helps to establish the interrelationship.

Invariance under a symmetry group would lead to exact model-independent relations between Green's functions i.e. W-T relations. These relations provide stringent restrictions in a consistent calculation of response functions. The energy of the collective states and the distribution of strength will depend on the detailed model for interaction among particles. The notion of spontaneous symmetry breaking establishes relationship in the static limit.

To find the dynamic response function, i.e. at finite q and $\omega$, a dynamical model for the interactions has to be defined. In such a case model dependent relationships between difference amplitudes can be established. In this paper we have restricted ourselves to the exact consequences of the symmetry of a system under an internal symmetry group and the spontaneous breakdown of such a symmetry. The constancy of quenching of the Gamow-Teller strength is again a consequence of PCAC and arises from the admixture with the $\Delta$-isobaric states into the nucleonic components.

It may be stressed that such relationship between the spontaneous-breaking of symmetry and N-G bosons has been extended[16] to the Elliot SU(3) group. In this case the rotation group, R(3), a subgroup of SU(3) is preserved. It has been shown that in a formalism that is an extension of the BCS theory i.e. a theory based on pairing, 6 N-G bosons are possible. Since rotational symmetry is preserved and the angular momentum is conserved providing orbital angular momentum as a good quantum number, the 6 N-G bosons can be expressed as components of an L=0 and an L=2 boson. The excitation modes have resemblance to the s- and d-boson excitations postulated in the Interacting Boson Model.

The concept of spontaneous-symmetry breaking and N-G bosons to a nuclear system treated as a field-theoretic system is highly attractive. It provides a unified description of several of the important excitations at low energies. A description within quantum mechanics based on pairing of two particles a la BCS theory is possible. Some consequences of the spontaneous breakdown of isospin symmetry have been discussed previously[17]. Implications for a consistent description of the response function will be dealt with elsewhere[14]. Numerical calculations will help to resolve the overall usefulness of the concept.

## ACKNOWLEDGEMENTS

This work was partially supported by the Natural Sciences and Engineering Research Council of Canada.

## REFERENCES

1. C. D. Goodman, et al., <u>Phys. Rev. Lett.</u>, 44:1755 (1980); C. D. Goodman, <u>in</u> "Spin Excitations in Nuclei," F. Petrovich et al., ed., Plenum, New York (1984); and C. D. Goodman, <u>in</u> these proceedings.
2. C. Gaarde et al., <u>Nucl. Phys.</u>, A369:258 (1981).
3. I. S. Towner and F. C. Khanna, <u>Nucl. Phys.</u>, A399:334 (1983); A. Arima and H. Hyuga, <u>in</u> "Mesons in Nuclei," M. Rho and D. W. Wilkinson, ed., North-Holland, Amsterdam (1979).
4. M. Rho, <u>Prog. Particle and Nucl. Phys.</u>, 1:105 (1978).
5. M. Rho, <u>Nucl. Phys.</u>, A466:678 (1985).
6. X. Q. Zhu, S. S. M. Wong, F. C. Khanna, Y. Takahashi and T. Toyoda, <u>Phys. Rev.</u>, C36:1968 (1987); X. Q. Zhu and S. S. M. Wong, <u>Nucl. Phys.</u>, A412:391 (1984).
7. J. C. Ward, <u>Phys. Rev.</u>, 78:182 (1950); Y. Takahashi, <u>Nuo. Cim.</u>, 6:370 (1957).

8.  S. L. Adler and Y. Dothan, <u>Phys. Rev.</u>, 151:1207 (1966).

9.  J. Goldstone, <u>Nuo. Cim.</u>, 19:154 (1961); Y. Nambu and G. Jona-Lasinio, <u>Phys. Rev.</u>, 122:345 (1961); <u>Phys. Rev.</u>, 124:246 (1961).

10. I.J.R. Aitchison, "An Informal Introduction to Gauge Field Theory", Cambridge University Press (1982).

11. H. Umezawa, M. Matsumoto and M. Tachiki, "Thermo Field Dynamics and Condensed States," North-Holland, Amsterdam (1982); S. Weinberg, <u>Prog. Theor. Phys.</u> Suppl. 86:43 (1986).

12. H. Matsumoto, H. Umezawa, S. Seki and M. Tachiki, <u>Phys. Rev.</u>, B17:2276 (1978); J. Whitehead, H. Matsumoto and H. Umezawa, <u>Phys. Rev.</u>, B25:4737 (1982).

13. J. D. Fox, C. F. Moore and D. Robson, <u>Phys. Rev. Lett.</u>, 12:198 (1964); D. Robson, G. M. Temmer and S. S. Hanna, <u>in</u> "Isospin in Nuclear Physics," D. H. Wilkinson, ed., North-Holland, Amsterdam (1969).

14. F. C. Khanna, H. X. He and H. Umezawa, to be published.

15. J. Wambach, <u>in</u> these proceedings.

16. X. Q. Zhu, F. C. Khanna and H. Umezawa, <u>Phys. Rev., C</u> (to be published).

17. M. D. Scadron, <u>Ann. Phys.</u> (N.Y.) 159:184 (1985).

# GROUND STATE GAMOW-TELLER STRENGTH IN $^{64}$Ni(n,p)$^{64}$Co

A. Ling, R.C. Haight, N.S.P. King, P.W. Lisowski,
D.S. Sorenson, and J.L. Ullmann

Los Alamos National Laboratory
Los Alamos, NM 87545

X. Aslanoglou, R.W. Finlay, B.K. Park, and J. Rapaport

Ohio University
Athens, Oh 45701

F.P. Brady and J.L. Romero

University of California
Davis, CA 95616

C.R. Howell and W. Tornow

Duke University
Durham, NC 27707

## INTRODUCTION

An important process occuring in presupernova stars is e⁻ capture on free protons and nuclei.[1,2] As e⁻ capture and the charge exchange reaction (n,p) between same initial and final states are both $T_o \rightarrow T_o + 1$ transitions involving the same nuclear matrix element, the (n,p) reaction can be used to provide the input required to calculate e⁻ capture rates. Specifically, the e⁻ capture rate $\lambda^{if}$ for a nucleus going from an initial state i to a final state f is proportional to the Gamow-Teller strength $B^{if}(GT)$[3]:

$$\lambda^{if} \quad \alpha \quad B^{if}(GT)$$

Where $B^{if}(GT)$ is given by[4]

$$B^{if}(GT) = \frac{1}{2J_i+1} \left| <f|\sigma\tau|i> \right|^2 .$$

Once $B^{if}(GT)$ for a particular transition is known, the corresponding $e^-$ capture rate $\lambda^{if}$ for that transition can be calculated.

As indicated in reference 5, the unit cross section $\hat{\sigma}_{GT}(E,A)$ relates $B^{if}(GT)$ to the zero degree differential cross section extrapolated to q=0 for (n,p) or (p,n) reactions:

$$\sigma(q=0) = \hat{\sigma}_{GT}(E,A)B^{if}(GT) .$$

As $\hat{\sigma}_{GT}(E,A)$ has been seen to have a smooth A dependence, measurements of zero degree (n,p) cross sections for (fp) shell nuclei can be used to obtain $B^{if}(GT)$, and thus $\lambda^{if}$, once a value for $\hat{\sigma}_{GT}(E,A)$ in this mass region is known. The data presented here allow a value for $\hat{\sigma}_{GT}$ in the (fp) shell to be calculated from the Gamow-Teller strength for the $\beta^-$ decay $^{64}Co(g.s.) \rightarrow$ $^{64}Ni(g.s.) + e^- + \bar{\nu}_e$. The $\beta^-$ strength is obtained from the relation[5]

$$B^{\beta^-}(GT) = \frac{6166}{(g_A/g_V)^2 ft} .$$

The constant 6166 reflects the choice of the vector coupling constant recommended by Wilkinson[6] and the value of 1.26 for the coupling constant ratio $g_A/g_V$ used here is that of Bopp[7]. A value for ft is calculated from the measured half-life and branching ratio. Reference 8 gives $t_{1/2} = 0.3 \pm 0.035s$ with a branching ratio of 90% while $\log(f) = 4.749 \pm 0.006$ was obtained from reference 9. Since $^{64}Co(g.s.)$ $\beta^-$ decay and the reaction $^{64}Ni(n,p)^{64}Co(g.s.)$ have opposite initial and final states, detailed balance can be used to relate the Gamow-Teller strengths for the two processes:

$$B^{np}(GT) = \frac{(2J_f + 1)}{(2J_i + 1)} B^{\beta^-}(GT) = 0.6228^{+.0328}_{-.0369}$$

It is this value of $B^{np}(GT)$ together with the differential cross section measurements presented here for the reaction $^{64}Ni(n,p)^{64}Co(g.s.)$ that allow $\hat{\sigma}_{GT}$ to be calculated and thus a calibration point in the (fp) shell is established.


EXPERIMENTAL METHOD

Differential cross sections were measured in the angular range $0°-10°$ for incident neutron energies 60-260 MeV using the WNR white neutron source at LAMPF[10]. A system of target chambers, drift chambers, CsI counters, and a DE scintillator were set up on the 15° left flight path about 90 meters from the neutron production target.[11] Events due to charged particles in the beam were tagged by two chambers placed in front of a target chamber assembly. The target chamber assembly consisted of four target chambers with a $^{64}$Ni (145.6 mg/cm$^2$) target of 97.93% enrichment placed in the second target location. A CH$_2$ (76.1 mg/cm$^2$) target was placed in the fourth target location for normalization purposes. Protons from the (n,p) reaction near 0° were swept by a 5 kG magnet into a rectangular calorimeter wall 11" tall by 18.5" wide consisting of 15 CsI counters. These counters gave the energy of the detected particle which was then corrected for energy loss of all material traversed. The energy resolution of the detected protons ranged from 1.5 MeV at low energies to 2.0 MeV at the higher energies. A large scintillator placed in front of the CsI array was used to obtain the incident neutron energy from time-of-flight and was also used as a DE detector. As the efficiency of the target chambers was ~ 95%, the excitation spectra for targets 2, 3 and 4 had to be corrected for misidentified events, i.e. events incorrectly identified has having come from a particular target. Up to 7% of the previous target excitation spectrum was subtracted out of the $^{64}$Ni excitation spectrum for each energy and angle range. The resulting corrected excitation spectra are shown in figure 1 for the angular range 0-4°. Of interest is the strong forward peak at 0 MeV excitation most prominantly seen in the 120-180 MeV spectrum. This peak contains the GT strength associated with the β$^-$ decay of $^{64}$Co. The GT peak region in each spectrum was fit with a $\chi^2$ minimization routine[12] to a gaussian representing the peak plus a quadratic term representing the part of the continuum under the peak to get the total number of counts in the peak. In most cases the centroid from the fit of the peak region agreed to within 2 keV of the Q value[13] of -6.525 ± 0.020 MeV for the reaction $^{64}$Ni(n,p)$^{64}$Co(g.s.) Background counts in the peak region obtained from runs with no targets in target locations 1, 2, and 3 were subtracted from the measured yield for the peak to give the correct yield used to calculate the $^{64}$Ni(n,p)$^{64}$Co(g.s.) cross section. The yield from the H(n,p) reaction of target 4, together with hydrogen cross sections calculated from Arndt's SP86 phase shifts,[14] gave the flux of neutrons impinging on the targets.

As the experimental resolution was no better than 1.5 MeV, the individual levels in $^{64}$Co could not be resolved. Very little is known about the spin and parity of the low-lying states in $^{64}$Co, but a recent paper[15] deduces a 1$^+$ assignment for the first clearly known excited state at 0.311 MeV. This state therefore corresponds to a GT transition in $^{64}$Ni(n,p) that is included in the ground state peak regions shown in figure 1. However, B(GT) for β$^-$ decay from the 0$^+$ ground state of $^{64}$Fe to the 0.311 MeV state of $^{64}$Co is 0.0489 compared to 0.6155 for β$^-$

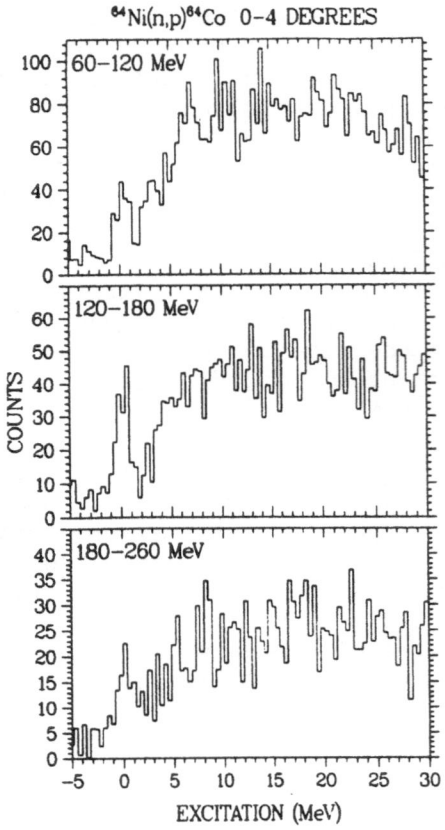

$^{64}$Ni(n,p)$^{64}$Co 0-4 DEGREES

**Figure 1.** Excitation spectra for the reaction $^{64}$Ni(n,p)$^{64}$Co from 0 to 4 degrees. The background from target empty runs have not been subtracted.

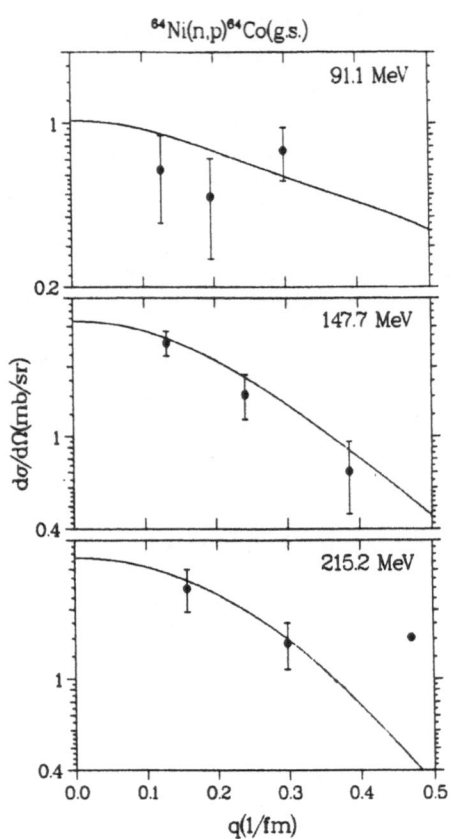

$^{64}$Ni(n,p)$^{64}$Co(g.s.)

**Figure 2.** Differential cross sections for the reaction $^{64}$Ni(n,p)$^{64}$Co(g.s.) as a function of q. The solid line is the DW81 calculation multiplied by $B^{np}(GT)$ normalized to the data. The normalization factors are 0.39, 0.69, and 0.64 for 91.1, 147.7, and 215.2 MeV respectively. The point without error bars is an upper limit and was not used to normalize the DWIA curve.

decay to the $^{64}$Co ground state. This suggests that the wavefunction for the 0.311 MeV state may have only a small overlap with the adjacent $0^+$ ground state, at about the 7% level. While the cross section quoted here for $^{64}$Ni(n,p)$^{64}$Co(g.s.) is an upper limit to the true value, the above considerations propose that the correction for the first excited state is likely to be at the 5% to 10% level.


RESULTS

Angular distributions for the reaction $^{64}Ni(n,p)^{64}Co(g.s.)$ are shown in figure 2 for mean neutron energies 91.1, 147.7, and 215.2 MeV. The angle binning was selected to be 0-4°, 4-6°, and 6-10° and the energy binning 60-120 MeV, 120-180 MeV, and 180-260 MeV. These ranges were chosen in order to evenly distribute statistics among the spectra. Each distribution was used to normalize a distorted wave impulse approximation curve generated by the program DW81 for that energy.[16] The code used the T matrices of Franey and Love[17] at 100, 140 and 210 MeV and the Schwandt optical model parameters.[18] Since Schwandt optical model parameters extrapolated out to 215.2 MeV are unrealistic, the DWIA calculation for 215.2 MeV used the Schwandt optical model parameters at 175 MeV. A five-particle one-hole $[\pi(f_{7/2})^{-1}$ $\nu(f_{5/2})^5]$ configuration was used with a Z coefficient of $(2J_i+1)^{-\frac{1}{2}} = 0.35355$. As the program DW81 generates DWIA angular distributions down to $\theta = 0°$, the curves had to be extrapolated to q=0 in order to obtain $\sigma(q=0)$. This was done by setting the Q value for the reaction equal to zero and running the program to get the shape of the curve from q=0 to $q(\theta=0°)$. The resulting DWIA curves extending to q=0 were then fit to the data according to the relation

$$\sigma^{data}(q) = N\, B^{np}(GT)\, \sigma^{dw81}(q)$$

with N the normalization factor obtained from the fit. The values of N for 91.1, 147.7, and 215.2 MeV were 0.39, 0.69, and 0.64 respectively. The resulting normalized DWIA curves are shown in figure 2. For the 215.2 MeV distribution, the point at the highest value of q without error bars is an upper limit and was not included in fitting the DWIA calculation. It is the value of the normalized DWIA curve at q=0 that is used to calculate the unit cross section.

Figure 3 shows the unit cross section as a function of energy for the $^{64}Ni(n,p)^{64}Co(g.s.)$ reaction using $\sigma(q=0)$ obtained from the angular distributions by the procedure discussed above. Also shown is a DWIA calculation performed at the energies of the existing Love-Franey t matrices, 100, 140, 175, and 210 MeV, normalized with the average value of N. The point at 210 MeV used the Schwandt optical model parameters at 175 MeV in order to give a more realistic behavior of the unit cross section with energy. Unit cross sections obtained in the (p,n) reaction for $^{58}Ni(p,n)$ are also shown for comparison.[5,19,20]

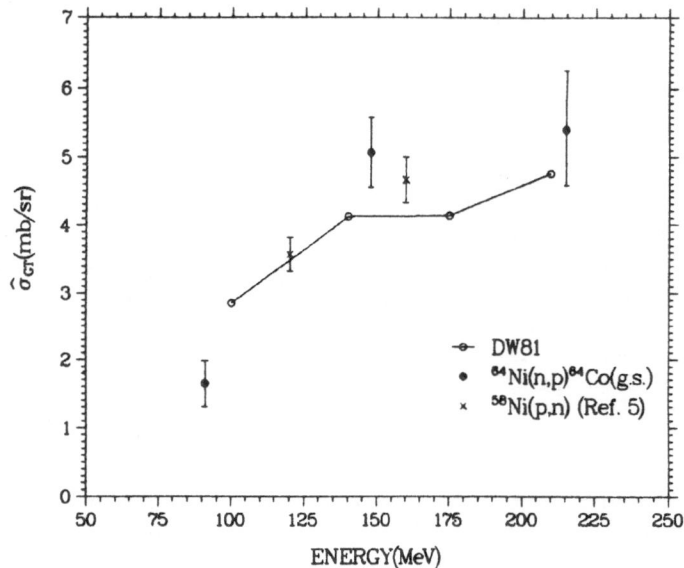

**Figure 3.** The unit cross section as a function of energy for (fp) shell nuclei. The errors shown for the reaction $^{64}Ni(n,p)^{64}Co(g.s.)$ are statistical in nature only and do not include the uncertainty in the measured ft value. Also shown are values for the reaction $^{58}Ni(p,n)$ from reference 5.

The work described here is the first absolute measurement of the unit cross section in the (n,p) reaction for an (fp) shell nucleus. It is seen that our measurements of $\hat{\sigma}_{GT}$ are in good agreement with values of $\hat{\sigma}_{GT}$ measured from (p,n) reactions on similar nuclei. Thus the trend seen in GT strength observed in p-shell nuclei ($^6Li$ and $^{12}C$) from (n,p) and (p,n) reactions of data taken at WNR[21], IUCF[5,22-26], and TRIUMF[5,27,28] is again seen in the (fp) shell, i.e $\hat{\sigma}_{GT}(n,p) \approx \hat{\sigma}_{GT}(p,n)$. Our measurement of $\hat{\sigma}_{GT}$ therefore allows calculations of $e^-$ capture rates of (fp) shell nuclei from measured (n,p) cross sections to be made with reasonable accuracy. A more precise theoretical description of the supernova phenomenon should then emerge.

REFERENCES

1. J. Cooperstein and J. Wambach, Nucl. Phys. **A420**, 591 (1984)
2. G. M. Fuller, W. A. Fowler, M. J. Newman, Ap. J. **252**, (1982)715
3. M. B. Aufderheide, G. E. Brown, T. T. S. Kuo, D.B. Stout, P. Vogel, to be published
4. A. Bohr and B. M. Mottelson, *Nuclear Structure, Vol. I*, W. A. Benjamin, 1969
5. T. N. Taddeucci et. al., Nucl. Phys. **A469**, 125 (1987)

6.  D. H. Wilkinson, Nucl. Phys. **A377**, 474 (1982)
7.  P. Bopp et. al., Phys. Rev. Lett. **56**, 919 (1986)
8.  V. Rahkonen and J. Kantele, Phys. Fenn. **9**, 103 (1974)
9.  Nuclear Data Center, Brookhaven
10. P. W. Lisowski, C. D. Bowman, G. J. Russell, S. A. Wender, Nucl. Sci. Eng **106**, 208 (1990)
11. J. L. Ullmann et. al., to be published
12. P. R. Bevington, *Data Reduction and Error Analysis for the Physical Sciences* (Mc Graw-Hill, New York,1969) p. 212
13. E. R. Flynn and J. D. Garrett, Phys. Lett. **42B**, 49 (1972)
14. R. A. Arndt and L. D. Roper, Scattering Analysis Interaction Dial-in (SAID) program, Virginia Polytechnic Institute, 1984 (unpublished)
15. E. Runte et. al., Nucl. Phys. **A441**, 237 (1985)
16. R. Schaeffer and J. Raynal, computer code DWBA70, Arizona State University, 1970 (unpublished); extended version: J. R. Comfort, computer code DW81, Arizona State University, 1984 (unpublished)
17. M. A. Franey and W. G. Love, Phys. Rev. C **31**, 488 (1985)
18. P. Schwandt et. al., Phys. Rev. C **26**, 55 (1982)
19. J. Rapaport et. al., Phys. Lett. **119B**, 61 (1982)
20. J. Rapaport et. al., Nucl. Phys. **A410**, 371 (1983)
21. D. S. Sorenson et. al., to be published
22. C. A. Goulding et. al., Nucl. Phys. **A331**, 29 (1979)
23. G. L. Moake et. al., Phys. Rev. C **21**, 2211 (1980)
24. J. Rapaport et. al., Phys. Rev. C **24**, 335 (1981)
25. J. Rapaport et. al., Phys. Rev. C **36**, 500 (1987)
26. K. Wang et. al., Phys. Rev. C **38**, 2478 (1988)
27. J. W. Watson et. al., Phys. Rev. C **40**, 22 (1989)
28. K. P. Jackson et. al., Phys. Lett. B **201**, 25 (1988)

# STRUCTURE OF THE NEUTRON RICH NUCLEUS $^{11}$Be

# STUDIED BY THE $^{11}$B($\vec{d}$,$^{2}$He) REACTION AT 70 MEV

H. Sakai, S. Ishida, Y. Nagai*, H. Okamura,
A. Okihana**, H. Okuno, H. Sagawa, K. Takeda*,
T. Toriyama*, and A. Yoshida*

University of Tokyo
Hongo 7-3-1, Bunkyo-ku, Tokyo 113, Japan
*Tokyo Institute of Technology
Ohokayama, Meguro, Tokyo 152, Japan
**Kyoto University of Education
Fukakusa, Fushimi-ku, Kyoto 612, Japan

## INTRODUCTION

Since the discovery of a neutron halo[1] in large neutron-excess nuclei, $^{11}$Li, $^{11}$Be and $^{14}$Be, a lot of interest has been placed on the nuclear structure of these nuclei. However such a study is hampered by experimental difficulties to excite those nuclei. Although the $^{11}$Be nucleus is the easiest to access among them, the spin and parity ($J^{\pi}$) is assigned for only few states[2]. The ground state is known to have an anomalous $J^{\pi}$ of $\frac{1}{2}^{+}$ which is very unexpected by the naive shell model predictions. The $J^{\pi}$ of the first excited state at $E_x$=0.32 MeV is also known to be $\frac{1}{2}^{-}$.

We studied the nuclear structure of $^{11}$Be by using the $^{11}$B($\vec{d}$,$^{2}$He) reaction. Here $^{2}$He indicates a proton-proton system coupled to the singlet S-state [$^{1}S_0$]. The ($d$,$^{2}$He) reaction has an excellent spin-isospin selectivity. It excites exclusively spin and isospinflip transitions e.g. Gamow-Teller (GT) type transition : $\Delta$ S=1, $\Delta$ L=0 and/or spinflip dipole (SFD) transition : $\Delta$ S=1, $\Delta$ L=1.

## EXPERIMENT

Present experiment was carried out by using a polarized deuteron beam of 70 MeV provided by the AVF cyclotron at the Research Center for Nuclear Physics, Osaka University. The $^{11}$Be target with a thickness of about 0.5 $mg/cm^2$ was made by evaporating isotopically enriched (> 95%) material onto a mylar film. The effect due to the mylar film was corrected for. The $^{2}$He "particle"s were detected with two sets of multicounter arrays. It was designed to have an optimum detection efficiency of two protons with small relative energy($^{2}$He). Each array was consisted with four sets of $\Delta E - E - E_{veto}$ Si counter telescopes. Owing to this configuration six different $p - p$ pair coincidences become available in each array and the detection efficiency of "²He" particles increases accordingly. Each telescope has a solid angle of 6.7 msr. Further details of the experimental procedures are described elsewhere[3].

RESULT

Figure 1 shows the typical energy spectrum of cross sections at $\vartheta_{Lab} = 35°$. Three prominent discrete peaks in the low excited region and a broad bump at an excitation energy around 10 MeV with a width of about 5 MeV are clearly observed. The ground state seems to be *not* excited by this reaction. Cross sections are extracted after subtracting a background due to the three-body phase space as indicated by the dashed curve in Fig.1. Angular distributions ($20° < \vartheta < 70°$) for the cross sections ($d\sigma/d\Omega$) as well as the vector and tensor analyzing powers (A$_y$ and A$_{yy}$) are displayed in Fig.2.

The J$^{\pi}$ of these observed peaks are deduced empirically by comparing observed angular distributions with those of known transitions in the $^{12}$C($d$,²He)$^{12}$B reaction at 70 MeV where the ground state transition is the GT-type ($\Delta$ S=1,$\Delta$ L=0) and the transition to the 2$^-$ state at $E_x = 4.5$ MeV is the SFD-type ($\Delta$ S=1,$\Delta$L=1). The solid and dashed curves in Fig.2 represent the $\Delta L = 0$ transition(GT-type) and $\Delta L$=1 transition(SFD-type) taken from Ref. [4,5], respectively.

As for the transitions to the low lying states($E_x = 0.3$, 2.7 and 3.8 MeV) the angular distributions for $d\sigma/d\Omega$, A$_y$ and A$_{yy}$ are very similar with those of the ground state transition in the $^{12}$C($d$,²He)$^{12}$B reaction. Therefore these peaks are considered to be due to the GT-type transitions with $\Delta L = 0$.

The angular distribution of the cross section for the broad bump observed at around $E_x$=10 MeV has a flatter shape suggesting a larger L transfer than those to the low lying discrete transitions. Moreover the angular distributions for $d\sigma/d\Omega$, A$_y$ and A$_{yy}$ are also similar to those of the transition to the 2$^-$ state at $E_x = 4.5$ MeV in the $^{12}$C($d$,²He)$^{12}$B reaction suggesting $\Delta L$=1 transition. These facts strongly indicate that this bump at $E_x$=10 MeV is mainly due to the spinflip dipole giant resonance.

COMPARISON WITH SHELL MODEL CALCULATION AND EFFECT DUE TO THE NEUTRON HALO

The shell model calculations with a neutron halo have been performed by Hoshino et al.[6] The effect of the neutron halo appears in the isovector SFD transition for the present reaction : The mean excitation energy $\overline{E}$ of SFD transitions is shifted down about 2 MeV and the SFD transition strength B(SFD) is enhanced about 40 % compared with the calculations(Millener-Kurath(MK) interactions) without the neutron halo[7]. In Fig.1 $\overline{E}$ of SFD transitions with(without) halo is indicated by the solid(dashed) arrow. The observed broad bump at around $E_x = 10$ MeV is closer to the shell model prediction with halo. The cross sections at 35° for the GT-transition (summed over three low lying peaks) and the SFD bump are listed in Table 1. The results for the $^{12}$C target taken under the same experimental condition are also listed in Table 1 for a comparison. The shell model estimations are also given. Cohen and Kurath interactions(CK)[8] are used for the GT-transitions and MK interactions with(without) halo for the SFD transitions. The unit cross section at 35° may be defined as $\hat{\sigma}_{\alpha} = \frac{d\sigma_{\alpha}}{d\Omega}/B(\alpha)$. $\hat{\sigma}_{GT}$ values for $^{12}$C and $^{11}$B are 20 and

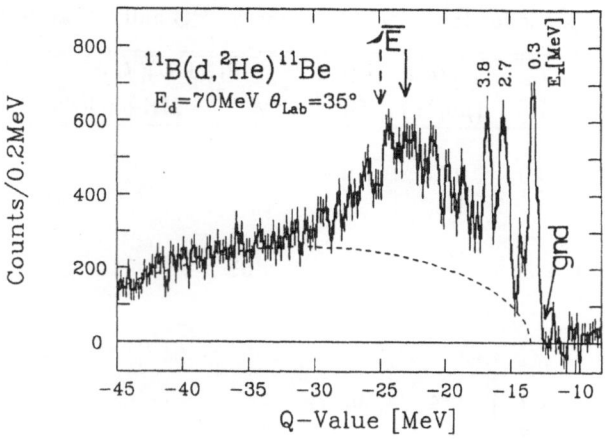

Figure 1. Energy spectrum for the $^{11}$B($\vec{d}$,$^2$He)$^{11}$Be reaction at 70 MeV and $\vartheta_{Lab} = 35°$.

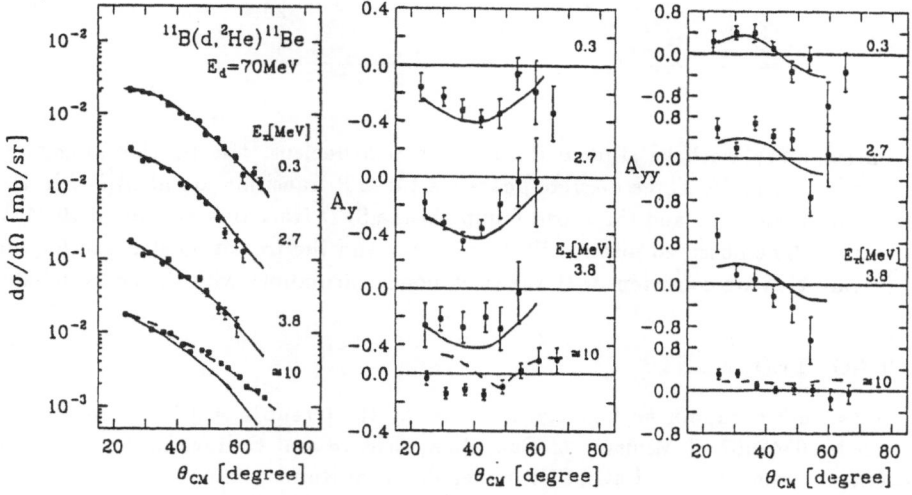

Figure 2. Cross sections and vector and tensor analyzing powers ($A_y$ and $A_{yy}$). Solid and dashed curves represent $\Delta L = 0$ transition(GT-type) and $\Delta L=1$ transition(SFD-type), respectively. See text for detail.

19 $\mu b/sr$, respectively. They are almost equal. Thus the proportionality relation seems to hold. As for the SFD transitions $\hat{\sigma}_{SFD}$ values of $^{11}$B are 1.1 or 1.5 $\mu b/sr$ depending on calculations with or without halo, respectively. These values should be compared to that of 1.02 $\mu b/sr$ of $^{12}$C. This result supports the shell model prediction of 40 % enhancement of SFD transitions due to the neutron halo if the same kind of proportionality holds for the SFD transitions. Note that $d\sigma/d\Omega$ for the SFD transition depends on an assumption of magnitudes of background by the three-body phase space. However both $^{11}$B and $^{12}$C data are analyzed in a consistent manner, the relative error in $d\sigma/d\Omega$ is estimated to be rather small ($\leq 20\%$).

Table 1. Cross sections for the $(d,{}^2\text{He})$ reaction on ${}^{12}\text{C}$ and ${}^{11}\text{B}$ at 70 MeV and 35°.

| Target | $J_i$ | $J_f$ | $E_x$ MeV | $d\sigma/d\Omega$ $(\mu b/sr)$ | $B(GT)$ CK[8] | $B(SFD)(fm^2)$ MK[7] | MK+halo[6] | $\hat{\sigma}_{GT}$ $(\mu b/sr)$ | $\hat{\sigma}_{SFD}$ $(\mu b/sr)$ |
|---|---|---|---|---|---|---|---|---|---|
| ${}^{12}\text{C}$ | $0^+$ | $1^+$ | 0.0 | 17.5 | 0.9 | | | 20. | |
| | $0^+$ | $0^-$<br>$1^-$<br>$2^-$ | 2-30 | 31.5 | | 31 | | | 1.02 |
| ${}^{11}\text{B}$ | $\frac{3}{2}^-$ | $\frac{1}{2}^-$<br>$(\frac{3}{2}^-)$<br>$(\frac{5}{2}^-)$ | 0.3<br>2.7<br>3.8 | 14 | 0.733 | | | 19 | |
| | $\frac{3}{2}^-$ | $\frac{1}{2}^+$<br>$\frac{5}{2}^+$<br>$\frac{7}{2}^+$<br>$\frac{9}{2}^+$ | 4-25 | 26 | | 17.5 | 24.4 | | 1.5<br>1.1 |

In summary we have studied the neutron rich nucleus ${}^{11}\text{Be}$ by the ${}^{11}\text{B}(\vec{d},{}^2\text{He})$ reaction at 70 MeV. Three discrete peaks via the GT transition are identified in the low excitation energy and the broad bump via the SFD transition at around 10 MeV excitation. The observed mean excitation energy and the transition strength for the SFD transition are consistent with the shell model calculations with the neutron halo.

## ACKNOWLEDGMENTS

This work is supported financially in part by the Grant-in-Aid for Scientific Research No.6342007 of Ministry of Education, Science and Culture of Japan. This experiment was performed at RCNP under Program Number 29A13.

## REFERENCES

1. I. Tanihata et al., Phys. Rev. Lett. **55** (1985) 2676.
2. F. Ajzenberg-Selove et al., Nucl. Phys. **A481** (1990) 1.
3. H. Okamura, PhD thesis, Kyoto University, 1989, unpublished.
4. T. Motobayashi et al., Phys. Rev. **C34** (1988) 2365.
5. T. Motobayashi et al., J. Phys. G: Nucl. Phys. **14** (1988) l137.
6. T. Hoshino, H. Sagawa and A. Arima, Nucl. Phys. in print and
   N. Fukunishi, H. Sagawa and T. Ohtsuka, private communication.
7. D.J. Millener and D. Kurath, Nucl. Phys. **A255** (1975) 315.
8. S. Cohen and D. Kurath, Nucl. Phys. **73** (1965) 1.

# DEVELOPMENT AND APPLICATION OF
# FULL–FOLDING OPTICAL POTENTIALS

C. Alvarez [*†], H. F. Arellano [*], F. A. Brieva [†] and W. G. Love [*]

*Department of Physics and Astronomy
University of Georgia, Athens, Georgia 30602, USA
†Departamento de Física
Facultad de Ciencias Físicas y Matemáticas
Universidad de Chile, Casilla 487–3, Santiago, Chile

## INTRODUCTION

The problem of describing nucleon–nucleus (NA) elastic scattering in a non–relativistic framework, starting from a realistic nucleon–nucleon (NN) force, has received considerable attention during the last few years. [1-6] The standard approach [7] is to reduce the (A+1)–body problem to a one–body problem where an average nuclear field, the optical potential, correctly describes the properties of the elastic channel. Thus, the primary focus is on the calculation of the nucleon–nucleus optical potential.

We can expect this non–relativistic approach to be reasonable for incident nucleon energies up to $\approx 400$ MeV. Indeed, one of the underlying assumptions in the formulation of the optical potential model is the existence of a NN potential which provides a good description of the existing data in the two–nucleon system. In practice, such NN potentials have been developed in the $0 \sim 400$ MeV energy range.

The involved structure of the NA optical potential has led to the development of several well–established, though limited, approximations. For incident nucleon energies above 200 MeV where medium corrections are assumed to be small, the NN effective interaction is approximated by the free NN $t$–matrix evaluated on the energy shell. This leads to a factorized $t\rho$ structure for the optical potential in momentum space. [8] The main advantage of this approach is that it may be extended well above 400 MeV through a knowledge of the NN phase shifts rather than the NN force. A more elaborate approach which also leads to a $t\rho$ form of the optical potential is the folding model based on the calculation of a complex, energy– and density– dependent effective NN interaction calculated from a realistic NN force. [2,3] As implemented, this model leads to a local optical potential in a coordinate representation. It has been used in the 30–300 MeV range, for nucleons on different nuclei, with varying degree of success when comparing the measured and calculated scattering observables.

Recent advances in the measurement of spin observables have exposed serious deficiencies in the simpler and more conventional non–relativistic approximations to the optical potential. Several authors have investigated the complicated computational structure of the theory by developing full– folding approaches to the optical potential. [4-6] The most flexible implementations of the theory are able to include explicitly the energy dependence and off–shell behavior of the NN effective interaction, and the associated knock–on exchange terms as well as the mixed–density of the target ground state.

In this work, we report some of the recent developments associated with the full–folding optical potential and its success for providing a satisfactory description of proton elastic scattering observables at intermediate energies within a non– relativistic framework. Some preliminary results give insight into the role of medium corrections in the context of the full– folding model.


THE OPTICAL POTENTIAL MODEL

The problem of describing the elastic scattering of a nucleon with energy $E$ from an A–nucleon target can be understood in terms of an effective one–body hamiltonian $h(E)$, [7]

$$h(E) = K_0 + U(E), \tag{1}$$

where $K_0$ is the projectile kinetic energy and $U(E)$ is the optical potential representing the average projectile–nucleus interaction in the elastic scattering channel. Following Refs.(4,7), the optical potential can be expressed, in a momentum representation, as the antisymmetrized matrix elements of a many–body transition operator $T(E)$,

$$U(\vec{k}', \vec{k}; E) = <\vec{k}'; \Phi_0|T(E + E_0)|\vec{k}; \Phi_0>_\mathcal{A}, \tag{2}$$

with $(E_0, \Phi_0)$ the energy and eigenstate of the target ground state. In this approach, the main problem is to relate $T(E)$ to the effective interaction $F$ between nucleon pairs in the medium. Considering single-particle processes predominant in a multiple scattering series expansion for $T$, [7] we can write

$$T(\omega) = \sum_{i=1}^{A} F_i(\omega) + \dots, \tag{3}$$

with $F_i$ the two–body effective interaction between the projectile and the $i^{th}$–nucleon in the target. Introducing Eq.(3) in Eq.(2) and assuming that only particle-particle propagation is important in intermediate states, the optical potential can be expressed as [9]

$$U(\vec{k}', \vec{k}; E) = \int_{-\infty}^{\epsilon_F} d\xi \int d\vec{p}d\vec{p}\,' < \vec{k}'\vec{p}\,'|F(E + \xi)|\vec{k}\vec{p}>_\mathcal{A} A(\vec{p}, \vec{p}\,'; \xi), \tag{4}$$

with $A(\xi)$ the spectral function for the target nucleus and $\epsilon_F$ the corresponding Fermi energy. Within the context of a single–particle model for the target with $\{\phi_\alpha, \epsilon_\alpha\}$ the single–particle wavefunctions and energies respectively, the optical potential reduces to [9]

$$U(\vec{k}', \vec{k}; E) = \sum_{\epsilon_\alpha \le \epsilon_F} \int d\vec{p}d\vec{p}\,'\phi_\alpha^\dagger(\vec{p}\,') < \vec{k}'\vec{p}\,'|F(E + \epsilon_\alpha)|\vec{k}\vec{p}>_\mathcal{A} \phi_\alpha(\vec{p}). \tag{5}$$

Eq.(5) shows the general folding structure for $U(E)$. Its main characteristic is the full off-shell sampling of the effective interaction $F$. Furthermore, Eq.(5) is the starting point from which the full-folding optical potential is calculated.

The choice of the effective interaction depends on the physics which is expected to dominate in the scattering process. Quite generally, the effective interaction $F$ satisfies, in the ladder approximation, the following integral equation, [9]

$$F(\omega) = V + V\Lambda(\omega)F(\omega), \tag{6}$$

where $V$ is the NN force and $\Lambda(\omega)$ is the two-body propagator in the nuclear medium. Different choices for $\Lambda(\omega)$ lead to alternative approximations for $F(\omega)$. If Pauli blocking and self-energy corrections are ignored in the calculation of $\Lambda$, namely

$$\Lambda(\omega) = \frac{1}{\omega - K_1 - K_2 + i\eta}, \tag{7}$$

with $K_i$ the kinetic energy of the interacting particle, then $F$ is identified as the free $t$-matrix. [4-6] When medium corrections are included in an infinite nuclear matter context (quasi-particle approximation), the $\Lambda$ propagator becomes

$$\Lambda(\omega; k_F) = \sum_{\vec{q}_1, \vec{q}_2} \frac{\theta[q_1 - k_F]\theta[q_2 - k_F]}{\omega - e(q_1) - e(q_2) + i\eta}, \tag{8}$$

and

$$e(q) = \frac{q^2}{2m} + U_{k_F}(q; e(q)), \tag{9}$$

the total energy for a nucleon of mass $m$ in nuclear matter at Fermi momentum $k_F$ and self-energy $U_{k_F}$. In this case, $F$ corresponds to the $g$-matrix. [1-3] Explicit calculations of effective forces in finite nuclei are not yet available.

The full-folding optical potential in the free $t$-matrix approximation has been discussed in Ref.(4). Defining the following momentum and energy variables,

$$\vec{\kappa}' = \frac{1}{2}(\vec{K} - \vec{P} - \vec{q}),$$

$$\vec{\kappa} = \frac{1}{2}(\vec{K} - \vec{P} + \vec{q}), \tag{10}$$

$$z_\alpha = E + \epsilon_\alpha - \frac{(\vec{P} + \vec{K})^2}{2M},$$

Eq.(5) reduces to

$$U(\vec{k}', \vec{k}; E) = \sum_{\epsilon_\alpha \leq \epsilon_F} \int d\vec{P} \phi_\alpha^\dagger(\vec{P} + \frac{\vec{q}}{2}) < \vec{\kappa}'|t_0[z_\alpha]|\vec{\kappa} >_\mathcal{A} \phi_\alpha(\vec{P} - \frac{\vec{q}}{2}), \tag{11}$$

where $\vec{P} = \frac{1}{2}(\vec{p} + \vec{p}')$ is the mean momentum of the struck nucleon, $\vec{q} = \vec{p}' - \vec{p} = \vec{k}' - \vec{k}$ is the momentum transfer to the struck nucleon, $\vec{K} = \frac{1}{2}(\vec{k} + \vec{k}')$ is the mean momentum of the scattered nucleon and $M$ is the total mass of the interacting nucleon pair. In Eq.(10), $t_0$ is a one-body operator satisfying

$$t_0(z) = V + V\frac{1}{z - K_r + i\eta}t_0(z), \tag{12}$$

with $K_r$ the kinetic energy operator of relative motion.

The primary feature which distinguishes the full–folding model, Eq.(10), from more traditional approaches is the consistent inclusion of off–shell effects associated with the NN $t$–matrix in conjunction with the mean–momentum $\vec{P}$ (or current) distribution of the target mixed density $\rho$. This can be made explicit when the single–particle energies of the bound nucleons are approximated by an average value $\epsilon$ ($\epsilon \approx -25$ MeV in the case of $^{40}$Ca ). In this case, the full–folding optical potential reduces to

$$U(\vec{k}', \vec{k}; E) = \int d\vec{P} \; \rho(\vec{P} + \frac{\vec{q}}{2}; \vec{P} - \frac{\vec{q}}{2}) < \vec{\kappa}' |t_0[z]| \vec{\kappa} >_{\mathcal{A}}, \qquad (13)$$

where the mixed ground–state density in momentum space ( in a single–particle model ) is given by

$$\rho(\vec{P} + \frac{\vec{q}}{2}; \vec{P} - \frac{\vec{q}}{2}) = \sum_{\alpha} \phi_{\alpha}^{\dagger}(\vec{P} + \frac{\vec{q}}{2})\phi_{\alpha}(\vec{P} - \frac{\vec{q}}{2}). \qquad (14)$$

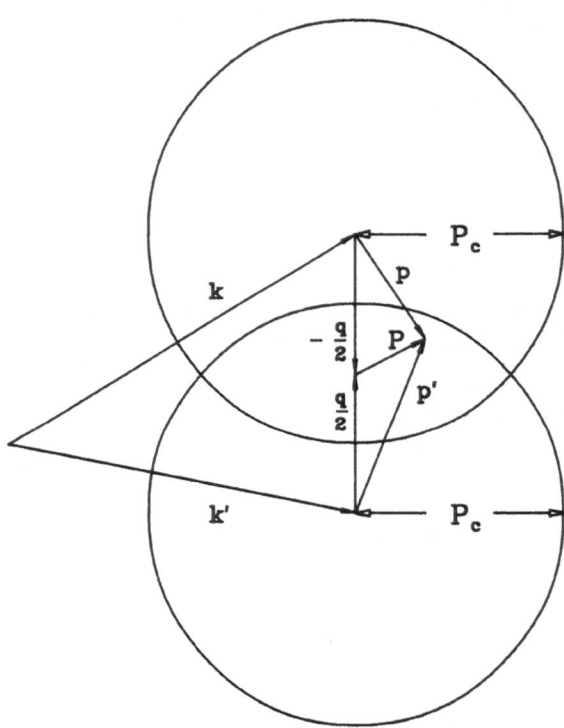

Fig. 1. Diagram of the overlap between the struck nucleon wavefunctions before and after the collision. $P_c$ represents the cut–off momentum described in the text.

In Fig.1 we show the geometry associated with the calculation of the full–folding model. In principle, the variation of $\vec{P}$ is unrestricted. However, the localized momentum distribution of the bound target nucleons [4] justifies limiting the integration over $\vec{P}$ to a relatively small volume. This is achieved by cutting–off the high Fourier components of the ground state wavefunctions at some momentum $P_c$ (typically of the order of $\sim 3 fm^{-1}$). The cut–off momentum defines an overlapping volume formed by the wavefunction of the struck nucleon before and after the collision. As observed from Fig.1, the overlap diminishes as the momentum transfer $\vec{q}$ increases. Thus, a reasonable choice for $P_c$ reduces considerably the time required to compute the optical potential matrix elements.

A comment is necessary regarding the use of the free $t$–matrix in the full–folding context. The model requires, in principle, the knowledge of the $t$–matrix at all NN center of mass energies below the projectile energy $E$ (Eq.(11) or (13)). This imposes a restriction in the sense that the calculations are only meaningful when the $t$–matrix elements required satisfy the condition $z_\alpha > 0$. We have studied this limiting situation and concluded that the full–folding model results are very reliable for $E \sim 300$ MeV and greater. [4]

## APPLICATIONS

In this section we show different results obtained for proton scattering from $^{40}$Ca at intermediate energies in the full–folding framework. We first calculate the fully off–shell free NN $t$–matrix ($t_0$) from Eq.(12), then the optical potential given by Eq.(13) and finally solve the Lippmann–Schwinger equation for the NA scattering amplitude. We have used the Paris NN potential [10] throughout unless stated otherwise.

First, we illustrate the role of off–energy–shell effects implicit in the full–folding model by comparing the full–folding results to those obtained in the simpler off–shell [5,11,12] and on–shell [7] $t\rho$ approximations to the optical potential, respectively.

In Fig.2 we show the calculated observables for p + $^{40}$Ca scattering at 200 and 300 MeV as a function of the momentum transfer $q$. The most pronounced differences between both $t\rho$ and the full–folding results occur at the lower energies. Thus we may conclude that neither factorized $t\rho$ approximation considered here accounts properly for all the off–shell effects included in the full–folding model. The agreement between the different approximations improves as the energy increases. Although this feature suggests that, in particular, the off–shell $t\rho$ model could be a suitable approximation to the full–folding model at higher energies, its validity needs to be assessed in the context of a broader class of NN forces.

In Fig.3 we show the measured and calculated observables for p + $^{40}$Ca elastic scattering for incident proton energies of 200, 300, 400 and 500 MeV. The data at 200 MeV were taken from Ref.(13). The cross section and analyzing power data at 300, 400 and 500 MeV were taken from Ref.(14). The spin–rotation data shown at 300 MeV were taken at 320 MeV and are from Ref.(15). The Q data at 500 MeV are from Ref.(16). The full–folding results are represented by the solid curves. The on–shell $t\rho$ results (dotted curves) are included as a reference.

The full–folding model provides a description of the data which is clearly superior to the on–shell $t\rho$ approximation at energies below 400 MeV. The deficiencies shown by the $t\rho$ model in describing the data are notably accounted for by the full–folding optical potential. In particular, the spin observables for p + $^{40}$Ca calculated with the full–folding model agree well with the data below $\approx 400$ MeV. The 500 MeV results should be regarded as exploratory; the clear deterioration of the full–folding results

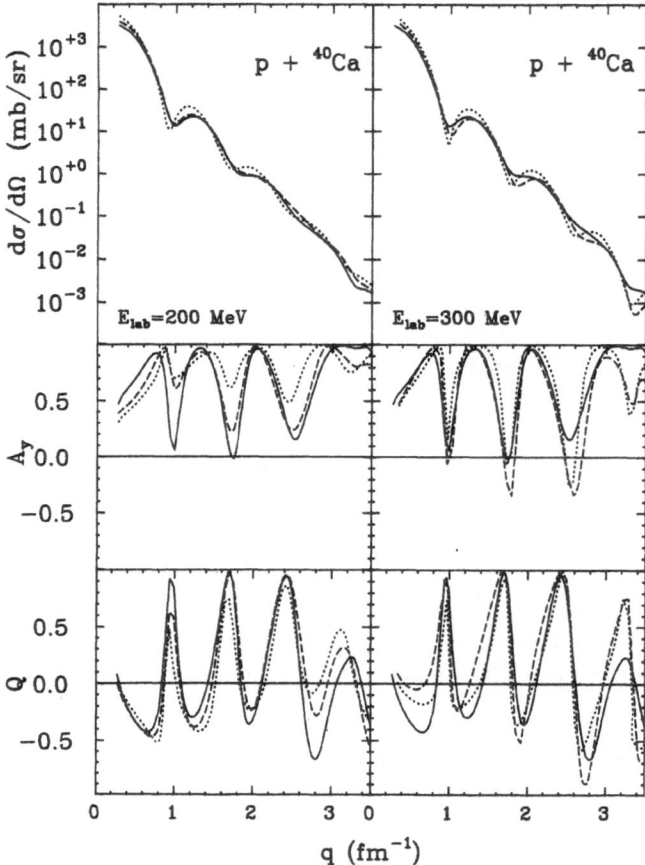

Fig. 2. Calculated elastic scattering observables for p + $^{40}$Ca at 200 and 300 MeV. Solid, dashed and dotted curves represent full–folding, off–shell $t\rho$ and on–shell $t\rho$ results respectively.

at this energy implies, among other factors, the need for an improved NN potential model above $\approx$ 400 MeV.

Similar behavior is observed for p + $^{16}$O elastic scattering when the full–folding and the on–shell $t\rho$ results are compared to the data. In general, we can conclude that the inclusion of off–shell effects together with the Fermi motion as prescribed by the full–folding model lead to a very reasonable description of the data.

The role that alternative NN potentials play in determining the off–shell properties of the free $t$–matrix may affect the full–folding results. We have investigated four different NN potentials which describe NN scattering data below $\approx$ 350 MeV. They are the Paris potential, [10] the momentum–space Bonn potential (OBEPQ), [17] the Hamada–Johnston (HJ) potential [18] and the Melbourne (M) potential. [19]

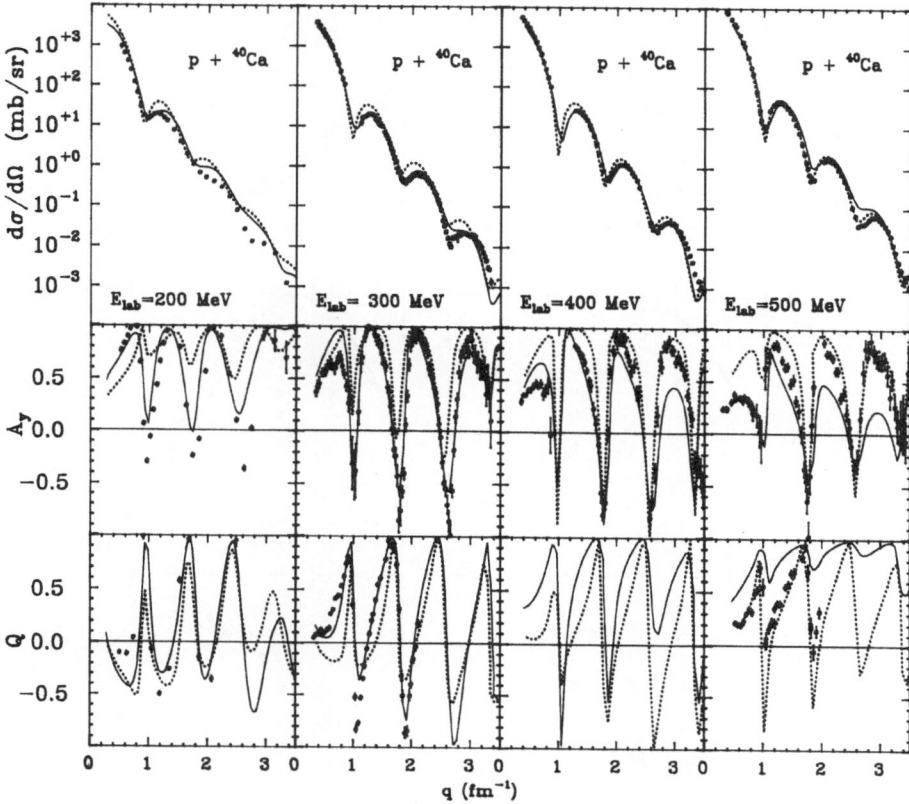

Fig. 3. Calculated and measured elastic scattering observables for p + $^{40}$Ca at 200, 300, 400 and 500 MeV. The data are from Refs.(13-16). Solid curves represent full-folding results. Dotted curves represent on-shell $t\rho$ results.

In order to illustrate the level of agreement between the different models in describing the free NN interaction we show in Fig.4 the scattering observables for NN scattering in the $\Delta T = 0$ "channel" at 210 and 325 MeV in the lab system. The same scattering "observables" were also calculated using Arndt's phase–shifts. [20] We observe that the potentials are distinguishable on–shell. This precludes making a detailed assessment of the different potentials based exclusively on their off–shell behavior. However, it is of interest to investigate the systematics of the full–folding model when considering alternative descriptions of the NN interaction based on realistic NN potential models. As a typical example of the results obtained, we show in Fig.5 the scattering observables at 300 MeV calculated from the Paris, Bonn, HJ and M potentials respectively. We notice differences between the observables predicted by the four underlying NN potential models considered.

In the case of the two meson–exchange based (Paris and Bonn) potentials the overall agreement with the data is similarly good, with the Bonn potential giving a slightly more diffractive structure in the cross section but a better description of the Q and $A_y$ observables than the Paris potential. A notably good agreement with

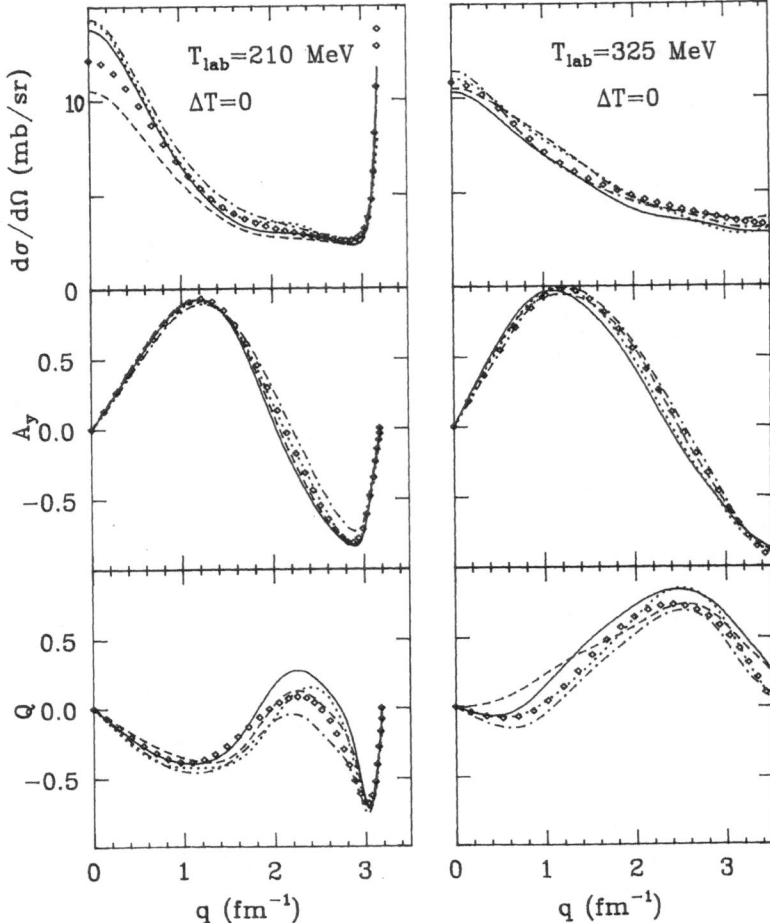

Fig. 4. Calculated cross section, analyzing power and spin rotation parameter for NN scattering for $\Delta T = 0$ from different NN potential models and phase shift analyses at 210 and 325 MeV lab energy. Solid curves correspond to the Paris potential, dashed curves to the Bonn potential, dotted curves to the HJ potential and dash–dotted curves to the M potential. The $\diamond$–symbol are the results obtained using Arndt's phase–shifts.

the data is also obtained with the HJ potential. This was not expected considering the presence of an infinitely strong repulsive core at short distances. However, the HJ potential has its long-range part constrained by the one–pion–exchange process and, in that sense, it belongs to the same class of potentials as the Paris and Bonn models. In contrast, it is interesting to note the difficulties that the M potential has describing the data, particularly the spin observables. We can conclude that the off–shell behavior of the NN $t$–matrix generated by the empirical Melbourne potential is unsatisfactory. More generally, our results emphasize that an adequate description of the on–shell NN interaction is not sufficient for describing a many-nucleon process such as NA scattering.

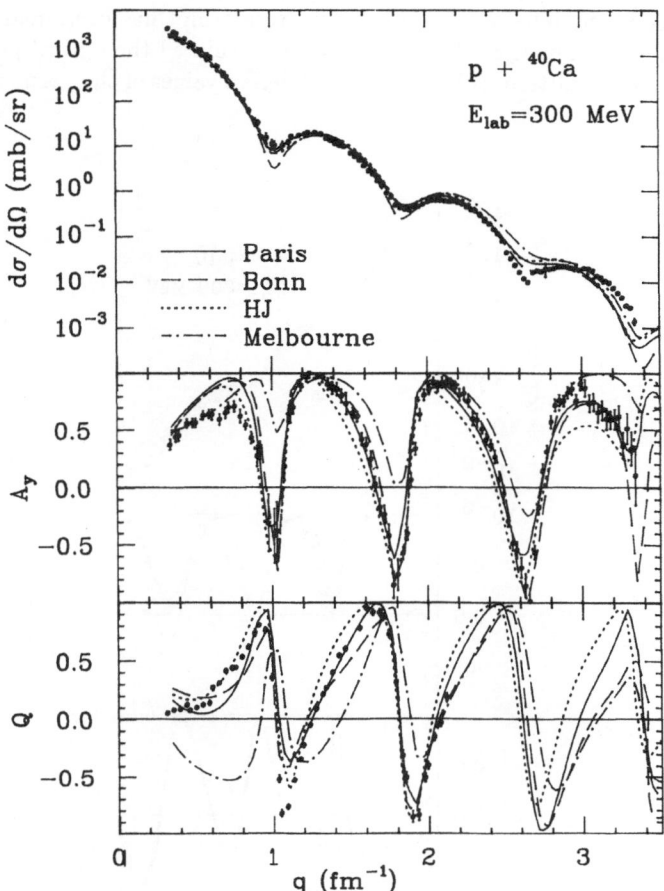

Fig. 5. Calculated elastic scattering observables for p + $^{40}$Ca at 300 MeV associated with the Paris, Bonn, HJ and M potentials.

One of the main limitations to the full–folding model results presented previously is the use of the free NN $t$–matrix as the effective interaction. Indeed, the agreement of the full–folding model results with the data in the absence of medium corrections, even at 200 MeV, seems questionable. In order to estimate the effect of Pauli blocking and self–energy corrections in the single–particle spectrum we have calculated the $g$–matrix effective interaction (Eqs.(6), (8–9)) in nuclear matter and have used this effective force in the full–folding framework. No attempt has been made to make a local density approximation. Instead, we have calculated the optical potential and the corresponding scattering observables at different values of the Fermi momentum, $k_F$.

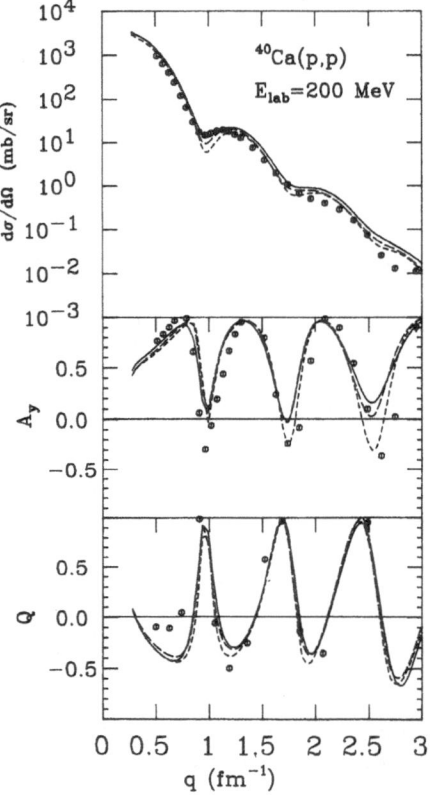

Fig. 6. Calculated elastic scattering observables for p + $^{40}$Ca at 200 MeV. The effective force is a $g$–matrix obtained from the Paris potential. The solid curve corresponds to $k_F = 0$, the long dashed curve to $k_F = 1.0\,fm^{-1}$ and the short dashed curve to $k_F = 1.36\,fm^{-1}$ respectively.

In Fig.6 we show the results for p + $^{40}$Ca elastic scattering at 200 MeV for values of $k_F$ between 0 (free $t$– matrix) and 1.36 $fm^{-1}$. Overall, these *prelimanary* results suggest that the density (or $k_F$) dependence is weak, the larger variations being observed for the analyzing power at momentum transfer $q > 2fm^{-1}$. This is an interesting feature of the full–folding model which differs from the results obtained by the standard folding model in a local density approximation. [2,3] At present, this difference is not fully understood.

## SUMMARY

The general full–folding structure of the non–relativistic microscopic optical potential has been developed and applied to proton elastic scattering at intermediate energies. Its main characteristic is that, for each NN collision, the complete dynamical variation of the (off–shell) NN $t$–matrix and the ground–state mixed density prescribed by the kinematics of the collision are included. The full–folding results show striking improvements in the description of nucleon–nucleus scattering observables relative to simpler models such as the on–shell $t\rho$ approximation. Furthermore, the full–folding model is able to discriminate among different classes of alternative NN potentials via their off–shell behavior.

Medium corrections associated with Pauli blocking in intermediate states and the inclusion of self–energy effects have been estimated to be small in the full–folding framework. This *preliminary* result is encouraging since it simplifies the understanding of medium effects in the calculation of the optical potential.

Although the success of the full–folding optical potential in reproducing the experimental data over a wider range of energies remains to be established, its implementation represents at present the most complete attempt towards obtaining a detailed microscopic, non–relativistic, description of NA elastic scattering.

## ACKNOWLEDGMENTS

We acknowledge Kanzo Nakayama for many helpful discussions and for providing the Bonn $t$–matrix elements. This work was supported in part by NSF Grant PHY–8903856. F.A.B. acknowledges partial support from FONDECYT Grant 1239–90. A grant for computing time provided by the University of Georgia is also appreciated.

## REFERENCES

1. J. P. Jeukenne, A. Lejeune and C. Mahaux, Phys. Rep. **25C**, 83(1976) ; Phys. Rev. **C16**, 80(1977).
2. F. A. Brieva and J. R. Rook, Nucl. Phys. **A291**, 317(1977); **A297**, 206(1978); **A307**, 493(1978).
3. L. Rikus and H. V. von Geramb, Nucl. Phys. **A426**, 496(1984).
4. H. F. Arellano, F. A. Brieva and W. G. Love, Phys. Rev. Lett. **63**, 605(1989); Phys. Rev. **C41**, 2188(1990); **C42**, 652(1990); **C43**, (1991) in press.
5. Ch. Elster and P. C. Tandy, Phys. Rev. **C40**, 881(1989); Ch. Elster, T. Cheon, E. Redish and P. C. Tandy, Phys. Rev. **C41**, 814( 1990).
6. R. Crespo, R. C. Johnson and J. A. Tostevin, Phys. Rev. **C41**, 2257( 1990).
7. K. M. Watson, Phys. Rev. **89**, 575(1953); G. Takeda and K. M. Watson, Phys. Rev **97**, 1336(1955); A. K. Kerman, H. McManus and R. M. Thaler, Ann. Phys. (N.Y.) **8**, 551(1959);

J. S. Bell and E. J. Squires, Phys. Rev. Lett. **3**, 96(1959);
H. Feshbach, Ann. Phys. (N.Y.) **5**, 357(1958); **19**, 287(1962).

8. L. Ray and G. W. Hoffmann, Phys. Rev. **C31**, 538(1985).

9. F. A. Brieva and M. A. Nagarajan, Nucl. Phys. **A452**, 221(1986);
N. Vinh-Mau and A.Bouyssy, Nucl. Phys. **A257**, 189(1978).

10. M. Lacombe, B. Loiseau, J. M. Richard, R. Vinh Mau, J. Côté, P. Pirès and R. de Tourreil, Phys. Rev. **C21**, 861(1980).

11. A. Picklesimer, P. C. Tandy, R. M. Thaler and D. H. Wolfe,
Phys. Rev. **C30**, 1861(1984).

12. H. F. Arellano, Ph. D. dissertation, University of Georgia (1990).

13. E. J. Stephenson, J. Phys. Jpn.(Suppl.) **55**, 316(1985).

14. D. A. Hutcheon *et al.*, Nucl. Phys. **A483**, 429(1988);
P. Schwandt, private communication.

15. E. Bleszynski *et al.*, Phys. Rev. **C37**, 1527(1988).

16. A. Rahbar *et al.*, Phys. Rev. Lett. **47**, 1811(1981).

17. R. Machleidt, K. Holinde and Ch. Elster, Phys. Rep. **149**, 1(1987);
R. Machleidt, in Advances in Nuclear Physics,
edited by J. Negele and E. Vogt (Plenum, N.Y., 1989), Vol. 19.

18. T. Hamada and I. D. Johnston, Nucl. Phys. **34**, 382(1962).

19. K. Amos, L. Berge, F. A. Brieva, A. Katsogiannis, L. Petris and L. Rikus, Phys. Rev. **C37**, 934(1988).

20. R. A. Arndt *et al.*, Phys. Rev. **D28**, 97(1983);
R. A. Arndt and L. D. Roper, unpublished.

# SPIN DEPENDENT EFFECTIVE INTERACTION FOR MEDIUM ENERGY NUCLEON-NUCLEUS SCATTERING

Lanny Ray

Department of Physics
The University of Texas
Austin, Texas 78712 USA

## INTRODUCTION

The study of nucleon-nucleus scattering at medium energies is one of the fundamental subjects in nuclear physics. As the incident nucleon energy becomes much greater than the Fermi energy of the nucleus, the interpretation of the scattering data in terms of the effective nucleon-nucleon (NN) interaction and nuclear structure should become more straightforward. Once the scattering process is well understood, analysis of the data might reveal new elements of nuclear structure and reaction dynamics.

During the 1980's interest in nucleon-nucleus scattering increased after it was demonstrated that theoretical descriptions of proton-nucleus (pA) elastic scattering data, based on the Dirac equation, were much more successful than the traditional nonrelativistic (NR) multiple scattering theories (as applied at the time).[1,2] The first relativistic nucleon-nucleus scattering model was the relativistic impulse approximation (RIA) - Dirac equation calculation. A later model, derived from covariant meson exchange theory (the IA2), has also been developed.[3] Although the relativistic models utilize a Lorentz invariant NN effective interaction, they lack a fundamental underpinning based on a relativistic field theory of fundamental constituents (*i.e.,* quarks and gluons). Such a description may eventually be required for the relativistic approaches, which depend critically on virtual $N\bar{N}$ pair propagation, which in turn, depends on very short distance scales.[4,5]

Comparisons between the relativistic and NR models have generally been made using only the lowest order term in the optical potentials of both theories.[1,2,6,7] At the time the RIA was introduced it was thought that the lowest order application of NR multiple scattering theory adequately represented the full calculation. Recent NR calculations have shown however, that this is not the case; contributions beyond the simplest version of the theory are important. These include off-shell, full-folding and medium effects.[8-12] Each of these corrections improves the NR model descriptions of the data; the overall success of NR predictions is now much better than it was a few years ago. For the NR approaches an essentially "complete" optical potential calculation will soon be available so that the success or failure of the traditional NR approach can be evaluated more fairly.

The IA2 relativistic model,[3] based on covariant meson exchange theory, provides a better overall description of proton-nucleus scattering data than the original RIA model, particularly for incident proton energies below 400 MeV. The IA2 and RIA descriptions of the 800 MeV data are very similar, however the IA2 fit to the 500 MeV data is inferior compared to the almost perfect fits obtained by the RIA.[7]

The validity of Lorentz invariance is not at issue in the comparison of NR and relativistic model predictions with each other and with data. Rather, the relevant question is what level are relativistic effects (*e.g.* virtual $N\bar{N}$ pairs, vacuum polarization, etc.) manifest in nuclear phenomenon. Two complementary approaches appear to be reasonable for approaching this subject. The first is to attempt to develope an effective Lagrangian for nuclear phenomenon which is derived from QCD and depends on effective nucleonic and mesonic degrees of freedom.[13,14] Effective interactions derived in these approximate field theories can be compared with phenomenological models. The second approach is to calculate the nonrelativistic many-body theory "completely" and then look to see if sizable discrepancies remain in the comparisons with experiment which might indicate the need for additional dynamics in the theory. Proton-nucleus scattering at medium energies provides an opportune area to study this problem.

In this talk I address the latter approach. After briefly reviewing the NR multiple scattering formalism, the recent improvements in the application of NR models will be discussed. These include the full-folding and off-shell calculations, medium corrections, correlations, and the electromagnetic spin-orbit (EMSO) potential.[15] The development of the NR density dependent (DD) effective interaction for energies up to 1 GeV will be discussed in some detail. Examples of calculations using the best available relativistic and NR models will be shown and compared with data. Results for even-even nuclei and the recent $\vec{p} + {}^{13}\vec{C}$ elastic scattering experiment are presented.

WATSON MULTIPLE SCATTERING FORMALISM

The NR calculations discussed here are based on the multiple scattering optical potential formalism of Watson[16] which provides a solution of the many-body Schrödinger equation for the projectile-nucleus system in terms of an expansion in quasi–two-body operators. The many-body Schrödinger equation is

$$[H_o + H_A + \sum_{i=1}^{A} v_i]\Psi = E\Psi, \qquad (1)$$

where $H_o$ is the kinetic energy operator acting on the relative projectile-nucleus coordinates, $H_A$ is the target nucleus Hamiltonian, $v_i$ is the two-body projectile - $i^{th}$ target nucleon constituent interaction potential (assumed to be the same in the nuclear medium as in free space where no three-body or higher $n$-body forces are considered), $A$ is the number of target nucleons, E is the energy parameter for the system in the pA center-of-momentum (COM) reference frame, and $\Psi$ is the total projectile-nucleus wave function.

The formal solution is given by

$$T = \sum_{i=1}^{A} v_i + \sum_{i=1}^{A} v_i G(E)T, \qquad (2)$$

where $T$ is the pA $t$ matrix and

$$G(E) = (E - H_o - H_A + i\delta)^{-1} \tag{3}$$

is the (A+1)-body propagator. The quasi–two-body $t$ matrix of the Watson approach is

$$t_i^W = v_i + v_i G(E) \mathcal{Q} t_i^W \tag{4}$$

where $\mathcal{Q}$ is the non-elastic channel projection operator. Eliminating dependence on $v_i$ in Eq. (2) in favor of $t_i^W$ and projecting the elastic scattering channel (using projection operator $\mathcal{P}$) yields

$$\mathcal{P}T\mathcal{P} = \mathcal{P}U^W\mathcal{P}(1 + G(E)\mathcal{P}T\mathcal{P}), \tag{5}$$

where $\mathcal{P}U^W\mathcal{P}$ is the Watson optical potential for elastic scattering given by

$$\mathcal{P}U^W\mathcal{P} = \sum_i \mathcal{P}t_i^W\mathcal{P} + A(A-1)\left[\frac{1}{A(A-1)}\sum_{i \neq j}\mathcal{P}t_i^W\frac{1}{\tilde{\alpha}_W}t_j^W\mathcal{P}\right.$$
$$\left. - \frac{1}{A}\sum_i\mathcal{P}t_i^W\mathcal{P}\frac{1}{\tilde{\alpha}_W}\frac{1}{A}\sum_j\mathcal{P}t_j^W\mathcal{P}\right], \tag{6}$$

where

$$\tilde{\alpha}_W \equiv G(E)^{-1} - \frac{A-1}{A}\sum_k \mathcal{Q}t_k^W\mathcal{Q}. \tag{7}$$

The first term on the right hand side (RHS) of the equation is the first-order optical potential; the second set of terms in square brackets is the second-order potential which is proportional to correlations in the nucleus. The utility of this particular organization of the many-body problem is that accurate approximations can be made at this point.

In general, even the first-order optical potential in Eq. (6) is very complicated. Contained in this term are medium corrections,[11,12,17] off-shell dynamics,[18] and dependence on the full ground state density matrix of the target.[8-10] A useful procedure is to expand the Watson $t$ matrix in terms of the free NN $t$ matrix, $t_i$, defined by

$$t_i = v_i + v_i g(\varepsilon)t_i, \tag{8}$$

where

$$g(\varepsilon) = (\varepsilon - h_o + i\delta)^{-1}, \tag{9}$$

$\varepsilon$ is the free two-body energy, and $h_o$ is the kinetic energy operator acting on the relative two-body coordinate. This expansion is

$$t_i^W = t_i + t_i(G(E)\mathcal{Q} - g(\varepsilon))t_i^W. \tag{10}$$

This expression for $t_i^W$ can be further expanded by separating out the static Coulomb contribution. Using the two-potential formula[19] the Watson quasi–two-body $t$-matrix may be written as

$$t_i^W = t_{i,C}^W + t_{i,CN,M}^W \tag{11}$$

where $t_{i,C}^W$ is the purely Coulombic scattering $t$-matrix given by

$$t_{i,C}^W = v_i^C + v_i^C G(E)\mathcal{Q}t_{i,C}^W \tag{12}$$

and $t_{i,CN,M}^W$ is the Coulomb distorted nuclear $t$-matrix including also magnetic scattering contributions. The pure nuclear Watson $t$-matrix, $t_{i,N}^W$, is

$$t_{i,N}^W = v_i^N + v_i^N G(E)\mathcal{Q}t_{i,N}^W, \tag{13}$$

where $v_i^N$ represents only the nuclear part of $v_i$. Similar expressions apply for the free projectile-nucleon $t$-matrices, $t_i$, $t_{i,C}$, $t_{i,N}$, and $t_{i,CN,M}$. Using Eqs. (8) - (13), $t_i^W$ is finally approximated by

$$t_i^W \cong v_i^C + t_{i,CN,M} + t_{i,N}(G(E)\mathcal{Q} - g(\varepsilon))t_{i,N}^W \tag{14}$$

where virtual Coulomb excitation[20] and Coulomb contributions to the $(G\mathcal{Q} - g)$ term were neglected. The first term in Eq. (14) leads to the usual Coulomb part of the optical potential.

## CALCULATION OF THE FREE $t$ MATRIX CONTRIBUTION, $\sum_i \mathcal{P}t_{i,CN,M}\mathcal{P}$

The leading term, $t_{i,CN,M}$, in the expansion of the Watson $t$ matrix should provide the dominant part of the optical potential and its contribution should be calculated as accurately as possible. The free NN energy parameter, $\varepsilon$, is arbitrary but should be chosen to minimize the $(G\mathcal{Q} - g)$ correction; thus the free $t$ matrix is evaluated using kinematics similar to that in $t_i^W$.

The most complete calculations of this contribution to the optical potential have been done by three independent groups,[8-10] resulting in the so-called full-folding form of the optical potential. The following simplifications are necessary to obtain this form of the optical potential: (1) the independent particle model for the nuclear wave function is assumed, (2) $1/A$ recoil kinematic corrections are neglected, and (3) an average, state independent energy for the struck, target nucleon is used. This full-folding form of the optical potential is given by (where $\mathcal{P}U^W\mathcal{P} \equiv \mathcal{U}_{oo}^W$)

$$\mathcal{U}_{oo}^{W(1)}(\vec{k}', \vec{k})_{FF} = \frac{1}{(2\pi)^3} \sum_\alpha \int d^3P u_\alpha^\dagger(\vec{p}')\eta\langle\frac{1}{2}(\vec{k}' - \vec{p}')|t_{CN,M}(\varepsilon)|\frac{1}{2}(\vec{k} - \vec{p})\rangle u_\alpha(\vec{p}), \tag{15}$$

where

$$\vec{p} = \vec{P} - \vec{q}/2, \qquad \vec{p}' = \vec{P} + \vec{q}/2, \qquad \vec{q} = \vec{k} - \vec{k}', \tag{16}$$

and

$$\varepsilon = T_{pA} + \varepsilon_N - \frac{\hbar^2(\vec{P} + \frac{1}{2}(\vec{k} + \vec{k}'))^2}{4m}. \tag{17}$$

In Eq. (15) $u_\alpha$ is the single particle eigenstate, $\langle t_{CN,M}\rangle$ is the fully off-shell, NN $t$ matrix for initial and final relative momentum $\frac{1}{2}(\vec{k} - \vec{p})$ and $\frac{1}{2}(\vec{k}' - \vec{p}')$, respectively, and $\eta$ is the Møller factor[18] which converts the NN $t$ matrix in the NN COM system to the pA COM system. In Eq. (17) $T_{pA}$ is the pA COM kinetic energy and $\varepsilon_N$ is the average single particle energy of the struck nucleon.

The nuclear structure in the integrand of Eq. (15) corresponds to the one-body density matrix. This function is strongly peaked at $P = 0$ (Ref. 21). This suggests that it should be reasonable to further approximate the optical potential by evaluating $\langle t_{CN,M}\rangle$ at $P = 0$ and factoring it out of the integral, resulting in the "off-shell $t\rho$" form given by

$$\mathcal{U}_{oo}^{W(1)}(\vec{k}', \vec{k})_{OFF} = \sum_{N=p,n} \tilde{\rho}_N(\vec{q})\eta\langle\frac{1}{2}(\vec{k}' - \vec{p}')|t_{CN,M}(\varepsilon)|\frac{1}{2}(\vec{k} - \vec{p})\rangle_{P=0}, \tag{18}$$

where $\tilde{\rho}_n(\vec{q})$ and $\tilde{\rho}_p(\vec{q})$ are the neutron and proton density form factors, respectively. Finally, the smooth dependence of $\langle t_{CN,M}\rangle$ on total momentum suggests a further simplification of the optical potential where the off-shell $t$ matrix, $\langle\vec{\kappa}'|t|\vec{\kappa}\rangle$, is evaluated using just the on-shell values according to,

$$\langle \vec{\kappa}' | t | \vec{\kappa} \rangle \cong t_{\text{on-shell}}(\vec{q} = \vec{\kappa} - \vec{\kappa}'). \tag{19}$$

Using this final approximation the simple, "on-shell $t\rho$" version of the optical potential is obtained where

$$\mathcal{U}_{oo}^{W(1)}(\vec{k}', \vec{k}) \cong \mathcal{U}_{oo}^{W(1)}(q)_{ON} = \sum_{N=p,n} \tilde{\rho}_N(q) t_{CN,M}(q)_{\text{on-shell}}. \tag{20}$$

Calculations representing all three of these forms of the first-order optical potential are shown in Fig. 8 of Ref. 8 for $p+^{40}$Ca at 200 and 300 MeV. These calculations show that the changes in the predictions due to the effects of full-folding are comparable in importance to that produced by including the off-shell dependence of the $t$ matrix.

## MEDIUM MODIFICATIONS

The basic premise of the multiple scattering optical potential description of proton-nucleus scattering and reactions is that the bulk of the pA interaction can be accounted for by a summation of the free, proton-nucleon $t$ matrix over the constituent nucleons of the target.[22] This is represented by the leading term in the optical potential in Eq. (6) and the first two leading terms in the expansion of the Watson operator, $t_i^W$, in Eq. (14). At high energies ($\sim$1 GeV), calculations based solely on this lowest order optical potential provide a reasonable, qualitative description of proton-nucleus elastic scattering data.[6,23]

Using just the free, two-body $t$ matrix in place of the full Watson operator, $t_i^W$, is referred to as the "impulse approximation" (IA). The basic requirement for the IA to be valid is that effects of the nuclear medium on the propagation of the two-body system in intermediate scattering states are negligible. For many years it was thought that the IA was valid throughout the intermediate energy range above a few hundred MeV. However, analyses of proton elastic scattering data at 400 and 500 MeV demonstrated the failure of nonrelativistic IA models at these energies.[24,25] Much work has since been done which shows the significant role played by the nuclear medium in modifying the NN effective interaction for energies from tens to many hundreds of MeV.[11,12]

In this section I discuss calculations of the contribution of the third term on the RHS of Eq. (14) to the pA optical potential. This is

$$\sum_i \mathcal{P} t_{i,N} (G(E)\mathcal{Q} - g(\varepsilon)) t_{i,N}^W \mathcal{P} = \sum_i \mathcal{P} (t_{i,N}^W - t_{i,N}) \mathcal{P}, \tag{21}$$

where a model estimate of the difference, $\sum_i \mathcal{P}(t_{i,N}^W - t_{i,N})\mathcal{P}$, is calculated. The basic approach involves replacing the many-body propagator $G(E)\mathcal{Q}$ in the definition of $t_i^W$ with that corresponding to two interacting nucleons in infinite nuclear matter with Fermi momentum $k_F$. In this limit the projection operator $\mathcal{Q}$ becomes the usual Pauli blocking factor.[17] In the energy denominator for the Watson $t$ matrix, binding potentials are applied only to the struck nucleon, not the projectile. This is the principal difference between the Watson $t$ matrix and the Brueckner $g$ matrix which has often been applied to scattering problems at lower energies.[12]

In order to provide estimates of medium corrections up to 1 GeV, a NN interaction model must be used which accounts for the large NN inelasticities. The NN coupled channels isobar model of Ref. 26 generates inelastic scattering by explicitly including $N\Delta$ and $NN^*$ channels. This model assumes the $\pi$, $2\pi$, $\rho$, $\eta$, and $\omega$ meson exchange potentials of Lomon and Feshbach[27] for $r > 0.735$ fm for the NN-NN channels. Phenomenological core and short range potentials are added to the meson exchange terms, while the coupling to isobar channels is included via one-pion exchange potentials (OPEP) and phenomenological interactions of reasonable range, assuming, in general, the coupling schemes of Lomon.[28,29] The two-pion exchange component of the Lomon-Feshbach interaction accounts for virtual $N\bar{N}$ pairs and nonresonant $NN\pi$ contributions in the two-body system, so that double counting of isobar effects does not occur. The form of the potential and the parameters are given in Refs. 11 and 26. The model provides good fits to the NN phase shifts[30] for partial wave states with $J \leq 6$, for $T = 0$ and 1, and for energies from 0 - 1000 MeV.

The corrections to the optical potential due to medium modifications to the free, two-body propagator are estimated by replacing the full (A+1)-body propagator $G(E)\mathcal{Q}$ in the definition of $t_i^W$ with $\hat{G}$ defined by

$$G(E)\mathcal{Q} \cong \hat{G} \equiv \frac{\hat{Q}}{\hat{e}} \tag{22}$$

where $\hat{Q}$ and $\hat{e}$ are the Pauli blocking factor and energy denominator for two propagating nucleons in infinite nuclear matter. For NN channels $\hat{Q}$ is determined by requiring

$$\hat{Q}_{NN}(\vec{k}_1', \vec{k}_2', k_F) = \begin{cases} 1 & \text{if } |\vec{k}_1'| > k_F \text{ and } |\vec{k}_2'| > k_F \\ 0 & \text{otherwise,} \end{cases} \tag{23}$$

where $\vec{k}_1'$ and $\vec{k}_2'$ are the intermediate momenta of the two nucleons in the rest frame of the nuclear medium (laboratory) and $k_F$ is the Fermi momentum. For $N\Delta$ channels $\hat{Q}$ is

$$\hat{Q}_{N\Delta}(\vec{k}_N', \vec{k}_\Delta', k_F) = \begin{cases} 1 & \text{if } |\vec{k}_N'| > k_F \\ 0 & \text{if } |\vec{k}_N'| \leq k_F. \end{cases} \tag{24}$$

Angle averaged values of $\hat{Q}$ using relativistic kinematics were used in Eq. (22). The full expressions are given in Appendix A of Ref. 11. The energy denominator $\hat{e}$ is

$$\hat{e} = \frac{\hbar^2 k_1^2}{2m_1} + \frac{\hbar^2 k_2^2}{2m_2} + U(k_2) - \frac{\hbar^2 k_1'^2}{2m_1} - \frac{\hbar^2 k_2'^2}{2m_2} - U(k_2') + i\delta, \tag{25}$$

where $\vec{k}_1$ and $\vec{k}_2$ are the initial nucleon momenta and $U$ is the target nucleon binding potential. Complete expressions for $\hat{e}$ are given in Appendix B of Ref. 11.

The approximate $t$ matrix to be evaluated is

$$\hat{t}_i^W = v_i + v_i \frac{\hat{Q}}{\hat{e}} \hat{t}_i^W. \tag{26}$$

The calculation is facilitated by introducing a wave function $\hat{\psi}$, defined by (omitting the target nucleon label)

$$\hat{t}^W \hat{\phi} = v\hat{\psi}, \tag{27}$$

where $\hat{\phi}$ is a NN plane wave and $v$ is the nucleon-isobar coupled channels model. The wave function $\hat{\psi}$ was obtained by solving the integral equation,

$$\hat{\psi} = \hat{\phi} + \frac{\hat{Q}}{\hat{e}} v\hat{\psi}, \tag{28}$$

in coordinate space using standard partial wave expansion techniques.[11,17]

Matrix elements of $\hat{t}^W$ in the NN COM system were obtained from Eq. (27) where

$$\langle \hat{t}^W \rangle = \langle \hat{\phi}_{\vec{\kappa}'} | v | \hat{\psi}_{\vec{\kappa}} \rangle, \tag{29}$$

and subscripts $\vec{\kappa}$ and $\vec{\kappa}'$ denote the initial and final relative momentum vectors in the NN COM system, respectively. The medium corrections or *density dependence* are therefore generated by the quantity $\langle \hat{t}^W(k_F) - \hat{t}^W(k_F = 0) \rangle_{Model}$ which corresponds to our model estimate of $\mathcal{P}(t^W_{i,N} - t_{i,N})\mathcal{P}$. For the "on-shell $t\rho$" calculation the density dependence correction is added to the on-momentum-shell free NN $t$ matrix, $\langle t_{CN,M} \rangle$, as in Eq. (14). Each of the $t$ matrix elements in Eq. (14) were computed using the full Lorentz transformation (Møller factor and Wigner rotation matrix[31]) of the NN $t$ matrix from the NN COM to the pA COM.[6] The final, on-momentum-shell $t$ matrix used in the calculations discussed here and in Ref. 11 is given by

$$\langle \hat{t}^W(k_F) \rangle_{Final} \equiv \langle t_{CN,M} \rangle + \langle \hat{t}^W(k_F) - \hat{t}^W(k_F = 0) \rangle_{Model}. \tag{30}$$

The binding potential $U$ in the energy denominator (Eq. (25)) is assumed to have the velocity dependent form $U(k'^2_2) = A + Bk'^2_2$. Parameters $A$ and $B$ for NN channels[11] were chosen to fit the interior well depths of the Schrödinger equivalent real, central optical potentials from Dirac phenomenological fits[32] to $p+{}^{40}Ca$ elastic scattering data over the appropriate range of momentum.[11] The binding potentials for the N$\Delta$ and NN* channels were assumed to be 0.6 times that of the NN channel. The reduction in strength is based on the position of the quasifree $\Delta$ peak at low momentum transfer in $(e,e')$ measurements[33] and should be regarded as only a crude estimate. Sensitivity to the $\Delta$ and N* nuclear binding potentials is minimal.[11] The intermediate binding potentials were assumed to be proportional to the nuclear matter density or to $k^3_F$.

The local form for the first-order, density dependent optical potential was calculated assuming the local density approximation according to

$$\mathcal{U}^{W(1)}_{oo}(r) = (2\pi)^{-3} \sum_{N=p,n} \left[ \int_0^\infty 4\pi q^2 dq t^a_{pN}(q, \rho(r)) \tilde{\rho}_N(q) j_0(qr) \right.$$
$$\left. + \frac{i}{r}\frac{\partial}{\partial r} \int_0^\infty 4\pi q^2 dq \tilde{t}^{c_1}_{pN}(q, \rho(r)) \tilde{\rho}_N(q) j_0(qr) \vec{\sigma} \cdot \vec{\ell} \right], \tag{31}$$

where

$$\tilde{t}^{c_1}_{pN}(q, \rho(r)) = \frac{t^{c_1}_{pN}(q, \rho(r))}{k^2_{pA} \sin(\theta_N)}, \tag{32}$$

subscript $N$ refers to target neutrons or protons, and $\theta_N$ is defined by $q = 2k_{pA}\sin(\theta_N/2)$. The quantities $t^a_{pN}$ and $t^{c_1}_{pN}$ are obtained from Eq. (30).

Medium modifications to the isoscalar spin independent ($t_0$) and spin-orbit ($t^{LS}_0$) $t$ matrices in the NN COM at 500, 650 and 800 MeV are shown in Fig. 1, where the solid, dotted and dashed lines correspond to $k_F = 1.4$, 0.7 and 0.0 fm$^{-1}$, respectively (from Ref. 11). The $t$ matrices represented in Fig. 1 were computed directly from the model and do not correspond to the scaled quantities in Eq. (30). The continued importance of medium corrections at 800 MeV is somewhat unexpected.

In Figs. 2 and 3 calculations from Ref. 11 for $p+{}^{40}Ca$ with (solid curves) and without (dashed curves) medium corrections are compared with each other and with

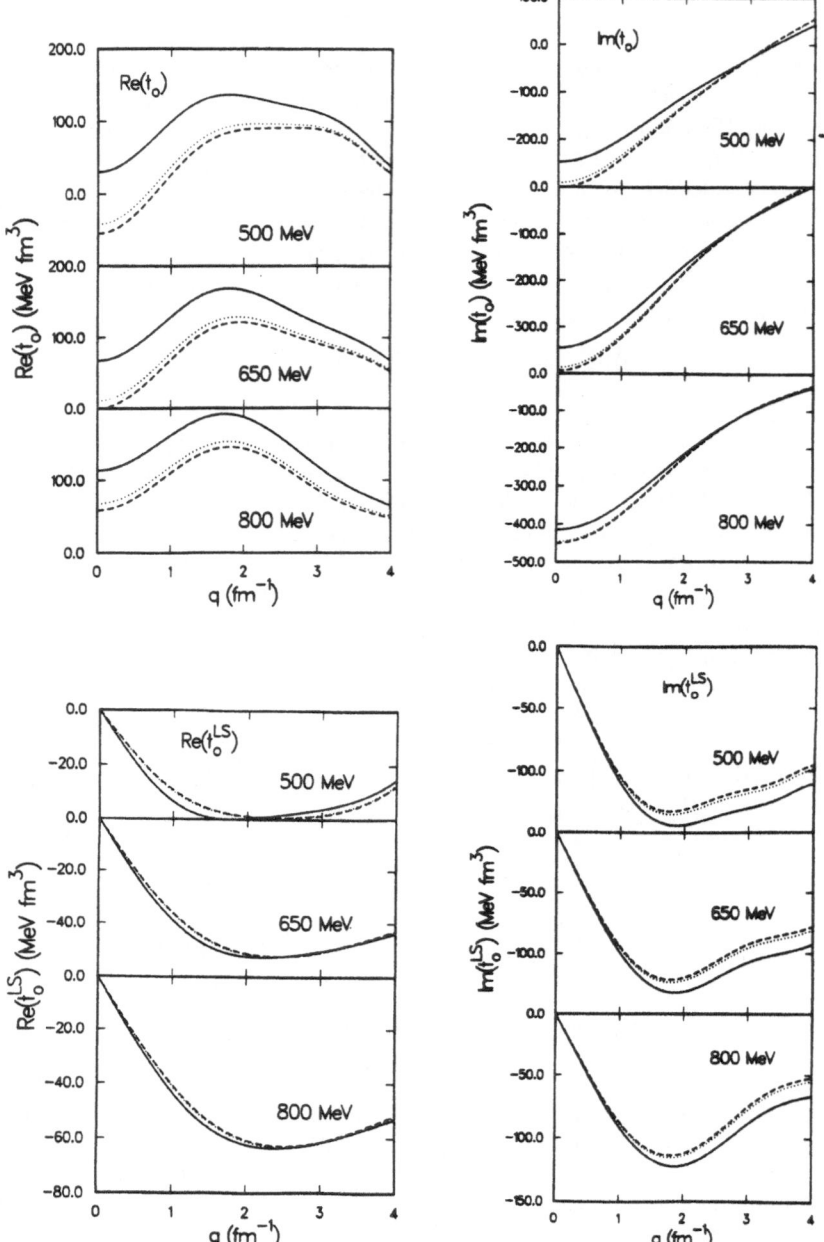

Figure 1. Density dependence of the real and imaginary parts of the spin-independent and spin-orbit NN $t$ matrices in the NN COM system. The solid, dotted, and dashed curves correspond to $k_F = 1.4$, 0.7, and 0.0 fm$^{-1}$, respectively.

data[25,34-40] for complete sets of elastic scattering observables including the differential cross section, analyzing power ($A_y$) and spin rotation (Q). The calculations use the "on-shell $t\rho$" form of the optical potential, assume the SP82 Arndt[30] amplitudes for $\langle t_{CN,M} \rangle$, include correlations (see the following), and use fixed nuclear densities obtained from electron scattering results.[41]

At 300, 400 and 500 MeV the differential cross section predictions including medium modifications do not display diffractive minima which are as deep as that predicted by calculations based on the impulse approximation and are in overall better agreement with the data. The general shapes of the experimental angular distributions are reproduced fairly well by the predictions which include medium modifications. At 300, 400 and 500 MeV quantitative descriptions of the analyzing power data are obtained with the density dependent model except at forward angles. Improved descriptions also result at 650 and 800 MeV although the effects due to medium modifications are smaller than at 500 MeV and below. Overall, calculations including medium corrections produce significantly better fits to the spin rotation data for each case. At 400 MeV the spin rotation angle $\beta$ is shown where $\sin\beta = Q/\sqrt{1 - A_y^2}$. The electromagnetic spin-orbit potential will be added to these calculations in the following sections where further improvements are shown to result.

At 300 MeV the changes in the observables due to Pauli blocking and binding energy corrections are similar in both magnitude and direction. At 500 MeV the effects due to binding energy are larger than those due to Pauli blocking while at 800 MeV almost all of the density dependent effects are due to the binding energy correction. Therefore the Pauli blocking contributions quickly diminish at higher energies as expected whereas significant binding energy effects persist even at 800 MeV.


## ELECTROMAGNETIC SPIN-ORBIT POTENTIAL AND CORRELATIONS

The relative motion between the projectile and the charged nucleus results in a magnetic dipole - electric current interaction that acts like a spin-orbit potential.[15] In $p+p$ scattering this effect contributes to the $t^c(\sigma_{1n} + \sigma_{2n})$ term (see Eq. (36)) which can be divided into two parts,

$$t_{pp}^c(q) = t_{pp}^{c,EM}(q) + t_{pp}^{c,CN}(q), \tag{33}$$

where $t_{pp}^{c,EM}(q)$ is the purely electromagnetic amplitude and $t_{pp}^{c,CN}(q)$ is the Coulomb-distorted nuclear amplitude. $t_{pp}^{c,EM}(q)$ has been calculated to order $\alpha^2$ (Ref. 42) including the full proton magnetic moment. This amplitude diverges like $q^{-1}$ at small momentum transfer, causing the spin-orbit potential (Eq. (31)) to asymptotically approach $(U_{EMSO}/r^3)\vec{\sigma} \cdot \vec{\ell}$, where $U_{EMSO}$ is a complex constant. The corresponding $p+n$ magnetic scattering amplitude does not contribute to the proton-nucleus optical potential for spin saturated, even-even nuclei. This long-range electromagnetic spin-orbit potential is usually omitted in NR calculations but has been included in a few analyses[15,34,43] of data by means of a convergence factor method.

Estimates of the second-order Watson optical potential, $\mathcal{U}_{oo}^{W(2)}(r)$, are given in Ref. 23 for the spin-independent and spin-orbit components of the NN $t$ matrix. For these parts of the NN interaction which are scalar with respect to the spins of the target nucleons, the second-order optical potential is proportional to two-body correlations in

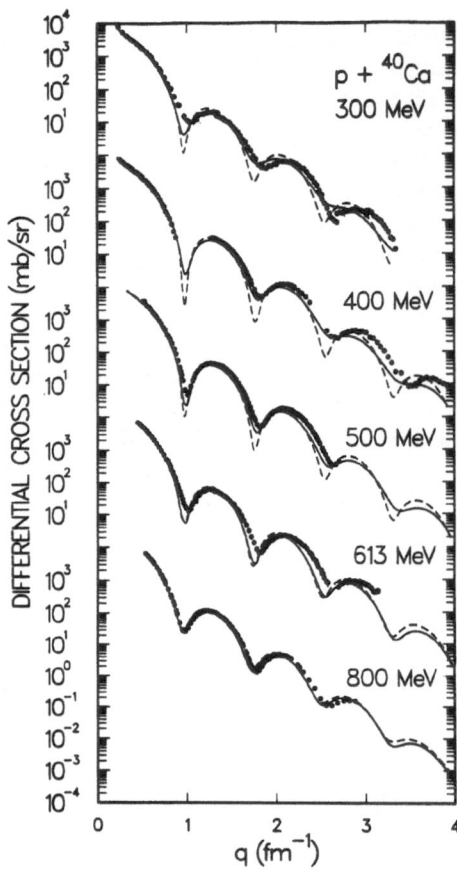

Figure 2. NR predictions for $p+^{40}Ca$ elastic scattering differential cross sections at various energies. The solid (dashed) curves correspond to NR calculations which do (do not) include medium corrections.

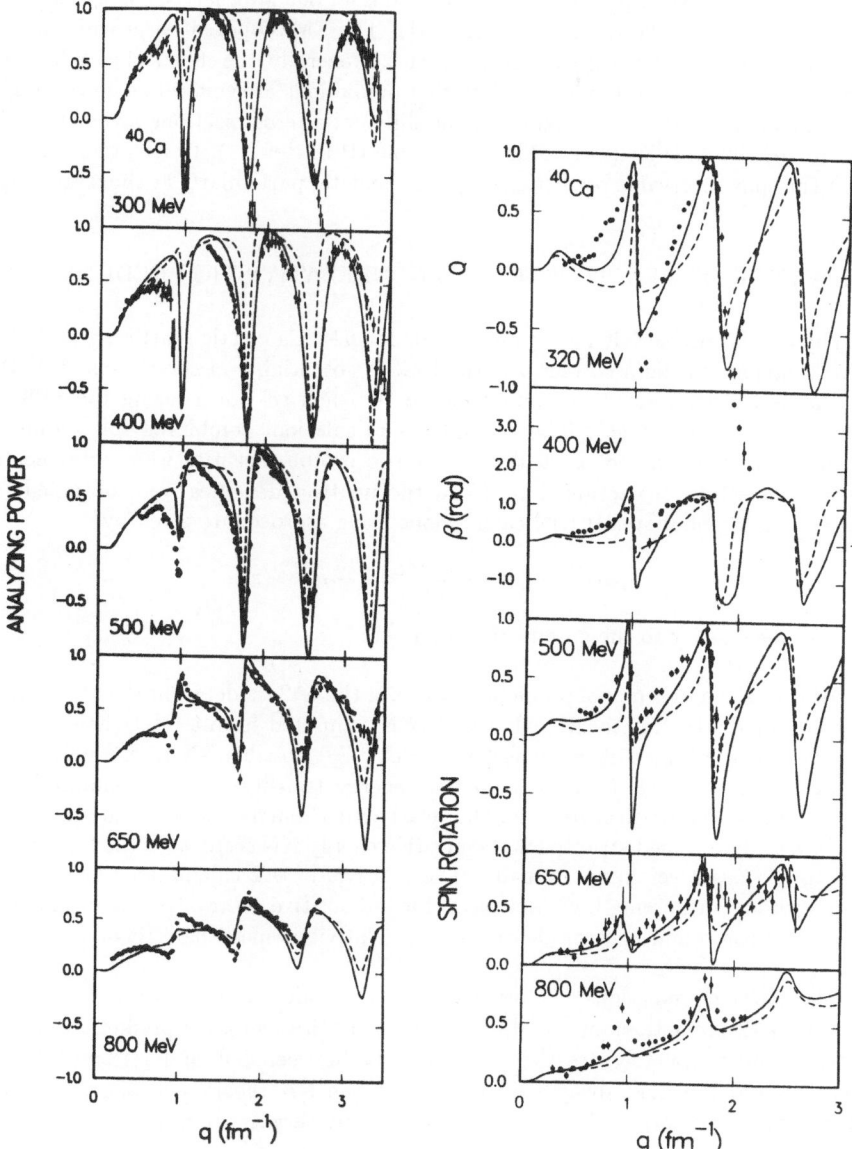

Figure 3. NR predictions for $p+{}^{40}\mathrm{Ca}$ elastic scattering analyzing power and spin rotation at various energies. The solid (dashed) curves correspond to NR calculations which do (do not) include medium corrections.

the nucleus. These correlations mainly arise from antisymmetrization of the nuclear wave function (Pauli correlations), the short-range repulsive NN force, and center-of-mass constraints.

Effects of the EMSO potential and correlations for 800 MeV $p+^{40}$Ca analyzing power and spin rotation are shown in Fig. 4 (Ref. 34). The EMSO potential provides much of the missing predicted structure at forward angles. Generally, the effects of correlations on intermediate energy proton-nucleus scattering predictions are small, but non-negligible. For example, the diffractive maxima in the differential cross sections are increased by about 10-20% when $\mathcal{U}_{oo}^{W(2)}(r)$ is added (see Table III in Ref. 23). Correlations also tend to shift the spin observable predictions toward the data, particularly at the larger angles.

## COMPARISON OF RELATIVISTIC AND NONRELATIVISTIC PREDICTIONS

The best available NR model predictions for $p+^{40}$Ca elastic scattering at 500 and 800 MeV include medium corrections, the EMSO potential, and correlations.[43,44] However, the first-order term is computed in the "on-shell $t\rho$" form, using the SP89 NN amplitudes of Arndt et al.[30] The "complete" calculation, combining full-folding and medium corrections has not been done yet. The proton densities were obtained from analyses of electron scattering data[41] and the neutron densities were obtained from Hartree-Fock-Bogoliubov (HFB)[45] calculations using a procedure given by[6]

$$\rho_n(r) = \rho_p(r) + [\rho_n^{HFB}(r) - \rho_p^{HFB}(r)] \tag{34}$$

to rescale the density to the empirical proton value.

For the relativistic model predictions those of the IA2 model from Ref. 7 are shown. In the IA2 model the relativistic NN $t$ matrix is computed from a relativistic covariant meson exchange, coupled channels isobar model.[46] All possible NN to NN initial - final spin transitions and $\pm E$ initial to $\pm E$ final energy transitions were computed. This model provides a theoretical basis for the relativistic $t$ matrix whereas the original RIA model[1,2] was determined strictly for the positive energy NN sector and the form extrapolated to the negative energy domain. The relativistic IA2 calculations use empirical NN amplitudes (in Lorentz invariant form) for the positive energy NN sector and scalar and vector, proton and neutron densities from relativistic mean field theory.[17]

The NRDD predictions[43,44] are shown by the solid curves in Figs. 5 and 6; the relativistic IA2 predictions by the dashed curves. Both theoretical predictions are very similar. Gone completely are the vast disparity between NR and relativistic model predictions and the huge disagreement between the NR theory and the data. More comparisons of this sort, including other energies and targets are in preparation.[44]

## POLARIZED NUCLEAR TARGETS

Analyses of proton elastic scattering from even-even targets are mainly sensitive to just two of the ten components of the NN effective interaction, these being the isoscalar spin-independent and spin-orbit amplitudes. The remaining parts of the effective interaction must be studied by means of other reactions, one of which is polarized proton elastic scattering from polarized, odd nuclear targets. The results from the recently

242

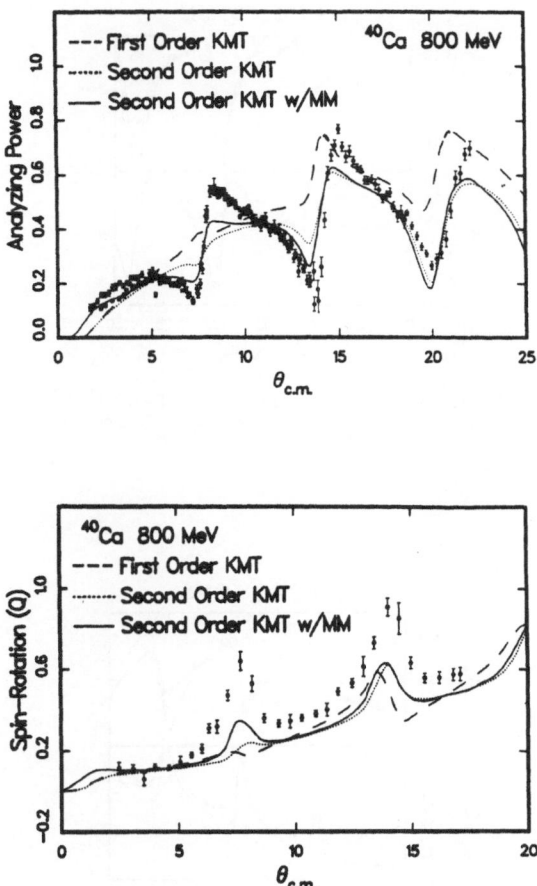

Figure 4. The 800 MeV $p+^{40}$Ca elastic scattering analyzing power and spin rotation data are compared with NRIA predictions. The calculations represented by the dashed, dotted, and solid curves correspond to first-order only, first-order plus correlations, and first-order plus correlations plus the EMSO potential, respectively.

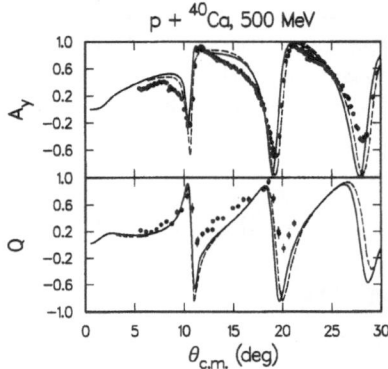

Figure 5. NRDD predictions including correlations and the EMSO potential are compared with the relativistic IA2 predictions of Ref. 7 for $p+^{40}$Ca elastic scattering at 500 MeV.

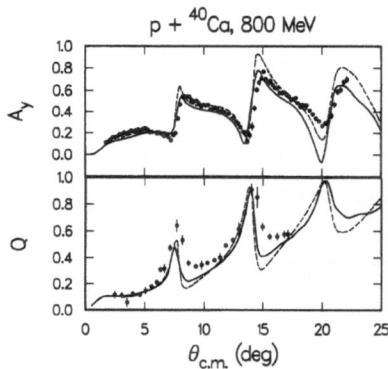

Figure 6. NRDD predictions including correlations and the EMSO potential are compared with the relativistic IA2 predictions of Ref. 7 for $p+^{40}$Ca elastic scattering at 800 MeV.

completed LAMPF experiment 955, $\vec{p}+^{13}\vec{C}$ at 500 MeV are discussed here.[48] In this experiment the target analyzing power, $A_{ooon}$, and spin correlation, $A_{oonn}$, were measured (see Ref. 49 for definition of spin observables).

The principal advantage of this approach is that it permits the spin-flip parts of the effective interaction to be studied in the elastic channel, rather than having to resort to a theoretically more complicated inelastic reaction. Another motivation for this type of work is that it provides a way to investigate possible relativistic effects in the nuclear wave function.[50] Finally, relativistic and nonrelativistic model predictions can be compared for completely different kinds of elastic scattering processes involving new spin degrees of freedom.

The optical potential for $p+^{13}$C contains many more spin dependent terms and multipoles than occur for $J^{\pi}=0^+$ targets; the exact solution of the equation of motion is much more difficult. A reasonable approximation for the scattering amplitude can be obtained by making use of the distorted wave Born approximation (DWBA). The relativistic DWBA model for $p+^{13}$C was developed in Refs. 50 and 51. The nonrelativistic DWBA scattering amplitude was used in Ref. 48. Here it is shown that the NR-DWBA amplitude can be related to the relativistic model in Ref. 50.

The DWBA for the $p+^{13}$C scattering amplitude is (suppressing spin labels, see Ref. 50)

$$f(\vec{k}',\vec{k}) \cong f_{core}(\vec{k}',\vec{k}) - \frac{\varepsilon_{pA}}{2\pi(\hbar c)^2}\langle \psi^{(-)}_{C,\vec{k}'} u_{1p_{\frac{1}{2}}} | \hat{t}^W | u_{1p_{\frac{1}{2}}} \psi^{(+)}_{C,\vec{k}}\rangle, \tag{35}$$

where $f_{core}$ and $\psi_C$ are the scattering amplitude and the distorted wave function for proton scattering from the core part of the optical potential, assumed to be 12 nucleons in the $s_{\frac{1}{2}}$, $p_{\frac{3}{2}}$ levels. The $1p_{\frac{1}{2}}$ valence neutron wave function is denoted by $u_{1p_{\frac{1}{2}}}$. The proton-nucleus reduced total energy is $\varepsilon_{pA}$. The $t$ matrix, $\hat{t}^W$, in the pA COM system assuming on-shell $t$ matrices, can be expressed as,

$$\langle \hat{t}^W \rangle = t^a + t^b \sigma_{1n}\sigma_{2n} + t^{c_1}\sigma_{1n} + t^{c_2}\sigma_{2n} + t^d \sigma_{1q}\sigma_{2q} + t^e \sigma_{1p}\sigma_{2p}, \tag{36}$$

where $\sigma_{1x} \equiv \vec{\sigma}_1 \cdot \hat{x}$, and in terms of the initial and final intermediate projectile momenta, $\vec{k}_1$ and $\vec{k}'_1$, $\hat{n} = (\vec{k}_1 \times \vec{k}'_1)/ | \vec{k}_1 \times \vec{k}'_1 |$, $\hat{p} = (\vec{k}_1 + \vec{k}'_1)/ | \vec{k}_1 + \vec{k}'_1 |$, and $\hat{q} = \hat{p} \times \hat{n}$.

Evaluation of the DWBA matrix element is greatly facilitated by converting from a two-component Pauli representation to a four-component Dirac representation for the spin wave functions and $t$ matrix operator. The advantage of this is that the interaction becomes strictly local; the momentum dependences associated with $\sigma_{1n}$, $\sigma_{2n}$, etc. are absorbed into the lower two components of the four-component wave functions. In this way evaluation of convection currents and composite spin currents becomes straightforward. We emphasize that this is simply a change of representation and does not mean that relativistic dynamics are entering the NR calculation.

In the partial wave expansions of the distorted waves and the bound state wave functions the two-component spinors are denoted by $\chi^{(1)}_{m_s}$ and $\chi^{(2)}_{\mu}$, respectively. To affect the change in representation we use the defining equation of the relativistic impulse approximation given by[31]

$$\chi^{(1)^\dagger}_{m'_s} \chi^{(2)^\dagger}_{\mu'} \langle \hat{t}^W \rangle \chi^{(1)}_{m_s}\chi^{(2)}_{\mu} \equiv \bar{u}^{(1)}_{m'_s}\bar{u}^{(2)}_{\mu'}\langle \Upsilon' \rangle u^{(1)}_{m_s}u^{(2)}_{\mu} \tag{37}$$

where $u_{m_s}^{(1)}$, etc. are the four-component, positive energy Dirac spinors, defined by

$$u_{m_s}^{(1)} = \sqrt{\frac{E_1 + m_1}{2m_1}} \left( \begin{array}{c} 1 \\ \frac{\vec{\sigma} \cdot \vec{k}_1}{E_1 + m_1} \end{array} \right) \chi_{m_s}, \tag{38}$$

where for particle (1), $E_1$ is the on-shell energy, $\vec{k}_1$ is the intermediate momentum, $\chi_{m_s}$ is the two-component Pauli spinor, and $\bar{u} = u^{\dagger} \gamma^0$. The operator $\langle \Upsilon' \rangle$ has the structure[31]

$$\langle \Upsilon' \rangle = \Upsilon_S + \Upsilon_P \gamma_1^5 \gamma_2^5 + \Upsilon_V \gamma_1^{\mu} \gamma_{2\mu} + \Upsilon_A \gamma_1^5 \gamma_1^{\mu} \gamma_2^5 \gamma_{2\mu} + \Upsilon_T \sigma_1^{\mu\nu} \sigma_{2\mu\nu}, \tag{39}$$

where $\gamma^{\mu}$, $\gamma^5$ and $\sigma^{\mu\nu}$ are the usual Dirac matrices. The Dirac spinors are recombined with the partial wave expansions to form a DWBA matrix element with the same form as the relativistic DWBA matrix element in Ref. 50. The nonrelativistic scattering amplitude for $p+^{13}C$ is then given by

$$f(\vec{k}', \vec{k}) \cong f_{core}(\vec{k}', \vec{k}) + \left( \frac{E_A}{E_p + E_A} \right) f_{sp}^{RDWBA}(\vec{k}', \vec{k}), \tag{40}$$

where the core scattering amplitude is calculated using the Schrödinger equivalent optical potential of the Dirac phenomenological fit to $p+^{12}C$ elastic scattering data at 500 MeV.[48,50] The $1p_{\frac{1}{2}}$ valence particle contribution to the scattering amplitude, computed as in Ref. 50, is $f_{sp}^{RDWBA}$. $E_p$ and $E_A$ are the total energies of the projectile proton and target nucleus in the pA COM, respectively.

To summarize, the $p+^{13}C$ NR elastic scattering amplitude is conveniently calculated using the relativistic DWBA formalism[50] where the following changes from the procedure described in Ref. 50 are carried out: (1) NR distorted waves and bound state wave functions are used for the upper components, (2) the lower components are set to the NR or free particle limit (i.e., no potential terms included in $(\vec{\sigma} \cdot \vec{k})/(E + m)$), (3) $\langle \Upsilon' \rangle$ corresponds to the choice of $\langle \hat{t}^W \rangle$ according to Eq. (37) and may be density dependent, and (4) the pA reduced energy factor, $E_A/(E_p + E_A)$, is included.

In Ref. 48 data were obtained for observables $A_{ooon}$ and $A_{oonn}$. In the relativistic models the predictions for these observables are dominated by the space-like vector interaction which in the Pauli representation corresponds to a combination of $t^{c_2}$, $t^b$ and $t^e$, with $t^{c_2}$ providing the dominant contribution at small q. In the results published in Ref. 48 a simple $1p_{\frac{1}{2}}$ valence neutron coupled to a $0^+$ $^{12}C$ core was assumed for the nuclear structure model. Also the IA was used in place of $\langle \hat{t}^W \rangle$ and the older, SP82 Arndt phase shift solutions were used.[30] The results shown in Fig. 1 of Ref. 48 indicate that both the NR and relativistic models are very similar and both qualitatively describe the data. This situation is in marked contrast to previous experiences when new spin observable data were first analyzed. For example, when the first proton-nucleus analyzing power data at 800 MeV (Ref. 52) and again at 500 MeV (Ref. 25) were initially analyzed the theoretical predictions were in gross disagreement with the data. Perhaps a calculable theory of the proton-nucleus system is finally emerging which has some predictability.

New $\vec{p}+^{13}\vec{C}$ NR-DWBA calculations were done in which SP89 NN amplitudes, shell model wave functions,[53] and density dependence were included. These results are shown in Fig. 7 where the dashed-dotted curves indicate the NR predictions from Ref. 48, the dotted curves use SP89 amplitudes, the dashed curves use the SP89 density dependent amplitudes, and the solid curves represent the best current NR prediction which uses

246

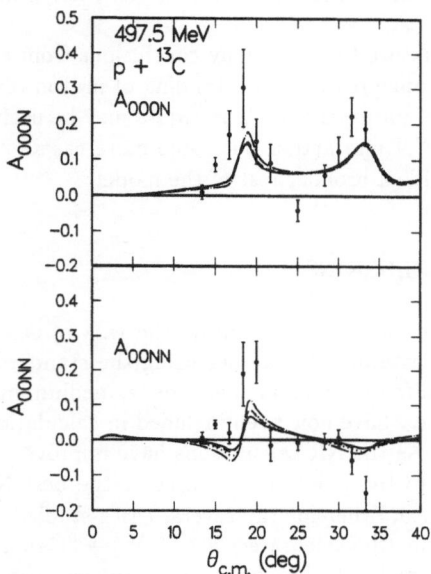

Figure 7. NR-DWBA predictions for $p+{}^{13}$C elastic scattering at 497.5 MeV for observables $A_{ooon}$ and $A_{oonn}$. The dashed-dotted, dotted, dashed, and solid curves represent calculations which used the SP82 IA NN amplitudes, the SP89 IA NN amplitudes, the SP89 DD amplitudes, and the SP89 DD amplitudes with the CK wave functions, respectively.

the SP89 density dependent amplitudes and Cohen and Kurath (CK)[53] wave functions. None of these effects are large for this case. This is not surprising since at 500 MeV the NN amplitudes have remained fairly stable for several years, the CK M1 elastic form factor is only different from that of the simple $1p_{\frac{1}{2}}$ valence wave function for $q > 1.5$ fm$^{-1}$ (Ref. 54), and the main effect of density dependence at this energy is in the isoscalar parts of $t^a$ and $t^{c_1}$, which is absorbed into the phenomenological $f_{core}$ and $\psi_C$.

Calculations have also been made for this case using a new coupled channels Dirac code[55] in which the full $p+{}^{13}$C RIA optical potential scattering solution is obtained without using the DWBA. The results indicate that the relativistic DWBA used in Ref. 50 is very accurate for this case. Further theoretical work related to these data to be done in the future includes an investigation of quadrupole excitation - deexcitation effects, a calculation using the IA2 interaction, and the consideration of magnetic moment effects.

It is certainly premature to draw many conclusions from analysis of this one data set, but it appears that our present understanding of proton scattering from polarized, odd nuclear targets is not unreasonable. These data may be useful for phenomenological studies of the effective NN interaction but some more cases (*e.g.* ${}^3$He, ${}^6$Li, ${}^{15}$N, etc.) need to be done to provide a broader test of the model.

## SUMMARY AND CONCLUSIONS

Renewed efforts are underway to "finish" the task of calculating medium energy proton-nucleus elastic scattering observables using the standard nonrelativistic optical potential formalism. Full-folding and off-shell effects, medium modifications, the EMSO potential, correlations, etc. have now been included in calculations and found to be significant for many cases. As the NR calculations have improved in their physics content so have the predictions to the point that at present the best NR and best relativistic predictions agree with each other for the several cases studied so far and qualitatively describe the data, even for the celebrated 500 MeV $p+{}^{40}$Ca analyzing power and spin rotation. Whether the "final" calculations, which simultaneously include all of the terms discussed here, will accurately describe all of the medium energy proton-nucleus scattering data remains to be seen. If the NR models are ultimately successful, one must then ask why the virtual $N\bar{N}$ pair process, which is large in the Dirac description, is not important. If significant discrepancies between these "final" NR predictions and data occur, then we should seriously examine exotic physical processes, such as vacuum polarization,[56] medium effects on the meson masses and couplings,[57] changes in nucleon sizes inside the nucleus,[58] etc. Whatever the outcome, an understanding of the results in terms of some approximation to QCD is a reasonable goal for the foreseeable future.

## REFERENCES

1. J. A. McNeil, J. Shepard, and S. J. Wallace, Phys. Rev. Lett. **50**, 1439, 1443 (1983).

2. B. C. Clark, S. Hama, R. L. Mercer, L. Ray, and B. D. Serot, Phys. Rev. Lett. **50**, 1644 (1983).

3. J. A. Tjon and S. J. Wallace, Phys. Rev. Lett. **54**, 1357 (1985); Phys. Rev. C **32**, 267 (1985); Phys. Rev. C **36**, 1085 (1987).

4. S. J. Brodsky, in *Short-Distance Phenomena in Nuclear Physics*, edited by D. H. Boal and R. M. Woloshyn (Plenum Pub. Corp., 1983), p. 141.

5. E. Bleszynski, M. Bleszynski, and T. Jaroszewicz, Phys. Rev. Lett. **59**, 423 (1987).

6. L. Ray and G. W. Hoffmann, Phys. Rev. C **31**, 538 (1985).

7. N. Ottenstein, S. J. Wallace and J. A. Tjon, Phys. Rev. C **38**, 2272, 2289 (1988).

8. H. F. Arellano, F. A. Brieva, and W. G. Love, Phys. Rev. C **41**, 2188 (1990).

9. R. Crespo, R. C. Johnson, and J. A. Tostevin, Phys. Rev. C **41**, 2257 (1990).

10. Ch. Elster, T. Cheon, E. F. Redish, and P. C. Tandy, Phys. Rev. C **41**, 814 (1990).

11. L. Ray, Phys. Rev. C **41**, 2816 (1990).

12. H. V. von Geramb, in *The Interaction Between Medium Energy Nucleons in Nuclei*, Proceedings of the Workshop on the Interactions Between Medium Energy Nucleons in Nuclei, AIP Conf. Proc. No. **97**, edited by H. O. Meyer (AIP, New York, 1983), p. 44; L. Rikus, K. Nakano, and H. V. von Geramb, Nucl. Phys. **A414**, 413 (1984); L. Rikus and H. V. von Geramb, Nucl. Phys. **A426**, 496 (1984).

13. S. Weinberg, University of Texas preprint, UTTG-31-90 (1990).

14. See the proceedings of the Workshop "From Fundamental Fields to Nuclear Phenomena," University of Colorado, Boulder, CO, Sept. 20 - 22, 1990; T. D. Cohen, R. J. Furnstahl, and D. K. Griegel, Univ. of Maryland preprint, UMPP No. 91-177 (1991).

15. G. W. Hoffmann *et al.*, Phys. Rev. C **24**, 541 (1981).

16. K. M. Watson, Phys. Rev. **89**, 575 (1953).

17. J.-P. Jeukenne, A. Lejeune, and C. Mahaux, Phys. Rev. C **10**, 1391 (1974).

18. A. Picklesimer, P. C. Tandy, R. M. Thaler, and D. H. Wolfe, Phys. Rev. C **30**, 1861 (1984).

19. L. S. Rodberg and R. M. Thaler, *Introduction to the Quantum Theory of Scattering*, (Academic Press, New York, 1967).

20. L. Ray, G. W. Hoffmann and R. M. Thaler, Phys. Rev. C **22**, 1454 (1980).

21. H. F. Arellano, F. A. Brieva, and W. G. Love, Phys. Rev. C **42**, 652 (1990).

22. G. F. Chew, Phys. Rev. **80**, 196 (1950).

23. L. Ray, Phys. Rev. C **19**, 1855 (1979).

24. D. A. Hutcheon *et al.*, Phys. Rev. Lett. **47**, 315 (1981).

25. G. W. Hoffmann *et al.*, Phys. Rev. Lett. **47**, 1436 (1981).

26. L. Ray, Phys. Rev. C **35**, 1072 (1987).

27. E. L. Lomon and H. Feshbach, Ann. Phys. (N. Y.) **48**, 94 (1968).

28. E. L. Lomon, Phys. Rev. D **26**, 576 (1982).

29. P. Gonzalez and E. L. Lomon, Phys. Rev. D **34**, 1351 (1986).

30. R. A. Arndt, J. S. Hyslop III, and L. D. Roper, Phys. Rev. D **35**, 128 (1987); Phase shift solutions obtained from the computer code SAID, R. A. Arndt, private communication.

31. J. A. McNeil, L. Ray, and S. J. Wallace, Phys. Rev. C **27**, 2123 (1983).

32. L. G. Arnold *et al.*, Phys. Rev. C **25**, 936 (1982).

33. R. M. Sealock *et al.*, Phys. Rev. Lett. **62**, 1350 (1989).

34. R. W. Fergerson *et al.*, Phys. Rev. C **33**, 239 (1986).

35. A. Rahbar *et al.*, Phys. Rev. Lett. **47**, 1811 (1981).

36. L. Ray *et al.*, Phys. Rev. C **23**, 828 (1981).

37. E. Bleszynski *et al.*, Phys. Rev. C **37**, 1527 (1988); C. A. Whitten, Jr., priv. comm.

38. D. Lopiano *et al.*, priv. comm.

39. D. A. Hutcheon *et al.*, Nucl. Phys. **A483**, 429 (1988).

40. G. Bruge, Saclay Report DPh-N/ME/78-1, 1978 (unpublished).

41. I. Sick *et al.*, Phys. Lett. **88B**, 245 (1979).

42. C. Lechanoine, F. Lehar, F. Perrot, and P. Winternitz, Nuovo Cimento **56**, 201 (1980).

43. W. R. Coker and L. Ray, Phys. Rev. C **42**, 659 (1990).

44. L. Ray, G. W. Hoffmann, and W. R. Coker, Phys. Reports, in preparation (1991).

45. J. Decharge and D. Gogny, Phys. Rev. C **21**, 1568 (1980); J. Decharge, M. Girod, D. Gogny, and B. Grammaticos, Nucl. Phys. **A358**, 203c (1981); J. Decharge, priv. comm.

46. E. van Faassen and J. A. Tjon, Phys. Rev. C **33**, 2105 (1986); C **30**, 285 (1984); Phys. Lett. **120B**, 39 (1983).

47. C. J. Horowitz and B. D. Serot, Nucl. Phys. **A368**, 503 (1981).

48. G. W. Hoffmann *et al.*, Phys. Rev. Lett. **65**, 3096 (1990).

49. J. Bystricky, F. Lehar and P. Winternitz, J. Phys. (Paris) **39**, 1 (1978).

50. L. Ray *et al.*, Phys. Rev. C **37**, 1169 (1988).

51. J. R. Shepard, E. Rost and J. Piekarewicz, Phys. Rev. C **30**, 1604 (1984).

52. G. W. Hoffmann *et al.,* Phys. Rev. Lett. **40**, 1256 (1978).

53. S. Cohen and D. Kurath, Nucl. Phys. **73**, 1 (1965).

54. R. S. Hicks, J. Dubach, R. A. Lindgren, B. Parker, and G. A. Peterson, Phys. Rev. C **26**, 339 (1982).

55. R. L. Mercer and B. C. Clark, private communication.

56. C. J. Horowitz and B. D. Serot, Phys. Lett. **140B**, 181 (1984).

57. G. E. Brown, A. Sethi, and N. M. Hintz, Univ. of Minnesota preprint (1990).

58. L. S. Celenza, A. Harindranath, and C. M. Shakin, Phys. Rev. C **32**, 2173 (1985).

# MEASUREMENTS OF POLARIZATION TRANSFER OBSERVABLES

## IN ($\vec{p}, \vec{p}'$) REACTIONS AT INTERMEDIATE ENERGIES

Scott W. Wissink

Indiana University Cyclotron Facility
Department of Physics
Bloomington, Indiana 47408

## INTRODUCTION

It has now been almost ten years since a charged-particle polarimeter was first coupled to a high resolution spectrometer, the HRS at Los Alamos National Laboratory, for use at intermediate energies. During this time, polarimeters of similar design and comparable performance have been built at TRIUMF and at IUCF. Yet despite this apparent potential for generating a wide variety of new data, this goal has not yet been fully realized. Despite the obvious advantages of determining, not just two, but up to eight independent parameters for a particular transition at each energy and angle, our progress in understanding these transitions has been slow. And despite some of the early claims that these types of measurements would allow for 'clean' isolation of individual terms in the nucleon-nucleon ($NN$) effective interaction, the waters today appear only slightly less muddy than they were ten years ago. Perhaps one of the greatest benefits derived from this early work was the optimistic fervor that prevailed for a few years in the mid 80's, when an impressive amount of theoretical effort was expended towards understanding these first results, and when experimentalists were measuring observables as rapidly as Advisory Committees would let them.

Some of these early successes are well known, and in fact the Telluride conferences have often served as the first sounding board for new and preliminary results in this field. In my talk today, however, I do not want to present an historical overview of these types of investigations, but would rather start by reminding you *why* these measurements are so interesting. I will then discuss some of the experimental efforts recently completed or currently underway at IUCF, mention briefly some of our future plans, and finally indicate why we feel these new studies may prove useful in resolving many of the problems noted above.

I should also point out that several of the particular physics issues I will address here will be discussed in more detail by other speakers. In particular, I would refer the reader to the talks presented at this conference by Jochem Wambach and Edward Stephenson.

# THE IUCF $(\vec{p}, \vec{p}\,')$ EXPERIMENTAL PROGRAM

Before discussing any specific measurements, a few general comments on the $(\vec{p}, \vec{p}\,')$ program at IUCF are in order. Though we are currently addressing a wide range of physics topics, our primary efforts to date have focussed on extracting information on some of the more poorly understood terms in the $NN$ effective interaction, particularly those that are spin-dependent. Much of our work has centered on investigations of excitation of narrow discrete states, exploiting the high momentum resolution ($\sim$25 keV at 200 MeV) obtainable with the K600 spectrometer. By measuring all three components of the polarization of the scattered proton (though not simultaneously, of course) for several different orientations of the polarization of the incident proton beam, we are able to deduce what I will call a "complete set" of spin transfer observables, the $D_{ij}$'s, for select nuclear transitions. As I will show below, this set consists in general of eight independent observables: the differential cross section $d\sigma/d\Omega$, the analyzing power $A_y$ (or $D_{n0}$), the induced polarization $P$ (or $D_{0n}$), the normal-component spin transfer coefficient $D_{NN'}$, and the four in-plane $D_{ij}$'s. The labelling conventions followed here are those that are fairly widely (though by no means universally!) accepted.

From a physics standpoint, the proton energies available at IUCF (near 200 MeV) lie in a particularly interesting regime. At these energies, the spin-dependent terms in the effective interaction, such as the spin-orbit and tensor pieces, are relatively strong, since the strength of the dominant central term is at a minimum. This should lead to an enhanced sensitivity to these components of the interaction. At the same time, the energies are sufficiently high that the impulse approximation should work reasonably well, thus providing the theoretical framework for connecting the nucleon-nucleus interaction with the more fundamental $NN$ interaction.

To most efficiently use the $(\vec{p}, \vec{p}\,')$ reaction as a probe of the effective $NN$ force, one must also exercise care in the selection of the excited states to be studied. Because we are primarily interested in investigating the spin-dependence of the interaction, we have concentrated thus far on transitions of unnatural parity (for which the spin transfer $\Delta S = 1$). More generally, we have chosen particular discrete states or regions of excitation for which the nuclear structure is believed to be fairly simple or well understood, such as the $1\hbar\omega$ "stretched" states we have already heard so much about. In order to study the dependence of the various interaction terms on the momentum transfer $q$, we have looked at transitions of both low spin and high spin, where the form factors tend to peak at low and high $q$, respectively (though this relationship becomes somewhat more tenuous when the effects of knock-on exchange are taken into account). Finally, because the $(p, p')$ reaction can effectively probe several different isospin transfers, we have often tried to study a set of states for which the nuclear structure is believed to be similar but which differ in isospin, *e.g.*, for stretched states, examining both T=0 and T=1 states in the light nuclei, or states that are predominantly either pure neutron or pure proton 1p-1h excitations in the heavier nuclei.

## Inelastic Proton Scattering Formalism

To illustrate more intuitively the physics content of these new spin transfer coefficients, the $D_{ij}$'s, it is useful to examine the relationship between these nucleon-nucleus ($NA$) observables and those that parameterize nucleon-nucleon ($NN$) scattering. Using the conventions of Kerman, McManus, and Thaler (KMT),[1] the $NN$ scattering amplitude can be expressed as

$$M_t(q) = A + B\sigma_{\hat{n}}^t\sigma_{\hat{n}} + C(\sigma_{\hat{n}}^t + \sigma_{\hat{n}}) + E\sigma_{\hat{q}}^t\sigma_{\hat{q}} + F\sigma_{\hat{p}}^t\sigma_{\hat{p}} \qquad (1)$$

with

$$\vec{q} = \vec{k}' - \vec{k} \qquad \vec{n} = \vec{k} \times \vec{k}' \qquad \hat{p} = \hat{q} \times \hat{n} \qquad (2)$$

where $\vec{k}$ ($\vec{k}'$) is the incident (scattered) proton momentum, the Pauli spin operators $\sigma$ that are $t$-superscripted (nonsuperscripted) are for the target (projectile) nucleon, and the complex amplitudes $A$, $B$, $C$, $E$, and $F$ are assumed to exhibit an isospin dependence of the form $A = A_1 + A_2(\vec{\tau}^t \cdot \vec{\tau})$, etc.

For a $0^+$ target excited via proton inelastic scattering to a state $J^\pi$, the impulse approximation allows us to express the $NA$ scattering amplitude in terms of the $NN$ amplitudes in the form

$$\overline{M}(q) = \langle J\mu | \sum_t M_t(q) e^{-i\vec{q}\cdot\vec{r}_t} | 0 \rangle \qquad (3)$$

where $M_t$ is the amplitude for scattering of the incident proton from the $t^{th}$ target nucleon. We can now *define* the polarization transfer coefficients $D_{ij}$ in terms of the $NA$ scattering amplitude $\overline{M}(q)$ and the nucleon spins, using the conventions of Wolfenstein[2]

$$I_0 D_{ij} \equiv \frac{1}{4}\text{Tr}[\overline{M}\sigma_i\overline{M}^\dagger\sigma_j] , \qquad (4)$$

where $\sigma_i$ ($\sigma_j$) is the Pauli spin matrix for the $i^{th}$ ($j^{th}$) component of the incident (scattered) nucleon polarization, and where

$$I_0 \equiv \frac{1}{4}\text{Tr}[\overline{M}\,\overline{M}^\dagger] \qquad (5)$$

is the usual unpolarized differential cross section. Written in this form, it is apparent that values for all of the reaction observables can be easily generated, given a model calculation for $\overline{M}$.

More intuitively, we see that the polarization transfer observable $D_{ij}$ relates the $i^{th}$ component of the incident proton polarization to the $j^{th}$ component of the outgoing proton polarization. Because $i$ and $j$ each run over four possible values (corresponding to the identity matrix and the three Pauli spin-matrices), 16 independent observables are required in principal to characterize a $0^+ \rightarrow J^\pi$ transition induced via inelastic scattering of a spin-1/2 particle. Invoking parity conservation and time-reversal invariance reduces the number of non-zero observables to eight, since there can be no mixing in the nuclear interaction between those components normal to the scattering plane and those that lie in the plane.[3] The relationship between the polarization components of the incident and scattered proton may therefore be expressed in the laboratory frame in the following matrix form:

$$\sigma \begin{bmatrix} 1 \\ p_{N'} \\ p_{L'} \\ p_{S'} \end{bmatrix} = \sigma_0 \begin{bmatrix} 1 & D_{N0} & 0 & 0 \\ D_{0N'} & D_{NN'} & 0 & 0 \\ 0 & 0 & D_{LL'} & D_{SL'} \\ 0 & 0 & D_{LS'} & D_{SS'} \end{bmatrix} \begin{bmatrix} 1 \\ p_N \\ p_L \\ p_S \end{bmatrix} \qquad (6)$$

where now $\sigma_0$ is the unpolarized cross section, $\sigma = \sigma_0(1 + D_{N0}p_N)$ is the familiar polarized cross section, and the polarization components for the incident (scattered) proton are measured in the unprimed (primed) coordinate system, defined by $(\hat{N}, \hat{L}, \hat{S})$ $\equiv (\hat{n}, \hat{k}, \hat{n} \times \hat{k})$ and $(\hat{N}', \hat{L}', \hat{S}') \equiv (\hat{n}, \hat{k}', \hat{n} \times \hat{k}')$.

EXPERIMENTAL TECHNIQUES AND FACILITIES

From the above discussion, it can be seen that the ability to orient the polarization vector of the incident proton beam in an arbitrary direction, coupled with the capability of completely determining the polarization of the proton after the nuclear scattering, yields eight independent pieces of information on the transition or region of excitation under study (only three if $J_f=0$, see below). I would now like to indicate very briefly how we perform such measurements at IUCF.

The Indiana University Cyclotron Facility provides polarized proton beams of up to several hundred nanoamps in intensity at energies between about 35 and 200 MeV. (These parameters will change dramatically, of course, if one injects into the Cooler Ring.) Polarization magnitudes of 0.75 are typical, with the orientation of the spin vector lying very close to vertical upon extraction from the main stage cyclotron. For measurements of differential cross sections or any of the three "normal-component" spin observables ($A_y$, $P$, or $D_{NN'}$), the beam may be transported directly to the spectrometer, since the nuclear scattering plane used is horizontal. For investigation of the four "in-plane" spin transfer observables, however, one must use an incident proton beam whose polarization vector has been rotated into the reaction plane at the desired orientation. This can be accomplished through the use of two independently adjustable solenoids, whose magnetic fields lie parallel to the proton's momentum, so long as there is an additional rotation of the proton's spin about the vertical axis in between the two solenoids. This rotation can be provided by a horizontally bending dipole magnet whose proton spin precession angle is close to 90°. The 45° energy analysis magnet at IUCF conveniently provides an in-plane spin precession angle of 97.9° for 200 MeV protons, a value which decreases relatively slowly with decreasing energy. Thus, by locating one magnetic solenoid immediately upstream of this magnet, and a second downstream, one can essentially "dial-in" the required currents to orient the final proton polarization vector in any desired direction in any of the various beamlines (assuming the bend angles in all subsequent magnets are known). Because precession angles as large as 90° may be required in these solenoids, those at IUCF have been designed to be capable of inducing up to 120° precession for 200 MeV protons. The solenoid coils themselves are superconducting.

In practise, though, the polarization vector may be tipped by as much as 15° away from the vertical during the acceleration process, and the solenoid settings selected must take this into account. More generally, one would *always* like to monitor the actual value of the three components of the beam polarization, rather than relying on calibrated values for appropriate currents in the solenoids. To provide this information, we have designed, installed, and recently calibrated a set of high-energy in-beam (transmission) polarimeters, based on proton–deuteron elastic scattering. Details as to the advantages and shortcomings of these devices can be found elsewhere,[4] along with a discussion of their physical layout and operating characteristics. Each polarimeter consists of a left/right pair of deuteron detectors (plus additional detectors for the co-incident proton) which provides information on the vertical component of the proton polarization, and an up/down pair of detectors for measurement of the sideways component. The longitudinal component of the beam polarization can not be measured directly, and one must exploit the same trick referred to in the preceding paragraph: use a horizontally bending dipole magnet to precess the beam spin about the vertical axis, then remeasure the sideways component. A bit of algebra will show that by placing both left/right and up/down polarimeter detector pairs just downstream of each

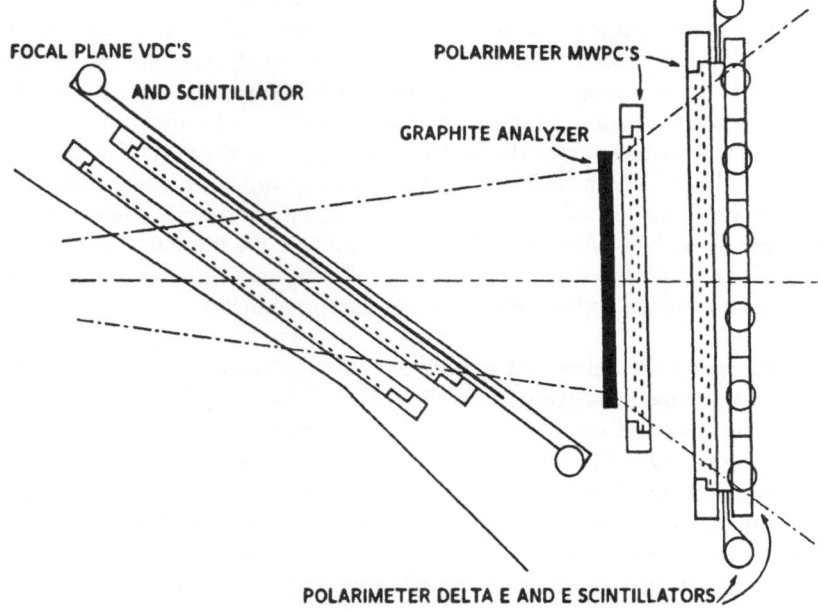

FOCAL PLANE VDC'S AND SCINTILLATOR

POLARIMETER MWPC'S

GRAPHITE ANALYZER

POLARIMETER DELTA E AND E SCINTILLATORS

Fig. 1   Top view of the IUCF K600 focal plane and polarimeter detector system

of the two solenoids, full information on the polarization of the proton beam may be obtained, for any set of solenoid currents, without any prior assumptions as to the polarization state of the beam extracted from the cyclotrons.

The spectrometer used for $(p, p')$ studies at IUCF is a horizontal K600 QDD configuration, with internal coils for kinematic (quadrupole) and hexapole aberration corrections.[5] The system can be rotated to cover laboratory scattering angles between –42° (scattering to beam right) and +82.5° (beam left). In the medium dispersion mode we have used, a momentum bite of about 7% is possible for the focal plane detectors and polarimeter. Energy resolutions of $\sim$ 20–25 keV are typical at 200 MeV, though this is often compromised by the use of thick targets for higher counting rates. Solid angles of up to about 2 msr have been used, but values of 0.5 to 1 msr are more common.

A schematic illustration of the focal plane and focal plane polarimeter detector geometry is presented in Fig. 1.. Descriptions of the focal plane detector system and its associated electronics are provided in Ref. 6, while the polarimeter is discussed in detail in Ref. 7. Only a brief description will be provided here.

The focal plane detection system consists of two vertical drift chambers (VDC's) and, for polarimeter operation, a single thin plastic scintillator. The two VDC's provide high resolution position and angle information on the scattered proton trajectory, while the scintillator is used for particle identification and fast timing. Note that the central momentum ray is inclined by approximately 55° from the normal through the focal plane detector system.

The polarimeter, located just downstream of the focal plane detectors and oriented so that the central momentum ray is nearly normal, is designed to detect and measure the polarization state of the scattered protons. The polarimeter consists of a 5 cm thick graphite block, used as the polarization analyzer, followed by two sets of paired $x$–$y$ multiwire proportional chambers (MWPC's) and two plastic scintillator planes. The two sets of wire chambers provide hit pattern information that is used to determine the scattering angle of the proton and verify that the event origin lies within the carbon analyzer. When multiple scattering effects are added in quadrature with the intrinsic resolution of the wire chamber geometry, an overall angular resolution of approximately 1.2° is achieved.[7] The two scintillator planes, 0.64 cm and 7.62 cm thick, provide additional particle identification information and ensure that the proton underwent an elastic scattering collision in the carbon analyzer.

The readout electronics used is fairly standard, although one aspect of the polarimeter hardware bears mentioning. We use the LeCroy[8] PCOS III system, which, in addition to encoding and compacting the full data set, also provides rapid ($< 500$ ns) encoding of the first wire hit in each chamber, with presentation of the encoded output at an ECLport. These data are used as input to a second-level trigger processor for online hardware rejection of events in which the detected proton did not scatter more than a few degrees in the carbon analyzer. By eliminating these "straight–through" events, we estimate that we increase our effective data aquisition rate by a factor of about eight to ten.

In our final analysis, we consider only those events in which the momentum-analyzed proton elastically scattered in the carbon analyzer by more than about 5°. By determining the yield asymmetry between those that scatter to the left and those that scatter right, we are able to deduce the vertical component of the proton polarization in the focal plane. Similar arguments applied to protons that scatter up or down allow extraction of the sideways component. While the vertical component measured is directly related to $p_{N'}$, the normal component of the scattered proton's polarization, the sideways component determined at the focal plane represents a linear combination of $p_{L'}$ and $p_{S'}$ due to the additional in-plane precession of the proton's spin through the K600 dipoles. Accurate knowledge of this precession angle is therefore required to convert a series of focal plane polarimeter asymmetry measurements into the desired set of spin transfer coefficients. These asymmetries must be measured on both beam left and right ($\pm\theta$) in order to independently determine $p_{L'}$ and $p_{S'}$.

One must also know the effective analyzing power $A_{FPP}$ of the focal plane polarimeter, *effective* in that it represents an average of several reaction analyzing powers weighted by the various cross sections and constraints of the polarimeter geometry and software requirements. The calibration procedures we followed exploited some of the simplifications in the form of the spin transfer matrix, Eq. 6, that result when one examines elastic scattering of a spin-1/2 particle from a spin-0 nucleus.[3] Details of these procedures, and quantitative discussions of the energy and focal plane position dependences that we measured, may be found in Ref. 7. Approximate values near 200 MeV, relevant for most of the work to be described below, are: overall polarimeter efficiency $\approx 2.0\%$; effective analyzing power $\approx 0.50$. As the proton energy at the focal plane decreases, the efficiency rises slightly while the effective analyzing power drops. The combined figure-of-merit for the focal plane polarimeter (essentially the efficiency times the square of the analyzing power) shows only a very gradual decrease as the energy is lowered, though operation at proton energies of less than 120 MeV introduces several additional complications.

# EXPERIMENTAL RESULTS – COMPLETED WORK

As an offshoot of some of the calibration work just mentioned, we realized we could exploit the high efficiency of the K600 Focal Plane Polarimeter to make precise determinations of reaction analyzing powers, using a technique previously unemployed at intermediate energies. Such determinations are useful, since (as I will show shortly) measurements of many spin observables can now be made with statistical accuracies significantly smaller than their corresponding systematic uncertainties, due largely to lack of precise knowledge of the incident beam polarization. Though much progress has been made in the development of in-beam high-energy polarimetry,[4] these devices must be calibrated against a well-determined analyzing power standard.

Our method relies on the quadratic relationship that exists among three of the polarization observables in $\frac{1}{2} + 0 \rightarrow \frac{1}{2} + 0$ spin configuration reactions,[3] namely

$$A_y^{\,2} + D_{LL'}^{\,2} + D_{SL'}^{\,2} = 1 \ . \tag{7}$$

(In this case, because there can be only three independent observables, it is also true that $P = A_y$, $D_{NN'} = 1$, $D_{LL'} = D_{SS'}$ and $D_{SL'} = -D_{LS'}$). By searching for appropriate energies and angles where $A_y$ *approaches* $\pm 1$, one can then *determine* $A_y$ through much less precise measurements of $D_{LL'}$ and $D_{SL'}$, which become very small. Because one is using a *null* method, i.e., searching for small asymmetries rather than large ones, accurate knowledge of the beam polarization and the effective analyzing power of the focal plane polarimeter become less important. Possible systematic errors arising from large spin-dependent dead-time corrections also disappear, as counting rates become comparable for all in-plane spin orientations. In order to make such precise analyzing power measurements most useful for beam polarization and cross-calibration checks, we have chosen targets that are easy to obtain and handle, and reactions with large cross sections in angle ranges where $A_y$ is known to be close to $\pm 1$.

Precise measurements of $D_{LL'}$ and $D_{SL'}$ for proton elastic scattering have now been completed at proton energies of 200, 190, and 180 MeV for $^{12}$C and at 180 MeV for $^4$He. At each energy, complete sets of measurements were made at three angles (in $1°$ steps) that spanned the region where $A_y$ was a maximum. Preliminary results for the $^{12}$C data are presented in Fig. 2.

The most interesting results were obtained for p + $^{12}$C scattering at 190 MeV, where optical model predictions[9] indicate that $A_y$ should come very close to 1. After several rounds of analysis of the data, our best determination of $A_y$ at $\theta_{cm} = 19.0°$ ($\theta_{lab} = 17.3°$) at $T_p = 188.9$ MeV yields a value of $A_y = 0.99963^{+0.00021}_{-0.00030}$. This error includes both statistical uncertainties and our best estimates of systematic contributions that arise due to lack of precise knowledge of the beam polarization magnitude and orientation and the effective analyzing power of the K600 focal plane polarimeter. Determinations of $A_y$ at other energies and angles have corresponding larger uncertainties as the analyzing power decreases from unity. In addition, the $D_{LL'}$ and $D_{SL'}$ data at 200 and 180 MeV are preliminary values for only two of the three angles measured, and represent only 30–40% of the total data collected. It is also significant that the energy dependence is rather smooth, as suggested by the solid lines in Fig. 2. Continuity arguments would therefore require that at an energy somewhat below 190 MeV the locus of $D_{LL'}$ and $D_{SL'}$ values will cross the origin; at that energy there will necessarily be an angle at which $A_y$ is identically equal to 1.

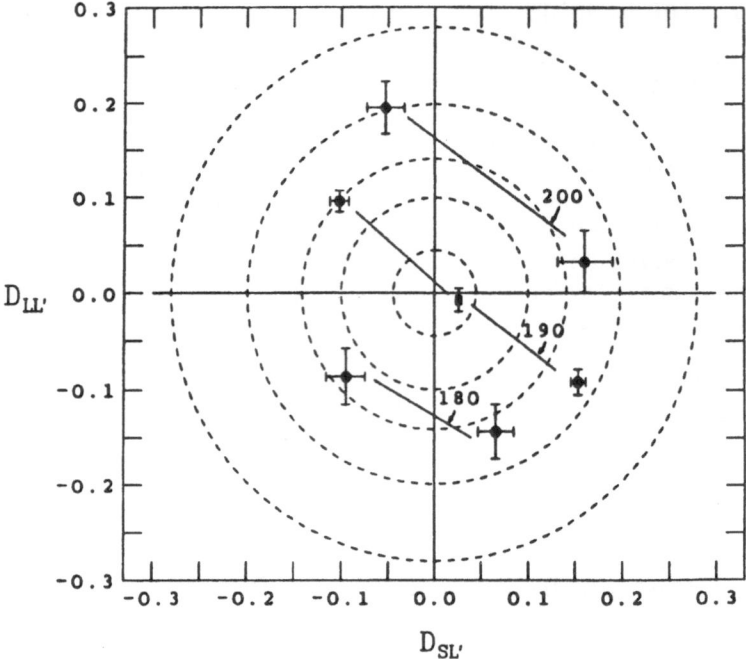

Fig. 2  Measurements of $D_{LL'}$ and $D_{SL'}$ for p + $^{12}$C elastic scattering at 200, 190, and 180 MeV. The solid lines are meant only to guide the eye to suggest the energy dependence of the curves. The dashed circles indicate contours of $A_y = 0.96$, 0.98, 0.99, 0.995, and 0.999 as one moves toward the origin.

Our eventual goal for this work is to map out a region in scattering angle and bombarding energy over which $A_y$ is known absolutely to within several tenths of a percent. This information can then be used to cross-calibrate other devices, *e.g.*, one could compare the left/right asymmetry measured in an in-beam polarimeter to the spin–up/spin–down asymmetry seen in the K600 spectrometer when positioned within this grid. The $^4$He data should prove useful for normalization of earlier pp analyzing power data[10] in which both helium and hydrogen gases were used during data acquisition.

Of perhaps more immediate interest to most participants at this conference, we have also recently finished measuring complete sets of polarization transfer observables ("complete" in the sense defined earlier) for several unnatural parity transitions in light closed-shell nuclei.[11] Specifically, we have measured the normal-component spin observables $D_{NN'}$, P, and $A_y$ at 200 MeV for the $1^+$, T=0 (12.71 MeV) and T=1 (15.11 MeV) states in $^{12}$C and for the $4^-$, T=0 (17.79 and 19.80 MeV) and T=1 (18.98 MeV) states in $^{16}$O. These spin observables were determined at five angles for the $^{12}$C $1^+$ transitions, corresponding to center-of-mass momentum transfers between 80 and 250 MeV/c, and at three angles for the $4^-$ transitions in $^{16}$O, at momentum transfers of 225 to 400 MeV/c. The angles were chosen to match those at which the in-plane spin transfer observables had been measured previously for these same transitions.[12] Typical excitation spectra are shown in Fig. 3.

The primary motivation for this work was to exploit the increased sensitivity pre-

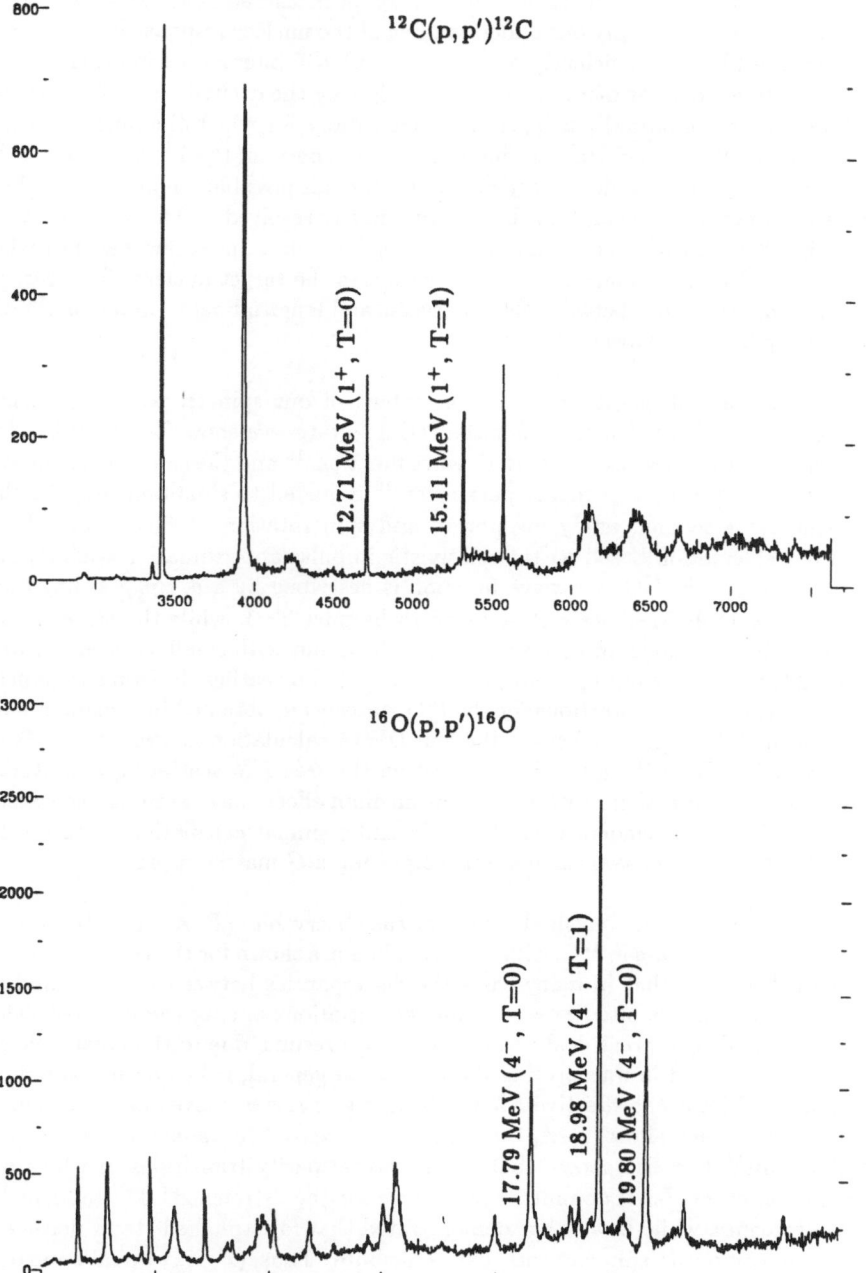

Fig. 3   Proton inelastic excitation spectra at 200 MeV on $^{12}$C at $\theta_{lab} = 16°$(*upper plot*) and on $^{16}$O at $\theta_{lab} = 28°$(*lower plot*).

dicted for these polarization transfer observables to some of the spin-dependent (and generally weaker) terms of the effective interaction used to describe nucleon-nucleus scattering at intermediate energies. Comparison of measured observables to their calculated values tests not only our understanding of the nuclear response for these transitions, but our ability to efficiently parameterize the $NN$ interaction and then modify it appropriately for nuclear medium effects. In selecting the excited states for this study, we chose strong, unnatural-parity, spin-flip transitions for which the nuclear structure has been carefully studied. In particular, different aspects of the $1^+$ states in $^{12}C$ have been investigated by a wide variety of probes, though possible ambiguities in the nuclear wave functions have still not been satisfactorily resolved.[13] The $4^-$ transitions in $^{16}O$, being "stretched" states, can only be excited (via a 1p–1h mechanism) when a single value of angular momentum is transferred to the target nucleus. This results in a simple proportionality between the transverse and longitudinal form factors,[14] which greatly simplifies the nuclear structure.

The statistical quality and angular extent of our spin transfer measurements are illustrated in Figs. 4 and 5. For the $^{12}C$ $1^+$ states we show four DWIA calculations, each of which uses Cohen-Kurath wave functions[15] and the same distorted waves, the latter based on optical model parameters[16] adjusted to simultaneously fit elastic scattering cross section, analyzing power, and spin rotation coefficient data.[17] Also shown is a calculation based on the relativistic impulse approximation code DREX.[18] For the $4^-$ states in $^{16}O$, the wave function is described by a $d_{5/2}p_{3/2}^{-1}$ configuration, which for the 18.98 MeV state is assumed to be pure T=1, while the states at 17.79 and 19.80 MeV are taken to be predominantly T=0 but with small T=1 admixtures of comparable magnitude but opposite sign, as suggested by earlier electron and pion scattering experiments.[19] Distortions for the $^{16}O$ states were obtained in a manner similar to that of the $^{12}C$ work. In all cases, the four DWIA calculations shown are the Franey-Love[20] and free Bonn[21] interactions, based on the free $NN$ scattering amplitudes, a modified Bonn interaction in which nuclear medium effects have been incorporated using a local density-dependent prescription,[22] and a similar calculation using the Paris interaction,[23] the latter two calculations employing a $G$-matrix approach.

Shown in Fig. 4 are our final values for the observables $(P-A_y)$ and $D_{NN'}$ for the strong isoscalar $1^+$ state in $^{12}C$, while $D_{NN'}$ values are shown for the two isoscalar transitions in $^{16}O$. Note that in many cases the discrepancies between the data and most of the calculations are often greater than the variations among the different calculations. I will not say a great deal about the $(P-A_y)$ results, due to the great difficulties we have encountered in interpreting these data. In general, it has been pointed out[24] that non-localities in the effective $NN$ interaction give rise to convection and composite spin currents, which allow $(P-A_y)$ to deviate from zero[25] (it vanishes in a direct-only PWIA). Moss[26] has also shown that for unnatural parity transitions in which a single combination of $[LSJ]$ dominates (as is true for the "stretched" $4^-$ states in $^{16}O$), $(P-A_y)$ vanishes, while Love[27] has demonstrated that for stretched states neither convection nor composite spin currents may contribute. Thus, $(P-A_y)$ would be expected to be small for the $^{16}O$ states (it is), and, though non-zero for the $1^+$ states in $^{12}C$, will require a much greater understanding of the nuclear structure of these states before any useful physics may be extracted.

Turning to $D_{NN'}$ for these isoscalar transitions, we note that the calculations using the Franey-Love $t$-matrix and the Bonn $G$-matrix (both the Free Bonn and the density-dependent Bonn) are quite similar to each other at all values of momentum

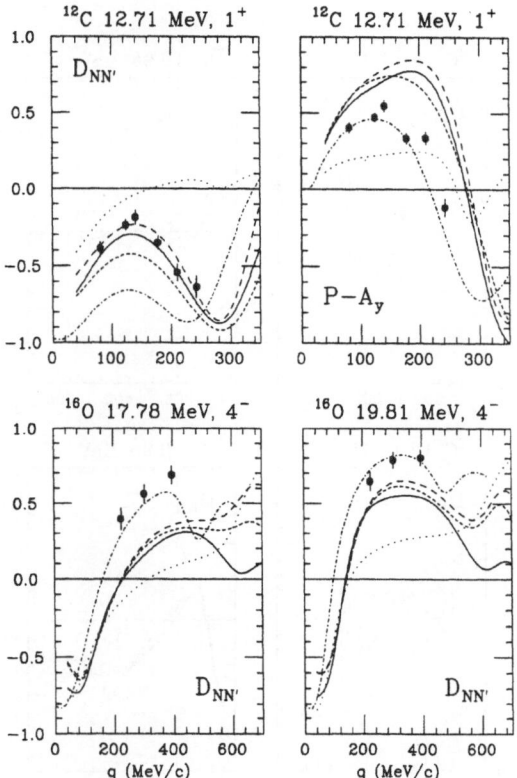

Fig. 4 Spin observables for the isoscalar transitions studied in $^{12}$C and $^{16}$O. In all cases, the solid curve is a Franey-Love interaction, the (short–)long–dashed curve is a (free) density-dependent Bonn interaction, the dotted curve is a Paris-Hamburg interaction, and the dot–dashed curve is a DREX calculation.

263

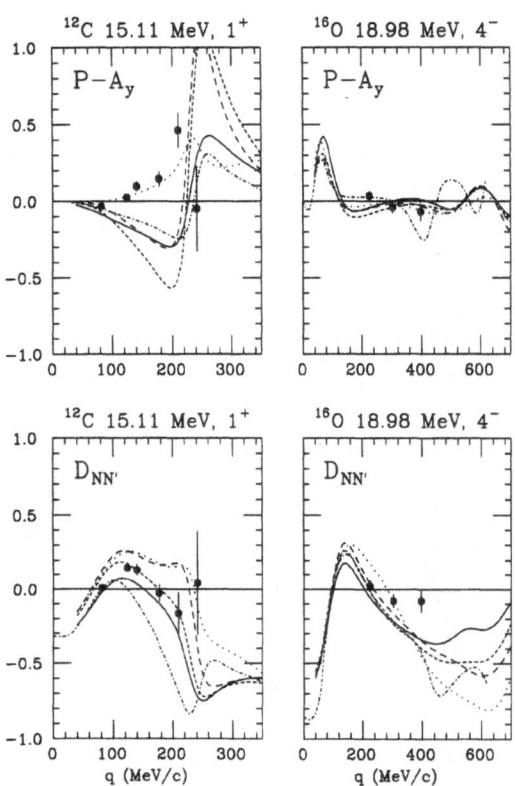

Fig. 5　Spin observables for the isovector transitions studied in $^{12}$C and $^{16}$O. The curves for the various interactions are described in the caption of Fig. 4.

transfer, while the calculations based on the Paris potential and DREX vary significantly from the other three, especially over the range where the data were measured. Inspection of the two Bonn calculations for the $^{12}$C state suggests a need to correct for medium effects for this transition, while the density-dependence is much weaker for the $^{16}$O states. This is consistent with the assumption that the $1^+$ state has a transition density that is peaked in the nuclear interior, while the "stretched" states are characterized by transition densities that are strongly surface peaked. It is also worth noting that for $^{16}$O the DWIA calculations are too small for both states, suggesting a problem with the isoscalar interaction. However, the *differences* between the calculations for the two states are due entirely to the small T=1 admixtures in the wave function. These differences are much larger than those indicated by the data, which implies either an error in the amount of isospin mixing introduced or a problem with the isovector interaction as well. Other experiments tend to rule out the former, and our own data support the latter, as will be shown below. We are currently looking into a means by which to place this idea on a more quantitative footing.

The final values obtained in this work for $(P-A_y)$ and $D_{NN'}$ for the two isovector transitions are shown in Fig. 5, along with the five calculations discussed above. As with the isoscalar states, $(P-A_y)$ is both measured and predicted to be close to zero for the $4^-$ stretched state in $^{16}$O. For the T=1 $1^+$ state in $^{12}$C, however, $(P-A_y)$ is significantly non-zero, but differs substantially from all calculations except the one based on the Paris interaction. The momentum transfer dependence is simply wrong for all other calculations. For $D_{NN'}$, the general agreement is not too bad at low values of $q$, but gets worse for momentum transfers greater than about 200 MeV/c. Although no single calculation fits the $^{12}$C data over the whole range of $q$, most follow the general trend of the data; the Free Bonn and Franey-Love calculations in particular track the data nicely over the entire first lobe of the form factor. The situation is quite different for the 18.98 MeV state in $^{16}$O, in that all of the calculations are quite similar to each other over the range of $q$ covered by the data, but all of the calculations pull away from the near-zero value of the data for momentum transfers above 300 MeV/c. Our interpretation of this discrepancy will be discussed shortly, after a brief digression into some interesting simplifications of the formalism presented earlier.

It was pointed out earlier, via Eq. 4, that values of all of the spin transfer observables can be generated if one is given a form for the nucleon-nucleus scattering amplitude $\overline{M}$, which is defined formally in Eq. 3. It has been shown by several groups[26,28] that if one works in a static, direct only, non-relativistic Plane Wave Impulse Approximation, the form for $\overline{M}$ simplifies considerably, to the extent that evaluation of the $D_{ij}$'s for unnatural parity, spin-flip transitions leads to expressions that are simple linear combinations of individual nuclear structure terms and bilinear products of the fundamental KMT $NN$ amplitudes. Explicitly, we find that, under the above assumptions, application of Eq. 4 yields directly

$$I_0 = (C^2 + B^2 + F^2)\chi_T^2 + E^2\chi_L^2$$

$$I_0 D_{nn} = (C^2 + B^2 - F^2)\chi_T^2 - E^2\chi_L^2$$

$$I_0 D_{pp} = (C^2 - B^2 + F^2)\chi_T^2 - E^2\chi_L^2$$

$$I_0 D_{qq} = (C^2 - B^2 - F^2)\chi_T^2 + E^2\chi_L^2$$

$$I_0 D_{n0} = I_0 D_{0n} = 2\chi_T^2 Re(BC^*)$$

$$I_0 D_{qp} = -I_0 D_{pq} = 2\chi_T^2 Im(BC^*). \qquad (8)$$

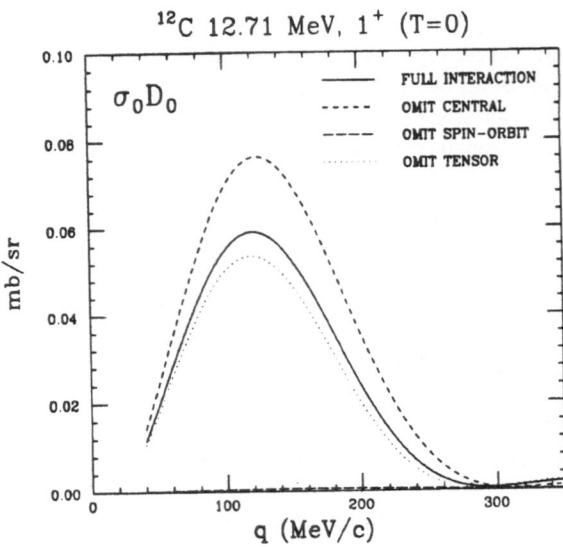

$^{12}$C 12.71 MeV, $1^+$ (T=0)

Fig. 6    Calculations of $\sigma_0 D_0$ for the 12.71 MeV (T=0) $1^+$ state in $^{12}$C from which individual terms of the interaction have been omitted. The calculations were made using the density-dependent Bonn interaction.[22]

From these expressions, it is clear that if a complete set of polarization transfer observables can be measured, these observables can be added and subtracted from one another to form appropriate combinations which isolate these interaction amplitudes ($A$, $B$, $C$, $E$, $F$) as well as the longitudinal and transverse nuclear form factors $\chi_L$ and $\chi_T$ (which are related to the two spin densities $\langle \vec{\sigma} \cdot \vec{q} \rangle$ and $\langle \vec{\sigma} \times \vec{q} \rangle$, respectively). Bleszynski and others[28] have defined such a set of linear combinations, labelled the $D_k$'s, which are given here in the lab frame:

$$D_0 \sigma_0 \equiv \sigma_0 [1 + (D_{SS'} + D_{LL'}) \cos\theta + D_{NN'} - (D_{LS'} - D_{SL'}) \sin\theta]/4 = C^2 X_T^2$$

$$D_1 \sigma_0 \equiv \sigma_0 [1 + D_{SS'} - D_{LL'} - D_{NN'}]/4 = E^2 X_L^2$$

$$D_2 \sigma_0 \equiv \sigma_0 [1 - (D_{SS'} + D_{LL'}) \cos\theta + D_{NN'} + (D_{LS'} - D_{SL'}) \sin\theta]/4 = B^2 X_T^2$$

$$D_3 \sigma_0 \equiv \sigma_0 [1 - D_{SS'} + D_{LL'} - D_{NN'}]/4 = F^2 X_T^2. \tag{9}$$

A similar reduction can also be performed in a relativistic PWIA, under the same set of assumptions, but this introduces additional nuclear structure dependences in the form of composite spin current terms.[29]

An obvious concern with this procedure lies in the assumptions that must be made to obtain these simple expressions. As a minimal check, we have used the best available calculation codes to test if the introduction of these approximations substantially alters the results of the calculation. A typical example is presented in Fig. 6, where we use the full density-dependent Bonn interaction to calculate the observable $\sigma_0 D_0$ for the T=0 $1^+$ state in $^{12}$C. This is a "complete" calculation, and does not involve any of the above simplifying assumptions. If Eq. 9 were exactly valid, omitting the spin-orbit interaction term $C$ should drive the calculation of $\sigma_0 D_0$ to zero, while omission of other terms should have little effect. As can be seen, these predictions are borne out reasonably well, in that removing the central and tensor terms individually changes the shape and magnitude of $\sigma_0 D_0$ only slightly, while removal of the spin-orbit term causes it to almost vanish. This and other similar studies[30] suggests that, in spite of the above caveats, the $D_k$'s should serve as a reliable guide as to which aspects of the $NN$ interaction or the nuclear structure may need modification to be brought into agreement with experiment.

Since there is not sufficient room to show all the $D_k$'s for all of the transitions we have studied, I will present only two sets. In Fig. 7 are shown the results for the $1^+$ isoscalar state in $^{12}$C. Despite the complexity of the nuclear structure for this state, the Franey-Love and Bonn interactions do a reasonable job of reproducing the full set of $D_k$'s. The momentum dependence in particular is tracked very well, though the Paris and DREX calculations seem to err in estimating the relative strengths of the different terms. In short, most of the comments made previously regarding $D_{NN'}$ for this transition apply to the $D_k$'s as well.

For the $4^-$ isovector state in $^{16}$O, however, the agreement is not quite as satisfying. Though the data for $D_0$ and $D_3$ are reproduced fairly well, serious deviations arise for the other two combinations, which are sensitive predominantly to the tensor interaction, including exchange contributions. Note that all five calculations are remarkably close in their predictions for $D_1$ and especially $D_2$, yet all overestimate the former and underpredict the latter as one goes out in momentum transfer. One interpretation would be that, for almost any reasonable parameterization of the $NA$ interaction, one can not predict the strength of the isovector tensor force *if* one is also constrained to fit the $NN$ database. Similar conclusions can be drawn for the two isoscalar $4^-$ states in $^{16}$O, though it is not clear how strongly correlated these two observations may be. We are continuing our investigations into possible means of resolving these problems, perhaps empirically, and I would refer the reader to the contribution by Ed Stephenson at this conference for details on some of our preliminary results along these lines.

Though our research efforts to date have concentrated primarily on transitions to discrete states, I should also mention some very interesting results[31] obtained recently on continuum studies in $^{208}$Pb. I will be brief, since Jochem Wambach will go into much more detail in his discussion of these same results later at this conference.

We have measured the spin-flip probability parameter $S_{NN'}$ (which is defined as $[1 - D_{NN'}]/2$) at 200 MeV on $^{208}$Pb for excitation energies between 2 and 24 MeV for momentum transfers from 0.45 to 0.84 fm$^{-1}$, as part of a larger program to investigate the spin response of the nuclear continuum. The nucleus $^{208}$Pb was chosen because of its narrow giant resonance widths, a property of heavier targets, and a lack of resonance splitting from nuclear deformation. Throughout the measurements, energy resolution in the K600 focal plane was kept at approximately 40 keV, so that fine structure could be studied. Spin-flip data were taken at four angles ($\theta_{lab} = 8°$, $10°$, $12°$, and $15°$). The smallest scattering angle corresponds to the peak for an $L = 2$ transfer, so we should be sensitive to spin quadrupole strength.

Our final results are presented in Fig. 9 for all four scattering angles. To improve our statistical precision for comparison with theory, we have collected the data into

$^{12}$C 12.71 MeV, $1^+$ (T=0)

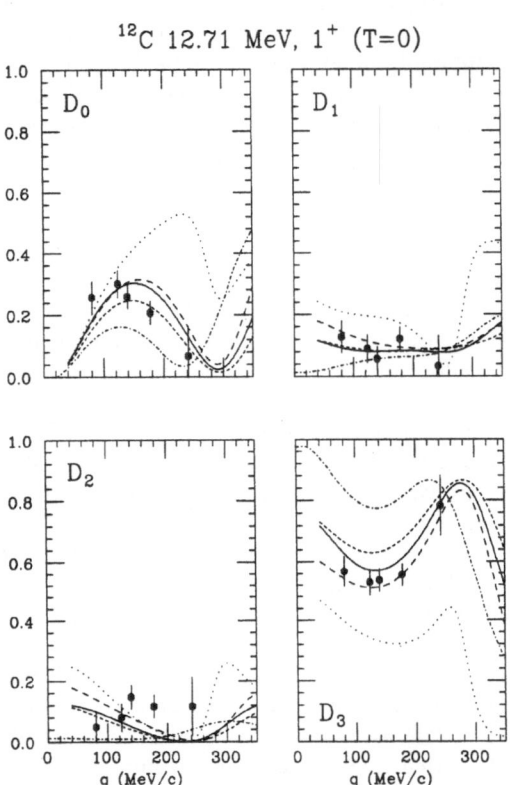

Fig. 7    Spin observable combinations for the $1^+$ isoscalar transition (12.71 MeV) in $^{12}$C. The curves for the various interactions are described in the caption of Fig. 4.

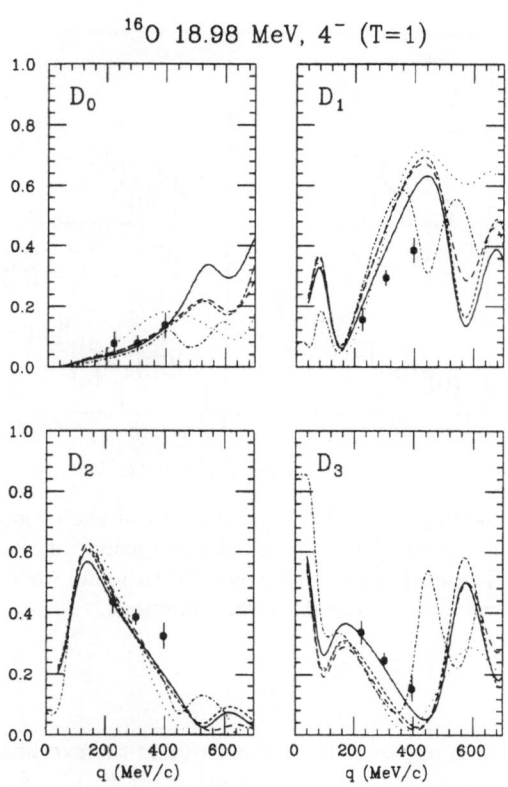

Fig. 8    Spin observable combinations for the $4^-$ isovector transition (18.98 MeV) in $^{16}$O. The curves for the various interactions are described in the caption of Fig. 4.

1 MeV bins. At all angles the data exhibit a broad structure centered near 7 MeV with a 2.5 to 3 MeV width. Since this energy range is known to have a high density of low angular momentum states,[32] this structure may result from a superposition of the spin-flip strength for these states. There is some suggestion of a narrower structure near 12 MeV, but the statistics are marginal. Fig. 9 also shows calculations by Unkelbach and Wambach,[31] based on 1p–1h RPA including 2p–2h damping into the continuum. Distortions are described by phenomenological optical potentials of Woods-Saxon type,[33] and both finite– and zero-range residual interactions were used.

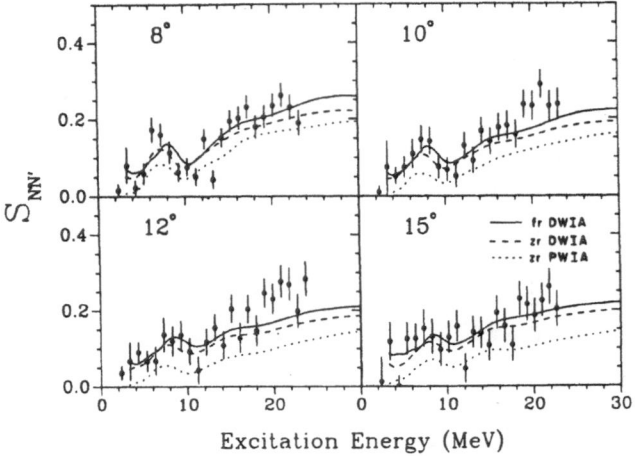

Fig. 9   Spin-flip probability for $^{208}$Pb as a function of excitation energy. The solid (dashed) curves are the result of DWIA calculations using a finite-range (zero-range) residual interaction. The dotted curves correspond to plane wave calculations with a zero-range force. All calculations use RPA wavefunctions.

The first conclusion apparent from this figure is the excellent overall agreement of the RPA calculations with the measured values of $S_{NN'}$. As Jochem will show, this agreement also extends to the calculated differential cross sections. A perhaps somewhat more surprising result, and also somewhat disturbing, is the strong sensitivity of this spin observable to distortions, as indicated by the difference between the DWIA and PWIA curves. This is believed to be due largely to the spin-orbit term in the nuclear distortions. This sensitivity will lead to problems in extracting spin-flip strength for the various multipoles if distortions are not properly taken into account, a fact often overlooked in earlier work. The dependence of $S_{NN'}$ on the residual interaction, illustrated by the differences between the finite- and zero-range DWIA calculations, is thought to arise from the tensor-exchange part of the interaction, which acts strongly in the isoscalar $\Delta S=1$ channel. Though not shown in the figure, a multipole decomposition of these calculations suggests[31] that in the continuum region the major contributors to the observed spin-flip strength are the natural parity excitations, with no obvious localization of strength for any spin and parity.

# EXPERIMENTAL RESULTS – WORK IN PROGRESS

I would now like to mention just briefly some of the other experiments for which data acquisition is essentially complete, but for which only preliminary results are available at this time.

In addition to the continuum studies described above, we have also exploited the unique features of the K600 spectrometer/polarimeter system to study the high-spin, particle-hole states that occur at slightly lower excitation energy in $^{208}$Pb. Since high-spin states with a stretched, maximum–$J$ configuration have a unique particle–hole description in a $1\hbar\omega$ basis, comparative studies of their excitation with several probes have been an important source of information about nuclear structure, components of the effective nucleon-nucleon interaction, and reaction models of inelastic scattering at intermediate energies. Comparative studies of stretched states in a doubly-closed-shell heavy nucleus offer the unique possibility of observing both proton and neutron particle-hole excitations that involve different shell-model orbitals.

We have concentrated our efforts on the known high-spin states in $^{208}$Pb at 6.10, 6.43, 6.74, and 7.05 MeV, with the configurations $(\nu i_{11/2}, \nu i_{13/2}{}^{-1})$ 12$^+$ (neutron), $(\nu j_{15/2}, \nu i_{13/2}{}^{-1})$ 12$^-$ (neutron), $(\nu j_{15/2}, \nu i_{13/2}{}^{-1})$ 14$^-$ (neutron), and $(\pi i_{13/2}, \pi h_{11/2}{}^{-1})$ 12$^-$ (proton), respectively. We have measured both differential cross sections and analyzing powers for the excitation of these states at 200 MeV with an overall resolution of 25 keV. Since the high-spin states of interest in $^{208}$Pb are preferentially excited at large momentum transfer (typically 1.8 to 2 fm$^{-1}$), our measurements focussed on the angular range 28° – 52°, corresponding to a range of momentum transfer from 1.5 to 3 fm$^{-1}$ at 200 MeV.

Fig. 10 shows the inelastic proton spectrum at a scattering angle of 36°, for a range of excitation energies from the first-excited 3$^-$ state at 2.615 MeV up to about 7.5 MeV. At this angle, the four high-spin states are excited about equally strongly, and are the dominant feature of the excitation region between 6 and 7 MeV. The 12$^+$ (neutron), 14$^-$ (neutron), and 12$^-$ (proton) states are easily resolved from nearby, weaker states of lower multipolarity, while the 12$^-$ (neutron) state at 6.43 MeV is not as cleanly separated from nearby levels. A more complete analysis, including both cross section and analyzing power angular distributions for over one hundred states in this energy range, is in progress. Measurements have been extended forward in angle to about 8° in order to provide additional information for the identification of lower-spin states in the spectrum. A more detailed comparison with recent high-resolution $(e, e')$ measurements[34] at Bates is underway.

Our measurements of the analyzing power angular distributions for these states indicate a sensitivity to the neutron and/or proton character of the nuclear excitation that is similar to predictions based on the DWIA. DWIA calculations of $D_{NN'}$, however, predict a considerably more pronounced sensitivity of this observable to the isospin character of the nuclear excitation. We have therefore recently completed the first measurements of $D_{NN'}$ and $P$ for these states at momentum transfers of 1.9 and 2.0 fm$^{-1}$ (34° and 38°), where the inelastic excitation of the 12$^-$ and 14$^-$ states reach their maximum values. We hope to extract $D_{NN'}$ values with statistical uncertainties of ±0.07. Online results suggest that there are significant differences among the values of $D_{NN'}$ for the high-spin states that correspond to predominantly neutron or proton particle-hole excitations.

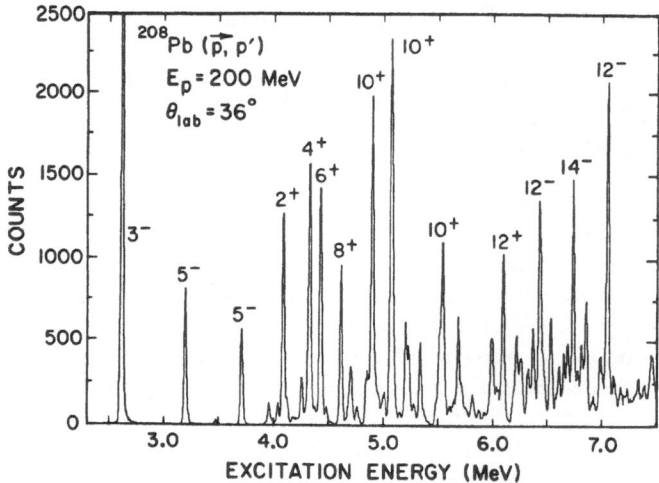

Fig. 10 Excitation spectrum of $^{208}$Pb at 200 MeV at 36°.

Another interesting case to investigate via proton inelastic scattering is that of $0^+ \rightarrow 0^-$ transitions. The constraints of zero angular momentum in both the initial and final nuclear states lead to simplifications not only in the nuclear structure ($\chi_T$ vanishes), but in the scattering matrix as well. Simple counting arguments reveal that there can be only three independent observables, so measurements of cross section, analyzing power, and an in-plane spin rotation parameter will therefore provide complete information on the transition amplitude (within an overall phase), and should place severe limits on the acceptability of different theoretical models. Because the spin-orbit component of the $NN$ effective interaction can not contribute, and the central component is weak, excitation of these states must occur primarily through the tensor component, one of the more poorly understood terms in the effective interaction. Moreover, these observables are expected to provide information specifically on the exchange terms, since it is this part of the interaction that yields non-zero values for some of the spin observables.[35]

To this end, we have performed high-resolution ($\sim$30 keV FWHM) studies of the $0^-$ states in $^{16}$O at 10.957 MeV (T=0) and 12.797 MeV (T=1). Cross section and analyzing power angular distributions for these two states were measured at 200 MeV for c.m. angles between 7.0° and 45.0° in 4° steps, providing the highest quality data currently available on any $0^+ \rightarrow 0^-$ transition. We will compare our results for the T=0 state to recent relativistic calculations, using the RIA-based code DREX,[18] and nonrelativistic calculations based on the Franey-Love $t$-matrix[20] and the code DW81.[36] Both approaches use a microscopic $t$-matrix fit to $NN$ phase shifts, and knock-on exchange is included explicitly. Woods-Saxon phenomenological Schrodinger optical potentials fitted to elastic cross section, analyzing power, and spin rotation data were employed in the DW81 calculations to correct for nuclear distortion effects, while the DREX calculations use the prescription of Horowitz.[37]

For the $0^-$ T=0 state, our analysis is nearly complete, and we expect final results for the cross section and analyzing power angular distributions to be available soon. For the much weaker T=1 state, on the other hand, better peak-fitting techniques need to be developed so that reliable values of the observables can be extracted, and such a program is currently underway. We have also completed the first polarization-transfer measurements for these transitions, from which the spin rotation parameter $Q$ will be determined. (This is truly a testament to the capabilities of a high-resolution device such as the K600 spectrometer coupled to a high-efficiency polarimeter!) This will allow for a complete decomposition of the transition matrix into its two non-vanishing complex amplitudes.

As one final illustration of some recent work, I would like discuss some measurements performed at 200 MeV using proton scattering on $^{10}$B at fairly low excitation energy. Cross section and analyzing power angular distributions for both elastic and inelastic scattering were measured between $\theta_{lab} = 7.5°$ and 80° in 2.5° steps, while lengthier runs were taken every 10° between 20° and 60° to accumulate the statistics necessary for precise measurements of the induced polarization $P$ and the polarization transfer coefficient $D_{NN'}$. All known transitions below 6.5 MeV were observed. We intend to extract information for the strongest and best isolated of these transitions, including elastic scattering.

Of particular interest is the $3^+ \rightarrow 0^+$ transition to the T=1 state at 1.74 MeV, a "stretched" $0\hbar\omega$ excitation that proceeds via spin recoupling within the $p_{3/2}$ shell. The form factor for this transition has been measured in $(e, e')$ studies[38] over a large range of momentum transfer, from 100 to 800 MeV/c. Unlike most stretched states, which involve $1\hbar\omega$ excitations, this state occurs sufficiently low in energy that there is essentially no underlying background, and the state can be cleanly observed over a comparable range of momentum transfers in proton inelastic scattering. This makes it a good candidate with which to test our knowledge of the reaction mechanism and the spin-dependent parts of the isovector effective interaction, if the distortions due to the highly deformed $3^+$ core can be properly handled.

Preliminary results for $A_y$ and $D_{NN'}$ are shown in Fig. 11. As can be seen, $A_y$ is nearly constant at about 0.25, while $D_{NN'}$ is very close to zero at all momentum transfers. The errors shown are purely statistical, and not all of the data has been included in all points. The calculation shown was generated using the DWIA code DW86,[39] assuming a pure $p_{3/2}$ $\Delta$T=1 transition. A reasonable fit is obtained for $A_y$ and for the differential cross section (not shown), but the calculation for $D_{NN'}$ is obviously far from the data. Between 200 and 400 MeV/c, this discrepancy is reminiscent of that observed for the $4^-$ isovector state in $^{16}$O discussed earlier (see Fig. 5). A common explanation in terms of medium modifications to the effective isovector tensor interaction is being sought.

As one final note, I would like to brag once more and point out that the final data point, taken near $q = 620$ MeV/c, corresponds to a cross section of only a few hundred nb/sr, and that the errors shown represent less than half of the total data, taken over only about one day.

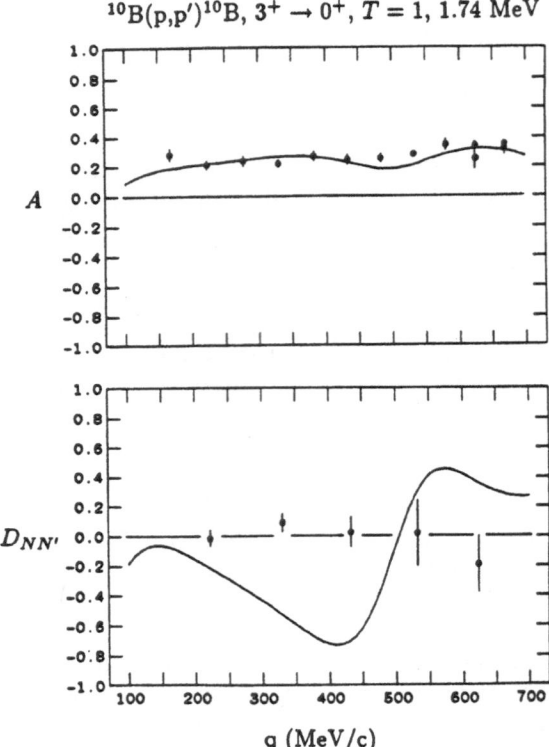

$^{10}$B(p,p′)$^{10}$B, $3^+ \rightarrow 0^+$, $T = 1$, 1.74 MeV

Fig. 11   Measurements of $A_y$ and $D_{NN'}$ for the $3^+ \rightarrow 0^+$, T=1 transition in $^{10}$B. The solid line is a DWIA calculation described in the text.

## FUTURE PLANS

To conclude this talk, I would like to mention quickly some of our "second generation" experiments, in order to convey some feeling for what we currently view as the most important set of measurements to be made. None of these experiments, however, fall into the speculative domain – indeed, most of the experiments I will discuss have already been approved by the IUCF Program Advisory Committee, and should have data acquisition well underway, if not completed, by next summer, or represent continuations of existing programs.

There are several obvious extensions of our current efforts. As I have stressed, a major thrust of the K600 program has been the investigation of discrete states of unnatural parity through proton inelastic scattering. The high resolution and count rate capabilities of the spectrometer system allow the study of states that are only weakly excited or fall in regions of high level density. By measuring a variety of spin observables for these transitions, one can use the quantum numbers of the nuclear excitation as a tool to enhance the effects of particular spin-dependent terms in the effective $NN$ interaction or to isolate specific pieces of the nuclear response.

The $D_{NN'}$ measurements on the $0\hbar\omega$ transition to the isovector $0^+$ state in $^{10}$B deviate significantly from standard DWIA calculations, for a state that has been well studied in electron scattering. To understand the source of this discrepancy, it may prove useful to measure several of the in-plane spin transfer coefficients as well, and we are presently running calculations to gain intuition as to which observables may best serve this purpose. Similar statements apply to our first measurements of $D_{NN'}$ and $P$ for the high-spin stretched states in $^{208}$Pb, although the existing data set requires much more detailed analysis before such a program would be considered.

We have also received approval for investigations of the T=0 and T=1 stretched $6^-$ states in both $^{32}$S and $^{28}$Si, which will require that high resolution be maintained for extended periods. We have very recently had our first run on $^{28}$Si, during which we obtained precise cross section and analyzing power data for elastic scattering at 200 MeV over a wide angle range, data necessary for eventual calculation of optical distortions for inelastic scattering from this nucleus. Some $D_{NN'}$ measurements on the $6^-$ states of primary interest were taken at $\theta_{lab} = 29°$, $35°$, and $41°$. It is important to point out that during these measurements data is simultaneously taken over a broad range of excitation, so that $D_{NN'}$ values may also be extracted for nearby $5^-$ natural parity states, as well as many others. Our specific goal is this work, however, remains an investigation into the tensor exchange terms in the effective $NN$ interaction as it operates within the nuclear medium, especially at momentum transfers above 200 MeV/c. This piece of the interaction appears to need substantial modification, for both T=0 and T=1, as evidenced first in the study of the $4^-$ states in $^{16}$O, and as later corroborated by study of the $0^+$ excitation in $^{10}$B that was discussed previously.

More generally, there is currently much theoretical interest in the distribution and fragmentation of the M6 response in the $sd$-shell. Though (p,n) and electron scattering studies have provided information on the location of the T=1 strength, the isoscalar strength is largely unexamined experimentally. Comparison of new data to the results of extended basis shell model calculations should prove very interesting. I would refer the reader to the contribution on exactly this topic presented at this conference by Jim Carr.

In addition to the studies on discrete states and continuum work already discussed, there are plans to investigate the quasifree region of the $(p, p')$ spectrum. The physics interest in such studies stems from various theoretical models[40] which suggest that the properties of nucleons bound in the nuclear medium differ from their corresponding free values, and that these effects should be apparent through study of inclusive, quasifree scattering (QFS) spin observables. A key assumption is that all of the proton scattering inclusive yield, at energy losses of $\omega \approx q^2/2m$, results from single-step QFS, $e.g.$, contributions from strong final-state interactions or more complicated multi-step scattering processes are not included. A program of $exclusive$ QFS measurements has been proposed to examine this assumption experimentally. The goal of this program, a collaborative effort between IUCF and the University of Maryland, is to compare the integrated exclusive QFS yield to the simultaneously measured inclusive $(p, p')$ yield for nuclei such as $^3$He, $^4$He, and $^{40}$Ca. By using the K600 spectrometer system as one detector arm, it will be possible to compare integrated exclusive to inclusive observables including the cross section, analyzing power, and $D_{NN'}$. The second proton will be detected in a large solid angle array of NaI scintillators, fronted by a series of wire chambers for trajectory information.

Finally, on a slightly longer timescale, there are several technical developments underway which should open up new possibilities for areas of study. Through installation of a septum magnet, we will be able to measure scattering angles as small as 4° while continuing to run the proton beam into a shielded dump (which enables the system to tolerate much higher beam currents). This will allow us, for example, to examine the distribution of M1 strength with much higher precision, since we will be able to observe fairly weak $1^+$ states and to track them over a larger fraction of the first lobe of the transition density. Measurements of this type have been proposed for several of the 4N even-even nuclei in the $sd$-shell.

A second new development, closely correlated with the septum magnet installation, will be the addition of several large solid angle $BaF_2$ arrays for use in $(\vec{p}, p'\gamma)$ studies, work that will be carried out in collaboration with Oak Ridge National Laboratory. Approval has been given for an investigation into the photon decay of $^{12}C$ following proton excitation to the T=1 $1^+$ state at 15.11 MeV, a $0^+ \to 1^+$ transition that has already been extensively studied at IUCF in $(\vec{p}, \vec{p}')$. We plan to position one detector pack, which consists of 19 individual $BaF_2$ scintillators, photomultiplier tubes, and bases, directly above the target. At this location, 90° out of the reaction plane, the coincident photon yield will provide an independent determination of $D_{NN'}$ and $(P-A_y)$, thus confirming (we hope!) our previous polarimeter-based measurements. However, if the detected photon lies $in$ the scattering plane, the experiment becomes sensitive to two new scattering amplitudes that play a major role in describing the spin dependence of knock-on exchange,[41] $A_{Kq}(\Sigma \cdot K)(\sigma \cdot q)$ and $A_{qK}(\Sigma \cdot q)(\sigma \cdot K)$. In a relativistic plane-wave impulse approximation, these terms are proportional to the composite spin convection currents $\langle \vec{\sigma} \times \vec{J} \rangle$ and $\langle \vec{\sigma} \cdot \vec{J} \rangle$, respectively.[29] Moreover, if one also measures the sideways and longitudinal analyzing powers, $i.e.$, the coincident proton yield asymmetry as one reverses the direction of the $in$-$plane$ proton polarization (a quantity that vanishes in a singles measurement due to parity conservation), it can be shown[41] that one obtains sufficient information to completely determine the $0^+ \to 1^+$ transition amplitude. This is possible due to the sensitivity of partially out-of-plane photon detection to the phases between the individual $NN$ amplitudes, not just the magnitude information provided by $(\vec{p}, \vec{p}')$ studies (see, for example, Eq. 8). Complete specification of any transition amplitude has never been accomplished other than for transitions of the form $0^+ \to 0^\pi$, for which only three observables are required. Though details are still a little rough, a schematic illustration of the experimental layout we will use is provided in Fig. 12.

This same system will also be used by the ORNL collaborators for nuclear structure studies, probing the Giant Dipole Resonance via photon decay techniques. The goal is to use the coincident photon requirement for a $(\vec{p}, p'\gamma)$ reaction to isolate the GDR cleanly from other nearby excitations, thus allowing for a quantitative study of the hadronic excitation of this very fundamental mode of collective nuclear behavior. The results should provide direct information on the isovector proton-nucleus interaction (symmetry potential) and shed light on the details of the GDR transition density and its variation (if it exists) as one moves across the resonance. The first nuclei to be studied will be $^{40}Ca$ and $^{208}Pb$.

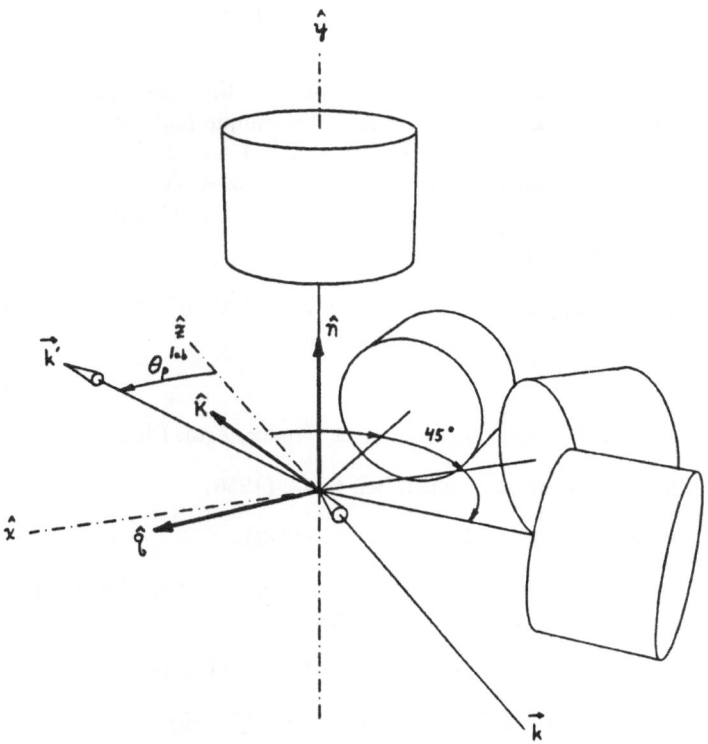

Fig. 12   Experimental configuration for $(\vec{p}, p'\gamma)$ studies showing the placement of $\gamma$ detectors both in and out of the reaction plane. Also shown are the proton and photon coordinate systems.


## SUMMARY AND OUTLOOK

I hope I have convinced you that spin transfer investigations using the $(\vec{p}, \vec{p}')$ reaction are alive and well, and that this technique can be used to address a number of important issues across the broad range covered today by nuclear physics. At IUCF, we are currently accumulating data of unprecedented statistical accuracy, and envision a healthy program that will continue for many more years, with several important technical advances already underway. I hope that at least a few of the measurements I have described strike you as interesting areas for continued investigation.

On the other hand, to use an old Telluride cliché, I feel it is important to emphasize that we are no longer hoping for the much sought after, but elusive, "smoking gun" experiment, that will clearly and unequivocally resolve some of the long-standing questions in this field, such as the 'need for relativity', or 'evidence for effective masses', phrases used so frequently over the last ten years. Progress will, unfortunately, be slow, and large databases need to be acquired before patterns will emerge from the morass. The importance of renewed and continuing theoretical interest in these studies can not be overemphasized, and if I have done nothing more than stimulate a few of you to go home and start thinking about these data, then I have succeeded today.

# ACKNOWLEDGEMENTS

It is a pleasure to acknowledge the efforts of the many people who over the past few years have been involved in making this unique piece of apparatus work and in taking and analyzing these data. In particular, I would like to thank Edward J. Stephenson and Jerry Lisantti, and my students Allena K. Opper and Steven P. Wells. The help of A.D. Bacher, S.M. Bowyer, S. Chang, J. Liu, C. Olmer, R. Sawafta, and T. Rinckel is also greatly appreciated.

This work was supported in part by the U.S. National Science Foundation.

# REFERENCES

1. A.K. Kerman, H. McManus, and R.M. Thaler, Ann. Phys. **8**, 551 (1959).

2. L. Wolfenstein, Ann. Rev. Nucl. Sci. **6**, 43, (1956).

3. G.G. Ohlsen, Rep. Prog. Phys. **35**, 717 (1972).

4. S.W. Wissink *et al.*, in *Proc. 7th Intl. Conf. on Polarization Phenomena in Nuclear Physics*, Paris (1990), p. C6-557.

5. G.P.A. Berg *et al.*, IUCF Scientific and Technical Report (1986) p. 152.

6. E.J. Stephenson *et al.*, Phys. Rev. C **42**, 2562 (1990).

7. S.W. Wissink *et al.*, to be submitted to Nucl. Inst. Meth.; see also Ref. 11.

8. LeCroy Research Corporation, 700 Chestnut Ridge Road, Chestnut Ridge, NY 10977.

9. P. Schwandt, private communication.

10. J. Sowinski *et al.*, IUCF Scientific and Technical Report (1986) p. 11.

11. A.K. Opper, Ph.D. thesis, Indiana University (1991).

12. C. Olmer, in *Antinucleon- and Nucleon-Nucleus Interactions*, ed. G.E. Walker, C.D. Goodman and C. Olmer, (Plenum Press, New York, 1985) p. 261.

13. J.B. McClelland *et al.*, Phys. Rev. Lett. **52**, 98 (1984).

14. R.A. Lindgren and F. Petrovich, in *Spin Excitations in Nuclei*, eds. F. Petrovich, G.E. Brown, G.T. Garvey, C.D. Goodman, R.A. Lindgren and W.G. Love, (Plenum Press, New York, 1982) p. 323.

15. S. Cohen and D. Kurath, Nucl. Phys. **A101**, 1 (1967).

16. P. Schwandt, computer code RUNT (unpublished).

17. E.J. Stephenson, in *Antinucleon- and Nucleon-Nucleus Interactions*, eds. G.E. Walker, C.D. Goodman and C. Olmer, (Plenum Press, New York, 1985) p. 299.

18. E. Rost and J.R. Shepard, computer code DREX (unpublished); E. Rost and J.R. Shepard, Phys. Rev. C **35**, 681 (1987).

19. D.B. Holtcamp *et al.*, Phys. Rev. Let. **45**, 420 (1980).

20. W.G. Love and M.A. Franey, Phys. Rev. C **24**, 1073 (1981); M.A. Franey and W.G. Love, Phys. Rev. C **31**, 488 (1985).

21. R. Machleidt *et al.*, Phys. Rep. **149**, 1 (1987); R. Machleidt, Adv. Nucl. Phys. **19**, 189 (1989).

22. K. Nakayama *et al.*, Nucl. Phys. **A431**, 419 (1984).

23. H.V. von Geramb, in *The Interaction Between Medium Energy Nucleons in Nuclei*, ed. H.O. Meyer, (A.I.P. New York, 1983) p. 44.

24. W.G. Love and A. Klein, in *Proc. 6$^{th}$ Intl. Symp. Polarization Phenomena in Nuclear Physics*, Osaka Japan (1985); W.G. Love and J.R. Comfort, Phys. Rev. C **29**, 2135 (1984).

25. J. Piekarewicz *et al.*, Phys. Rev. C **32**, 949 (1985).

26. J.M. Moss, Phys. Rev. C **26**, 727 (1982).

27. W.G. Love, in *Spin Excitations in Nuclei*, eds. F. Petrovich, G.E. Brown, G.T. Garvey, C.D. Goodman, R.A. Lindgren and W.G. Love, (Plenum Press, New York, 1982) p. 205.

28. E. Bleszynski, M. Bleszynski, and C.A. Whitten Jr., Phys. Rev. C **26**, 2063 (1982).

29. J.R. Shepard, E. Rost, and J.A. McNeil, Phys. Rev. C **33**, 634 (1986).

30. T. Carey, in *Proc. 6$^{th}$ Intl. Symp. Polarization Phenomena in Nuclear Physics*, Osaka Japan (1985); S. Seestrom-Morris *et al.*, Phys. Rev. C **26**, 2131 (1982).

31. J. Lisantti *et al.*, accepted for publication in Phys. Rev. C.

32. M.J. Martin, Nuclear Data Sheets **47**, 797 (1986).

33. F.E. Bertrand *et al.*, Phys. Rev. C **34**, 45 (1986).

34. J.P. Connelly *et al.*, Bull. Am. Phys. Soc. **34**, 1152 (1989).

35. J. Piekarewicz, Phys. Rev. C **35**, 675 (1987).

36. J. Comfort, code DW81 (unpublished).

37. C.J. Horowitz, Phys. Rev. C **31**, 1340 (1985).

38. R. S. Hicks *et al.*, Phys. Rev. Lett. **60**, 905 (1988).

39. R. Schaeffer and J. Raynel, program DW70, as modified by S. M. Austin, W. G. Love, J. R. Comfort, and C. Olmer.

40. C.J. Horowitz and M.J. Iqbal, Phys. Rev. C **33**, 2059 (1986); C.J. Horowitz and D.P. Murdock, Phys. Rev. C **37**, 2032 (1988).

41. J. Piekarewicz, E. Rost, and J.R. Shepard, Phys. Rev. C **41**, 2277 (1990).

# MODIFICATIONS OF THE EFFECTIVE ISOVECTOR INTERACTION
# FROM STUDIES OF $(\vec{p}, \vec{p}')$ POLARIZATION TRANSFER

Edward J. Stephenson and Jeffrey A. Tostevin†

Indiana University Cyclotron Facility
Bloomington, IN 47405

†University of Surrey
Guildford, Surrey GU2 5XH, United Kingdom

## INTRODUCTION

A complete set of polarization transfer coefficients $(D_{ij})$, as well as differential cross section, analyzing power $(A_y)$, and induced polarization $(P)$, are now available[1,2] for the $4^-$ "stretched" $T = 0$ and $T = 1$ transitions in $^{16}O(\vec{p}, \vec{p}')^{16}O$ at $E_p = 200$ MeV. These transitions at 17.79 and 19.80 MeV $(T = 0)$ and at 18.98 MeV $(T = 1)$ can be described within the framework of the distorted wave impulse approximation, which models the transition with an effective $t$-matrix based on NN scattering. The spin transfer $(\Delta S = 1)$ required by the dominant $1p_{3/2}^{-1}1d_{5/2}$ character of these transitions emphasizes their sensitivity to the spin-orbit and tensor parts of the $t$-matrix. Various interactions[3,4] based on free NN scattering (phase shifts or potentials) often agree with each other but not with the measurements, giving several systematic discrepancies[1,2] with the polarization transfer observables. Because the "stretched" transitions occur predominantly in the low density of the nuclear surface, interactions[4] that correct for Pauli blocking in the nuclear medium have little effect on these calculations.[2]

In a recent article, Brown and Rho[5] have suggested additional modifications to the isovector interaction in the nuclear medium. These may be characterized as an increase in the $\rho$-meson coupling and a decrease in the $\rho$-meson mass in proportion to the change in the nucleon effective mass (from $m$ to $m^*$). In this contribution, we will illustrate a change to the effective isovector interaction that removes the discrepancies for the $T = 1$ transition and compare this change quantitatively with the Brown and Rho suggestion.

## MODIFICATION TO THE EFFECTIVE ISOVECTOR INTERACTION

The calculations reported in the contribution were made with the impulse approximation program DW86.[6] The distorted waves in the entrance and exit channels were calculated using an Dirac optical potential adjusted to reproduce the cross section, analyzing power, and spin rotation data for the elastic scattering of 200-MeV protons from $^{16}$O (ref. 7). The potential parameter optimization was made using the program RUNT,[8] and the Schrödinger equivalent potential[9] used in DW86. The free interaction taken as the basis for this study was from Franey and Love.[3]

The output from DW86 was arranged to provide observables for separate pieces of the interaction as well as their real and imaginary cross terms. These sets of observables were combined in a second program to yield a final set equivalent to the altered interaction $t'$,

$$t' = a_0 t + a_1 \delta t_1 + a_2 \delta t_2 + \dots \; , \tag{1}$$

where $t$ was the original interaction, $\delta t_i$ modifications to it, and $a_i$ adjustable coefficients. This program compared the observables with the data for the 18.98 MeV, $T = 1$ transition, adjusting the coefficients to best reproduce the data. Comparison with the calculation was always made to either the cross section or the product of the cross section and a polarization observable. During the adjustment, $a_0$ was real while the remaining coefficients were allowed to be complex.

Modifications to $\rho$-meson exchange should appear in the isovector term of the tensor interaction, $t_\tau^T$, and the vector-isovector term of the central interaction, $t_{\sigma\tau}^C$. Using a model that will be discussed in the next section, forms were chosen for $\delta t_1$ and $\delta t_2$ that modelled the $q$-dependence of the $\rho$-meson contribution in the tensor and central interactions, respectively. Using the effective range expansion and values of $R$ from among those used in ref. 3

$$\delta t_1 = \delta t_\tau^T = 5 \times 10^4 \; 32\pi \; \frac{q^2 R^7}{[1 + (qR)^2]^3} \; , \qquad R = 0.15 \text{ fm} \tag{2}$$

$$\delta t_2 = \delta t_{\sigma\tau}^C = 50 \; 4\pi \; \frac{R^3}{1 + (qR)^2} \; , \qquad R = 0.25 \text{ fm} \tag{3}$$

where the normalization was chosen to give values of $a_i$ near unit magnitude in the adjustment process.

Final values of the $a_i$ coefficients are shown in Table I. The errors include contributions from both the statistical precision of the measurements as well as the quality of agreement in the final calculation. The value of $a_0$ was close to one; departures from one may reflect experimental normalization errors as well as true changes in the interaction. The signs of the real parts of $a_1$ and $a_2$ are consistent with a stronger $\rho$-meson contribution in the medium. While imaginary pieces are crucial in the parameter adjustment, there is no guidance from the simple idea of increased $\rho$ coupling for an interpretation of these values.

Figure 1 shows each observable. The dashed curves are the calculations based on the free interaction of Franey and Love. The solid curves show the effect of altering

Table I. Coefficients for the Altered Interaction

|  | real | imaginary |
|---|---|---|
| $a_0$ | $1.08 \pm 0.05$ |  |
| $a_1$ (tensor) | $-0.32 \pm 0.09$ | $-0.52 \pm 0.14$ |
| $a_2$ (central) | $0.12 \pm 0.05$ | $-0.12 \pm 0.09$ |

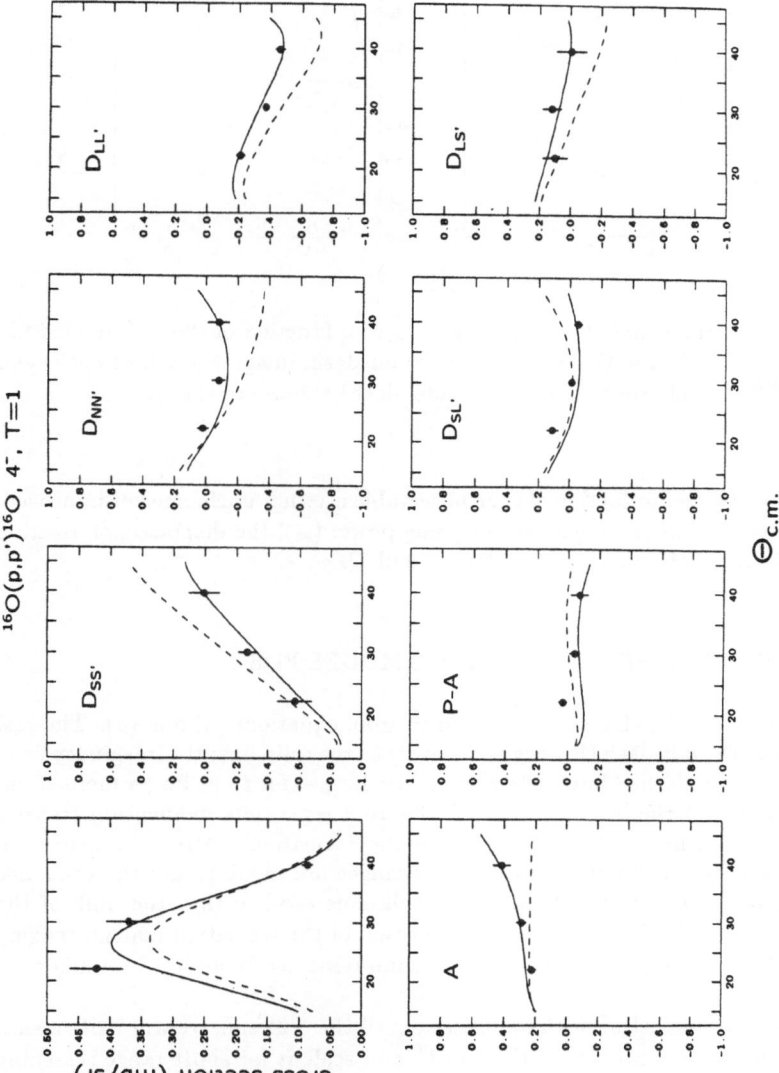

Fig. 1. Measurements of the cross section and polarization observables for the $^{16}\mathrm{O}(p,p')^{16}\mathrm{O}$ reaction to the $4^-$, $T = 1$ state at 18.98 MeV. The dashed curve represents calculations made with the free Franey-Love interaction; the solid curve includes modifications to the isovector terms.

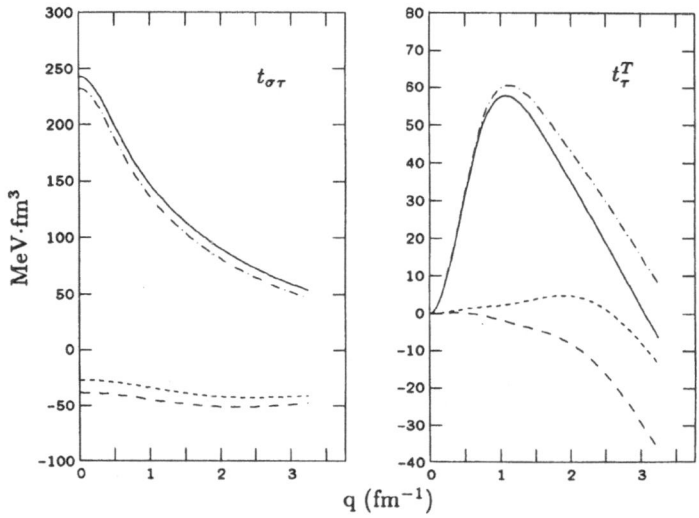

Fig. 2. Values of the $t$-matrix terms $t_{\sigma\tau}^C$ and $t_\tau^T$ as a function of the NN momentum transfer. Both the Franey-Love (real=dot-dash, imaginary=short dash) and modified (real=solid, imaginary=long dash) values are shown.

the interaction. The reproduction of each observable is either unchanged or improved, with the best results obtained for the analyzing power ($A$), the diagonal polarization transfer coefficients ($D_{NN'}$, $D_{LL'}$ and $D_{SS'}$), and $D_{LS'}$.

## COMPARISON WITH MESON EXCHANGE MODEL FOR $t$

Figure 2 shows the changes to $t_{\sigma\tau}^C$ and $t_\tau^T$ from equations (2) and (3). The real part moves from the dot-dash (Franey and Love[3]) to the solid line; the imaginary from the short to the long dashed line. The changes are largest for $t_\tau^T$ at larger momentum transfer. Noting that the horizontal axis in figure 2 represents momentum transfer in the NN center-of-mass, the direct part of the transition matrix is sensitive to momentum transfers near 1 fm$^{-1}$ where the changes are modest, and the exchange part is sensitive to values near 3 fm$^{-1}$. Calculations confirm that the bulk of the changes in figure 1 arise from the exchange portion of the transition matrix, making formulations[10] based on plane-wave, direct assumptions nearly useless as a guide.

To make the connection with modifications of the $\rho$-meson, we will write a simple meson-exchange potential[11] for $t_{\sigma\tau}^C$ and $t_\tau^T$ and scale it for short-range screening (using coefficients $\alpha$ and $\beta$) until it matches the Franey and Love[3] interaction in figure 2. The reproduction will not be perfect, since we are assuming in effect that the phenomenological $t$-matrix contains contributions only from $\pi$ and $\rho$ exchange. Nevertheless, we will compare the *changes* in the empirical $t$-matrix with the *changes* in the potential model as a function of $m^*/m$. Because of the ranges selected in equations (2) and (3), the reproduction of the $q$-dependence of these changes is excellent. The size of the screening coefficients $\alpha$ and $\beta$ sets the scale for interpreting the changes in terms of the model parameter $m^*/m$.

The tensor potential is

$$V_T = \alpha \frac{1}{3}\left(\frac{f_\pi^2}{m_\pi^2}\frac{q^2}{q^2 + m_\pi^2} - \frac{f_\rho^2}{m_\rho^2}\frac{q^2}{q^2 + m_\rho^2}\right) S_{12}(\hat{q})\, \tau_1 \cdot \tau_2, \qquad \alpha = 0.7 \qquad (4)$$

and the central potential is

$$V_C = \alpha \frac{1}{3}\left(\frac{f_\pi^2}{m_\pi^2}\frac{m_\pi^2}{q^2 + m_\pi^2} + \beta\, 2\frac{f_\rho^2}{m_\rho^2}\frac{m_\rho^2}{q^2 + m_\rho^2}\right) \sigma_1 \cdot \sigma_2 \tau_1 \cdot \tau_2, \quad \alpha = 0.7, \quad \beta = 0.3 . \quad (5)$$

The equation for $V_C$ has the contact term removed. A similar value for $\beta$ may be found in Speth.[12] The coupling constants were chosen following Brown and Rho[5] to be $f_\pi^2/m_\pi^2 = 0.08$ and $f_\rho^2/m_\rho^2 = 2\, f_\pi^2/m_\pi^2$. Since we match to the empirical $t$-matrix, more recent values of the coupling constants[13] will only change the screening coefficients.

Since the values of $a_1$ and $a_2$ were determined independently, each may become the basis for a value of $m^*/m$, thus giving a check in the internal consistency of the model. For this comparison, we chose to vary only the $\rho$-meson coupling constant and not the $\rho$ mass, in accord with the calculations discussed in Brown and Rho.[5] The results are:

$$\frac{m^*}{m} = 0.954 \pm 0.021 \ (V_C), \quad = 0.915 \pm 0.025 \ (V_T), \quad = 0.938 \pm 0.016 \ (\text{avg.}). \quad (6)$$

We find that the model parameter $m^*/m$ is determined consistently from the changes to $t_{\sigma\tau}^C$ and $t_\tau^T$. Choosing to vary the $\rho$-meson mass as well would have resulted in values closer to one. Our values are larger than the Brown and Rho estimate of 0.75 because we are observing a "stretched" surface transition in a light nucleus, and the density is much less than would be expected for nuclear matter.

Through exchange, we expect that these changes will influence calculations for $\Delta T = 0$. An examination of these transitions in $^{16}$O(p,p')$^{16}$O is underway.

## CONCLUSIONS

We have examined a full set of polarization observables for the $4^-$, $T = 1$ transition in $^{16}$O($\vec{p},\vec{p}'$)$^{16}$O and find that they can be fully explained by adjusting the short-range contribution to $t_{\sigma\tau}^C$ and $t_\tau^T$. The real parts of this modification are consistent with an increase in the $\rho$-meson coupling in the medium corresponding to $m^*/m = 0.938 \pm 0.016$.

## ACKNOWLEDGMENTS

We wish to acknowledge helpful conversations with Gerald E. Brown, Norton Hintz, W. Gary Love, Anil Sethi, and Jochen Wambach. Support for this work has come from the US National Science Foundation, the UK Science and Engineering Research Council (GR/F/4105.1 and GR/F/1080.6), and the NATO International Scientific Exchange Programme.

REFERENCES

1. C. Olmer, in "Antinucleon- and Nucleon-Nucleus Interactions," eds. G. E. Walker, C. D. Goodman, and C. Olmer (Plenum, New York, 1985) p. 261.

2. A. K. Opper, S. W. Wissink, A. D. Bacher, J. Lisantti, C. Olmer, R. Sawafta, E. J. Stephenson, and S. P. Wells, in "7th International Conference on Polarization Phenomena in Nuclear Physics, eds. A. Boudard and Y. Terrien (Les Éditions de Physique, Les Ulis, 1990) p. C6-607; and private communication.

3. M. A. Franey and W. G. Love, Phys. Rev. C 31, 488 (1985).

4. K. Nakayama, S. Krewald, J. Speth, and W. G. Love, Nucl. Phys. A431, 419 (1984).

5. G. E. Brown and Manneque Rho, Phys. Lett. B237, 3 (1990).

6. R. Schaeffer and J. Raynal, program DWBA, as modified by S. M. Austin, W. G. Love, J. R. Comfort, and C. Olmer, private communication.

7. C. Olmer, private communication, based on the measurements of C. W. Glover, P. Schwandt, H. O. Meyer, W. W. Jacobs, J. R. Hall, M. D. Kaitchuck, and R. P. deVito, Phys. Rev. C 31, 1 (1985), with corrections to the laboratory scattering angle, and E. J. Stephenson, A. D. Bacher, J. D. Brown, M. S. Cantrell, J. R. Comfort, V. R. Cupps, D. L. Friesel, J. A. Gering, W. P. Jones, D. A. Low, R. S. Moore, C. Olmer, A. K. Opper, P. Schwandt, J. W. Seubert, A. Sinha, and S. W. Wissink, J. Phys. Soc. Jpn. 55 (1986) Suppl. p. 926.

8. E. D. Cooper, program RUNT, private communication.

9. L. G. Arnold, B. C. Clark, R. L. Mercer, and P. Schwandt, Phys. Rev. C 23, 1949 (1981).

10. E. Blezynski, M. Blezynski, and C. A. Whitten, Jr., Phys. Rev. C 26, 2063 (1982).

11. T. E. O. Ericson and W. Weise, *Pions and Nuclei* (Clarendon Press, Oxford, 1988).

12. J. Speth, V. Klemt, J. Wambach, and G. E. Brown, Nucl. Phys. A343, 382 (1980); see also M. R. Anastasio and G. E. Brown, Nucl. Phys. A285, 516 (1977).

13. R. Machleidt and F. Sammarruca, Phys. Rev. Lett. 66, 564 (1991) and references therein.

# MEDIUM MODIFICATIONS OF THE NN INTERACTION AND (p,p') SCATTERING

N.M. Hintz and A. Sethi

School of Physics, University of Minnesota
Minneapolis, MN 55455

A.M. Lallena

Departamento de Fisica Moderna, Universidad de Granada
E-18071 Granada, Spain

## I. Introduction

We are exploring the consequences of "non standard" medium modifications in the NN interaction used in the non relativistic DWIA for
proton-nucleus scattering at intermediate energy, and in the corresponding
nuclear structure calculation. The modifications being considered are
those resulting from the reduction, in medium, of the masses of scalar and
vector mesons and nucleons as suggested by G. Brown and co-workers[1-3].
They present arguments for a universal scaling, $m_s^*/m_s \simeq m_v^*/m_v \simeq M_N^*/M_N$ with
$m^*/m = 1 - \lambda/2\, \rho/\rho_0$ where $\rho$ is the local nuclear density and $\rho_0$ is the
density of infinite nuclear matter, and $\lambda \sim 0.3 - 0.5$.

Some of the expected consequences for proton scattering and nuclear
structure calculations are:

1.    Elastic optical potential, dominated by spin-independent central
interaction $(V_0)$ at $q \lesssim 0.5$ fm$^{-1}$. $V_0$ is expected to scale as $(m/m^*)^2$
leading to an enhancement at the origin of the real central potential by $\sim$
$(1-\lambda)^{-1}$ and a "shrinking" of the effective radius by $R' = R - \lambda a$, where a
is the surface diffuseness[1]. This modification with $\lambda \simeq 0.3 - 0.4$
eliminates the radius discrepancy seen even in fancy DWIA calculations[4],
and puts theory in phase with experiment.

2.    Spin-Orbit interaction, $V_{LS}$
$V_{LS}$ is expected to scale approximately[3] as $(m/m^*)^3 \simeq 1.4$ (in the
nuclear surface).

3.     Isovector spin (V$_{\sigma\tau}$) and Tensor (V$_{T\tau}$) interactions

In the OBE ($\pi+\rho$) model the repulsive $\rho$ contribution should be enhanced[2] by $\sim (m_\rho/m_\rho^*)^2$ leading to an overall reduction of V$_{T\tau}$ and an enhancement of V$_{\sigma\tau}$.

Some evidence which supports the need for the above m$^*$ modifications comes from the analysis of p-shell[5] and s-d shell[6] matrix elements, from spin observable (D$_{ij}$) data[7,8], from the quenching of R$_L$ in (p,p')[2], and from empirically derived effective interactions[9].

## II.  Mixing of stretched states in $^{208}$Pb

In this talk we present a new analysis of (e,e')[10] and (p,p')[11] excitation (at T$_p$ = 318 Mev) of the 12$_1^-$, 12$_2^-$ and 14$^-$ states in $^{208}$Pb.  The dominant 1p-1h components of these states are $\nu(j_{15/2}, i_{13/2}^{-1})$, $\pi(i_{13/2}, h_{11/2}^{-1})$ and $\nu(j_{15/2}, i_{13/2}^{-1})$ respectively.  The excitation of these states involves the same spin density in (e,e') and (p,p') and so the quenching factors, Q = $\sigma_{exp}/\sigma_{theo}$ are directly comparable.  In the previous analysis[11] it was shown that the quenching factors were nearly the same in (ee') and (p,p') for the very pure 14$^-$ neutron state, but that a small mixing of the $\nu$ and $\pi$ configurations in the 12$^-$ states was required to get Q$_e$ = Q$_p$.

The new analysis, which we present here, was made to see the effect of m$^*$ modifications of Re V$_{T\tau}$ and Re V$_{LS}$ (2 and 3 above), and to compare the results with the large basis RPA calculations[12] by one of us (A.L.). In the (e,e') and (p,p') calculations shown here we have used a simple two-component model for the 12$^-$ states (since these still dominate in the RPA calculations):

$$|12_1^- > = (1-a^2)^{1/2} \mid \nu(j_{15/2}, i_{13/2}^{-1})> + a \mid \pi (i_{13/2}, h_{11/2}^{-1}>$$

$$|12_1^- > = -a \mid \nu(j_{15/2}, i_{13/2}^{-1})> + (1-a^2)^{1/2} \mid \pi (i_{13/2}, h_{11/2}^{-1}> \qquad (1)$$

In the (p,p') calculation the Re V$_{T\tau}$ and Re V$_{LS}$ components of the L-F interaction[13] were modified.  The V$_{T\tau}$ modification was based on the OBE model:

$$V_{T\tau}^{OBE}(q) = G_\pi \frac{q^2}{q^2 + m_\pi^2} - G_\rho \frac{q^2}{q^2 + m_\rho^2} , \quad G_i = \frac{4\pi\hbar c}{3} (\frac{f_i}{m_i})^2 \qquad (2)$$

The procedure was first to fit the OBE potential to the unmodified Re V$_{T\tau}$ of L-F by adjusting G$_\pi$ and G$_\rho$.  G$_\rho$ was then increased by $(m_\rho/m_\rho^*)^2$ = 1.563 (for m$_\rho^*$/m$_\rho$ = 0.8) and m$_\rho$ replaced by m$_\rho^*$ = .8.  The coefficients in the free L-F coordinate space t-matrix were then modified to reproduce the modified OBE potential.  These steps are shown in Figure 1.  It can be seen that the modified Re V$_{T\tau}$ is about half that of the unmodified, around q = 2fm$^{-1}$, the region effective for the direct terms in (p,p') to the

288

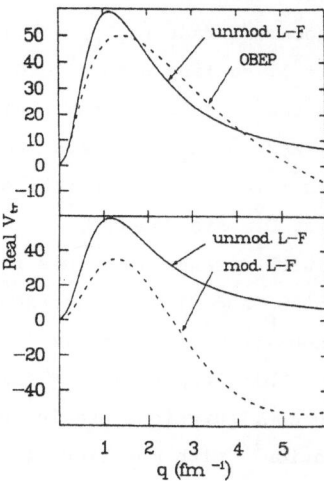

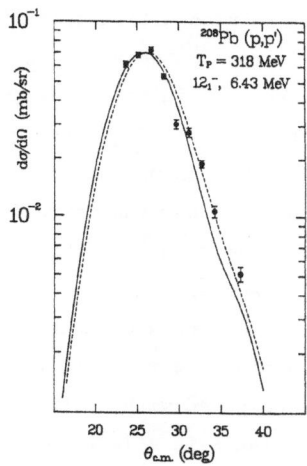

Fig. 1. Upper: Unmodified Love-Franey (L-F) Re $V_{Tr}$ and OBEP fit with $G_\pi$=80 MeV fm$^3$ $G_\rho$=122 MeV fm$^3$ (the "free" values are $G_\pi$= 135 and $G_\rho$=270). Lower: Modified and un-modified L-F interactions. The modified L-F is a fit to the modi-fied OBEP with $G_\pi$ = 80 MeV fm$^3$ and $G_\rho$=190 MeV fm$^3$.

Fig. 2. Cross section for $^{208}$Pb (p,p') to the $12_1^-$ (6.43 MeV) state. The curves are DWIA predictions using the unmodified L-F interaction (solid) and the modified interaction as described in the text (dashed). Both are calculated assuming a pure neutron configuration (a = 0) and are normalized to the data with $Q_p$ = 0.82 and 0.67 respectively.

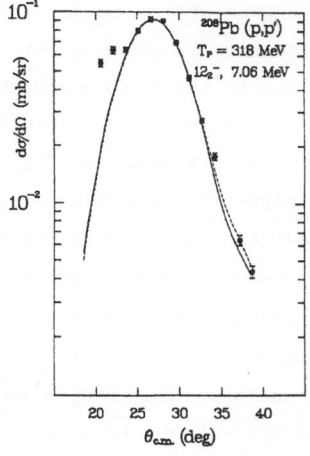

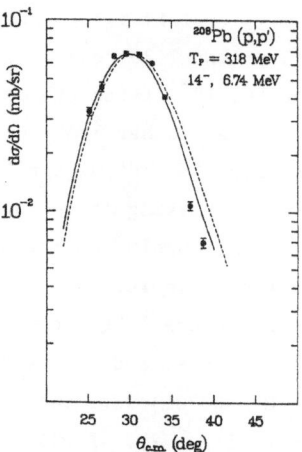

Fig. 3. Same as Fig. 2 but for the $12_2^-$ "proton" state. Here $Q_p$ = 0.39 (unmod.) and $Q_p$ = 0.36 (mod.).

Fig. 4. Same as Fig. 2 but for the $14^-$ neutron state. Here $Q_p$ = 0.63 (unmod.) and $Q_p$ = 0.47 (mod.).

stretched states. The 14⁻ state was used to fix Re $V_{LS}$ so as to obtain $Q_p$ = $Q_e$ = 0.47, where $Q_e$ has been corrected for meson exchange current contributions. This condition determined the enhancement factor, $F_{LS}$ = 1.4 for Re $V_{LS}$, close to the value expected for $(m/m^*)^3$ scaling with $m^*$ = 0.9, which is reasonable in the surface region where the transition densities peak. The resulting (p,p') predictions, normalized to the data, are shown in Figs. 2-4. The shapes of $\sigma(\theta)$ using modified forces are close to those of the unmodified, but the normalization factors, $Q_p^2$, are reduced. Calculations were then made for the 12⁻ states with a range of values for the mixing parameter a. The resulting values of $Q_p$ and $Q_e$ vs. a are shown in Fig. 5. It can be seen that $Q_e$ and $Q_p$ are consistent for both 12⁻ states for a mixing parameter, a = 0.06 ± 0.01. This "experimental" value of a can be compared with the mixing of $\pi$ and $\nu$ configurations predicted for the 12⁻ states in a large basis RPA calculation[12] with modified ($\pi$ + $\rho$) OBE interactions. The (unmodified) residual interaction was taken as

$$V_{res} = V_{mig} + V_{\sigma\tau}^{\pi} + V_{T\tau}^{\pi} + V_{\sigma\tau}^{\rho} + V_{T\tau}^{\rho} \tag{3}$$

where $V_{mig}$ is the usual Migdal zero range interaction[12)]

$$V_{mig} = C_0 [ f_0 + f_0' \tau_1 \cdot \tau_2 + g_0 \sigma_1 \cdot \sigma_2 + g_0' \sigma_1 \cdot \sigma_2 \tau_1 \cdot \tau_2 ] \delta(\vec{r}_1 - \vec{r}_2) \tag{4}$$

and the $V^{\pi}$ and $V^{\rho}$ are the OBE potentials (with contact terms omitted).

Calculations were made varying $V_{res}$ in three ways:

1)    $V_{res} = V_{mig} + \alpha (V_{\sigma\tau}^{\pi} + V_{T\tau}^{\pi} + V_{\sigma\tau}^{\rho} + V_{T\tau}^{\rho})$

2)    $V_{res} = V_{mig} + V_{\sigma\tau}^{\pi} + V_{\sigma\tau}^{\rho} + \beta (V_{T\tau}^{\pi} + V_{T\tau}^{\rho})$

3)    $V_{res} = V_{mig} + V_{\sigma\tau}^{\pi} + V_{T\tau}^{\pi} + \epsilon (V_{\sigma\tau}^{\rho} + V_{T\tau}^{\rho})$    (5)

The parameters $\alpha$, $\beta$, $\epsilon$ were then varied, adjusting the $g_0$ and $g_0'$ parameters of $V_{mig}$ (mainly $g_0$) to keep fixed the energies of the two 1⁺ states in $^{208}$Pb (5.85 and 7.30 MeV). The dominant components in the RPA calculations for the 12⁻ states are the same as we used in the two-component model above. (All other RPA components have $X \lesssim .02-.05$). The results for the dominant effective amplitudes, $A_{eff}$ = X + Y, are shown in Fig. 6-8. The observed mixing (a = 0.06) is obtained either for $\alpha$ = 0.6, or $\beta$ = 0.4 or $\epsilon$ ~ 2.1, roughly consistent with the modifications used in the (p,p') reaction calculations. The quantitative results at this point are tentative as the small RPA components were not included in the (p,p') and (e,e') calculations, and in the RPA calculations only the overall strengths ($G_{\pi}$ and $G_{\rho}$) were varied, and not the corresponding ranges ($m_i^2$ in denominators of Eq. 2). Preliminary (p,p') and RPA calculations indicate that the results will not change qualitatively when the DWIA (p,p') and the RPA interactions are treated consistently. Thus our main conclusion is that a net reduction of the tensor and enhancement of the spin-orbit interactions, in medium, is necessary for an understanding of the excitation of the 12⁻ states in (p,p') and (e,e').

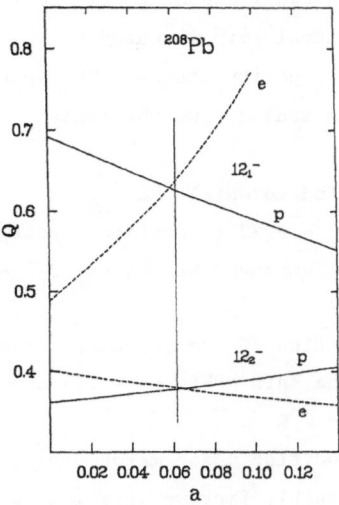

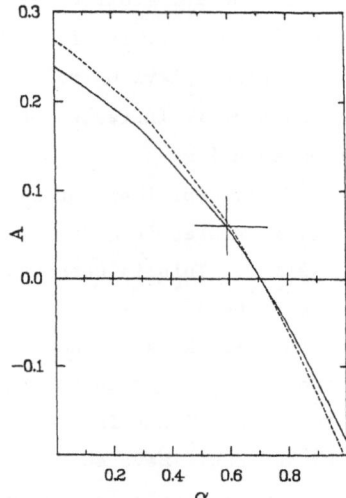

Fig. 5. Quenching factors, Q = $\sigma_{exp}/\sigma_{theo}$, for (e,e') (dashed) and (pp') (solid) excitation of the $12^-_1$ (upper) and $12^-_2$ (lower) states of $^{208}$Pb vs the mixing parameter, a, in the two component model (Eq. 1).

Fig. 6. RPA effective amplitudes, $A_\pi = X + Y$, for $\pi$ ($i_{13/2}$, $h^{-1}_{11/2}$) in the $12^-_1$ state (solid) and $-A_\nu$ for $\nu(j_{15/2}, i^{-1}_{13/2})$ in the $12^-_2$ state (dashed) vs. strength parameter $\alpha$, in eq. 5.

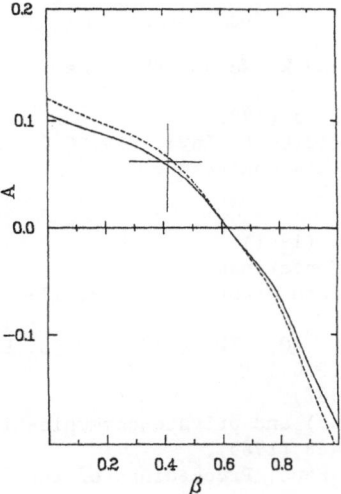

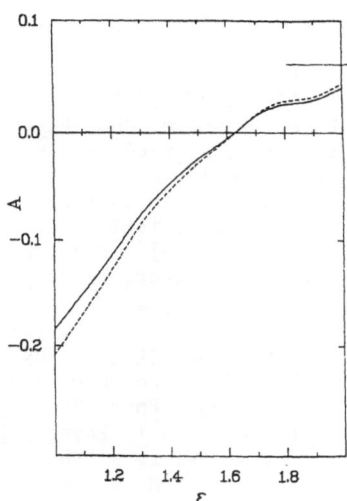

Fig. 7. Same as Fig. 6 but vs the strength parameter $\beta$.

Fig. 8. Same as Fig. 6 but vs. the strength parameter $\varepsilon$.

Finally, some comparisons can be made with medium modifications found in other analyses:

1) "Standard" medium modifications[14] due to Pauli blocking, two-body correlations, etc., mainly effect $V_0$, the central spin independent interaction, which plays no role for unnatural parity states. The spin-orbit interaction is increased only 10-15% in medium, and the tensor interaction even less.

2) In the Kelly empirical interaction[9] (fitted to data), Re $V_{LS}$ (isoscalar) increases in medium by $\sim 1.3$ at $\rho = \rho_0/2$ ($\sim$ surface density) around $q = 2fm^{-1}$. This is consistent with our enhancement factor, $F_{LS} = 1.4$ discussed above.

3) Zheng and Zamick[5] have shown that a reduction of the residual tensor interaction by $\sim 0.5$, <u>or</u> an enhancement of the spin-orbit by $\sim 1.5$ can explain the small GT matrix element for $^{14}C \rightarrow ^{14}N$.

4) Hosaka and Toki[6] have shown that using modified meson masses ($m^*/m \sim 0.8$), in a G-matrix calculation, can significantly improve agreement with 2s-1d shell matrix elements determined empirically by Brown, <u>et al</u>.[15]

ACKNOWLEDGEMENT

We are grateful to Gerry Brown for inspiring and educating us as we progressed in this work.

REFERENCES

1. G. E. Brown, C. B. Dover, P. B. Siegal and W. Weise, Phys. Rev. Lett. <u>60</u>, 2723 (1988).
2. G. E. Brown and M. Rho, Phys. Lett. <u>B237</u>, 3 (1990).
3. G. E. Brown, A. Sethi and N. Hintz, submitted to Phys. Rev. <u>C</u>.
4. See for example, L. Ray, invited talk, this Conference.
5. D. C. Zheng and L. Zamick, preprint, 1990.
6. A. Hosaka and H. Toki, preprint, 1990.
7. E. Donoghue, <u>et al</u>., Phys. Rev. <u>C43</u>, 213 (1991).
8. E. Stephenson, contributed paper, this Conference.
9. J. J. Kelly, <u>et al</u>., University of Maryland preprint PP # 91-119, submitted to Phys. Rev. <u>C</u> (1991).
10. J. Lichtenstadt, <u>et al</u>., Phys. Rev. Lett. <u>40</u>, 1127 (1978); Phys. Rev. <u>C20</u>, 497 (1979) and Ph.D. thesis, MIT (1980).
11. D. Cook, <u>et al</u>., Phys. Rev. <u>C35</u>, 456 (1987).
12. A. M. Lallena, Nucl. Phys. <u>A489</u>, 70 (1988) and private communication.
13. M. Franey and W. Love, Phys. Rev. <u>C31</u>, 488 (1985).
14. See for example, H. V. Geramb, and K. Nakano, Proceedings of the Workshop on the Interaction Between Medium Energy Nucleons in Nuclei, AIP Conference Proceedings No. 97, ed. by H. O. Meyer (AIP, New York, 1983), p.44.
15. B. A. Brown, W. A. Richter, R. E. Julies, and B. H. Wildenthal, Ann. Phys. <u>182</u>, 191 (1988).

# EXPERIMENTAL FOUNDATION FOR NN INTERACTIONS: A COMMENT

James A. Carr

Supercomputer Computations Research Institute
The Florida State University, B-186
Tallahassee, FL 32306-4052

## ABSTRACT

It is noted that the NN data set, upon which phenomenological NN interactions are based, is lacking quality spin-dependent measurements at the energies of interest at IUCF. Data to constrain potential and t-matrix models are needed now.

## INTRODUCTION

The combination of an excellent spectrometer and a focal-plane polarimeter at IUCF has opened up some fascinating areas of research involving a wide variety of spin-transfer measurements, some of which we have seen at this meeting.[1] Reliable use of these results to extract nuclear structure information requires that the relevant parts of the nucleon-nucleon (NN) interaction be calibrated accurately. Some useful steps are being taken along these lines with the new nucleon-nucleus data.[2] However, this effort assumes that the NN interaction is already known from fits to the NN data taken over the past 3 decades. I have serious doubts that this is the case.

For some time I have had nagging doubts concerning various unexpected differences between inelastic-scattering cross sections calculated with interactions obtained from potentials and from phase shifts, both fitted to the NN data. At this meeting we saw another example, in Figures 4 and 5 of Ref. 1. These interactions each purport to be from a best fit to the *same set* of NN data, yet they produce large differences in the predicted value for the spin observable. Something must be wrong.

## OVERVIEW OF THE DATA SET

During a visit to Los Alamos, I learned from Mike McNaughton that the SAID program[3] contained a complete and accessible description of the NN data set used to determine the interactions. I pulled out the data between 100 MeV and 250 MeV lab

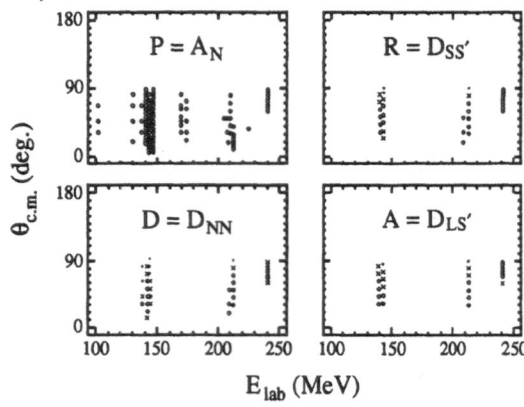

Figure 1. Extent of the data for the p+p system, where a circle denotes errors < 0.05, an X denotes errors between 0.05 and
0.10, and a small symbol denotes errors > 0.10.

energy for *every* observable and devised a way to display the quality and kinematic
range of the data. Selected observables from the p + p data set are shown in Fig. 1.
The figure indicates that data for the p+p system, which is the iso-triplet scattering
channel, are sparse; many also date from the 1960's. Data for the n+p system,
which is a combination of iso-singlet and iso-triplet amplitudes, are fewer and of
much poorer quality (except for the recent $C_{NN}$ data sets). Cross-section data are
generally in good shape, with the n+p data set being most complete and accurate.
More extensive graphs and discussion are in a recent preprint.[4]

## CONCLUSION

The present data cannot precisely constrain the NN interaction over the entire
range of energy and momentum transfer being explored at IUCF. The interactions are
interpolated between data sets of varying accuracy, age, and reliability – yet the p+p
system should be easy to do and the n+p setup exists at IUCF. New measurements
of complete spin observables over a range of energies and angles should be made to
determine the interaction precisely at the energy where it will be used.

I wish to acknowledge the hospitality of the MP division at LANL, where some
of these ideas came into focus. This work supported by the F.S.U. Supercomputer
Computations Research Insitute which is partially funded by the U.S. Department of
Energy through Contract No. DE-FC05-85ER250000.

## REFERENCES

1. S. W. Wissink, these proceedings.
2. E. J. Stephenson and J. A. Tostevin, these proceedings.
3. R. A. Arndt and L. O. Roper, SAID phase shift analysis program.
4. J. A. Carr, FSU-SCRI preprint.

# SPIN-LONGITUDINAL CORRELATIONS

C. Gaarde

The Niels Bohr Institute
University of Copenhagen
Blegdamsvej 17, Copenhagen Ø

## 1    INTRODUCTION

We shall discuss the nuclear spin response as measured in charge exchange reactions at intermediate energies. We shall especially be interested in the spin-longitudinal part of the response. This is where the largest effects from particle-hole correlations at larger momentum transfer are expected. The attraction in the spin-isospin channel as seen in $NN$ scattering and dominantly coming from one pion exchange would give very dramatic effects inside a nucleus. That is, if we just took the particle-hole interaction equal to the free $NN$-interaction in this $\vec{\sigma} \cdot \vec{q}$ spin-longitudinal channel. Similar dramatic effects could be expected in the $\Delta$-region. The question is then, what are the effective particle-hole or $\Delta$-hole interactions at finite momentum transfers.

The experimental challenge is to separate the response into a spin-longitudinal and spin-transverse part. This is attempted in two different ways.

We have used composite projectile-ejectile systems with different formfactors for the spin components. The $(^3He, t)$ and $(d, 2p[^1S_0])$ reactions are examples of very strong enhancements of the spin-longitudinal formfactor at finite momentum transfer q. In both cases caused by the presence of the d-state in the wavefunctions, in turn due to tensor components in the elementary interaction. For the $(^{12}C, ^{12}Ng.s.)$ reaction the spin-transverse component is completely dominant at small q's, whereas the opposite is the case at larger q. This is demonstrated in fig. 1, where the squares of formfactors are given for four different reactions [1]. We shall discuss results obtained with these reactions in the following.

The other method to separate the spin response is to use polarized beams and get spin observables. The simplest reaction to analyse and interprete is then the $(\vec{p}, \vec{n})$ reaction and results obtained at Los Alamos have been presented at this conference (T. Taddeucci). We have used the $(\vec{d}, 2p[^1S_0])$ [2,3] and $(^6\vec{Li}, ^6He)$ reactions, where in both cases tensor polarized spin 1 projectiles, charge exchange to spin zero ejectiles. A measurement of the spin of the outgoing particle is therefore not necessary (as in $(\vec{p}, \vec{n})$) and only relative yields from the different polarization states of the beam is needed, to obtain similar information. The problem with these reactions is, that distortion effects complicate the interpretation of the data.

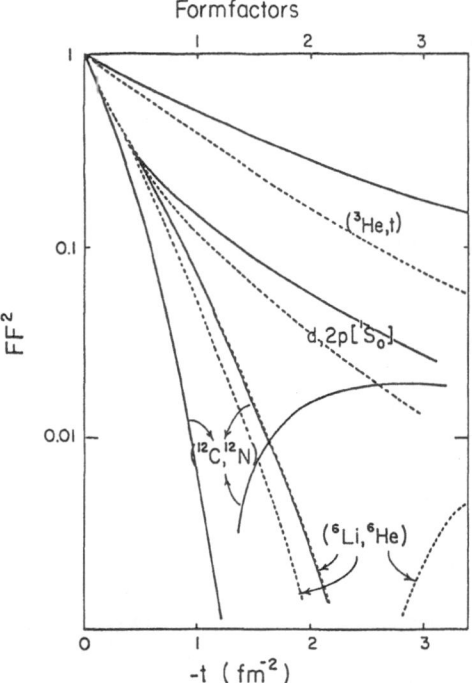

Figure 1. The spin-longitudinal (full drawn) and spin transverse (dashed) formfactors (squared) are given for a number of projectile-ejectile systems. The spin-longitudinal formfactor for $^6Li - ^6He$, happens to almost coincide with the transverse formfactor for $^{12}C - ^{12}N$. Only the larger is shown beyond the first minimum for these two reactions.

In this paper we briefly discuss three cases where spin-longitudinal correlations are essential.

i) The $0^-$ collective state.
This is the only truly pionic mode, a pure spin-longitudinal state. The basic question is to what extent, this state is quenched by coupling to the pion-field. The coupling could be so strong, that the effective particle-hole interaction becomes too small to form a real collective state.

ii) Spin response in the quasielastic region.
The challenge here is to understand the medium effects on the spin dependent interactions through a determination of the effective particle-hole interactions in the different spin channels.

iii) The $\Delta$-region.
The situation is very similar to the one in the quasielastic region. What are the effective $\Delta$-hole interactions? Do we see genuine correlation effects?

We have used the $(^6\vec{Li},^6He)$ reaction to examine the spin response at small momentum transfer. We shall very briefly show some data obtained with this reaction.

## 2.1    The $(^6\vec{Li},^6He)$ reaction

Data have been obtained at Laboratoire National Saturne at 2 bombarding energies, 200 and 750 $MeV$ per nucleon. Tensor polarized beams were used with an intensity of about $10^9$ particles per sec. The scattered particles were analysed in a magnetic spectrometer, Spes4, equipped with drift-chambers and scintillators in the focal plane. The $^6He$-ejectiles are very stiff particles and only tritons have a similar rigidity. The tritons are however easily rejected by the energy loss signal, and very clean spectra can usually be obtained in the $(^6Li,^6He)$ reaction.

The simplest reaction is the one on the deuteron-target. The cross section and tensor analysing power is given in fig.2. The curves correspond to calculated quantities, assuming a $n \rightarrow p$ transition as for a free nucleon. We have written e.g. the cross section as

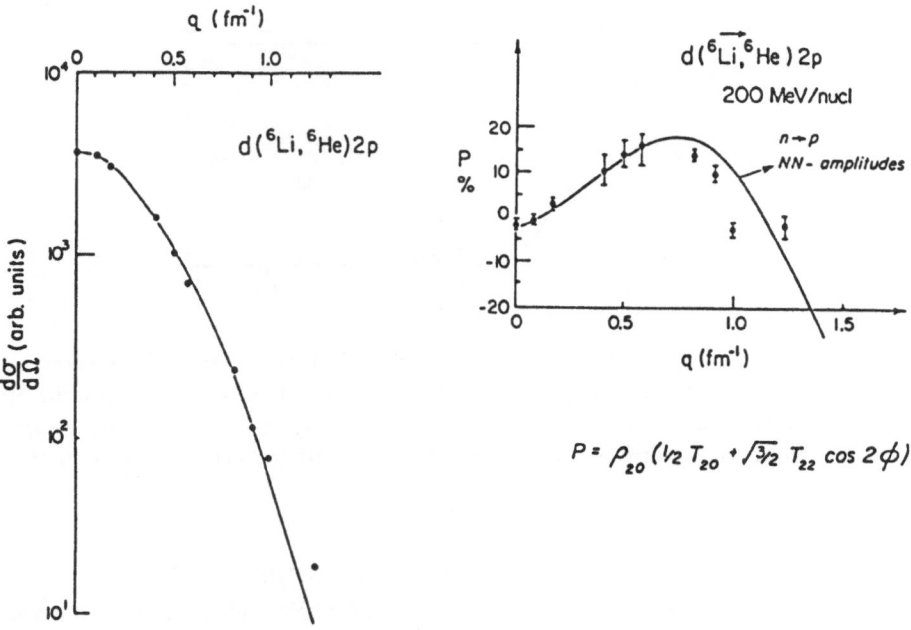

Figure 2. Cross section and tensoranalysing power for the $(^6Li,^6He)$ reaction on the deuteron at 1200 $MeV$ bombarding energy. The curves correspond to a plane wave approximation for the reaction on the neutron i.e. the $n \rightarrow p$ transition with $NN$-amplitudes taken from ref. [4]. The beam polarization $\rho_{20}$ was 0.60 and $\cos 2\phi = 1$, i.e. the beam polarization axis is normal to the scattering plane.

$$\frac{d\sigma}{d\Omega}(q) = D\{FF_T^2 \frac{d\sigma}{d\Omega_T}(q) + FF_L^2 \frac{d\sigma}{d\Omega_L}(q)\}$$

a product of a distortion factor D (independent of q), formfactors for the $(^6\vec{Li}, ^6He)$ reaction and cross sections for the elementary $n \to p$ transition. The contributions from the spin-longitudinal and transverse amplitudes are added incoherently as in a plane wave impulse approximation.

We see that such an approach describes the data quite well out to 1 $fm^{-1}$.

In fig. 3 the spectra for the reaction on $^{12}C$ are shown at 2 angles. The zero degree spectrum is dominated by the Gamow-Teller transition to the g.s. of $^{12}N$. At the larger angle the $l = 1$ structures at 4.1 and 7.2 $MeV$ excitation energy show up more clearly. The spectra are very similar to those seen in the $(p, n)$ and $(^3He, t)$ reactions at the same energy per nucleon [5].

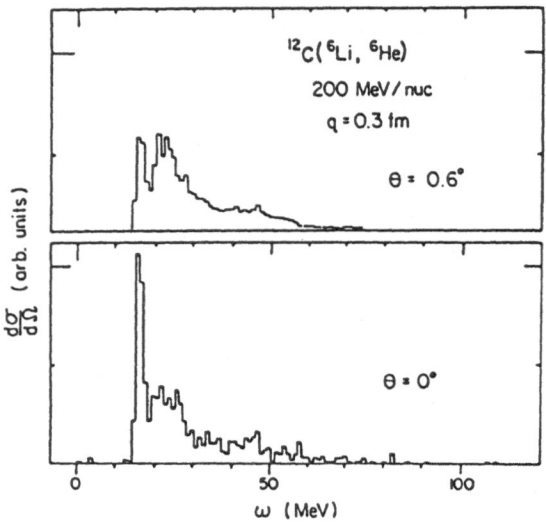

Figure 3. Spectra for the $(^6Li, ^6He)$ reaction on $^{12}C$ at 1200 $MeV$. The spectra are obtained with an aperture in the spectrometer of $0.2^0 * 1.6^0$. The two spectra are therefore measured simultaneously and separated in angle by a ray-tracing procedure using the information from the two driftchambers. The effective aperture is then chosen as $0.2^0 * 0.2^0$ for both spectra.

In fig. 4 we compare zero degree spectra for $^{12}C$ and $^{208}Pb$.

In fig. 5 is shown $(p, n)$ and $(^3He, t)$ spectra for $^{208}Pb$ and a striking contrast to the $(^6Li, ^6He)$ spectrum is observed. In ref. [5] we have discussed distortion effects for the $(^3He, t)$ reaction and seen that transitions with angular momentum transfers of 1 and 2 have larger zero degree cross sections than for $(p, n)$. This is especially so for heavy target nuclei. We see the same effect for the $^{208}Pb(^6Li, ^6He)$ reaction. The Gamow-Teller resonance is strongly excited at zero degree, but so are the $l = 1$ and 2 transitions. We have no reasons to suspect, that the $(^6Li, ^6He)$ reaction is a complicated reaction. The distortion effects are however very important for heavier target nuclei and the uncertainties on multipole decompositions of the spectra at different angles would be larger for $(^6Li, ^6He)$ than for $(p, n)$ spectra.

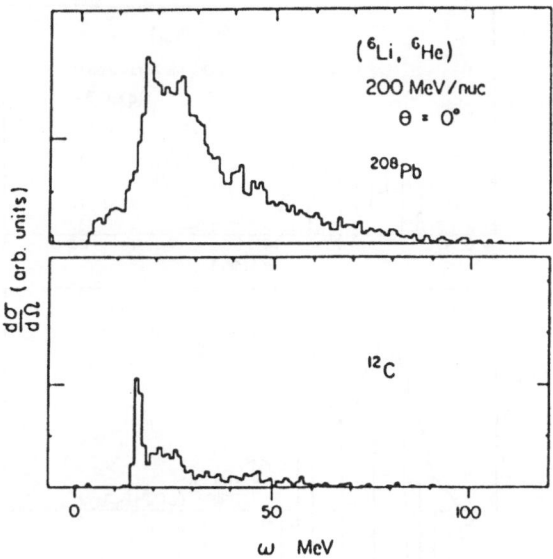

Figure 4. Zero degree spectra for $^{12}C$ and $^{208}Pb$ at 1200 $MeV$. The relative cross section scale is correct. The spectra are given versus energy loss $\omega = T(^6Li) - T(^6He)$.

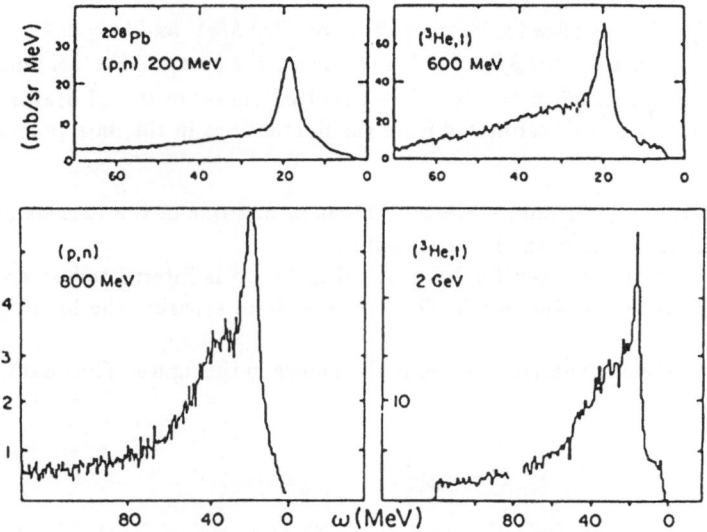

Figure 5. $(^3He, t)$ and $(p, n)$ spectra for $^{208}Pb$ are compared at 2 energies at $\Theta = 0^0$ [5,6,7].

## 2.2 Spin dipole strength

In fig. 6 we show spectra for $^{40}Ca$ and $^{90}Zr$ at a finite angle where the cross sections for $l = 1$ transitions have their maxima. The spectra are very similar to those found for the $(p, n)$ and $(^3He, t)$ reactions. The signal to background ratio is in fact somewhat better for the $(^6Li,^6 He)$ reaction in contrast to the zero degree spectra. The formfactor as given in fig. 1 is so steep that the transitions with $l = 2$ and larger have smaller cross

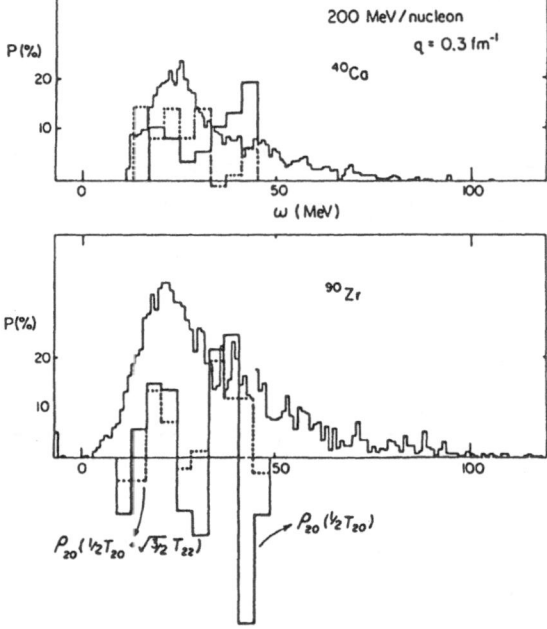

Figure 6. $(^6Li, {}^6He)$ spectra for $^{40}Ca$ and $^{90}Zr$ at $1200\ MeV$ and $\Theta = 0.6^0$ (corresponding to a momentum transfer of $0.3 fm^{-1}$). Also given are the measured tensor analysing power $P = \rho_{20}(\frac{1}{2}T_{20} + \sqrt{\frac{3}{2}}cos2\phi)$ with $cos\ 2\ \phi = 1$ (dotted curve) or 0 (full drawn). Error bars are not given, but can be estimated from the fluctuations in the data points.

sections relative to $(p, n)$ and $(^3He, t)$. This is in fact one of the reasons for using this reaction, that it enhances the $l = 1$ transitions.

The broad structure seen for both nuclei in fig. 6 is interpreted as an envelope of $2^-, 1^-$ and $0^-$ collective states [6]. The challenge is to separate the broad peak into its components.

Data on the tensor analysing power is also shown in the figure. The quantity measured is

$$P = \rho_{20}(\frac{1}{2}T_{20} + \sqrt{\frac{3}{2}}T_{22}cos2\phi)$$

where $\rho_{20}$ is the beam polarization (around 60% in this experiment). The $T_{\lambda\mu}$ are tensor-analysing powers in the reaction, and $\phi$ is the angle between the normal to the scattering plane and the beam polarization.

A superconducting solenoid placed in front of the target and with a maximum field strength of 11 Tesla-meter allows the rotation of the spin of $^6Li$-projectiles i.e. changing the angle $\phi$.

In fig. 6 data are shown with $cos2\phi = 1$ and 0, and it is therefore possible to get $T_{20}$ and $T_{22}$ separately. This is principle enough information to separate the $2^-, 1^-$ and $0^-$ states. In a plane wave approximation the cross sections and tensoranalysing powers can be expressed in terms of squares of NN-spin amplitudes. The transitions to $1^-$ states are

300

purely spin-transverse and purely spin-longitudinal for $0^-$ transitions. In other words: three quantities are measured (cross section, $T_{20}$ and $T_{22}$) to determine 3 amplitudes (2 spin-transverse and 1 spin-longitudinal)

The statistical uncertainties in the data in the present experiment are however too large that a meaningful separation can be made. We shall therefore only indicate trends in the results. In a plane wave approximation the value for P, in going from the situation with the solenoide off $(P = \rho_{20}(\frac{1}{2}T_{20} + \sqrt{\frac{3}{2}}T_{22}))$ to the case with $cos 2\phi = 0$, would

i) increase for a $0^-$-state

ii) decrease for a $1^-$-state

iii) small effect for a transition to a $2^-$-state.

With these rules it is tempting to claim that the three states can be identified from the data in fig. 6. The real problem is however to get the strength of the $0^-$-state.

## 3    THE QUASIELASTIC RESPONSE

The $(d, 2p[^1S_0])$ and $(^3He, t)$ reactions have been used to study the response in the quasielastic region [8,9,10]. The strong enhancement of the spin-longitudinal formfactor, as shown in fig. 1, make there reactions interesting as possible probes for correlations at finite momentum transfers. Both reactions show well developed quasielastic peaks from $q \sim 1.2 fm^{-1}$ and as far and as they are studied ($\sim 3 fm^{-1}$). The spectra are almost identical in shape to the ones seen in the $(p, n)$ reaction (at the same q) but shifted in energy. This shift in energy could possibly be due to correlations, but also distortion effects could play a role. In this paper we shall not discuss this aspect, but concentrate on relative cross sections. In fig. 7 $(p, n)$ spectra obtained at LAMPF are shown and in fig. 8 a $(d, 2p)$ spectrum at $q \sim 1.3 fm^{-1}$ corresponding to one of the $(p, n)$ spectra. The spectra are very similar. This is also true for $(^3He, t)$ spectra at the same momentum transfers.

In fig. 9 ratio of cross sections are given for $(d, 2p)$ and $(^3He, t)$ reactions relative to $(p, n)$ reactions.

The two upper curves refer to the elementary reactions e.g. $p(d, 2p[^1S_0])n$ is compared to $p(n, p)n$. The crosses refer to data whereas the curve is calculated from $NN$ data and the formfactor for $d - 2p[^1S_0]$. For all the cases in figure 9 the curves or data have been divided by the squares of the transverse formfactors. This is for practical reasons to avoid the strong q dependence of the formfactors and cross sections. The curve for the ratio of cross sections for $n(^3He, t)p$ and $n(p, n)p$ is calculated from $NN$ amplitudes and the $^3He - t$ formfactor.

The curves for the ratio for the elementary reactions are just different ways of showing the enhancement of the spin longitudinal formfactors. The curves do however show this effect rather nicely.

The two lower curves in figure 9 are drawn to guide the eye through the experimentally determined ratios of cross sections for the reactions on $^{12}C$. The ratios are for the peak of the cross sections of the quasielastic response. Since the spectra have very similar shape at the same momentum transfer, these ratios are well defined quantities with small relative errorbars. The uncertainties on the absolute values could be as large as 20%.

The ratios for the reactions on $^{12}C$ have a similar q-dependence as the ratio for the

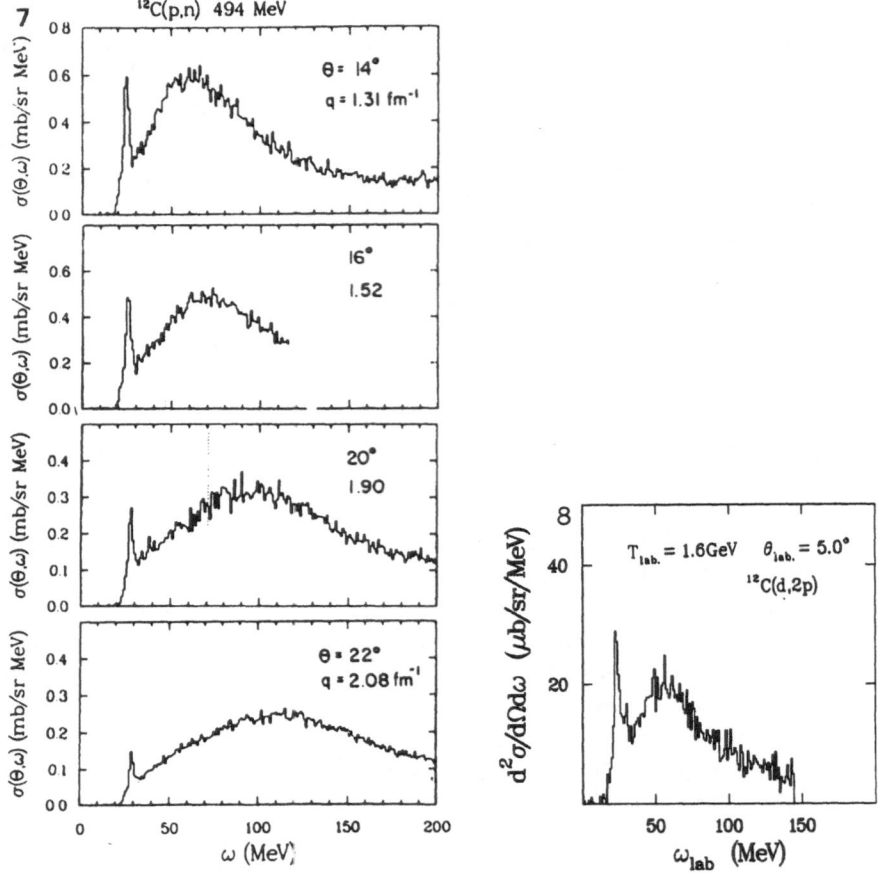

Figure 7. $(p, n)$ spectra at 4 angles at 500 $MeV$ bombarding energy obtained at LAMPF. The sharp peak in the spectra is the $4^-$-state in $^{12}N$, with $d_{5/2}p_{3/2}^{-1}$ as the dominant shell model configuration.

Figure 8. A $(d, 2p[^1S_0])$ spectrum at 800 $MeV$ per nucleon and $\Theta = 5.0^0(q = 1.3fm^-)$, to be compared to the $(p, n)$ spectrum at the same q as given in fig. 7.

elementary reactions with a possible exception for the $(d, 2p)$ data at the largest value for q.

The three reactions therefore seem to have the same response. The conclusion would then be that possible spin-longitudinal correlations are weak. This is based on the fact that when we probe the $^{12}C$ nucleus with reactions that should be sensitive to enhancements in this channel we see rather small effects.

A much more detailed analysis is of course needed to be quantitative about such a conclusion. We refer to the work of Thomas Sams presented at this meeting on distortion effects for the $(\vec{d}, 2p)$ reaction. The absolute cross sections at the larger moentum transfers seem not accounted for in a one step process. The same conclusion is reached by M. Ichimura and coworkers for the $(p, n)$ reaction [11]. It is therefore interesting to see that the ratio of cross sections as given in figure 9 have such a simple q-dependence. The distortion effects could be rather different in the three reactions, but the data seem to show that the response is similar.

The study of isobar-excitations in nuclei with charge exchange reactions has recently been discussed in a review article [1]. A very general result is that the Δ-peak for nuclear targets is lower in energy than for a proton target. Part of the shift can be explained as effects from Fermi-motion, but a genuine shift of the energy of Δ in the nuclear medium seems to be necessary to explain the data. Detailed calculations of $(p, n)$ and $(^3He, t)$ spectra have been performed with Δ-hole interactions based on one-pion exchange potential. Such interactions are therefore attractive in the spin-longitudinal channel.

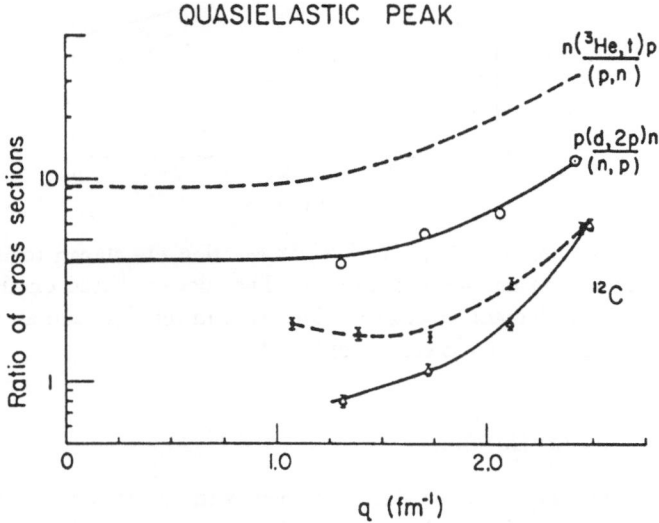

Figure 9. Ratio of cross sections divided by $FF_T^2$, the square of the transverse formfactor for $^3He - t$ or $d - 2p[^1S_0]$. The upper curve (dashed) is calculated from $NN$-amplitudes and the formfactors as given in fig. 1 and refers to the ratio $\frac{(^3He,t)}{(p,n)} * \frac{1}{FF_T^2}$ on the nucleon target, the neutron in this case. The next (full drawn) curve is the calculated ratio $\frac{(d,2p)}{(n,p)} * \frac{1}{FF_T^2}$ and this can now be compared to data on the proton target. The two lower curves are drawn to guide the eye through the experimentally determined ratios on the $^{12}C$ target for $(^3He, t)$ (dashed) and $(d, 2p)$ (full drawn) relative to the $(p, n)$ cross sections (both divided by the $FF_T^2$ used for the nucleon targets).

Here we note that for Δ-excitations with photons such a shift for nuclear targets is not observed, only a broadening of the resonance. This is however consistent with the Δ-hole model mentioned above. Photons probe the spin-transverse response and not the pionic modes.

The Δ-hole model is then rather specific on the origin of the energy shift of the Δ-peak, and effects should show up in the spin-longitudinal response. The two methods mentioned above have been used to study such effects in the inclusive spectra. A recent Δ-decay experiment, presented by B. Ramstein at this meeting could possibly be even more specific on the properties of the Δ-isobar in the nuclear medium.

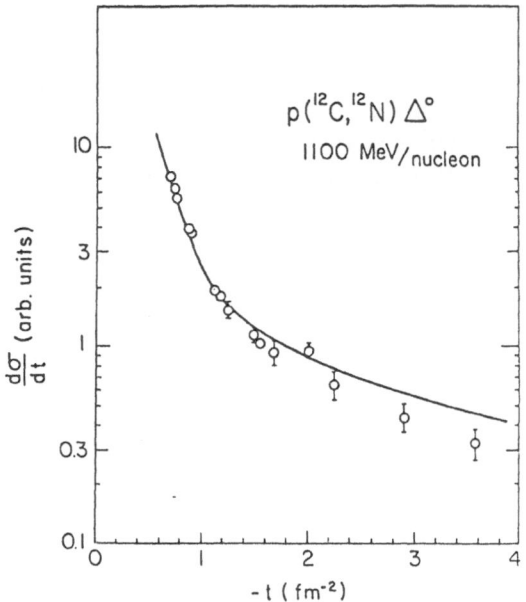

Figure 10. Cross sections for the $p(^{12}C,^{12}N)\Delta^0$ reaction are shown together with the calculated dependence on momentum transfer. The curve is based on the formfactors given in fig. 1 and the elementary cross section given in ref. [12] and an assumed ratio of transverse to longitudinal cross section of 2 to 1.

### 4.1 $(^{12}C,^{12}N)$ and $(^6Li,^6He)$ data

Here we shall briefly discuss some data obtained with the $(^6Li,^6He)$ and $(^{12}C,^{12}N)$ reactions. In fig. 10 is shown the cross section versus momentum transfer for the $p(^{12}C,^{12}N)\Delta^0$ reaction. The curve is the calculated dependence on four-momentum transfer in a plane wave approximation. The expression for the cross section is in fact allready given above for the $d(^6Li,^6He)2p$ reaction. In the $p(^{12}C,^{12}N)\Delta^0$ case the elementary $NN \rightarrow N\Delta$ cross section is taken as one-pion exchange from ref. [12], but more importantly an expression that describes the $p + p \rightarrow n + \Delta^{++}$ cross section data. The ratio between longitudinal and transverse cross section is taken as 1:2, which is certainly not OPE.

This ratio is taken from our $p(\vec{d},2p[^1S_0]\Delta^0$ data. [3] The formfactor for the $^{12}C - ^{12}N$ g.s. transition is given in fig. 1 and basically corresponds to the Cohen-Kurath wave functions for these states. The impressive agreement demonstrated in fig. 10 shows, that the simple approach is adequate. The figure also shows that for -t larger than $2fm^{-2}$ the $N \rightarrow \Delta$ transition is induced by a pure spin-longitudinal driving force. The $(^{12}C,^{12}N)$ reaction could therefore be an interesting probe also for nuclear targets. Data on $^{208}Pb$ are being analyzed. The problem will be to handle the distortion effects.

In fig. 11 is shown a zero degree spectrum for the $(^6Li,^6He)$ reaction on $^{12}C$ at 4.5 $GeV$. The ground state is strongly excited, whereas the $\Delta$-resonance is so weakly excited, that it is difficult to separate the $\Delta$-part from the 1p - 1h and 2p - 2h excitations. This is due to the very steep formfactor for the $^6Li - ^6He$ transition as given in fig. 1.

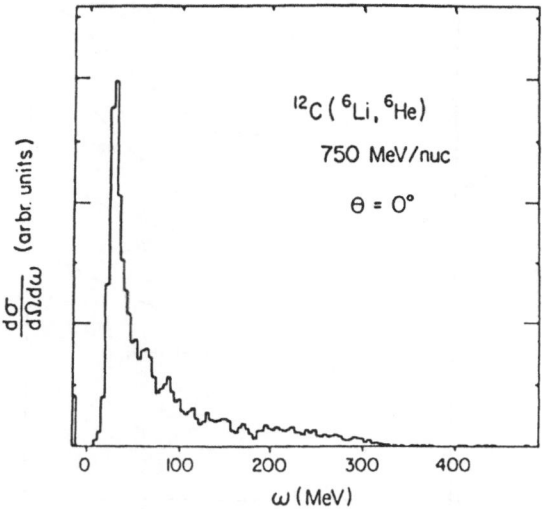

Figure 11. Zero degree spectrum for the $^{12}C(^6Li,^6He)$ reaction at 4.5 $GeV$ bombarding energy.

This also gives a very large shift of the energy of the $\Delta$-resonance is these inclusive spectra. The resonance is therefore pushed down in energy into the dip-region dominated by 2p - 2h states.

We also have data on the spin observables in the $\Delta$-region, but shall not discuss them here.

The $(^6Li,^6He)$ reaction with a tensor polarized beam is a very interesting and potentially useful reaction. In the $\Delta$-region the minimal four-momentum transfer is however so large that the formfactor in the reaction is small. This gives small cross sections and distortion effects could become important. This could be compensated by going up in bombarding energy. The limitation at Saturne has been the bending of the $^6He$-ejectiles. The maximal field in Spes4 corresponds to a momentum 4.2 $GeV/C$ close to 4.5 $GeV$ doubly charged $^6He$-particles.

## 4.2  $(\vec{d},2p[^1S_0])$ data

For the $(d,2p)$ reaction we have data on the spin observables in the $\Delta$-region [3]. The inclusive spectra show an energy shift of the $\Delta$-peak between $^{12}C$ and the proton. An analysis based on quasifree $\Delta$ excitation and an eikonal approximation for the distortion effects can only account for half of the shift. This shift should therefore be accompanied by an enhancement of the spin-longitudinal rresponse. This is not the immediate conclusion from the data on the tensor analysing power for $^{12}C$. The data seem to show the opposite, namely that the ratio of spin-transverse to spin-longitudinal cross section is larger for $^{12}C$ than for the proton target. We do however ascribe this to a distortion effect. We have seen that the distortion effects on the spin observables on the quasielastic peak is quite large for the $(\vec{d},2p)$ reaction. A detailed analysis of the effect in the $\Delta$-region has not yet been performed.

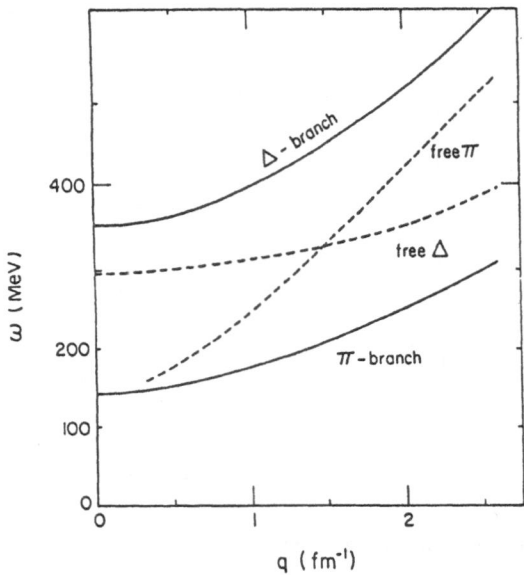

Figure 12. The two-level model by Guichon and Delorme is illustrated with the Landau-Migdal parameter g' = 0.5. The figure is adapted from ref. [15].

## 4.3  Δ-decay

The most detailed information on Δ-properties in the nuclear medium will probably come from a recent experiment on the Δ-decay following excitation by a $(^3He, t)$ reaction. The data has been shown by B. Ramstein at this meeting and I shall only very briefly point to some of the results

i) The $(p + \pi^+)$-decay channel for $^{12}C$ follows the decay from the proton or the deuteron, targets.

ii) The 2p-channel in $^{12}C$ is shifted almost 100 $MeV$ down in excitation energy relative to the $(p + \pi^+)$ channel.

iii) The missing mass spectrum corresponding to the 2p-decay have a very small (if any) contribution in the pion-mass region. That means that projectile excitation do not contribute in this 2p-channel. The shift in energy is therefore not due to projectile excitation.

The shift of the 2p-decay channel is the most direct evidence for a medium effect. The only question left is whether a dramatic broadening of the width of the Δ-isobar in the nucleus would show up like that. On the low energy side of such a broadened resonance the $(p + \pi)$ decay would we hindered because it would be closer to threshold for this channel. This threshold effect is further emphasized by the energy dependence of the width. The dependence is because the Δ is a p-wave resonance in the $\pi + p$ system. Detailed calculations on this decay data are underway. These are cascade calculations, based on a classical picture, but also quantal calculations are being done.

We shall end this section by pointing to an instructive picture developed by the Lyon group [13,14,15]. In fig. 12 are shown curves in a $\omega$ versus q plot i.e. energy-loss versus three momentum transfer for the free pion and Δ-hole states.

In this model the pion in the medium and the Δ-hole state is treated as a two level system. These levels would cross at some q and any interaction repulsive or attractive between the two, would make them repel each other and the two states would mix. In the crossing region they would be an equal mixture of a pion and isobar-hole state, (sometimes refered to as a pisobar). We also see from the figure, that we can get an enormeous downward shift. The figure refers to infinite nuclear matter, but many of the basic results survive in a treatment, where the finiteness of the nucleus is included. The Δ-hole model is a specific case of such a treatment.

The question is then whether the decay experiment finally has found the pisobar.

## 5   SUMMARY

We have very briefly discussed data from charge exchange reactions studied at Laboratoire National Saturne. These reactions are used as probes of the spin response and we have emphasized that the spin-longitudinal response can only be studied this way. The electromagnetic probes measure the spin transverse response.

The $0^-$-strength function is the most onteresting part of the spin-longitudinal response at small momentum transfer. Several attempts have been made to find the collective $0^-$-state, so far without success. We have shown data from the $(^6\vec{Li}, ^6He)$ reaction with a tensor polarized beam and the results look promising as to allow for a separation of the spin-dipole resonance in its components. Data with better statistics could hopefully also give the strength with sufficient accuracy so that the expected quenching of the strength can be examined.

We have shown data from the quasielastic region. Data obtained with the $(p, n)$, $(d, 2p[^1S_0])$ and $(^3He, t)$ reactions were compared and the ratios of cross sections at the same momentum transfers were found to follow the ratios for the reactions on the nucleon. Both $(d, 2p)$ and $(^3He, t)$ are reactions with a strong emphasis of the spin-longitudinal component. The results of the ratios of cross sections is therefore a very strong indication for small effects from correlations on the spin response in the quasielestic region. The $(\vec{p}, \vec{n})$ data from Los Alamos demonstrate this at $q \sim 1.75 fm^{-1}$, but the comparison of the data from the three reactions seem to show, that this is also so at larger momentum transfers.

In the Δ-region charge exchange reactions have consistently shown a shift of the Δ-peak for nuclear targets. This shift is ascribed to correlations in the spin-longitudinal channel. This is therefore in contrast to the findings from the quasielastic region, where the correlations seem to be weak. Various attempts have been made to find other signals from the Δ-hole interactions. The spin response with the $(\vec{d}, 2p)$ reaction does not seem to support the conclusion from the shift of the Δ-peak. Distortion effects could however be important in this case and a more detailed analysis is needed.

The most direct information on medium effects on the Δ-properties is coming from an experiment presented at this conference. The decay of the Δ-isobar is studied in a $4\pi$-detector following the formation in a $(^3He, t)$ reaction. The decay into the 2p channel shows a large broadening of the Δ width in the medium, and possibly also a downward shift. Only a more detailed analysis of the decay data can help clarify the effects from Δ-hole correlations.

In conclusion: Little is known about spin-longitudinal correlations in nuclei, but a number of recent experiments have given valuable hints on how to proceed. Over the next few years we will learn more and a coherent description of the data can be hoped for. A crucial parameter is the density. Can we develop experiments, where the density dependence can be studied?

## 6    ACKNOWLEDGEMENTS

The material presented in this paper is the result of the work of many people. I want especially to thank the collaborators around the activities at Saturne, D. Bachelier, J.L. Boyard, C. Ellegaard, T. Hennino, J.C. Jourdain, J.S. Larsen, M. Østerlund, P. Radvanyi, M. Roy-Stephan, P. Zupranski, B. Ramstein and T. Sams. The two last have presented other results from the collaboration at this meeting.

## REFERENCES

[1]  C. Gaarde, Ann. Rev. of Nuclear and Particle Science **41** (1991) 187.

[2]  C. Ellegaard, C. Gaarde, T.S. Jørgensen, J.S. Larsen, C. Goodman, I. Bergqvist, A. Brockstedt, P. Ekström, M. Bedjidian, D. Contardo, J.Y. Grossiord, A. Guichard, D. Bachelier, J.L. Boyard, T. Hennino, J.C. Jourdain, M. Roy-Stephan, P. Radvanyi, and J. Tinsley, Phys. Rev. Lett. **59** (1987) 974.

[3]  C. Ellegaard, C. Gaarde, T.S. Jørgensen, J.S. Larsen, B. Million, C. Goodman, A. Brockstedt, P. Ekström, M.Österlund, M. Bedjiidian, D. Contardo, D. Bachelier, J.L. Boyard, T. Hennino, J.C. Jourdain, M. Roy-Stephan, P. Radvanyi, and P. Zupranski, Phys. Lett **231B** (1989) 365.

[4]  D.V. Bugg., private comm.

[5]  A. Brockstedt, I.Bergquist, L. Carlén, L.P. Ekström, B. Jakobsson, C. Ellegaard, C. Gaarde, J.S. Larsen, C. Goodman, M. Bedjidian, D. Contardo, J.Y. Grossiord, A. Guichard, J.R. Pizzi, D. Bachelier, J.L. Boyard, T. Hennino, J.C. Jourdain, M. Roy-Stephan, M. Boivin, T. Hasegawa, and P. Radvanyi, Nucl. Phys. **A** (1991) in print.

[6]  C. Gaarde, J. Rapaport, T.N. Taddeucci, C.D. Goodman, C.C. Foster, D.E. Bainum, C.A. Goulding, M.B. Greenfield, D.J. Horen, and E. Sugarbaker, Nucl. Phys. **A369** (1981) 258.

[7]  R.G. Jeppesen., private comm.

[8]  C. Gaarde, Nucl. Phys. **A507** (1990) 79c.

[9]  C. Gaarde, Physica Scripta **T32** (1990) 14.

[10]  I. Bergqvist, A. Brockstedt, L. Carlén, L.P. Ekström, B. Jakobsson, C. Ellegaard, C. Gaarde, J.S. Larsen, C. Goodman, M. Bedjidian, D. Contardo, J.Y. Grossiord, A. Guichard, R. Haroutunian, J.R. Pizzi, D. Bachelier, J.L. Boyard, T. Hennino, J.C. Jourdain, M. Roy-Stephan, M. Boivin, and P. Radvanyi, Nucl. Phys. **A469** (1987) 648.

[11] M. Ichimura, K. Kawahigashi, T. Jørgensen, and C. Gaarde, Phys. Rev. **C39** (1989) 1446.

[12] V. Dmitriev, O. Sushkov, and C. Gaarde, Nucl. Phys. **A459** (1986) 503.

[13] P.A.M. Guichon and J. Delorme, *Journees d'etudes Saturne, Piriac*, in , p. 53, 1989.

[14] G. Chanfray, Ann. de Phys. **(to be published)** (1991) .

[15] M. Ericson, Nucl. Phys. **A518** (1990) 116.

# THE CONTINUUM SPIN RESPONSE TO INTERMEDIATE ENERGY

# PROTONS AT LOW MOMENTUM TRANSFER

F.T. Baker

Physics Department, University of Georgia, Athens, GA 30602

and

C. Glashausser

Physics Department, Rutgers University, New Brunswick, NJ 08903
and
Institut de Physique Nucleaire, University of Paris, F91406 Orsay, France

Let us start with some interesting new data. Shown in Fig. 1 are the spin-longitudinal ($S_L$) and spin-transverse ($S_T$) spin-flip probabilities for proton scattering from $^{40}$Ca at 500 MeV and 580 MeV. They are the thesis data of Andrew Green. The momentum transfer $q$ here is small, about 100 MeV/c. The excitation energies $\omega$, while fairly large, are still in the region where nuclear structure, nuclear collectivity, is expected to be important. The quasielastic peak is not seen at these momentum transfers because it would appear in the region of the giant resonances or even below. But the same one-step scattering mechanism which excites one- particle one-hole states in first order and yields the quasielastic peak at high $q$ is responsible for continuum excitation, including the giant resonances, at low q. The data of Fig. 1 show that the probe plus the nucleus yield approximately equal strengths in the longitudinal and transverse channels. This may not seem surprising. But, if we divide the values of $S_L/S_T$ from Fig. 1 by the values of $S_L/S_T$ from isospin averaged NN scattering amplitudes,[2] we get the ratios of around 4.0 shown in Fig. 2. These may seem startling, particularly when they are compared to values slightly less than unity measured by Rees et al. at much higher momentum transfer (350 MeV/c) for a very similar ratio.[1] Is the long sought pionic (ie, spin-longitudinal) enhancement in nuclei to be found at low $q$ instead of high $q$? The answer is no, of course, but it will take awhile, even into the next talk by Jochen Wambach, to see why. We first have to review the measurements and interpretation that led us to measure $S_L$ and $S_T$.

Kevin Jones opened the last Telluride Conference with a presentation of $S_{nn}$ measurements for inelastic scattering around 300 MeV.[3] (The sum of $S_L$ and $S_T$ mentioned above is $S_{nn}$, the spin-flip probability relative to the the normal (n) to the reaction plane.) For $^{40}$Ca and several other medium weight nuclei, he pointed out that a strong enhancement of $S_{nn}$ relative to free NN scattering is observed for $\omega$ near 40 MeV and $q$ around 100 MeV/c.[4] With several assumptions discussed below, these values of $S_{nn}$ were translated into a relative nuclear spin response which was roughly 80% S=1 in this $(q,\omega)$ region. Several new examples of this behavior are shown in Figs. 3 and 4. The data for $^{12}$C (Fig. 3) at 7° (100 MeV/c) taken at LAMPF at 319 MeV rise up sharply from very small values at low $\omega$ (the isolated states below 10 MeV yield essentially zero for $S_{nn}$) to values above 0.4 at high excitation. This is almost twice as high as the isospin averaged free value of $S_{nn}$ at a laboratory scattering angle of 7°, shown by the solid horizontal line in the figure. At 12° (q=178 MeV/c), $S_{nn}$ is only slightly higher than the free value at $\omega$ near 40 MeV, and at 18° (q=258 MeV/c) it barely reaches the free value. There is considerable structure in the 10-20 MeV region in $^{12}$C, due to strong individual unnatural parity states; this structure is almost entirely

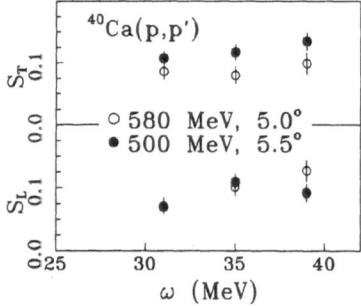

Figure 1. Spin-longitudinal $(S_L)$ and spin-transverse $(S_T)$ spin-flip probabilities for inelastic excitation of $^{40}$Ca at incident energies of 500 and 580 MeV. The momentum transfer is about 100 MeV/c at each energy.

absent from the other nuclei we have examined. The TRIUMF data for $^{208}$Pb at 200 MeV incident energy in Fig. 4 show a similar but somewhat weaker enhancement of $S_{nn}$ at 40 MeV. Because of the lower bombarding energy, the $q$ at even 10.7° is only 135 MeV/c, and so we have no data comparable to the 12° or 18° data in $^{12}$C. Compared to the free NN data shown by the solid line, the enhancement is hardly visible at 5°, but it becomes significant when compared to the dashed line where the effects of fermi motion have been included via the optimal frame. This distinction is also important at 7.7°, and will be discussed below. The enhanced values of $S_{nn}$ in the 40 MeV region for low $q$ scattering thus appear over the entire periodic table.[5] This is true also of the general rise from small values at low $\omega$ to large values at high $\omega$. But details of the $S_{nn}$ spectrum, such as the 10-20 MeV structure in $^{12}$C and the rather large values in the 5-10 MeV region in $^{208}$Pb, do vary from one nucleus to another, or perhaps from one region of the periodic table to another.

Inelastic proton scattering produces a mix of isospin transfers zero and one in the excitation of the continuum. In an attempt to see directly the T=0 spin response, work has begun at Saturne to measure spin transfer in the inelastic scattering of

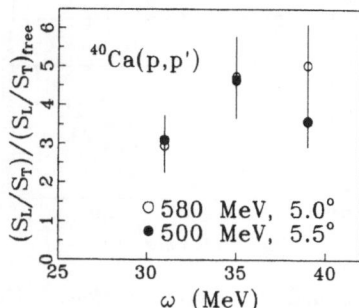

Figure 2. The ratio of the values of $S_L$ to $S_T$ for $^{40}$Ca divided by the isospin-averaged values of this same ratio for free nucleon-nucleon scattering. The scattering amplitudes of Franey and Love (Ref. 2) were used.

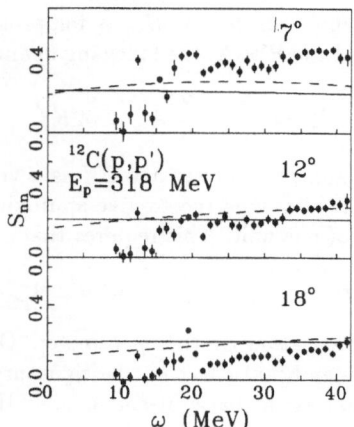

Figure 3. Values of the spin-flip probability $S_{nn}$ for inelastic scattering from $^{12}$C at 318 MeV. Free NN values of $S_{nn}$ are shown by the solid lines; fermi-motion corrections to these values have been included in the dashed lines.

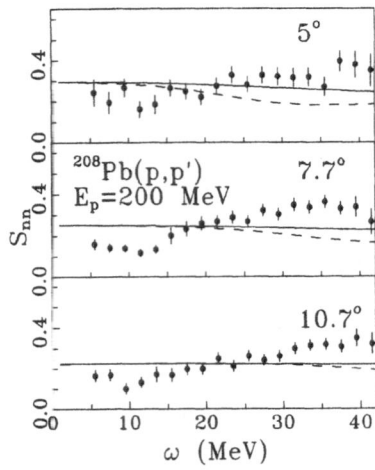

Figure 4. Values of $S_{nn}$ for inelastic scattering from $^{208}$Pb at 200 MeV. Free NN values of $S_{nn}$ are shown by the solid lines; fermi-motion corrections to these values have been included in the dashed lines.

polarized deuterons.[6] In the absence of a convenient analyzer of the tensor polarization of the scattered deuterons at this time, the initial experiments have measured only the vector polarization with a focal plane polarimeter only slightly modified from a standard proton polarimeter. In this situation there is no observable which is rigorously a signature of spin transfer, as $S_{nn}$ is for protons. However, assuming time-reversal invariance and the PWIA, the following quantity is expected to be zero for $\Delta S=0$ transitions:

$$S_d^y = \frac{4}{3} + \frac{2}{3} A_{yy} - 2K_y^{y'}.$$

Here $A_{yy}$ is the tensor analyzing power and $K_y^{y'}$ is a vector polarization transfer parameter. This quantity is also expected to give approximately the probability for a change in spin direction of one unit. This requires that

$$P_y = A_y \quad \text{and} \quad K_{yy}^{y'y'} = 2 - A_{yy},$$

where $K_{yy}^{y'y'}$ is a tensor polarization transfer parameter. Only large deviations from these assumptions have a significant effect on the signature. In addition, it assumes a single step NN interaction as the basic mechanism, so that spin two transfers are forbidden.

To determine whether this signature can indeed separate S=0 and S=1 excitations in deuteron scattering, measurements were taken first on a $^{12}$C target where individual states of both types can be easily resolved.[6] Fig. 5a shows the standard unpolarized yield of 400 MeV deuteron scattering from $^{12}$C at 4°; this yield multiplied by the signature $S_d^y$ is illustrated in Fig. 5b. In the latter spectrum, the strong natural parity S=0 states are greatly reduced and the unnatural parity states at 12.71 and 18.3 MeV stand out nicely. It is interesting to note the lack of strength in the 15 MeV region in this spectrum, a region where the $1^+$ T=1 is very strong in proton scattering. Preliminary results from a measurement with a $^{40}$Ca target are shown in Fig 6. The signatures are noticeably smaller than the $S_{nn}$ values for proton scattering at 300 MeV, but this is not unexpected because of the relative weakness of the T=0 S=1 NN interaction. In fact, the shape of the curve does bear some resemblance to the shape of the proton data at 800 MeV where the relative strengths of S=1 and S=0 interactions are more comparable to the deuteron case. Much better statistics expected in a forthcoming run will be necessary to establish whether significant T=0 S=1 strength is indeed being observed. Theoretical work will also be important to determine the validity of the signature and to try to obtain a quantitative measure of the strength. It may be that tensor polarization measurements will prove to be essential in obtaining reasonably precise determinations of the T=0 S=1 response.

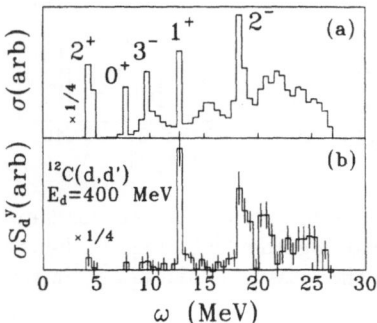

Figure 5. Results of polarization transfer measurements with 400 MeV deuterons for inelastic scattering from $^{12}$C at 4°. Fig. 5a shows the normal unpolarized spectrum. Fig. 5b shows the same spectrum multiplied by the measured values of the spin-transfer signature $S_d^y$ described in the text.

For $S_{nn}$ spectra in proton scattering, Jones[3] described how we attempt to take out the effects of the incident probe, so that we can determine the actual nuclear response. We label the relative S=1/(S=0 + S=1) response $R_s$. If the nucleus were a gas of non-interacting fermions, then $R_s$ should be 0.5. If the many-body effect of the nucleus tends to enhance S=1 excitations over S=0 excitations in a particular excitation region, then $R_s$ should be greater than 0.5. Conversely, if S=0 correlations dominate, then $R_s$ should be less than 0.5. An example of a recent determination of $R_s$ is shown in Fig. 7, where the values for $^{48}$Ca at 7° are illustrated; the $S_{nn}$ data were taken at 319 MeV.[7] The shape of $R_s$ is very similar to the shape of the $S_{nn}$ spectrum with a change of scale. Note that the nuclear response appears to be about 80% S=0 at low $\omega$ and just the reverse at high $\omega$; even in the giant resonance region where strong S=0 correlations are manifest in the giant resonances, $R_s$ is not much less than 50%.

But the $R_s$ we determine depends on many assumptions, and certainly they are not all precisely true. The basic assumption is that $S_{nn}$ in this energy region is robust,[4] that it is strong enough to maintain its free value in the face of possible attacks from effects unrelated to the intrinsic nuclear structure we are trying to understand, such as

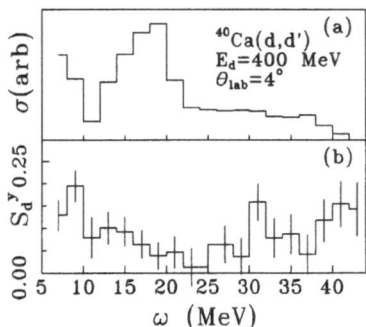

Figure 6. Preliminary results of polarization transfer measurements with 400 MeV deuterons for inelastic scattering from $^{40}$Ca. Fig. 6a shows the standard unpolarized spectrum. Fig. 6b shows the measured values of $S_d^y$ in this region, averaged over 2 MeV bins.

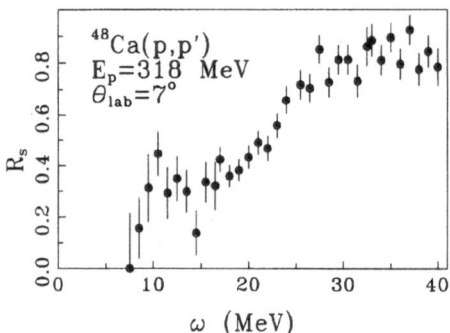

Figure 7. Values of the relative nuclear spin response $R_s$ determined from 318 MeV measurements of $S_{nn}$ on $^{48}$Ca at 7°.

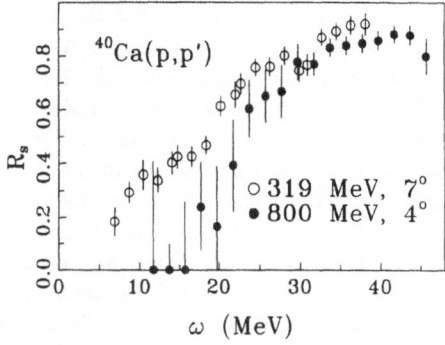

Figure 8. Comparison of the values of $R_s$ for $^{40}$Ca at 100 MeV/c determined from $S_{nn}$ measurements at 319 MeV and 800 MeV.

distortions, two-step processes, relativistic effects in the nucleon-nucleus interaction, fermi motion effects, and the like. We also assume for the values we have published that the value of $S_{nn}$ for pure S=1 processes inside the nucleus is the same as its free space value. We label this quantity alpha; it is typically around 0.5. Nuclear correlations, however, may well enhance a particular spin component at some excitation energies, particularly in the giant resonance region, and modify this value of alpha. An RPA/IA calculation with particular attention to the treatment of the two-body spin-orbit term is necessary to obtain a better value of alpha. All of the assumptions about the cleanliness of the reaction mechanism have been checked theoretically at some level, as described previously,[4] but we have also carried out an experimental check, as reported recently.[5,8] We redid the original $^{40}$Ca measurement at 800 MeV. Essentially all the reaction mechanism effects likely to be significant are very different at 800 and 300 MeV. The free value of $S_{nn}$ itself is reduced by a factor of 4 or so at 800 MeV. If $R_s$ is indeed a measure of the intrinsic nuclear response, it should stay unchanged as we change the incident energy. The results of our measurements, shown in Fig. 8, reveal wonderful agreement between the values of $R_s$ determined over essentially the entire spectrum. This is true at the momentum transfer where the enhancement of $S_{nn}$ is strongest.

All these remarks suggest that we should be tempted to take the results of the $S_L/S_T$ measurement discussed in the beginning of this talk literally, without worrying about interaction mechanism effects. If we did, we would have some problems with the tentative explanation of the $S_{nn}$ enhancement that we have previously put forward. We (and our theoretical colleagues, one of whom is giving the next talk) have remarked that, at a qualitative level, the enhancement of $S_{nn}$ at high excitation is not hard to understand.[5] The residual S=0 force in nuclei is attractive, and it thus pulls the S=0 resonances down well below their uncorrelated 1p-1h positions. The reverse is true for S=1 resonances. Thus, in $^{40}$Ca, eg, at about 100 MeV/c, the L=2 giant resonance should be an important component of the cross section and it occurs at about 16 MeV excitation. The octupole resonance does not become really important until about 12°. Thus there is little S=0 strength at high excitation at 7°. On the other hand, the L=2 spin resonance is expected at around 35 MeV. We have seen evidence for it in this region, as well as evidence for the L=1 spin-dipole at about 18 MeV.[9] The latter helps explain the fact that $S_{nn}$ is not much smaller than the free value even in the giant resonance region, while the former would help explain the enhancement in the high excitation energy region. Indeed, calculations based on the schematic model, as well as calculations based on the PWIA and RPA with only the T=1 channel included, tend to confirm these arguments.[5] Such an explanation of the enhanced $S_{nn}$ strength would not suggest that we should see a ratio of $S_L/S_T$ in nuclei very different from the free value of this ratio. The exhaustion of strength in the S=0 channel around 40 MeV would affect both $S_L$ and $S_T$ equally, since the L and T terms contribute to the S=1 channel only. But even the extra strength in the S=1 channel at high $\omega$ is expected to be roughly equally divided between L and T components, because the strength of the residual interaction in the two channels is expected to be comparable at low q. While uncertainties in the residual interaction might allow some enhancement in the L or T channel, and nuclear structure effects could enhance L or T in a small region, certainly the factor of four enhancement in the L channel over a 10 MeV region could not be explained this way. No exotic mechanisms–like the pionic enhancements searched for in the experiment by Rees et al.[1]– were expected. We were looking for a straightforward verification of models of collectivity in nuclei.

This means we must look more closely at our assumptions about the simplicity of the reaction mechanism. Even if $S_{nn}$ is robust, it doesn't guarantee that $S_L$ and $S_T$ are robust. And $S_{nn}$ itself? Even in the figures we have shown here, there is evidence that fermi motion has some effect on $S_{nn}$. The solid and dashed lines in Fig. 4 compare the free value of $S_{nn}$ with the values calculated in the optimal frame, as described by R.D. Smith at the last Telluride meeting.[10] The effect is significant at 200 MeV; it arises from the energy dependence of the scattering amplitudes in this region. The effect is much smaller at 319 MeV, as evidenced by the difference between the solid and dashed curves in Fig. 3. Around 580 MeV, the spin-transverse amplitude is changing rather rapidly with energy, and this gives rise to a significant effect on $S_T$ as shown in Fig. 9; the effect on $S_L$ is very small. The impact of such changes can be included rather simply in the formulas we have used previously; the free NN values are just replaced by the fermi-corrected values. Now, however, we have an even bigger problem than before with the ratio of $S_L/S_T$. We already had too large a ratio of $S_L/S_T$ compared to the free value; now we've increased the effective free value of $S_T$ without affecting the free value of $S_L$.

The effect of distortions on $S_{nn}$ has also been examined by R.D. Smith[10] in the context of the slab model with a zero range interaction. At 319 MeV, the effects were significant only at very forward angles like 3°. On the other hand, DWIA calculations of the modification of the strengths of individual shell-model particle-hole transitions show that spin-longitudinal and spin-transverse components are reduced from their plane wave values by different amounts.[9] The transverse components are much more strongly affected than the longitudinal ones. This is reminiscent of the arguments adduced by Klein, Love, and Franey[11] to suggest that the strong absorption in $N\bar{N}$ collisions should yield a strong preference for spin longitudinal excitations relative to spin transverse excitations, because of the long range of the $\pi$ exchange force relative to $\rho$ exchange. Preliminary calculations by Castel of the effects of distortion on $S_L$ and $S_T$ using the schematic model do in fact suggest that $S_L/S_T$ is significantly increased, ie, that distortions may be able to account for the apparent enhancement of $S_L$ in our measurements.[12] Note finally, however, that the effects of distortions on

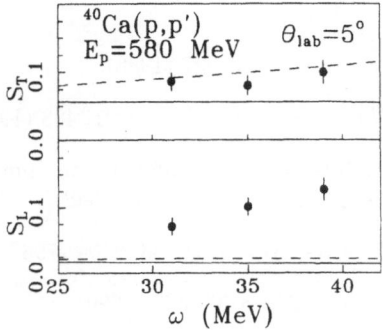

Figure 9. Values of $S_L$ and $S_T$ calculated for NN scattering with fermi-motion corrections included via the optimal frame described by Smith (Ref. 10) are shown by the dashed lines. Uncorrected values are shown by the solid lines.

$S_L/S_T$ at large $q$ go just the other way. When RPA correlations are included in the calculations by Ichimura et al.[13] for the data of Rees et al.,[1] $S_L/S_T$ is dramatically decreased by distortions.

The stage is now set for the theoretical discussion by Jochen Wambach. The experimental data reveal an enhancement of $S_{nn}$ relative to the free values over the entire periodic table in the 40 MeV excitation energy region at momentum transfers around 100 MeV/c. This region is thus dominated by S=1 correlations while the excitation energy region below about 20 MeV is dominated by S=0 correlations. Work has begun on trying to separate the T=0 and T=1 isospin components of this excitation by measuring polarization transfer with deuterons. Assumptions about the simplicity of the reaction mechanism are corroborated by the agreement between the 300 MeV and 800 MeV values for the relative nuclear spin response, but various possible complications should be considered in more detail. It is expected that the results should be reasonably explained in terms of a complete RPA calculation with good descriptions of S=0 and S=1 collectivity in nuclei. In fact, explanation of these results, including cross sections and spin-flip cross sections as well as $S_{nn}$, should be a sensitive test of the accuracy of contemporary RPA calculations, and thus a test of our understanding of S=0 and S=1 collectivity in nuclei. Recent data on the spin longitudinal and spin transverse separation of the $^{40}$Ca response in the interesting region pose an apparent challenge to simple models. But it is just here that complications of the nuclear reaction might be expected to distort our view of the nuclear structure.

## ACKNOWLEDGMENTS

The work reported here is the result of the concerted effort of a large number of collaborators over several years at LAMPF, TRIUMF, and SATURNE. We cite here particularly L. Bimbot, R.W. Fergerson, A. Green, O. Hausser, K. Jones, C.A. Miller, M. Morlet, S.K. Nanda, and M. Vetterli, but many others, co-authors on the papers mentioned above, have also made important contributions. We are grateful also to our theoretical colleagues who have helped us enormously, especially B. Castel, R.D. Smith, J. Wambach, and J. Unkelbach. Our work is partially supported by the Department of Energy and the National Science Foundation.

## REFERENCES

1. L. B. Rees et al., Phys. Rev. C34, 627 (1986).

2. M. A. Franey and W.G. Love, Phys. Rev. C31, 488 (1985).

3. K. Jones, Proceedings International Conference on Spin Observables of Nuclear Probes, Telluride, Plenum Press, New York (1988).

4. C. Glashausser et al., Phys. Rev. Lett. 58, 2404 (1987).

5. F. T. Baker et al., Phys. Lett. B237, 337 (1990).

6. M. Morlet et al., Phys. Lett. B247, 228 (1990).

7. F. T. Baker et al., submitted to Phys. Rev. C. (1991).

8. L. Bimbot et al., Phys. Rev. C42, 2367 (1990).

9. F. T. Baker et al., Phys. Rev. C40, 1877 (1989).

10. R. D. Smith, Proceedings International Conference on Spin Observables of Nuclear Probes, Telluride, Plenum Press, New York (1988).

11. A. Klein, W.G. Love, and M. Franey, Phys. Rev. Lett. 56, 700 (1986).

12. B. Castel, private communication.

13. M. Ichimura et al., Phys. Rev. C39, 1449 (1989).

# THE NUCLEAR SPIN RESPONSE IN EXTENDED RPA THEORIES

J. Wambach† and W. Unkelbach

Institut für Kernphysik
Forschungszentrum Jülich
D-5170 Jülich, FRG

## 1. INTRODUCTION

The study of the spin-isospin response of nuclei induced by medium energy nucleon-nucleus scattering is of great interest, both, in theory and experiment. Major issues are: (1) the existence of spin resonances excited at forward angles and small energy loss, (2) the relative enhancement of the spin response as one approaches the nuclear continuum and (3) spin-longitudinal collectivity in the quasielastic region. The latter is related to the enhancement of the virtual pion field in the nucleus which is of great importance in the mesonic interpretation of the EMC-effect and recent Drell-Yan measurements of the quark sea in the nuclear medium. Theoretically, the main uncertainty is the role of short-range nucleon correlations in screening the pion field. In a somewhat simplified fashion, such screening effects are summarized by the Fermi liquid parameter $g'$. On the experimental side, the field has witnessed major advances with the availability of polarized beams, coupled with scattered particle detection by magnetic spectrometers and polarization analysis with focal plane polarimeters. Such measurements have resulted in the measurement of polarization transfer coefficients $D_{ij'}$ which give detailed information on the momentum-spin correlations (spin-longitudinal or spin-transverse).

In this talk we discuss calculations of polarized proton-nucleus reactions in the forward direction and for energy loss up to 40 MeV. In this kinematical region, collective modes are predominantly excited and one begins to see the onset of quasielastic scattering. We will concentrate on the spin-flip probability $S_{nn'}$ which singles out the $\Delta S = 1$ component of the cross section. $S_{nn'}$ is determined by measuring the normal-component spin-transfer coefficient $D_{nn'}$ $[S_{nn'}=(1-D_{nn'})/2]$. Measurements are now available over a large range of excitation energies and on a variety of targets. This body of data has stimulated considerable theoretical interest. To assess the global features of the spin response one of the earliest models used a semi-infinite slab picture of the nucleus[1]. This model qualitatively describes both cross section[2] and spin-flip data[3,4,5] The calculations were extended in ref.[6] pointing out the importance of 2p2h damping. To account for finite nucleus effect in a more realistic way, Boucher et al.[7] have used a schematic model that is based on energy-weighted sum rules with coupling to 2p2h states. They have also performed calculations using an RPA model for the nuclear structure[8]. In both cases their calculations also qualitatively describe the data. Ichimura et al.[9] have performed Distorted-Wave-Impulse-Approximation (DWIA) calculations within the RPA. For the first time the effects of distortions on spin observables were pointed out. Here we will present new DWIA calculations for (p,p') reactions in $^{40}$Ca and $^{208}$Pb using the "second RPA" (SRPA) which includes 1p1h- as well as 2p2h excitations. The results will be compared to cross section and spin-flip measurements.

---

† also: Institut für Kernphysik, Universität Bonn
D-5300 Bonn, Fed. Rep. Germany

*Spin and Isospin in Nuclear Interactions*
Edited by S.W. Wissink *et al.*, Plenum Press, New York, 1991

## 2. THEORY

### 2.1 CROSS SECTIONS AND POLARIZATION TRANSFER COEFFICIENTS

The inelastic nucleon-nucleus scattering process is described by the transition amplitude

$$T_{if} = \langle \phi_f \Phi_f | \hat{T} | \phi_i \Phi_i \rangle \tag{1}$$

from an initial state $|\phi_i \Phi_i\rangle$ to a final state $|\phi_f \Phi_f\rangle$ were $|\phi\rangle_i (|\phi\rangle_f)$ and $|\Phi\rangle_i (|\Phi\rangle_f)$ denote the initial (final) state of the projectile and target respectively. Knowing $T_{if}$ we obtain the scattering observable.

The double differential cross section is given by

$$\frac{d^2\sigma}{d\Omega dE} = \left[ \frac{E_{cm}^2}{4\pi} \right]^2 \sum_f \frac{k_f}{k_i} |T_{if}|^2 \delta(E - E_{if}) \tag{2}$$

where $k_{i(f)}$ are the initial (final) energy and three-momentum of the projectile, $E_{cm} = \sqrt{m^2 + k_i^2}$ the $cm$-energy, and $E_{if} = E_i - E_f$ the energy loss in the reaction which equals the excitation energy in the target.

To determine the polarization transfer coefficients $D_{ij}$ we introduce coordinate systems $(\hat{x}, \hat{y}, \hat{z})$ and $(\hat{x}', \hat{y}', \hat{z}')$ before and after the collision, according the Madison convention[10]

$$
\begin{aligned}
\hat{l} &= \mathbf{k}_i / |\mathbf{k}_i| & \hat{l}' &= \mathbf{k}_f / |\mathbf{k}_f| \\
\hat{n} &= \frac{\mathbf{k}_i \times \mathbf{k}_f}{|\mathbf{k}_i \times \mathbf{k}_f|} & \hat{n}' &= \hat{y} \\
\hat{s} &= \hat{n} \times \hat{l} & \hat{s}' &= \hat{n}' \times \hat{l}'
\end{aligned}
\tag{3}
$$

The scattering plane is defined by $(\hat{s}, \hat{l})$ or $(\hat{s}', \hat{l}')$, and $\hat{n} = \hat{n}'$ is perpendicular to the plane. Denoting the polarization vector for the incoming (outgoing) projectile as $\vec{P} = (P_s, P_n, P_l)$ $(\vec{P}' = (P_{s'}', P_{n'}', P_{l'}'))$, where $P_i = \langle \sigma_i \rangle$ is the expectation value of the spin operator in a given direction, these are related as[11] save

$$
\begin{pmatrix} P_{s'}' \\ P_{n'}' \\ P_{l'}' \end{pmatrix} = c(\theta) \left[ \begin{pmatrix} 0 \\ D_{0n'} \\ 0 \end{pmatrix} + \begin{pmatrix} D_{ss'} & 0 & D_{ls'} \\ 0 & D_{nn'} & 0 \\ D_{sl'} & 0 & D_{ll'} \end{pmatrix} \begin{pmatrix} P_s \\ P_n \\ P_l \end{pmatrix} \right] \tag{4}
$$

with $c(\theta)$ depends only on the scattering angle and

$$D_{nm'} = \frac{Tr\left(T\sigma_n T^\dagger \sigma_{m'}\right)}{Tr\left(TT^\dagger\right)} \qquad , n, m = 0, s, n, l \quad . \tag{5}$$

The coefficients $D_{ij'} (i = s, n, l; j' = s', n', l')$ determine the probability for rotating the spin from direction $i$ to direction $i'$. As mentioned the coefficient $D_{nn'}$ is related to the spin-flip probability

$$S_{nn'} = \frac{\sigma_{\uparrow\downarrow}}{\sigma_{\uparrow\uparrow} + \sigma_{\uparrow\downarrow}} \tag{6}$$

as $S_{nn'} = \frac{1}{2}(1 - D_{nn'})$. Another combination of $D_{ij}$'s yields the longitudinal and transverse spin-flip probabilities, $S_L$ $S_T$, as

$$S_L = \tfrac{1}{4}\left[1 - D_{nn'} + (D_{ss'} - D_{ll'})\right]$$
$$S_T = \tfrac{1}{4}\left[1 - D_{nn'} - (D_{ss'} - D_{ll'})\right] \tag{7}$$

In general, $\hat{T}$ is a many-body operator and the evaluation of the transition amplitude $T_{if}$ is difficult. At incident energies above 100 MeV and for small scattering angles and moderate energy loss the nucleon-nucleus reaction is, however, predominantly a one-step process. Then the problem separates into a reaction part involving the projectile and a structure part which describes the target dynamics. The reaction can be treated in DWIA and

$$T_{if} = \langle \Phi_f | \hat{T}_{if} | \Phi_i \rangle \tag{8}$$

where the transition operator $\hat{T}_{if}$ involves the optical model distorted waves $\chi_{\mathbf{k}_i}^{(+)}(\chi_{(\mathbf{k}_f)}^{(-)})$ for the initial (final) scattered nucleon and the free nucleon-nucleon T-matrix in the nucleon-nucleus $cm$-frame

$$\hat{T}_{if} = \sum_{j=1}^{A} \int d^3 r_p \chi_{\mathbf{k}_i}^{+}(\mathbf{r}_p) \chi_{\mathbf{k}_f}^{-*}(\mathbf{r}_p) t(\mathbf{r}_p, \mathbf{r}_j; E). \tag{9}$$

The projectile coordinate $\mathbf{r}_p$ is integrated over such that $\hat{T}_{if}$ is a one-body operator in the target coordinates $\mathbf{r}_j$.

In the applications we will use the T-matrix parameterization by Franey and Love[12] which is local

$$t(\mathbf{r}_p, \mathbf{r}_j; E) = \int d^3 q\, e^{i\mathbf{q}\cdot\mathbf{r}} t(\mathbf{q}, E); \qquad \mathbf{r} = \mathbf{r}_p - \mathbf{r}_j. \tag{10}$$

The $(q, E)$-dependence of the various spin-isospin components is expanded in Yukawa functions with energy-dependent, complex coupling strengths. The largest component is spin-scalar, isospin-scalar which implies that isoscalar non-spin flip modes will be predominantly excited in (p,p') reactions. The next most important contribution is the spin-isospin component which couples to isovector spin-flip excitations. In the region of interest it is about a factor of two smaller than the dominant term which means that spin-flip excitations will make up roughly one quarter of the total cross.

For completeness we give the DWIA expressions for the $D_{ij}$'s. Using the angular momentum coupled representation for $T_{if}$

$$T_{if} = \sum_{LSJ} \frac{1}{\sqrt{2J+1}} \langle J_i M_i J M | J_f M_f \rangle \, \beta_{LSJ}^{M m_f m_i}(\vec{k}_i, \vec{k}_f) \tag{11}$$

we obtain in terms of the reduced amplitudes

$$D_{ss} \sum |\beta_{LSJ}^{M m_f m_i}|^2 = \sum Re(\beta_{LSJ}^{M m_f m_i} \beta_{LSJ}^{M -m_f -m_i *})$$
$$\times \{\delta_{m_i,1/2}\delta_{m_f,1/2} + \delta_{m_i,1/2}\delta_{m_f,-1/2} + \delta_{m_i,-1/2}\delta_{m_f,1/2} + \delta_{m_i,-1/2}\delta_{m_f,-1/2}\}$$
$$D_{sl} \sum |\beta_{LSJ}^{M m_f m_i}|^2 = \sum Re(\beta_{LSJ}^{M m_f m_i} \beta_{LSJ}^{M m_f -m_i *})$$
$$\times \{\delta_{m_i,1/2}\delta_{m_f,1/2} - \delta_{m_i,1/2}\delta_{m_f,-1/2} + \delta_{m_i,-1/2}\delta_{m_f,1/2} - \delta_{m_i,-1/2}\delta_{m_f,-1/2}\}$$
$$D_{nn} \sum |\beta_{LSJ}^{M m_f m_i}|^2 = \sum Re(\beta_{LSJ}^{M m_f m_i} \beta_{LSJ}^{M -m_f -m_i *})$$
$$\times \{\delta_{m_i,1/2}\delta_{m_f,1/2} - \delta_{m_i,1/2}\delta_{m_f,-1/2} - \delta_{m_i,-1/2}\delta_{m_f,1/2} + \delta_{m_i,-1/2}\delta_{m_f,-1/2}\}$$
$$D_{ll} \sum |\beta_{LSJ}^{M m_f m_i}|^2 = \sum Re(\beta_{LSJ}^{M m_f m_i} \beta_{LSJ}^{M m_f m_i *})$$
$$\times \{\delta_{m_i,1/2}\delta_{m_f,1/2} - \delta_{m_i,1/2}\delta_{m_f,-1/2} - \delta_{m_i,-1/2}\delta_{m_f,1/2} + \delta_{m_i,-1/2}\delta_{m_f,-1/2}\}$$
$$D_{ls} \sum |\beta_{LSJ}^{M m_f m_i}|^2 = \sum Re(\beta_{LSJ}^{M m_f m_i} \beta_{LSJ}^{M -m_f m_i *})$$
$$\times \{\delta_{m_i,1/2}\delta_{m_f,1/2} + \delta_{m_i,1/2}\delta_{m_f,-1/2} - \delta_{m_i,-1/2}\delta_{m_f,1/2} - \delta_{m_i,-1/2}\delta_{m_f,-1/2}\} \tag{12}$$

323

where the summation is for all $m$-quantum numbers as well as all $L$, $S$ and $J$. These equations are valid if the incoming and outgoing polarization vectors are described in the same coordinate system. In the Madison convention one then has the following transformation:

$$
\begin{pmatrix} D_{ss'} \\ D_{sl'} \\ D_{nn'} \\ D_{ll'} \\ D_{ls'} \end{pmatrix} = \begin{pmatrix} \cos\theta & -\sin\theta & 0 & 0 & 0 \\ \sin\theta & \cos\theta & 0 & 0 & 0 \\ 0 & 0 & 1 & 0 & 0 \\ 0 & 0 & 0 & \cos\theta & -\sin\theta \\ 0 & 0 & 0 & \sin\theta & \cos\theta \end{pmatrix} \begin{pmatrix} D_{ss} \\ D_{sl} \\ D_{nn} \\ D_{ll} \\ D_{ls} \end{pmatrix}
\tag{13}
$$

where $\theta$ is the scattering angle.

## 2.2 TARGET TRANSITIONS

The target wave functions entering in the transition amplitude $T_{if} = \langle \Phi_f | \hat{T}_{if} | \Phi_i \rangle$ we shall describe in the SRPA which includes 1p1h and 2p2h excitations on the correlated ground state $|\Phi_i\rangle$. Since the DWIA transition operator $\hat{T}_{if}$ is a one-body operator in the target space[13] the scattering amplitude is a superposition of 1p1h excitations

$$
T_{if} = \sum_{ph} t^{if}_{ph} X^{if}_{ph} + t^{if}_{hp} Y^{if}_{ph}
\tag{14}
$$

where $t^{if}_{ph} = \langle h | \hat{T}_{if} | p \rangle$ are the ph-transition matrix elements. The forward (backward) $X^{if}_{ph}$ ($Y^{if}_{ph}$) amplitudes obey the matrix equation

$$
\begin{pmatrix} \tilde{A}(E_{if}) & B \\ B^* & \tilde{A}^*(E_{if}) \end{pmatrix} \begin{pmatrix} X^{if} \\ Y^{if} \end{pmatrix} = E_{if} \begin{pmatrix} 1 & 0 \\ 0 & -1 \end{pmatrix} \begin{pmatrix} X^{if} \\ Y^{if} \end{pmatrix}
\tag{15}
$$

which formally looks like the usual RPA equations. The effect of the 2p2h states is included in the energy-dependent A-matrix

$$
\begin{aligned}
\tilde{A}_{ph,p'h'}(E) &= A_{ph,p'h'} + \Sigma_{ph,p'h'}(E) \\
\Sigma_{ph,p'h'}(E) &\equiv \sum_{\substack{p_1 p_2 h_1 h_2 \\ p_1' p_2' h_1' h_2'}} A^*_{ph,p_1 p_2 h_1 h_2} \left(E - A_{p_1 p_2 h_1 h_2, p_1' p_2' h_1' h_2'}\right)^{-1} A_{p_1' p_2' h_1' h_2', p'h'} \cdot
\end{aligned}
\tag{16}
$$

and the RPA is recovered by setting the ph self energy $\Sigma_{ph,p'h'}(E)$ equal to zero. The solutions determine the transition amplitudes $T_{if}$, as well as the target excitation energies $E_{if}$. It should be noted that, due to the poles in the ph self energy $\Sigma$, there are many more eigenvalues $E_{if}$ in SRPA than in RPA. This leads to a fragmentation of the RPA transition strength. The amount of fragmentation is determined by the magnitude of imaginary part of $\Sigma_{ph,p'h'}$ or the "spreading width" $\Gamma^\downarrow$

$$
\Gamma^\downarrow_{ph,p'h'}(E) = 2\mathrm{Im}\,\tilde{\Sigma}_{ph,p'h'}(E).
\tag{17}
$$

$\Gamma^\downarrow$ is not diagonal in the ph-indices. The off-diagonal terms arise from interference between the particle- and hole decay amplitudes[13]. Two comments should be added:

- At $E = 0$ the spreading width vanishes which ensures stability of the SRPA ground state.

324

– There is a dispersion relation which relates the real part of $\Sigma$ to its imaginary part

$$Re\, \Sigma_{ph,ph'}(E) = \frac{1}{2\pi}\mathcal{P}\int dE' \frac{\Gamma^{\downarrow}_{ph,p'h'}(E')}{E'-E} \tag{18}$$

This "Kramers-Kronig" relation ensures proper normalization of excited state wave functions.

Because of the large number of 2p2h states, needed to correctly describe the cross sections and spin observables at the excitation energies of interest, the 2p2h propagator in the ph self energy (eq. (16)) involves the inversion a matrix of very large dimension (typically of the order of $10^3$-$10^4$). For all practical purposes it is prohibitive to invert such matrices. To proceed nonetheless with meaningful calculations, one has to rely on certain approximations for the energy-dependent $A$-matrix whose validity has to be judged by comparison with experiment and by other arguments of consistency. In the present work we use a semi empirical model which relies on optical potential information and is very easy to handle numerically[6]. We make the simplifying assumption that $\Sigma_{ph,p'h'}$ does not depend on the ph indices, i.e.

$$\Sigma_{ph,p'h'}(E) \approx \Sigma(E) = \Delta(E) + i\Gamma^{\downarrow}(E)/2. \tag{19}$$

Such an approximation is justified if many 1p1h excitations contribute to the strength function. The imaginary part determines a global, energy-dependent spreading width $\Gamma^{\downarrow}(E)$ while the real part $\Delta(E)$ gives an energy shift. According the eq. (18), both are not independent but related via a dispersion relation. It is therefore sufficient to know $\Gamma^{\downarrow}(E)$.

Within the approximation (19) there is a simple expression for the SRPA cross section[6]

$$\frac{d^2\sigma^{(SRPA)}}{d\Omega dE} = \sum_f |T^{(RPA)}_{if}|^2[\rho(E^{(RPA)}_{if}, E) + \rho(E^{(RPA)}_{if}, -E)] \tag{20}$$

where

$$\rho(E', E) = \frac{1}{2\pi}\frac{\Gamma^{\downarrow}(E)}{(E'-E-\Delta(E))^2 + \Gamma^{\downarrow}(E)^2/4} \tag{21}$$

i.e. the $\delta$-function in the RPA expression is replaced by the sum two Breit-Wigner functions (the second term accounts for ground state correlations). It can be easily shown that the total strength, as well as the total energy-weighted strength, is the same as in the RPA, in agreement with sum rule expectations.

We model the energy dependence of $\Gamma^{\downarrow}(E)$ by using information from the decay width of hole states, $\gamma_h$ and the imaginary part of the empirical optical potential which specifies the decay width of particle states, $\gamma_p$. Such data have been complied by Mahaux et al.[14], for instance. In the energy region of interest, the widths are nearly symmetrical around the Fermi energy $\epsilon_F$, i.e. $\gamma_h(-\epsilon) \approx \gamma_p(\epsilon) \equiv \gamma(\epsilon)$ and a smooth function of single-particle energy $\epsilon$. An adequate parameterization is given by

$$\gamma(\epsilon) = \alpha_\gamma \left[\frac{\epsilon^2}{\epsilon^2 + \epsilon_0^2}\right]\left[\frac{\epsilon_1^2}{\epsilon^2 + \epsilon_1^2}\right] \tag{22}$$

with $\alpha_\gamma = 10.75$ MeV, $\epsilon_0 = 18$ MeV and $\epsilon_1 = 110$ MeV.

$\Gamma^{\downarrow}(E)$ is then obtained as an energy-conserving average over $\gamma_p$ and $\gamma_h$ as

$$\Gamma^{\downarrow}(E) = \frac{\alpha_\Gamma}{E}\int_0^E d\epsilon\,[\gamma_p(\epsilon) + \gamma_h(\epsilon - E)]. \tag{23}$$

It contains a parameter $\alpha_\Gamma$ which we have adjusted to reproduce the known width of the isoscalar giant quadrupole resonance in $^{208}$Pb. This yields $\alpha_\Gamma = 0.5$. As a consequence of the strong energy dependence of $\gamma_p$ and $\gamma_h$ near the Fermi surface the spreading width $\Gamma^\downarrow$ is also strongly energy dependent, especially at low excitation energies. Once $\Gamma^\downarrow(E)$ is determined, the real part $\Delta(E)$ can be obtained from the dispersion relation (18).

## 2.3 RESIDUAL INTERACTION

The target wave functions are determined by the mean field and the residual ph interaction $\mathcal{F}_{ph}$. While the mean field was taken in standard Woods-Saxon form with parameters given in[15] we have employed two different forms of the residual interaction in the present calculations. The first one is a standard Landau-Migdal interaction[16]

$$\mathcal{F}_{ph} = \frac{4\pi f_\pi^2}{m_\pi^2}(f + f'\boldsymbol{\tau}\cdot\boldsymbol{\tau}' + g\boldsymbol{\sigma}\cdot\boldsymbol{\sigma}' + g'\boldsymbol{\sigma}\cdot\boldsymbol{\sigma}'\boldsymbol{\tau}\cdot\boldsymbol{\tau}') \tag{24}$$

where $f$ is strongly density dependent. The parameters were taken from[17]. The second interaction is an extended Landau-Migdal interaction[18] in which the spin-isospin channel is replaced by correlated $\pi + \rho$-exchange

$$\mathcal{F}_{ph} = \frac{4\pi f_\pi^2}{m_\pi^2}(f + f'\boldsymbol{\tau}\cdot\boldsymbol{\tau}' + g\boldsymbol{\sigma}\cdot\boldsymbol{\sigma}') + \tilde{V}_\pi + \tilde{V}_\rho \tag{25}$$

where

$$\tilde{V}_{\pi(\rho)}(\mathbf{q}) = \frac{1}{(2\pi)^3}\int d^3k\, g(\mathbf{k}+\mathbf{q})V_{\pi(\rho)}(\mathbf{k}) \tag{26}$$

and $g(\mathbf{q}) = (2\pi)^3\delta(q) - 2\pi^2/q_c^2\delta(|\mathbf{q}| - q_c)$ is a two-body correlation function obtained from a $G$-matrix with $q_c = 3.93$ fm$^{-1}$. The bare interactions $V_{\pi(\rho)}$ are of the usual form

$$
\begin{aligned}
V_\pi &= -\frac{4\pi f_\pi^2}{m_\pi^2}\Gamma_\pi^2\frac{\boldsymbol{\sigma}\cdot\mathbf{q}\boldsymbol{\sigma}'\cdot\mathbf{q}}{q^2 + m_\pi^2}\boldsymbol{\tau}\cdot\boldsymbol{\tau}' \\
V_\rho &= -\frac{4\pi f_\rho^2}{m_\rho^2}\Gamma_\rho^2\frac{\boldsymbol{\sigma}\times\mathbf{q}\boldsymbol{\sigma}'\times\mathbf{q}}{q^2 + m_\rho^2}\boldsymbol{\tau}\cdot\boldsymbol{\tau}'
\end{aligned}
\tag{27}
$$

with $f_{\pi(\rho)}^2 = 0.08(4.86)$ and where $\Gamma_{\pi(\rho)}$ denote the $\pi NN(\rho NN)$ formfactors.

It is well known, that the G-matrix yields too small a value for $g'$. With the parameters given above we obtain $g' = 0.57$ as compared to the empirical value $g' = 0.70 \pm 0.05$ deduced from the systematics of Gamow-Teller resonances. There is an interesting new possibility to explain this discrepancy which we briefly comment on. It relates to the role of the $\rho$-meson and its medium modifications. To a large extent the $\rho$-meson can be considered as a resonance in the $J^\pi = 1^-, T = 1$ $\pi\pi$-scattering channel (the time-like electromagnetic form factor of the pion is well this way[19]). In the nuclear medium, however, the pion propagation is modified, due to large coupling to nucleon-hole and $\Delta$-hole states. This leads to a density-dependent reshaping of the $\rho$-strength function. Theoretical estimates for uniform nuclear matter are shown in Fig. 1.

For our discussion, the most interesting feature is a softening near threshold, which becomes very pronounced as the density increases. Qualitatively we can understand this behavior as an increase of the pion effective mass (near saturation density $m_\pi^{eff} \approx 2 - 3m_\pi$). In a one-boson exchange picture of the NN-interaction, the modification of the $\rho$-strength distribution can be included by introducing a light $\rho$-meson, $\rho'$, with 2.5 to 3.0 $m_\pi$ whose coupling constant, $f_{\rho'}$, increases with density. The coupling constant $f_\rho$, itself, decreases accordingly. The partitioning of the coupling strengths can be obtained from the calculated $\pi\pi$-strength distribution[21]. Including this effect in the $\rho$-component of the spin-isospin interaction, as we have done in the present calculations, leads to a strong density dependence of $g'$. At half density one obtains $g' = 0.72$, close to the empirical value. At saturation density $g' = 0.91$.

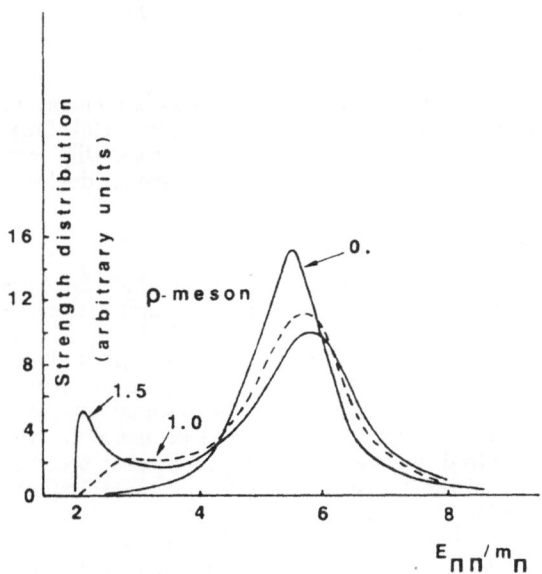

Fig. 1. Theoretical strength distribution of the $J^\pi = 1^-, T = 1$ $\pi\pi$-scattering channel (the $\rho$-channel) as a function of nuclear density. The density is quoted relative to the saturation density, $\rho_o$=0.16 fm$^{-3}$. The calculations were taken from[20].

## 3. RESULTS

### 3.1 (p,p') CROSS SECTIONS

Given the transition operator $\hat{T}_{if}$ (eq. (9)) the SRPA cross sections can be calculated according to eqs. (20) and (21). A comparison between theory and experiment is given in Fig. 2 which displays cross section measurements at 319 MeV incident energy in $^{40}$Ca[22,23,24] and at scattering angles between 5° and 12°, corresponding to momentum transfers ranging from 0.35 fm$^{-1}$ to 0.84 fm$^{-1}$.

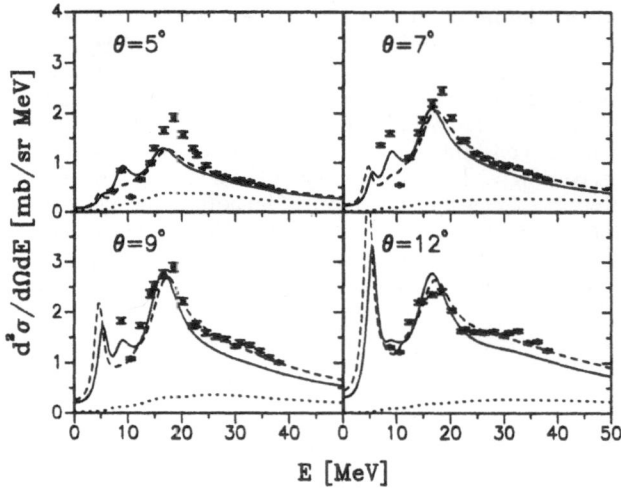

Fig. 2. Double differential cross sections for intermediate energy proton-nucleus scattering $^{40}$Ca for 319 MeV incident energy. A finite-range (full lines) and a zero-range (dashed lines) residual interaction have been used. The spin-flip component of the cross section is displayed by the dotted lines for the zero-range residual interaction . The data were taken from [22,23,24].

The theoretical description of the data is quite good which indicates that the reaction mechanism is predominantly of one-step nature. The underestimate of the cross section at 5°, is due to Coulomb excitation of the $GDR$ which has not been included in the calculations. The cross sections are dominated by the $GQ_0R$ at about 17 MeV excitation energy. At larger scattering angles (9°, 12°) the low-lying collective 3$^-$-state is strongly excited. The calculations using the finite-range residual interaction (full lines) and the zero-range residual interaction (dashed lines) do not differ very much. Spin-flip transitions ($\Delta S$=1) (dotted lines) are generally suppressed because of the small spin-dependent components of the T-matrix. They become more important with increasing excitation energy E above the giant resonance region.

Fig. 3 shows the double differential cross section for the reactions $^{48}$Ca (p,p') at 319 MeV and $^{208}$Pb (p,p') at 200 MeV incident energy, again at four different scattering angles. The $^{208}$Pb data correspond to momentum transfers between 0.45 fm$^{-1}$ and 0.84 fm$^{-1}$. The $^{48}$Ca curves show similar features as in the case of $^{40}$Ca for both, the total cross sections and the spin-flip components. For $^{208}$Pb, at about 11 MeV excitation energy the $GQ_0R$ can be identified clearly. At 2.6 MeV and 4.5 MeV the low-lying collective 3$^-$- and 2$^+$-states are excited. The peak at about 13 MeV in the 8° experimental spectrum is the Coulomb-excited $GDR$, which has not been included in the calculation. The description in the giant resonance region is quite good. In the continuum, above the $GQ_0R$, we underestimate the cross sections. This could be due to two-step processes becoming more important at higher excitation-energies. The spin-flip cross sections, again, are suppressed strongly, but become more important above the giant resonance region.

## 3.2 SPIN OBSERVABLES

At low momentum transfer, the spin-flip probability $S_{nn'}$ is of particular importance since it allows extraction of the $\Delta S = 1$-part of the total inelastic cross section. This is quite small in (p,p')-scattering (see Figs. 2 and 3). The calculated spin-flip probabilities are compared to the measurements in Fig. 4. Again the agreement is satisfactory.

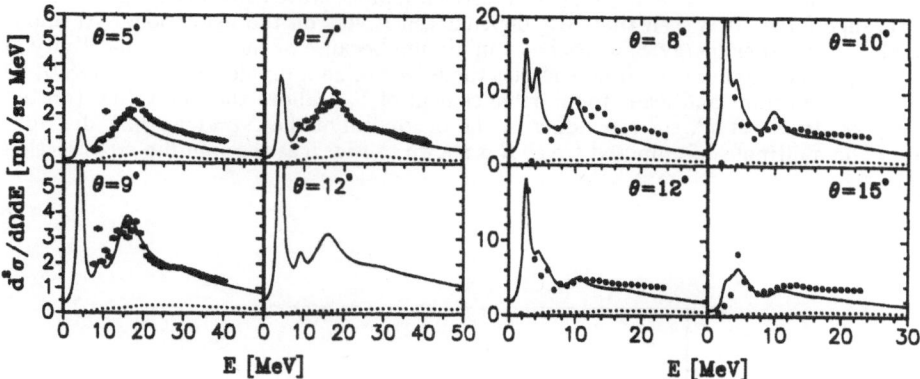

Fig. 3. Double differential cross sections for the reactions $^{48}$Ca (p,p') at 319 MeV (right-hand side) and $^{208}$Pb (p,p') at 200 MeV incident energy (left-hand side) at four scattering angles. The total cross sections (full lines) as well as the spin-flip components (dotted lines) are displayed. The results were obtained with a zero-range residual interaction. The data were taken from refs.[25,26], respectively.

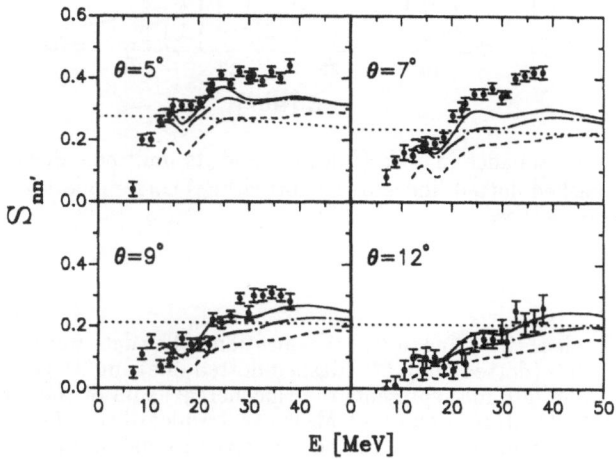

Fig. 4. The spin-flip probability $S_{nn'}$ from polarized (p,p')-scattering in $^{40}$Ca at 319 MeV at different scattering angles. The DWIA-SRPA calculations are compared with the data of ref.[24]. The full and dashed-dotted lines denote the distorted-wave calculations with finite-range and zero-range residual interaction, respectively. The dashed lines correspond to a plane-wave calculation (zero-range), and the dotted lines denote the free NN-scattering (Fermi gas).

The full lines denote the SRPA-DWIA calculation with the finite-range residual interaction. They result in larger values for $S_{nn'}$ than a calculation using the zero-range residual interaction (dashed-dotted lines). This is due to the tensor exchange-contributions of the finite-range force. The dashed lines represent a plane-wave calculation, based on the zero-range force. It is seen that $S_{nn'}$ depends sensitively on the distortions. At about 17 MeV excitation energy $S_{nn'}$ reaches a minimum because of the $\Delta S=0$-collectivity in the $GQ_0R$. For higher excitation energies the isovector spin-dipole resonance ($GSD_1R$) becomes important and leads to an enhancement of $S_{nn'}$ above the free values (dotted lines), especially at 7°, corresponding to the maximum of the $\Delta L=1$ angular distribution. The multipole decomposed $GSD_1R$ is shown in Fig. 5 as the spin-flip cross section $\sigma \cdot S_{nn'}$.

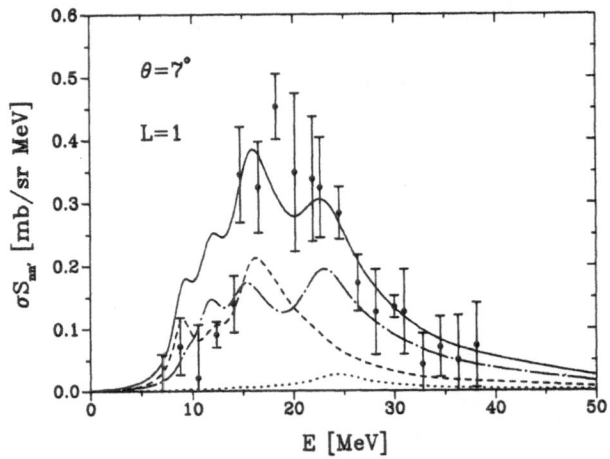

Fig. 5. Spin-dipole resonance in $^{40}$Ca (full line) and its multipole decomposition. The dashed line, the dashed-dotted line and the dotted lines correspond to the $2^-$-, $1^-$- and $0^-$-contributions respectively.

The product $\sigma \cdot S_{nn'}$ is entirely due to $\Delta S=1$-transitions. The data were taken from ref.[24]. To this resonance, $0^-$- (dotted line), $1^-$- (dashed-dotted line) and $2^-$-transitions (dashed line) contribute. The full line represents the (incoherent) sum of these multipolarities. A broad resonance-structure around 20 MeV can be identified. This corresponds to $1\hbar\omega$ transitions, pushed up in energy by the repulsive residual interaction in the vector-isovector channel ($1\hbar\omega \approx 14$ MeV). The inclusion of 2p-2h damping is important in introducing strong spreading of $\Delta S=1$-strength to higher excitation energies E. Fig. 6 shows $S_{nn'}$ for $^{48}$Ca and $^{208}$Pb at 319 MeV and 200 MeV incident energy, respectively. The calculations are based on the finite-range residual interaction. The full lines denote the distorted-wave calculations and the dashed lines correspond to plane-wave calculations. The data were taken from refs.[25,27], respectively. Again the agreement is satisfactory. The features for $^{48}$Ca are similar to those in $^{40}$Ca. For $^{208}$Pb, there is a minimum at 10 MeV due to the $GQ_0R$. The isovector spin-quadrupole resonance ($GSQ_1R$) and the isovector spin-octupole resonance ($GSO_1R$) lead to an enhancement of $S_{nn'}$ at higher excitation energies.

The sensitivity of $S_{nn'}$ to distortions is somehow surprising, since the main effect is absorption due to the imaginary part of the central optical potential. Scattering is localized at the nuclear surface and the effective number of participating nucleons is reduced. The spin observables are then insensitive to distortions. However, the real part of the central optical potential and the $\mathbf{L} \cdot \mathbf{S}$- part can influence $S_{nn'}$. To study further the effect of distortions, the longitudinal and transverse spin-flip probabilities $S_L$ and $S_T$ (see eq. (7)), have also been investigated. Besides $D_{nn'}$, the spin observables $D_{ss'}$ and $D_{ll'}$ have to be known. These have recently been measured by A. Green *et al.* at LAMPF[28].

Fig. 7 shows $S_L$ and $S_T$ for the reaction $^{40}$Ca (p,p') at 580 MeV incident energy. For free scattering (dotted lines) and in the plane-wave calculation (dashed lines) the ratio $S_L/S_T$ is about 1/2, reflecting the existence of one longitudinal and two transverse directions. In the distorted-wave calculation (full lines) this ratio is larger than one for higher excitation energies (> 30 MeV). Distortions increase $S_L$ and decrease $S_T$. The influence on $S_L$ is larger than on $S_T$, reflecting the longer range of the pion as compared to the $\rho$-meson. A comparison with the distorted-wave calculation ignoring $\mathbf{L} \cdot \mathbf{S}$-distortions (dashed-dotted lines) shows that this type of distortions is mainly responsible for the increase of $S_L/S_T$. No indication for the existence of a pion enhancement can be found.

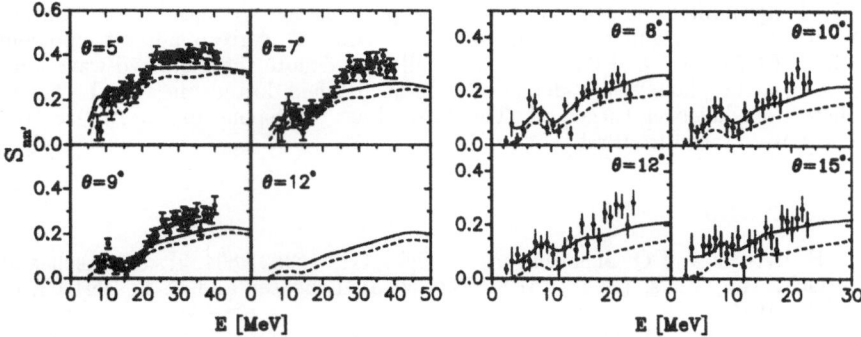

Fig. 6. The spin-flip probability, $S_{nn'}$, for the reactions $^{48}$Ca (p,p') at 319 MeV and $^{208}$Pb (p,p') at 200 MeV incident energy. Distorted-wave calculations (full lines) and plane-wave calculations (dashed lines) are displayed. The data were taken from refs.[25,27], respectively.

## 4. SUMMARY

We have shown that SRPA-DWIA calculations account well for intermediate energy (p,p') reactions at forward angles and for small energy loss. The reaction mechanism is predominantly of one-step nature. Two-step contributions may become important at lower incident energy and at higher energy loss. Double differential cross sections and spin-flip probabilities $S_{nn'}$ can be reproduced quite well. These spectra show moderate sensitivity to the residual interaction. The enhancement of $S_{nn'}$ at excitation energies above the giant resonance region is due to the repulsive residual interaction in the $\Delta S=1/\Delta T=1$-channel which leads to collective spin modes. The inclusion of distortions, especially of the $\mathbf{L}\cdot\mathbf{S}$-parts, are important for describing $S_{nn'}$ properly. The ratio $S_L/S_T$ of longitudinal and transverse spin-flip probability is increased by more than a factor of 2 after inclusion of distortions.

ACKNOWLEDGEMENT

We thank C. Glasshausser and J. Lisantti for helpful discussions and suggestions. This work was supported by NSF grant PHY89-21025 and NATO grant RG 85/093.

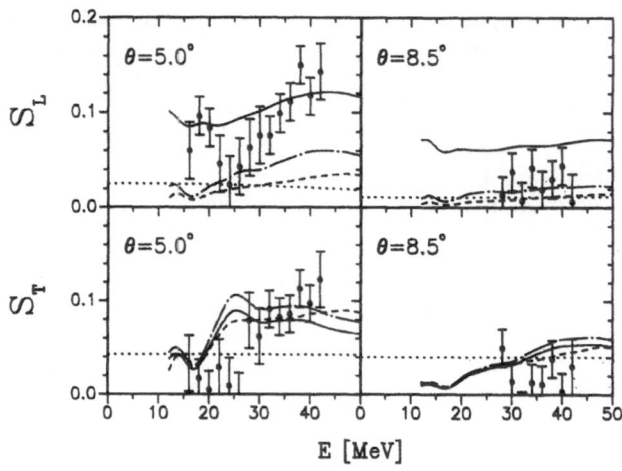

Fig. 7. Longitudinal and transverse spin-flip probabilities $S_L$ and $S_T$ for $^{40}$Ca at 580 MeV incident energy and scattering angles of 5° and 8.5°, corresponding to momentum transfers of 0.53 fm$^{-1}$ and 0.9 fm$^{-1}$. The full lines denote the distorted-wave and the dashed lines the plane-wave calculations. In the dashed-dotted lines the $\mathbf{L} \cdot \mathbf{S}$-part of the distortions has been turned off. The dotted lines correspond to free $NN$-scattering. The data were taken from ref.[28].

## REFERENCES

1. G. F. Bertsch and O. Scholten, Phys. Rev. C25 (1982) 804; H. Esbensen and G. F. Bertsch, Ann. Phys. (N.Y.) 157 (1984) 255; H. Esbensen and G. F. Bertsch, Phys. Rev. C34 (1986) 1419
2. K. W. Jones, Spin Observables of Nuclear Probes, (Plenum Press, New York) p. 1
3. J. M. Moss et al., Phys. Rev. Lett. 48 (1982) 789
4. C. Glashausser et al., Phys. Rev. Lett. 58 (1987) 2404
5. O. Häusser et al. Phys. Rev. C43 (1991) 230
6. R. D. Smith and J. Wambach, Phys. Rev. C38 (1988) 100
7. P. M. Boucher at al., Phys. Rev. C37 (1988) 907
8. P. M. Boucher et al., Z. Phys. A333 )1989) 137
9. M Ichimura et al., Phys. Rev. C39 (1989) 1446
10. H. H. Barschall and W. Haeberli (ed.), Proc. 3rd Int. Symp. on 'Polarization Phenomena in Nuclear Reactions', 1970
11. G. G. Ohlsen, Rep. Prog. Phys. 35 (1872) 717
12. M. A. Franey and W. G. Love, Phys. Rev. C31 (1985) 488
13. S. Drożdż, S. Nishizaki, J Speth and J. Wambach, Phys. Rep. 191 (1991) 1
14. C. Mahaux and H. Ngô, Nucl. Phys. A378 (1982) 205
15. W. Unkelbach, Ph.D.-thesis, University of Bonn
16. L. D. Landau, JETP 3 (1957) 920; 5 (1957) 101; 8 (1959) 655; A. B. Migdal, Nuclear Physics 13 (1959) 655
17. G. A. Rinker and J. Speth, Nuclear Physics A366 (1978) 306
18. J. Speth, V. Klemt, J. Wambach and G. E. Brown, Nucl. Phys. A343 (1980) 382
19. J. Nitschkowski, Diploma-thesis, University of Bonn
20. G. Chanfray et al., Phys. Lett. B256 (1991) 323
21. V. Mull and J. Wambach, in preparation
22. C. Glashausser et al., Phys. Rev. Lett. 58 (1987) 2404
23. F. T. Baker et al., Phys. Rev. C37 (1988) 1350
24. F. T. Baker et al., Phys. Rev. C40 (1989) 1877
25. F. T. Baker, private communication
26. J. Lisantti, private communication
27. J. Lisantti et al., to be published
28. A. Green, private communication

# FRAGMENTATION OF THE M6 RESPONSE

James A. Carr

Supercomputer Computations Research Institute
The Florida State University, B-186
Tallahassee, FL 32306-4052

## ABSTRACT

Calculations have been made to examine the effect of configuration mixing on the spectral distribution of "stretched" M6 strength in $sd$-shell nuclei. These calculations explain much of the observed reduction in strength of the lowest $6^-$ states, relative to single-particle estimates, and predict that some of the missing strength should be found within a few MeV of the strong yrast state. Recent measurements with the $^{28}$Si(p,n) reaction are in surprising agreement with these predictions, as are the qualitative features of spectra from the $^{32}$S(e,e') and (p,n) reactions. Experiments to study the isoscalar M6 response would be most valuable and interesting.

## INTRODUCTION

"Stretched" particle-hole states are $1\ \hbar\omega$ excitations with negative parity and total angular momentum transfer $J$ that is the fully aligned sum of the maximum possible particle and hole angular momenta. For the $sd$-shell nuclei considered in this work, these are M6 transitions to the "stretched configuration" $f_{7/2}d_{5/2}^{-1}$. This configuration is unique in a $1\ \hbar\omega$ shell-model basis and provides the only means to excite a $6^-$ state via a one-step reaction. As a consequence, a single spin transition density $\rho_{JL}^{S}$ (where $L=5$, $S=1$, and $J=6$) describes electron, nucleon, and pion scattering within a quasi-local single-scattering model.[1,2] The plane wave expression[3]

$$\frac{d\sigma}{d\Omega}^{\mathrm{PWBA}} = 4\pi\left(\frac{\mu}{2\pi\hbar^2}\right)(2J+1)\ |\bar{t}_s|^2\ |\rho_{JL}^{S}|^2 \qquad (1)$$

gives a schematic picture of this relationship, the essence of which is preserved in full DWA scattering calculations.[1] Strictly, non-localities can modify this relation.[4]

The fact that a single transition density drives the inelastic transition, regardless of the probe used, provides us with a very powerful tool. Because magnetic electron scattering is mainly (by a factor of 28) sensitive to isovector excitations, we can deduce the strength of $6^-$, T=1 states with great precision. This information can be used to calibrate the reaction model for hadronic probes, which subsequently can be used to deduce the strength distribution of T=0 states. Research over the past decade has established that the calibration of the various probes is well understood (at the 10% level), and these probes have been used with great effectiveness to identify the primary features of the lowest-lying stretched states.[5,6]

The difficulty is that, in striking contrast to early expectations that the stretched configuration would be found concentrated in a single state, experiments found the spectra to be dominated by a single state (for each isospin) that contained only a small fraction of the expected M6 strength.[2] The situation for isovector amplitudes as of the Telluride meeting in 1985 (Ref. 5) is summarized in Table 1. In this table, we make use of the convenient fact that the transition strength can be represented by a single number, the square of the spectroscopic amplitude

$$Z_T = \langle \Psi_{6^-,T} \| A^6_{T,0}(f_{7/2}, d_{5/2}) \| \Psi_{g.s.} \rangle, \qquad (2)$$

that (along with a radial size parameter) is sufficient to specify $\rho^S_{JL}$. The table compares the strength $Z^2$ for isovector transitions to the extreme single-particle-hole model (ESPHM), a model that places all of the strength in a single state built from a simple closed-shell ground state. Notice that we observe only 30-50% of the ESPHM, and only about 60% of what theory predicts, for a wide range of nuclei.

We can state this problem of the missing stretched strength another way: only a small fraction (about a third) of the wave function of a so-called "stretched state" consists of the *stretched configuration*. Since the remainder must come from admixtures of other possible configurations with the same spin (one must not forget that there are many thousands of ways to make a $6^-$ state in the 1 $\hbar\omega$ space for $sd$-shell nuclei, it is just that the M6 operator only connects *one* of them to the ground state), our problem is to understand how it is that the stretched configuration has become mixed with them and fragmented across many states.

Our approach is to examine the stretched $6^-$ states in $sd$-shell nuclei within the largest practical shell model basis. Our expectation was that we would improve upon previous theoretical efforts if we could perform our calculation in a basis with enough degrees of freedom. We restricted our study to $N = Z$ nuclei where neutron-proton mixing does not introduce any additional complications. Although we will not discuss the additional effect of core polarization,[7] we note in passing that our premise is that it should provide a small, perturbative correction to the spectrum obtained in the 1 $\hbar\omega$ space. We emphasize that it is the *location* of the missing fragments of M6 strength that is most crucial to understanding the physics of these states.

Table 1. Isovector stretched strength as reported by Lindgren and Petrovich[5] at the 1985 Telluride meeting.

| Nucleus | $E_x$ | $J^\pi, T$ | $Z^2_{exp}$/ESPHM | $Z^2_{thy}$/ESPHM |
|---------|-------|------------|--------------------|--------------------|
| $^{12}$C | 19.50 | $4^-, 1$ | $0.37 \pm 0.04$ | 0.60 |
| $^{16}$O | 18.98 | $4^-, 1$ | $0.41 \pm 0.02$ | 0.71 |
| $^{24}$Mg | 15.05 | $6^-, 1$ | $0.27 \pm 0.02$ | 0.47 |
| $^{28}$Si | 14.36 | $6^-, 1$ | $0.31 \pm 0.01$ | 0.55 |
| $^{54}$Fe | 13.26 | $8^-, 2$ | $0.51 \pm 0.02$ | 0.72 |

## METHODS

In this study we encounter some cases where there are more than 80 000 $6^-$ states, described by over 400 000 Slater determinants, in the vector space where we need to diagonalize the Hamiltonian. Such problems are not easily done with conventional methods. The Lanczos iterative technique, outlined below, turns out to be extremely well suited to them. It provides rapid convergence for the lowest states, where the wavefunction is of particular interest, and it has an additional feature we can exploit to get an accurate spectrum of the remaining strength at no additional cost.

### Lanczos Procedure

The Lanczos procedure starts with the Hamiltonian $H$ in some simple basis of Slater-determinant shell-model wavefunctions and iteratively transforms $H$ into a tri-diagonal matrix $T$ (with the same eigenvalues as $H$ of course) in a new basis $\{\,|L_i\rangle\,\}$, the so-called Lanczos vectors, as follows:

$$
\begin{aligned}
H|L_1\rangle &= \alpha_1|L_1\rangle + \beta_2|L_2\rangle \\
H|L_2\rangle &= \beta_2|L_1\rangle + \alpha_2|L_2\rangle + \beta_3|L_3\rangle \\
H|L_3\rangle &= \beta_3|L_2\rangle + \alpha_3|L_3\rangle + \beta_4|L_4\rangle \\
H|L_4\rangle &= \beta_4|L_3\rangle + etc.
\end{aligned}
\tag{3}
$$

After $N$ iterations we have an approximation $T_N$ to $T$,

$$
T_N =
\begin{bmatrix}
\alpha_1 & \beta_2 & & & & \\
\beta_2 & \alpha_2 & \beta_3 & & & \\
 & \beta_3 & \alpha_3 & \beta_4 & & \\
 & & \beta_4 & \cdot & \cdot & \\
 & & & \cdot & \cdot & \beta_N \\
 & & & & \beta_N & \alpha_N
\end{bmatrix}.
\tag{4}
$$

When the iterations are complete, the Lanczos vectors span the sub-space of the states being considered and the eigenvalues are found by diagonalizing $T$. The power of the Lanczos method is that the lowest (and highest) eigenvalues of $T_N$ converge rapidly and monotonically to the corresponding eigenvalues of $H$. (One needs fewer than 24 iterations to determine accurately the properties of the lowest state for a space of 53637 T=0 $6^-$ states in $^{28}$Si.) Further, the spectrum describing the fragmentation of the first Lanczos vector amongst the eigenstates is described more accurately than one might expect: at iteration $N$ the lowest $E^{2N}$ moments of the spectrum are exact.[8] Hence, a careful choice of $|L_1\rangle$ can simplify the calculations.

### Collective Vector Method

This last feature of the Lanczos method was noticed by Stew Bloom at Livermore and applied to calculations of the GT strength distribution.[9] We have found a variant of his technique to be even more effective in the stretched state problem. The basic idea is that, since you wish to compute the distribution of strength associated with some operator, you should form the first Lanczos vector by operating on the ground state with that operator. A "collective vector" is defined by

$$
|\chi_T\rangle = A_{T,0}^{6,6}\,|\Psi_{g.s.}\rangle
\tag{5}
$$

335

which contains all of the M6 strength in this basis ($\Sigma = \langle \chi | \chi \rangle$ is a sum rule) and is used as the first Lanczos vector

$$|L_1\rangle \;=\; |\chi_T\rangle \; \langle \chi_T | \chi_T \rangle^{-1/2} \tag{6}$$

in a calculation with the Livermore shell model code. It is now trivial to follow the fragmentation of the strength: When we find the *pseudo*-eigenstates

$$|E_i\rangle \;=\; \sum_{j=1}^{N} a_j(i) \, |L_j\rangle \tag{7}$$

after $N$ iterations of the Lanczos algorithm, we can easily compute the approximate spectrum from

$$
\begin{aligned}
Z_T^2(i) &= \langle E_i | A_{T,0}^{6,6} | \Psi_{g.s.} \rangle^2 \;=\; \langle E_i | \chi_T \rangle^2 \\
&= |a_1(i)|^2 \; \langle \chi_T | \chi_T \rangle \;=\; |a_1(i)|^2 \, \Sigma,
\end{aligned}
\tag{8}
$$

where the eigenenergies come from the diagonalization, and the widths

$$\langle \, ( \, H - \langle H \rangle )^2 \, \rangle^{1/2}$$

are due to incomplete convergence. The theorem mentioned earlier guarantees that the important lowest moments of this approximate spectrum are the same as the "exact" spectrum computed with $H$.

What makes the collective vector so powerful when studying stretched states is that our operator builds just a simple, single particle-hole configuration on the many-particle ground state. This means our calculation, *and the measured M6 spectrum*, is showing us how this particular configuration is being fragmented and mixed among all of the $6^-$ states in a given nucleus. It is a nice example of one of the classic problems in nuclear structure physics.

RESULTS

Calculations were made for $^{20}$Ne, $^{24}$Mg, $^{28}$Si, $^{32}$S, and $^{36}$Ar. The Hamiltonian was composed of an $sd$-shell part, taken from Wildenthal's fits,[10] and a part that connects the $sd$ and $fp$ shells, taken to be a particular Shiffer-True interaction[11] employed in previous work in Mg and Si.[12] The space we used allowed any combination of particles in the $sd$ shell but was truncated to permit only one particle in the $f_{7/2}$ shell. The single particle energy for the latter was fixed by the energy of the yrast $6^-$ T=0 state in $^{28}$Si. A full description of this work has been prepared for publication;[13] two Letters presented preliminary results for $^{28}$Si (Ref. 14) and $^{32}$S (Ref. 15) in a smaller basis.

A summary of our results for prominent isovector transitions is given in Table 2. The table is limited to information relevant for comparison to the results presented in Table 1. One notices that there is now data for $^{32}$S and that the data now include values (denoted by $\int$) for the M6 strength integrated over a region of $\sim$ 6 MeV excitation energy about the yrast level. The new theoretical results are much closer to the data than those in Table 1. The differences are sometimes less than 20%. It is of particular interest that the integrated strength also appears to be described well. Other properties of these states [transfer reaction spectroscopic factors and B(M1) rates] are given reasonably well.[12,14,13]

Table 2. Comparison of new theoretical results to data for *sd*-shell nuclei.

| Nucleus | $J^\pi, T$ | ESPHM | $\Sigma$ | Theory | Experiment | Ref. |
|---------|-----------|-------|----------|--------|------------|------|
| $^{24}$Mg | $6^-, 1$ | 0.66 | 0.50 | $Z^2 = 0.32$ | $0.18 \pm 0.01$ | 16,2 |
|  |  |  |  | $\int = 0.37$ | $0.41 \pm 0.04$ | 17 |
| $^{28}$Si | $6^-, 1$ | 1.0 | 0.77 | $Z^2 = 0.37$ | $0.31 \pm 0.01$ | 18,2 |
|  |  |  |  | $\int = 0.70$ | $0.43 \pm 0.04$ | 17 |
| $^{32}$S | $6^-, 1$ | 1.0 | 0.90 | $\int = 0.77$ | $0.75 \pm 0.01$ | 15 |
|  |  |  |  |  | $0.54 \pm 0.05$ | 17 |

Although the pattern of agreement noted in Table 2 looks quite promising, in this talk I want to emphasize the importance of some particular features of the M6 spectrum that we predicted[13] and which have now been seen in experiment. This will also give me an opportunity to show the progress made on the experimental side, since there are no other talks at this conference on the subject of stretched states.

## SPECTRUM OF THE M6 RESPONSE

Recall that we actually measure the distribution of a *particular configuration* rather than some set of combinations of configurations as is the case for GT transitions. This a very special feature of stretched transitions. Experiment and theory both examine fragmentation via configuration mixing of a particular particle-hole state in a particular nucleus, which means we should pay as much attention to the strength distribution as we do to the strengths of particular states. Our calculations, such as the one for $^{28}$Si shown with the solid curve in Fig. 1, make a rather specific prediction for the location of the "missing" strength in a bump about 3.5 MeV above the strongest state and in a long tail extending to high excitation energy.

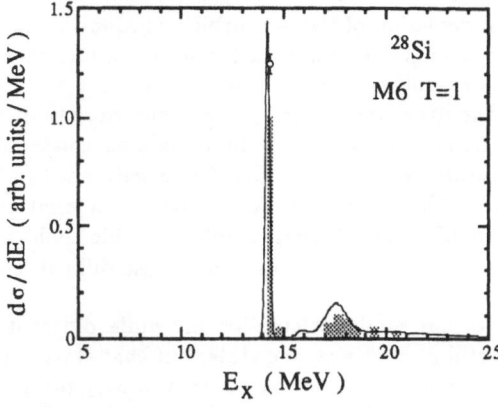

Figure 1.  Calculated T=1 spectrum (solid curve) for $^{28}$Si compared to the original electron scattering measurement[16,2,5] (data point) and a recent (p,n) experiment[17] (shaded area).

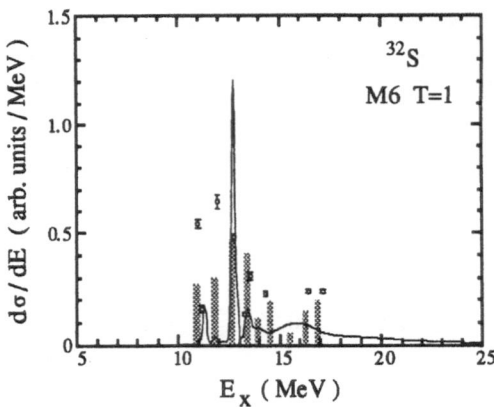

Figure 2.  Calculated T=1 spectrum (solid curve) for $^{32}$S compared to recent inelastic electron scattering[15] (data point) and (p,n) reaction[17] (shaded area) experiments.

The data point in Fig. 1 shows the strength measured in the original electron scattering experiment;[16,2,5] it is about 20% larger than the strength measured in a recent (p,n) experiment[17] shown with a shaded bar. (The bar and "data point" are constructed so that their heights are equal to the height of a gaussian with a 250 keV FWHM; one can compare their heights to the top of the theory curve or the area of the bar to the area under our curve.) This difference between the electron scattering and (p,n) measurements could be attributable to the need for meson-exchange corrections in the electron-scattering analysis[5] or for calibration of the nucleon-nucleon interaction used in the (p,n) analysis, but from the perspective of this report the difference merely sets the scale for systematic errors in the comparison of theory and data. What is most important is the nearly quantitative agreement between the spectrum measured in the (p,n) experiment[17] and our prediction that a noticeable part of the M6 strength in $^{28}$Si might be located about 3-4 MeV above the strong T=1 state.

The details are interesting. The total strength $\Sigma$ is about 77% of the ESPHM (see Table 2) as a result of depletion of the $d_{5/2}$ orbit by ground-state correlations. About half of the surviving strength is placed in the lowest state; much of the remainder is located in a region extending about 6 MeV above that state with about 10% at higher energies. On the experimental side, somewhat more than half of the observed T=1 strength is in the lowest state with the remainder clustered in a region about 3-6 MeV above that state. It is not possible for experiment to identify the long tail of weak M6 strength at high energy because of limits on what can be isolated from the background. It would be nice if some spin-observable could be found that would increase our sensitivity to stretched states under these difficult circumstances.

The predictions of our model calculation are quite different for the case of $^{32}$S shown in Fig. 2. We still predict a strong state, but that state is no longer the lowest T=1 state. There is much less depletion of the the $d_{5/2}$ orbit, so $\Sigma$ is 90% of the ESPHM. We predict that most (about 85%) of this total should be seen in the 7 MeV region of excitation energy studied by the experiments. This prediction is in quite good agreement with the electron scattering result,[15] as seen in Table 2, although we

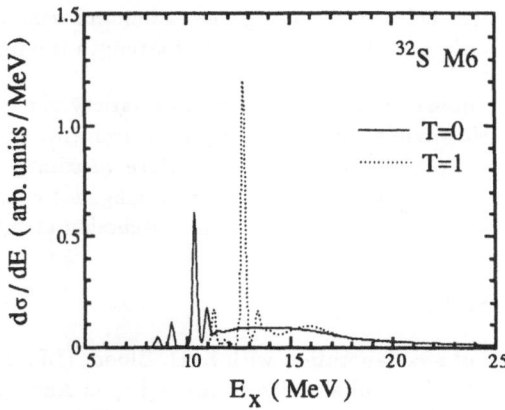

Figure 3. Predicted spectrum for the T=0 response (solid curve) in
$^{32}$S compared to the T=1 response (dotted curve).

do not agree on the exact distribution of strength, as is clear from Fig. 2. Theory is doing a good job on the lowest moments of the strength distribution.

The (p,n) experimental result[17] is about 30% lower than the (e,e') measurement in integrated strength, but both agree on the size of the strongest state in the spectrum. The reason for the disagreement on the weaker states is not clear. It may be (see below) that isospin mixing adds some T=0 strength to the strength measured with electron scattering. Qualitatively, however, the two probes agree that the number of observable fragments has increased dramatically as we start filling the orbit above the one where the hole is formed, and theory predicts a similar effect.

Experimental effort has been centered on the isovector probes, magnetic electron scattering and the (p,n) reaction, in recent years. However, I wish to point out that the T=0 spectrum can be as rich and interesting as what has been presented so far. Fig. 3 shows a comparsion of the isoscalar M6 response (solid curve) in $^{32}$S to the isovector M6 response (dotted curve) we displayed in Fig. 2. The isoscalar spectrum is complex, suggesting experiments would be interesting. There is also a region where strong T=0 states are close enough to isospin mix with the lower T=1 states.

## CONCLUSIONS

Configuration mixing, in a sufficiently large basis, seems to give a good indication of where stretched strength will be distributed in $sd$-shell nuclei. However, we do not have a quantitative description of the spectra currently known from experiment. It is important to explore the role of the remainder of the full $1\hbar\omega$ basis, particularly that of the $f_{5/2}$ spin-orbit partner, and our choice for $H_{sd-fp}$. (Better $H$'s exist.)

There are some other, broader, open questions. We are fortunate that $\Delta$-hole contributions to the fragmentation of stretched strength are expected to be small.[19] This allows us to focus our attention on some of the more conventional, but no less challenging, issues discussed in this talk. However, important effects due to collective correlations in nuclei are not treated very efficiently in the shell model. Some are surely incorporated into $H_{eff}$, but the corresponding change in the effective

M6 operator has not been calculated. Core polarization and weak mixing calculations can help to further clarify the physics of stretched strength fragmentation.

Measurements of more complete M6 spectra in a variety of nuclei would be useful for comparison to such models. In particular, measurements of the T=0 strength distribution would be especially valuable. The close proximity of T=0 and T=1 states in the $^{32}$S spectrum could lead to isospin mixing, for example. Challenging theoretical and experimental problems remain for discussion at future meetings.

## ACKNOWLEDGEMENTS

This work is part of a collaboration with S. D. Bloom (LLNL) and F. Petrovich and R. J. Philpott (FSU). I would also like to thank Bryon Anderson for allowing me timely access to the (p,n) data of the Kent State group. This research is supported by the F.S.U. Supercomputer Computations Research Institute which is partially funded by the U.S. Department of Energy through Contract No. DE-FC05-85ER250000.

## REFERENCES

1. F. Petrovich, in *The (p,n) Reaction and the Nucleon-Nucleon Force*, eds. C. D. Goodman et al. (Plenum, New York, 1980), pg. 115.
2. R. A. Lindgren et al, *Phys. Rev. Lett.* **42**:1524 (1979).
3. F. Petrovich, J. A. Carr, and H. McManus, *Ann. Rev. Nucl. Part. Sci.* **36**:29 (1986).
4. F. Petrovich, J. A. Carr, R. J. Philpott, and A. W. Carpenter, *Phys. Lett. B* **207**:1 (1988).
5. R. A. Lindgren and F. Petrovich, in *Spin Excitations in Nuclei*, eds. F. Petrovich et al. (Plenum, New York, 1984), pg. 323.
6. R. A. Lindgren, *Can. J. Phys.* **65**:666 (1987).
7. A. Yokoyama and H. Horie, *Phys. Rev. C* **36**:1657(RC) (1987).
8. R. R. Whitehead, in *Moment Methods in Many-Fermion Systems*, eds. B. J. Dalton et al. (Plenum, New York, 1980), pg. 235.
9. S. D. Bloom, C. D. Goodman, S. M. Grimes, and R. F. Hausman, Jr., *Phys. Lett.* **107b**:336 (1981).
10. B. H. Wildenthal, *Prog. Part. Nucl. Phys.* **11**:5 (1984).
11. J. P. Schiffer and W. W. True, *Rev. Mod. Phys.* **48**:191 (1976).
12. A. Amusa and R. D. Lawson, *Phys. Rev. Lett.* **51**:103 (1983).
13. J. A. Carr, S. D. Bloom, F. Petrovich, and R. J. Philpott, FSU-SCRI preprint.
14. J. A. Carr, S. D. Bloom, F. Petrovich, and R. J. Philpott, *Phys. Rev. Lett.* **62**:2249 and **63**:918(E) (1989).
15. B. L. Clausen et al, *Phys. Rev. Lett.* **65**:547 (1990).
16. H. Zarek et al, *Phys. Rev. Lett.* **38**:750 (1977);
    H. Zarek et al, *Phys. Rev. C* **29**:1664 (1984).
17. N. Tamimi et al, Kent State preprint.
18. S. Yen et al, *Phys. Lett.* **93B**:250 (1980).
19. T. Susuki, S. Krewald, and J. Speth, *Phys. Lett.* **107B**:9 (1981).

# MEASURING THE QUARK CONTRIBUTION TO THE

# PROTON SPIN THROUGH $\nu p \to \nu p$

W. C. Louis for the LSND Collaboration

Los Alamos National Laboratory
Medium Energy Physics Division
Los Alamos, NM 87545, U.S.A.

## INTRODUCTION

The LSND (Liquid Scintillator Neutrino Detector) experiment will be performed at LAMPF in the next several years. The main goal of the experiment is to search for $\nu_\mu$-$\nu_e$ oscillations with high sensitivity; however, an increasingly important by-product of this search is to measure $\nu p \to \nu p$ elastic scattering and determine the strange quark contribution, $\Delta s$, to the spin of the proton. With the 800-MeV proton energy of LAMPF, neutrinos are produced from pion decay-in-flight with an average energy of about 150 MeV. This energy is sufficiently high so that the $\nu p \to \nu p$ cross section is large and is sufficiently low so that the low $Q^2$ approximation ($Q^2 \ll m_p^2$) is valid and the cross section can be expressed in a simple form dependent upon $\Delta s$ as the only unknown. LAMPF with its 1-mA proton intensity is, therefore, an ideal accelerator to perform this measurement.

## THE LSND EXPERIMENT

### Detector and Beam

The Liquid Scintillator Neutrino Detector (LSND) experiment[1] is a collaboration of groups from the University of California at Irvine, University of California at Riverside, Embry-Riddle Aeronautical University, Los Alamos National Laboratory, Louisiana State University, University of New Mexico, University of Pennsylvania, Temple University, and Valparaiso University. The proposed detector is shown in Fig. 1 and consists of a cylindrical tank of dilute mineral-oil-based liquid scintillator approximately 6 m in diameter by 9 m long with an active mass of 200 tons. The tank will reside inside an existing veto shield, which is located 28 m downstream of the A6 proton beam stop and is at an angle of approximately 13° to the beam direction. The proton kinetic energy entering the beam stop is 780 MeV and the typical proton current is 800 $\mu$A. The A6 beam stop area consists of a 30-cm long water production target followed by an array of isotope-production target stringers and a water-cooled

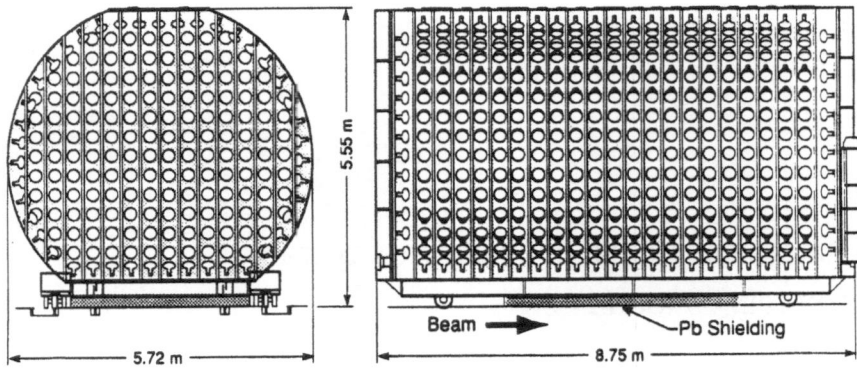

Fig. 1. A schematic view of the LSND detector, consisting of a 6-m diameter by 9-m long cylindrical tank of dilute liquid scintillator with 1224 8″ photomultiplier tubes covering about 25% of the surface area of the tank.

Cu beam dump. There exists about 8.5 m of Fe-equivalent between the Cu beam dump and the detector. Approximately 97.5% of the pions produced in the beam stop decay at rest and emit 30-MeV muon-neutrinos, while 2.5% of the pions decay in flight and emit a muon-neutrino beam with an average energy of about 150 MeV. An upgraded A6 beam stop is now being considered, which would have bending and focusing magnets and a larger separation between the water production target and the Cu beam dump. Such an upgraded beam stop would produce an order of magnitude more decay-in-flight neutrinos and, therefore, an order of magnitude more $\nu p \rightarrow \nu p$ events.

The tank will be made of 1-cm thick steel and have 1224 8″ diameter, low time jitter photomultiplier tubes mounted uniformly over the inside surface (25% coverage). These tubes are manufactured by Hamamatsu (R1408) and have good timing resolution, 3.8-ns FWHM for single photons and full-face illumination. They also have single photoelectron separation and typical noise rates < 5 kHz. We plan to use a dilute concentration of scintillator such that about 20% of the total light output will be Čerenkov light, 80% scintillation light, and 20 photoelectrons will be observed per MeV of deposited energy. The Čerenkov light will enable us to distinguish electrons from protons and to reconstruct the directions of electrons. The liquid scintillator consists mostly of mineral oil (> 99.99%) with the addition of about 0.03 g/ℓ of b-PBD. Mineral oil has advantages relative to water for the detection of Čerenkov light as it has a higher index of refraction, a lower density so that electrons travel farther before stopping, and a sharper Čerenkov ring due to the longer radiation length.

Electrons and protons have been reconstructed by a Monte Carlo simulation of the detector. The simulation is based on EGS4 and includes scintillator absorption, Rayleigh scattering, reflections off phototube surfaces, phototube noise counts, etc. For 45-MeV protons produced inside the detector, we expect an average reconstructed position error of 10 cm, an average reconstructed time error of 0.5 ns, and an energy resolution of 5%. For 45-MeV electrons produced inside the detector, we expect the same spatial, timing, and energy resolutions, and an average angular error of 12°.

Liquid Scintillator Tests

The liquid scintillator we plan to use in LSND consists of mineral oil with 0.03 g/ℓ of b-PBD. The mineral oil comes from Petroleum Specialties International (BRITOL 6NF HP) and has an attenuation length which increases from 20 m at a wavelength of 400 nm to 50 m at a 500-nm wavelength. The mineral oil is also very low in radioactivity and has a Th content of less than one part per trillion. The b-PBD scintillator additive has the advantages of a low wavelength absorption band ($\lambda < 360$ nm), a short fluorescence lifetime ($\tau = 1.2$ ns), and a fluorescence emission maximum ($\lambda = 365$ nm) which matches well the response of our Hamamatsu phototubes.

A series of measurements were performed in the LAMPF test beam during the summer of 1990 to determine the properties of our liquid scintillator mixture. A 1.5-m long, 5-cm diameter PVC pipe was filled with different liquids and placed in the test beam. A 2″ phototube with a photocathode response similar to the 8″ Hamamatsu tubes was mounted at one end of the pipe and at a distance of 1 m from the beam crossing point. The angle of the pipe with respect to the beam was varied from 25 to 90°. Figure 2 is a histogram of the relative number of photoelectrons measured in the phototube as a function of the angle with respect to a beam of 100 MeV/$c$ positrons for pure mineral oil and mineral oil with 0.25 g/ℓ of b-PBD. With pure mineral oil there is a pronounced peak at the 47° Čerenkov angle and very little light outside the

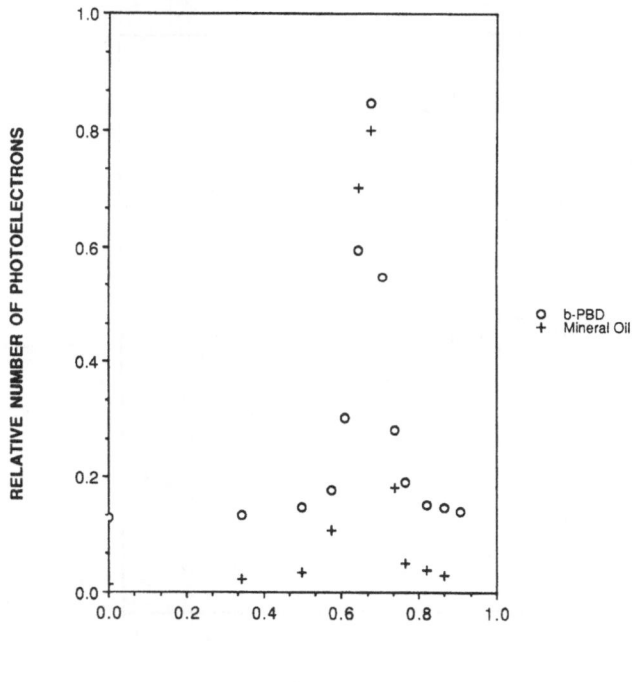

Fig. 2. Relative number of photoelectrons vs. the angle of the test cylinder with respect to the positron beam for pure mineral oil and mineral oil with 0.25 g/ℓ of b-PBD scintillator.

peak, while with the b-PBD mixture there is the Čerenkov peak and a flat distribution of scintillation light. The Čerenkov peaks in the two plots are almost identical, which shows that very little Čerenkov light ($< 15\%$) is lost by adding the b-PBD. The scintillation light time distribution for $\beta = 1$ positrons and $\beta = 0.5$ protons is shown in Fig. 3 for the pipe at 90 degrees to the beam so that there is no Čerenkov light background. The number of photoelectrons for the two beam particles is normalized to be the same, and as can be seen in the figure, the proton scintillation light has a larger slow component than the positron light. This pulse shape discrimination (PSD) is unexpected and will be a way to perform particle identification in addition to fitting the Čerenkov cone.

Physics Objectives

The main physics objective is to search for $\nu_\mu \rightarrow \nu_e$ and $\bar{\nu}_\mu \rightarrow \bar{\nu}_e$ oscillations with high sensitivity in two independent ways by using neutrinos from $\pi^+$ decay in flight and $\mu^+$ decay at rest, respectively. Shown in Fig. 4 is the two-parameter space, $\Delta m^2 \equiv m_1^2 - m_2^2$ (eV$^2$) and $\sin^2 2\theta$, that customarily describes the neutrino oscillation parameters, in which are plotted (i) the resulting limiting curve from the reactor (Gosgen)[2] disappearance ($\bar{\nu}_e \rightarrow \bar{\nu}_x$) experiments, (ii) the BNL (E734)[3] limiting curve for $\nu_\mu \rightarrow \nu_e$, (iii) the LAMPF (E645)[4] limit for $\bar{\nu}_\mu \rightarrow \bar{\nu}_e$, and (iv) the LAMPF (E764)[5] limiting curve for $\nu_\mu \rightarrow \nu_e$. Also indicated in Fig. 4 are the limiting curves expected

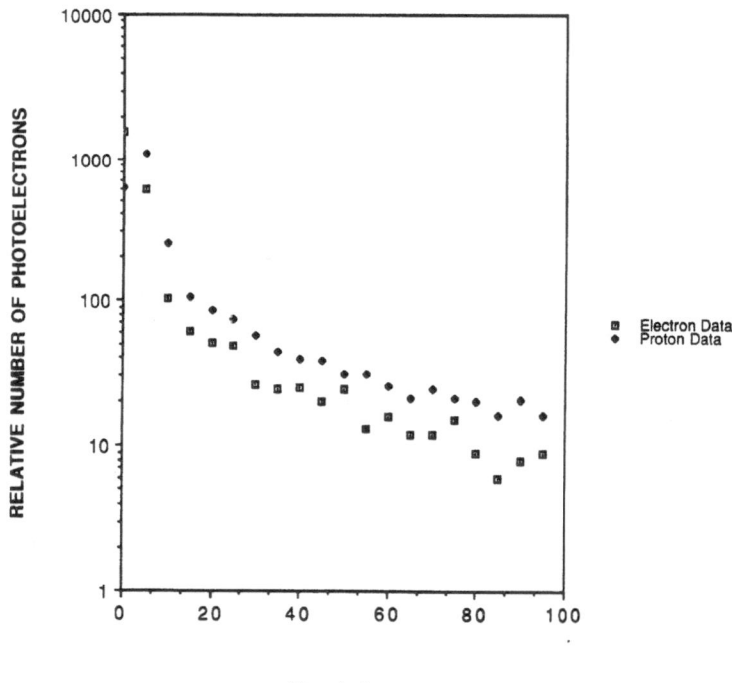

Fig. 3. The time distribution of scintillation light for $\beta = 1$ positrons and $\beta = 0.5$ protons.

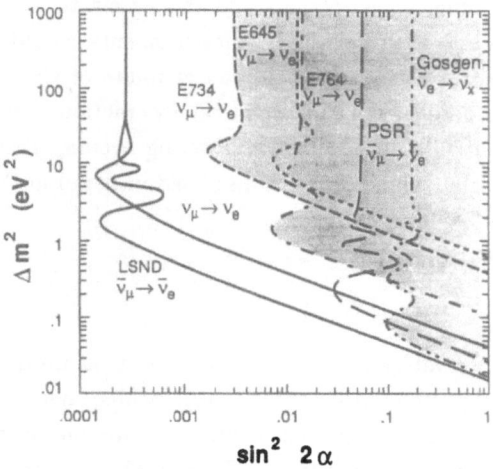

Fig. 4. The present $\bar{\nu}_\mu \to \bar{\nu}_e$ oscillation limits from reactor (Gosgen), BNL (E734), and LAMPF (E645, E764) experiments expressed in terms of the $\Delta m^2$ vs. $\sin^2 2\theta$ two parameter space. Also shown are the limiting curves expected from LSND with neutrinos from $\mu^+$ decay at rest and $\pi^+$ decay in flight.

from our two independent oscillation searches. After two years of data collection the limit on $\sin^2 2\theta$ of approximately $3 \times 10^{-4}$ for all $\Delta m^2 > 1$ eV$^2$ can be achieved, and the limit on $\Delta m^2$ of approximately $1.7 \times 10^{-2}$ eV$^2$ will be obtained for $\sin^2 2\theta = 1$. An additional search for $\bar{\nu}_\mu \to \bar{\nu}_e$ oscillations can be made with neutrinos from the Proton Storage Ring (PSR). Again, it will be done concurrently with the measurements above, and distinguished by timing alone. The sensitivity of this measurement is principally in the region of large $\sin^2 2\theta$, as shown in Fig. 4.

There are many other physics objectives that can be pursued with the liquid scintillator detector. The LSND data acquisition system (DAQ) will have the capability of triggering on hundreds of events per second and transferring up to 4 MB/s of data to a farm of UNIX workstations, where the events will be fully reconstructed, so that many different physics topics can be pursued simultaneously. Searching for $\bar{\nu}_\mu \to \bar{\nu}_e$ oscillations is equivalent to searching for the lepton number violating decay $\mu^+ \to e^+ \nu_\mu \bar{\nu}_e$. Limits on this decay are directly comparable to limits on neutrino oscillations, so that we should be sensitive to branching ratios as low as $10^{-4}$. The $\nu C \to \nu C^*$ (15.11-MeV $\gamma$) neutral-current reaction, one of the only neutrino-nuclear neutral-current reactions that can be easily observed, will be measured to approximately 10% accuracy and will be recognized by the detection of the 15.11-MeV $\gamma$ emitted by the excited carbon nucleus when it decays to the ground state. The rare decays $\pi^0 \to \nu\bar{\nu}$ and $\eta \to \nu\bar{\nu}$, followed by $\nu_e C \to e^- N$, can be searched to sensitivities of about $10^{-8}$ and $10^{-4}$, respectively, with very little background because the neutrinos from these decays are extremely energetic. These decays are forbidden for massless Weyl neutrinos and can proceed only if neutrino states of both chiralities exist or if lepton number is not conserved.[6] We shall also measure the $\nu_e C$ and $\nu_\mu C$

charged-current scattering cross sections and the $\nu e$ elastic scattering cross section to approximately 10–15% accuracy. These measurements should test present theories of neutrino-nucleus scattering and provide an estimate of $\sin^2\theta_W$, the fundamental parameter of the standard model of electroweak interactions. Finally, we will obtain a large sample of neutrino-proton elastic-scattering events, where the neutrinos are from pion decay in flight. This process is discussed in detail in the next sections.

## $\nu p \rightarrow \nu p$ FORMALISM

### Observables

The only two observables in $\nu p \rightarrow \nu p$ are the recoil proton kinetic energy and the angle of the recoil proton with respect to the incident neutrino. (We assume that the recoil neutrino and the recoil proton polarization are not measurable in a practical way.) The proton kinetic energy is given by $T_p = Q^2/2m_p$, where $Q^2$ is the momentum transfer squared of the reaction and $m_p$ is the proton mass. The angle of the recoil proton in the lab frame can be expressed as $\cos\theta = (1 + m_p/E_\nu)/(1 + 2m_p/T_p)^{1/2}$, where $E_\nu$ is the energy of the incident neutrino. In LSND we are unable to measure the proton angle, so that the proton kinetic energy, $T_p$, is the only signature.

It is worth giving a heuristic argument that shows how $\nu p \rightarrow \nu p$ is analogous to polarized deep inelastic scattering (DIS). If we consider 180° scattering in the center of mass frame, then as neutrinos always have negative helicity, protons with positive helicity are forbidden as the scattering process is $s$-wave. Therefore, at 180° the protons which interact must have negative helicity, even though the protons in the target are unpolarized, because of the fixed helicity of the neutrinos.

### Neutrino-Proton Elastic Scattering Form Factors

Fortunately, all three form factors relevant to $\nu p \rightarrow \nu p$ are well known at $Q^2 = 0$, due to measurements with electromagnetic probes, except for the strange quark contributions. Two of the form factors, $F_1$ and $F_2$, are vector and the third, $G_1$, is axial vector. $F_1(0) = 0.034 - F_1^s/2$; however, $F_1^s = 0$ at $Q^2 = 0$ because $F_1^s$ is the strange charge radius. $F_2(0) = 1.017 - F_2^s/2$, where $F_2^s$ is the strange magnetism form factor which will be measured by the SAMPLE experiment at BATES.[7] Finally, $G_1(0) = -g_A/2 + G_1^s/2$, where $g_A = 1.26$ as known from neutron decay and $G_1^s = \Delta s$, the strange quark contribution to the proton spin. Other form factors, $F_3$ and $G_2$, are zero because they are second class, and the $G_3$ contribution is zero because the neutrino mass is approximately zero. The $Q^2$ dependences of the form factors are assumed to have the dipole forms $F_2(Q^2) = F_2(0)/(1 + \tau)(1 + Q^2/M_V^2)^2$, $F_1(Q^2) = \tau F_2(Q^2) + F_1(0)/(1 + Q^2/M_V^2)^2$, and $G_1(Q^2) = G_1(0)/(1 + Q^2/M_A^2)^2$, where $M_A \simeq 1.0$ GeV/$c^2$, $M_V = 0.84$ GeV/$c^2$, and $\tau = Q^2/4m_p^2$.

### Neutrino-Proton Elastic Scattering Cross Section

The general form of the $\nu p \rightarrow \nu p$ cross section is given by[8,9]

$$\frac{d\sigma}{dQ^2} = \frac{G_F^2}{8\pi} \frac{Q^2}{E_\nu^2} \Big[ G_1^1(1+\tau) - F_1^2(1-\tau) + F_2^2\tau(1-\tau) + 4\tau F_1 F_2$$

$$+ 4\left(\frac{E_\nu}{m_p} - \tau\right) G_1(F_1 + F_2) + \left(\frac{4E_\nu^2}{Q^2} - \frac{2E_\nu}{m_p} + \tau\right)(G_1^2 + F_1^2 + \tau F_2^2) \Big] \ .$$

This complicated expression simplifies considerably at low $Q^2$ ($Q^2 \ll m_p^2$) to the form

$$\frac{d\sigma}{dQ^2} \sim \frac{G_F^2}{2\pi} \left[ G_1^2\left(1 + \frac{Q^2}{4E_\nu^2}\right) + F_1^2\left(1 - \frac{Q^2}{4E_\nu^2}\right) \right] \ .$$

Substituting the values of the form factors at $Q^2 = 0$, we obtain

$$\frac{d\sigma}{dQ^2} \sim \frac{G_F^2}{2\pi} \left[ \left(-0.63 + \frac{G_1^s}{2}\right)^2 \left(1 + \frac{Q^2}{4E_\nu^2}\right) + (0.034)^2 \left(1 - \frac{Q^2}{4E_\nu^2}\right) \right] \ ,$$

where the only unknown term is $G_1^s = \Delta s$, the strange quark contribution to the proton spin. Note that the form factor $F_2$ drops out at low $Q^2$, so that there is little uncertainty due to $F_2^s$, the strange magnetism term.

Relation to Polarized DIS Experiments

As shown above, a measurement of $\nu p \to \nu p$ at low $Q^2$ determines $\Delta s$. (We use the convention that $\Delta q$ is the q quark contribution to the proton spin.) The best measurement of this cross section was made by the BNL E734 experiment,[8] from which a value of $\Delta s = -0.15 \pm 0.08$ was determined.[9] E734, however, was performed at a higher neutrino energy ($Q^2 \sim 1$ GeV$^2/c^2$) so that the low $Q^2$ approximation is not valid and there are additional uncertainties, such as the form factors $F_1^s$ and $F_2^s$ and the masses $M_V$ and $M_A$. Polarized DIS experiments measure the integral

$$\int g_1^p(x)dx = \frac{4}{9}\Delta u + \frac{1}{9}\Delta d + \frac{1}{9}\Delta s = \frac{5}{18}(\Delta u + \Delta d) + \frac{1}{9}\Delta s + \frac{1}{6}g_A \ ,$$

and three years ago the EMC experiment measured[10] this integral to be $\int g_1^p(x)dx = 0.126 \pm 0.010 \pm 0.015$, in disagreement with the Ellis-Jaffe sum rule[11] prediction of $0.189 \pm 0.005$. (In the next several years there should be much better measurements of $\int g_1^p(x)dx$ from the SMC experiment at CERN and the HERMES experiment at DESY.) Many people have invoked SU(3) symmetry to interpret this result to mean that $\Delta s$ is nonzero and that $\Sigma\Delta q \sim 0$. However, Manohar and Lipkin have recently argued[12] that SU(3) symmetry is suspect and that it may be invalid for SU(3) symmetry to relate matrix elements of charged and neutral axial currents. Therefore, $\nu p \to \nu p$ and polarized DIS experiments are very complementary. If one assumes SU(3) symmetry to be valid, then both measurements determine $\Delta u$, $\Delta d$, and $\Delta s$ separately in independent ways. If one does not assume SU(3) symmetry to be valid, then $\nu p \to \nu p$ will measure $\Delta s$, polarized DIS will measure $\frac{5}{18}(\Delta u + \Delta d) + \frac{1}{9}\Delta s + \frac{1}{6}g_A$, and combined, the experiments will determine $\Sigma\Delta q = \Delta u + \Delta d + \Delta s$.

Proton Energy Distribution in LSND

Figure 5 shows the $\nu p \to \nu p$ elastic scattering cross section as a function of neutrino energy. At LAMPF the average decay-in-flight neutrino energy is about 150 MeV, which corresponds to a cross section of $3 \times 10^{-40}$ cm$^2$ and a maximum $Q^2$ of 0.07 GeV$^2/c^2$. The neutrino energies available at LAMPF are, therefore, optimal; the energy is sufficiently high that the cross section is almost at saturation but sufficiently low that the low $Q^2$ approximation, $Q^2 \ll m_p^2$, is valid. (For a neutrino of 150 MeV, the low $Q^2$ cross section approximation discussed in the previous section differs by 5% from the general cross section expression.) The recoil proton kinetic energy distribution from $\nu p \to \nu p$ for 150-MeV incident neutrinos is shown in Fig. 6. At this low energy the cross section rises almost linearly with $Q^2 = 2m_p T_p$. Folding in the neutrino decay-in-flight energy spectrum, we finally obtain the overall recoil proton kinetic energy distribution as shown in Fig. 7. The average proton energy is 15 MeV and about 27% of the protons have $T_p > 20$ MeV.

Event Rates and Backgrounds

We calculate the number of events per year based on 130 days of running or about 3000 hours of actual beam, which at the LAMPF A6 beam stop means an average proton intensity of 0.8 mA at an energy of 780 MeV. Table 1 gives the $\nu p \to \nu p$ event rate and background rates for recoil protons with $T_p > 20$ MeV. As seen in the table, we expect 2000 such $\nu p \to \nu p$ events per year with the present beam stop, and by upgrading the beam stop, as mentioned in the second section, this rate should increase by an order of magnitude to 20,000 events per year. These rates are doubled by using a 10-MeV recoil proton threshold; however, we will be conservative and use the 20-MeV threshold, where we are more confident with the background rate, for estimating event rates.

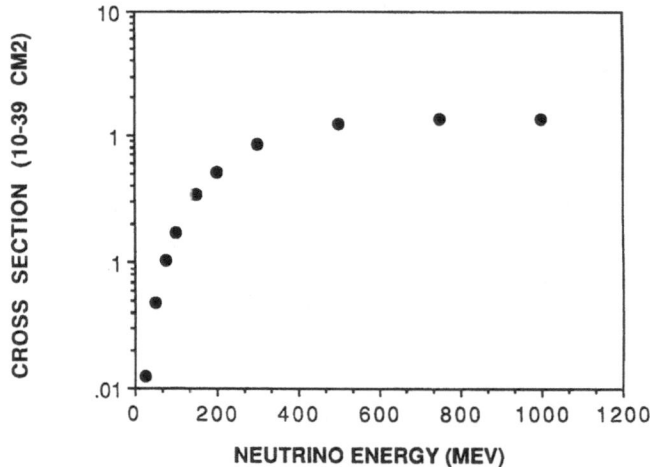

Fig. 5. The $\nu p \to \nu p$ elastic scattering cross section as a function of neutrino energy.

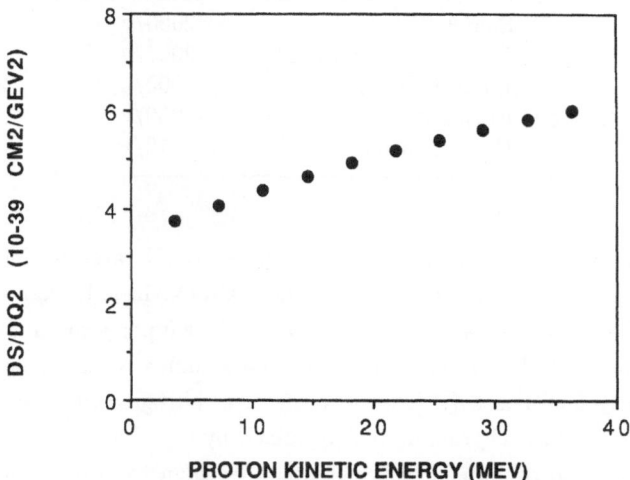

Fig. 6. The recoil proton kinetic energy distribution from $\nu p \to \nu p$ for 150-MeV incident neutrinos.

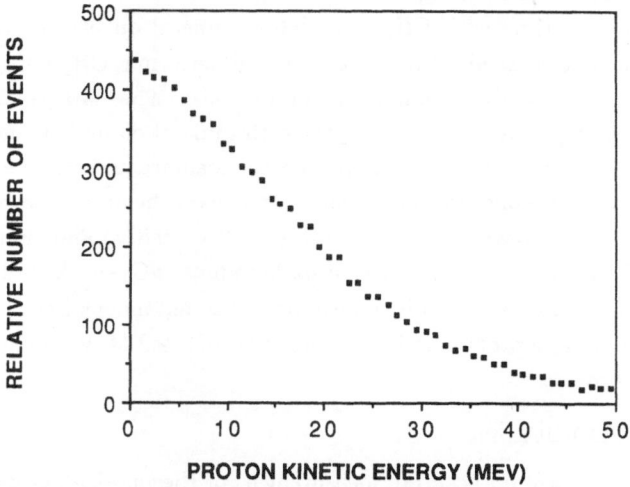

Fig. 7. The overall recoil proton kinetic energy distribution from $\nu p \to \nu p$ after folding in the LAMPF neutrino energy spectrum.

**Table 1.** The expected $\nu p \to \nu p$ event rate, background rate, and normalization rate per year (8678 C of protons) of data collection for the present beam stop and for an upgraded beam stop.

| Process | Type | Present Beam | Upgraded Beam |
|---|---|---|---|
| $\nu p \to \nu p$ | Signal | 2000/y | 20,000/y |
| $np \to np$ | Extrinsic Background | 2000/y | 2000/y |
| $\nu C \to pX$ | Intrinsic Background | 2000/y | 20,000/y |
| $\nu\,{}^{12}C \to \mu^-\,{}^{12}N(\text{g.s.})$ | Normalization | 250/y | 2500/y |
| $\nu e \to \nu e$ | Normalization | 10/y | 100/y |

Also shown in Table 1 are the $\nu p \to \nu p$ background rates, which are divided into two categories: extrinsic and intrinsic. The main extrinsic background is due to cosmic-ray induced neutrons at an estimated rate of 2000 per year. This background is large but can be easily handled with a beam-on minus beam-off subtraction, as the LAMPF duty factor is 6%. The beam-neutron background is less than 5% of the cosmic ray neutron background, as determined by a previous experiment (E645), and can be eliminated by fiducial volume cuts. (We plan to have a nominal fiducial volume that is 50 cm inside the photocathode surfaces.) The main intrinsic background is $\nu C \to pX$, neutrino scattering off a bound proton, which is estimated to occur at a rate comparable to $\nu p \to \nu p$. It is possible that this background becomes negligible at low $Q^2$ or low recoil proton energies, although it is difficult to make a precise estimate. Therefore, our most definite method for dealing with this bound proton background is to run half of the time with $CH_2$ scintillator (mineral oil based), the other half with $CH_{1.3}$ scintillator (pseudo-cumene based), and perform a $CH_2$-$CH_{1.3}$ subtraction to extract the neutrino cross section off free protons. The data collection can be performed in two ways: we can run one year with mineral oil and the next year with pseudo-cumene, or we can construct a transparent membrane down the center of the tank with mineral oil on one side and pseudo-cumene on the other. Note that mineral oil and pseudo-cumene have similar densities (0.858 vs. 0.877) and similar indices of refraction (1.47 vs. 1.51). Other intrinsic backgrounds, $\nu C \to nX$ and $\nu C \to \mu^-X$, are not problems because we can distinguish neutrons and muons from protons due to neutron capture on free protons (with the emission of a 2.2-MeV $\gamma$) and muon decay to a Michel electron.

Neutrino Flux Normalization

In order to perform an accurate measurement of the $\nu p \to \nu p$ cross section, it is imperative to obtain an accurate neutrino flux normalization. We have three possible ways to normalize: dead reckoning with our beam Monte Carlo simulation (and measuring pion production with an incident proton beam to improve the input to the Monte Carlo), the reaction $\nu\,{}^{12}C \to \mu^-\,{}^{12}N(\text{g.s.})$, and the reaction $\nu e \to \nu e$. The first method, using the Monte Carlo to infer the neutrino flux from the incident proton beam, should have an uncertainty of less than 10%; however, as we prefer to use a more direct technique to obtain the neutrino flux, this method will only be used as a check on our normalization. The second method is to use the exclusive charged-current reaction $\nu\ {}^{12}C \rightarrow \mu^-\ {}^{12}N(g.s.)$, which has a very clean threefold signature due to the muon, the decay electron, and the ${}^{12}N$ beta decay. All of the form factors in the cross section expression for this reaction have been determined at $Q^2 = 0$, and we estimate a systematic uncertainty to the cross section of 7.5% for a 150 MeV neutrino energy. Part of this uncertainty comes from the error in the form factor determination at $Q^2 = 0$, and the other part comes from the uncertainty in the extrapolation to nonzero $Q^2$. The third method, using the reaction $\nu e \rightarrow \nu e$, has essentially no systematic error and only suffers from statistics. As shown in Table 1, this reaction becomes useful if the beam is upgraded with an order of magnitude increase in the decay-in-flight neutrino flux.

Projected Sensitivity

The projected sensitivity for measuring the $\nu p \rightarrow \nu p$ cross section after two years of data collection is shown in Table 2. With the present beam stop and with a 20-MeV recoil proton energy threshold, we estimate that the $CH_2$-$CH_{1.3}$ subtraction will yield $700 \pm 85 \pm 63$ events, where the first error is the statistical error due to the different liquid scintillator subtraction and the second error is the statistical error due to the beam-on minus beam-off subtraction. With a normalization error of $\pm 6\%$ (statistical) and $\pm 7.5\%$ (systematic), we expect a total error on the cross section of $\pm 16\% \pm 7.5\% = \pm 18\%$, which corresponds to a $\pm 0.11$ error on $\Delta s$. For an upgraded beam stop with the same 20-MeV recoil proton energy threshold, the total error on the cross section is estimated to be $\pm 4.5\% \pm 5.2\% = \pm 6.8\%$, corresponding to a $\pm 0.04$ error on $\Delta s$.

CONCLUSION

LAMPF is an ideal accelerator for mounting an experiment to measure $\nu p \rightarrow \nu p$ due to its high intensity and optimal energy for producing neutrinos which are sufficiently energetic that the cross section is near saturation but sufficiently low in energy that the low $Q^2$ approximation, $Q^2 \ll m_p^2$, is valid. The LSND experiment, which should begin data collection in 1993, will measure about 2000 $\nu p \rightarrow \nu p$ events per year with the present beam stop (20,000 events per year with an upgraded beam stop), and after two years LSND should measure the $\nu p \rightarrow \nu p$ cross section to an accuracy of $\pm 18\%$ ($\pm 7\%$). Such a cross section measurement will determine the strange quark

Table 2. The projected sensitivity for measuring the $\nu p \rightarrow \nu p$ cross section after two years of data collection.

| Estimate | Present Beam Stop | Upgraded Beam Stop |
|---|---|---|
| $CH_2$-$CH_{1.3}$ Events | $700 \pm 85 \pm 63$ | $7000 \pm 270 \pm 63$ |
| Normalization Error | $\pm 6\% \pm 7.5\%$ | $\pm 2\% \pm 5.2\%$ |
| Total Cross Section Error | $\pm 18\%$ | $\pm 6.8\%$ |
| Error in $\Delta s$ | $\pm 0.11$ | $\pm 0.04$ |

contribution to the proton spin, $\Delta s$, to an error of $\pm 0.11$ ($\pm 0.04$ with the upgraded beam stop) and will be complementary to future polarized DIS experiments, such as the SMC experiment at CERN and the HERMES experiment at DESY.

## REFERENCES

1. LSND Proposal, Los Alamos National Laboratory report LA-UR-89-3764 (1989). The LSND collaboration consists of: X-Q. Lu, G. Yodh (University of California, Irvine); S. Y. Fung, J. H. Kang, B. C. Shen, W. Strossman, G. J. VanDalen (University of California, Riverside); D. Smith (Embry-Riddle Aeronautical University); J. Amann, H. Baer, A. Band, R. Bolton, R. Burman, J. Donahue, W. Foreman, G. T. Garvey, M. Hoehn, T. Kozlowski, D. M. Lee, W. C. Louis, J. Margulies, J. McClelland, M. Oothoudt, V. Sandberg, M. Schillaci, N. Thompson, R. Werbeck, D. H. White, D. Whitehouse (Los Alamos National Laboratory); A. Fazely (Louisiana State University); M. Brooks, B. B. Dieterle, C. P. Leavitt, R. Reeder (University of New Mexico); M. Albert, A. K. Mann (University of Pennsylvania); L. B. Auerbach, W. K. McFarlane, D. Works (Temple University); D. D. Koetke, R. Manweiler (Valparaiso University).

2. V. Zacek et al., Phys. Lett. 164B:193 (1985).

3. L. A. Ahrens et al., Phys. Rev. D31:2732 (1985).

4. L. S. Durkin et al., Phys. Rev. Lett. 61:1811 (1988); James J. Napolitano et al., Nucl. Instrum. Methods A274:152 (1989).

5. T. Dombeck et al., Phys. Lett. 194B:591 (1987).

6. Cyrus M. Hoffman, Phys. Lett. 208B:149 (1988); Peter Herczeg, in "Proceedings of the Paris Workshop on Production and Decay of Light Mesons," ed. Patrick Fleury, World Scientific (1988), p. 16; P. Herczeg and C. M. Hoffman, Phys. Lett. 100B:347 (1981); 102B:445(E) (1981).

7. Bates Proposal #89-06, R. D. McKeown and D. H. Beck, contact people.

8. L. A. Ahrens et al., Phys. Rev. D35:785 (1987).

9. David B. Kaplan and Aneesh Manohar, Nucl. Phys. B310:527 (1988).

10. J. Ashman et al., Phys. Lett. 206B:364 (1988).

11. J. Ellis and R. L. Jaffe, Phys. Rev. D9:1444 (1974); D10:1669(E) (1974).

12. Aneesh Manohar, talk presented at the Workshop on Accelerator-Based Low-Energy Neutrino Physics at Los Alamos, January 1991; Harry J. Lipkin, Phys. Lett. 256B:284 (1991).

# $^{37}$Ca $\beta^+$-DECAY: IS THE $GT$ STRENGTH REALLY 'QUENCHED'?

A. García, E.G. Adelberger, P.V. Magnus, H.E. Swanson
and D.P. Wells

University of Washington, Seattle, Washington

O. Tengblad and The Isolde Collaboration

Isolde-CERN, Geneva, Switzerland

D.M. Moltz

Lawrence Berkeley Laboratory, Berkeley, California

## 1  Motivation

### 1.1  The efficiency of the $^{37}$Cl $\nu$-detector

The efficiency of the Homestake Mine $^{37}$Cl solar $\nu$-detector is calibrated[1] using transition rates measured in the $\beta^+$ decay of $^{37}$Ca.

The captures of $\nu$'s in $^{37}$Cl to levels in $^{37}$Ar are 'isobaric reflections' of the $\beta^+$ decays of $^{37}$Ca to levels in $^{37}$K (see fig. 1). Isospin non-conservation effects are expected to produce differences of only 1 or 2 % [2] between the $GT$ strength of the two processes.

One can also estimate the $GT$ strength of the A=37 system by using the $(p,n)$ reaction near zero degrees[3]. However, previous measurements[4, 5] of the $GT$ strength using these two methods show strong disagreements which, as shown in fig. 2, is significant at all excitation energies. In particular, at excitation energies of $\sim$ 7.5 MeV the $(p,n)$ experiment shows a large concentration of $GT$ strength that seems to be missing in the $\beta^+$ decay measurement.

Adelberger and Haxton[6] proposed an explanation for the disagreement between the measurements of Rapaport et al.[5] and Sextro et al.[4] In order to understand it let us first notice that the phase space for the $\beta$-decay drops very rapidly with excitation energy as can be seen in fig. 3, so that the prominent peak in the Gamow-Teller strength distribution, observed at high excitation energies ($\approx$ 7.5 MeV) in the $(p,n)$ measurement, would show up only as a few counts in the $\beta$-decay measurement. Since Sextro et al. could not discriminate between decays from $^{37}$K that leave $^{36}$Ar in different final states they assumed that all decays were mainly to the $^{36}$Ar(0$^+$) ground state. Adelberger and Haxton made a shell model calculation and observed a rather good agreement with the $(p,n)$ data. They also used the calculation to estimate branching ratios of levels in $^{37}$K to the first excited 2$^+$ level in $^{36}$Ar and observed that a group of levels in $^{37}$K lying at excitation energies of about 7 to 9 MeV had a significant probability of decaying to $^{36}$Ar(2$^+$) which is very

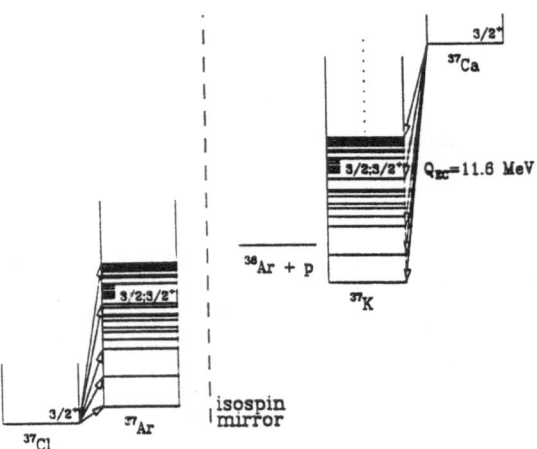

Figure 1. The $\beta^+$-decays of $^{37}$Ca are the isobaric analogs of the $\nu$ captures in $^{37}$Cl.

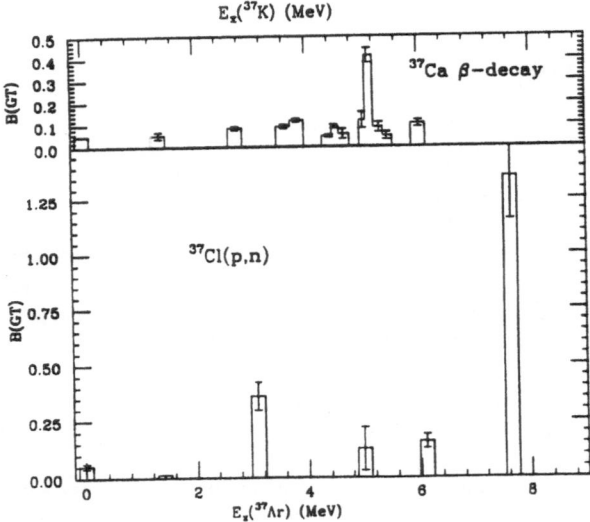

Figure 2. Comparison of the $B(GT)$ for A=37 obtained by previous experiments: The $\beta^+$-decay of $^{37}$Ca and the $^{37}$Cl$(p, n)$ reaction near zero degrees. The disagreement was the main motivation for this work.

reasonable considering that some of them are $J^\pi = 5/2^+$ and could have an s-wave decay to that excited state. Assuming the shell model as the truth, they then deduced the kind of intensity distribution they expected Sextro et al. to have observed and found agreement with the actual measurement of these authors. **It was then clear that a measurement of protons in coincidence with $\gamma$-rays that would 'tag' decays leaving $^{36}$Ar in the first excited state was essential to correctly estimate the B(GT) distribution.** One should bear in mind that, due to the drastic change in phase space, even though the effect should have an important consequence for the $B(GT)$, the total number of decays leaving $^{36}$Ar in the $2^+$ state, is expected to be only a small fraction ($\approx 2\%$) of the total.

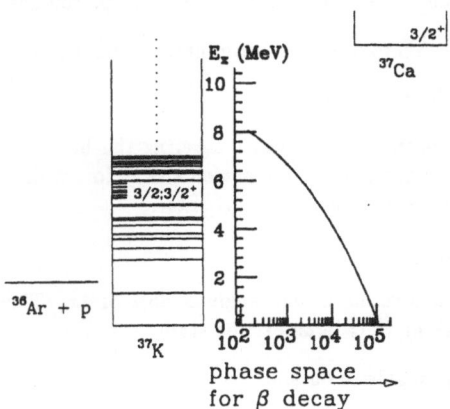

Figure 3. Phase space for $^{37}$Ca $\beta^+$-decay.

## 1.2  Quenching of the Gamow-Teller strength

Brown and Wildenthal[7] have performed a systematic comparison of the observed $GT$ strength with predictions using complete-$sd$-shell effective-interaction wave-functions and found that the shell model calculation systematically overestimates the $GT$ strength. The magnitude of the effect is expressed by indicating that, for nuclei, the axial vector constant is, on average, 'quenched', *i.e.* $|g_A/g_V| \approx 1$ instead of $|g_A/g_V| = 1.26$ which is the value for free nucleons (see also [8]). Brown and Wildenthal went further: based on the comparison, they were able to show that the origin of the problem is apparently two-fold:

- configuration mixing: second order effects due to mixing with other shells.

- substructure of nucleons: meson exchange currents with, for example, $\Delta$'s.

However, the average excitation energy of the transitions used in Brown and Wildenthal's comparison is only 3.5 MeV and the question may be raised of whether the observed quenching is not merely due to a failure of the shell-model to correctly distribute the $GT$ strength. In that sense the $\beta$-decay of $^{37}$Ca provides a special opportunity to test the quenching of the $GT$ strength since the Q-value for the $\beta$-decay is very large ($Q_{EC} =11.6$ MeV). In addition, our experiment provides an ideal benchmark for testing the accuracy of the $(p,n)$ reaction as an estimator of the $GT$ strength.

## 2 The experiment

A beam of six $^{37}$Ca atoms/s was produced at the Isolde-3 separator at CERN. The radioactivity was produced by bombarding a Ti-foil target with 600 MeV protons. The target was heated to 1800 °C to diffuse the $^{37}$Ca($\tau_{1/2}$ = 175 ms) to a hot ionizing surface. The source was elevated to 60 kV to extract the ions and a system of 90° plus 60° magnets separated the appropriate mass with a resolution of $\Delta m/m$= 1/7000. This resolution was sufficient to produce a beam of $^{37}$Ca with very little contamination of $^{37}$K which is ionized much more readily than $^{37}$Ca. This was critical for the succesful measurement of proton-$\gamma$ coincidences since significant amounts of $^{37}$K would have produced an intolerable background in our $\gamma$-ray counters.

The beam of $^{37}$Ca was implanted in the window of our detection system which consisted of a telescope surounded by $\gamma$-ray detectors. The telescope consisted of:

- An ultra-thin gas $\Delta$E counter to discriminate $\beta$'s from protons.

- A silicon surface-barrier E counter.

- A silicon surface-barrier Veto counter covering the back of the E counter to avoid satellite peaks from summing of the proton energy loss with the $\approx$ 150 keV that high energy coincident $\beta$'s deposit in the E counter.

We detected $\gamma$-rays using:

- two NaI detectors covering a solid angle of 65% of $4\pi$. These gave a signature for proton decays leaving $^{36}$Ar in an excited state.

With this system we were able to obtain:

- excellent particle ID: almost complete $\beta$ rejection.

- very good energy resolution: FWHM=16 keV for $\approx$ 1 MeV protons.

- reliable signature for decays leaving $^{36}$Ar in an excited state.

Fig. 4 shows the proton spectrum obtained with this detection system. For each peak in the spectrum, we obtained a ratio of coincidences to singles defined by:

$$R = \frac{Area\ of\ protons\ in\ coinc.\ with\ 1.97\ MeV\ \gamma - rays}{Area\ of\ protons\ in\ singles\ \times\ effc.\ for\ 1.97\ MeV\ \gamma's} \qquad (1)$$

which is shown in fig. 5

## 3 Extraction of the axial vector strength

With the information obtained above we were able to extract the $GT$ strength using the definition:

$$\frac{K}{f(E_i)\,t_i} = B_i(F) + B_i(GT)\ , \qquad (2)$$

where $K$ = 6170±4 s, $t_i$ and $E_i$ are the partial half-life and the energy release for decay to the $i^{th}$ level of $^{37}$K, $B_i(F)$ and $B_i(GT)$ are the Fermi and Gamow-Teller reduced transition strengths defined in ref. [9], and the statistical function $f(E_i)$ is computed using the prescription of Wilkinson and Macefield[10].

In fig. 6 we compare our results to the $(p,n)$ data of Rapaport et al.. Note that we have not only found the strength that was missing in the previous experiment of Sextro et al., but we saw 50% more $GT$ strength than in the $(p,n)$ measurement. The problem is not

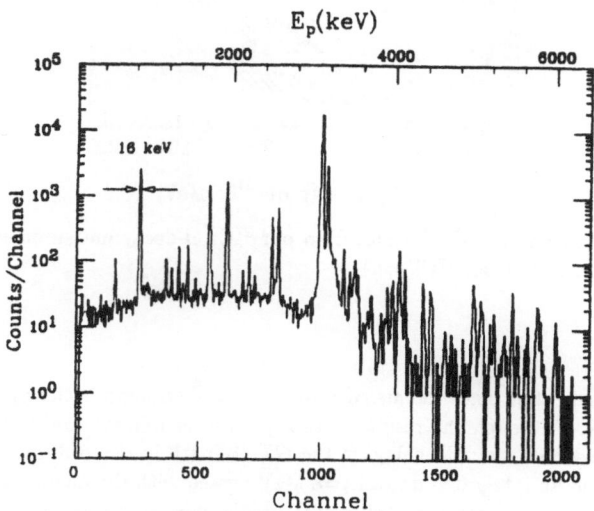

Figure 4. Singles proton spectrum from the $\beta$-decay of $^{37}$Ca.

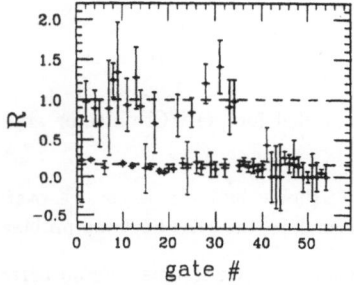

Figure 5. Coincidence ratio $R$ as a function of the peak number defined by order of appearance in the proton spectrum. $R=1$ gives a signature for a decay leaving $^{36}$Ar in its first excited state. The decays leaving $^{36}$Ar in the ground state have a value of $R$ slightly greater than zero because of detection of brehmstrahlung or $\beta$'s in the NaI counters.

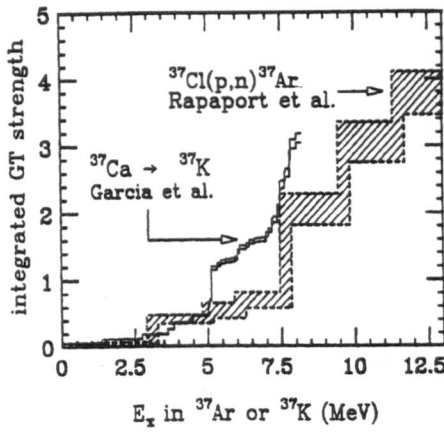

Figure 6. Comparison of $B(GT)$ values from our $^{37}$Ca $\beta$-decay measurement to those deduced by Rapaport $et~al.$ for $^{37}$Cl$(p,n)$

just one of normalization since the *distribution* of the $GT$ strength is clearly different in the two experiments. Although the results of the $(p,n)$ measurement are in good agreement with the hypothesis of the 'quenching' of the $GT$ strength in our $\beta$-decay experiment the summed $GT$ strength integrated up to $E_x=8$ MeV agreed with the unquenched shell-model prediction. Fig. 7 compares the shell-model predictions to our results and shows that the $\beta$-decay experiment sees all the predicted strength the experimental excitation energy window. The shell-model, however, fails to reproduce the *distribution* of $GT$ strength -a similar result was seen in the decay of $^{33}$Ar[11], which also probes a wide range of excitation energies ($Q_{EC} = 11.6$ MeV).

## 4 Conclusions

The neutrino cross-sections needed for the $^{37}$Cl detector are now on a secure basis. In particular:

- The efficiency for $^8$B $\nu$'s should be increased by 8% over the standard value[1] (this goes in the direction of making the solar neutrino problem worse).

- The efficiency for supernova $\nu$'s (we assume a Fermi-Dirac distribution with $T \approx 5.5$ MeV) should be increased by 25% over the standard value.

- Our experiment will improve the accuracy of a proposed direct test[12] of the $^{37}$Cl detection scheme that uses $\nu_e$'s from $\mu^+$ decay that have energies up to $E_{\nu_e} \approx 50$ MeV.

Most disturbing are our observations regarding the 'quenching' of the $GT$ strength:

- The $B(GT)$ values deduced from the $^{37}$Cl $(p,n)$ measurement do not agree with the results of our high sensitivity $\beta$-decay experiment and the problem is not just one of normalization. The fact that the $(p,n)$ measurement misses approximately 1/3 of

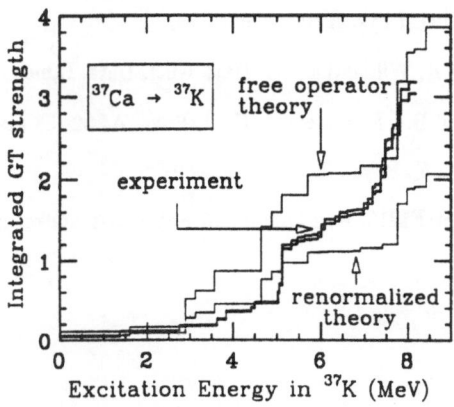

Figure 7. $B(GT)$ in the decay of $^{37}$Ca: comparison between our results and the shell-model calculation of Brown and Wildenthal.

the strength seen in our $\beta$-decay work leads us to question the evidence for 'quenching' from the former kind of measurements. Because the A=37 system presents an ideal opportunity to compare $(p, n)$ measurements to $\beta$-decays, the experiment on $^{37}$Cl$(p, n)$ done in 1981, certainly deserves a re-measurement after a decade of remarkable progress in this area.

- The most systematic support for the 'quenching' hypothesis comes from a comparison of the shell-model calculation to $\beta$-decay measurements of transitions to levels at rather low excitation energies. This shell-model, however, fails to reproduce the distribution of $GT$ strength measured in the decay of $^{37}$Ca in which transitions to levels in a broad excitation energy range were observed. This indicates problems with the $0\hbar\omega$ shell-model -the relevance of $2\hbar\omega$ excitations could be tested by measuring $^{37}$Cl$(n, p)$ which in the $0\hbar\omega$ model should show no $GT$ strength.

We must conclude then that there is no compelling evidence for the 'quenching' of the $GT$ strength.

# References

[1]  J.N. Bahcall and R.K. Ulrich, Rev. Mod. Phys. **60**, 297 (1988).

[2]  W.E. Ormand and B.A. Brown, Nucl. Phys. **A491**, 1 (1989); and B.A. Brown, private communication.

[3]  C.D. Goodman, these proceedings.

[4]  R.G. Sextro, R.A. Gough, and J. Cerny, Nucl. Phys. **A234**, 130 (1974).

[5]  J. Rapaport *et al.*, Phys. Rev. Lett., **47**, 1518(1981).

[6]  E.G. Adelberger and W.C. Haxton, Phys. Rev. C **36**, 879 (1987).

[7]  B.A. Brown and B.H. Wildenthal, Ann. Rev. Nucl. Part. Sci. **38**, 29 (1988).

[8]  D.H. Wilkinson, Nucl. Phys. **A209**, 470(1973).

[9]  B.A. Brown and B.H. Wildenthal, At. Data Nucl. Data Tables **33**, 347(1985).

[10] D.H. Wilkinson and B.E.F. Macefield, Nucl. Phys. **A232**, 58 (1974).

[11] M.J.G. Borge *et al.*, Z. Phys., **A332**, 413(1989).

[12] LAMPF experiment E1213, Ken Lande and Ray Davis, spokesmen.

# ISOSPIN AND QUARKS IN NUCLEAR BETA-DECAY

D.H. Wilkinson

TRIUMF, 4004 Wesbrook Mall, Vancouver, B.C., Canada V6T 2A3
and
University of Sussex, Brighton BN1 9QH, England

## 1. PURPOSE

The purpose of the present paper is to expose in some detail the technical problems relating to the extraction of the vector coupling constant from the beta decay of complex nuclei. The paper also considers the extraction of the axial coupling constant from the beta-decay of the neutron. The internal consistency of all data relating to beta-decay, including that of the muon, is also examined, within the standard model, with a view to the possible intervention of $W_R$.

## 2. INTRODUCTION

Ideally, allowed vector (Fermi) beta-decay just spirits away charge since the leptons are in a mutual $j = 0$ state and, furthermore, do not carry orbital angular momentum away from the nucleus; the nuclear wave function is unchanged; the nucleus is simply tilted in isospin space. Allowed axial (Gamow-Teller) beta-decay, on the other hand, flips nucleon spin with the leptons being in a mutual $j = 1$ state so that such decay may change the $J$-value of the nucleus, or may change the nuclear wave function, leaving $J$ unchanged, or may leave the nuclear wave function unchanged in which case it simply tilts the nucleus in ordinary space as well as in isospin space. Evidently, if the transition is between states of $J = 0$ without change of parity only vector decay is possible and that only if the states are members of the same isomultiplet $T$. Furthermore, if we accept the hypothesis of the conserved vector current (CVC), at least in its weak form, we have:

$$ft = \frac{K}{G_V^2 \mid M \mid^2} \tag{1}$$

where

$$
\begin{aligned}
K &= 2\pi^3 \ ln2 \ \hbar^7/(m_e^5 c^4) \\
&= (8.120270 \pm 0.000012) \times 10^{-7} \mathrm{GeV}^{-4} \mathrm{sec}
\end{aligned}
\tag{2}
$$

*Spin and Isospin in Nuclear Interactions*
Edited by S.W. Wissink *et al.*, Plenum Press, New York, 1991

and where

$$| M |^2 = T(T+1) - T_Z(T_Z+1) \qquad (3)$$

for the transition $T_Z \rightarrow T_Z + 1$.

## 3.  THE REAL WORLD

Equation (1) and its concomitant $G_V$ apply in an ideal, completely charge-independent, world, $G_V$ there being the primitive constant of vector coupling for the nucleon. In practice, de facto charge dependences of several kinds subvert Eq. (1) at various levels. That such effects are of major importance is demonstrated in Fig. 1(a) where the experimental $(ft)_e$[1] of the eight accurately-measured $J^\pi = 0^+ \rightarrow 0^+$ transitions within $T = 1$ isomultiplets* (for which $| M |^2 = 2$ in Eq. (3)), as listed in Table 1, are displayed; these $(ft)_e$-values have been corrected for electron capture

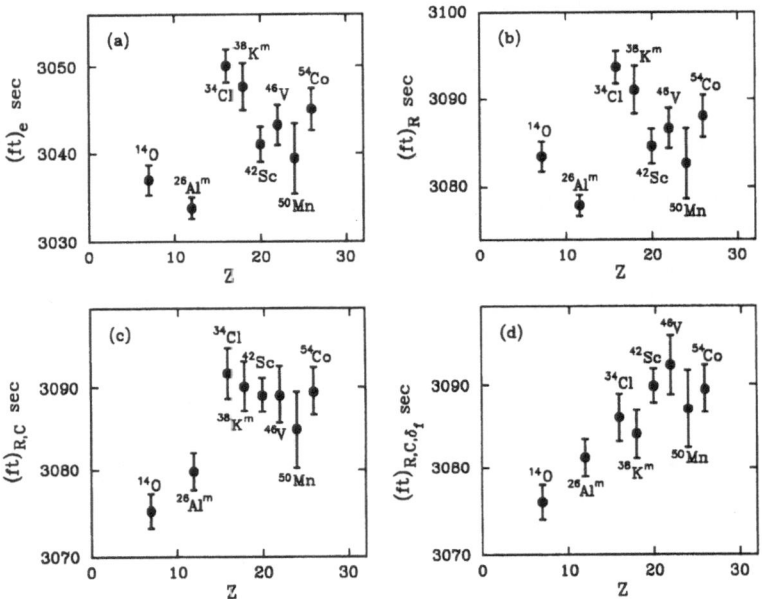

Fig. 1(a) Experimental $(ft)_e$ from the Chalk River compilation[1]; (b) $(ft)_e$ corrected for outer radiative corrections: $(ft)_R = (ft)_e(1 + \delta_R)$; (c) As (b) with additional correction for the nuclear-structure-dependent part, $C_{NS}$, of the inner radiative correction: $(ft)_{R,C} = (ft)_e(1+\delta_R)(1+\frac{\alpha}{\pi}C_{NS})$; (d) As (c) with additional correction for the case-to-case fluctuations, $\delta_{cf}$, of the nuclear mismatch: $(ft)_{R,C,\delta_f} = (ft)_e(1 + \delta_R)(1 + \frac{\alpha}{\pi}C_{NS})(1 - \delta_{cf})$.

---

*It is amusing to note that the availability, for accurate study, of all eight members of this set is due to various "accidents": in the case of $^{14}$O because the (Gamow-Teller) transition to the ground state of $^{14}$N, which might have been expected to overwhelm that of the Fermi decay in question, is highly suppressed (cf. the analogue $^{14}$C); in the cases of $^{26}$Al$^m$ and $^{38}$K$^m$ because, although the $J^\pi = 0^+, T = 1$ states of those bodies, from which the vector decay of Fig. 1(a) takes place, are excited states they are separated from the $T = 0$ ground states of high spin ($J = 5$ and 3 respectively) by such small energy differences (0.23 and 0.13 MeV respectively) that their gamma-decay is much slower than their beta-decay; in the remaining cases because although the nuclei are self-conjugate their ground states are not of $T = 0$ but of $T = 1, J^\pi = 0^+$.

Table 1. The accurately-measured vector transitions.

| Body | $(ft)_e$ | $\delta_f^{\mathrm{THH}}$ | $\delta_f^{\mathrm{QB}}$ | $\delta_1$ | $\delta_2$ | $\delta_3$ | $C_{NS}$ | $(ft)^*$ | $N_q$ |
|------|----------|---------|---------|----------|----------|----------|----------|----------|-------|
| $^{14}$O | $3037.1 \pm 1.6$ | -0.02 | -0.04 | 1.29 | 0.22 | 0.01 | $-1.13 \pm 0.17$ | $3150.3 \pm 2.0$ | $3.1 \pm 0.5$ |
| $^{26}$Al$^m$ | $3033.9 \pm 1.2$ | -0.06 | -0.03 | 1.11 | 0.32 | 0.02 | $0.26 \pm 0.26$ | $3155.1 \pm 2.2$ | $2.9 \pm 0.5$ |
| $^{34}$Cl | $3049.9 \pm 1.9$ | 0.19 | 0.18 | 1.01 | 0.39 | 0.03 | $-0.26 \pm 0.26$ | $3159.9 \pm 2.8$ | $3.2 \pm 0.7$ |
| $^{38}$K$^m$ | $3047.7 \pm 2.7$ | 0.22 | 0.17 | 0.97 | 0.41 | 0.04 | $-0.14 \pm 0.00$ | $3157.8 \pm 2.8$ | $2.6 \pm 0.5$ |
| $^{42}$Sc | $3041.1 \pm 2.0$ | -0.12 | 0.06 | 0.94 | 0.45 | 0.04 | $0.60 \pm 0.10$ | $3163.7 \pm 2.2$ | $3.4 \pm 0.6$ |
| $^{46}$V | $3043.1 \pm 2.2$ | -0.08 | -0.13 | 0.91 | 0.47 | 0.05 | $0.32 \pm 0.32$ | $3166.1 \pm 3.3$ | $3.9 \pm 1.2$ |
| $^{50}$Mn | $3039.4 \pm 3.9$ | -0.05 | -0.07 | 0.88 | 0.49 | 0.05 | $0.32 \pm 0.32$ | $3160.5 \pm 4.6$ | $2.5 \pm 0.7$ |
| $^{54}$Co | $3044.9 \pm 2.3$ | 0.01 | -0.01 | 0.86 | 0.50 | 0.06 | $0.18 \pm 0.18$ | $3163.2 \pm 2.8$ | $2.9 \pm 0.6$ |

The $ft$ columns are in s; the $\delta$ are in %.

(of some 0.1%) and for branching (everywhere greater than 99% and known to $\pm 0.01\%$); the phase space factors $f$ incorporate the gross effect of the Coulomb field of the daughter nucleus $Z$ on the departing positron through the Fermi function $F(Z, W)$ as precisely calculated – see section 3.1: the best-fit quadratic in $Z$ has $\chi^2/\nu \simeq 11$.

We now discuss the ways in which Eq. (1) is affected by charge dependences bearing in mind that, as seen from Fig. 1(a) and Table 1, the experimental accuracy on the $(ft)_e$-values is a few times 0.01% which provides the measure for the confidence with which the various charge-dependent effects must be estimated.

## 3.1 The $f$-value

The phase space is profoundly affected by the Coulomb field of the daughter nucleus acting upon the departing positron; this introduces a factor of order 2 for the heaviest bodies of present concern in Table 1. At the level of precision here sought it is not adequate to use the usual analytical point-nucleus solution for the positron wave function: we must consider the effect of an extended charge distribution. This move from point to finite nucleus results in a further correction of up to 2% or so for our practical cases on top of that due to the use of the point-nucleus $F(Z, W)$ evaluated at $R = \sqrt{5/3} R_{\mathrm{rms}}$ where $R_{\mathrm{rms}}$ refers to the extended charge distribution.[2] This finite-size correction is dominated by $R_{\mathrm{rms}}$ and is only weakly sensitive to the form of the charge distribution; however, differences of up to about 0.1% arise in the present cases as between a uniform spherical charge distribution and one of more realistic form of the same $R_{\mathrm{rms}}$ so this point must be taken.[3] An accurate approximation to this use of a more realistic form for the charge distribution in the present cases[4] is to multiply $f$ by $1 + A - BW_o/2$ where $A = 1.8 \times 10^{-5} \times |Z|^{1.36}$, $B = 2.4 \times 10^{-6} \times |Z|$ and $W_o$ is the total end-point energy in natural units. (The finite-size correction is not itself strongly sensitive to $R_{\mathrm{rms}}$ in our region of practical interest: knowledge of the nuclear size to about 2% is needed to determine the correction to 0.01%.[2]) Similarly, it is not fully adequate in effecting the lepton-nucleon convolution to use for the nucleons wave functions uniform through the nuclear volume: use of more realistic single-nucleon wave functions can entrain changes of order 0.1%.[5] (Note that we are not here concerned with charge-dependent differences between the initial proton wave function and that of the neutron into which the proton transforms itself, such as will be considered in section 3.3 in the context of the nuclear mismatch, the charge-dependent modification of $|M|^2$, but only with the role of the nucleonic wave function in weighting the convolution of the positron and neutrino wave functions.)

Two further corrections to the $f$-value relate not to the size of the nucleus but to its mass. The first recognizes that the nucleus recoils from the resultant of the lepton momenta; this converts the two body phase space for a nucleus of infinite mass into three body phase space and is worth up to 0.01% in the present context.[2] The second recognizes that the Coulomb field that affects the positron's wave function is not fixed in space but recoils with the nucleus; this correction is here less than $10^{-3}\%$ and so may be ignored.[6]

Analytical expressions have been presented[2,7] that permit the evaluation, with an accuracy that significantly betters 0.01%, of the fundamental $F(Z,W)$ and its associated point-nucleus $f$-value together with the finite size and mass effects with the exception of those relating to the forms of the charge distribution and of the single-nucleonic wave functions; a numerical parameterization of similar content that betters 0.1% is also available[8] but without the desirable facility for change of $R$ for a given $Z$ that the analytical treatment has. However, even apart from effects associated with the forms of the charge distribution and of the single-nucleonic wave functions it must be recognized that tractable analytical expressions for the positron wave functions even of great elaboration and refinement,[9] are necessarily approximate in the degree to which they can accommodate higher terms in the radial behaviour and may not be wholly adequate at the level of precision to which we must currently aspire. Thus, two alternative analytical expressions that have been suggested[9,10] differ in the $f$-values that they generate by up to 0.26% in the present context[2]; this significantly exceeds the experimental error. It seems, therefore, that if we are to seek confidence in the $f$-value at the 0.01% level we must resort to direct numerical integration of the Dirac equation in the field of a charge distribution to represent the daughter nucleus that is synthesized from appropriate proton orbitals derived from some suitable Saxon-Woods or Hartree-Fock prescription that correctly reproduces the nuclear size and then to convolute the resultant positron wave functions with exact neutrino wave functions weighting that convolution by the appropriate single-nucleonic wave functions deriving from that same prescription. This is done in preparation of the Chalk River data base of Table 1.[1] It is then perfectly adequate to use the numerical parameterization[8] to adjust for possible subsequent changes to $Q$-values and the analytical expressions[2,7] for possible changes in nuclear dimensions. This procedure omits only the relativistic term that is written $\int \alpha.r$ in traditional notation; its evaluation is not unambiguous but its magnitude is very small; it occasions no significant anxiety.

In all computations of the $f$-value allowance must be made for the effect of the atomic electrons in partially screening the nuclear charge. For the decays in question this correction is only a little over 0.1% in all cases (see e.g. Ref.[11]) and may be made with great accuracy.

We may also note that the slowing-down of the beta-decay due to the rearrangement of the atomic electrons as between initial and final states is very slight: the correction is about $3.2 \times 10^{-4} \times |Z|^{-0.54}$, viz. 0.01% or less for the cases in question.[4]

Before leaving the $f$-value we should remark that we have so far considered only the electrostatic interaction of the positron with the charge of the daughter nucleus;

but there is also a magnetostatic interaction which is strong but of short range: it is incorporated into the term $C$ of the inner radiative correction to be discussed in section 3.2.2. Overall, the $f$-value evaluation, for a given $Q$-value, is probably reliable to ± 0.01%.

## 3.2 The Radiative Corrections

The virtual photon exchanges that build up the effective electrostatic positron-nucleus potential, within which the Dirac equation is solved in deriving the $f$-value as above, represent the greater part of the electromagnetic intercourse involved in nuclear beta-decay. There are, however, other terms, both nucleon-dependent and nucleus-dependent that involve additional photons, both real (inner bremsstrahlung) and virtual and that also involve the mechanism of the weak interaction and that therefore entrain the $W$ and $Z$ bosons that are the vehicles of that interaction. These additional photonic and specifically electro-weak effects are the radiative corrections; they separate rather accurately, although only to an approximation to be discussed below, into two parts: inner and outer. Within this initial approximation the outer correction is decay-energy-dependent and $Z$-dependent and, slightly, nuclear-size-dependent but does not depend upon nucleon structure nor upon the anatomy of the weak interaction process; the inner correction is concerned with nucleon structure and the weak mechanism but not with the energy release nor, in lowest order, with the nuclear context within which the decaying nucleon is immersed (although, as we discuss in section 3.2.2, it does contain a very important small term of that nature) and so, to a first approximation, may be treated as a renormalization of the vector coupling constant converting the primitive $G_V$ into the operational $G_V^*$.

3.2.1 The outer corrections. The outer radiative corrections are of various orders $Z^n \alpha^m$; they must be carefully defined so as to avoid double counting with $F(Z, W)$. Naive vertex counting involving $n$ virtual photons would suggest terms of order $(Z^2 \alpha)^n$ which would be disastrous but such terms, in fact, vanish: the lowest that survive are of the form $Z^n \alpha^m$ where $m \geq n$; furthermore, those remaining of order $(Z\alpha)^n$ are negligible following the construction of $F(Z, W)$ which is itself a function of $Z\alpha$.[12]

The correction of order $\alpha$, relating to a single nucleon, is known exactly; call it $\delta_1 = \frac{\alpha}{2\pi} \overline{g(W, W_o)}$ where the bar indicates integration of the explicit function $g(W, W_o)$[13] over the positron spectrum to its end point $W_o$. The corrections of order $Z\alpha^2(\delta_2)$[14,15] and $Z^2\alpha^3(\delta_3)$[14] have recently been stabilized and may be written down explicitly in various orders of approximation; $\delta_2$ depends significantly upon the nuclear size as well as upon $Z$; $\delta_3$ is available only in lowest approximation, also involving the nuclear size. $\delta_1, \delta_2$ and $\delta_3$ are listed in Table 1; all speed the decay and, being designed to be combined in this way (and with $F(Z, W)$), total $\delta_R = \delta_1 + \delta_2 + \delta_3$. $\delta_R$, about 1.4–1.5%, is huge in relation to the experimental errors; however, $\delta_1$ is "exact", $\delta_2$ is probably secure to ±0.02% or better and $\delta_3$ to ±0.01% or so. Higher terms are probably negligible and we conclude that the outer radiative corrections are under good control in relation to the experimental errors. $\delta_R$ is traditionally applied to the experimental lifetime leading to the outer-radiative-corrected:

$$(ft)_R = (ft)_e(1 + \delta_R). \tag{4}$$

Figure 1(b) displays the $(ft)_R$ which has $\chi^2/\nu \simeq 13$ for the best quadratic fit in $Z$ viz. slightly poorer than for the raw $(ft)_e$ of Fig. 1(a).

3.2.2 The inner correction. For a point 4-fermion interaction the inner radiative correction is $\frac{\alpha}{2\pi}3ln\frac{\Lambda}{m_p}$ where $\Lambda$ is a cut-off that must be imposed to eliminate the ultra-violet divergence; with the introduction of just the $W$-boson to mediate the interaction $\Lambda$ is replaced by $m_W$ but divergences remain elsewhere in the calculation; with the full electro-weak unification the residual divergences disappear[16] and we may define an inner radiative correction $\Delta^*$ for the nucleon (subsuming that for the muon):

$$G_V^{*2} = G_V^2(1 + \Delta^*) \qquad (5)$$

$$\Delta^* = \frac{\alpha}{2\pi}\{3ln\frac{m_Z}{m_p} + 6\overline{Q}ln\frac{m_Z}{m_A} + 2C + \mathcal{A}\}. \qquad (6)$$

In Eq. (6) the first, vector, term is exact by CVC; the second term is axial and is brought in by photon exchange, the infra-red divergence of the $\gamma W$ box diagram being cut off at some "axial mass" $m_A$ which is usually taken to be that of the $a_1$-meson[†] viz. $(1.262 \pm 0.023)$ GeV which is close to the empirical mass, $(1.032 \pm 0.036)$ GeV, implied by the dipole parameterization of the axial form factor in $\nu N$ interactions at low energy.[17] $\overline{Q}$ is the mean charge of the fundamental constituents of the nucleon i.e. $\overline{Q} = \frac{1}{2}$ for fundamental neutrons and protons and $\overline{Q} = \frac{1}{2N_q}$ for $N_q$ quarks per nucleon. $C$ is an asymptotic long-distance axial correction (which reaches out of the nucleon and which therefore entrains consideration of nuclear structure[18] and means that the inner radiative correction in the nuclear context cannot be thought of purely as a renormalization of the coupling constant according to Eq. (5)). $\mathcal{A}$ is a small QCD correction of magnitude -0.34.[19]

The total radiative correction is now $(1 + \delta_R)(1 + \Delta^*)$ which amounts to some 3.7%.

The importance of the nuclear-structure-dependent part of the asymptotic term $C$ may be seen by writing $C = C_{\text{BORN}} + C_{\text{NS}}$ where

$$C_{\text{BORN}} = 0.788\lambda(\mu_p + \mu_n) = 0.885 \qquad (7)$$

where $\lambda$ is as defined in Eq. (30) and where the nuclear-structure-dependent $C_{\text{NS}}$ is as listed in Table 1[‡] and then removing $C_{\text{NS}}$ from the inner radiative correction $\Delta^*$ and applying it as a correction to the individual transitions, defining:

$$(ft)_{R,C} = (ft)_e(1 + \delta_R)(1 + \frac{\alpha}{\pi}C_{\text{NS}}). \qquad (8)$$

Figure 1(c) shows the substantial effect this correction has in bringing the data into smoother concordance: for the best quadratic fit in $Z$ we now have $\chi^2/\nu \simeq 2.4$.

Previous analyses have not had regard for $C_{\text{NS}}$ and it has been customary to use for $C$ its nuclear-structure-independent part $C_{\text{BORN}}$ which yields $\Delta^* = 0.02258$; this preserves the neat separation of inner and outer radiative corrections and one writes:

$$\mathcal{F}t = (ft)_R(1 - \delta_c) \qquad (9)$$

[†]Some uncertainty should properly attach to this assumption but it is difficult to say what it might be.
[‡]The errors on the $C_{\text{NS}}$-values as listed in Table 1 reflect varying degrees of confidence in the wave functions used in making the estimates[18]; the use of better wave functions, as are now available, will result in smaller uncertainties.

where $\delta_c$ is the purely nuclear mismatch between initial and final nuclear wave functions induced by the various charge dependences that we shall consider in section 3.3. $\mathcal{F}t$, which should be the same for all eight transitions of Table 1, if weak CVC holds, now takes the place of $ft$ in Eq. (1), $G_V$ being there similarly replaced by $G_V^*$. However, this treatment relies on absolute theoretical estimates of the mismatch $\delta_c$ which, as will be discussed at length in section 3.3, we must regard with some reserve and is also no longer adequate in respect of the inner radiative correction which must be revised to incorporate: (i) higher power of $\alpha$; (ii) the running QED coupling constant appropriate to the masses of all the real and virtual particles involved; (iii) the nuclear-structure-dependent part, $C_{NS}$, of $C$ the importance of which we have seen in Fig. 1(c).

The above revisions lead to the replacement of $(1+\delta_R)(1+\Delta^*)$ of the traditional approach by $(1 + \Delta + \delta_2 + \delta_3)$ where[19]:

$$1 + \Delta = \{1 + \frac{\alpha}{2\pi} \left[ ln\frac{m_p}{m_A} + 2C \right] + \frac{\alpha(m_p)}{2\pi}(\overline{g(W, W_o)} + \mathcal{A})\}S(m_p, m_Z). \quad (10)$$

$S(m_p, m_Z)$ is derived by renormalization group methods which sum up all leading terms of the form $(\alpha\ lnm_Z)^n$ and so constitutes an approximation to the incorporation of all powers of $\alpha$. Explicitly[19,20]:

For $m_t < m_W$:

$$S(m_p, m_Z) = \left[\frac{\alpha(m_c)}{\alpha(m_p)}\right]^{\frac{3}{4}} \left[\frac{\alpha(m_\tau)}{\alpha(m_c)}\right]^{\frac{9}{16}} \left[\frac{\alpha(m_b)}{\alpha(m_\tau)}\right]^{\frac{9}{19}} \left[\frac{\alpha(m_t)}{\alpha(m_b)}\right]^{\frac{9}{20}} \left[\frac{\alpha(m_W)}{\alpha(m_t)}\right]^{\frac{3}{8}} \left[\frac{\alpha(m_Z)}{\alpha(m_W)}\right]^{\frac{12}{11}} \quad (11a)$$

For $m_W < m_t < m_Z$:

$$S(m_p, m_Z) = \left[\frac{\alpha(m_c)}{\alpha(m_p)}\right]^{\frac{3}{4}} \left[\frac{\alpha(m_\tau)}{\alpha(m_c)}\right]^{\frac{9}{16}} \left[\frac{\alpha(m_b)}{\alpha(m_\tau)}\right]^{\frac{9}{19}} \left[\frac{\alpha(m_W)}{\alpha(m_b)}\right]^{\frac{9}{20}} \left[\frac{\alpha(m_t)}{\alpha(m_W)}\right]^{\frac{36}{17}} \left[\frac{\alpha(m_Z)}{\alpha(m_t)}\right]^{\frac{12}{11}} \quad (11b)$$

For $m_Z < m_t$:

$$S(m_p, m_Z) = \left[\frac{\alpha(m_c)}{\alpha(m_p)}\right]^{\frac{3}{4}} \left[\frac{\alpha(m_\tau)}{\alpha(m_c)}\right]^{\frac{9}{16}} \left[\frac{\alpha(m_b)}{\alpha(m_\tau)}\right]^{\frac{9}{19}} \left[\frac{\alpha(m_W)}{\alpha(m_b)}\right]^{\frac{9}{20}} \left[\frac{\alpha(m_Z)}{\alpha(m_W)}\right]^{\frac{36}{17}} \quad (11c)$$

The QED running coupling constants $\alpha(\mu)$ are defined by modified minimal subtraction:

$$\alpha^{-1}(\mu) = \alpha^{-1}(0) - \frac{2}{3\pi}\sum_f Q_f^2 ln\frac{\mu}{m_f} + \frac{7}{2\pi}ln\frac{\mu}{m_W} \quad (11d)$$

$$\alpha^{-1}(0) = \alpha^{-1} + \frac{1}{6\pi} = 137.089 \quad (11e)$$

In Eq. (11d) $f$ refers to all elementary fermions, the second term on the RHS being evaluated for $m_f < \mu$ and the third for $\mu > m_W$.

$S(m_p, m_Z)$, evaluated using $m_Z = 91.16$ GeV and $m_W = 80.6$ GeV[21] (and $m_{u,d,s} = 0.07$ GeV[19]; $m_c = 1.35$ GeV; $m_b = 5$ GeV[21]) has the value 1.0225 almost independently of $m_t$ in the experimentally-permitted range $m_t > 80$ GeV.[22]

The full radiative corrections are now expressed through the factors:

$$\text{Rad}_{OLD} = (1 + \delta_R)(1 + \Delta^*)$$
$$\text{Rad}_{NEW} = 1 + \Delta + \delta_2 + \delta_3$$

Rad$_{NEW}$ - Rad$_{OLD}$ ranges from 0.00152 for $^{14}$O to 0.00134 for $^{54}$Co and so it is important to use the full treatment represented by Rad$_{NEW}$ since the change that it entrains is significant in relation to experimental accuracy.

## 3.3 The Nuclear Mismatch

There remains to be discussed $\delta_c$, the nuclear mismatch due to charge-dependences, which causes the square of the matrix element in the denominator of Eq. (1) to fall below its value given in Eq. (3) by the factor $(1 - \delta_c)$. Since, as will be seen, $\delta_c$ is of the order of ten times the experimental error on individual decays its adequate treatment is a matter of high concern. Unfortunately there is no unambiguous direct procedure through which $\delta_c$ may be confidently calculated. One reason for this is that the overlap that defines $\delta_c$ involves all the nucleons, neutrons and protons, of the entire nucleus and not just the nucleon nominally responsible for making the beta-transition so that, for example, a misattribution of only 0.001% to the contribution of individual nucleons to $\delta_c$ would, in the heavier bodies of our concern, result in an error in the overall $\delta_c$ equal to the experimental error in the data to be analyzed. However, as will be seen, uncertainties in the estimation of $\delta_c$ go beyond that just indicated and make the matter additionally unsure in respect of dead reckoning.

The nuclear mismatch $\delta_c$ can be estimated via global considerations relating to the general behaviour of nucleons in charge-dependent potentials[23] or via detailed microscopic shell model wave functions case by case. It is clear from Fig. 1(c) that sufficient case-to-case fluctuations remain in $(ft)_{R,C}$ to demand consideration by the latter method.

It is usual, in shell model calculations of $\delta_c$, to divide the effect into two parts: $\delta_c = \delta_{c1} + \delta_{c2}$. $\delta_{c1}$ is due to the $T_Z$-dependence of the configurational mix out of which the detailed wave functions are constructed (which contains a component that can be described as isospin mixing although it is not very useful to use that term in this context). $\delta_{c2}$ is due to the fact that the change of $T_Z$ between initial and final states gives rise to different binding energies for the "decaying" proton and the "resultant" neutron; this, together with the Coulomb effect on the proton, results in different wave functions $\psi_p$ and $\psi_n$ for the proton and neutron states respectively so that their radial overlap:

$$\Omega_\pi = \int_o^\infty \psi_{p\pi}\psi_{n\pi} dr \qquad (12)$$

is less than unity. Here the subscripts $\pi$ remind us that when we deal with many-particle shell model wave functions the decaying proton does not observe the parent nucleus of the $A - 1$ system solely, if at all, in its ground state, to which the nominal binding energy is referred, but rather in an extensive spectrum of parent states $\pi$ in relative abundances described by the respective spectroscopic factors $S_\pi$ and with isospins $\pi_< = T - \frac{1}{2}$ and $\pi_> = T + \frac{1}{2}$; and similarly for the resultant neutron. At this point one introduces the conventional and dubious fiction that the effective binding energies, to be used in generating the $\psi_{p\pi}$, $\psi_{n\pi}$ via some fancied overall nuclear potential, are the nominal binding energies, as referred to the ground states of the respective $A - 1$ nuclei, plus the excitation energies of the parent states $\pi$ in the respective $A - 1$ nuclei; these excitation energies are taken from experiment if available, and if sufficiently confident association can be made between the experimental states and those of the model, or from the theoretical shell model wave functions if not. The associated spectroscopic factors $S_\pi$ are similarly derived from theory or experiment following which we have[24]:

$$(1 - \delta_{c2})^{\frac{1}{2}} = \frac{1}{2} \left[ \frac{1}{T} \sum_{\pi<} S_\pi \Omega_\pi - \frac{1}{T+1} \sum_{\pi>} S_\pi \Omega_\pi \right]. \tag{13}$$

Obviously, the more reliable the shell model wave functions, from the point of view of the parentage spectrum with its associated $S_\pi$ and excitation energies, the more confidence we may repose in the $\delta_{c2}$ so computed. However, we still have to reckon with: (i) the uncertainties in the generation of the $\psi_{N\pi}$, which include the effects of the deformation, as well as the form, of the optical model potential; (ii) most particularly the uncertainty associated with the effective $\pi$-dependent binding energy fiction to which reference has been made; (iii) the fact that Eq. (13) concerns itself only with the mismatch between the initial proton and final neutron states of the transforming nucleon whereas there will also be mismatch as between the parent states themselves, represented by the remainder of the valence nucleons of the shell model wave functions, in the initial and final nuclei; (iv) the fact that the core nucleons, that are not involved in the generation of the shell model wave functions, being outside the basis of the shell model calculation, will also suffer mismatch as between initial and final nuclei. It is not possible, at this time, accurately to quantify these uncertainties and so direct computation of $\delta_{c2}$ must be viewed with appropriate reserve.

Some orientation into the dependence of the calculated $\delta_c$ on the details of its generation is given by comparing the two most recent computations due to Towner, Hardy and Harvey (THH)[1,25] and to Ormand and Brown (OB)[26]; the former use single-nucleonic $\psi_N$ generated in a standard Saxon-Woods potential while the latter use a Hartree-Fock mean field approach. Table 2 lists the respective $\delta_{c1}, \delta_{c2}$ and $\delta = \delta_{c1} + \delta_{c2}$. (The "errors" quoted for the $\delta_{c2}$ are somewhat impressionistic figures thought by the authors to be reasonable reflections of uncertainties in their respective procedures.) It is seen: (i) that the $\delta_{c1}$ are relatively small and although the differences are considerable as between THH and OB those differences are, for the most part, not too worrying in relation to the experimental uncertainties although in some cases comparable with them; (ii) that the $\delta_{c2}$ are considerably larger than the $\delta_{c1}$ and differ unacceptably as between THH and OB in relation to the experimental errors: the mean value of $\delta_c^{THH} - \delta_c^{OB}$ is 0.16% (as is that of the modulus of that quantity).

In regard to the mismatch of the core: (i) if an estimate of the core contribution to $\delta_{c2}$ is made in the literal Saxon-Woods spirit of THH a very substantial figure, of the percentage order, is obtained; (ii) the Hartree-Fock approach of OB suffers from the fact that such wave functions do not respect isospin even when the two-body force is an isospin scalar and so cannot be trusted for a direct evaluation of the core mismatch. OB overcame this difficulty by an ingenious if ad hoc stratagem following which the core mismatch was estimated to be negligible. However, the ad hoc nature of the remedy must leave some doubt as to the reliability of the conclusion.

The upshot of this discussion is that present direct computation of $\delta_c$ cannot be relied upon to better than the difference, 0.2% or so, between the $\delta_c^{THH}$ and $\delta_c^{OB}$ of Table 2 and that, lurking behind this figure, there are the additional uncertainties associated with the core nucleons and other considerations to which reference has been made.

However, there is an alternative approach to the assessment of $\delta_c$[27] that seeks to by-pass much of the above uncertainty by looking to the experimental data themselves to help reveal the mismatch. This approach considers that $\delta_c$ will have two

Table 2. Calculated and inferred nuclear mismatch in %.

| Body | THH | | | OB | | | |
|---|---|---|---|---|---|---|---|
| | $\delta_{c1}$ | $\delta_{c2}$ | $\delta_c$ | $\delta_{c1}$ | $\delta_{c2}$ | $\delta_c$ | $\delta_{cINF}$ |
| $^{14}$O | 0.00 | 0.28 ± 0.03 | 0.28 | 0.01 | 0.18 ± 0.05 | 0.19 | 0.32 |
| $^{26}$Al$^m$ | 0.06 | 0.27 ± 0.04 | 0.33 | 0.01 | 0.23 ± 0.06 | 0.24 | 0.48 |
| $^{34}$Cl | 0.02 | 0.62 ± 0.07 | 0.64 | 0.06 | 0.42 ± 0.07 | 0.48 | 0.84 |
| $^{38}$K$^m$ | 0.16 | 0.54 ± 0.07 | 0.70 | 0.11 | 0.38 ± 0.12 | 0.49 | 0.90 |
| $^{42}$Sc | 0.04 | 0.35 ± 0.06 | 0.39 | 0.11 | 0.28 ± 0.05 | 0.39 | 0.71 |
| $^{46}$V | 0.09 | 0.36 ± 0.06 | 0.45 | 0.01 | 0.20 ± 0.06 | 0.21 | 0.65 |
| $^{50}$Mn | 0.10 | 0.40 ± 0.09 | 0.50 | 0.004 | 0.28 ± 0.06 | 0.28 | 0.72 |
| $^{54}$Co | 0.03 | 0.56 ± 0.06 | 0.59 | 0.005 | 0.34 ± 0.06 | 0.35 | 0.79 |

components: $\delta_c = \delta_{cu} + \delta_{cf}$; $\delta_{cu}$ is a smooth (monotonically increasing) function of $Z$, the underlying mismatch, whose form and magnitude are unknown; $\delta_{cf}$ are the case-to-case fluctuations about $\delta_{cu}$ associated with explicit shell model effects, case-to-case irregularities in binding energies and so on. All that we know, a priori, is that $\delta_{cu}$ and $\delta_{cf}$ must, presumably, separately go to zero as $Z$ goes to zero. We might now hope that although different direct computations of $\delta_c$ may differ considerably in magnitude, and might be afflicted by the various uncertainties that we have considered, yet they might be more reliable in respect of, and agree better with each other in respect of, the case-to-case fluctuations $\delta_{cf}$ since these are more nearly due just to the valence nucleons of the shell model wave functions and to the de facto binding energy fluctuations that are taken into account in the computational procedures. Indeed[27] if we extract the $\delta_{cf}$ from the THH and OB calculations by separately best-fitting the arbitrary function $AZ^B$ to each set of $\delta_c$ taken from Table 2 (the fitting form chosen matters little and the parameters have no significance) we find the $\delta_{cf}$-values listed in Table 1 from which it is seen that the mean value of $\delta_{cf}^{THH} - \delta_{cf}^{OB}$ is -0.005% and that of $| \delta_{cf}^{THH} - \delta_{cf}^{OB} |$ is only 0.05%: $\delta_c^{THH}$ and $\delta_c^{OB}$ tend strongly to fluctuate up and down synchronously.

If we now apply this case-to-case correction $\delta_{cf}$ to the experimental data (using for $\delta_{cf}$ the mean of $\delta_{cf}^{THH}$ and $\delta_{cf}^{OB}$), defining:

$$(ft)_{R,C,\delta_f} = (ft)_e (1 + \delta_R)(1 + \frac{\alpha}{\pi} C_{NS})(1 - \delta_{cf}) \qquad (14)$$

we should be left with data requiring further correction only for the smooth underlying $\delta_{cu}$, which the data themselves should then reveal, such that the appropriate extrapoation of $(ft)_{R,C,\delta_f}$ to $Z = 0$ should define $(ft)_o$ which would then yield:

$$G_V^{*2} = K/2(ft)_o \qquad (15)$$

Figure 1(d) shows the $(ft)_{R,C,\delta_f}$, the best-fitting of which to a quadratic in $Z$ has $\chi^2/\nu \simeq 0.7$ which is wholly satisfactory and justifies our procedure.

## 4. ANALYSIS OF THE FERMI TRANSITIONS

The preceding discussion has led to the point at which, from Fig. 1(d), we could extract $(ft)_o$, hence derive $G_V^*$ by Eq. (15), hence $G_V$ by Eq. (5) using $\Delta^*$ from Eq. (6) (setting $C = C_{\text{BORN}}$). However, as has been emphasized, such a procedure is no longer adequate in view of the development of the superior $\Delta$ of Eq. (10). The better plan is to construct from the experimental data:

$$(ft)^* = (ft)_e(1 - \delta_{cf})(1 + \Delta + \delta_2 + \delta_3) \tag{16}$$

and extrapolate $(ft)^*$ to $(ft)_o^*$ at $Z = 0$ then deriving:

$$G_V^2 = K/2(ft)_o^*. \tag{17}$$

This construction of $(ft)^*$ from Eq. (16) is given in Fig. 2 following the listing in Table 1 (where the increase in the errors of the $(ft)^*$ above those of the $(ft)_e$ is due to the theoretical uncertainties listed for the $C_{\text{NS}}$). We find by extrapolation quadratic in $Z$[§]:

$$(ft)_o^* = (3139.1 \pm 2.6)\text{sec} \tag{18}$$

hence:

$$G_V^2 = (1.2934 \pm 0.0011) \times 10^{-10}\text{GeV}^{-4}. \tag{19}$$

Before proceding to further discussion note that the best way to extract the operational $G_V^*$ is to start from the $G_V$ of Eq. (19), to note the equivalence:

$$(1 + \Delta^*)(1 + \delta_1) \simeq 1 + \Delta \tag{20}$$

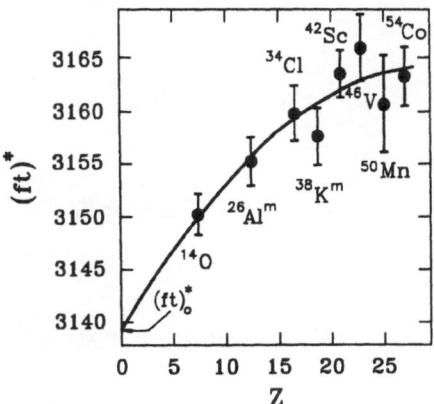

Fig. 2. $(ft)^* = (ft)_e(1 - \delta_{cf})(1 + \Delta + \delta_2 + \delta_3)$ as a function of $Z$ extrapolated quadratically to $(ft)_o^*$ at $Z = 0$.

---

[§]No theoretical justification attaches to this fitting; it is simply the lowest-order adequate polynomial in $Z$.

from which derive $\Delta^*$ on a case-to-case basis and average them to gain the best effective $\Delta^*$ (it is 0.02397) and then use Eq. (5) to find:

$$G_V^* = (1.1508 \pm 0.0005) \times 10^{-5}\text{GeV}^{-2}. \tag{21}$$

It is also of interest to use the $(ft)^*$-fit of Fig. 2 to derive the $\delta_{cuINF}$ inferred from our present procedure, then to combine them with the $\delta_{cf}$-values that we have used in obtaining the $(ft)^*$ thereby gaining the overall inferred $\delta_{cINF} = \delta_{cuINF} + \delta_{cf}$. Figure 2 lists these $\delta_{cINF}$; it is seen that they are broadly of the same order as the $\delta_c$ of the direct computations but are, as we anticipated, somewhat larger: by a factor, on average of 1.4 than $\delta_c^{THH}$ and of 2.1 than $\delta_c^{OB}$.

## 5. UNITARITY OF THE CABIBBO-KOBAYASHI-MASKAWA MATRIX

An immediate use for the $G_V$ of Eq. (19) is in a test for the unitarity condition relating to the first row of the Cabibbo-Kobayashi-Maskawa (CKM) matrix:

$$| V_{ud} |^2 + | V_{us} |^2 + | V_{ub} |^2 = 1. \tag{22}$$

From muon decay we have:

$$G_\mu^2 = (1.36046 \pm 0.00005) \times 10^{-10}\text{GeV}^{-4} \tag{23}$$

(where the outer radiative correction, here also extended to higher orders in $\alpha$ by renormalization group methods,[28] has been applied to the experimental data[21] and where we remember that the muon's inner radiative correction has been subsumed into $\Delta$).

With:

$$| V_{ud} |^2 = (G_V/G_\mu)^2 \tag{24}$$

we have:

$$| V_{ud} |^2 = 0.9507 \pm 0.0008. \tag{25}$$

This we may now combine with[29]:

$$| V_{us} |^2 = 0.0481 \pm 0.0008 \tag{26}$$

and[30]:

$$| V_{ub} |^2 \simeq 3 \times 10^{-5} \tag{27}$$

to find:

$$| V_{ud} |^2 + | V_{us} |^2 + | V_{ub} |^2 = 0.9989 \pm 0.0012 \tag{28}$$

which constitutes a satisfactory test.

# 6. THE NUMBER OF QUARKS PER NUCLEON

For an amusement in passing we can:

(i) Assume CKM unitarity;

(ii) Use $|V_{us}|$ (Eq. (26)) and $|V_{ub}|$ (Eq. (27)) to infer $|V_{ud}|$;

(iii) Use $G_\mu$ (Eq. (23)) and $|V_{ud}|$ to find $G_V$ by Eq. (24);

(iv) Use $G_V$ from (iii) to derive $\Delta^-$ (namely $\Delta$ from which $C_{NS}$ has been extracted) on a case-by-case basis from:

$$K/2G_V^2 = (ft)_e(1 - \delta_{cINF})(1 + \frac{\alpha}{\pi}C_{NS})(1 + \Delta^- + \delta_2 + \delta_3)$$

(v) Use the equivalence of Eq. (20) to infer $\Delta^{*-}$ on a case-by-case basis from the $\Delta^-$ from (iv);

(vi) Infer $\overline{Q}$ (viz. $N_q$) on a case-by-case basis from the $\Delta^{*-}$ from (v) using $C = C_{\text{BORN}}$ in Eq. (6).

When this is done[31] we find the $N_q$-values listed in Table 1 which average $N_q = 3.02 \pm 0.20$ without regard for the error in $|V_{us}|$ (Eq. (26)); when that is included we find $N_q = 3.0 \pm 0.6$.

# 7. NEUTRON DECAY

From the $G_V^*$ of Eq. (21) and the best "post 1986" (directly measured) neutron lifetime of[32,33]:

$$\tau_m = (888.8 \pm 2.4)\text{sec} \tag{29}$$

we derive[6]:

$$G_V^{*2} + 3G_A^{*2} = (7.687 \pm 0.021) \times 10^{-10}\text{GeV}^{-4} \tag{30}$$

($G_A^*$ here being the operational axial coupling constant analogous to the vector $G_V^*$).

Another important quantity is $A_o$, the asymmetry parameter for the beta-decay of polarized neutrons which is, in lowest order, given by:

$$A_o = -2\frac{\lambda(\lambda - 1)}{1 + 3\lambda^2} \tag{31}$$

where:

$$\lambda = |G_A^*/G_V^*|. \tag{32}$$

Two recent accurate measurements of $A_o$ exist yielding, after appropriate correction for weak magnetism (strong CVC), recoil and Coulomb effects[6] and after assurance that radiative corrections are negligible in their effect upon $\lambda$[34]:

$$G_A^*/G_V^* = -1.262 \pm 0.004 \quad (\text{Ref. } 32, 35) \tag{33}$$

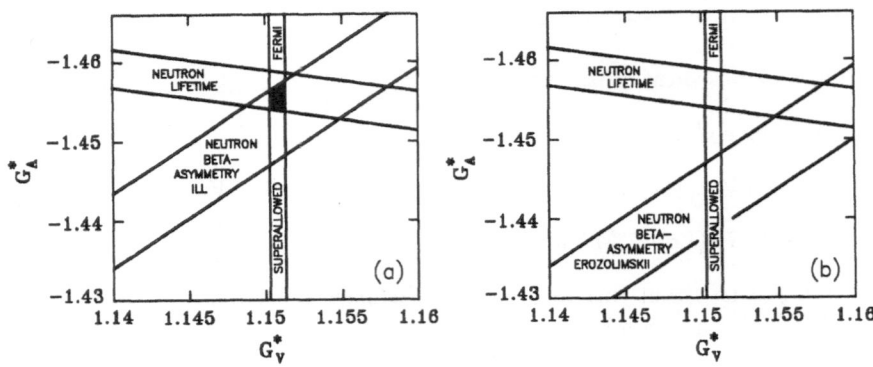

Fig. 3(a) $G_V^*$ derived from vector (superallowed Fermi) decay (Eq. (21)); $G_V^{*2} + 3G_A^{*2}$ derived from the neutron lifetime (Eq. (30)) and $G_A^*/G_V^*$ derived from the ILL measurement of neutron decay asymmetry (Eq. (33)); (b) As (a) but with the $G_A^*/G_V^*$ measurement of Erozolimskii et al., (Eq. (34)) replacing the ILL measurement.

$$G_A^*/G_V^* = -1.254 \pm 0.004 \quad (\text{Ref. 36}). \tag{34}$$

We may now confront our values for $G_V^*, G_V^{*2} + 3G_A^{*2}$ and $G_A^*/G_V^*$ which we do in Fig. 3(a) using Eqs. (21), (30) and (33) for these three quantities respectively and in Fig. 3(b) where we replace Eq. (33) by Eq. (34). Figure 3(a) shows good mutual accord and permits us to derive the overall:

$$G_A^* = -(1.4554 \pm 0.0016) \times 10^{-5} \text{GeV}^{-2} \tag{35}$$

or:

$$G_A^*/G_V^* = -1.2647 \pm 0.0014. \tag{36}$$

If we choose to derive $G_V$ via unitarity of the CKM matrix as indicated in section 5 followed by allowance for inner radiative corrections via $\Delta$ and $\Delta^*$ as indicated in section 3 to gain $G_V^*$ we should find:

$$G_A^* = -(1.4557 \pm 0.0020) \times 10^{-5} \text{GeV}^{-2} \tag{37}$$

$$G_A^*/G_V^* = -1.2649 \pm 0.0017. \tag{38}$$

Figure 3(b) shows no area of mutual consistency so does not provide a fit within the standard model.

## 8.  LEFT-RIGHT SYMMETRIC MODELS

The lack of internal consistency of Fig. 3(b) suggests that we should explore the consequences of the intervention of right hand currents as a natural extension of the standard model[37] in the form of the introduction of a right handed partner, $W_R$, to the $W_L$ of the standard model with the mass parameter $\delta = (m_L/m_R)^2$ and with the mixing angle $\zeta$ between the $L$ and $R$ mass eigenstates.

Such $L - R$ symmetry has a number of consequences for beta-decay that we now examine:

(i) The beta-decay asymmetry of polarized neutrons becomes[38]:

$$A_o = -2\frac{\lambda_p(\lambda_p - 1) - \lambda_p y(\lambda_p y - x)}{1 + 3\lambda_p^2 + x^2 + 3\lambda_p^2 y^2} \tag{39}$$

where, very nearly, $x = \delta - \zeta$ and $y = \delta + \zeta$ and where $\lambda_p$ must be determined from parity-conserving variables only, viz. $G_V^*$ and $\tau_m$ which, via Eqs. (21) and (29), yield:

$$\lambda_p = 1.2655 \pm 0.0022. \tag{40}$$

This analysis, using the $A_o$-value that lies behind Eq. (33) namely

$$A_o = -0.1146 \pm 0.0015 \tag{41}$$

gives only a useful constraint in the $\delta, \zeta$-plane such that the permitted area lies, at 90% CL, to the left of the line in Fig. 4(a) labelled "neutron/Fermi". Using the $A_o$-value that lies behind Eq. (34), namely

$$A_o = -0.1114 \pm 0.0015 \tag{42}$$

we find, at 90% CL, the allowed band of Fig. 4(b).

(ii) The longitudinal polarization of beta-particles will manifestly be affected by the presence of right hand currents: (a) absolutely, as has been sought in the case of Gamow-Teller transitions,[39] which yield the 90% CL upper constraint labelled "Gamow-Teller" in Fig. 4(a); (b) in the relative polarization in Fermi and Gamow-Teller transitions[40] which prescribe the 90% CL region between the curves of Fig. 4(a) labelled "F/GT".

(iii) Powerful constraints derive from muon decay: (a) via the Michel parameter[21] which, at 90% CL, limits the area available to that between the lines of Fig. 4(a) labelled "Michel"; (b) via the combinations of decay parameters $P_\mu \xi \delta / \rho$[41] and $P_\mu \xi$[42] which, at 90% CL, limit the allowed regions to those below their respectively-labelled curves in Fig. 4(a).

(iv) If we assume CKM unitarity the existence of $W_R$ induces a shortfall of $2\zeta$ in the test applied to the first row of the matrix[43] as quoted in Eq. (22) and reported in Eq.

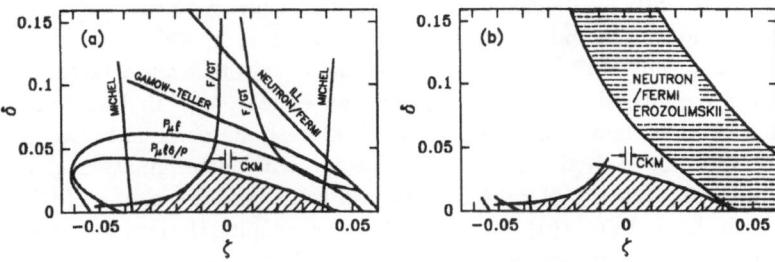

Fig. 4(a) constraints at 90% CL on the $\delta, \zeta$ parameters of the left-right symmetric model derived as described in the text; the neutron/Fermi constraint is that deriving from the ILL measurement of $A_o$ (Eq. (41)); (b) The shaded area is transferred from (a); the shaded band derives from the $A_o$-measurement of Erozolimskii *et al.*, (Eq. (42)).

(28) so that we should, from the latter, infer, at 90% CL, $\zeta = 0.0006 \pm 0.0010$; these limits are shown in Fig. 4(a).

The ensemble of these constraints gathered in Fig. 4(a) limit the allowed region of $\delta, \zeta$-space to the shaded area at 90% CL without reference to the sharp constraint in $\zeta$ afforded by the CKM test of (iv) above. We derive $m_R/m_L > 5.5$ at 90% CL from the shaded area, i.e. a $W_R$ mass of greater than 440 GeV, or $m_R/m_L > 5.8$ at 90% CL if the CKM constraint is accepted, i.e. a $W_R$ mass of greater than 470 GeV.

We now transfer the shaded allowed area of Fig. 4(a) to Fig. 4(b) there to confront the shaded band that represents the $\delta, \zeta$-region permitted at 90% CL by the $A_o$-value of Eq. (42). We see that there is only marginal compatibility between the area and the band without admitting the CKM constraint, which is also shown in Fig. 4(b), and none at all if that constraint is accepted.

## 9. OTHER TESTS

We note that the method of analysis of vector beta-decay presented in sections 2 and 3 is implicitly based not only upon the assumption of weak CVC, which therefore cannot simultaneously be tested, but also upon the assumption of the absence of scalar and induced-scalar couplings for which the data could be analyzed were we to adopt directly-computed $\delta_c$ and to construct the $\mathcal{F}t$ of Eq. (8)[1,44]; a similar remark applies to testing for the possible intervention of massive neutrinos.[1,45] In the latter context, however, we may note a curious feature of Fig. 2 namely that the inferred underlying $\delta_{cu}$ of the nuclear mismatch increases approximately linearly with $Z$, indeed the best quadratic fit in $Z$, as seen, increases more slowly than linearly, whereas, offhand, we might have expected, following the Behrends-Sirlin-Ademollo-Gatto theorem,[46,47] a mismatch going approximately as $Z^2$ as has also been remarked elsewhere[48]; the sense of the difference is as would be expected from the intervention of massive neutrinos.[49]

## 10. $G_V^*$: FUTURE DEVELOPMENTS

### 10.1 $^{10}$C Decay

The analysis presented here depends critically upon extrapolation of the experimental data, after allowance for radiative corrections (including $C_{NS}$) and for fluctuations in the nuclear mismatch, to $Z = 0$. Obviously any further $(ft)_e$ at lower $Z$ than those presently available, and of comparable accuracy to the present set, would be most precious. The only such body is $^{10}$C, at $Z = 5$, which has a sufficiently-accurately determined lifetime and $Q$-value and decays by a pure vector transition, of the type considered here, to the second excited state of $^{10}$B. Unfortunately this desired transition shows only a 1.5% branch against the dominant (Gamow-Teller) decay to the first excited state of $^{10}$B. The present[50] determination of the Fermi branch with an accuracy of $\pm 0.4\%$ of its own value, which yields an $(ft)_e$-value in accord with those of Fig. 1(a) but with an error enveloping the entirety of that figure, is already a considerable experimental tour de force; it seems unreasonable to look for the further improvement by an order of magnitude such as would be necessary to constitute a significant contribution to the sharpening of our analysis.

## 10.2 Neutron Decay

As we have seen, the problems of confidently analyzing the vector decays of complex nuclei are formidable despite the high precision of the experimental data. If the neutron lifetime could be measured with adequate precision and also, by $A_o$-measurement, the $G_A^*/G_V^*$ ratio $\lambda$ (both quantities now being known, as reported above, to about $\pm 0.3\%$) then we could access $G_V^*$ independently of all the uncertainties entrained by the use of complex nuclei; this would be highly desirable. (It would also, of course, depend on the assumption of the standard, $W_L$-only, model.) The precision to which we should aspire in the extraction of $G_V^*$ must match that ($\pm 0.05\%$) reported above from complex nuclei. This demands measurements of neutron lifetime and of $\lambda$ whose mutual accuracy lies within an ellipse with semi-axes of $\pm 0.06\%$ on the $\lambda$-axis and $\pm 0.1\%$ on the $\tau_m$-axis; this is very hard.

## 10.3 Pion Beta-Decay

Pion beta-decay, $\pi^\pm \rightarrow \pi^o + e^\pm + \overset{(-)}{\nu}_e$, is the archetypal test of weak CVC, directly yielding $G_V$ (after application of the appropriate inner and outer radiative corrections both of which are, in lowest order, the same as for nucleon-decay[16]) without primary concern for structural effects (but see section 11). The present lifetime measurement for this branch of $10^{-8}$ against $\pi \rightarrow \mu + \nu_\mu$ has an accuracy of $\pm 3\%$[21] so that an improvement of more than an order of magnitude is needed. (Present uncertainty in the $\pi^\pm - \pi^o$ mass difference[21] corresponds to a $\pm 0.03\%$ uncertainty in the extracted $G_V$.)

## 11. PARTICLE STRUCTURE

We have tacitly assumed that the $G_V$ of the nucleon or of the pion, arrived at after confident removal of the inner radiative corrections (which are themselves, as we have seen, particle-structure-sensitive), relate directly to the $V_{ud}$ of the CKM matrix by Eq. (24) viz. reflect directly the fundamental weak quark-quark couplings. However, this is not the case: just as the vector decay of a complex nucleus is affected by charge-dependent effects within its structure so that of a "fundamental" particle such as a pion or a nucleon is affected by the fact that it is not a truly fundamental particle, such as quarks and leptons putatively are, but is itself structured and that that structure entrains the charge dependences, such as the mass differences, inherent to the truly fundamental elements of that structure, namely the quarks, so that, for example, hadrons are not pure in isospin, although that is only part of the problem. The magnitude of this effect, perhaps to be ascribed specifically to $\rho - \omega$ mixing, is a matter of contention it being argued: (i) that it could have a reflection on $| V_{ud} |$ as large as 0.2%[51]; (ii) that, as a consequence of the Behrends-Sirlin-Ademollo-Gatto theorem,[47] namely that renormalization of vector coupling constants goes only as the square of the mass splittings between the de facto (complex) particles involved so that there is here, for example, no first-order term in $m_u - m_d$, the effect is less than 0.001%.[52] But it is clear that, at some level, we shall have to wrestle with particle structure just as we are now wrestling with nuclear structure.

# REFERENCES

1. J.C. Hardy et al., Nucl. Phys. A509:429 (1990).
2. D.H. Wilkinson, Nucl. Instrum. Methods A290:509 (1990).
3. H. Behrens and W. Bühring, Nucl. Phys. A150:481 (1970); A179:297 (1972); D.H. Wilkinson, A. Gallmann and D.E. Alburger, Phys. Rev. C18:401 (1978).
4. D.H. Wilkinson, Nuclear Physics with Heavy Ions and Mesons, Les Houches, Session XXX 1977 (North-Holland, Amsterdam, 1978) p. 877.
5. D.H. Wilkinson, Nucl. Phys. A232:93 (1974).
6. D.H. Wilkinson, Nucl. Phys. A377:474 (1982).
7. D.H. Wilkinson, Nucl. Instrum. Methods A275:378 (1989).
8. D.H. Wilkinson and B.E.F. Macefield, Nucl. Phys. A232:93 (1974).
9. H. Behrens and W. Bühring, Electron Radial Wave Functions and Nuclear Beta-Decay (Clarendon, Oxford, 1982).
10. H. Behrens and J. Jänecke, Landolt-Börnstein Tables, Gruppe I, Band 4 (Springer, Berlin, 1969).
11. D.H. Wilkinson, Nucl. Phys. A150:478 (1970).
12. M.A.B. Bég, J. Bernstein and A. Sirlin, Phys. Rev. D6:2597 (1972).
13. A. Sirlin, Phys. Rev. 164:1767 (1967); G. Källén, Nucl. Phys. B1:225 (1967).
14. W. Jaus and G. Rasche, Phys. Rev. D35:3420 (1987).
15. A. Sirlin, Phys. Rev. D35:3423 (1987).
16. A. Sirlin, Rev. Mod. Phys. 50:573 (1978).
17. U. Amaldi, et al., Phys. Rev. D36:1385 (1987).
18. W. Jaus and G. Rasche, Phys. Rev. D41:166 (1990).
19. W.J. Marciano and A. Sirlin, Phys. Rev. Lett. 56:22 (1986).
20. A. Sirlin, private communication.
21. J.J. Hernandez et al., Review of Particle Properties, Phys. Lett. B239 (1990).
22. F. Abe et al., Phys. Rev. Lett. 64:142 (1990).
23. J. Damgaard, Nucl. Phys. A130:233 (1969); A.M. Lane and A.Z. Mekjian, Advances in Nuclear Physics, vol. 7 (Plenum, New York, 1973) p. 97.
24. D.H. Wilkinson, Phys. Lett. B65:9 (1976).
25. I.S. Towner, J.C. Hardy and M. Harvey, Nucl. Phys. A284:269 (1977).
26. W.E. Ormand and B.A. Brown, Phys. Rev. Lett. 62:866 (1989).
27. D.H. Wilkinson, Nucl. Phys. A511:301 (1990).
28. W.J. Marciano and A. Sirlin, Phys. Rev. Lett. 61:1815 (1988).
29. J.F. Donoghue et al., Phys. Rev. D35:934 (1987); T. Yamaguchi et al., Nucl. Phys. A500:429 (1989); D.H. Wilkinson, Nuclear Weak Processes and Nuclear Structure, Osaka (World Scientific, 1989) p. 1.
30. R. Fulton et al., Phys. Rev. Lett. 64:16 (1990).
31. D.H. Wilkinson, Nucl. Phys. A518:138 (1990).
32. S. Freedman, Comments Nucl. Part. Phys. 19:209 (1990).
33. W. Mampe et al., Phys. Rev. Lett. 63:593 (1989); J. Byrne et al., Phys. Rev. Lett. 65:289 (1990).
34. Y. Yokoo and M. Morita, Prog. Theor. Phys. Suppl. 60:37 (1976); R.T. Shann, Nuovo Cimento 5A:591 (1971); A. Garcia and M. Maya, Phys. Rev. D17:1376 (1978).
35. P. Bopp et al., Phys. Rev. Lett. 56:919 (1986); E. Klemt et al., Z. Phys. C37:179 (1988).
36. B.G. Erozolimskii et al., Annual Session of USSR Academy of Sciences, Moscow (Nuclear Division) 23-25 Jan. 1990; See also: Yu.V. Gaponov et al., IAE-5032/2, 1990 (Moscow).

37. J.C. Pati and A. Salam, Phys. Rev. Lett. 31:661 (1973); M.A.B. Bég *et al.*, Phys. Rev. Lett. 38:1252 (1977).
38. A.-S. Carnoy *et al.*, Phys. Rev. D38:1636 (1988).
39. J. van Klinken *et al.*, Phys. Lett. B79:199 (1978).
40. V.A. Wichers *et al.*, Phys. Rev. Lett. 58:1821 (1987); A.-S. Carnoy *et al.*, WEIN '89, Gif-sur-Yvette (Editions Frontières, 1989) p. 581.
41. A. Jodido *et al.*, Phys. Rev. D34:1967 (1986) and Phys. Rev. D37:237 (1988).
42. I. Beltrami *et al.*, Phys. Lett. B194:326 (1987).
43. B.R. Holstein and S.B. Treiman, Phys. Rev. D16:2369 (1977); J. Deutsch, Acta Physica Hungarica, May 1989.
44. W.E. Ormand *et al.*, Phys. Rev. C40:2914 (1989).
45. V.I. Isakov and M.I. Strikman, Phys. Lett B181:195 (1986); J. Deutsch and R. Prieels, Nucl. Phys. A478:725c (1988); J. Deutsch, M. Lebrun and R. Prieels, Nucl. Phys. A518:149 (1990).
46. D.H. Wilkinson and D.E. Alburger, Phys. Rev. C13:2517 (1976).
47. R.E. Behrends and A. Sirlin, Phys. Rev. Lett. 4:186 (1960); M. Ademollo and R. Gatto, Phys. Rev. Lett. 13:264 (1964).
48. G. Rasche and W.S. Woolcock, Mod. Phys. Lett. A5:1273 (1990).
49. D.H. Wilkinson, Proc. XIII PANIC Conf., 1990, Nucl. Phys. A, in press.
50. Y. Nagai *et al.*, Nuclear Weak Processes and Nuclear Structure, Osaka (World Scientific, 1989) p. 64.
51. G. Lopez-Castro and J. Pestiau, Phys. Lett. B203:315 (1988).
52. J.F. Donoghue and D. Wyler, Phys. Lett. B241:243 (1990).

# SCATTERING OF PROTONS AND PIONS

# FROM POLARIZED $^3$He

Otto Häusser

Simon Fraser University, Burnaby, B.C., Canada, V5A 1S6
and
TRIUMF, 4004 Wesbrook Mall, Vancouver, B. C., Canada, V6T 2A3

## INTRODUCTION

The subject of this paper involves, directly or implicitly, spin-spin interactions at the atomic, nuclear, and subnuclear levels. The production of polarized $^3$He targets by the optical pumping spin exchange method (see Chupp et al.[1]) proceeds first by absorption of a circularly polarized photon by an alkali atom (Rubidium). A constant flux of photons from a laser is required to compensate for alkali spin destruction, either in alkali–alkali collisions or, as will be shown below, in three–body collisions leading to the formation of alkali–$^3$He molecules. The collisional transfer of alkali polarization to $^3$He involves the Fermi contact hyperfine interaction between the angular momentum of the alkali atom and the $^3$He nuclear spin.

At the nucleonic level the main component of the $^3$He wave function[2] consists of a spatially symmetric S state, with the two protons in a spin singlet state, and with the neutron carrying the $^3$He spin. However, the spin-dependent nucleon-nucleon interaction introduces small components into the $^3$He wave function which can be calculated by the Faddeev method. The results depend on whether, in addition to two-body forces between nucleon pairs, three-body forces, or meson degrees of freedom and nucleon substructure, or relativistic effects are included in the calculations. Consistency tests of the A=3 wave function are provided by experimental binding energies, charge radii, magnetic moments, electric and magnetic form factors. The $^3$He wave function has recently been investigated in a novel way by nucleon knockout reactions from polarized $^3$He (see Rahav et al.[3]). A comparison of target-related spin observables in a nucleon knockout experiment with those for NN scattering can provide information on spin-momentum correlations of nucleons in the polarized target nucleus. The power of the method depends critically on the importance of rescattering corrections which are difficult to calculate.

The question to what extent polarized $^3$He is equivalent to a polarized neutron is of interest for investigations of the subnucleonic structure of the neutron. Quasielastic scattering of longitudinally polarized electrons from polarized $^3$He is sensitive to interference terms between longitudinal and transverse scattering (see Blankleider and Woloshyn[4]) and shows promise[5,6] for measurements of the electric form factor of the neutron at large momentum transfers. The non-vanishing electric form factor and an expected[7,8] non-zero quark contribution to the neutron spin are manifestations of quark spin-spin interactions which arise in a QCD description of nucleon structure.

In the following sections we report progress made recently at TRIUMF to improve the density and volume of polarized $^3$He targets. Quantitative estimates of the laser power required for Rb optical pumping at various Rb densities, $^3$He pressures and target cell geometries will be given. First nuclear physics results obtained with the TRIUMF target will then be presented. Spin observables in proton–induced nucleon knockout reactions are shown to be sensitive probes of the nucleonic spin structure of $^3$He. Experimental results on spin observables and asymmetries in elastic scattering of protons and pions from polarized $^3$He determine previously unexplored spin-dependent amplitudes.

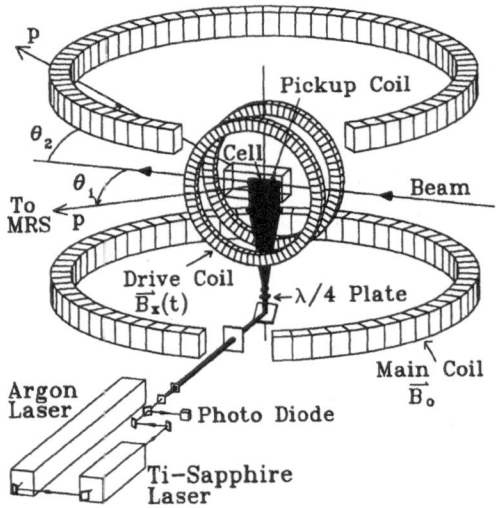

Fig. 1. Schematic layout of the polarized $^3$He target system used at TRIUMF.

## THE TRIUMF POLARIZED $^3$He TARGET

The polarized $^3$He target setup used in the experiments is shown schematically in Fig. 1. A full description of the technical details can be found in a recent paper by Larson *et al.*[9] We discuss here some of the new features which differ from earlier work[1]. The $5S_{1/2} \rightarrow 5P_{1/2}$ D1 transition in Rb is optically pumped by circularly polarized photons from a frequency tuneable Ti:Sapphire laser pumped by all visible lines from an Argon ion laser. With two such systems 8–9 Watts of power at 795 nm is obtained. The expanded, parallel light beam illuminates the target from below. The target glass cell is inside an oven to select the desired Rb density (typically $4 \times 10^{14}$ cm$^{-3}$ at 453 K). The target cell contains about 120 Torr of $N_2$ for collisional quenching of fluorescent Rb light, and $^3$He. A cryogenic setup using cold helium gas has allowed us to produce cells at relative $^3$He densities $p$ from 3 to 12 ($p = 1$ corresponds to 760 Torr at 273 K).

Transmission measurements with continuous, linearly ($\sigma_o$) or circularly polarized ($\sigma_\pm$) light of wavelengths between 790–800 nm were carried out[9] to determine the laser power required for optical pumping of a certain Rb density, [Rb]. The Rb D1 lineshapes are strongly asymmetric and exhibit shifts and broadening approximately linear in the $^3$He pressure. In contrast to $\sigma_o$ light which has a typical range of 82 $p$ $\mu$m at [Rb] = $4 \times 10^{14}$ cm$^{-3}$, $\sigma_\pm$ light has a much longer range in polarized Rb since it can only be absorbed by the depleted magnetic substate. The transmission scans with circularly polarized light determine the spin destruction rate per Rb atom, $\Gamma_{SD}$, and an approximately 40 $\mu$m thick unpolarized Rb layer at the glass boundary.

The strong increase of $\Gamma_{SD}$ versus $^3$He pressure (see Fig. 2) was unexpected and implies that Rb spin is lost not only by Rb–Rb collisions (constant contribution to $\Gamma_{SD}$) but also in 3-body collisions of Rb and two $^3$He atoms (quadratic contribution). The 3-body collisions may lead to the formation of Rb-$^3$He molecules, with the third partner required to carry away the molecular binding energy. It is our conjecture that Rb spin is lost by the spin-rotation interaction between the Rb atomic spin and the rotational angular momentum of the molecule. The molecules are expected to have a short lifetime because of the high binary collision rate ($\approx 2\ p\ 10^{10}$ Hz), however a sizeable spin rotation angle may still accumulate stochastically between absorption of subsequent photons. The strong increase of $\Gamma_{SD}$ versus $p$ makes it unfavorable to optically pump Rb at high $^3$He pressures. The lower limit for the laser power required on resonance to obtain 96% Rb polarization, at [Rb] = $4 \times 10^{14}$ cm$^{-3}$, and for various cell geometries and $^3$He pressures, is shown in Fig. 3.

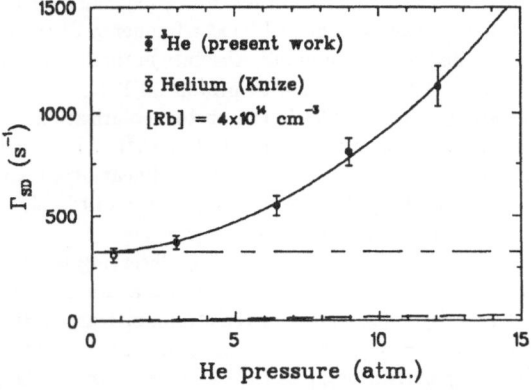

Fig. 2. Dependence of the Rb spin destruction rate on the $^3$He pressure. The solid curve represents a fit to the total measured spin destruction rate, the dotted line is the assumed constant Rb-Rb contribution, the dashed line is the spin destruction calculated for Rb–$^3$He spin exchange.

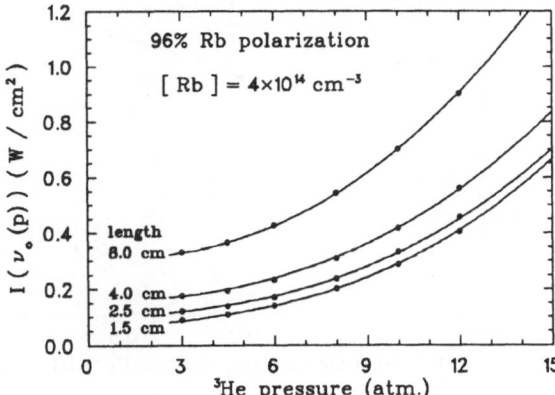

Fig. 3. Calculations of the laser power required to obtain 96% average Rb polarization in cells of various thicknesses. The curves represent an empirical fit.

The rate of Rb-$^3$He spin exchange was determined from the decay of $^3$He polarization at various Rb densities following laser irradiation for at least 12 hours. At [Rb] = 4 × 10$^{14}$ cm$^{-3}$ we find a rate of (10.6 hr)$^{-1}$ independent of $^3$He pressure. This corresponds to a velocity-averaged spin-exchange cross section, $< \sigma_{SE} v > = (6.1 \pm 0.2) \times 10^{-20}$ cm$^3$ s$^{-1}$.

Two different target sizes have been used. For achromatic proton beams typical cell volumes are 17 cm$^3$; the inner cell diameter of 1.7 cm is sufficient to avoid beam interactions with the side walls. For pion–induced reactions cell diameters are as large as 2.5 cm (35 cm$^3$ volume) although this does not prevent a small fraction of the pion beam from striking the cell walls. During proton experiments tracking wire chambers are used to separate events from the gas volume from those originating in the ≈.1 mm thick glass end windows. For pion experiments full vertex reconstruction is necessary using tracking wire chambers for both incoming beam and scattered particles.

## DETERMINATIONS OF THE $^3$He TARGET POLARIZATION

The bulk $^3$He polarization is analysed and reversed using the nuclear magnetic resonance (NMR) technique of adiabatic fast passage (AFP) at a frequency of 100 kHz. Our NMR setup is similar to the one described by Chupp et al.[1] Absolute normalization factors were obtained by comparing the $^3$He AFP NMR signals with proton AFP NMR signals from water-filled cells of the same dimension. Since at 293 K the proton polarization in water is 8.2 × 10$^{-9}$ the $^3$He signal is larger by a factor of ($p\, P(^3\text{He})$ 3.71 × 10$^4$). Thus for measurements of the absolute target polarization the NMR system has to be linear over six orders of magnitude. Furthermore, the temperature dependence of the impedance of the pickup coil circuit has to be accurately known between 293 K and 460 K.

We have recently tested the AFP NMR method by observing the reaction $^3\vec{\text{He}}(\vec{p}, \pi^+)^4\text{He}$ which samples the $^3$He polarization only in the beam interaction region. For reactions of the spin type $\frac{1}{2} + \frac{1}{2} \rightarrow 0 + 0$ parity conservation in the strong interaction implies the equalities[10] $A = A_{no} = \mp A_{on}$ and $A_{nn} = \mp 1$, where the first and second subscripts refer to beam polarization $p_b$ and target polarization $p_t = P(^3\text{He})$, respectively. The lower sign applies if there is a parity change in the reaction. At a proton energy of 416 MeV and a laboratory scattering angle of 28° we have observed a clean $\pi^+$ peak from the reaction and measured

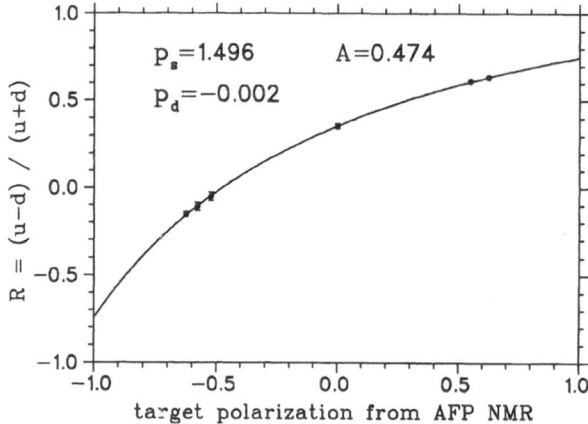

Fig. 4. Beam-related asymmetry R for the $^3$He(p,$\pi^+$)$^4$He reaction versus the target polarization determined by AFP NMR.

the beam-related analyzing power, $A = 0.474 \pm 0.018$, using an unpolarized target. Unknown target polarizations can then be determined by a single measurement of the beam-related yield asymmetry R = (u-d)/(u+d) via

$$p_t = \frac{2R - p_s A + p_d R A}{p_s - 2rA - p_d R}$$

where $p_s = p_b^u - p_b^d$ and $p_d = p_b^u + p_b^d$. The values of the beam-related asymmetries, R, measured during pump-up of the target are plotted in Fig. 4 at the values of the target polarization $p_t$ measured simultaneously by AFP NMR. The results for $p_t$ from the reaction method are found to be in good agreement with AFP NMR. The reaction method is more direct and less susceptible to systematic errors than the NMR method. The largest $^3$He polarizations observed during bench tests were 79% and 72% for cells at p=7 and p=9, respectively. Typical values during experiments are in the range from 0.45 to 0.70.

## NUCLEON KNOCKOUT FROM POLARIZED $^3$He

We have carried out (p,2p) and (p,pn) two-arm coincidence experiments at incident proton energies of 290 MeV[3] and, very recently, at 220 MeV[11]. Only results from the 290 MeV experiment will be reported here. Directions and momenta of both the scattered proton and the knock-on nucleons were determined. The leading proton was detected in a magnetic spectrometer (MRS), whereas the knock-on protons or neutrons were identified by two separate two-scintillator telescope arrays. The momentum of the knock-on nucleon was inferred from the nucleon time-of-flight.

If one assumes plane waves for all particles (PWIA) the reaction can be described as follows: the incident polarized proton (four-momentum $[k_o, \vec{k}]$, spin $s$, mass $M$) scatters off a bound nucleon ($[E(p), \vec{p}]$, $S$, $M$) in the polarized target ($[M_A, 0]$, $p_A$, $M_A$). Detection of the scattered proton ($[E(k'), \vec{k'}]$, $s_1$, $M$) determines the energy transfer $\omega = E(k) - E(k')$ and momentum transfer $\vec{q} = \vec{k}_o - \vec{k'_o}$ to the ejected nucleon ($[E(p'), \vec{p'}]$, $s_2$, $M$). Four-momentum conservation yields for the residual (A-1) system ($[E(P_{A-1}), \vec{p}_{A-1}] = [\omega + M_A - E(p'), \vec{q} - \vec{p'}]$). In the PWIA the recoil momentum $\vec{p}_{A-1}$ is a direct measure of $\vec{p}$, the initial momentum of the struck nucleon.

The kinematics were chosen to emphasize low momenta of the struck nucleon. The invariant missing energy, $E_m = E^2(P_{A-1}) - \vec{P}_{A-1}^2 + M - M_A$, is then confined to a narrow region corresponding to either a deuteron [in (p,2p)] or a slightly unbound two-nucleon pair. Since the time-of-flight resolution was insufficient to separate d and pn residual final states we have summed over missing masses in the time-of-flight spectra for each $\omega$ bin measured by the MRS. The comparison with the PWIA was then made using the closure approximation by replacing $E(P_{A-1})$ by $\{(\vec{P}_{A-1})^2 + \overline{M}_{A-1}^2\}^{1/2}$, where $\overline{M}_{A-1}$ was fixed at twice the nucleon mass. The spin dependent cross section can then be written[3,4]

$$\frac{d\sigma(s, S_A)}{d\Omega_{k'} d\Omega_{P'} d\omega} = \int \frac{M^4 \cdot k' \cdot p'^2 \cdot dp'}{(2\pi)^2 \cdot |\vec{k}| \cdot E \cdot E(P')}$$
$$\times \delta(E(k) + M_A - E(k') - E(p') - \{\vec{P}_{A-1}^2 + \overline{M}_{A-1}^2\}^{1/2})$$
$$\times \sum_{s_1', s_2'} \sum_{S, S'} M_{NN}^\dagger(p, S') \cdot M_{NN}(p, S) \cdot D^{SS'}(p) .$$

The "momentum distribution"

$$D^{SS'}(p) = \sum_{(A-1)\text{final-states}} g_{A-1,A}^\dagger(p, S') \cdot g_{A-1,A}(p, S) \qquad (1)$$

depends only on the target wave function, with the quantity $g_{A-1,A}(p, S)$ the probability amplitude for finding a nucleon in a plane wave state of momentum p and spin S with the rest of the nucleus in a state $S_{A-1}$ and invariant mass $\overline{M}_{A-1}$. The sum over final states includes all final state quantum numbers not otherwise indicated. The invariant nucleon-nucleon (NN) amplitude $M_{NN}(p,S)$ was constructed from the SM86 phase shift solution[12]. For the spin dependent momentum distribution the Afnan and Birrell[13] wave function was used as discussed in Ref. 4.

The experiment had to satisfy several, sometimes conflicting, requirements. The NN amplitudes had to be well known and large for both pp and pn scattering. This implied beam and target polarization normal to the scattering plane, and incident proton energies of 300 MeV or less. The angle for the leading proton ($\theta_1 = 27.5°$) had to be large enough to allow vertex reconstruction to the target, yet sufficiently small to retain a tolerably large NN cross section. The angles for the scintillator arrays ($\theta_2 = 58°$ and 71°) and the $\omega$ range of the MRS (50–120 MeV) had to encompass overlapping nucleon momenta ranging from 0–110 MeV/c for the 58° array, to 70–160 MeV/c for the 71° array.

For beam and target polarized normal to the scattering plane, the spin-dependent yield can be written as

$$\sigma = \sigma_o \left(1 + A_{no} p_b + A_{on} p_t + A_{nn} p_b p_t\right) \tag{2}$$

where the subscripts (bt) in $A_{no}$, $A_{on}$ and $A_{nn}$ refer to the direction of the projectile and target

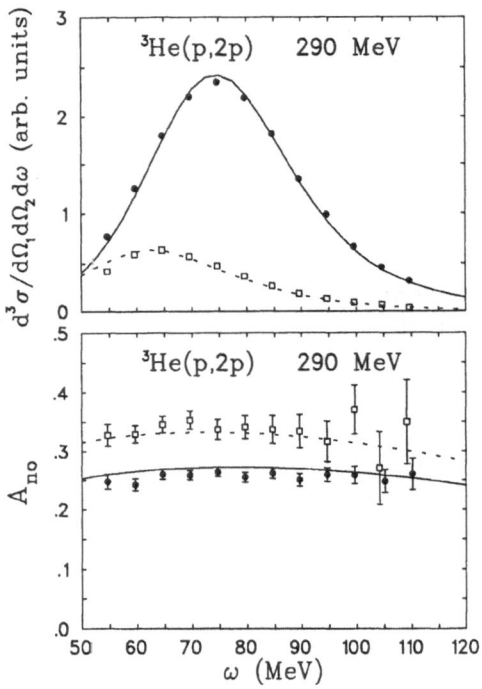

Fig. 5. Differential cross section (top) and beam-related analyzing power $A_{no}$ (bottom) for the $^3\vec{\text{He}}(\vec{p},2p)$ proton knockout reaction at 290 MeV and PWIA predictions. The filled circles and solid lines refer to recoil protons detected in the 58° array, the open squares and dotted lines to those in the 71° array.

polarization and $p_b$ and $p_t$ are the known polarizations of beam and target, respectively. The results from a least-squares fit of four (p,2p) yield combinations for opposite beam and target polarizations are shown in Figs. 5 and 6 together with PWIA calculations. Corrections for energy loss, spectrometer acceptance, random coincidences, and (negligible) background from $N_2$ in the target have been applied. The differential cross sections and beam-related analyzing powers $A_{no}$ in Fig. 5 are shown separately for the 58° and the 71° arrays used to detect the recoil protons. The same normalization factor was used for both the 58° and the 71° data. These data, together with previously obtained absolute spectral functions from (e,e′p)[14] and (p,2p) reactions[15] and spin correlation parameters $A_{nn}$ for the $pd \to ppn$ reaction[16] would seem to support the validity of the PWIA and the smallness of rescattering corrections provided $q < 160$ MeV/c.

The spin correlation parameter $A_{nn}$ shown in Fig. 6 is three times more sensitive to proton spin-momentum correlations than $A_{on}$, and is less susceptible to instrumental effects arising from the infrequent reversal of the target spin (several hours). The (p,2p) $A_{nn}$ results differ strongly from those for pp scattering ($A_{nn}(pp) \sim 0.82$) and reflect the spin-momentum correlation of protons in $^3$He. At $q = 0$ about 14% of the protons have their spin antiparallel to the $^3$He spin, whereas for protons with $q \sim 90$ MeV/c the momentum distributions $D^{SS'}(p)$ are spin independent. This is in good agreement with the PWIA calculations which, furthermore, show that the spin-momentum correlation effects are predominantly caused by the weak mixed-symmetry $S'$ states. Our measurement provides evidence that the $S'$ states contribute about 1.5% to the $^3$He wave function in agreement with Faddeev calculations, e.g. those of Afnan and Birrell[13].

Whereas the (p,2p) results are in reasonable agreement with the PWIA the (p,pn) results are not (see Fig. 7). The data sample for neutron knockout is only about 6% of that for proton knockout as a consequence of the three times smaller cross sections and of the poor neutron detection efficiencies of 15-20%. Although a few % proton contamination of the neutron sample could explain the observed discrepancy no such misidentified 'neutrons' were observed in elastic PP scattering runs which test the proton veto efficiency over about 30% of the active area of the 58° array. While no instrumental explanation has been found, a failure of the PWIA specific to the (p,pn) reaction seems surprising. In an effort to clarify the (p,pn) situation a new experiment[11] has been performed, with an order of magnitude improvement in rate per incident beam particle, with additional veto wire chambers in front of the scintillator array, and at 220 MeV where the sensitivity to neutron spin-momentum correlations is a factor of two larger than at 290 MeV. This experiment is presently being analyzed.

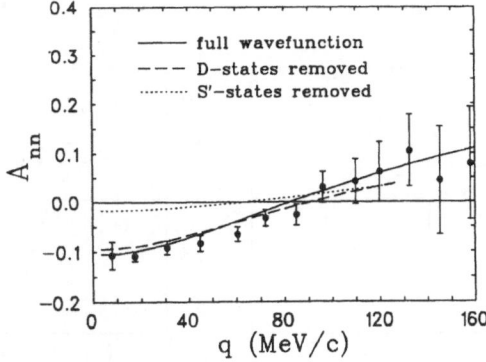

Fig. 6. Spin correlation parameter $A_{nn}$ for the $^3$He(p,2p) reaction at 290 MeV versus the momentum of the struck proton. The curves represent PWIA calculations referred to in the main text.

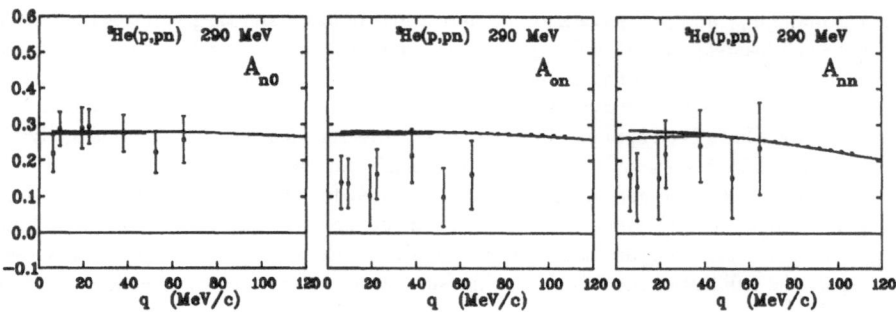

Fig. 7. Spin observables for (p,pn) neutron knockout from polarized $^3$He at 290 MeV. The curves are PWIA predictions.

## ELASTIC PROTON SCATTERING FROM POLARIZED $^3$He

Spin observables $A_{no}$, $A_{on}$ and $A_{nn}$ for proton elastic scattering have been measured at incident energies of 200, 290, 400 and 500 MeV. A preliminary summary of these results can be found elsewhere[17]. The results for 500 MeV are shown in Fig. 8.

Full four-body calculations of elastic p-$^3$He scattering at intermediate energies are not available at present. Landau and collaborators[18] have developed a momentum-space optical

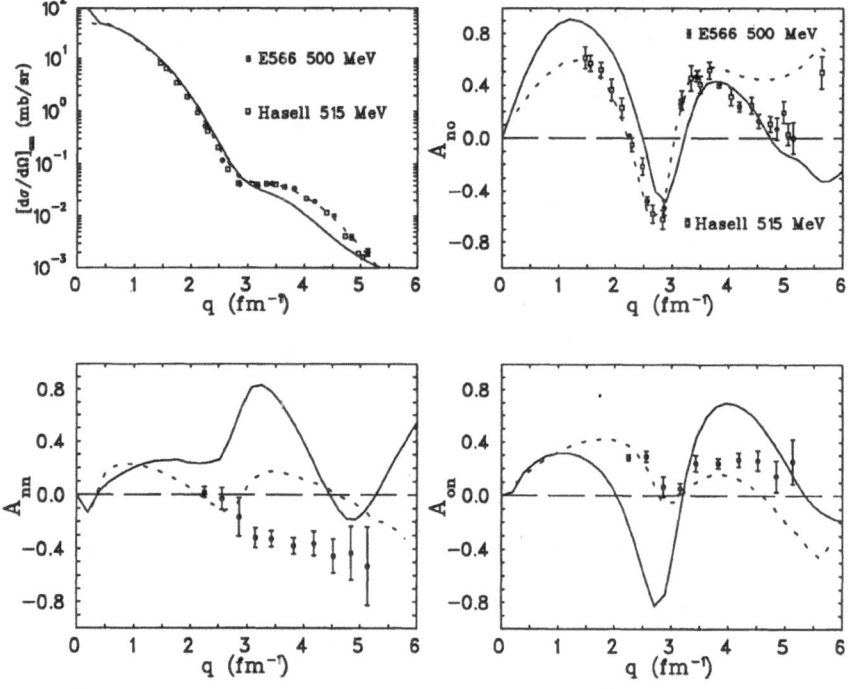

Fig. 8. Cross sections (top left), beam and target related analyzing powers $A_{no}$ (top right) and $A_{on}$ (lower left), and spin correlation parameter $A_{nn}$ (lower right) for $^3$He(p,p) elastic scattering at 500 MeV. The solid lines represent the parameter-free calculations of ref. 18, the dashed lines are from a preliminary calculation by L. Ray[20].

potential using a number of assumptions but without adjustable parameters (solid lines in Fig. 8). The dependence of the observables in the spin 1/2 × spin 1/2 system on a total of six complex amplitudes is as follows:

$$\sigma = (|a|^2 + |b|^2 + |c|^2 + |d|^2 + |e|^2 + |f|^2)/2$$
$$A_{no} = Re(a^*e + b^*f)/\sigma$$
$$A_{on} = Re(a^*e - b^*f)/\sigma$$
$$A_{nn} = (|a|^2 - |b|^2 - |c|^2 + |d|^2 + |e|^2 - |f|^2)/2\sigma$$

The importance of the $f$ term in the nuclear amplitude which is parity forbidden in the NN system is evident from the difference between $A_{no}$ and $A_{on}$ in the data of Fig. 8. The theoretical calculations[18] clearly fail to treat the dependence of these observables on the $b^*f$ term correctly. For $A_{nn}$ the calculations bear little resemblance to the data. The small measured values imply that the large amplitudes $a$, $b$, $e$, and $f$ are interfering destructively.

Very recently, preliminary DWBA calculations have been carried out by L. Ray[20] who used the relativistic impulse approximation (RIA, see Ref. 21; dashed lines in Fig. 8). Parameters in this calculation were adjusted to reproduce cross sections and analyzing powers. Although this calculation does better it appears that the new spin observables are sensitive to parts of the scattering amplitude not tested by $\sigma$ and $A_y$ measurements and that these parts are not calculated well enough.

## ELASTIC PION SCATTERING FROM POLARIZED $^3$He

The elastic scattering of pions from a spin 1/2 nucleus is described by two complex amplitudes $f(\theta)$ (non-spin-flip) and $g(\theta)$ (spin-flip). The differential cross section

$$\sigma = |f(\theta)|^2 + |g(\theta)|^2$$

corresponds to the incoherent sum of these amplitudes, whereas the asymmetry

$$A_y = 2Im[f(\theta)g^*(\theta)]/\sigma$$

is determined by the interference of the two amplitudes and is thus more sensitive to the weak spin-flip amplitude. Several momentum-space optical potentials have been constructed which imply substantial asymmetries near the cross section minima[22,23].

The first asymmetries in elastic pion scattering from polarized $^3$He were obtained recently at TRIUMF's M11 channel[24]. Wire chambers in the beam line allowed tracking of the 100 MeV $\pi^+$ beam at a rate of 5 MHz into the target cell. The scattered pions were momentum analyzed in the QQD magnetic spectrometer. Position information from two additional wire chambers between target and QQD resulted in satisfactory vertex reconstruction and in clean identification of scattering events originating inside the glass cell. Normalized momentum spectra observed at $\theta_{lab}$=80° for target spin up and down are shown in Fig. 9. The elastic $^3$He peak has a large analyzing power whereas the background to both sides of the peak which is ascribed to excited states of $N_2$ in the target has negligible analyzing power. The broad peak at large energy losses ($> 20$ MeV) has a contribution from quasielastic inclusive scattering from $^3$He.

The analyzing power at three angles is shown in Fig. 10 together with two optical model predictions by Landau[22]. The asymmetries are extraordinarily sensitive to the spin density in $^3$He derived from the magnetic form factor for $^3$He, $F_{mag}(q)$. The solid (dashed) lines correspond to the lower (upper) bound for fits of $F_{mag}(q)$ from early data by Collard et al.[25] It appears worthwhile to repeat the calculation[22] using recent more accurate data[26] for $F_{mag}(q)$ which appear to favor the lower bound (solid line) in agreement with the $A_y$ data. It would also be of interest to obtain $\pi^-$ data which are predicted to be much less sensitive to $F_{mag}(q)$.

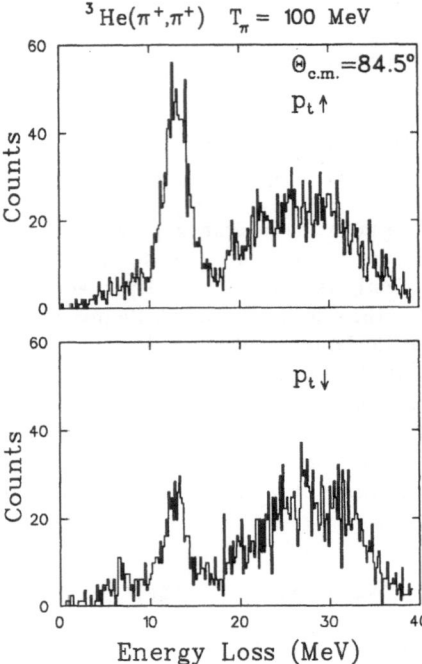

Fig. 9. Momentum spectra for elastic scattering of 100 MeV $\pi^+$ from $^3$He. The upper spectrum is for target spin up, the lower spectrum for target spin down. The smooth, continuous background is ascribed to inelastic scattering from $N_2$ and to quasielastic scattering from $^3$He.

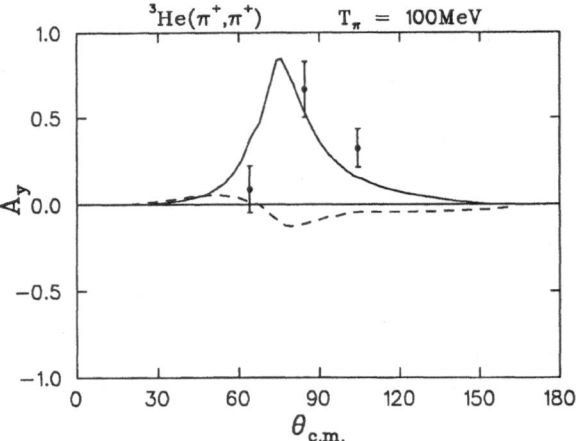

Fig. 10. Asymmetries for elastic scattering of 100 MeV $\pi^+$ from polarized $^3$He. The theoretical curves were calculated with a microscopic momentum space optical potential[22]. The solid and dashed curves are for lower and upper limits to fits of the magnetic form factor from an early experiment[25].

The analyzing power observed at $\theta_{c.m.} = 84.5°$ is much larger than any previously observed polarization effect in pion scattering from nuclei, i.e. from $^6$Li (Ref. 27), $^{13}$C (Ref. 28), or $^{15}$N (Ref. 27). As a very rough guideline, the maximum asymmetries appear to decrease with the number of nucleons in the valence shell.

## ACKNOWLEDGMENT

The experiments described here were the result of a team effort. Special thanks are due to P. Delheij, B. Larson, P. Levy, D. Thiessen and D. Whittal for their contributions to the target development, and to R. Woloshyn for theoretical support. Major contributions to the experiments and their analysis were also made by E. Brash, R. Henderson, A. Miller, A. Rahav, and M. Vetterli. This work is supported by grants from the National Science and Engineering Research Council of Canada.

## REFERENCES

1. T.E. Chupp et al., Phys. Rev. C36, 2244 (1987).

2. J.L. Friar et al., Phys. Rev. C42, 2310 (1991).

3. A. Rahav et al., to be published.

4. B. Blankleider and R.M. Woloshyn, Phys. Rev. C29, 538 (1984).

5. C.E. Woodward et al., Phys. Rev. Lett. 65, 698 (1990).

6. A.K. Tompson et al., preprint.

7. J. Ashman et al., Phys. Lett. B206, 364 (1988).

8. F.E. Close and A.W. Thomas, Phys. Lett. B212, 227 (1988).

9. B. Larson et al., submitted to Phys. Rev. A.

10. G.G. Ohlsen, Rep. Prog. Phys. 35, 717 (1972).

11. TRIUMF experiment E616; and E.J. Brash et al., to be published.

12. R.A. Arndt, and L.D. Roper, Scattering Analysis Dial-In (SAID) program, unpublished.

13. I.R. Afnan, and N.D. Birrell, Phys. Rev. C16, 823 (1977).

14. E. Jans et al., Nucl. Phys. A475, 687 (1987).

15. M.B. Epstein et al., Phys. Rev. C32, 967 (1985).

16. M. Garçon, Coll. de Physique 51, Suppl. C6-61 (1990).

17. O. Häusser, Coll. de Physique 51, Suppl. C6-99 (1990).

18. R.H. Landau, M. Sagen, and G. He, Phys. Rev. C41, 50 (1990).

19. D.K. Hasell et al., Phys. Rev. C34, 236 (1986).

20. L. Ray, private communication.

21. L. Ray et al., Phys. Rev. C37 1169 (1988).

22. R.H. Landau, Phys. Rev. C15, 2127 (1977); and R.H. Landau, Int. Workshop on Pion Single Charge Exchange, Los Alamos, 1979, p. 150 ff.

23. F.M.M. Geffen et al., Nucl. Phys. A468, 683 (1987).

24. B. Larson et al., to be published.

25. H. Collard et al., Phys. Rev. 138, B57 (1965).

26. P.C. Dunn et al., Phys. Rev. 27, 71 (1983); D. Beck et al., Phys. Rev. Lett. 59, 1537 (1987).

27. R. Tacik et al., Phys. Rev. Lett. 63, 1784 (1989).

28. Yi-Fen Yen et al., Phys. Rev. Lett. 66, 1959 (1991).

# POLARIZATION TRANSFER IN (p,n) REACTIONS AT 495 MeV

T.N. Taddeucci

Los Alamos National Laboratory
Los Alamos, NM 87545

## ABSTRACT

Polarization transfer observables have been measured with the NTOF facility at LAMPF for (p,n) reactions at 495 MeV. Measurements of the longitudinal polarization transfer parameter $D_{LL}$ for transitions to discrete states at 0° show convincing evidence for tensor interaction effects. Complete sets of polarization transfer observables have been measured for quasifree (p,n) reactions on $^2$H, $^{12}$C, and $^{40}$Ca at a scattering angle of 18°. These measurements show no evidence for an enhancement in the isovector spin longitudinal response.

## INTRODUCTION

The spin response probed by quasifree (p,n) reactions is of particular interest because of its relationship to the strength and momentum-transfer dependence of the residual isovector particle-hole interaction. Collectivity induced by this interaction is expected to produce significant differences between the isovector spin longitudinal ($\vec{\sigma} \cdot \vec{q}$) and spin transverse ($\vec{\sigma} \times \vec{q}$) responses at a momentum transfer of about 1.75 fm$^{-1}$.[1] In principle, the two responses can be experimentally distinguished by measuring complete sets of polarization transfer (PT) observables.[2]

The first measurement of the collective spin response induced by proton scattering involved (p,p') quasifree scattering at 500 MeV and 18.5°.[2] This measurement found no evidence for the expected enhancement of the longitudinal spin response with respect to the transverse spin response.

However, a lingering source of uncertainty in interpreting the data is the mixed isoscalar and isovector nature of the (p,p') reaction. Ideally, this uncertainty is removed by repeating the experiment using the pure isovector (p,n) reaction.

Even with a pure isovector probe, however, additional uncertainties remain. One effect of great interest is the nature of the nucleon-nucleon interaction when embedded in the nuclear medium. Horowitz, Murdock, and Iqbal have used a Fermi-gas model of the nucleus to calculate the signatures for relativistic modifications of the interaction in quasifree scattering.[3] These signatures involve differences between the PT observables for quasifree scattering and those for free scattering. Collectivity in the nuclear response will also alter the PT observables, however. An obvious difficulty, then, is deciding whether to attribute observed differences between free and quasifree observables to a medium modification of the interaction or to collectivity in the nuclear response.

An alternate strategy for investigating the effective interaction is to make use of transitions to discrete states as a nuclear filter to isolate specific components of the interaction. In many cases spin and momentum transfer are the only important parameters involved and simple expressions can be derived for PT observables in terms of basic nucleon-nucleon amplitudes.[4]

In this article I will present new polarization transfer data for both discrete (p,n) transitions and for quasifree (p,n) scattering for a bombarding energy of 495 MeV. The data for discrete transitions involve the longitudinal polarization transfer parameter $D_{LL}(0°)$. For quasifree scattering a complete set of PT observables has been obtained for $^2$H, C, and Ca. This latter set of measurements was obtained at the same bombarding energy and momentum transfer as the earlier (p,p') data[2] and provides the first look at the purely isovector spin responses induced by nucleon scattering.

EXPERIMENTAL TECHNIQUE

The data presented here were obtained with the Neutron Time-of-Flight Facility (NTOF) at the Clinton P. Anderson Meson Physics Facility (LAMPF) in Los Alamos. The NTOF facility has been described briefly in previous conference proceedings.[5] Cross section and analyzing power data for (p,n) reactions were first obtained with this facility in 1987 with polarized beam provided by a Lamb-shift source. The recent commissioning of a new optically-pumped polarized ion source[6] (OPPIS) that can provide intense (100 nA) chopped (200 ns pulse spacing) polarized beam has made possible the first measurements of complete sets of polarization transfer observables for (p,n) reactions.

High – resolution measurements of polarization transfer in (p,n) reactions commenced at the Indiana University Cyclotron Facility (IUCF) in 1982, and many of the techniques developed there[7] have been scaled up and applied to the NTOF facility at LAMPF. The NTOF detector/polarimeter is illustrated schematically in Fig. 1. The detector consists of four parallel "planes" oriented perpendicular to the incident neutron flux: three stainless steel tanks filled with liquid scintillator (BC–517s, H:C=1.7) and a fourth set of ten plastic scintillators (BC–408). The liquid scintillator tanks are each subdivided into ten optically-isolated cells with dimensions of 10 cm × 10 cm × 107 cm. The plastic scintillator cells have the same dimensions. Thin plastic scintillators in front of and between the front and back pairs of neutron detectors are used to tag charged particles.

The front pair of planes serve as neutron polarization analyzers. Time, position, and pulse-height information from front and back planes are used to kinematically select n+p interactions. Neutron polarization is determined from the azimuthal intensity distribution of these events. Elastic $^1$H(n,n) and charge-exchange $^1$H(n,p) events are identified and sorted separately. Because of energy-resolution limitations in the interplane timing, quasifree n+C events also contribute to both of these reaction channels.

Incident neutron energy is determined by time-of-flight (TOF) to the front detector planes with respect to an rf stop signal derived from the linac. The neutron flight path varies according to resolution and count rate requirements. The range of possible values is 170 m to 620 m. Calibration measurements employing $^{14}$C were made on a 400 m-flight path with an average energy resolution of about 0.75 MeV at a bombarding energy of 495 MeV. The best resolution obtained at this flight path was 0.6 MeV. The measurements of quasifree polarization transfer were made on a flight path of 200 m. In both cases time and energy spread in the beam were minimized with a rebunching technique that employs nonaccelerating rf cavities in the linac.[8]

The beam intensity and polarization for the present measurements were typically 70 nA and 55%. Nominal beam energy was 495 MeV. With the thick targets (1 g/cm$^2$) used in the quasifree measurements, approximately one day

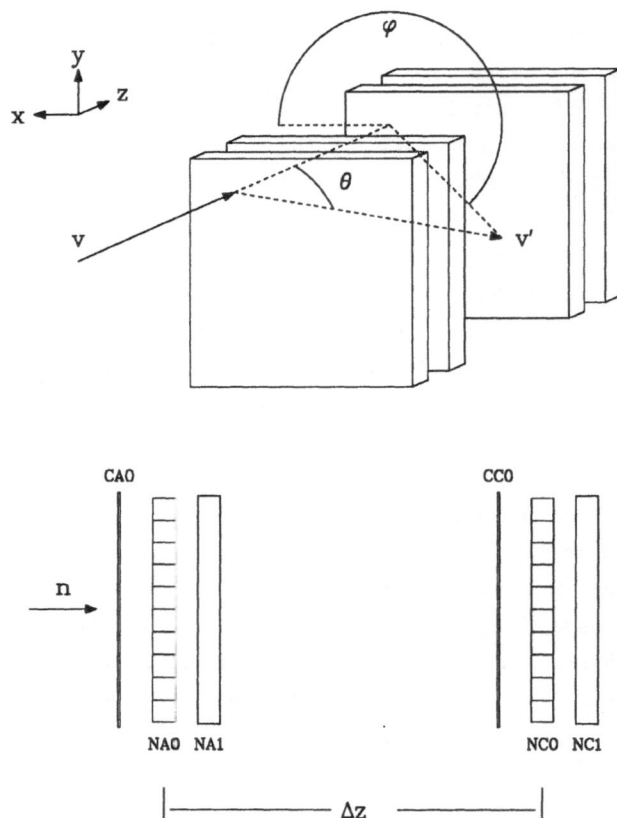

Fig. 1.  The NTOF neutron polarimeter.  The front detector planes
(NA0,NA1) serve as polarization analyzers and are separated from
the two back planes (NC0,NC1) by an average separation of
$\Delta z=1.7$ m.  Two sets of thin scintillators (CA0,CC0) are used to
tag charged particles.

of beam on target was required for each incident polarization state per target.

The effective analyzing power for each reaction channel in the polarimeter has been determined by observing neutrons produced by the $^{14}C(p,n)^{14}N(2.31-MeV)$ reaction. For this $0^+ \rightarrow 0^+$ transition, outgoing neutrons at a scattering angle of $0°$ have the same polarization as the incident proton beam. A spectrum for the $^{14}C(p,n)$ reaction at 495 MeV is shown in Fig. 2. This measurement employed longitudinally polarized beam. Horizontal and vertical dipole magnets were used to precess the outgoing L-type polarization into equal N and S components at the detector.

Two different precession schemes are used to measure polarization transfer observables at nonzero angles. For sideways (S) and longitudinally (L) polarized beams, a horizontal dipole field is used to precess the

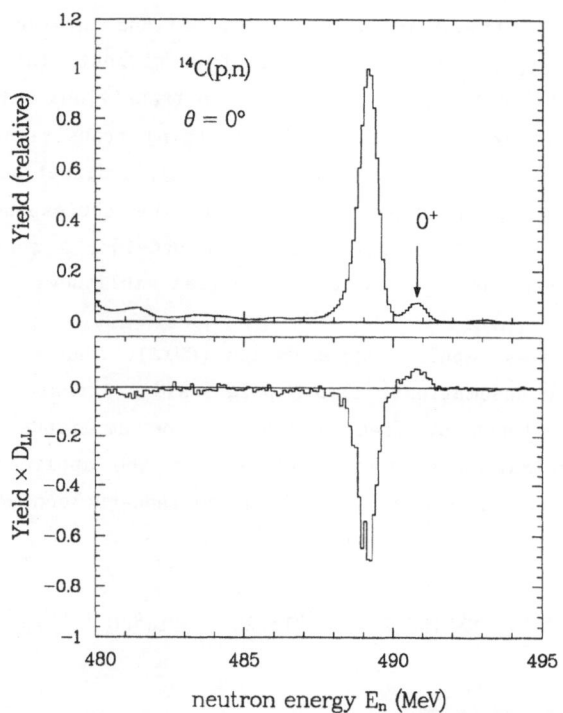

Fig. 2.    Spectrum for the $^{14}C(p,n)$ reaction at 495 MeV. Neutrons from the transition to the $0^+$ isobaric analog state at 2.31 MeV are used to calibrate the polarimeter.

outgoing L component into N (normal to reaction plane) polarization at the detector. The outgoing S polarization is unaffected by this field, while the induced N-type polarization is precessed into L polarization at the detector. Because the induced N-type polarization is now unobservable, reversal of the proton polarization also reverses the observable components of neutron polarization incident on the detector. This reversal allows cancellation of instrumental asymmetries. For measurements with N-type beam, the horizontal dipole field is reduced for minimum precession effect (it cannot be turned off because this magnet also functions as a sweep magnet) and a superconducting solenoid is used to precess the outgoing N type polarization alternately by ±90°. These reversals are again necessary for cancellation of instrumental asymmetries.

ZERO-DEGREE SCATTERING AND THE TENSOR EXCHANGE INTERACTION

At sufficiently high bombarding energy ($E_p$>100 MeV) and at low momentum transfer ($\theta$=0°), the spectrum of final states excited in (p,n) reactions is dominated by $\Delta J^\pi$=1$^+$ transitions. These transitions involve the same matrix elements that apply in beta decay, but without the binding-energy restrictions that limit actual beta decays to transitions between only a few low-lying levels. To the extent that the (p,n) transitions are pure $\Delta L$=0 and are mediated by the central interaction only, a direct connection can be made between the (p,n) cross sections and the corresponding beta-decay transition strengths. Polarization transfer provides a sensitive means of testing for the presence of $\Delta L$>0 and noncentral amplitudes.

The plane-waves impulse approximation (PWIA), coupled with simplifying nuclear-structure assumptions, allows polarization transfer coefficients to be expressed in terms of free nucleon-nucleon amplitudes.[4] Much of the utility of polarization transfer derives from the applicability of these simple expressions. A common form for the nucleon-nucleon scattering matrix is

$$M = A + C(\sigma_{1n}+\sigma_{2n}) + B\sigma_{1n}\sigma_{2n} + E\sigma_{1q}\sigma_{2q} + F\sigma_{1p}\sigma_{2p} \tag{1}$$

or, after rearranging terms

$$M = A + C(\sigma_{1n}+\sigma_{2n}) + \frac{1}{3}(B+E+F)\vec{\sigma}_1\cdot\vec{\sigma}_2 + \frac{1}{3}(E-B)S_{12}(\hat{q}) + \frac{1}{3}(F-B)S_{12}(\hat{p}). \tag{2}$$

Here, the coordinates $\hat{q} = \hat{k}_f - \hat{k}_i$, $\hat{n} = \hat{k}_i \times \hat{k}_f$, and $\hat{p} = \hat{q} \times \hat{n}$ are defined in terms of the initial and final momenta $k_i$ and $k_f$; $S_{12}$ is the tensor operator. The coefficients in Eqs. (1) and (2) can be subdivided into isoscalar and isovector components. In the following discussion, isovector terms are assumed.

The first four combinations of amplitudes in Eq. (2) can be identified with the familiar t-matrix interaction amplitudes $V_\tau$, $V_{LS\tau}$, $V_{\sigma\tau}$, and $V_{T\tau}$. The fifth term in Eq. (2) has no direct analog in local t-matrix parametrizations; it is, instead, approximated by the tensor **exchange** amplitude.

At a scattering angle of $0°$ there are only two unique polarization transfer coefficients: $D_{NN}$ and $D_{LL}$. Furthermore, if the spin and parity of the transition is well-defined these two coefficients are not independent but are related by the expression

$$D_{NN} = \pm \frac{1}{2}(1 + D_{LL}), \qquad (3)$$

where the plus sign applies for natural parity transitions ($0^+, 1^-, \ldots$) and the minus sign applies for unnatural parity transitions ($0^-, 1^+, \ldots$). The N–N scattering matrix simplifies at $\theta = 0°$, where E=B and C=0. The general expression for the polarization transfer coefficient $D_{NN}$ for pure $\Delta L = 0$, $1^+$ transitions then becomes[4]

$$D_{NN}(0°, 1^+) = \frac{-F^2}{2B^2 + F^2}. \qquad (4)$$

If the exchange tensor amplitude is negligible, B=F, then

$$D_{NN}(0°, 1^+, \text{central}) = -\frac{1}{3}. \qquad (5)$$

Deviations from this "ideal" value therefore signal the presence of noncentral or $\Delta L > 0$ amplitudes. Many $1^+$ transitions have been observed at IUCF in the energy range 120-200 MeV. Individual data points are displayed in Fig. 6 of Ref. 9. The average empirical value for $1^+$ transitions in this energy range is $D_{NN}(0°) = -0.33 \pm 0.05$.

There is no fundamental reason for the exchange tensor amplitude to be negligible at small momentum transfer. Indeed, it is interesting to employ Eq. (4) and the free N-N amplitudes to map out the expected value of $D_{NN}$ as a function of bombarding energy. This is done in Fig. 3. The solid data points plotted in this figure represent measured values for the $^{14}C(p,n)^{14}N(3.95-MeV)$ transition. The 120 MeV and 160 MeV points ($D_{NN}$) were obtained at IUCF. The 495 MeV point ($D_{LL}$ converted to $D_{NN}$) was obtained recently at LAMPF. The thick hashed line represents the prediction of Eq. (4). Amplitudes were obtained from the SM89 phase-shift solution of Arndt.[11] The thin lines represent plane-waves calculations for several p-shell single-particle transitions in carbon, calculated with the code DWBA80[11] and the t-matrix interaction of Franey and Love.[12]

The t-matrix calculations and Eq. (4) both give the same general energy trend. In the IUCF energy range the t-matrix values are comfortably close to the measured values, while the N-N amplitude values are too negative. Near 500 MeV, the $^{14}C$ measurement is better described by the simple N-N amplitude expression. Comparisons such as those in Fig. 3 can emphasize specific deficiencies in the interaction parameters, in this case the tensor (exchange) interaction, or alternately, indicate energy regions in which simple impulse approximation expressions such as Eq. (4) are less accurate than reaction-model calculations that better incorporate the off-shell nature of the interaction in the nuclear medium.

The $^{2}H(p,n)2p$ reaction represents a case for which there is minimal nuclear-medium modification of the interaction. In this reaction, at 0° the two residual protons are forced by the Pauli principle to be in a relative S-state ($^{1}S_{0}$) when the momentum transfer is small. For small energy loss, the transition can therefore be characterized as $1^{+} \rightarrow 0^{+}$, and Eq. (4) will apply. Two measured values for the polarization transfer for this reaction (open boxes) are displayed in Fig. 3. The 160 MeV datum is a $D_{NN}$ measurement.[13] The 495-MeV datum is $D_{LL}$ converted to $D_{NN}$. Both of these data points correspond to a narrow interval (2-4 MeV wide) centered on the peak of the 0° distribution. The agreement between the $^{2}H(p,n)$ measurements and the phase-shift predictions is very good. The 0° spectrum for this reaction at 495 MeV is displayed in Fig. 4. This figure shows that the polarization transfer varies with energy loss. This dependence arises from the contribution of $\Delta L=1$ ($^{3}P_{0,1,2}$) amplitudes to the zero degree scattering. The narrow energy interval employed for the above comparison was deliberately chosen to minimize these contributions.

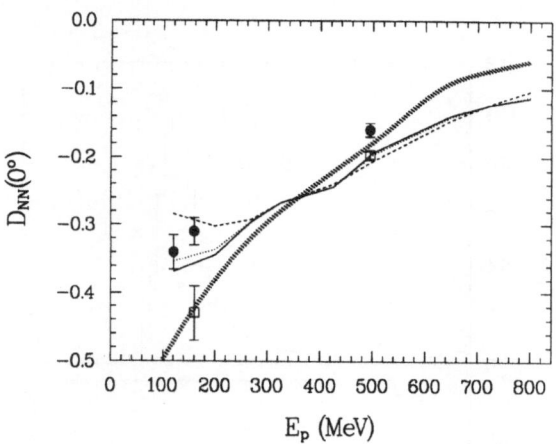

Fig. 3. Polarization transfer coefficient $D_{NN}(0°)$ for $\Delta J^\pi=1^+$ transitions. The solid data points are for the $^{14}C(p,n)^{14}N(3.95\text{-MeV})$ transition. The open boxes correspond to $^2H(p,n)$. The broad hashed line is calculated from N-N amplitudes and Eq. (4). The thin lines correspond to different single-particle transitions in carbon, calculated with a distorted-waves reaction code with the optical potential set to zero.

The data and calculations presented here show quite clearly that $\Delta J^{\pi}=1^+$ transitions do not have a unique energy-independent polarization transfer signature. Rather, the observed value of $D_{NN}(0°)$ [or $D_{LL}(0°)$] depends sensitively on the relative strengths of the central and tensor interactions. In the energy region accessible at IUCF (100–200 MeV) the effective tensor exchange contribution is very small in target nuclei heavier than $^2$H and $1^+$ transitions have a PT signature characteristic of a purely central interaction. At LAMPF energies (200–800 MeV), however, the tensor exchange interaction has a significant effect on PT observables at

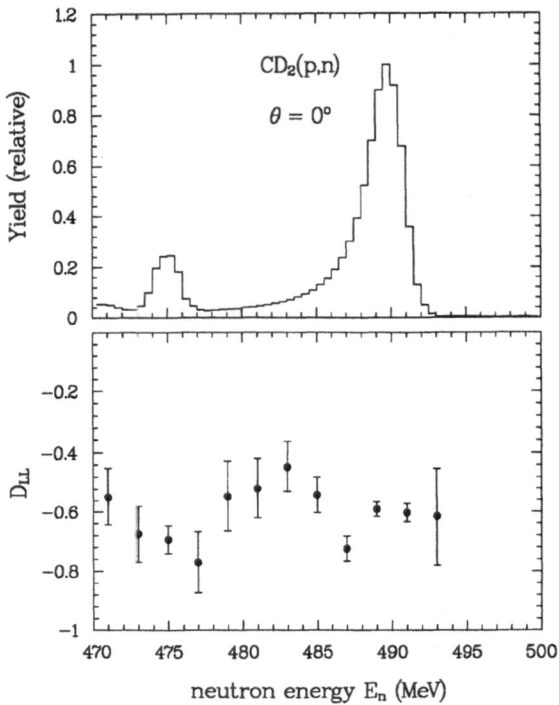

Fig. 4. Spectrum for $^2$H(p,n) at 495 MeV. The peak at $E_n$=475 MeV is the $^{12}$C(p,n)$^{12}$N(g.s.) transition from the carbon content of the $CD_2$ target.

0°. A related question that has not yet been systematically addressed is how much of an effect this interaction has on the 0° cross sections for the higher bombarding-energy region. The answer to this question has obvious implications with regard to the choice of bombarding energy for studying $1^+$ strength distributions.

In the simplest model of quasifree scattering the projectile nucleon undergoes a single hard collision with a target nucleon and ejects it into the continuum. The remaining A-1 target nucleons act as spectators and do not participate in the reaction. The observables for this process should look very much like those for free scattering, except for differences arising from the Fermi motion of the struck nucleon.

Absorptive corrections to this simple model localize the interaction region to near the nuclear surface. Additional corrections include nuclear-medium modification of the interaction between the projectile and target nucleon, and inclusion of the nuclear collective response. In the latter case the remaining A-1 target nucleons are not simple spectators (i.e., a free Fermi gas), but rather participate in the reaction through the influence of the residual particle-hole interaction.

The simplest observable that can be calculated in the simple model is the position of the quasifree "peak". Using nonrelativistic kinematics, it is easy to show that the energy loss $\omega$ of the projectile is given by

$$\omega = \frac{q^2}{2m} + \frac{\vec{p} \cdot \vec{q}}{m} + \frac{A}{A-1} \frac{p^2}{2m} - Q + \langle E_x \rangle \qquad (6)$$

where q is the momentum transfer, p is the Fermi momentum of the struck target nucleon, Q is the reaction Q-value for ejecting a nucleon into the continuum, and $\langle E_x \rangle$ is the average excitation energy of the residual A-1 nucleons. The first term is the energy loss for free scattering, the second term describes the width of the distribution arising from the Fermi motion of the struck nucleon, and the remaining terms represent a binding-energy shift in the position of this distribution. This shift should typically be of the order of 10-20 MeV. For example, the reaction Q-values for the ejection of one neutron or one proton from $^{12}$C are $Q_{pn}$ = -17.6 MeV and $Q_{pp}$ = -15.7 MeV, respectively.

A good example of the (p,n) quasifree distribution is presented in Fig. 5. This figure shows the spectrum for $^{12}$C(p,n) at $E_p$ = 795 MeV and several different scattering angles. As expected, the width of the distribution increases with angle (increasing momentum transfer) and the centroid of the distribution is shifted by about 25 MeV with respect to the energy loss for free scattering (marked by vertical dashed lines).

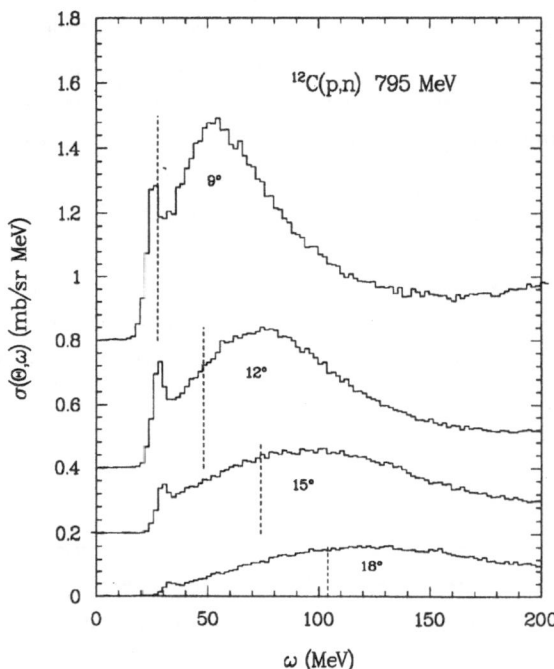

Fig. 5. Spectra for $^{12}C(p,n)$ at 795 MeV. The dashed vertical lines mark the energy loss for free scattering. Cross section normalization is uncertain by about 20%.

A similar picture for (p,p') quasifree scattering is presented in Fig. 6. Here the data are from the work of Chrien et al.,[14] and show $^{12}$C(p,p') spectra for $E_p$ = 795 MeV and several angles. An immediate difference is evident when compared to the (p,n) spectra in Fig. 5. The location of the (p,p') quasifree peak is very nearly consistent with the energy loss for free scattering, and therefore inconsistent with Eq. (6) and the (p,n) data. A possible explanation for this difference is collectivity in the isoscalar response excited by the (p,p') reaction. Collectivity in the isoscalar channel [absent in (p,n)] would enhance the response at low energy loss[15,16] and shift the centroid of the quasifree distribution toward low $\omega$. It should be noted that corrections for the isoscalar component in the (p,p') PT transfer data explicitly assumed no collectivity in the isoscalar channel.[2]

The simplest spin observable to measure is the analyzing power. An additional difference between (p,n) and (p,p') quasifree scattering can be seen in the data for this observable. Figure 7 presents cross section and analyzing power spectra for (p,n) quasifree scattering on $^2$H, $^{12}$C, and $^{40}$Ca

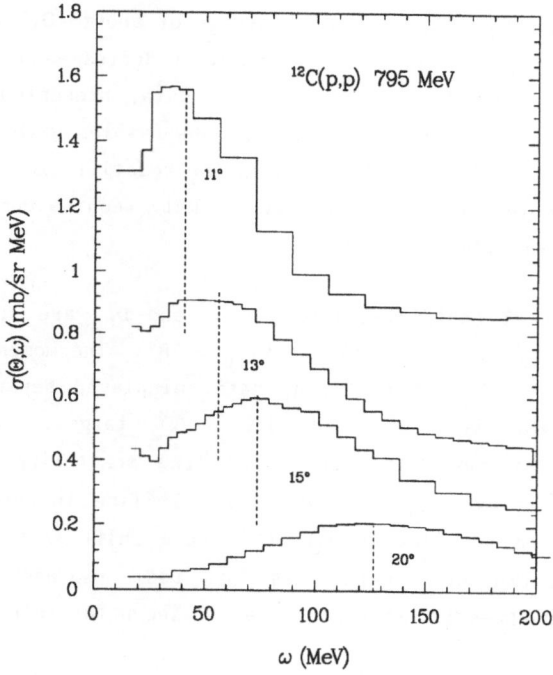

Fig. 6.  Spectra for $^{12}$C(p,p') at 795 MeV. The dashed vertical lines are positioned at the energy loss for free scattering. Data are from the tables in Ref. 14.

for $E_p$ = 495 MeV and $\theta_{lab}$ = 18°. The analyzing power for $^2$H(p,n) is consistent with the free value, which is indicated by the solid horizontal line. The $^{12}$C and $^{40}$Ca data are either consistent with or slightly larger than the free value.

It is well known that the analyzing power for (p,p') quasifree scattering is substantially reduced with respect to free scattering.[3,17] This reduction has been attributed to a sensitive cancellation between the large scalar and vector potentials that apply in relativistic models of this reaction.[3] For (p,n) quasifree scattering there is only a single dominant term in the relativistic parametrization of the interaction. According to whether this term is parametrized as a pseudoscalar or pseudovector invariant, the analyzing power for (p,n) is predicted to be slightly smaller or larger than the free value, respectively.[3] The data presented in Fig. 7 would seem to favor the pseudovector parametrization, but the differences are small enough that distortion effects must also be considered.

The relativistic effects predicted for PT observables for (p,n) quasifree scattering at 500 MeV and $\theta$ = 18° are rather small.[3] If the pseudovector parametrization is assumed, the largest differences will occur in the coefficients $D_{NN}$ and $D_{SS}$. An increase of about +0.1 with respect to the free value is predicted for $D_{NN}$, while a decrease of about the same amount is predicted for $D_{SS}$. The new (p,n) data, presented in Fig. 8, do not show any significant changes in the PT observables with respect to free scattering. Because of the ambiguities in the relativistic parametrization, a more conclusive test of this model will likely require comparison to data at several momentum transfers.

The diagonal PT coefficients $D_{SS}$, $D_{NN}$, and $D_{LL}$ are shown if Fig. 8. The data were obtained at 495 MeV and $\theta_{lab}$ = 18°. The momentum transfer is approximately 1.72 fm$^{-1}$. The $^2$H(p,n) data, displayed separately from the $^{12}$C and $^{40}$Ca data, were obtained with a $CD_2$ target. A careful $CD_2$-C subtraction yielded the $^2$H results. The free scattering value for each coefficient is displayed as a solid horizontal line in each panel. These values are obtained from the recent SM91 phase-shift solution of Arndt.[10] The $^2$H data should be regarded as the best representation of free scattering; the phase-shift values are merely shown for reference.

It is clear from Fig. 8 that there is little difference between the quasifree polarization transfer for $^{12}$C and $^{40}$Ca and that for free scattering, as represented by the $^2$H(p,n) data. With regard to collectivity in the nuclear spin response, the sought after signature is a ratio of

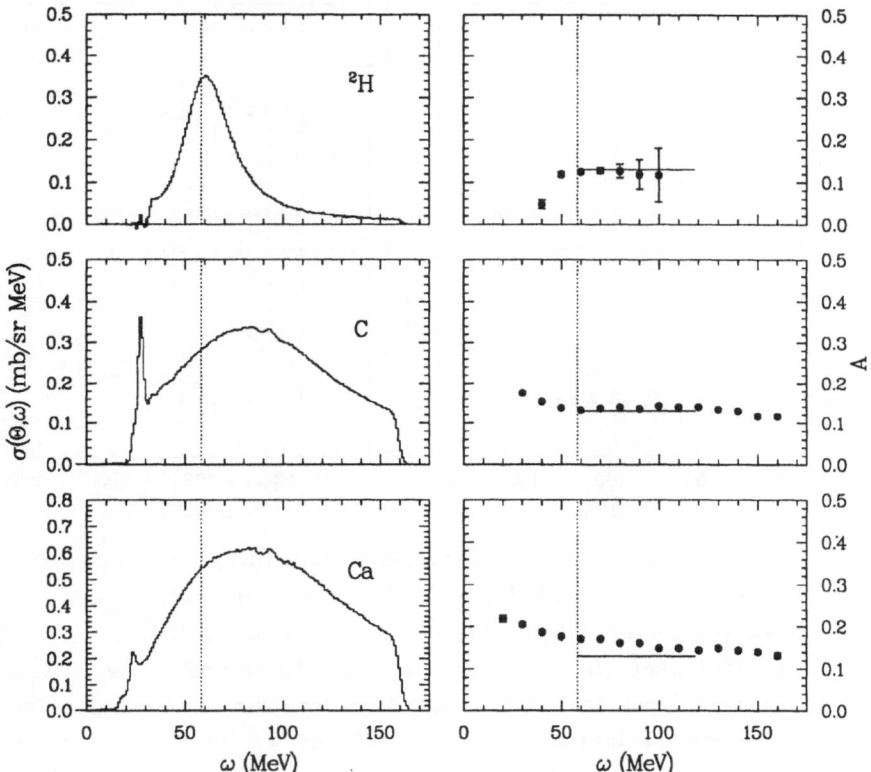

Fig. 7.   Cross section and analyzing power for (p,n) reactions on $^2$H, $^{12}$C, and $^{40}$Ca at 495 MeV and $\theta_{lab}$ = 18°.  The dashed vertical line in each panel marks the energy loss for free scattering.  The analyzing power for free scattering (from N-N phase shifts) is plotted as a solid horizontal line.  A vestige of the two-proton final-state interaction can be seen at approximately 31 MeV in the $^2$H spectrum.

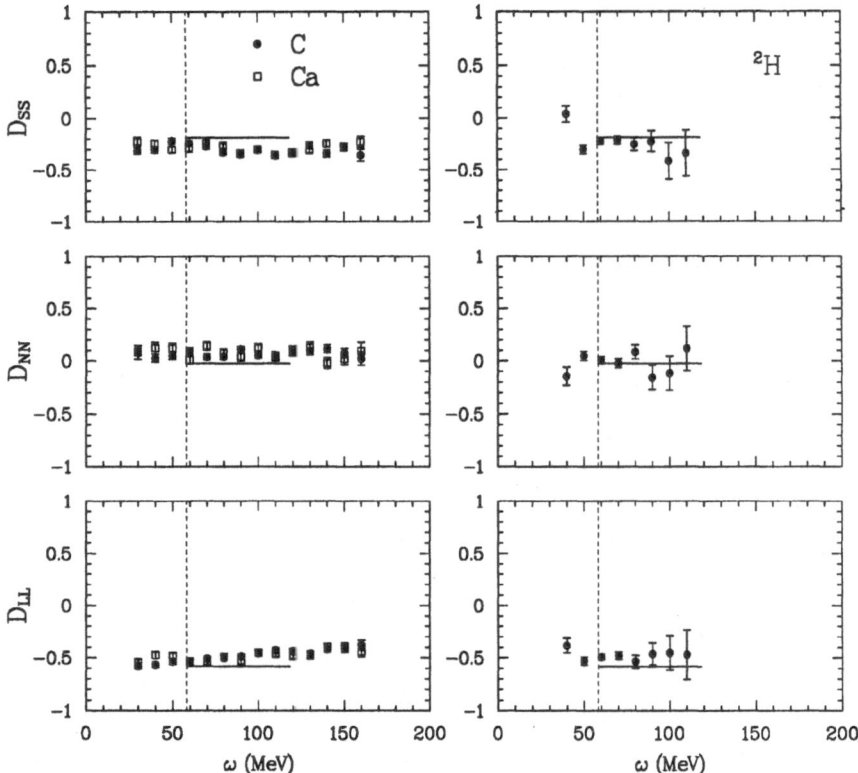

Fig. 8. Polarization transfer coefficients for (p,n) reactions on $^2$H (right) and $^{12}$C and $^{40}$Ca (left) at 495 MeV and $\Theta_{lab}$ = 18°. Free-scattering values obtained from the Arndt SM91 phase-shift solution (Ref. 10) are shown as solid horizontal lines. Dashed vertical lines mark the energy loss for free scattering. These data are preliminary and are subject to an energy-dependent normalization uncertainty of about 10%.

longitudinal to transverse spin-flip probabilities that is different from the ratio for free scattering. If the longitudinal response is collectively enhanced relative to the transverse response, this ratio should be larger than unity. The longitudinal and transverse spin-flip probabilities are defined by[2]

$$S_L = \frac{1}{4}[1 - D_{NN} + (D_{SS} - D_{LL})\sec\Theta_{lab}] \qquad (7)$$

and

$$S_T = \frac{1}{4}[1 - D_{NN} - (D_{SS} - D_{LL})\sec\Theta_{lab}]. \qquad (8)$$

For free scattering, the ratio of these two quantities gives the ratio of the longitudinal and transverse amplitudes [Eq. (1)],

$$\frac{S_L}{S_T} = \frac{E^2}{F^2}.$$  (9)

In the static PWIA, this ratio for quasifree scattering yields a ratio of effective nucleon-nucleon amplitudes times the ratio of longitudinal ($\vec{\sigma}\cdot\vec{q}$) and transverse ($\vec{\sigma}\times\vec{q}$) responses:

$$\frac{S_L}{S_T} = \frac{E^2}{F^2}\frac{R_L}{R_T}.$$  (10)

If the ratio of effective amplitudes is the same as the ratio of free amplitudes, then Eq. (10) divided by Eq. (9) yields the ratio of nuclear spin responses:

$$\frac{(S_L/S_T)_A}{(S_L/S_T)_D} = \frac{R_L}{R_T},$$  (11)

where A refers to the quasifree ratio for nuclide A and D refers to the deuterium ratio.

The ratio of longitudinal and transverse spinflip probabilities for deuterium is presented in Fig. 9. The solid horizontal line represents the free nucleon-nucleon value from the SM91 phase-shift solution. As an indication of how stable the phase-shift solutions have been, the SM86 value is plotted on this figure as a dotted line, and the SM90 value as a dashed line.

The super ratio defined in Eq. (11) is plotted for $^{12}$C and $^{40}$Ca in Fig. 10. The (p,n) results obtained here are very similar to the earlier (p,p') results. The ratio is everywhere consistent with unity or somewhat smaller. The expected longitudinal enhancement is not seen. The results for $^{12}$C and $^{40}$Ca are very similar and suggest that choice of target nuclide is not very important. This similarity is not too surprising in view of the surface-peaked nature of the reaction.[2]

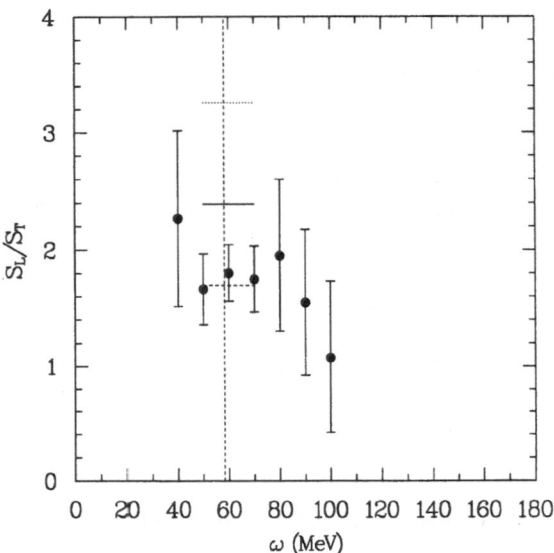

Fig. 9. Ratio of the longitudinal and transverse spinflip probabilities for $^2$H(p,n) at 495 MeV and $\Theta_{lab} = 18°$. The ratios obtained from N-N phase shifts (Ref. 10) are: SM86 (dotted), SM90 (dash), SM91 (solid).

Several effects may conspire to suppress the expected longitudinal/transverse enhancement as seen in the experimental ratio. In the surface region, decreased nuclear density and mixing between the longitudinal and transverse modes will both diminish the expected signature.

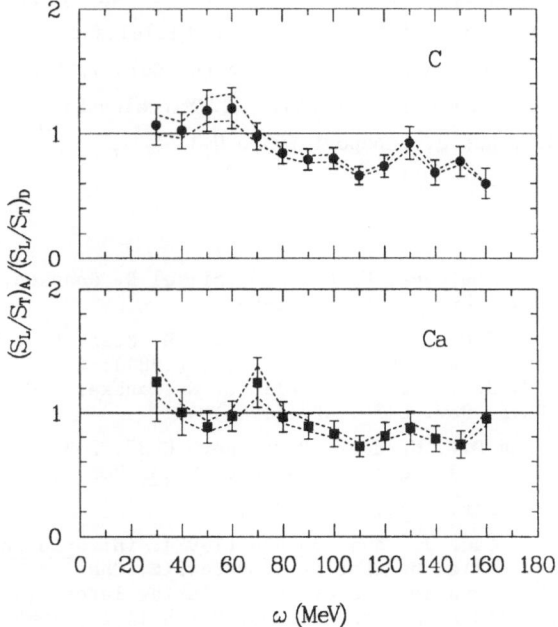

Fig. 10. Longitudinal/transverse spinflip ratios for carbon and calcium divided by the ratio for deuterium. This "super-ratio" should closely approximate the ratio of longitudinal and transverse spin responses. The dashed lines represent the error bounds on the data corresponding to a ±10% normalization uncertainty in the polarization transfer measurements.

Distortion effects must also be considered. What is needed is a calculation that combines both nuclear structure and reaction dynamics in a realistic way. Ichimura et al. have performed distorted-waves RPA calculations in a study of the (p,p') data,[18] and are currently producing new calculations for the (p,n) case.[19] The preliminary conclusion to be drawn from the new calculations is that the theoretical ratio is still too large in the low-$\omega$ region, and best consistency with the data is obtained when the RPA correlations are turned off. This is a surprising result, and may indicate a problem with the assumed form of the residual particle-hole interaction.

Additional experimental data at different momentum transfers will help to understand this problem. Such measurements are already planned.


ACKNOWLEDGEMENTS

Many people have contributed to the effort that produced the data in this article. They are: T.A. Carey, J.B. McClelland, and L.J. Rybarcyk (LANL), D. Mercer, X.Y. Chen, and D. Prout (U. Colo.), C.D. Goodman (IUCF), E. Gülmez and C.A. Whitten (UCLA), D. Marchlenski, B. Luther, and E. Sugarbaker (OSU), and J. Rapaport (Ohio U.)

REFERENCES

1. W.M. Alberico, A. De Pace, M. Ericson, Mikkel B. Johnson, and A. Molinari, Phys. Rev. C 38 109 (1988).

2. T.A. Carey, K.W. Jones, J.B. McClelland, J.M. Moss, L.B. Rees, N. Tanaka, and A.D. Bacher, Phys. Rev. Lett. 53, 144 (1984); L.B. Rees, J.M. Moss, T.A. Carey, K.W. Jones, J.B. McClelland, N. Tanaka, A.D. Bacher, and H. Esbensen, Phys. Rev. C 34, 627 (1986).

3. C.J. Horowitz and D.P. Murdock, Phys. Rev. C 37, 2032 (1988); C.J. Horowitz and M.J. Iqbal, Phys. Rev. C 33, 2059 (1986).

4. J.M. Moss, Phys. Rev. C 26, 727 (1982).

5. J.B. McClelland, Can. J. Phys. 65 633 (1987); in: Spin Observables of Nuclear Probes, edited by Charles J. Horowitz, Charles D. Goodman, and George E. Walker, Proceedings of the Telluride International Conference on Spin Observables of Nuclear Probes, March 14-17, 1988, Telluride, Colorado (Plenum, New York, 1988) pp. 183-193.

6. R.L. York, O.B. van Dyck, D.R. Swenson, D. Tupa, in Proceedings of the International Workshop on Polarized Ion Sources and Polarized Gas Jets, February 12-17, 1990, Tsukuba, Japan.

7. T.N. Taddeucci, C.D. Goodman, R.C. Byrd, T.A. Carey, D.J. Horen, J. Rapaport, and E. Sugarbaker, Nucl. Instrum. Methods A241, 448 (1985).

8. J.B. McClelland, D.A. Clark, J.L. Davis, R.C. Haight, R.W. Johnson, N.S.P. King, G.L. Morgan, L.J. Rybarcyk, J. Ullmann, P. Lisowski, W.R. Smythe, D.A. Lind, C.D. Zafiratos, and J. Rapaport, Nucl. Instrum. Methods A 276, 35 (1989).

9. T.N. Taddeucci, Can. J. Phys. 65, 557 (1987).

10. R.A. Arndt and L.D. Roper, Scattering Analyses Interactive Dial-in (SAID) program, Virginia Polytechnic Institute and State University (unpublished).

11. Program DWBA70, R. Schaeffer and J. Raynal (unpublished); extended version DW81 by J.R. Comfort (unpublished).

12. M.A. Franey and W.G. Love, Phys. Rev. C 31, 488 (1985).

13. H. Sakai, T.A. Carey, J.B. McClelland, T.N. Taddeucci, R.C. Byrd, C.D. Goodman, D. Krofcheck, L.J. Rybarcyk, E. Sugarbaker, A.J. Wagner, and J. Rapaport, Phys. Rev. C 35, 344 (1987).

14. R.E. Chrien, T.J. Krieger, R.J. Sutter, M. May, H. Palevsky, R.L. Stearns, T. Kozlowski, and T. Bauer, Phys. Rev. C 21, 1014 (1980).

15. G.F. Bertsch and O. Scholten, Phys. Rev. C 25 (1982) 804; H. Esbensen and G.F. Bertsch, Ann. Phys. 157, 255 (1984); H. Esbensen and G.F. Bertsch, Phys. Rev. C 32, 553 (1985).

16. R.D. Smith, Spin Observables of Nuclear Probes, edited by Charles J. Horowitz, Charles D. Goodman, and George E. Walker, Proceedings of the Telluride International Conference on Spin Observables of Nuclear Probes, March 14–17, 1988, Telluride, Colorado (Plenum, New York, 1988) pp. 15–51.

17. X.Y. Chen, L.W. Swenson, F. Farzanpay, D.K. McDaniels, Z. Tang, Z. Xu, D.M. Drake, I. Bergqvist, A. Brockstedt, F.E. Bertrand, D.J. Horen, J. Lisantti, K. Hicks, M. Vetterli, and M.J. Iqbal, Phys. Lett. B205, 436 (1988); O. Häusser, R. Abegg, R.G. Jeppesen, R. Sawafta, A. Celler, A. Green, R.L. Helmer, R. Henderson, K. Hicks, K.P. Jackson, J. Mildenberger, C.A. Miller, M.C. Vetterli, S. Yen, M.J. Iqbal, and R.D. Smith, Phys. Rev. Lett. 61, 822 (1988); R. Fergerson, J. McGill, C. Glashausser, K. Jones, S. Nanda, Sun Zuxun, M. Barlett, G. Hoffmann, J. Marshall, and J. McClelland, Phys. Rev. C 38, 2193 (1988).

18. M. Ichimura, K. Kawahigashi, T.S. Jorgensen, and C. Gaarde, Phys. Rev. C 39, 1446 (1989).

19. M. Ichimura, private communication.

# RELATIVISTIC EFFECTS ON SPIN OBSERVABLES

Charles J. Horowitz

Nuclear Theory Center
Indiana University
Bloomington, Indiana 47405

## ABSTRACT

For the last ten years there has been considerable work on relativistic descriptions of proton scattering. By now, many of the basic ideas have been elucidated with detailed calculations. Therefore it is a good time to review relativistic effects on proton nucleus spin observables.

First, a brief introduction is presented in Section I which motivates relativistic approaches. Next, Section II lists the basic relativistic ideas and provides a very short summary of their present status. Section III reviews the present status of the relativistic impulse approximation for elastic scattering. Section IV discusses quasielastic scattering. This provides a direct test of relativistic changes in the NN interaction. We consider both $(p, p')$ and $(p, n)$ reactions. The new $(p, n)$ data provide some of the first tests of relativistic effects on the isovector NN interaction. Conclusions are contained in Section V.

## I. INTRODUCTION

We motivate and characterize relativistic approaches by answering a few questions.

### Why Relativity?

There has long been a close connection between relativistic effects and spin. For example, relativistic effects give rise to the atomic spin-orbit potential. Presumably, the nuclear spin-orbit force (both one- and two-body) is also a relativistic effect. Furthermore, this spin-orbit force is very large. This suggests that relativistic effects could be important.

## Why is This Interesting?

Studying relativity in nuclei is interesting for several reasons. For example, a small relativistic effect seen in a detailed laboratory spin observable may have crucial implications for nuclear matter under extreme (clearly relativistic) conditions. These conditions may arise in astrophysics (early universe, supernovas, neutron stars) or in relativistic heavy ion collisions. Indeed, relativistic work on proton scattering has already had an impact on theoretical models of heavy ion collisions.

Furthermore, many high momentum transfer experiments are planed at CEBAF and elsewhere. These will require the development of relativistic reaction theories. It may be difficult and misleading to simply extend nonrelativistic reaction theories to describe momentum transfers greater than the nucleon mass.

## What does one mean by Relativity?

There is much confusion over what one means by relativistic effects. Most (so called) nonrelativistic models include relativistic kinematics. However, there is no such thing as just relativistic kinematics for strongly interacting particles. Instead one must always specify the relativistic *dynamics*. Often what one means by just relativistic kinematics is minimal relativity. Here, it is arbitrarily assumed that the nucleon optical potential transforms like (the time component of) a vector under a Lorentz boost. This assumption is not based on any calculations, but is made for simplicity.

However, a number of calculations suggest that the optical potential $U_{opt}$ has large Lorentz scalar $S$ and vector $V$ components,

$$U_{opt} = S + \gamma_0 V. \tag{1}$$

These strong potentials modify the four component nucleon spinors in the medium. The free positive energy spinors $U$ are replaced by $U^*$ spinors with enhanced lower components.

For uniform nuclear matter these spinors are,

$$U^*(p) = N \left[ \begin{matrix} 1 \\ \frac{\boldsymbol{\sigma} \cdot \boldsymbol{p}}{E^* + M^*} \end{matrix} \right] \chi_\lambda. \tag{2}$$

Here $\chi_\lambda$ is a two component Pauli spinor of spin projection $\lambda$ and $N$ is a normalization constant. The upper components are of order one while the lower components are of order the momentum $p$ over the effective mass $M^*$ ($E^* = \sqrt{p^2 + M^{*2}}$),

$$M^* = M + S. \tag{3}$$

Note, $M^*$ may be significantly smaller than the free mass $M$. *In this paper, relativistic effects refer to those effects from the change in four component Dirac spinors* $U \to U^*$.

## Why include negative energy states?

Given the substructure in a nucleon, there is debate over the role of negative energy states. First, the size of negative energy components depends on the basis. For example, Eq. (2) corresponds to a 100 percent positive energy state of a nucleon in the potential, Eq. (1), with energy $E = E^* + V > 0$. However, if one arbitrarily expands Eq. (2) in a *free* basis,

$$U^* = \alpha U + \beta V, \tag{4}$$

then the expansion coefficient $\beta$ for free negative energy spinors $V$ is nonzero.

These virtual (free) negative energy components are indicated pictorially in the "Z graph" of Fig. 1. People have argued that form factors at the vertices in Fig. 1 should greatly reduce the contributions of the Z graphs. However, the form factors are evaluated at very small momentum transfers where they are close to one (see below). [Of course, the intermediate negative energy state is quite far off mass shell as a result.] Therefore, the Z graph contributions should not be reduced just because the nucleon has substructure.

Indeed, negative energy states (or antiparticles) are added to a theory for symmetry reasons which are independent of whether the nucleon is "fundamental" or not. If one introduces a theory with effective coordinates then it is natural to also introduce "effective anticoordinates" to insure that the theory is Lorentz covariant and gauge invariant.

As an example, consider Compton scattering of a low energy photon from a proton as shown in Fig. 1. The amplitude for this process follows directly from gauge invariance and is simply proportional to the square of the proton's charge. However, this amplitude comes completely from the Z graph with a virtual negative energy state. If one neglects the Z graph by projecting onto only positive energy intermediate states then gauge invariance is violated and the amplitude will be zero.

Therefore, one must keep (at least some) negative energy states to insure gauge invariance. This is true for a meson-nucleon field theory. The only alternative to negative energy states is to give up locality and use a relativistic potential model. However, then one must deal with arbitrary and complicated many-body forces which are not simply related to physical processes such as particle exchange.

## II. BASIC RELATIVISTIC IDEAS

We now list some of the basic relativistic approaches and provide a few words as to their present status.

### A.) Relativistic Mean Field Theory

Several authors [1,2] have performed relativistic mean field calculations for the ground states of nuclei. Here, Dirac nucleons interact with classical meson fields. Most models have large (several hundred MeV) Lorentz scalar, $S$, and vector, $V$, meson fields. The strength of these fields follows directly from the known (at least in order of magnitude) meson-nucleon couplings. Relativistic effects in these fields

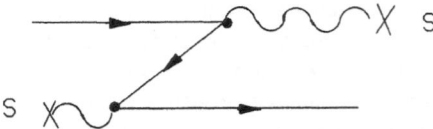

Figure 1a. Z graph contribution to nucleon-nucleus scattering.

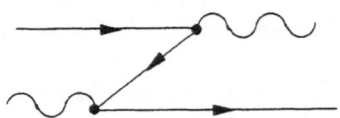

Figure 1b. Compton scattering from a proton for small momentum transfers.

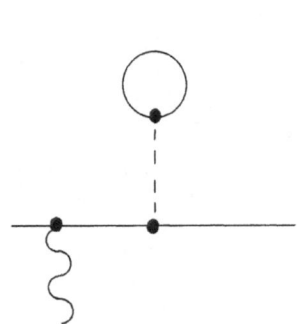

Figure 2a. Photon coupling to a valence nucleon in the strong field of the core nucleons.

Figure 2b. Core and vacuum polarization correction to Fig. 2a.

give rise to the single particle spin-orbit splittings of the shell model. Calculations vary in how they treat scalar meson self-interactions and negative energy Dirac sea nucleons.

**Present Status:** There are now many calculations for a variety of differently shaped nuclei (spherical, axially deformed, odd A ...) [3]. Agreement on a variety of ground state properties is often comparable to the best nonrelativistic mean field models. Thus, relativistic mean field theory (MFT) provides a good phenomenological description of nuclei.

**B.) Quantum Hadrodynamics (QHD).** Walecka, Serot and others have attempted to use a full relativistic quantum field theory (involving nucleon and meson coordinates) to calculate corrections to the MFT [4]. For example, relativistic Brueckner theory [5] calculates NN correlations and solves for the NN interaction in the medium. These correlation corrections reduce the strong scalar mean field by only about 10 percent, while $V$ is reduced by about 20 percent. Thus, the existence of strong fields is not changed by NN correlations. Furthermore, the change in spinor $U \rightarrow U^*$ reduces the scalar attraction in the NN interaction with respect to the vector repulsion. As a result, relativistic Brueckner calculations saturate nuclear matter at a lower density and binding energy than nonrelativistic calculations. (This is interesting because nonrelativistic calculations, with only two body potentials, may saturate at too high a density.) This basic change in the NN interaction leads to a very stiff equation of state for high density nuclear and neutron matter. Below, we search for more direct experimental evidence of this change.

**Present status:** Many corrections to the MFT have now been calculated. However, there are fundamental computational problems in dealing with a strong coupling quantum field theory (see for example [6]). These greatly limit our ability to make hard predictions from a given Lagrangian.

## C.) Dirac Phenomenology

Clark et al. [7], Cooper [8] and others have simply fit nucleon-nucleus elastic scattering data with a Dirac equation and appropriately parameterized (say Wood-Saxon) scalar and vector potentials. They obtain good fits to cross section, analyzing power and spin rotation data with strong S and V potentials.

**Present Status:** There now exist global fits of elastic data over large ranges of energy, E, and target mass, A [9]. Here the potentials are simple analytic functions of E and A. (One can contact E. D. Cooper for parameter sets [10].)

## D.) Relativistic Impulse Approximation (RIA)

The relativistic impulse approximation, folds a relativistic form of the NN interaction with target densities to produce an optical potential. This will be discussed in the next section.

## E.) Relativistic RPA or Linear Response

Several authors have realized that relativistic RPA or linear response theory is the correct way to calculate MFT predictions for a variety of reactions. Indeed, this was a hot topic at the 1988 Telluride conference [11]. Relativistic linear response theory is interesting because the probe can induce mixing between positive and negative energy states. As an example, consider the magnetic moment of a particle outside a closed shell. The strong scalar field from the core nucleons changes the effective mass, Eq. (3), of the valence particle as shown in Fig. 2a. As a result it has a higher velocity and convection current when probed by a photon. This effect modifies the old Schmidt magnetic moments and (alone) is a phenomenalogical disaster.

However, gauge invariance requires one to also consider Fig. 2b where the photon couples to a core nucleon which has been polarized by the valence particle. Core polarization is, of course, a well known feature of nonrelativistic nuclear structure. In addition, relativistic models involve an important mixing between particle- hole and particle-antiparticle excitations (i.e. between vacuum polarization and core polarization). Indeed, the Pauli blocking of vacuum polarization excitations (which are present in free space but no longer possible in the nucleus) are crucial for restoring the isoscalar Schmidt magnetic moments. Thus Fig. 2b cancels the enhancement from Fig. 2a. Both diagrams can be consistently included using relativistic RPA.

Relativistic RPA has several implications for nuclear structure calculations. First, one needs a detailed model of what gives rise to the strong S and V (in order to know how to 'attach' the probe). Second, self-consistency is very important. One must use the same interaction in the ground state and in RPA. Finally, one needs complete relativistic nuclear structure models because conventional nuclear structure effects such as core polarization mix with relativistic effects such as vacuum polarization. One can not build relativistic reaction models on top of nonrelativistic nuclear structure wave functions.

**Present Status:** There have been many relativistic RPA calculations for electron scattering. One now needs RPA calculations for proton scattering. However, these are somewhat complicated because of the relativistic proton distortions.

## III. THE RELATIVISTIC IMPULSE APPROXIMATION (RIA)

The relativistic impulse approximation (RIA) is do to McNeil, Shepard and Wallace [12]. In the RIA, a model of the NN amplitudes $\hat{F}$, is folded with mean field target densities to get an optical potential. In this section, we briefly list the present status of the RIA applied to elastic scattering. Quasielastic scattering is discussed in the next section.

**Present status for elastic scattering.** The RIA provides an excellent description of elastic spin observables both $A_y$ and $Q$ over a wide range of energies and nuclei. However, (a) elastic scattering only tests the isoscalar, Lorentz scalar and vector parts of the NN amplitudes. Relativistic effects in other amplitudes are essentially unconstrained.

(b) Relativistic multiple scattering theory does not exist. It was originally claimed that the RIA was the first term in a multiple scattering expansion. Indeed, one can try and construct such an expansion in analogy to the nonrelativistic version [13]. However, relativistic theories have qualitatively more coordinates than nonrelativistic theories since they involve both meson and particle-antiparticle production. Indeed, even the very first step in a multiple scattering expansion, dividing the hamiltonian into target and projectile pieces, is problematic. One might say the following: The target hamiltonian describes a nucleon interacting with an identical meson if and only if the meson "came from" a target rather than a projectile nucleon. This is nonsense. Therefore, the RIA is not the first term of any controlled approximation scheme. At this time, it is not clear what the RIA is an approximation to (aside from the data). Nevertheless, the RIA is still a very good phenomenology that relates NN to nucleon-nucleus observables.

(c) There exist important ambiguities in the relativistic form of the NN amplitudes. The RIA needs the matrix elements of $\hat{F}$ between $U^*$ spinors in the medium. However, only *free* spinor matrix elements of $\hat{F}$ are determined by NN phase shifts. There is no direct experimental information to determine how the matrix elements of $\hat{F}$ will change as the spinors $U \rightarrow U^*$ change. Originally, one arbitrarily postulated the Dirac operator form for $\hat{F}$ shown in Table 1. This form is simple and local. (There are no factors of momentum dotted into gamma matrices.)

Table 1.

| $\hat{F} = \sum_i F^i(q, E)\lambda^i_{(1)}\lambda^i_{(2)}$ | |
|:---:|:---:|
| $i$ | $\lambda^i$ |
| $S$ (scalar) | 1 |
| $V$ (vector) | $\gamma_\mu$ |
| $P$ (pseudoscalar) | $\gamma_5$ |
| $A$ (axial vector) | $\gamma_5\gamma_\mu$ |
| $T$ (tensor) | $\sigma_{\mu\nu}$ |

However, Table 1 is arbitrary. One can make other choices for $\hat{F}$ which will have the same free spinor matrix elements but different $U^*$ matrix elements. These different forms will lead to different nucleon-nucleus observables. The biggest ambiguity in $\hat{F}$ concerns the pseudoscalar invariant. Indeed, chiral symmetry and other arguments suggest that the $\gamma_5$ in Table 1 should be replaced by a pseudovector form $\gamma^\mu q_\mu \gamma_5/2M$. The exchange contribution of this pseudovector form will decrease the scalar and vector optical potentials at low energies, around 200 MeV (but not very much at 500 MeV or above) [14].

In addition to the pseudovector, there may be other important ambiguities in the form of $\hat{F}$. Furthermore, elastic scattering is only sensitive to the scalar and vector invariants. Therefore, many of these ambiguities could have gone undetected.

Wallace and Tjon [15] have calculated the full matrix structure of $\hat{F}$ within a relativistic one boson exchange model. This is a step in the correct direction and further work along these lines should be pursued. However, there is nothing sacred about a relativistic one boson exchange model. It may not be a good answer for the $U^*$ matrix elements of $\hat{F}$. After all one boson exchange models are only fit to the phase shifts with free $U$ spinors. Indeed, the correct answer for the matrix structure of $\hat{F}$ may not exist in nature. At best it can only be calculated within a model and there may be more than one 'correct' model.

**Example of RIA:** As an example of the RIA for elastic scattering we show results from Murdock et al. [14]. Here relativistic mean field densities [16] are folded with a simple direct plus exchange model of the NN amplitudes [17]. Next corrections from Pauli blocking (based on relativistic Brueckner results in nuclear matter) and a pseudovector form of the NN amplitudes are made. The calculations reproduce all measured spin observables for elastic scattering from closed shell nuclei at energies of 200 MeV and above. (Except for some very large angle $A_y$ data.)

All measured spin rotation function, Q, data for closed shell nuclei at 200 MeV are shown in Fig. 3. The calculation reproduces all of the data. Similar good results are obtained for the analyzing power. Note, Q at small angles in heavy nuclei (Zr and Pb) is sensitive to the details of the calculation. Also, there are "1/A" recoil corrections which are not treated correctly in the RIA. These may have a small effect in $^{16}O$.

This calculation essentially reduces to the original RIA for energies of 500 MeV and above (where the original RIA does very well). Interestingly, above 800 MeV there is little difference between the RIA and nonrelativistic descriptions. This is because the vector interaction dominates at high energy so there is no longer a sensitive cancellation between S and V. Finally, the computer codes for this calculation (including TIMORA.FOR for ground state densities) have now been published [18] and are available on request.

**Inelastic Scattering to Bound States:** There has been some work on relativistic descriptions of inelastic proton scattering in a distorted wave impulse approximation (RDWIA) [19]. The goals of this work are (a) to isolate changes in the NN interaction in the medium from relativistic effects and (b) to search for relativistic effects on nuclear structure.

The density dependence of the NN interaction is very interesting. It is just as "fundamental" as the density dependence of the single nucleon form factor. A measured density dependence will have important implications for the underlying QCD model of the NN interaction. Density dependence comes form (a) conventional sources such as Pauli blocking and binding energy corrections, (b) the change of spinors $U \rightarrow U^*$ and (c) other more exotic sources such as density dependent meson masses.

RDWIA calculations have shown that certain observables which are sensitive to nonlocalities, such as the difference between the polarization and analyzing power or those for $0^+$ to $0^-$ transitions, may show relativistic effects. However, inelastic calculations, to date, suffer from two problems. First, they use inconsistent nuclear wave functions. For example, many calculations have used nonrelativistic Cohen and Kurath particle-hole amplitudes. Instead one needs to perform consistent relativistic RPA calculations. Second, all of the calculations have used the simple form of the NN amplitudes shown in Table 1 or a similar direct plus exchange model [17]. Therefore, the inelastic results may suffer from ambiguities in the NN amplitudes.

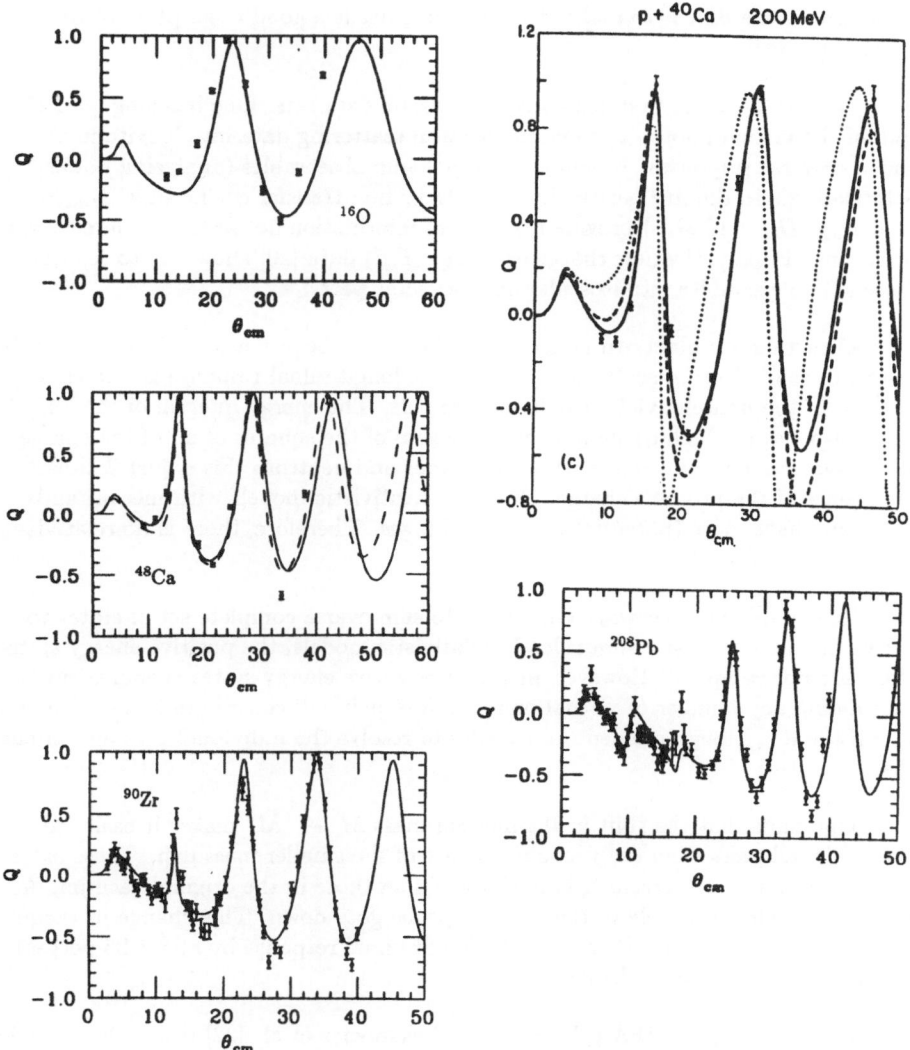

Figure 3. Spin rotation function for elastic proton scattering at 200 MeV. The solid curves are relativistic RIA calculations of ref [14], see also [18].

## IV. QUASIELASTIC SPIN OBSERVABLES

The quasielastic response is a fundamental property of nuclear matter. At these high excitation energies one nucleus looks like another. One is probing intrinsic properties of nuclear matter rather than details of the structure of a given nucleus. Furthermore, the reactive content of the RIA is single nucleon knockout. Therefore, if the RIA is to make any sense, one needs a good description of the quasielastic region.

We now have the first complete quasielastic data sets. One has long used the spin of the virtual photon to separate electron scattering data into longitudinal and transverse responses. In addition, $(\vec{p}, \vec{p}')$ spin observables (analyzing power, polarization and the five parity allowed polarization transfer coefficients: $D_{nn}$, $D_{ll'}$, $D_{ss'}$, $D_{ls'}$ and $D_{sl'}$) provide a wealth of information not accessible in electron scattering. Finally, the new charge exchange $(\vec{p}, \vec{n})$ data [20] allow one to separate these spin observables into isoscalar and isovector parts.

**Quasielastic electron scattering:** There has been much work on relativistic descriptions of electron scattering [21,22]. The longitudinal response is interesting because of the nonrelativistic Coulomb sum rule. The energy integral of the longitudinal response (at high q) should equal the sum of the squares of all of the charges in the system. For a nucleus with only protons and neutrons this is just Z times the square of the proton's charge. However, relativistic models with mesons and $N\bar{N}$ pairs have more (potentially visible) charges. Therefore, there is no relativistic Coulomb sum rule.

To say this another way: One needs to sum over a complete set of states to derive the nonrelativistic sum rule. In relativistic models, the positive energy states alone are not complete. However, including negative energy states is equivalent to considering any number of virtual pairs. A $N\bar{N}$ pair will contribute to the Coulomb sum if there is enough momentum transfer to resolve the individual plus and minus charges.

One finds that the shift in the nucleon mass $M \rightarrow M^*$ makes it easier to excite virtual pairs from the vacuum because of the smaller mass gap. These extra pairs do a better job screening bare charges than those in the original vacuum. As a result, the charge visible to the electron probe goes down. This change in vacuum polarization with $M \rightarrow M^*$ reduces the longitudinal response by about 25 percent in several relativistic RPA calculations.

For example, the RPA calculation of Piekarewicz et al. [22] treats the particle-hole response exactly with a full finite nucleus RPA. Vacuum polarization is treated in a local density approximation. The calculation consistently uses the same interaction for the mean field ground state and the RPA. As a result, current is explicitly conserved. The reduction in the Longitudinal response is in good agreement with Saclay data. Furthermore, this change in vacuum polarization is isoscalar ($M^*$ is the same for neutrons and protons). The transverse response, which is dominated by the isovector anomalous moment, is almost unchanged.

**Present status for electron scattering:** Because of computational difficulties with QHD it hard to improve on the RPA calculations of the vacuum response. Nevertheless, the negative energy states are crucial for the consistency of the theory (i.e. gauge invariance, completeness...). Therefore, it is bad to arbitrarily neglect the vacuum contributions.

**Quasielastic proton scattering:** The spin observables in proton scattering provide a direct test of the NN interaction in the medium. This is true for several reasons. First, there is a clear bench mark: one can directly compare to the free NN spin observables. Second, there are only small corrections from $L \cdot S$ distortions. (Of course, most of the effects from central distortions cancel in the spin observables.)

Likewise, corrections from Fermi motion averaging and multiple scattering are predicted to be small. This can be directly tested because there are a number of "null" spin observables which are measured to be close to their free values. If there were large corrections from multiple scattering, for example, it would take accidental cancellations to reproduce the free values. This provides an important check on the quasifree reaction mechanism.

If the reaction mechanism is under control, any observed changes in spin observables could be either a change in the NN interaction or a change in the nuclear structure. However, it may be possible to separate these two effects since they should have a different dependence on target mass and kinematics. Nuclear structure effects should be largest for heavy nuclei at relatively low excitation energies. For example, the analyzing power is predicted to be enhanced at low excitation energies because of nuclear structure effects. This is because the isoscalar response (which has a large analyzing power) is shifted down in RPA. In contrast, relativistic effects on the NN interaction are predicted to reduce the analyzing power for a large range of momentum transfers and excitation energies (see below). Experimental data for many nuclei and momentum transfers, clearly show both effects.

Unfortunately, there have not been many relativistic calculations for quasielastic proton scattering. Clearly, more calculations would be very useful. As an example, we show the plane wave impulse approximation calculations of Murdock et al. [23]. These calculations use a relativistic Fermi gas model for the nuclear structure. This should be a good approximation for momentum transfers, q, and energies E, such that E and $q^2/2M$ are significantly above giant resonance energies. However we emphasize, the Fermi gas model is a bad approximation for lower excitation energies or smaller momentum transfers.

The calculation includes the change in the spin dependence of the NN amplitudes when the spinors change,

$$\bar{U}_1 \bar{U}_2 \hat{F} U_1 U_2 \rightarrow \bar{U}_1^* \bar{U}_2^* \hat{F} U_1^* U_2^*. \tag{5}$$

Since the reaction is somewhat surfaced peaked, a simple eikonal estimate of the average density is used [23]. This yields a Fermi momentum of about $1\ \mathrm{Fm}^{-1}$. At this density, $M^*$ is about 0.8 to 0.9. Finally, old Arndt phase shifts SM86 from the summer of 1986 are used for the NN amplitudes. The pp phase shifts have not changed much since then. However, the np phase shifts have. Therefore, these old phases should be all right for $(p, p')$ but not for $(p, n)$.

Figure 4 shows results compared to the Carey et al. $(\vec{p}, \vec{p}')$ data on $^{40}$Ca at 500 MeV and a scattering angle of 18.5 degrees. The "nonrelativistic" curves

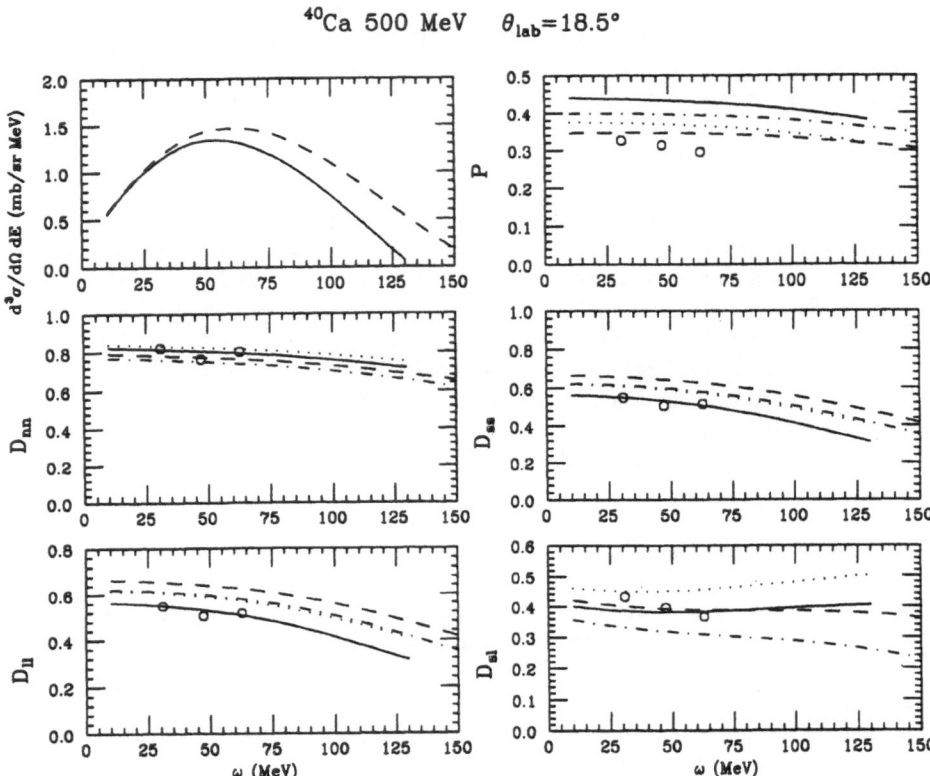

Figure 4. Relativistic plane wave impulse approximation calculations [23] for $(\vec{p}, \vec{p}')$ scattering from $^{40}$Ca at 500 MeV and 18.5 degrees. The dashed curves use relativistic $U^*$ spinors while the solid curves are for free $U$ spinors. The data is from ref. 24.

correspond to Fermi motion averaging over NN amplitudes in Eq. (5) evaluated with free spinors $U$. Fermi motion averaging is a small effect. Therefore, these curves are close to the free NN ones. The relativistic curves are for a calculation with $U^*$ spinors.

The data is quite striking. All of the polarization transfer coefficents are close to the free values and agree well with the nonrelativistic calculation. However, the polarization is substantially reduced from its free value. Taken together, this is strong evidence that something nontrivial is going on with the NN interaction in the medium. Standard sources of background such as multiple scattering or distortion effects would be expected to also change the $D_{ij'}$s.

The largest predicted relativistic effect is a reduction in the analyzing power or polarization. Furthermore, this reduction has now been seen over a large range of nuclei and energies. The analyzing power is dominated by the S and V amplitudes (Table 1). At small angles $A_y$ is proportional to

$$A_y \sim \mathrm{Im}(F_s F_v)/|M^* F_s + E^* F_v|^2. \tag{6}$$

Here the $M^*$ and $E^*$ factors are from appropriate matrix elements of the $U^*$ spinors. The denominator involves sensitive cancelations between the S and V amplitudes. As $M^*$ decreases the scalar attraction is reduced w.r.t. the vector repulsion. This increases the denominator and decreases $A_y$. This is directly related to the relativistic effect on nuclear matter saturation which was discussed in Section II. This reduction in $A_y$ may be the clearist relativistic signature found to date. At this time, there has been no alternative nonrelativistic explanation. All nonrelativistic calculations predict $A_y$ close to the free value.

Unfortunately, the polarization transfer coefficients involve all of the amplitudes in a more complicated way than Eq. 6. They depend on the axial, tensor and pseudoscalar invariants of Table 1 which have not been probed in elastic scattering. Relativistic effects are correctly predicted to be small for $D_{nn}$ and $D_{sl'}$. However, enhancements in $D_{ss'}$ and $D_{ll'}$ are predicted which are not seen in the data. This suggests that there may be some problems in these other amplitudes and that the form in Table 1 is too simple.

The TRIUMF $(\vec{p}, \vec{p}')$ data [25] from $^{54}$Fe at 290 MeV and 20 degrees is shown in Fig. 5. Also shown is a nonrelativistic RPA calculation of Smith [26] which shows some nuclear structure effects. These are larger at this lower momentum transfer. Again the analyzing power is seen to be below the free value. However, the enhanced slope in $A_y$ can be explained as a nuclear structure effect. Presumably a relativistic RPA calculation will reproduce both the slope and magnitude of $A_y$.

Relativistic enhancements in $D_{ll'}$ and $D_{ss'}$ are in reasonable agreement with the data. However, $D_{sl'}$ and $D_{ls'}$ data are between nonrelativistic and relativistic values. This suggests that there still may be some problems with the other NN amplitudes at 290 MeV but they are not as large as at 500 MeV. Overall, the simple relativistic calculation does a better job than the nonrelativistic version.

**Quasielastic $(\vec{p}, \vec{n})$ charge exchange:** The isovector NN amplitudes are quite different from the isoscalar ones. Now, there is no longer a large cancelation between S and V. Instead, the pseudoscalar amplitude from pion exchange domi-

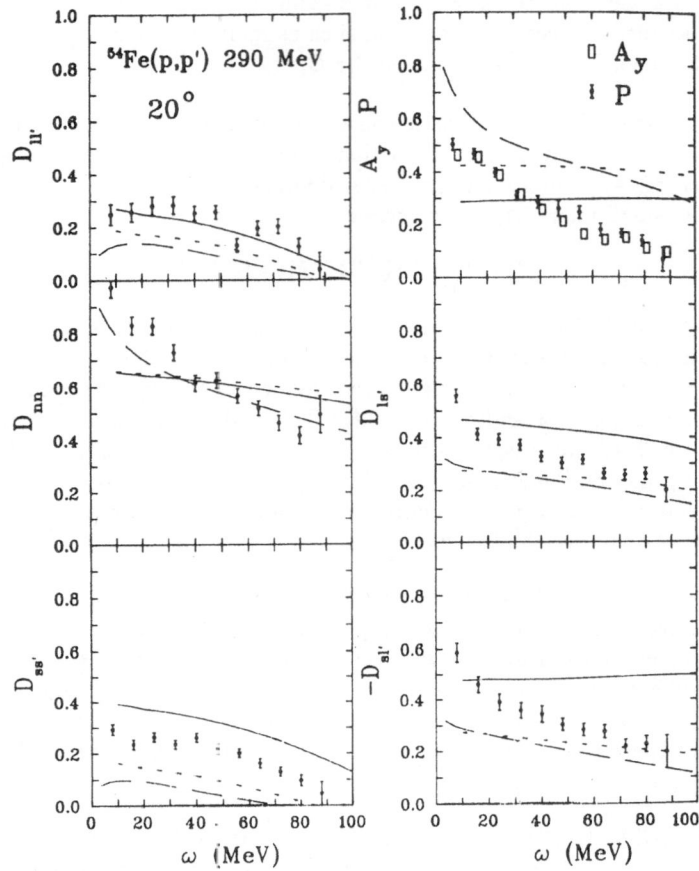

Figure 5. Relativistic PWIA calculations [23] for $(\vec{p}, \vec{p}\,')$ scattering from $^{54}$Fe at 290 MeV and 20 degrees. The solid curves use relativisitic $U^*$ spinors while the light dashed curves use free $U$ spinors. Also shown are nonrelativistic RPA calculations of Smith [26] (long dashed curves). The data is from ref [25].

nates. As a result, relativistic effects for charge exchange reactions are predicted to be quite different from those for $(p, p')$.

Results are shown in Fig. 6 for $(\vec{p}, \vec{n})$ from $^{12}C$ at 500 MeV and 18.5 degrees. [We caution that this figure involves the old SM86 phase shifts.] Pion exchange is very important for these kinematics. As a result, there is a large difference between a pseudoscalar and pseudovector form for the NN amplitudes. (Note, this difference is small for $(p, p')$.) On theoretical grounds the pseudovector form is preferred. Indeed, the pseudoscalar form predicts a large drop in $D_{nn}$ which is not seen in Taddeucci's data [20]. Instead, the pseudovector form predicts that $D_{nn}$ will be about 0.1 above the free value. This is in good agreement with data if one compares the measurement to the 1990 Arndt phase shifts.

Because there is no longer a large cancelation between S and V the analyzing power is not predicted to decrease. Indeed there is even a very small increase for a pseudovector invariant. This agrees with data.

Decreases in $D_{ss'}$ and $D_{ll'}$ are predicted for a pseudovector form. Again, the data also appears to be below the new phase shift values. Overall, the psuedovector calculations do a reasonable job predicting changes from free NN values.

**Future work:** One needs relativistic RPA calculations of quasielastic scattering with more sophisticated models of the NN interaction. In addition, more $(\vec{p}, \vec{n})$ data for different kinematics will help separate nuclear structure from NN amplitude effects. Finally, nonrelativistic calculations of the analyzing power in quasielastic scattering would be useful. To date, nonrelativistic calculations have been completely unable to explain the data. When relativistic RPA nuclear structure models and the model of the NN amplitudes are under control in quasielastic calculations then it is appropriate to return to relativistic DWIA calculations for bound states.

## V. SUMMARY

Relativistic effects are predicted from the large scalar and vector potentials. These enhance the lower components of the Dirac spinors $U \rightarrow U^*$. The relativistic impulse approximation based on this provides an excellent description of elastic spin observables.

These relativistic effects may have a large influence on nuclear matter under extreme conditions. Relativistic calculations predict a stiff high density equation of state. This is because the scalar attraction is reduced with respect to the vector repulsion. This change in the NN interaction is the most important relativistic effect.

Spin observables in quasielastic scattering provide a good way to look at the NN interaction in the medium. One directly measures several $(\vec{p}, \vec{p}')$ observables which are close to free values. This suggests that corrections to simple quasifree knockout are small. The analyzing power is substantially below the free value. This may provide the best relativistic signature to date and is directly related to the relativistic change in the NN interaction.

All relativistic effects seen in both proton and electron scattering are isoscalar. This is because the effective mass of the neutron is predicted to be close to that of the proton. Therefore the analyzing power is not predicted to decrease in $(\vec{p}, n)$. The present and future $(\vec{p}, \vec{n})$ data will allow one to study relativistic effects in isovector amplitudes.

Unfortunately, there are important computational difficulties converting relativistic mean field or RIA phenomenologies into multiple scattering or quantum hadrodynamics theories. This has important limitations for the field. In the future it is important to work on these difficulties and to develop a relativistic nuclear reaction theory. This will insure consistency among nuclear structure, reaction mechanism and coupling of the probe. The development of a relativistic reaction theory is crucial to analyze the many high momentum transfer experiments to be done at CEBAF and elsewhere.

## REFERENCES

1. J. D. Walecka, *Ann. Phys.* (N.Y.) **83** (1974) 491.

2. L. D. Miller and A. E. S. Green, *Phys. Rev.* **C5** (1972) 241.

3. C. E. Price and G. E. Walker, *Phys. Rev.* **C36** (1987) 354; R. J. Furnstahl, C. E. Price and G. E. Walker, *Phys. Rev.* **C36** (1987) 2590; W. Pannest, P. Ring and J. Boguta, *Phys. Rev. Lett.* **59** (1988) 2420; R. J. Furnstahl and C. E. Price, *Phys. Rev.* **C40** (1989) 1398; R. J. Furnstahl and C. E. Price, *Phys. Rev.* **C41** (1990) 1792.

4. B. D. Serot and J. D. Walecka in "Advances in Nuclear Physics" ed. J. W. Negele and E. Vogt (Plenum, N.Y. 1986) Vol. 16.

5. C. J. Horowitz and Brian D. Serot, *Nucl. Phys.* **A464** (1987) 613.

6. R. J. Furnstahl, R. J. Perry and B. D. Serot, *Phys. Rev.* **C40** (1989) 321.

7. B. C. Clark, S. Hama and R. L. Mercer, in "The Interaction Between Medium Energy Nucleons in Nuclei", *AIP Conf. Proc.* No. 97, ed. H. O. Meyer (AIP, N.Y. 1983).

8. E. D. Cooper, Ph.D. Thesis, Univ. of Alberta, unpublished.

9. S. Hama, B. C. Clark, E. D. Cooper, H. S. Sherif and R. L. Mercer, *Phys. Rev.* **C41** (1990) 2737.

10. Bitnet: Cooper @ Ohstpy.

11. See talks by R. J. Furnstahl, J. R. Shepard and C. J. Horowitz in "Spin Observables of Nuclear Probes", ed. C. J. Horowitz et al., (Plenum, N.Y. 1989).

12. J. A. McNeil, J. R. Shepard and S. J. Wallace, *Phys. Rev. Lett.* **50** (1983) 1439; J. R. Shepard, J. A. McNeil and S. J. Wallace, *Phys. Rev. Lett.* **50** (1983) 1443.

13. J. D. Lumpe and L. Ray, *Phys. Rev.* **C35** (1987) 1040.

14. D. P. Murdock and C. J. Horowitz *Phys. Rev.* **C35** (1987) 1442.

15. J. A. Tjon and S. J. Wallace *Phys. Rev.* **C36** (1987) 1085.

16. C. J. Horowitz and Brian D. Serot, *Nucl. Phys.* **A368** (1981) 501.

17. C. J. Horowitz, *Phys. Rev.* **C31** (1985) 1340.

18. C. J. Horowitz, D. P. Murdock and Brian D. Serot in "Computational Nuclear Physics" ed. S. E. Koonin et al. (Plenum, N.Y. 1991).

19. E. Rost and J. R. Shepard, *Phys. Rev.* **C35** (1987) 681; J. R. Shepard, E. Rost and J. Piekarewicz *Phys. Rev.* **C30** (1986) 1604.

20. T. Taddeucci in these proceedings, ed. S. Wissink et al.

21. H. Kurasana and T. Suzuki, *Nucl. Phys.* **A490** (1988) 571; C. J. Horowitz, *Phys. Lett.* **B208** (1988) 8.

22. C. J. Horowitz and J. Piekarewicz, *Phys. Rev. Lett.* **62** (1989) 391; *Nucl. Phys.* **A511** (1990) 461.

23. C. J. Horowitz and D. P. Murdock, *Phys. Rev.* **C37** (1988) 2032.

24. T. Carey et al., *Phys. Rev. Lett.* **53** (1984) 144 and private communication.

25. O. Hausser et al., *Phys. Rev. Lett.* **61** (1988) 822.

26. R. Smith in "Spin Observables of Nuclear Probes", ed. C. J. Horowitz et al. (Plenum, N.Y. 1989).

# RECENT RESULTS OF (n,p) STUDIES AT INTERMEDIATE ENERGIES [+]

J. Rapaport

Ohio University

## INTRODUCTION

In recent years charge–exchange reactions from nuclei at intermediate energies have been the subject of intensive studies. At these energies, isovector spin excitations are preferentially induced in nuclei making these reactions extremely sensitive to these aspects of nuclear structure. At the last Telluride conference, Jackson and Celler (1) reported on the (n,p) reaction as a probe of nuclear structure. New results will be presented here, mainly those obtained with the recently implemented white neutron source at LAMPF–WNR.

Two very distinct empirical features (fig. 1) appear in the (n,p) reactions that represent two different regions of the nuclear spin–isospin response: (a) the nuclear sector and (b) the quasifree region. In the nuclear sector, at small momentum transfer $q \lesssim 1$ fm$^{-1}$, and low energy loss $\omega \lesssim 25$ MeV, the spectra are characterized by giant resonances of low multipolarity, superimposed on a continuous background. The Gamow–Teller (GT) resonance, the dipole resonance (GDR) and especially the spin dipole resonance (GSDR) are the main features in the nuclear sector that will be discussed in the present work. At larger momentum transfers $1 \lesssim q \lesssim 3$ fm$^{-1}$, the inclusive spectra exhibit a broad peak at an energy loss $\omega \approx q^2/2m$ that follows the kinematics for NN scattering from a nucleon at rest. This is called the quasifree region; some empirical results will be compared with a model calculation.

The (n,p) reaction for nuclei with $T_0 \neq 0$ has a different but complementary role than the (p,n) reaction. The latter excites final states with $T_f = T_0-1$, $T_0$ and $T_0+1$. The cross section which is proportional to the corresponding isospin Clebsch–Gordon coefficient decreases with increasing $T_f$ values. Thus, excitation of $T_0+1$ states are very weak and they are in a high density region of $T_0-1$ and $T_0$ states. On the other hand, for the (n,p) reaction, the targets are 100% polarized in isospin space, given rise only to excitation of $T_0+1$ states. The obvious consequence is that (n,p) reactions do not excite Fermi transitions and thus cross normalizations of GT and F strength as done in (p,n) reactions (2) cannot be performed.

---

[+]    Supported in part by the National Science Foundation.

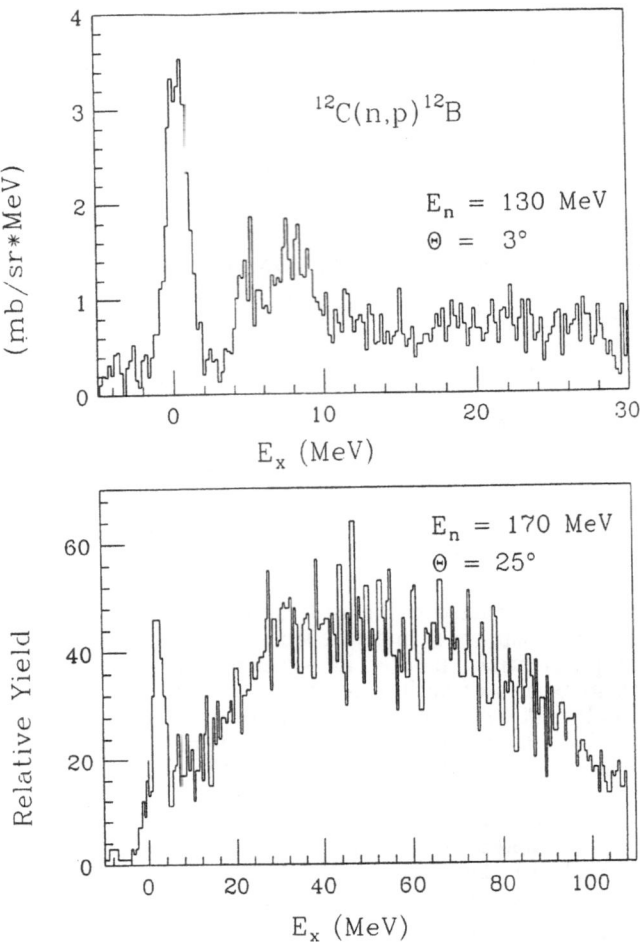

Fig. 1    Spectra for the $^{12}C(n,p)^{12}B$ reaction at $\theta_L = 3°$ and $E_n = 130$ MeV (top) and at $\theta_L = 25°$ and $E_n = 170$ MeV (bottom). In the top, the gs is a GT transition while the dipole region is observed below 10 MeV excitation. The bottom panel shows the excitation of the quasifree region while the prominent peak corresponds to the $4^-$ state in $^{12}B$ at 4.5 MeV excitation energy.

1.    The LAMPF–WNR White Neutron (n,p) Facility

A complete report of this facility will be published (3), thus only a few main details are presented here. The 800 MeV LAMPF pulsed beam is focussed on a 7.5 cm thick tungsten target. Spallation neutrons (4) emitted at 15° are collimated into a 10×10 cm target located 90 m from the source. A flux of about 3×10³ n/sec–MeV are available in the target area in the 50–300 MeV energy range. A set of four targets, each one followed by a charged particle tagging wire chamber similar to the segmented target chamber reported by Henderson et al. (5), is located in front of the spectrometer. The latter consist of four drift chambers, a thin plastic scintillator as a $\Delta E$ detector and a CsI(Tl) "wall" as a calorimeter. A 5 kG dipole field is used to study the small angle region (0°–15°). The magnet is removed to study reactions at larger angles. The thickness (15 cm) of the CsI detectors was selected to stop up to 250 MeV incident protons. An overall proton resolution of about 1% has been achieved in the 60–250 MeV energy interval. This represents a combination of target thickness, neutron beam binning resolution (1 nsec) and detectors resolution. The present facility spans a neutron energy range not studied until now, but its main strength that makes it a unique facility is in its continuous energy character. This novel feature allows, therefore, the study of the energy dependence of nuclear reaction mechanisms because all data in this energy range are taken simultaneously. In fig. 2 we present data showing the $\theta_L = 3°$ energy dependence of the $^{32}S(n,p)^{32}P$

reaction for neutron energies between 75 and 190 MeV. The energy dependence of GT states is clearly shown. Also shown is the dominance of GDR cross section at lower incident energy that decreases with energy given rise to the excitation of the GSDR.

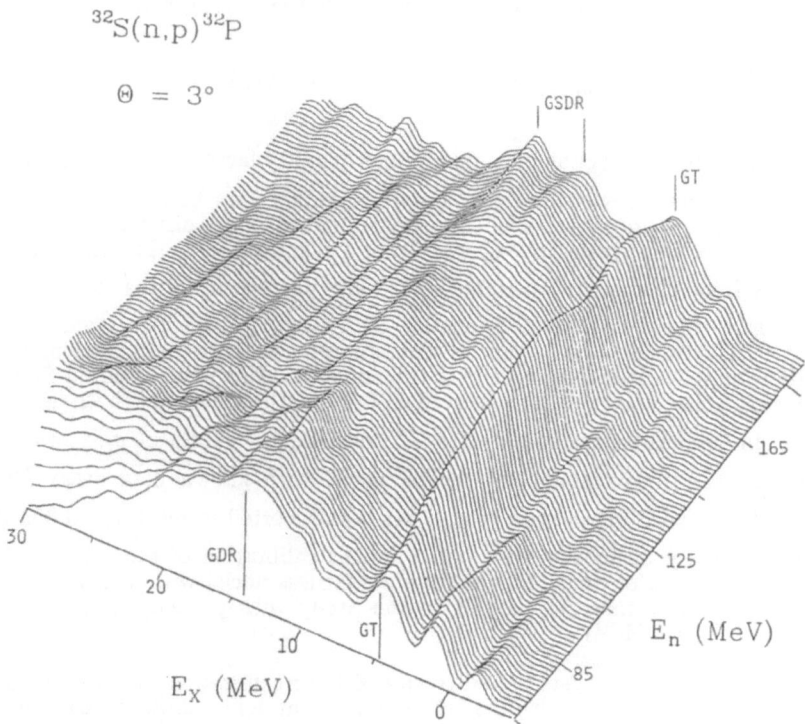

Fig. 2    Data for the $^{32}S(n,p)^{32}P$ reaction at $\theta_L = 3°$ and $E_n$ between 75 and 190 MeV. The excitation region up to 30 MeV is shown, indicating the energy dependence of the GT, GDR and GSDR peaks.

In the following sections, we will present recent results obtained with this facility.

2.    Nuclear Sector—Giant Resonances

2.1    Gamow–Teller Resonance

At intermediate energies, typical zero–degree (p,n) or (n,p) spectra are seen to be dominated by strong peaks which are characterized by $\Delta L=0$, $\Delta S=1$

transfers ($\Delta J^{\pi}=1^+$) which are interpreted as GT transitions (fig. 1). The differential charge–exchange cross section at small q is usually factorized (2) as follows:

$$\sigma_{GT}(q,\omega) = K(E_i,\omega)\, N^D_{GT}\, |J_{\sigma\tau}(q)|^2\, B(q) \qquad (1)$$

where $K(E_i,\omega)$ is a kinematic factor, $N^D_{GT}$ is a distortion factor defined by the ratio of plane waves and distorted waves cross sections, $J_{\sigma\tau}(q)$ represents the volume integral of the spin–isospin component of the nucleon–nucleus interaction, and $B(q)$ is a nuclear structure coefficient that becomes the GT strength for the transition in the limit q = 0. Explicit in the derivation of eq. (1) is the assumption that tensor interactions and spin quadrupole ($\Delta L=2$, $\Delta S=1$) transition densities may be neglected for small q.

We use the notation of ref. 2 to define the unit GT cross section $\hat{\sigma}_{GT}$ as:

$$\sigma_{GT}(E,q=\omega=0) = \hat{\sigma}_{GT}(E,A)\, B(GT) . \qquad (2)$$

The "unit cross section" $\hat{\sigma}_{GT}$ is a proportionality factor that depends both in nuclide and incident nucleon energy. Empirical values may be obtained by taking the ratio of the extrapolated q=0 measured forward angle differential GT cross section and the corresponding Gamow–Teller strength B(GT), known in specific transitions from beta decay measurements. The purpose of evaluating $\hat{\sigma}_{GT}(A,E)$ values, which should be a smooth function of A, is to use them in transitions for which the GT strength is not known. Thus charge–exchange cross sections to GT transitions may be calibrated in units of GT strength. This is an essential first step in the evaluation of the total observed GT strength which has been used, for instance, in (p,n) reactions to evaluate GT quenching. Values of $\hat{\sigma}_{GT}$ have been obtained in the (p,n) reaction for several nuclei with known beta decay ($\beta^+$) ft values. The A and E dependence of $\hat{\sigma}_{GT}$ has been reported in ref. 2 up to 200 MeV and more recently up to 800 MeV (ref. 6). The calibration of the probe in (n,p) reactions is not that obvious because there are far less nuclei with well known beta decay ($\beta^-$) ft values that are suitable to this study with present (n,p) achievable resolutions (about 1 MeV).

Among the few targets presently amenable to calibration of the (n,p) probe are $^6$Li, $^{12}$C, $^{13}$C and $^{64}$Ni. The first three with accurately known ft values for the $\beta^-$decay of the gs of $^6$He, $^{12}$B and $^{13}$B to the ground state of the daughter nuclei provides ideal cases for this comparison. This has been reported by Jackson and Celler (1) at $E_n$ = 198 MeV. The $^6$Li and $^{12}$C nuclei are T = 0 targets and thus similar unit cross section values are expected for the (p,n) and (n,p) reaction. This observation has been verified at $E_n$ = 198 MeV (1). We have extended this comparison for $^{32}$S (7). We also report on the $^{64}$Ni(n,p)$^{64}$Co (gs) reaction (8) in the 100–250 MeV energy range.

In fig. 3 we present spectra obtained by Sorenson et al. (9) for the (n,p) reaction on $^6$Li, $^{12}$C and $^{13}$C targets. Following the procedure indicated above, unit cross section values have been obtained for the gs transitions. The energy dependence of $\hat{\sigma}_{GT}$ is shown in fig. 4. In the same figure we present results for $^{32}$S and for $^{64}$Ni. Values of $\hat{\sigma}$ obtained in the $^6$Li(p,n)$^6$Be (gs) reaction (10) at energies between

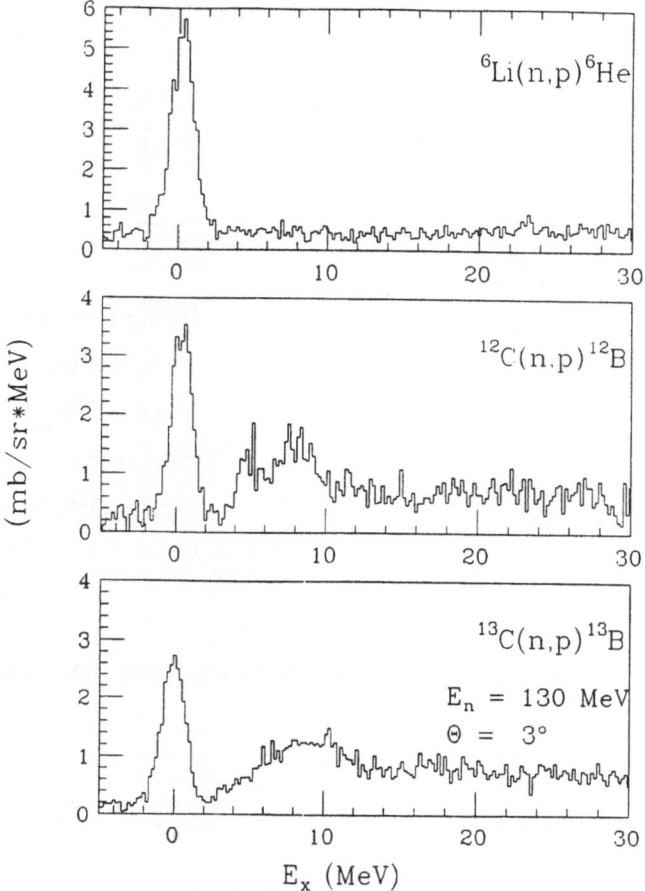

Fig. 3    Spectra for the indicated reactions at $\theta_L = 3°$ and $E_n = 130$ MeV. In all cases the gs is a GT transition.

100 and 200 MeV and values of $\hat{\sigma}$ for the $^{12}C(p,n)^{12}N$ (gs) reaction (2,11) in the 60–300 MeV energy range agree very well with $\hat{\sigma}$ values reported here. However, values of $\hat{\sigma}$ reported for the $^{13}C(p,n)^{13}N$ (gs) transition (2) are about 40% larger than the present results. What is more puzzling is that a recent measurement (12) of the $^{13}C(p,n)^{13}N$ (15.1 MeV) GT transition reports a $\hat{\sigma}$ value in very good agreement with the value $\hat{\sigma}$ obtained in the $^{13}C(n,p)^{13}B$ (gs) transition.

The unit cross section $\hat{\sigma}_{GT}(E,A)$ may be approximately described (see eqs. 1 and 2) as products of the kinematic factor $K(E,A)$, a distortion factor $N_{GT}^D$ and $|J_{\sigma\tau}|$ such that

$$\hat{\sigma}_{GT}(E,A) = K(E,A)\, N_{GT}^D(E,A)\, |J_{\sigma\tau}(E,q=0)|^2 .$$

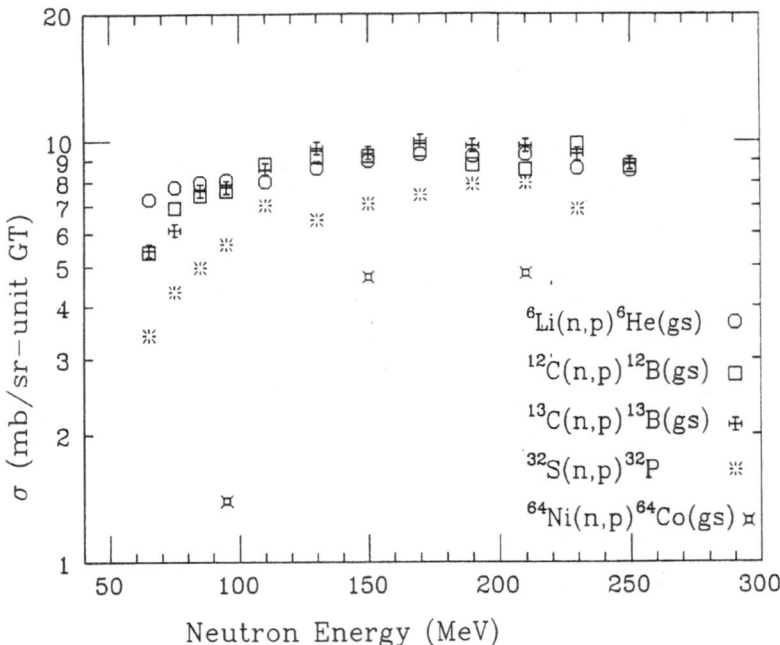

Fig. 4    Unit GT cross sections between 60 MeV and 250 MeV for the indicated reactions.

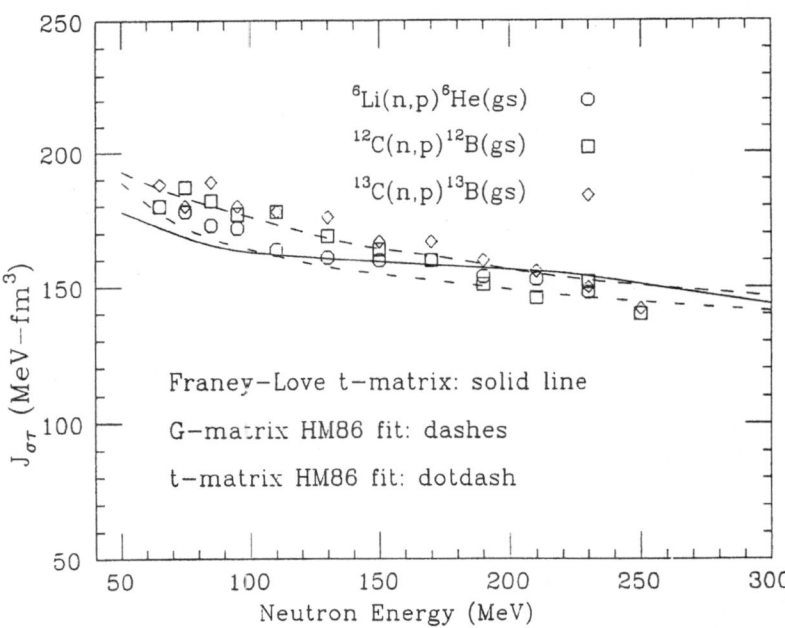

Fig. 5    Energy dependence of the spin–isospin term of the effective interaction at q = 0 obtained from data in fig. 4.  The curves are predictions from ref. 13.

The quantities K(E,A) and $N_{GT}^D$(E,A) may be calculated to obtain $J_{\sigma\tau}$(E,q=0) values. This has been done in ref. 9 and the results are shown in fig. 5. The curves are theoretical evaluations reported by Nakayama and Love (13).

The $^{32}$S(p,n)$^{32}$Cl reaction has been studied at 135 MeV by Anderson et al. (14). We present in fig. 6 (bottom panel) the reported zero degree cross sections for GT transitions distributed in 1 MeV width Gaussian distributions. A slice at 130 MeV from the $^{32}$S(n,p)$^{32}$P data shown in fig. 2 is shown at the top panel of fig. 6. The middle frame of fig. 6 presents results of a shell model calculation using the code OXBASH (14). An excellent agreement among the three spectra is observed. The calculated ΣGT strength normalized by 0.8 agrees very well with the reported ΣGT in the (p,n) experiment (14). We have used the GT strength observed in the states at about 5 MeV excitation to calculate the unit cross section values shown in fig. 3.

The $^{64}$Ni(n,p)$^{64}$Co reaction has also been studied (8) to obtain the unit cross section for the gs transition. The reported values are also shown in fig. 3. (See contribution by A. Ling in these proceedings.)

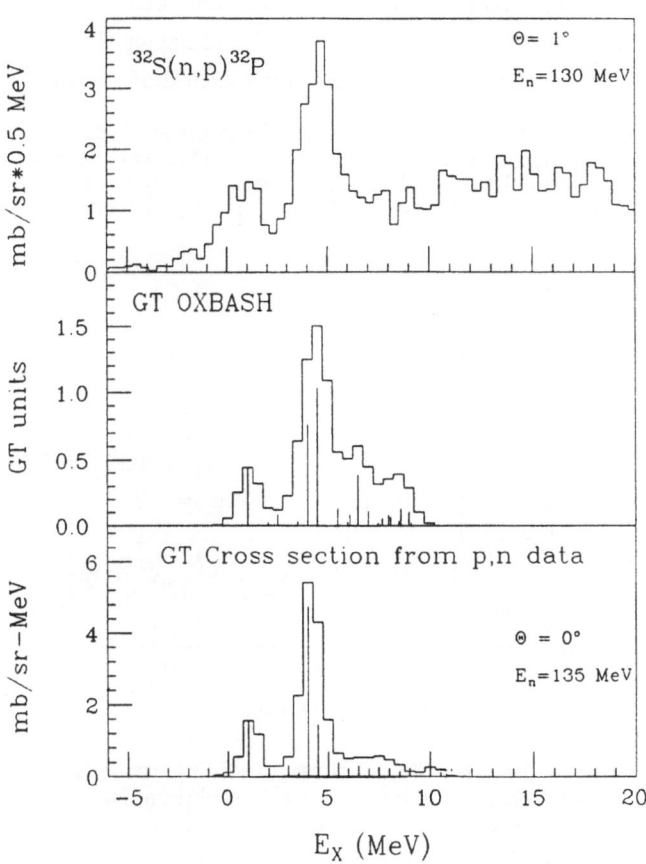

Fig. 6       Comparison of the GT strength distribution observed in the $^{32}$S(p,n)$^{32}$Cl reaction (ref. 14) (bottom panel), a calculated GT strength distribution (middle panel), and data for the $^{32}$S(n,p)$^{32}$P at $\theta_L = 1°$ and $E_n = 130$ MeV.

## 2.2  Dipole and Spin Dipole Resonances

The X(n,p) reaction is probably one of the best probes to study GDR and GSDR in nuclei.  Only $(T_0+1)$ final states are excited.  The strong energy dependence of the non–spin transfer isovector effective interaction makes it also plausible to recognize these two important transitions just from the (n,p) cross section data.  At low energy (E ~ 55 MeV) the ratio of the isovector spin transfer to non–spin transfer effective interaction is close to 1, but increases rapidly with energy to a value close to 12 at about 200 MeV (2).

These giant dipole resonances, macroscopically interpreted in terms of collective oscillations of protons moving away from neutrons may be microscopically evaluated in terms of particle hole excitations.  In the next paragraph, we present (n,p) data that are compared with results of simple 1p–1h calculations that illustrate the main features of the observed cross sections.

The dipole resonances are all characterized via $\Delta L=1$.  Those mediated via the $J_\tau$ term of the effective interaction corresponds to $\Delta J = 1^-$ transfers and make up the GDR.  Those mediated via the $J_{\sigma\tau}$ term lead to $\Delta J^\pi = 0^-$, $1^-$ and $2^-$ transitions and form the GSDR.  Self–consistent Hartree–Fock RPA calculations for nuclei with N > Z (15) indicate that the $J^\pi$ splitting, which has an origin in the spin–orbit interaction, results in $\Delta J^\pi = 0^-$ transitions having the highest excitation energy, while $\Delta J^\pi = 2^-$ transitions have the lowest excitation energy.

The dipole and spin dipole in $^{12}$N and $^{12}$B have been studied by Gaarde et al. (16) and by Millener (17).  These authors use shell model calculations derived from the Cohen–Kurath model extended in a $1\hbar\omega$ space to include negative parity state as excitations from the p into the sd shell.  For $^{12}$B, the simple 1p–1h  model

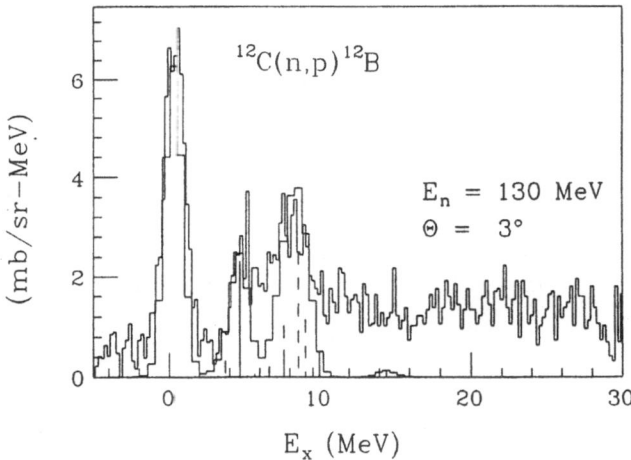

Fig. 7    Spectrum for the $^{12}$C(n,p)$^{12}$B reaction compared with 1p–1h calculated cross sections, assuming a 1 MeV resolution (wide bars histogram). The gs is a GT transition, while the others are dipole states.  The location of $J^\pi = 2^-$, $1^-$ and $0^-$ states are represented with solid vertical lines, dash lines and dots.  The latter is the group at about 14 MeV excitation energy.

predicts three $J^\pi = 0^-$ states, five $J^\pi = 1^-$ states and five $J^\pi = 2^-$ states below 10 MeV and a fourth $J^\pi = 0^-$ state at 14 MeV excitation. We have used the reported (17) OBDME in a DWIA calculation to obtain the differential cross sections. Empirical optical model potential parameters (18) and the Franey–Love (19) t–matrix interaction are used in these calculations. The obtained cross sections corrected by the solid angle acceptance and distributed in 1 MeV width Gaussian distributions (wide bar histogram) are compared with data in fig. 7. There is very good agreement in the magnitude and shape of the two spectra.

We also have recently reported (20) on the $^{16}O(n,p)^{16}N$ reaction at $E_n = 298$ MeV. A multipole decomposition of the observed angular distributions indicate that the main feature of the forward angle cross section is the excitation of the GSDR. We present in fig. 8 a comparison of data and DWIA calculated 1p–1h cross sections. The main features of data and calculation seem to agree. Only about 60% of the calculated cross section is experimentally observed.

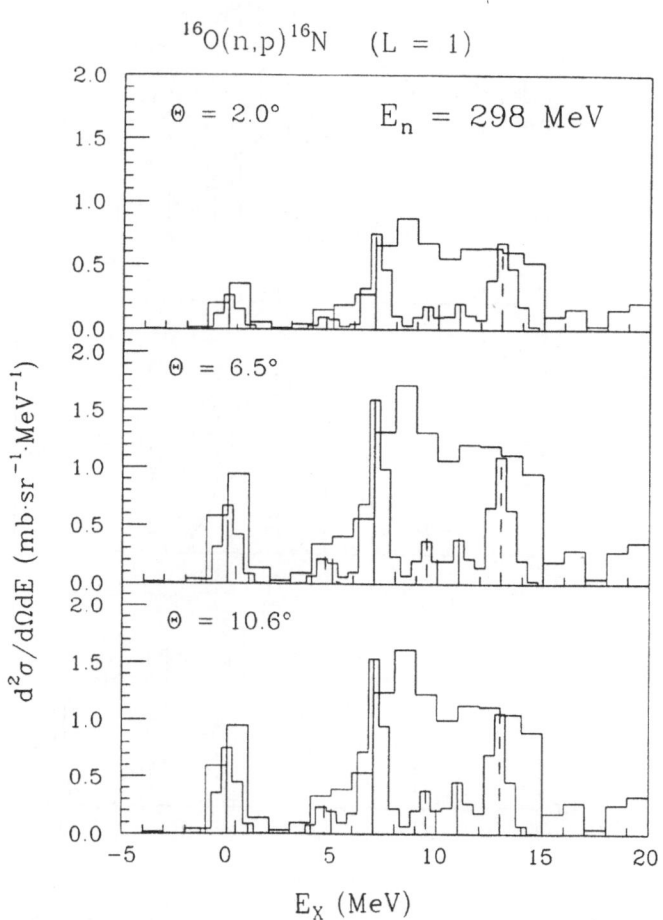

Fig. 8    Comparison of L = 1 cross section obtained in the $^{16}O(n,p)^{16}N$ reaction at 298 MeV (ref. 20) (1 MeV histogram) and 1p–1h predicted cross section (0.5 MeV histogram). The location of $J^\pi = 2^-$, $1^-$ and $0^-$ states are represented with solid vertical lines, dash lines and vertical dots.

In fig. 9 we present data for the $^{40}$Ca(n,p)$^{40}$K cross section at 75, 150 and 210 MeV. The nucleus $^{40}$Ca is a spin saturated nucleus. Thus, the simple shell model predicts no GT strength. Correlations in the $^{40}$Ca ground state wave function contribute to some GT strength, but it is rather weak, thus making $^{40}$Ca a perfect nucleus to study the GDR and GSDR. Superimposed on the lower panel of fig. 9 is the shape of the observed GDR in $^{40}$Ca (21). The top panel has super-imposed the GSDR as obtained from the $^{40}$Ca(p,p′) at $E_p = 316$ MeV (22). In both cases the $^{40}$Ca(n,p)$^{40}$K data and the dipole shapes are in good agreement. The middle panel has a calculated 1p–1h cross section with the main features of the observed excitation function. A more sophisticated RPA calculation (see Wambach in these proceedings), including a 2p–2h spreading width, predicts quite well the GSDR observed by Baker et al. (22). This calculation has similar features (i.e., location of multipoles) than the present calculation.

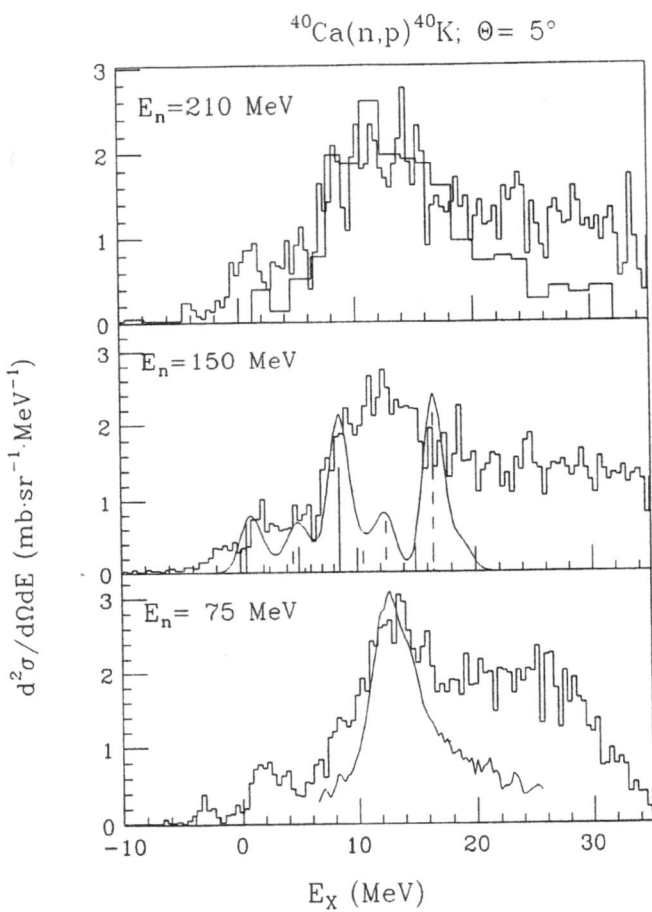

Fig. 9    Data for the $^{40}$Ca(n,p)$^{40}$K cross section at 75, 150 and 210 MeV and $\theta_L = 5°$. The bottom panel has overlayed the shape of the GDR, the middle panel a calculated 1p–1h cross section and the top panel a GSDR histogram shape obtained for ref. 22.

## 3.    The Quasifree Region

The LAMPF/WNR (n,p) spectrometer was moved to larger angles ($15° < \theta < 35°$) and we have preliminary data for the $^{12}C(n,p)$ reaction, shown in fig. 1. These results compare quite well with recent $^{12}C(p,n)^{12}N$ data obtained at $E_p = 186$ MeV and at the same angles. In both spectra the main features are the QF peak and the sharp 4$^-$ state at $E_x \sim 4.5$ MeV. In recent years both relativistic plane wave calculations (23) and non–relativistic full–folding calculations (24) have been published. One of the fundamental questions to answer when comparing calculations and data is to whether the nuclear response in this region may be represented

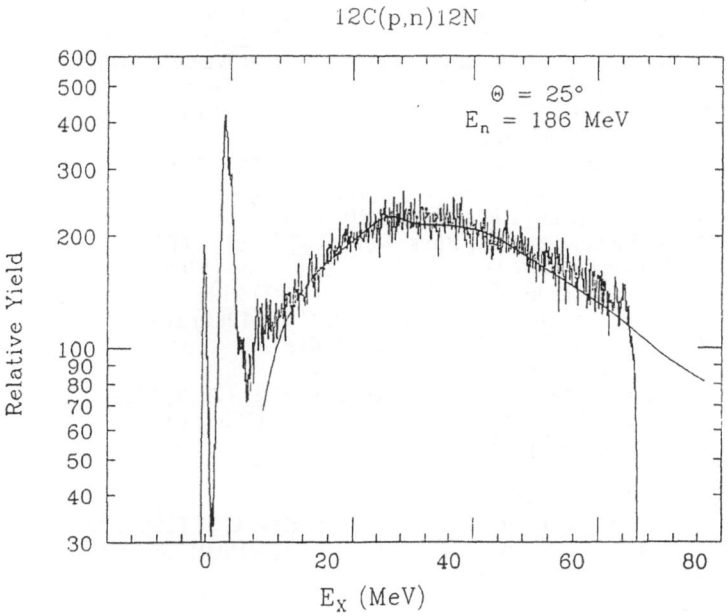

Fig. 10    The $^{12}C(p,n)^{12}N$ reaction at $\theta_L = 25°$ and $E_p = 186$ MeV. The two peaks at low excitation correspond to 2$^+$ and 4$^-$ states in $^{12}N$ while the upper region corresponds to the QF. The solid line is an interactive Fermi gas calculation (24).

by a Fermi gas or whether many body effects are important in this (q,$\omega$) region. We present in fig. 10 a comparison of a calculation for the QF region using an interactive Fermi gas model (24) with data for the $^{12}C(p,n)^{12}N$ reaction. The experimental and calculated shape agree quite well. A shift of about 26 MeV is noted for the location of the QF peak with respect to the value $\frac{q^2}{2M}$. This shift is similar to the value noted at higher energy (p,n) data and in other nuclei. (See paper by DePace for a discussion of this subject.)

## 4. Conclusions

The study of X(n,p) reactions complement similar X(p,n) studies, providing unique information on $(T_0+1)$ transitions. The LAMPF/WNR white neutron source provides a unique opportunity to study the energy dependence of effective interactions. Future studies will be extended to 800 MeV neutrons.

## REFERENCES

1. K.P. Jackson and A. Celler, in "Spin Observables of Nuclear Probes", ed. by C.J. Horowitz, C.D. Goodman and C.E. Walker (Plenum, New York, 1988), p. 139.
2. T.N. Taddeucci et al., Nucl. Phys. A469 (1987) 125.
3. J. Ullmann et al., to be published.
4. P.W. Lisowski et al., Nucl. Sci. and Eng. 106 (1990) 208.
5. R.S. Henderson et al., Nucl. Inst. and Meth. A257 (1987) 97.
6. J. Rapaport, in Fundamental Symmetries and Nuclear Structure, ed. by J.N. Ginocchio and S.P. Rosen, World Scientific, 1989, p. 186.
7. B.K. Park et al., to be published, BAPS 35 (1990) 1650.
8. A. Ling et al., to be published.
9. Sorenson et al., to be published, BAPS 34 (1989) 1834.
10. Rapaport et al., Phys. Rev. C 41 (1990) 1920.
11. J.W. Watson et al., Phys. Rev. C 40 (1989) 22.
12. J.M. Mildenberger et al., Phys. Rev. C 43 (1991) 1777.
13. K. Nakayama and W.G. Love, Phys. Rev. C 38 (1988) 51.
14. B.D. Anderson et al., Phys. Rev. C 36 (1987) 2195.
15. N. Auerbach and A. Klein, Phys. Rev. C 30 (1984) 1032.
16. C. Gaarde et al., Nucl. Phys. A422 (1984) 189.
17. F.P. Brady et al., Phys. Rev. C 31 (1991).
18. J.R. Comfort and B.C. Karp, Phys. Rev. C 21 (1980) 2162.
19. M. Franey and G.A. Love, Phys. Rev. C 31 (1985) 488.
20. K. Hicks et al., Phys. Rev. C, to be published.
21. J. Ahrens et al., Nucl. Phys. A251 (1975) 479.
22. T. Baker et al., Phys. Rev. C 40 (1989) 1877.
23. C.J. Horowitz and D.P. Murdock, Phys. Rev. C 37 (1988) 2032.
24. F. Brieva and G. Love, Phys. Rev. C 42 (1990) 2573; F. Brieva (private communication).

# THEORY OF INCLUSIVE (p,n) AND ($^3$He,t) REACTIONS ON NUCLEI IN THE QUASIELASTIC AND Δ-ISOBAR EXCITATION REGION

Yu. L. Ratis, E. A. Strokovsky
JINR, Dubna, USSR

F. A. Gareev,[*] J. S. Vaagen
SENTEF, Institute of Physics, University of Bergen, Norway

## 1 INTRODUCTION

Starting from a formalism of effective numbers for the inclusive integral cross sections of (p,n) with formation of quasielastic and Δ-isobar nuclear excitation we obtain, without free parameters, a satisfactory quantitative description of the existing experimental data. The theory gives a factorized form which allows for a useful separation of the reaction and the structure of the light and heavy reaction partners. We have now been successful in using this formalism with only small modifications for ($^3He, t$) reactions at intermediate energies. We argue from the similarity of the cross sections for the (p,n) and ($^3He, t$) reactions obtained by removing the form factors of ($^3He, t$) from the cross sections and bringing the experimental data for different energies to the same energy. It is shown that the reaction mechanism for the processes (p,n) and ($^3He, t$) is in principle essentially the same.

## 2 PRECIS

Inclusive reactions of the type (e,e'), (p,p'), ($\pi, \pi'$), (p,n), ($^3He, t$), ($^6Li, ^6He$),... provide a basic source of information on effective NN and NΔ interactions, reaction mechanisms, and on the nuclear structure at intermediate energies. The knowledge of spin and isospin components of NN and NΔ interactions is still somewhat poor in spite of a substantial amount of works (see for example the review talks in Ref.[1]).

Thus, the theoretical understanding of these reactions is far from satisfactory [2]. Specifically, since large energy and momentum transfer is involved, the traditional methods have to be re-examined to see if approximations that were made at low $q$ and $\omega$ are still valid. New attention has to be given to how one may correctly separate the nuclear structure effects from the mechanisms of the reactions at intermediate energies. The analysis of the processes is significantly complicated by collective excitations of the type

---

[*]Permanent address JINR, Dubna, USSR

$(\Delta N^{-1})$, i.e. $\Delta$-isobar plus a nucleon hole [3]. The renormalization of the interaction $NN - N\Delta$ in nuclei requires further investigations [4]. Furthermore, proper calculations of the distortions in the entrance and exit channels including the "exchange" term (the Pauli principle between the projectile nucleons and the bound target nucleons) are not clear cut [5,6,7]. And finally, a main subject of investigation and a major part of this paper is the connection between the mechanisms of charge-exchange on a nuclear target compared to a free proton target [8].

It is necessary to keep in mind that a characteristic feature of the inclusive treatment of cross sections is the lack of sensitivity to details of the structure of the target nuclei. Following the terminology of Refs. [5,7] we may refer to this as the "universality" of inclusive cross sections. The "universality" in the theoretical calculations displays itself as an essential independence of the results to the choise of the basic shell model functions. For example, wave functions of the Saxon-Woods potential, oscillator wave functions, quasiclassical or even the Fermi-gas wave functions give the same cross sections within an accuracy of about 5-10%. This "universality" manifests itself in the cross sections of the charge-exchange reactions, which contain all information about the $NN - N\Delta$-interactions in nuclei and of the reaction mechanism, being smooth functions of arguments such as the mass number $A$ and the energy of the projectile.

In [8] (see also [9]) we showed that under definite conditions one can establish a link between the cross sections of the (p,n) charge-exchange reactions on nuclei and on the free proton (referred to as an "elementary" process)

$$\frac{d\sigma[A(p,n)_\Delta B]}{d\Omega_n} = \overline{N}\frac{d\sigma[p + p \to n + \Delta^{++}]}{d\Omega_n}. \tag{1}$$

We will recall our arguments in Section 3. Here $\overline{N} = (Z + \frac{1}{3}N) < f^2 >$ is the effective number of nucleons participating in the process $(p,n)_\Delta$. The quantity $< f^2 >$ is a folding of the target density with the Glauber absorption factor and leads to an effective (reduced) number of active nucleons, comparision with data and an extention of the theory are given in Sections 4 and 5.

A main aim of this contribution is also to discuss common properties of integral cross sections of inclusive $(^3He, t)$ and (p,n) reactions for $T > 0.6$ GeV/A in the $\Delta$-isobar excitation region, on the basis of the effective-number approximation. Aside from the form factor we will, in Section 6, argue that these reactions have the same underlying reaction mechanism and that the cross section can be given a factorized form in terms like(1) of the nuclear structure of the participants in the reaction.

## 3 RELATION BETWEEN THE INCLUSIVE A(p,n)$_\Delta$B REACTION AND THE ELEMENTARY PROCESS

In Refs.[8,10] it has been argued that the $A(p, n)_\Delta B$ reaction can be described in the framework of distorted waves. In this approach the invariant cross section is given by (c=ħ=1)

$$d\sigma = \frac{2E_i E_A}{\lambda^{1/2}(s_{NA}, m_N^2, M_A^2)}\frac{1}{2}\frac{1}{2J_i + 1}\sum(2\pi)^4\delta^{(4)}(P_i + P_A - P_n - P_f) \mid T^{np}_{B+\Delta,A} \mid^2 d\vec{P}_n, \tag{2}$$

where sums are over the angular momentum projection quantum numbers and the final channels f. In first order Born approximation the matrix element of the T operator can be written as (spin coordinates are suppressed)

$$T_{B+\Delta,A}^{np} = < \hat{A}\{\chi_n^{(-)}(\vec{k}_n, \vec{r})\Psi_{B+\Delta}(\vec{r}_1, ..., \vec{r}_A)\} \mid \sum_{j=1}^{A} V_{N\Delta}(\vec{r}_j, \vec{r}) \mid$$

$$\hat{A}\{\chi_p^{(+)}(\vec{k}_p, \vec{r})\Psi_{\alpha_i J_i M_i}(\vec{r}_1, ..., \vec{r}_A)\} > . \tag{3}$$

In (2) and in the following we use a notations where: $E_i = (\vec{P}_i^2 + m_N^2)^{1/2}$ is the energy of the incoming proton with impuls $\vec{P}_i$ and mass $m_N$, $E_A = (\vec{P}_A^2 + M_A^2)^{1/2}$ is the energy of the target-nucleus A, $s_{NA}$ is the square of the invariant mass of the system p+A, $\lambda(x, y, z) = (x - y - z)^2 - 4yz$ is the kinematic or triangular function [11]. The quatities $P_i, P_A, P_n$ and $P_f$ are the four momenta of the projectile-proton, target-nucleus, registered neutron and nonregistered final fragments (for example, f=B+$\Delta$), respectively. Formula (3) is given in the CM of p+A. It contains the wave function $\Psi_{\alpha_i J_i M_i}$ of the target-nucleus A with the spin $J_i$ and projection $M_i$ and other quantum numbers $\alpha$, in overlap with the wave function $\Psi_{B+\Delta}$ of the unobserved system. Futhermore, $\chi_p^{(+)}(\chi_n^{(-)})$ are the distorted waves of the incident and exit channels, $V_{N\Delta}$ is the transition operator for NN$\rightarrow$ N$\Delta$ and $\hat{A}$ the antisymmetrizer. The factor $[2(2J_i + 1)]^{-1}$ in formula (2) is due to averaging/summing over the projections of projectile and target spins.

In Refs. [5,6,7] it was shown that the usual method of evaluation of distorted waves $\chi_p^{(+)}$ ($\chi_n^{(-)}$) in the framework of the optical model underestimates the cross sections because this method ignores the contributions of processes due to incoherent rescattering of protons (neutrons) on nucleons in the target nucleus (residual nucleus $B + \Delta$). This effect can be taken into account using the distorted waves in Glauber approximation [5]

$$\chi_p^{(+)}(\vec{k}_p, \vec{r}) = (2\pi)^{-3/2} exp(i\vec{k}_p \cdot \vec{r}) \prod_{j=1}^{A}[1 - \Gamma(\vec{b} - \vec{b}_j)\theta(z_j - z)]\chi_m(\vec{\sigma}), \tag{4}$$

$$\chi_n^{(-)}(\vec{k}_n, \vec{r}) = (2\pi)^{-3/2} exp(i\vec{k}_n \cdot \vec{r}) \prod_{j=1}^{A-1}[1 - \Gamma(\vec{b} - \vec{b}_j)\theta(z - z_j)]\chi_m(\vec{\sigma}), \tag{5}$$

where $\Gamma(\vec{b})$ is the profile function,

$$\Gamma(\vec{b}) = \frac{1}{2\pi i k} \int e^{i(\vec{q}\vec{b})} A_{NN}(\vec{q}) d^2q. \tag{6}$$

Here $\vec{b}$ is the impact parameter, $\vec{q}$ is the transferred momentum, $A_{NN}(\vec{q})$ the nucleon-nucleon scattering amplitude, $\chi_m(\vec{\sigma})$ the spin function of a nucleon with the spin $\vec{\sigma}$ and projection m, while $\theta(z)$ is the step function

$$\theta(z) = \begin{cases} 1 & \text{if } z \geq 0 \\ 0 & \text{otherwise} \end{cases}$$

Using the completeness of the states of the nonregistered fragments the inclusive cross section of the reaction $A(p, n)_\Delta B$ can be written as [5,7]

$$\frac{d\sigma_{A(p,n)_\Delta B}}{d\Omega_n} = \int d\vec{Q}[\Phi_N^A(\vec{Q})]^2 \frac{d\sigma_{p+p\rightarrow n+\Delta++}(\vec{P}_i, \vec{Q})}{d\Omega_n}, \tag{7}$$

where $[\Phi_N^A(\vec{Q})]^2$ is the momentum distribution of the nucleons in nucleus A, participating in the charge-exchange reaction.

The total momentum distribution function $[\Phi_N^A(\vec{Q})]^2$ can be expressed through the partial distributions of protons $[\Phi_p^A(\vec{Q})]^2$ and neutrons $[\Phi_n^A(\vec{Q})]^2$,

$$[\Phi_N^A(\vec{Q})]^2 = [\Phi_p^A(\vec{Q})]^2 + \frac{1}{3}[\Phi_n^A(\vec{Q})]^2, \tag{8}$$

where $\frac{1}{3}$ is the isotopic weight factor for the creation of a $\Delta^+$- isobar on the neutron. The effective number of protons (neutrons) participating in the process $A(p,n)_\Delta B$ is defined by the integrated impuls distributions,

$$\overline{N}_N^A = \int d\vec{Q}[\Phi_N^A(\vec{Q})]^2, \tag{9}$$

with distorted waves folded in (see eq. (13) below). In the plane wave approximation the effective numbers are equal to $\tilde{N}_p^A(PW) = Z$, $(\tilde{N}_n^A(PW) = \frac{1}{3}N)$, where Z(N) is the number of the protons (neutrons) in nucleus A.

The presence of the momentum of an intranuclear nucleon $\vec{Q}$ among the arguments of the cross section of charge-exchange on a nucleon in the nuclear medium $d\sigma_{p+p\to n+\Delta^{++}}(\vec{P}_i, \vec{Q})/d\Omega_n$ points to the possible necessity to include of effects of going off mass shell. However, in the investigated region of energies, $T_p > 0.6$ GeV, the influence of the off-mass-shell effects can be neglected as the momentum of the incident nucleon $\vec{P}_i$ and the transferred momentum $\vec{q}$ obey the conditions $\mid \vec{P}_i \mid \gg P_F$ and $\mid \vec{q} \mid \gg P_F$, where $P_F$ is the Fermi momentum. The point is that the strength of the interaction depends on the momentum of the incident proton $\vec{P}_i$ and on that of an intranuclear nucleon $\vec{Q}$ by $(P_i^2 + Q^2)^{1/2}$, so that the total correction for the Fermi motion of the nucleons and for their binding does not exceed 3-5%. In addition, the off-mass-shell effects influence the A-dependence of the integral inclusive cross section of $A(p,n)_\Delta B$, that we are interested in, rather weakly. These considerations allow us to use the following approximation

$$\frac{d\sigma_{p+p\to n+\Delta^{++}}(\vec{P}_i, \vec{Q})}{d\Omega_n} \approx \frac{d\sigma_{p+p\to n+\Delta^{++}}(\vec{P}_i, \vec{Q} = 0)}{d\Omega_n} \equiv \frac{d\sigma_{p+p\to n+\Delta^{++}}(\vec{P}_i)}{d\Omega_n}\mid_{free}. \tag{10}$$

As we intended to show, in this approximation (7) becomes factorized, as follows

$$\frac{d\sigma_{A(p,n)_\Delta B}}{d\Omega_n} = \overline{N}\frac{d\sigma_{p+p\to n+\Delta^{++}}(\vec{P}_i)}{d\Omega_n}\mid_{free}, \tag{11}$$

where the quantity $\overline{N}$ (we now drop sub (super) scripts N (A)) has the simple expression,

$$\overline{N} = \int d\vec{Q}[\Phi_N^A(\vec{Q})]^2 = (Z + \frac{1}{3}N)\int d\vec{r}\rho(\vec{r})f^2(b,z) = (Z + \frac{1}{3}N) < f^2 > . \tag{12}$$

Here $< f^2 >$ is the effective factor of absorption, $\rho(\vec{r})$ is the one-nucleon density with the normalization $\int d\vec{r}\rho(\vec{r}) = 1$ and $f^2(b,z)$ the Glauber factor of absorption

$$f^2(b,z) = [(1 - \frac{\sigma_{pN}^{tot} - \sigma_{pN}^{el}}{A}T_-(b,z))(1 - \frac{\sigma_{nN}^{tot} - \sigma_{nN}^{el}}{A}T_+(b,z))]^A \equiv \sum_{\lambda_p=0}^{A}\sum_{\lambda_n=0}^{A}f_{\lambda_p\lambda_n}^2(b,z)$$

$$f_{\lambda_p\lambda_n}^2(b,z) = \sum_{\lambda_p=0}^{A}\sum_{\lambda_n=0}^{A}\binom{A}{\lambda_p}(1 - \frac{\sigma_{pN}^{tot}}{A}T_-(b,z))^{A-\lambda_p}(\frac{\sigma_{pN}^{el}}{A}T_-(b,z))^{\lambda_p}$$

$$\binom{A}{\lambda_n} (1 - \frac{\sigma_{nN}^{tot}}{A} T_+(b, z))^{A-\lambda_n} (\frac{\sigma_{nN}^{el}}{A} T_+(b, z))^{\lambda_n}. \tag{13}$$

Here $\sigma_{pN}^{tot}$ ($\sigma_{nN}^{tot}$) is the total cross section of proton (neutron)-nucleon scattering and $\sigma_{pN}^{el}$ ($\sigma_{nN}^{el}$) the corresponding elastic scattering cross section. The thickness functions $T_\pm$ are given by the standard forms

$$T_+(b, z) = A \int_z^\infty d\xi \rho([b^2 + \xi^2]^{1/2}), \tag{14}$$

$$T_-(b, z) = A \int_{-\infty}^z d\xi \rho([b^2 + \xi^2]^{1/2}). \tag{15}$$

Formula (13) represents the decomposition of the absorption factor over the number of quasielastic collisions $\lambda_p$ ($\lambda_n$) of ejectile-protons (outgoing neutrons) with the nucleons of nucleus A (B). It allows us to represent the effective numbers in the physically transparent form

$$\overline{N} = (Z + \frac{1}{3}N) \int d\vec{r} \rho(\vec{r}) \sum_{\lambda_p=0}^{A} \sum_{\lambda_n=0}^{A} f_{\lambda_p \lambda_n}^2 (b, z). \tag{16}$$

$$\overline{N}_{\nu_p \nu_n} = (Z + \frac{1}{3}N) \int d\vec{r} \rho(\vec{r}) \sum_{\lambda_p=0}^{\nu_p} \sum_{\lambda_n=0}^{\nu_n} f_{\lambda_p \lambda_n}^2 (b, z), \tag{16a}$$

where the latter will be referred to as partial effective numbers. Each of the partial sums describes the contribution to the total cross section from definite groups of final states of the system $B + \Delta$. For example, $\overline{N}_{00}$ corresponds to generating the state $(\Delta N^{-1})$ in the nucleus B in the reaction; the $\overline{N}_{10}$ corresponds to the process where the incident proton excites at the beginning a state (1p-1h) in nucleus A and only after the charge-exchange process the state $(\Delta N^{-1})$ is generated. Generally if $\nu_p + \nu_N = i$ this means that in the reaction $A(p, n)_\Delta B$ one generates the excited state (ip-ih)+$(\Delta N^{-1})$.

Assuming charge symmetry allows us to simplify expression (13) somewhat since it implies equality of the elementary cross sections $\sigma_{pN} = \sigma_{nN} = \sigma_{NN}$, and consequently instead of two functions $T_\pm$ it is possible to introduce one function only

$$T(b) = T_-(b, z) + T_+(b, z) = A \int_{-\infty}^\infty d\xi \rho([b^2 + \xi^2]^{1/2}). \tag{14a}$$

In the case of large mass numbers $A \gg 1$ we obtain the well known eikonal approximation

$$f_{eik}^2 = exp[-(\sigma_{NN}^{tot} - \sigma_{NN}^{el})]T(b). \tag{17}$$

In obtaining (17) from (13) we have used the smooth energy dependence of the cross sections $\sigma_{pN}(T_p)$ and $\sigma_{nN}(T_n)$, because, strictkly speaking $\sigma_{pN}(T_p) = \sigma_{nN}(T_n)$ only at $T_p = T_n$.

If the nucleus is registered in the ground state as was for example done in [12], than in the absorption factors (13) and (17) we have formally to put $\sigma_{NN}^{el} = 0$. In that case, formula (17) corresponds to the eikonal approximation for the elastic scattering in the optical model.

By inserting (17) in (12) we can write the effective absorption factor $< f^2 >$ in the eikonal approximation in the following form

$$< f^2 >= \frac{2\pi}{A} \int dbbT(b)e^{-\sigma T(b)}, \tag{18}$$

Table 1. The A-dependence of effective numbers for the reaction $A(p,n)_\Delta B$ at $T_p = 6$ GeV. The index "opt" means that the calculations was carried out in the framework of the optical model for elastic scattering. The meaning of the other indices is described in the main text.

| A | $N_{opt}$ | $N_{00}$ | $N_{11}$ | $N_{22}$ | $N_{33}$ | $N_{eik}$ |
|-----|-----|-----|-----|-----|-----|-----|
| 12 | 1.79 | 1.69 | 2.23 | 2.32 | 2.34 | 2.43 |
| 16 | 1.79 | 1.71 | 2.28 | 2.40 | 2.42 | 2.51 |
| 27 | 2.66 | 2.58 | 3.46 | 3.65 | 3.68 | 3.78 |
| 40 | 3.07 | 3.00 | 4.08 | 4.33 | 4.38 | 4.47 |
| 58 | 3.38 | 3.32 | 4.54 | 4.85 | 4.92 | 5.02 |
| 118 | 3.90 | 3.86 | 5.31 | 5.69 | 5.74 | 5.88 |
| 208 | 4.35 | 4.33 | 5.98 | 6.43 | 6.55 | 6.63 |

Table 2. The $T_p$-dependence of effective numbers for the reaction $^{12}C(p,n)_\Delta B$.

| $T_p$(GeV) | $N_{opt}$ | $N_{00}$ | $N_{11}$ | $N_{22}$ | $N_{33}$ | $N_{eik}$ |
|-----|-----|-----|-----|-----|-----|-----|
| 1 | 1.79 | 1.69 | 2.67 | 2.98 | 3.06 | 3.12 |
| 6 | 1.79 | 1.69 | 2.23 | 2.32 | 2.34 | 2.43 |
| 10 | 1.83 | 1.74 | 2.24 | 2.33 | 2.34 | 2.43 |
| 14 | 1.83 | 1.74 | 2.30 | 2.40 | 2.41 | 2.51 |
| 20 | 1.93 | 1.84 | 2.27 | 2.33 | 2.34 | 2.43 |

where $\sigma = \sigma_{NN}^{tot}$ for the exclusive and $\sigma = \sigma_{NN}^{tot} - \sigma_{NN}^{el}$ for inclusive reactions, respectively. The integral (18) can be estimated by the saddle point method

$$< f^2 >= [(2\pi)^{3/2} b_0]/[A\sigma^2 e \mid T'(b_0) \mid], \qquad (19)$$

where $b_0$ is root of the equation

$$T(b_0) = \sigma^{-1}. \qquad (20)$$

At the energies $T_p > 0.6$ GeV the values of $\sigma$ lie in the interval $\sigma \approx 20\text{-}40$ mbarn ($\sigma_{NN}^{tot} \approx 40$ mbarn, $\sigma_{NN}^{el} \approx 10\text{-}20$ mbarn). In this case $b_0 \approx R_A$, $\mid T'(b_0) \mid \approx 1/(\sigma a)$, where $R_A$ is the nuclear radius, a the diffuseness of the surface. An approximate expression for $< f^2 >$ may now be written as

$$< f^2 > \approx \frac{(2\pi)^{3/2} R_A a}{\sigma A e}. \qquad (21)$$

From (21) we obtain immediately that $< f^2 > \propto A^{-2/3}$ and consequently, that the A-dependence $\overline{N}$ goes as

$$\overline{N} \propto A^{1/3}, \qquad (22)$$

The A- and $T_p$-dependences of $\overline{N}$ as well as of the partial ones calculated according to eqs. (13) and (16) are given in Tables 1 and 2. Assuming an A-dependence $\overline{N}_{ij} = \kappa_1 A^{\alpha_{ij}}$ ($\kappa_1 = 50$) we find from the tables that $\alpha$ is a smoothly increasing function of $i = \nu_p$ and $j = \nu_n$: $\alpha_{00} = 0.31$ for $\overline{N}_{00}$ and $\alpha_{33} = 0.35$ for $\overline{N}_{33}$. Thus, the series (16) converges sufficienly fast, about 90% of the final value $\overline{N}$ is given by the partial sum $\overline{N}_{11}$. This

**Table 3.** The A- and $\theta$-dependences of experimental cross sections $d\sigma[A(p,n)_\Delta B]/d\Omega_n$ [mb/sr] $\equiv \sigma(\theta)$ (in lab. system) and effective numbers $\overline{N}^{exp}$ at $T_p = 1$ GeV [14]. The values $\overline{N}^T$ are the results of calculations at $\theta_n = 0°$.

| Target | $\overline{N}^T$ | $\sigma(4°)$ | $\sigma(7.5°)$ | $\sigma(11.3°)$ | $\overline{N}^{exp}(4°)$ | $\overline{N}^{exp}(7.5°)$ | $\overline{N}^{exp}(11.3°)$ |
|---|---|---|---|---|---|---|---|
| $H$ | 1.00 | 42.7 | 31.0 | 20.7 | 1.00 | 1.00 | 1.00 |
| $D$ | 0.61 | 52.1 | 44.7 | 29.6 | 1.22 | 1.44 | 1.43 |
| $^7Li$ | 2.15 | 123.8 | 90.5 | — | 2.90 | 2.92 | — |
| $^9Be$ | 2.50 | 177.1 | 132.4 | 81.2 | 4.15 | 4.27 | 3.92 |
| $^{10}B$ | 2.77 | 165.8 | 118.3 | 73.2 | 3.88 | 3.82 | 3.54 |
| $^{11}B$ | 2.74 | 159.8 | 117.5 | 74.3 | 3.74 | 3.79 | 3.59 |
| $^{12}C$ | 2.96 | 162.3 | 122.9 | 81.3 | 3.80 | 3.96 | 3.92 |
| $^{16}O$ | 3.15 | 220.6 | 159.9 | 118.0 | 5.17 | 5.16 | 5.07 |
| $^{19}F$ | 3.11 | 246.6 | 169.4 | 117.9 | 5.78 | 5.46 | 5.91 |
| $^{24}Mg$ | 4.65 | 254.3 | 153.6 | 128.4 | 5.96 | 4.95 | 5.91 |
| $^{25}Mg$ | 4.65 | 243.6 | 179.6 | 122.5 | 5.70 | 5.79 | 5.92 |
| $^{26}Mg$ | 4.65 | 263.6 | 207.1 | 128.8 | 6.17 | 6.68 | 6.22 |
| $^{27}Al$ | 4.84 | 255.8 | 191.1 | 133.7 | 5.99 | 6.16 | 6.46 |
| $^{40}Ca$ | 5.90 | 331.5 | 228.2 | 162.1 | 7.76 | 7.36 | 7.83 |
| $^{44}Ca$ | 5.86 | 344.0 | 245.1 | — | 8.06 | 7.90 | — |
| $Cu$ | 6.79 | 400.2 | 297.9 | 209.9 | 9.37 | 9.61 | 10.14 |
| $^{116}Sn$ | 8.13 | 554.7 | — | — | 12.99 | — | — |
| $^{124}Sn$ | 8.06 | 533.2 | — | — | 12.49 | — | — |
| $^{181}Ta$ | 9.02 | 611.5 | 451.8 | 303.3 | 14.32 | 14.67 | 14.66 |
| $Pb$ | 9.27 | 588.4 | 481.6 | 324.5 | 13.78 | 15.53 | 15.68 |

result justifies using the completeness approximation to obtain (7) and gives credability to expressions (21,22). From Tables 1 and 2 we also conclude that the optical model limit underestimates the cross sections ($N_{opt} < \overline{N}_{ij}$, i and j=1,2,3). For the calculations of the $T_p$-dependence of the values $\overline{N}$ and $<f^2>$ we used experimental cross sections $\sigma_{NN}^{tot}$ and $\sigma_{NN}^{el}$ from [13]. From Table 2 we conclude that in the considered region of energies the $T_p$-dependence of the effective numbers is smooth which substantiate the approximation $\sigma_{pN}(T_p) \approx \sigma_{nN}(T_n)$ employed above.

## 4 COMPARISON WITH DATA FOR THE REACTION A(p,n)$_\Delta$B

We now turn to an analysis of the experimental data on the reaction $A(p,n)_\Delta B$ in the termenology of effective numbers and with reference to eq.(11). Table 3 contains the $\theta_n$- dependence (three angles) of effective numbers $\overline{N}^{exp}$, extracted according to formula (11) on the basis of experimental cross sections [14] for a number of nuclei and for the elementary process at $T_p = 1$ GeV. From the table it is seen that within the experimental errors the effective number approximation describes the angular dependence of cross sections of charge-exchange. The observed anomalies, the value $\overline{N}^{exp}$=4.95 for the nucleus $^{24}Mg$ ($\theta_n = 7.5°$) and also $\overline{N}^{exp}$=13.78 for Pb ($\theta_n = 4°$), most likely reflect

Table 4. The A-dependence of cross sections $d\sigma[A(p,n)_\Delta B]/d\Omega_n \equiv \sigma(\theta_n)$ [mb/sr] and effective numbers for reaction $A(p,n)_\Delta B$ at $T_p = 0.8$ GeV [15,16] at $\theta = 0°$. The values $\overline{N}^T$ are our results.

| Target | A | Z | $\sigma(0°)$ | $\overline{N}^{exp}$ | $\overline{N}^T$ | $\Delta N = \overline{N}^{exp} - \overline{N}^T$ | $K = \frac{\overline{N}^{exp}}{\overline{N}^T}$ | $<f^2_{33}>$ |
|---|---|---|---|---|---|---|---|---|
| $H$ | 1 | 1 | $33.0 \mp 3.0$ | 1.0 | 1.0 | 0 | 1.00 | — |
| $Al$ | 27 | 13 | $271.4 \mp 2.0$ | 8.2 | 6.0 | 2.2 | 1.36 | 0.34 |
| $Ti$ | 47.9 | 22 | $372.1 \mp 2.7$ | 11.3 | 7.9 | 3.4 | 1.43 | 0.26 |
| $Cu$ | 63.5 | 29 | $425.0 \mp 3.2$ | 12.9 | 8.9 | 4.0 | 1.45 | 0.22 |
| $W$ | 183.9 | 74 | $695.5 \mp 5.6$ | 21.1 | 12.3 | 8.8 | 1.71 | 0.11 |
| $Pb$ | 207.2 | 82 | $695.4 \mp 5.5$ | 21.1 | 12.6 | 8.5 | 1.70 | 0.10 |
| $U$ | 238 | 92 | $767.9 \mp 6.3$ | 23.3 | 13.0 | 10.3 | 1.80 | 0.09 |

Table 5. The same as in table 4 but for $T_p = 1$ Gev at $\theta_n = 4°$ [14]

| Target | $\sigma(4°)$ | $\overline{N}^{exp}$ | $\overline{N}^T$ | $\Delta N = \overline{N}^{exp} - \overline{N}^T$ | $K = \frac{\overline{N}^{exp}}{\overline{N}^T}$ | $<f^2_{33}>$ |
|---|---|---|---|---|---|---|
| $H$ | $42.7 \mp 4.3$ | 1.0 | 1.0 | 0 | 1.00 | — |
| $D$ | $52.1 \mp 5.2$ | 1.2 | 0.6 | 0.6 | 2.0 | 0.45 |
| $^7Li$ | $123.8 \mp 3.8$ | 2.9 | 2.2 | 0.7 | 1.35 | 0.50 |
| $^9Be$ | $177.1 \mp 5.3$ | 4.2 | 2.5 | 1.7 | 1.66 | 0.44 |
| $^{10}B$ | $165.8 \mp 5.0$ | 3.9 | 2.8 | 1.1 | 1.40 | 0.42 |
| $^{11}B$ | $159.8 \mp 4.7$ | 3.7 | 2.7 | 1.0 | 1.36 | 0.39 |
| $^{12}C$ | $162.3 \mp 4.8$ | 3.8 | 3.0 | 0.8 | 1.28 | 0.37 |
| $^{16}O$ | $220.6 \mp 11.0$ | 5.2 | 3.2 | 2.0 | 1.64 | 0.30 |
| $^{19}F$ | $246.6 \mp 12.4$ | 5.8 | 3.1 | 2.7 | 1.86 | 0.25 |
| $^{24}Mg$ | $254.3 \mp 17.5$ | 6.0 | 4.7 | 1.3 | 1.29 | 0.29 |
| $^{25}Mg$ | $243.6 \mp 17.0$ | 5.7 | 4.7 | 1.0 | 1.23 | 0.28 |
| $^{26}Mg$ | $263.6 \mp 17.6$ | 6.2 | 4.7 | 1.5 | 1.33 | 0.28 |
| $^{27}Al$ | $255.8 \mp 7.5$ | 6.0 | 4.8 | 1.2 | 1.24 | 0.27 |
| $^{40}Ca$ | $331.5 \mp 23.2$ | 7.8 | 5.9 | 1.9 | 1.38 | 0.22 |
| $^{44}Ca$ | $344.0 \mp 24.2$ | 8.1 | 5.9 | 2.2 | 1.38 | 0.21 |
| $Cu$ | $400.2 \mp 16.2$ | 9.4 | 6.8 | 2.6 | 1.38 | 0.17 |
| $^{116}Sn$ | $554.7 \mp 28.0$ | 13.0 | 8.1 | 4.9 | 1.60 | 0.11 |
| $^{124}Sn$ | $533.2 \mp 26.9$ | 12.5 | 8.1 | 4.4 | 1.55 | 0.11 |
| $^{181}Ta$ | $611.5 \mp 26.1$ | 14.3 | 9.0 | 5.3 | 1.59 | 0.08 |
| $Pb$ | $588.4 \mp 23.8$ | 13.8 | 9.3 | 4.5 | 1.49 | 0.08 |

statistical hoist of the corresponding experimental data more than a presence of dynamical or structural factors in the reaction mechanism. An analogous situations was described in Refs.[5,7] in the analysis of the inclusive process $(p, pd)$ (The experimental number of the deuterons in Pb, $\overline{N}^{exp} = 19 \pm 2.5$, measured at $T_p = 1$ GeV, was excluded because it diviated from the systematics). We conclude that $\overline{N}^{exp}(\theta_n)$, within experimental uncertainties, does not depend on $\theta_n$. Thus, the angular dependence of cross sections $d\sigma[A(p,n)_\Delta B]/d\Omega_n$ testifies in favour of of the approximation of effective numbers. Physically, this means that the process takes place on the nuclear periphery, i.e. in the region where the density of nucleons is small, and consequently, all NN and $\Delta N$ interactions in the nuclear are close to free interactions. While the angular behaviour seems to be accounted for by the factorized form (11), the magnitude of the extracted effective numbers $\overline{N}^{exp}$ deviates from what is obtained theoretically, $\overline{N}^T$, by the approach described above. In Tables 4 and 5 we list the results of data processing performed in [15,16] (at $T_p = 0.8$ GeV at $\theta = 0°$) and in [14] (at $T_p = 1$ GeV at $\theta = 4°$). From the tables it is seen that $\overline{N}^{exp}$ systematically exceeds $\overline{N}^T$ by about a factor 1.5, which clearly points to the insufficiency of the approximation of effective numbers for the description of integral cross section of the reaction $A(p,n)_\Delta B$. The employment of "optical" absorption factors instead of Glauber's gives even a further deterioration of the description. In this case the ratio $\overline{K} = \overline{N}^{exp}/\overline{N}^T$ becomes about 2.

## 5 EXTENSION OF THE THEORY

This situation may be qualitatively understood if we consider the A-dependence of the quantity $\Delta N = \overline{N}^{exp} - \overline{N}^T$. It follows from Tables 4 and 5 that $\Delta N$ grows, on the whole, according to the law $A^\alpha$, where $\alpha$=0.6-0.8. If we assume (remember a previous conclusion) that the effect of going off mass-shell in the cross section $d\sigma_{p+p\to n+\Delta^{++}}(\vec{P_i}, \vec{Q})/d\Omega_n$ is not very large, we may consider expression (11) to be valid for description of that part of the cross section of the reaction $A(p,n)_\Delta B$ where a realistic $\Delta^{++}$ or a $\Delta^+$-isobar is generated (see diagram 1, Fig.1a). By definition, diagram 1 contains only the direct charge-exchange process, but in the presense of the nuclear medium, implying some renormalization of the 2-body interaction. However, within the developed formalism, the cross section of diagram 1 includes both the direct and exchange (Pauli principle) terms. The latter term corresponds physically to excitation of the $\Delta$-isobar in the incident particle and plays an important role. A similar comment can be made for diagram 2 (see Fig.1b) discussed below. Taken this into account the observed cross section should be written as the sum

$$\frac{d\sigma_{A(p,n)_\Delta B}}{d\Omega_n} = \frac{d^{(1)}\sigma_{A(p,n)_\Delta B}}{d\Omega_n} + \frac{d^{(2)}\sigma_{A(p,n)_\Delta B}}{d\Omega_n}. \tag{23}$$

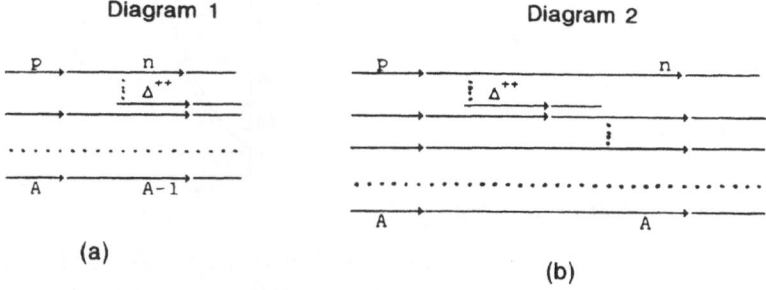

Fig.1. (a)This process will in shorthand notation be denoted by $(p, n)_\Delta^{\pi N}$ with emphasis on reality- principal observability- of a produced $\Delta$-isobar whose de-excitation takes place through decay into a pion and a nucleon. (b)This process may be referred to as mesonless $\Delta$-isobar de-excitation [17]. We will denote it by $(p, n)_\Delta^{NN}$, thus emphazising that a virtual $\Delta$-isobar is involved through charge-exchange on one of the intranuclear nucleons.

The first term corresponds to the approximation of effective numbers (11); the second term is connected with the process of exchange of virtual mesons, in which a virtual $\Delta$-isobar also is present (see diagram 2), bringing in effects from 3-body and higher order forces.

The cross section corresponding to diagram 2 can be written as

$$\frac{d^2\sigma_{A(p,n)\Delta B}}{d\Omega_n} \approx \int d\vec{P} \int d\vec{Q}' \int d\vec{Q}\phi(\vec{P},\vec{Q},\vec{Q}')[\Phi_N^{A-1}(\vec{Q}')]^2 \frac{d\sigma_{p+p\rightarrow n+\Delta++}(\vec{P_i},\vec{Q}')}{d\vec{P}}$$

$$\mid G(E_\Delta)\mid^2 [\Phi_N^A(\vec{Q})]^2 \frac{d\sigma_{p+p\rightarrow n+\Delta++}(\vec{P_i},\vec{Q})}{d\Omega_n}. \tag{24}$$

Here $\phi(\vec{P},\vec{Q},\vec{Q}')$ is the kinetic factor, $G(E_\Delta)$ is the Green's function describing the propagation of the $\Delta$-isobar with energy $E_\Delta$ in the intermediate state, $\vec{P}$ the impuls of the fast nucleon created as result of the second act of the charge-exchange. $\vec{Q}$ and $\vec{Q}'$ are the momenta (Fermi motion) of nucleons in the target-nucleus involved in the first and second act of charge-exchange, respectively. To obtain expression (24) we have carried out averaging over the spin projection of the $\Delta$ and used the approximation of completeness. As in the case of the single-step charge-exchange, the inequality $\mid \vec{P_i} \mid \gg P_F$ allows one to ignore the $\vec{Q}$-dependence of cross section $d\sigma_{p+p\rightarrow n+\Delta++}(\vec{P_i},\vec{Q})/d\Omega_n$. Thus, the cross section of the process $(p,n)_\Delta^{NN}$ is also factorized,

$$\frac{d^2\sigma_{A(p,n)\Delta B}}{d\Omega_n} = \Delta\overline{N}\frac{d\sigma_{p+p\rightarrow n+\Delta++}(\vec{P_i},\vec{Q})}{d\Omega_n}\mid_{free}, \tag{25}$$

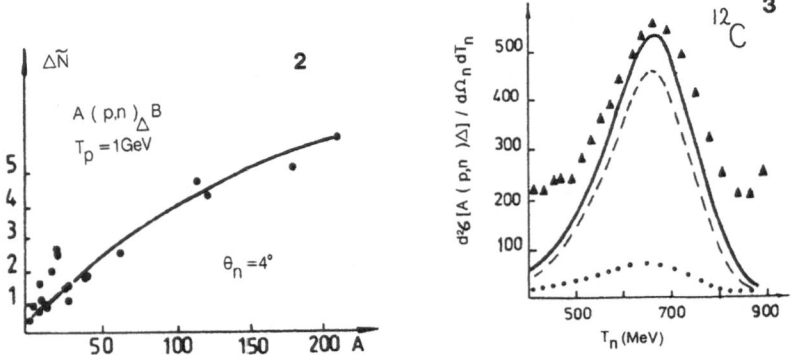

Fig.2 Mass number (A) dependence of the effective number $\Delta\overline{N}$ (solid line). The dots are the differences between experimental data [14] and calculated by the formula (11).

Fig.3 The energy spectrum of neutrons from the reaction $^{12}C(p,n)_\Delta$ at $T_p$=1 GeV. The experimental data [14] at $\theta_n = 4°$ are denoted by triangles. The dashed curve is the contribution of diagram 1; the dotted, diagram 2; and solid line their sum. The theoretical spectra were calculated at $\theta_n = 0°$.

$$\Delta \overline{N} = \int d\vec{P} \int d\vec{Q}' \int d\vec{Q} \phi(\vec{P}, \vec{Q}, \vec{Q}')[\Phi_N^{A-1}(\vec{Q}')]^2 \frac{d\sigma_{p+p\to n+\Delta^{++}}(\vec{P_i}, \vec{Q}')}{d\vec{P}}$$

$$|\, G(E_\Delta)\,|^2\,[\Phi_N^A(\vec{Q})]^2. \tag{26}$$

To neglect of the off-mass-shell effects in formula (26) is inadmissible because the momenta $|\vec{P}|$ and $P_F$ are commensurable quantities. The region of integration in (24) and (26) is determined by the width of the $\Delta$-peak and the Fermi momentum, by the properties of the Green's function and momentum distribution $[\Phi_N^A(\vec{Q})]^2$. As final result we have found a modified factorized form for (23),

$$\frac{d\sigma_{A(p,n)\Delta B}}{d\Omega_n} = [\overline{N} + \Delta\overline{N}]\frac{d\sigma_{p+p\to n+\Delta^{++}}(\vec{P_i}, \vec{Q})}{d\Omega_n}\,|_{free}, \tag{27}$$

$$\overline{N} + \Delta\overline{N} = \kappa_1 A^{1/3} + \kappa_2 A^{2/3}. \tag{27a}$$

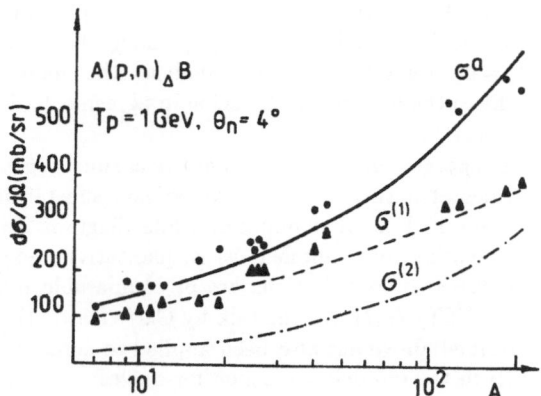

Fig.4 Partial contributions $\sigma^{(1)}$ and $\sigma^{(2)}$ from the processes $(p,n)_\Delta^{\pi N}$ and $(p,n)_\Delta^{NN}$ pictered in diagram 1 and 2 of Fig.1, $\sigma^{(a)}$ is their sum. The full dots are experimental data from Ref.[14]. The partial contributions $\sigma^{(1)}$ and $\sigma^{(2)}$ were obtained by a $\chi^2$ fit using the formula $d\sigma/d\Omega_n = d\sigma^{(1)}/d\Omega_n + d\sigma^{(2)}/d\Omega_n = \overline{\kappa}_1 A^\alpha + \overline{\kappa}_2 A^{2\alpha}$, ($\overline{\kappa}_1=50$, $\overline{\kappa}_1=5.1$, $\alpha=0.38$), where the first (second) term corresponds to diagram 1 (2).

Expression (27) indicates that the angular spectra of neutrons from the charge-exchange channel $(p,n)_\Delta^{NN}$ coincide in form with analogous spectra of neutrons from the charge-exchange channel $(p,n)_\Delta^{\pi N}$. This result allows us to understand the experimental data from Ref. [14] and it also points to the impossibility in principle to separate the contribution of the channel $\Delta \to \pi N$ to the charge-exchange cross section from the contribution of the channel $\Delta N \to NN$ on the basis of merely angular spectra of neutrons.

Table 6. The A-dependence of effective numbers $\overline{N}_{QCE}^{exp}$ and $\overline{N}_{QCE}^{T}$ for the reaction A(p,n)B in the quasielastic region at $T_p$=1 GeV at $\theta_n = 4°$ [14].

| Target | $\overline{N}_{QCE}^{exp}$ | $\overline{N}_{QCE}^{T}$ |
|---|---|---|
| $^{12}C$ | 2.23 | 1.80 |
| $^{16}O$ | 2.72 | 1.88 |
| $^{27}Al$ | 3.22 | 3.02 |
| $^{40}Ca$ | 3.39 | 3.41 |
| $^{116}Sn$ | 6.62 | 5.53 |
| Pb | 9.79 | 6.92 |

The leading A-dependence of the contribution from diagram 2 follows directly from the 2-step character of the process. Direct calculation in the plane wave approximation shows that the functions $\phi(\vec{P}, \vec{Q}, \vec{Q}')$ and $\mid G(E_\Delta) \mid^2$ depend little on A, while $[\Phi_N^A(\vec{Q})]^2 \propto A^{1/3}$ as discussed previously. Thus, $\Delta N \propto A^{1/3} A^{1/3} = A^{2/3}$ which is in full agreement with the experimental data [14,15,16] (see also expression (27a) and Fig. 2). In Fig. 3 we present the energy spectrum of neutrons from $d\sigma^{(2)}[^{12}C(p,n)_\Delta]/d\Omega_n dT_n$ at $T_p$=1 GeV compared with theory (we used the plane wave approximation for the fast nucleon created as result of the second act of charge-exchange). The theoretical spectra were calculated for $\theta_n = 0°$. As follows from our analysis, the channel described by diagram 2 improves the fit to the data in the $\Delta$-region in agreement with the conclusions of Ref. [18] devoted to $(e, e')_\Delta$.

In Fig.4 we present separate contributions from the channels $(p, n)_\Delta^{\pi N}$ and $(p, n)_\Delta^{NN}$ extracted from experimental data [14]. The comparision substantiates that for low mass numbers A diagram 1 (Fig.1a) dominates, while diagram 2 (Fig.1b) plays an increasing role with increasing mass number, as is qualitativley expected. For large mass number ($\geq 100$) the two contributions are of comparable magnitude. For an additional discussion for $^{12}C(^3He, t)$, see the talk by Carl Gaarde [19] in this volume.

The formalism described above has also been applied to compute the cross section of reaction the A(p,n)B in the quasielastic region for nuclear excitation, i.e. in a high momentum region of the neutron spectrum in which an incident proton either knocks out a neutron or suffers a charge-exchange on target-nucleons without excitation of the $\Delta$-isobar. From Table 6 it is seen (for details see [8]) that also in this region of excitation of a nucleus, the proposed approach describes the integral cross section of reaction A(p,n)B with reasonable accuracy.

# 6 COMPARATIVE ANALYSIS OF (p,n) AND ($^3$He,t) REACTIONS ON NUCLEI

The experimental information on the inclusive cross sections of the $(p, n)$ reactions on nuclei for the quasielastic and the $\Delta$-isobar excitation region is far from sufficient. The most comprehensive tabulation is that of Ref.[14] which presents experimental data on neutron production in 1 GeV proton interactions with different nuclei at angles $4°, 7.5°$ and $11.3°$. Neutron spectra at $\theta = 0°$ from p-p and p-d collisions at $T_p$=647 and 800 MeV incident energies were measured in Ref.[15], and systematics of $\theta = 0°$ neutron production by 800 MeV protons on targets with $27\leq A \leq238$ have been reported in

Ref. [16]. Neutron spectra at $\theta = 0°$ from p-p collisions have also been measured for $T_p$=647, 771 and 805 MeV [20], we have employed these results above.

A systematic experimental study of $\Delta$-isobar excitations in nuclei started with experiments of the inclusive type $(^3He, t)$ in Dubna [17] at beam energies $T_{^3He}$ from 800 MeV/A up to 5.23 GeV/A and in Saclay [21] at energies 500, 667 and 767 GeV/A, near the threshold of the $\Delta$-isobar production. In these experiment one measured differential cross sections of the charge- exchange reactions on free proton targets and nuclear targets as functions of the energy $Q = (E_{^3He} - E_t)$ transferred to the target at an angle of $\theta \approx 0°$ for the outgoing tritons.

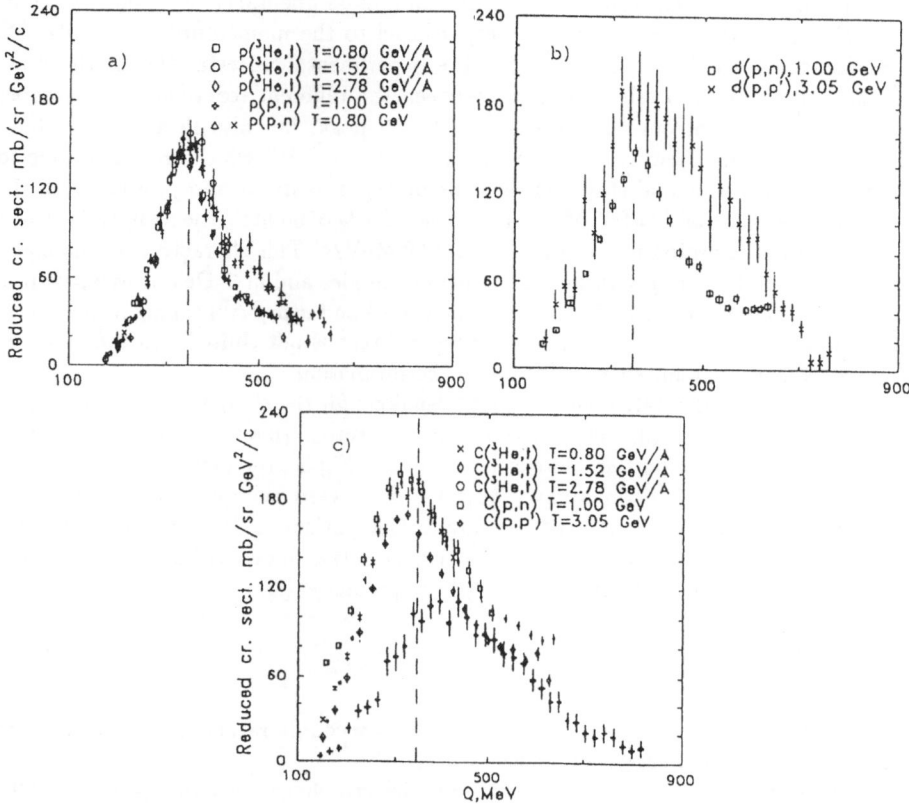

Fig.5 (a)"Reduced" cross sections of the inclusive p(p,n) and $p(^3He, t)$ charge exchange reactions, when the effects of the $(^3He, t)$ formfactor and the energy dependence of the "elementary" reaction cross section are removed as explained in the text. The dashed line marks the $\Delta$-peak position in the p(p,n) data. The $p(^3He, t)$ data are taken from Ref.[17], p(p,n) data are from Refs.[14,20,23]. (b)"Reduced" cross sections of the inclusive d(p,n) charge exchange [14] and exclusive $d(p, p'\Delta^0)$ reaction [22]. The isospin factor for the latter case is taken into account. (c)The same as for Figs. a) and b) for charge exchange reactions on the C-nucleus. In addition to the factors discussed above, the difference in the absorption factors of the projectile-ejectile passing through the target nucleus is taken into account. A clear shift to lower Q is evident in the inclusive data. A shift to higher Q is present in the quasifree $\Delta^0$-production data of KEK [22].

We have calculated the quantity

$$\bar{\sigma} = \frac{C(p_i, p_f)}{3F(t)} \frac{d^2\sigma[A(^3He, t)B]}{p\,dQ\,d\Omega}, \tag{28}$$

where $F(t) = \exp(-27.736 \mid t \mid)$ is the square of magnetic transition formfactor for $(^3He, t)$ and the factor $C(p_i, p_f) = \sigma(pp \to pn\pi^+) \mid_{p_f} /\sigma(pp \to pn\pi^+) \mid_{p_i}$ is introduced to compensate the energy dependence of the cross sections and to bring the experimental data at different energies to the same one, which is choosen to be 800 MeV/A. Here $\sigma(pp \to pn\pi^+) \mid_{p_f}$ $(\sigma(pp \to pn\pi^+) \mid_{p_i})$ is the cross section of the elementary process at impuls $p_f(p_i)$, and $d^2\sigma A(^3He, t)B/p\,dQ\,d\Omega$ the invariant cross section of charge-exchange reactions on nuclei. Fig.5a shows the corresponding experimental cross sections for (p,n) and $(^3He, t)$ reactions on a proton-target, reduced to the momentum 1.47 GeV/c/A, as explained before. Taking into account the experimental uncertainties of the cross sections there is remarkable agreement between the different reactions for a number of beam energies, not only in the shape of the $\Delta$-peak, but also in absolute values. Some diviations are, however, present at high Q$\approx$ 500 MeV, which are partly due to the interaction in the final state between the detected neutrons and protons from the reaction $pp \to pn\pi^+$ at $E_p$=800 MeV. In that case for Q$\approx$500 MeV the neutron is at rest in the CM and the proton momentum is about 30 MeV/c. This characteristic kinematic region moves to higher Q with increasing beam energies and at 1 GeV it escapes from the region of interest for us. Fig.5b shows the d(p,n) and $d(p, p'\Delta^0)$ taking into account the isospin factor. The $\Delta$-peak in the deuteron-target is not shifted and the width of $\Delta$-isobar is increasing for small value due to Fermi-motion.

An analogous comparative analysis has been done for the charge-exchange reactions on $^{12}C$. In this case we take into account that the protons (neutrons) and $^3He(t)$ have different absorption properties. As can be seen from Fig.5c, the inclusive experimental data for the different reactions have the same downward energy shifts, about 30-40 MeV, relative to the peak for the proton target. Comparision of the reaction $C(p, p'\Delta^0)$ with quasifree $\Delta$-production leads to the conclusion that its contribution to the total inclusive cross section is about 50% as has been predicted [8].

## 7 SUMMARY

We have outlined basic features of a theory which accounts rather well for (p,n) data at $T_p$=1 GeV on a wide range of nuclei:

1) The A-dependence of the peak for the quasielastic charge-exchange process (QCE) is

$$\{\frac{d\sigma A(p, n)B}{d\Omega_n}\}_{QCE} \sim A^{\frac{1}{3}}, \tag{29}$$

which argues in favor of a one-step reaction mechanism and indicates that the periph-erical region gives the main contribution to the cross section due to a strong volume absorption. The maximum is shifted to lower Q values compared with elastic scattering of protons on a neutron (that is on a deuteron, experimental data for the free p+n $\to$ n+p scattering are not available, only data for the inverse reaction n+p $\to$ p+n exist). This shift is trivially associated with the binding energy of the knock-out neutrons in the target nucleus. The width of the quasielastic peak is almost completely associated with the Fermi-motion of the target nucleons. The angular distributions of the reaction $A(p,n)B_{QCE}$ are practically independent of the target nucleus.

2) The A-dependence of the charge-exchange reaction $A(p,n)_\Delta B$ in the $\Delta$-excitation region is

$$\{\frac{d\sigma A(p,n)_\Delta B}{d\Omega_n}\} \sim aA^{\frac{1}{3}} + bA^{\frac{2}{3}}. \tag{30}$$

The first term is again associated with the single step reaction mechanism while the second is due to a multi-step (but "direct") mechanism. The maximum is shifted by about 30-40 MeV toward the region of high neutron momentum which indicates the presence of some reaction mechanism which compensates the binding effects of nucleons in the nuclei. The width of $\Delta$ in nuclei is about 1.5 times larger than the width of the decay of the free $\Delta$-isobar. It increases strongly with increasing A and cannot be explained fully by the Fermi-motion of nucleons.

Our main conclusion is that both the inclusive cross sections of the reactions (p,n) on a number of targets and $(^3He,t)$ on C are found to be proportional to the cross section of the corresponding processes on a free proton;

$$\frac{d^2\sigma[A(a,b)_\Delta B]}{pdQd\Omega} = N_{eff}F(t)\frac{d^2\sigma(p+p \to n+\Delta^{++})}{pdQd\Omega}, \tag{31}$$

where (a,b) is (p,n) or $(^3He,t)$. This is a theoretical result, supported by availalable data. The increased complexity of the projectile (ejectile) is contained in the extra factor F(t). This suggests that all NN and $\Delta$N interactions in the nucleus are close to free interactions. The shift of the $\Delta$-peak in C relative to the free nucleon target is the same in both reactions and is equal to about 30-40 MeV independent of the type and energy of the projectile. It means that the reaction mechanism for the (p,n) and $(^3He,t)$ is in principle almost the same, the process takes place on the nuclear periphery and all NN ana $\Delta$N interactions in the nucleus are close to free interactions.

We expect that the same conclusion applies to other types of light ion charge-exchange reactions. If that is true possibilities open up for extraction of information about formfactors for more exotic nuclei such as the radioactive nuclei $(^6He, ^6Li)$, $(^{11}Li, ^{11}Be)$.

More experimental data for charge-exchange reactions induced by different projectiles is desirable, in particular of the exclusive type. Prespectivies of performing exclusive studies of charge-exchange reactions are challenging, including spin observables, and constitute the next logical step in understanding the basic mechanism of such reaction.

## REFERENCES

[1] Proc.of the Telluride Int. Conf.on "Spin Observables of Nuclear Probes", 14-17 March, (1988), in Telluride, Colorado.

[2] F.A.Gareev, S.N.Ershov, N.I. Pyatov and S.A.Fayans, Particles and Nuclei 19(1988)864.

[3] S.-W.Hong, F.Osterfeld and T.Udagawa, Phys.Lett. B245(1990)1.

[4] A.B.Migdal, E.E.Saperstein, M.A.Troitsky and D.N.Voskresensky, Physics Reports 192(1990)179.

[5] S.G.Kadmensky and Yu.L.Ratis, Jad. Fis. 38(1983)1325.

[6] R.D.Smith and S.J.Wallace, Phys. Rev.C32(1985)1654.

[7] S.G.Kadmensky and V.I.Furman, "Alpha-decay", Moscow, Atomizdat (1985).

[8] F.A.Gareev and Yu.L.Ratis, Communications JINR P2-89-805, Dubna, (1989); Preprint JINR E2-89-876, Dubna,(1989); Hirschegg (1990); F.A.Gareev, Yu.L.Ratis and J.S.Vaagen (in preparation for Nucl. Phys. A).

[9] V.G.Neudatchin,Yu.F.Smirnov and N.F.Golovanova, Adv. Nucl. Phys. 11(1979)1.

[10] C.Gaarde, Nucl. Phys., A476(1988)475c.

[11] E.Byckling and K.Kajantie, "Particle kinematics", ed. John Wiley and sons, (1973).

[12] T.Hennino et al., Phys. Rev, Lett. 48(1982)997.

[13] B.S.Baraschenkov "Sechenja vsaimodestvja elementarnex chactiz", Moscow (1966).

[14] V.N.Baturin et al., JETP Lett. 30(1979)86; Preprint LIJAPH, N 1322, (1987).

[15] C.W.Bjork et al., Phys.Lett. B63(1976)31.

[16] B.E.Bonner et al., Phys. Rev. C18(1978)1418.

[17] V.G.Ableev et al., Pis'ma ZHETF, 40(1984)35; Jad. Fiz. 46(1987)549; 48(1988)27.

[18] C.R.Chen and T.-S.H.Lee, Phys.Rev.C38(1988)2187.

[19] C.Gaarde, Telluride Talk (1991).

[20] G.Glass et al., Phys. Rev. D15(1977)36.

[21] C.Ellegaard et al., Phys. Rev. Lett. 50(1983)1745; Phys. Lett. B154(1985)110; Phys. Rev. Lett. 59(1987)974; D.Contardo et al., Phys. Lett. B168(1986)331.

[22] T. Nagae et al., Phys. Lett. B191(1987)31.

[23] T.Rupp et al., Phys. Rev. C28(1983)1696.

# RECENT DEVELOPMENTS IN LOW-ENERGY NUCLEON-NUCLEON
# INTERACTION STUDIES

Werner Tornow

Department of Physics, Duke University and Triangle Universities
Nuclear Laboratory
Durham, NC 27706, USA

## INTRODUCTION

Even after decades of intensive work on nucleon-nucleon (NN) scattering, there are still
unanswered, basic questions about subtle details of the NN interaction at low energies. In
this talk "low energy" refers to nucleon laboratory energies of less than 100 MeV. The
answers to these questions are important since they are related to the fate of modern,
meson-exchange based NN potential models. Of course, any significant new
developments at low energies will also have an impact on the parameterization and
understanding of the NN interaction at higher energies.

I will discuss the questions of possible charge-independence and charge-symmetry
breaking of the $^3P$ NN interactions and review recent progress in determining the strength
of the on-shell NN tensor force at low energies. As is known since a long time, the 2N
systems provide insufficient information about some details of the NN force. We will
show that certain three-nucleon (3N) observables display a larger sensitivity to specific
parts of the NN interaction. Therefore, we will, whenever possible, include in our
discussion results obtained from 3N studies.

## CHARGE-INDEPENDENCE BREAKING STUDIES IN $^3P$ NN STATES

Charge-independence breaking (CIB) effects in the NN interaction are expected from the
up-down quark mass difference and from electromagnetic interactions among quarks. The
difference between the proton-proton (pp) scattering length $a_{pp} = -17.3 \pm 0.3$ fm and the
neutron-proton (np) scattering length $a_{np} = -23.715 \pm 0.015$ fm is a measure of CIB in the
$^1S_0$ state.[1] Experimental information on CIB in higher angular momentum NN states is

sparse. Arndt's[2] global NN phase-shift analysis SP89 includes very small CIB effects in the $^3P_{0,1,2}$ NN partial waves. Langacker and Sparrow[3] calculated the differences between the pp and np P-waves from 0 to 350 MeV using an isospin violating potential based on quark mass differences and both pseudoscalar $\pi$-$\eta$, $\pi$-$\eta'$ and vector $\rho$-$\omega$ mixing. Their calculated differences between the pp and np $^3P_0$ and $^3P_1$ phase shifts agree well with Arndt's phase-shift analysis results.

## The $\pi$NN Coupling Constant

Renewed interest in the question of CIB in $^3P$ NN states was triggered by the Nijmegen group in 1987. Bergervoet *et al.*[4] determined the neutral $\pi^\circ$NN coupling constant $g_0^2 = 13.1 \pm 0.1$ from a model-directed pp phase-shift analysis in the 0-350 MeV energy range. A refined analysis with the magnetic moment interaction included yields a somewhat larger result:[5] $g_0^2 = 13.53 \pm 0.14$. This value implies a surprisingly large charge-dependence of the $\pi$NN coupling since it is smaller than the charged $\pi^\pm$NN coupling constant $g_c^2 = 14.28 \pm 0.18$ obtained from $\pi^\pm$p scattering by Koch and Pietarinen[6] in 1980.

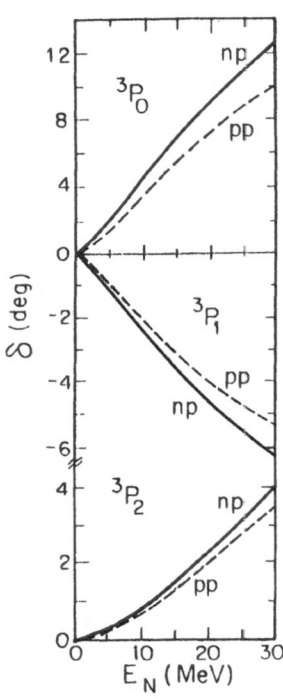

Fig. 1. Nijmegen's neutron-proton (np, solid curves) and proton-proton (pp, dashed curves)$^3P_0$, $^3P_1$, and $^3P_2$ phase shifts (Ref. 7).

The Nijmegen group incorporated their findings into a np phase-shift analysis[7] in the 0-30 MeV energy range and calculated $^3P_{0,1,2}$ np phase shifts which are considerably stronger than the corresponding pp $^3P_{0,1,2}$ phase shifts. Fig. 1 shows the results of Ref. 7. At 30 MeV the CIB obtained for $^3P_0$ is about a factor of ten larger than calculated by Langacker and Sparrow.

Although the Nijmegen work has been criticized (see Ref. 8), it prompted a new analysis of $\pi^\pm$p scattering data below 2 GeV by Arndt *et al.*[9] at VPI. This analysis, based essentially on data taken in the 1980's, yields a value for the charged $\pi^\pm$NN coupling constant of $g_c^2 = 13.31 \pm 0.27$, in agreement with Nijmegen's result for $g_0^2$ and in clear disagreement with the value of Koch and Pietarinen for $g_c^2$. If we assume that both the Nijmegen result for $g_0^2$ and the VPI result for $g_c^2$ are correct, then CIB effects in the $^3P_{0,1,2}$ interactions are very small or nonexistent. However, very recently, Machleidt and Sammarruca[10] examined the deuteron properties using the new value $g^2 = 13.3 \pm 0.3$ for the charged and neutral $\pi$NN coupling constant.

They find that the deuteron properties cannot be described in meson-exchange based NN potential models with such a small value for the pion coupling if the strong empirical $\rho$ coupling[11] is used. The smallest value for $g^2$ that still gives a reasonable value for the deuteron quadrupole moment is $g^2 = 13.9$.

In order to resolve the present confusion and uncertainty concerning the $\pi$NN coupling constants, $g_c^2$ must be determined in an independent way, i.e., from np scattering rather than from $\pi^{\pm}$p scattering alone. Such an analysis is currently underway at VPI and Nijmegen. In addition, the model dependence of the Nijmegen result for $g_0^2$ must be studied carefully.

## The $^3P_{0,1,2}$ Phase Shifts from np Scattering

As was shown by the Nijmegen[7] group, any charge dependence of the $\pi$NN coupling constant results in differences between the pp and np $^3P_{0,1,2}$ phase shifts. Therefore, if we assume that the pp $^3P_{0,1,2}$ phase shifts are accurately known from pp scattering, the question of charge dependence of the pion coupling is tied to the accurate determination of the np $^3P_{0,1,2}$ phase shifts.

In the past, as a first approximation, the $^3P_{0,1,2}$ phase shifts for np scattering were simply taken from pp scattering. They were then slightly modified as required to fit the existing np data base. As we have shown recently, among np observables, only the analyzing power $A_y(\theta)$ is sufficiently sensitive to allow an accurate determination of the $^3P_{0,1,2}$ phase shifts.[12] However, the accuracy required for CIB studies appears to be out of reach. Fig. 2 shows $A_y(\theta)$ at 10 MeV calculated with Nijmegen's np phase shifts (solid curve) and Nijmegen's pp phase shifts (dashed curve).[7] The difference between the two curves is smaller than the accuracy of the available np $A_y(\theta)$ data[13] at this energy. Note also the small magnitude of $A_y(\theta)$. Furthermore, one should also keep in mind that the three $^3P$ phase shifts cannot be determined unambiguously from data obtained for one observable only.

## The $^3P_{0,1,2}$ NN Phase Shifts from a Combined Analysis of 2N and 3N Data

The analyzing power $A_y(\theta)$ for elastic nucleon-deuteron (Nd) scattering is one order of magnitude larger than $A_y(\theta)$ for np and two orders of magnitude larger than $A_y(\theta)$ for pp scattering, thereby providing superior sensitivity to the $^3P_{0,1,2}$ phase shifts than that obtainable in the NN system. Fig. 3 demonstrates the large sensitivity of the nd $A_y(\theta)$ to the individual $^3P_{0,1,2}$ phase shifts.[14] In order to take full advantage of this feature, one needs exact solutions of the 3N Faddeev equations in the continuum using realistic NN potentials (for example Paris[15] and Bonn[16]). Such calculations are available since 1986.[17] Details of the calculational method used to solve the Faddeev equations in momentum space are given in Ref. 18. The numerical accuracy of these so-called "rigorous" calculations is

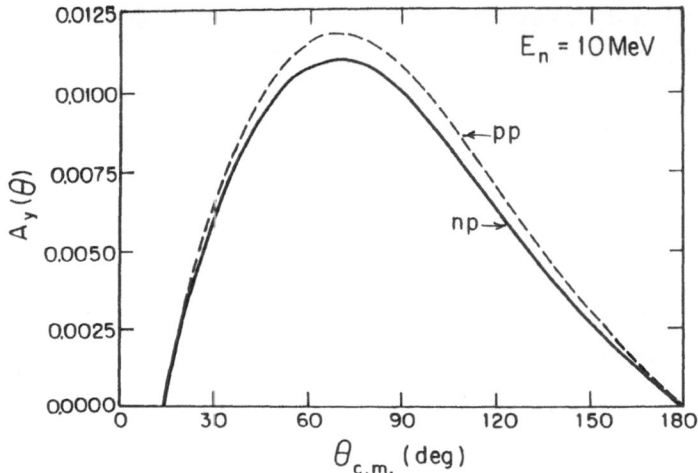

Fig. 2 Neutron-proton analyzing power $A_y(\theta)$ at 10 MeV calculated with Nijmegen's np (solid curve) and pp ${}^3P_{0,1,2}$ phase shifts (dashed curve).

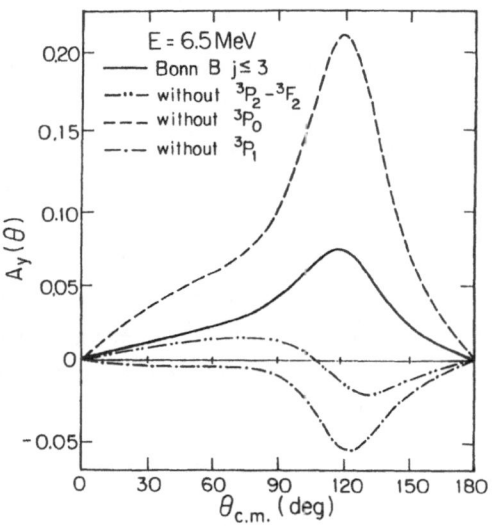

Fig. 3
Sensitivity of $A_y(\theta)$ in elastic neutron-deuteron scattering at $E_n$=6.5 MeV to the ${}^3P_0$, ${}^3P_1$ and ${}^3P_2$ - ${}^3F_2$ Bonn B NN phase shifts (Ref. 14).

about 1%.[19] Fig. 4 shows, as an example, nd differential cross-section $\sigma(\theta)$ data[20] at 14.1 MeV in comparison with a 3N calculation[21] using as the NN interaction the Bonn B[22] potential. The agreement between data and calculation is very good. The rigorous calculations are restricted to the nd or dn systems since the inclusion of the long-range Coulomb interaction in the case of pd or dp scattering creates difficulties which have not been solved for incident nucleon laboratory energies above the deuteron breakup threshold of 3.4 MeV. Nevertheless, Fig. 5 shows the comparison between dp tensor analyzing

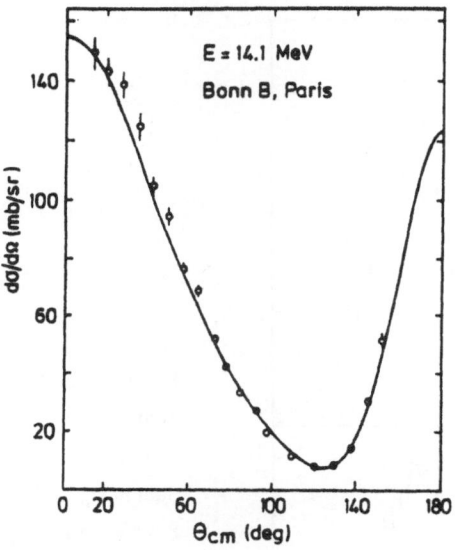

Fig. 4 Comparison between neutron-deuteron elastic differential cross-section data[20] at 14.1 MeV and rigorous calculations performed by the Bochum group.[21] The Bonn B and Paris NN potential model predictions are identical.

powers $T_{20}(\theta)$ and $T_{22}(\theta)$ data[23] and dn calculations.[24] As can be seen, except for the very forward angular range where Rutherford scattering dominates, the calculations are in excellent agreement with the data. The only nd observable which is not well described by the calculations using either the Paris or Bonn B potential is $A_y(\theta)$. Here, as displayed in Fig. 6, the calculated $A_y(\theta)$ between 5 and 8.5 MeV is too small by about 25% in comparison with the data in the angular region of the maximum of $A_y(\theta)$ which occurs near 120° c.m.[14] The same type of discrepancy was already observed earlier in the 10-14 MeV

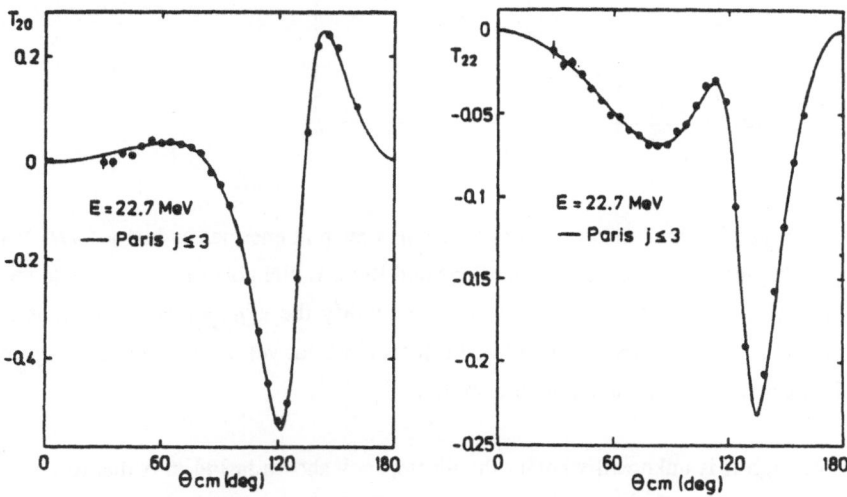

Fig. 5 Comparison between deuteron-proton tensor analyzing power data[23] $T_{20}$ (left side) and $T_{22}$ (right side) at 45.4 MeV and rigorous deuteron-neutron calculations (Ref.24) using the Paris NN potential.

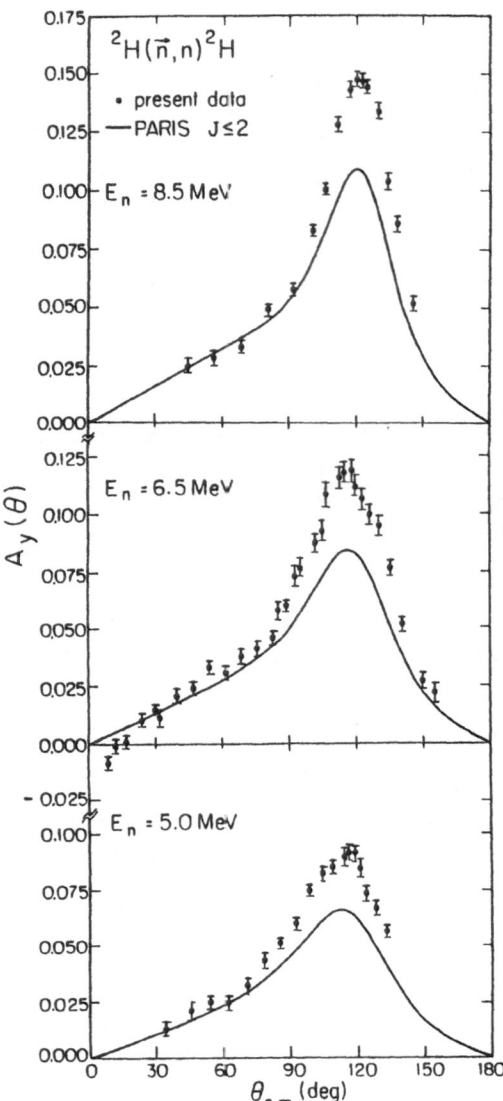

Fig. 6 Neutron-deuteron elastic analyzing power $A_y(\theta)$ data at $E_n$=5.0, 6.5 and 8.5 MeV in comparison with rigorous calculations of the Bochum group using the Paris NN potential (Ref. 14).

energy range.[21] Fig. 6 clearly demonstrates that even at energies as low as 5 MeV the $^3P_{0,1,2}$ NN phase shifts as used in the Paris and Bonn B NN potential are inadequate, or alternatively, that three-nucleon force effects modify the $^3P_{0,1,2}$ NN interactions to a substantial amount. Three-nucleon force effects have not yet been included in rigorous Faddeev calculations of scattering observables.

Although it is unknown whether the discrepancy shown in Fig. 6 is due to on-shell deficiencies of the $^3P_{0,1,2}$ NN interactions or off-shell (three-body force) effects, it is interesting to investigate what kind of modifications to the $^3P_{0,1,2}$ phase shifts are required to bring the calculated n-d $A_y(\theta)$ in closer agreement with the data. Very recently, such an analysis was performed by Witala and Glöckle.[25] Starting from the Bonn B potential

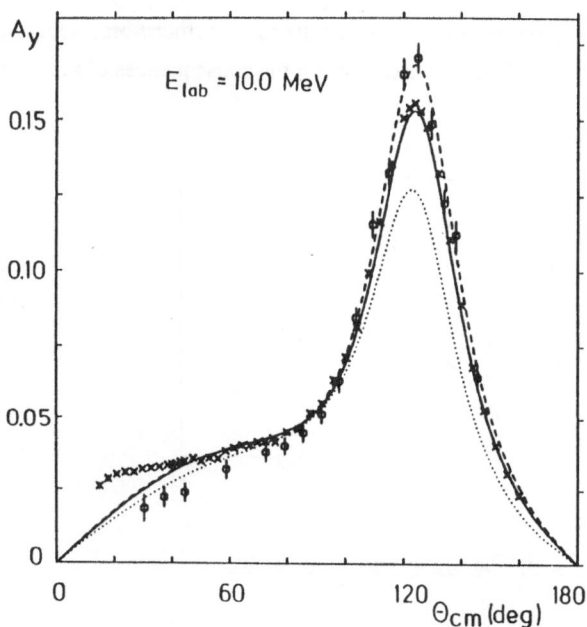

Fig. 7 Neutron-deuteron[26] (open circles) and proton-deuteron[23] (crosses) analyzing power data at 10 MeV in comparison with the Bonn B prediction (dotted curve). The solid (proton-deuteron) and dashed (neutron-deuteron) curves are based on charge-independence breaking (CIB) $^3P_{0,1,2}$ NN phase shifts (Ref. 25).

parameters the $^3P_0$, $^3P_1$, $^3P_2$, $\varepsilon_2$ and $^3F_2$ phase parameters were varied by searching on 2N <u>and</u> 3N observables simultaneously. Results of this novel approach are shown in Fig. 7. The magnitude of $A_y(\theta)$ in nd (dashed curve) and pd (solid curve) scattering at 10 MeV is in excellent agreement with the data in the critical angular range near 120° c.m.[26] Only the nd data forward of 60° c.m. are not described very well by the calculation. It has to be seen whether a readjustment of the D-wave NN interactions is possible in order to improve the fit in this angular region. Since the calculation does not include the Coulomb interaction it is not surprising that the solid curve does not fit the pd data at small angles. It should be pointed out that the fits to all 2N data are as good as those of the original Bonn B potential (for pp scattering the Bonn B $^1S_0$ phase shift was replaced by that of the Paris potential). Finally, Fig. 8 shows the $^3P_0$ phase shifts obtained in the work of Witala and Glöckle. In agreement with the Nijmegen 2N analyses[6,7] the solid and dashed curves clearly display a sizeable CIB in the $^3P_0$ phase shift in this simultaneous analysis of 2N and 3N data. However, as seen from Table 1, the $^3P_1$ and $^3P_2$ phase shifts are essentially charge independent.[25] Presently, it is not clear whether or not the surprising results of Ref. 25 require a complete departure from the, until recently, well-accepted parameterization of the $^3P_0$ NN phase shifts. Sensitivity studies at higher energies (>100 MeV) are necessary to investigate the impact of the present findings. Furthermore, accurate NN and ND analyzing power $A_y(\theta)$ data are needed to test the consequences of the work of Witala and Glöckle.

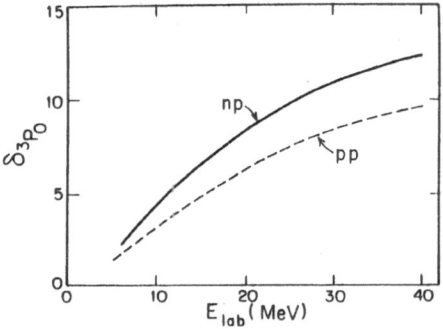

Fig. 8 Neutron-proton (solid curve) and proton-proton (dashed curve) $^3P_0$ NN phase shifts (Ref. 25).

<u>Charge-Symmetry Breaking Studies in $^3P$ NN States</u>

The recommended values for the nn scattering length $a_{nn} = -18.5 \pm 0.5$ fm and the Coulomb corrected pp scattering length $a_{pp} = -17.3 \pm 0.3$ fm are a measure of charge-symmetry breaking (CSB) in the $^1S_0$ NN state.[1] Due to the lack of a suitable free neutron target, the determination of $a_{nn}$ is not trivial and still subject of intensive experimental efforts. Nevertheless, the value given above for $a_{nn}$ now appears well established. Similar to the role of 3N <u>scattering</u> calculations in determining the $^3P$ NN interactions, 3N <u>bound state</u> calculations played an important role in the determination of $a_{nn}$. Since the Coulomb force accounts for only 680 keV of the 760 keV $^3H$-$^3He$ binding energy difference, the remaining 80 keV[27] require $a_{nn}$ to be more negative than $a_{pp}$ by about 1 fm.

Table 1. $^3P$ NN phase shifts in degree.

| $E_N$(MeV) | $^3P_0$ | $^3P_1$ | $^3P_2$ | $\varepsilon_2$ | $^3F_2$ | NN |
|---|---|---|---|---|---|---|
|  | 1.99 | -1.13 | 0.26 | -0.06 | 0.002 | np |
| 5 | 1.59 | -1.12 | 0.26 | -0.06 | 0.002 | pp |
|  | 1.36 | -1.14 | 0.26 | -0.06 | 0.002 | nn |
|  | 5.36 | -2.95 | 0.97 | -0.31 | 0.023 | np |
| 12 | 4.27 | -2.90 | 0.96 | -0.30 | 0.023 | pp |
|  | 3.64 | -2.97 | 0.97 | -0.31 | 0.023 | nn |
|  | 9.80 | -5.64 | 2.73 | -0.89 | 0.111 | np |
| 25 | 7.80 | -5.55 | 2.70 | -0.88 | 0.110 | pp |
|  | 6.65 | -5.68 | 2.73 | -0.89 | 0.111 | nn |
|  | 12.81 | -9.26 | 6.41 | -1.88 | 0.349 | np |
| 50 | 10.28 | -9.12 | 6.32 | -1.86 | 0.345 | pp |
|  | 8.79 | -9.33 | 6.41 | -1.88 | 0.349 | nn |

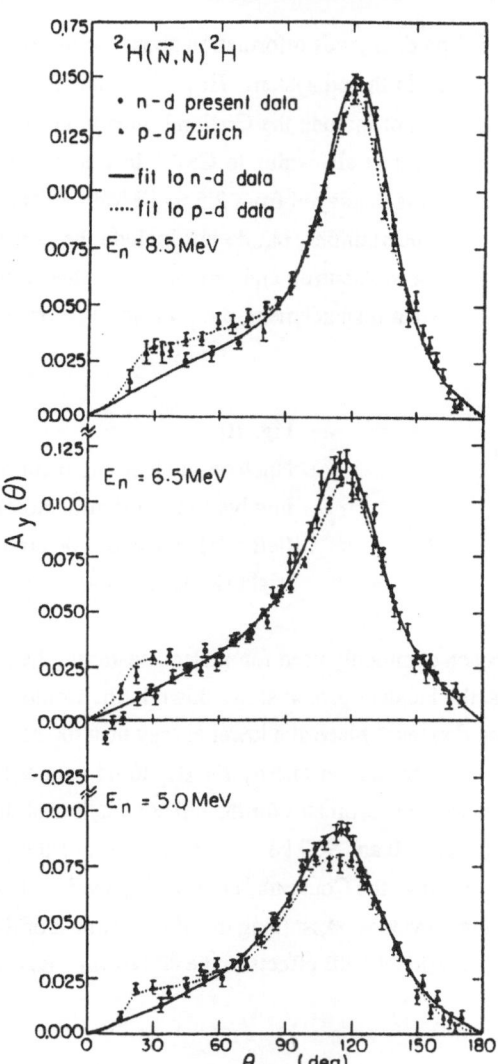

Fig. 9  Comparison of neutron-deuteron[14] and proton-deuteron[23] analyzing power data at 5.0, 6.5 and 8.5 MeV.  The solid (neutron-deuteron) and dotted (proton-deuteron) curves are based on Legendre polynomial fits to the product $\sigma(\theta)\, A_y(\theta)$ (Ref. 29).

The 3N binding energy calculations are not sensitive to small changes in the higher partial wave NN interactions.  Considering the experimental problem mentioned above, we conclude that 2N scattering and 3N bound state data are not well suited to yield information about CSB effects in the $^3P$ NN states.  According to Ref. 3 only small CSB effects are expected for $^3P_0$ and $^3P_1$.  However, Table 1 reveals a sizeable difference between the $^3P_0$ pp and nn phase shifts.  This apparent large Class III[28] CSB is intimately tied to differences between the pd and nd $A_y(\theta)$ data near 120° c.m. as shown in Fig. 7.  To further document this small difference Fig. 9 displays pd[23] and nd[14] $A_y(\theta)$ data at lower energies between 5 and 8.5 MeV.  Here, the solid (nd) and dotted (pd) curves are based on fitting the product $\sigma(\theta)A_y(\theta)$ using associated Legendre polynomials.[29]  Besides the difference at very forward angles which is due to the Coulomb interaction, the difference at $A_y(max)$, the maximum of $A_y(\theta)$ near 120° c.m. is clearly noticeable.  Fig. 10 indicates that, in principle, the comparison of nd and pd data gives information about the nn force acting in the nd system and the pp force acting in the pd system. However, since the 3N calculation shown in Fig. 7 for pd $A_y(\theta)$ does not include the Coulomb interaction it is unclear how much of the observed difference, if at all, is due to CSB. In Fig. 11, the energy dependence of $A_y(max)$, for nd and pd is displayed from 2.5 to 14 MeV.[29] The width of the bands represents the associated uncertainties, but do not include the "scale uncertainty" associated with the polarization of the neutron and proton beams that were used in the measurements. Except for a constant displacement, the pd and nd $A_y(max)$ curves are nearly identical.

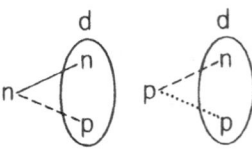

Fig. 10
Nucleon-nucleon interactions involved in neutron-deuteron (left side) and proton-deuteron (right side) scattering.

In the spirit of an optical-model approach commonly used for nucleon scattering from, of course, heavier nuclei than deuterons, the incident proton slows down in the Coulomb field of the nucleus and the nuclear interaction takes place at a lower energy than the beam energy. Consequently, one should compare pd data at energy $E+\Delta E_c$ to nd data at E, where $\Delta E_c$ is a "Coulomb shift." Using reasonable estimates for the effective radius of the deuteron one obtains values for $\Delta E_c$ between 400 and 900 keV. In fact, a shift of the pd values shown in Fig. 11 by $\Delta E_c = 650$ keV brings the Coulomb "corrected" pd values into excellent agreement with the corresponding nd values. Assuming that this treatment of the Coulomb effect is correct to first order, the residual CSB effects in the $^3P$ NN interactions appear to be very small.

To further illustrate the basis for this conclusion, in Fig. 12 we plot the relative difference between $A_y(max)$ for nd and pd.[29] The dashed error bars display the effect of merely adding the contribution of the scale uncertainty in the neutron and proton beam polarizations. The point plotted at 2.5 MeV was obtained from the theoretical work of Berthold et al.[30] Their calculation includes, for the first time, the Coulomb interaction below the deuteron breakup threshold in an exact way in pd Faddeev calculations using a realistic NN interaction; no CSB effects were incorporated in the $^3P_{0,1,2}$ NN interactions. Figure 12 shows that this point is consistent with the trend of the measurements at higher energies. The solid curve is based on an optical-model calculation for Nd scattering.[29] In this calculation the nd data at 10 MeV were fitted. In order to predict pd scattering results, the Coulomb interaction (assuming a uniformly charged sphere) was added without any modification of the nuclear part of the interaction. We used an energy dependence for the real part of the nuclear potential that is common for heavier nuclei $(V_0 - 0.3E_N)$ and obtained the solid curve. Although this calculation yields larger relative differences it compares favorably with the experimental trend.

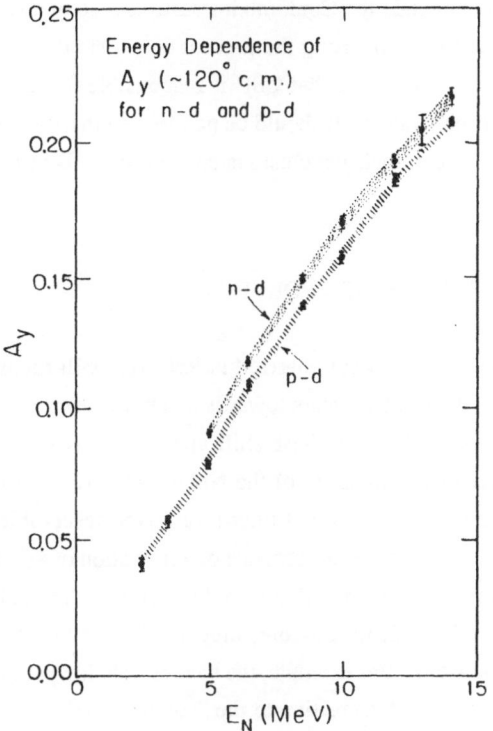

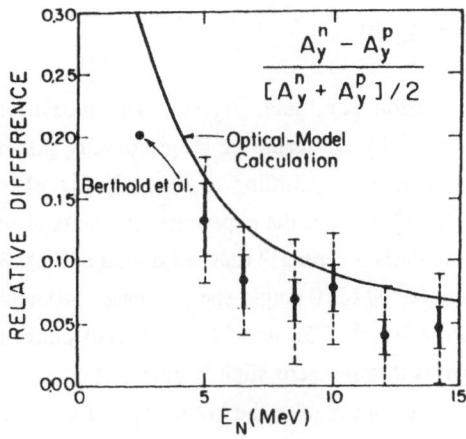

Fig. 11
Energy dependence of
$A_y$(max), the maximum
analyzing power near 120°
c.m. for proton-deuteron (pd)
and neutron-deuteron (nd)
scattering (Ref. 29).

Fig. 12
Relative difference between
neutron-deuteron and proton-
deuteron $A_y$(max) near 120° c.m.
Error bars are described in the
text. The point at 2.5 MeV was
obtained from Ref. 30. The solid
curve is an optical-model
calculation (Ref. 29).

We conclude from Figs. 11 and 12 that it appears that the major part of the observed differences between nd and pd $A_y$(max) data can be accounted for by the Coulomb interaction. It might well be that CSB effects account for only 10% of the discussed differences, similar to what was found for the 3N bound state binding energy difference. However, rigorous pd calculations that include the Coulomb interaction exactly above the deuteron breakup threshold are needed for supporting the present conclusion. As stated earlier, the work of Witala and Glöckle[25] assumes that CSB is responsible for the entire difference between the nd and pd $A_y$(max) values. It should be pointed out that even if this assumption should turn out to be incorrect, their conclusions concerning CIB in the $^3P_0$ NN states remain unchanged.

## THE ON-SHELL NUCLEON-NUCLEON TENSOR FORCE

Since the experimental situation was already summarized in Ref. 31, I will focus here only on the progress achieved during the last two years towards a better determination of the on-shell NN tensor force at low energies. In phase-shift analyses of NN data, the isoscalar $^3S_1$ - $^3D_1$ mixing parameter $\varepsilon_1$ is a measure of the NN tensor force at positive energies. It has been known for quite some time that only a few NN observables are sufficiently sensitive to allow, at least in principle, an accurate determination of $\varepsilon_1$. These observables involve the measurement of the polarization of at least two of the nucleons. Not only are these observables very difficult to measure, they are, unfortunately, also sensitive to other NN phase-shift parameters. For example, the spin-correlation parameters $A_{yy}(\theta)$ and $A_{zz}(\theta)$ are, except at 90° c.m., also sensitive to the "controversial" $^1P_1$ phase shift. As a consequence, $\varepsilon_1$ cannot be determined accurately from $A_{yy}(\theta)$ or $A_{zz}(\theta)$ measurements via NN phase-shift analyses unless $^1P_1$ is known very well.

### Neutron-Proton Spin-Correlation Parameter $A_{zz}(\theta)$

Very recently, the np spin-correlation parameter $A_{zz}(\theta)$ was measured at $E_n = 67.5$ MeV at PSI by the Basel group[32] by scattering of longitudinally polarized neutrons from longitudinally polarized protons. According to Ref. 33, $A_{zz}(\theta)$ is the observable that is most sensitive to $\varepsilon_1$. Fig. 13 displays the experimental results obtained by the Basel group in comparison with the Paris potential[15] (dashed-dotted curve), Bonn potential[16] (full Bonn, dashed curve), Arndt's[2] 50 MeV single-energy phase-shift analysis "C50 89" (dotted curve) predictions and a Basel "C50 new" (solid curve) phase-shift result.[32] As clearly noticeable, $A_{zz}$ crosses through zero slightly above 150° c.m. The determination of the zero-crossing angle does not require the knowledge of the neutron beam and proton target polarization. Therefore, in principle, the zero-crossing method is a very powerful tool for determining $\varepsilon_1$. As can be seen from Fig. 13 the Bonn (full Bonn) and Paris potential predictions are almost identical in the entire angular range studied, while Arndt's "C50 89" analysis does not even cross through zero. More importantly, the data in the zero-crossing region are in excellent agreement with the potential model predictions. Inspection of Fig. 14, which gives $\varepsilon_1$ as a function of neutron energy, suggests that the Basel measurement must support a value of about 2° for $\varepsilon_1$. However, the Basel group quoted $\varepsilon_1 = 2.9° \pm 0.3°$. This surprising result is intimately tied to the value of -9.2° used for the $^1P_1$ np phase shift. This feature is illustrated in Fig. 15. Near 50 MeV the Bonn (full Bonn) and the Paris potential have a $^1P_1$ value close to -10.5°. The solid curve in Fig. 15 is based on $\varepsilon_1 = 2°$ and $^1P_1 = -10.5°$. The dashed curve was calculated with the same value for $\varepsilon_1$, but with $^1P_1 = -9.2°$. Fig. 15 clearly demonstrates the shift in the zero-crossing angle. In order to move the zero-crossing angle back to smaller angles, $\varepsilon_1$ has to be increased. In fact, the dashed-dotted curve was obtained with $^1P_1 = -9.2°$ and $\varepsilon_1 = 3°$. This example convincingly displays the correlation between $\varepsilon_1$ and $^1P_1$. Fig. 15 also shows that $A_{zz}$ is independent of $^1P_1$ at 90° c.m. Furthermore, the entire forward angular range does not depend much on $^1P_1$. Even more importantly, this angular region is very sensitive to the value of $\varepsilon_1$. Of course, to take full advantage of these features, the neutron and proton polarizations must be measured accurately. In addition, since the recoil-proton detection method used by the Basel group is not applicable for small neutron c.m. angles, one has to detect the scattered neutrons. This is not a trivial task. The polarized proton target contains a large number of nuclei other than hydrogen from which neutrons can scatter, thus making the correct identification of np scattering events difficult.

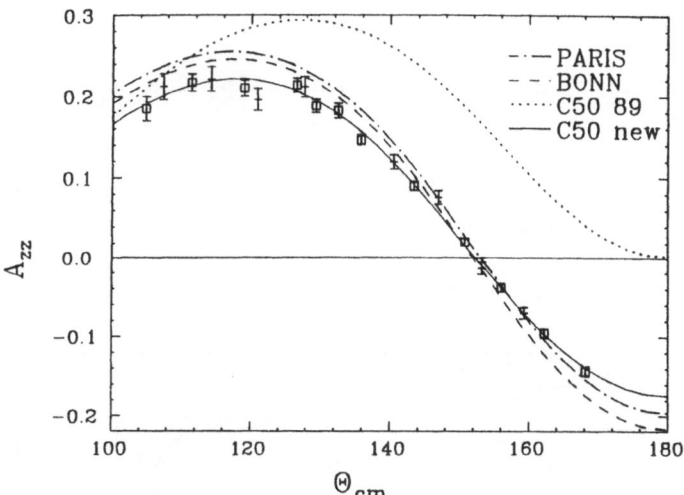

Fig. 13    Data for the spin-correlation parameter $A_{zz}(\theta)$ at $E_n = 67.5$ MeV (Ref. 32) in comparison with NN potential model and phase-shift predictions.

Returning to the Basel experiment at $E_n = 67.5$ MeV, we notice that the result for $\varepsilon_1$ is given at 50 MeV. This is due to a combined analysis of the Basel $A_{zz}(\theta)$ result and $A_{yy}(\theta)$[34] and differential cross-section $\sigma(\theta)$ data[35] obtained at Karlsruhe at 50 MeV. It is well known[33] that $^1P_1$ can be determined from accurate $\sigma(\theta)$ data. In addition, the spin-correlation parameters $A_{zz}(\theta)$ and $A_{yy}(\theta)$ represent complementary observables in the sense that the sign of the correlation between $\varepsilon_1$ and $^1P_1$ is opposite.[32] Therefore, using both data sets helps to reduce ambiguities with respect to $^1P_1$. Fig. 14 shows the new Basel result for $\varepsilon_1$ at 50 MeV. If confirmed, this result will have far reaching consequences, since modern, meson-exchange based NN potential models cannot yield such a large value for $\varepsilon_1$ at 50 MeV.[36] Therefore, measurements of $A_{zz}(\theta)$ at the same energy at forward angles would be extremely desirable.

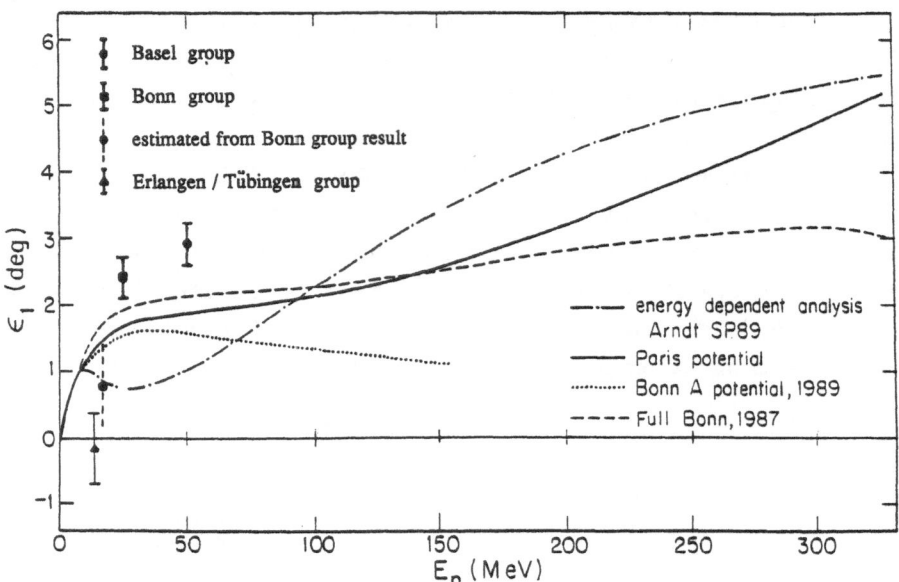

Fig. 14 Recent data for the $^3S_1$ - $^3D_1$ mixing parameter $\varepsilon_1$ and NN potential model and phase-shift predictions.

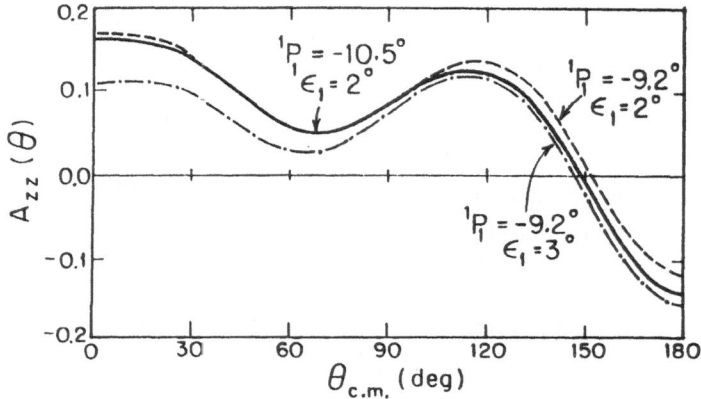

Fig. 15 Sensitivity calculations for $A_{zz}(\theta)$ using different values for $^1P_1$ and $\varepsilon_1$.

Recently, Ockenfels *et al.*[37,38] at Bonn measured the np spin-transfer parameter $K_y^{y'}(\theta)$ at 25.8 and 17.4 MeV near 130° c.m.  In this experiment, transverse polarized neutrons were scattered from an unpolarized proton target and subsequently the transverse polarization of the recoil protons was determined.  The observable $K_y^{y'}(\theta)$ displays only a very weak sensitivity to the $^1P_1$ phase shift.  Therefore, in principle, $K_y^{y'}(\theta)$ measurements are very well suited for determining $\varepsilon_1$.  Of course, in this "triple scattering experiment," the magnitude of the neutron and recoil proton polarization must be measured very accurately.  Fig. 16 shows the result obtained by the Bonn group in comparison with the Paris (dashed curve) and Bonn (full Bonn, solid curve) potential model predictions as well as with Arndt's global phase-shift analysis result SP89 (dotted curve).  Also given is a curve (long dashes) where the sensitivity of $K_y^{y'}(\theta)$ is displayed by adding 3° to the $\varepsilon_1$ value of the Bonn (full Bonn) potential.  As can be seen from Fig. 16, the result of Ockenfels *et al.*[37] is slightly larger than the potential model predictions and considerably larger than Arndt's $K_y^{y'}$.

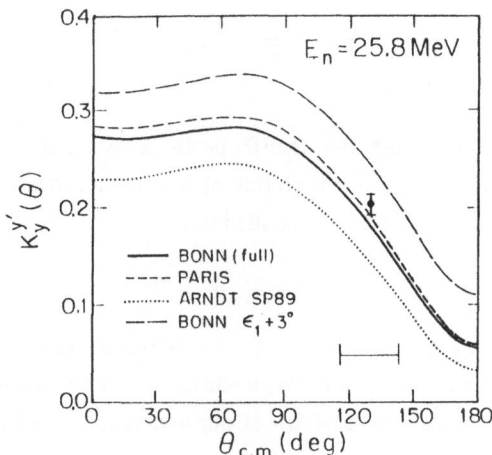

Fig. 16

Datum for the neutron-proton spin-transfer parameter $K_y^{y'}$ at $\theta_{c.m.} = 130°$ and $E_n = 25.8$ MeV from Ref. 37 in comparison with NN potential model and phase-shift predictions. The horizontal bar indicates the angular resolution of the experiment.

As mentioned above, $K_y^{y'}(\theta)$ is practically insensitive to $^1P_1$. Like $A_{zz}$, the spin-transfer coefficient is sensitive to $^1S_0$. In addition, $K_y^{y'}$ depends on the $^3S_1$ and $^3P_{0,1,2}$ phase shifts. Therefore, a sophisticated phase-shift analysis is required in order to extract a reliable result for $\varepsilon_1$. The Bonn group obtained $\varepsilon_1 = 2.8° \pm 0.4°$, a value which is also very large compared to the potential model predictions (see Fig. 14). Again, if confirmed, this result would support the observation that modern NN potential models do not predict $\varepsilon_1$ correctly, a result concluded already from the Basel $A_{zz}(\theta)$ measurement. The Bonn group performed in addition a combined analysis of their $K_y^{y'}$ measurement and the Karlsruhe data for $A_{yy}(\theta)$ at 25 MeV.[34] The result of this combined analysis, $\varepsilon_1 = 2.4° \pm 0.3°$, is plotted in Fig. 14.

Fig. 17 displays the result of the Bonn group for $K_y^{y'}(\theta)$ at 17.4 MeV.[38] Contrary to the case at 25.8 MeV, $K_y^{y'}$ is here lower than all realistic potential model predictions and even slightly lower than Arndt's global NN phase-shift analysis result. Although $\varepsilon_1$ has not yet been determined by the Bonn group, it seems that $\varepsilon_1$ will turn out to be near 0.8°, a result which is about a factor of 2 smaller than that of any potential model predictions (see Fig. 14). With this new result at 17.4 MeV, the value of the Erlangen / Tübingen group[39] at 13.7 MeV (see Fig. 14) does not appear so unphysical anymore.

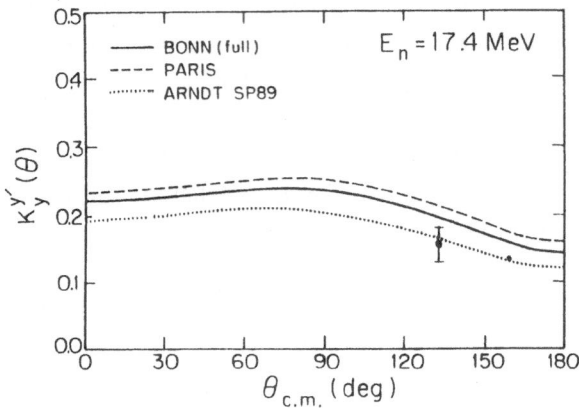

Fig. 17     Datum for the neutron-proton spin-transfer parameter $K_y^{y'}$ near $\theta_{c.m.} = 130°$ and $E_n = 17.4$ MeV from Ref. 38 in comparison with NN potential model and phase-shift predictions.

Inspection of Fig. 14 reveals strong indications that the very recent "experimental" determinations of $\varepsilon_1$ are not in agreement with NN potential model predictions. If confirmed, this would result in a dramatic breakdown of our understanding of the NN interaction in terms of meson exchange. Since the consequences are so severe, the present data call for a critical and reliable determination of the overall uncertainty of the extracted $\varepsilon_1$ values. For this purpose, realistic uncertainties have to be assigned to all phase-shift parameters involved in the determination of $\varepsilon_1$.

Nucleon-Deuteron Spin-Transfer Parameter $K_y^{y'}$

In NN potential models the deuteron D-state probability $P_D$ is a measure of the tensor force on and off the energy shell. As summarized in Table 2, modern NN potential models are characterized by quite different values for $P_D$. A strong correlation exists between $P_D$ and the calculated triton binding energy. The Bonn A potential with its low $P_D$ of 4.4% almost correctly predicts the triton binding energy while the Paris potential with $P_D = 5.8\%$

underbinds the triton by about 1 MeV.[22] The $\varepsilon_1$ values of Bonn A are given in Fig. 14 by the dotted curve. It should be noted that triton binding energy calculations are insensitive to $\varepsilon_1$ at energies above 100 MeV.[44]

Table 2.   Deuteron D-state probability $P_D$ of NN potential models

| Potential | $P_D$ (%) |
|---|---|
| Bonn 1987 (full Bonn)[16] | 4.25 |
| Bonn A[22] | 4.4 |
| Bonn B[22] | 5.0 |
| Nijmegen[40] | 5.4 |
| Bonn C[22] | 5.6 |
| Paris[15] | 5.8 |
| de Tourreil, Rouben, Sprung[41] | 5.9 |
| Argonne[42] | 6.1 |
| Reid soft core[43] | 6.5 |

In the previous sections of our tensor force discussion we frequently referred to the Bonn (full Bonn) NN potential. To be more specific, we referred to calculations based on the one-boson exchange approximation OBEPQ of the full Bonn potential in momentum space.[16] The difference between OBEPQ and Bonn A[22] is not only the slightly larger value for $P_D$ (see Table 2) of Bonn A, but more importantly, the enhanced $\omega$NN coupling constant used in Bonn A. The larger $\omega$NN coupling constant increases the spin-orbit strength as is required to fit np analyzing power data more accurately.[8] Bonn B and Bonn C differ from Bonn A essentially only in their values for $P_D$.[22]

Sensitivity calculations performed by Witala et al.[45] revealed that, among 3N scattering observables, the nd polarization transfer parameter $K_y^{y'}(\theta)$ is very sensitive to the magnitude of $P_D$. Fig. 18 displays pd $K_y^{y'}(\theta)$ data at $E_p = 22.7$ MeV obtained by Clajus et al.[45] in comparison to the nd Paris (dashed curve), Bonn A (solid curve) and Bonn B (dashed-dotted curve) NN potential model predictions. The data clearly favor the Bonn A calculation with its low $P_D$ over the Paris or Bonn B predictions with their larger $P_D$ values. Of course, the Coulomb interaction is not included in the present calculations. However, a sizeable influence of the Coulomb force is not expected in the angular region of interest. Therefore, the $K_y^{y'}(\theta)$ data appear to support the small $P_D$ value of the Bonn A potential, i.e., a weak tensor force. This result is in clear disagreement with both the $\varepsilon_1$ value obtained from the Bonn np $K_y^{y'}(\theta)$ experiment at 25.8 MeV and the Basel np $A_{zz}(\theta)$ experiment at 67.5 MeV.

We conclude this section by referring to the 1989 Long Range Plan[46] by the DOE/NSF Nuclear Science Advisory Committee: "At low and intermediate energies, the effects of the tensor force have not been well determined, owing to the difficulty of carrying out double spin-flip experiments. The recent development of polarized-beam polarized-target technology has made such experiments more feasible. The next few years should see high-precision measurements that will lead to a much better determination of the tensor force." As correctly predicted, high-precision measurements were performed during the last two years. However, it appears that the present polarized-beam polarized-target experiments are not sufficient for an accurate determination of the tensor force. This result is by no means surprising. It was probably too embarrassing to the Nuclear Science Advisory Committee to openly support, after more than 50 years of 2N interaction studies, an accurate determination, for example, of the np differential cross section, or more generally, of the low-energy NN phase-shift parameters, a prerequisite towards progress in understanding the role of mesons and quarks in nuclear physics.

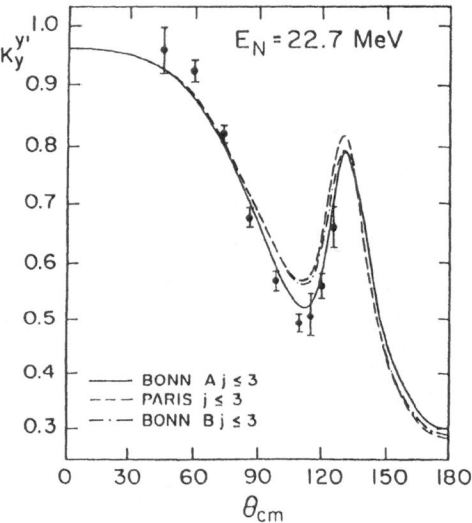

Fig. 18    Data for the spin-transfer parameter $K_y^{y'}(\theta)$ in pd scattering at 22.7 MeV in comparison with nd calculation using NN potentials with different tensor strength.

## Polarized Neutron-Polarized Proton Total Cross-Section Differences $\Delta\sigma_L$ and $\Delta\sigma_T$

As we have shown in the previous sections, the accurate determination of $\varepsilon_1$ from measured 2N observables like $A_{zz}$ and $K_y^{y'}$ requires sophisticated NN phase-shift analyses. In general, the mixing parameter $\varepsilon_1$ cannot be determined without the accurate knowledge of other NN phase-shift parameters. Since <u>all</u> phase-shifts are known at zero energy, it appears that $\varepsilon_1$ can be determined accurately at very low energies, say below 10 MeV, where only a few phase-shift parameters are non-zero. Fig. 19 shows that potential model predictions for $\varepsilon_1$ agree with each other in this energy range. Arndt's phase-shift analyses solutions SP89 and FA90 are also in close agreement with the potential models below 10 MeV. This observation is related to the deuteron properties: the quardupole moment of the deuteron determines the slope of $\varepsilon_1$ at very low energies.[47] At energies above 10 MeV the potential model predictions start to deviate from each other. Fig. 19 displays that NN potentials with a large value for $P_D$ do not necessarily have a large mixing parameter $\varepsilon_1$. Although $P_D$ of the Paris potential is 5.8% (see Table 2) the associated $\varepsilon_1$ is smaller than that of the Bonn B, Nijmegen and Bonn C potentials, which all have smaller $P_D$ values. This surprising observation is due to the fact, that $\varepsilon_1$ and the $^3S_1$ phase shift are related via the equation[16]

$$\tan 2\varepsilon_1 = 2R_{DS}^1/(R_{SS}^1 - R_{DD}^1).  \tag{1}$$

For example, an increase of $^3S_1$ (given by the matrix element $R_{SS}^1$) will lower $\varepsilon_1$, although the matrix element $R_{DS}^1$, which represents the strength of the tensor, is kept unchanged.

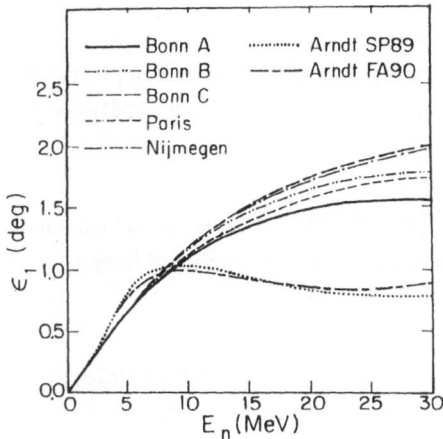

Fig. 19 Mixing parameter $\varepsilon_1$ between 0 and 30 MeV in comparison to NN potential model and phase-shift analysis predictions.

As was stated already (see Fig. 14), the recent experimental determinations of $\varepsilon_1$ at 13.7, 17.4, 25.8 and 50 MeV appear to disagree with potential model predictions. Although one may question the validity of the assigned uncertainties, it seems appropriate to investigate the discrepancy more carefully, especially at energies below 10 MeV where no data exist. Only if the validity of the potential model predictions for $\varepsilon_1$ has been verified experimentally in this energy range, can one rule out the experimental results in the 10 to 30 MeV energy range, where correlations between $\varepsilon_1$ and other phase-shift parameters prevent an accurate determination of $\varepsilon_1$.

It was shown in Ref. 31 that the longitudinal and transverse np total cross-section differences $\Delta\sigma_L$ and $\Delta\sigma_T$ are very sensitive to $\varepsilon_1$ at low energies. The spin-dependent total cross-section differences are defined as the differences in the total cross-sections with neutrons and protons polarized parallel and anti-parallel to each other and either longitudinal to the beam direction (L) or transverse to the beam direction (T):

$$\Delta\sigma_L = \sigma(\rightrightarrows) - \sigma(\rightleftarrows) \quad \text{and}$$
$$\Delta\sigma_T = \sigma(\uparrow\uparrow) - \sigma(\uparrow\downarrow), \tag{2}$$

where the top (first) arrow refers to the proton and the bottom (second) arrow to the neutron spin orientation. Fig. 20 shows the sensitivity of $\Delta\sigma_L$ to a $\pm1°$ change of $\varepsilon_1$ and the $^1P_1$ phase shift. As can be seen , $\Delta\sigma_L$ is extremely sensitive to $\varepsilon_1$ and only slightly sensitive to $^1P_1$ at energies below 30 MeV. This surprising sensitivity to $\varepsilon_1$ is due to a cancellations between the individual phase shifts that contribute to $\Delta\sigma_L$. The following equations express $\Delta\sigma_L$ in terms of phase-shift parameters:

$$\Delta\sigma_L = \frac{\pi}{k^2}\,x \tag{3}$$

with wave number k and

$$x = 2 - \cos2\delta_{1S_0} + \cos2\delta_{3S_1} - 3\cos2\delta_{1P_1} - \cos2\delta_{3P_0} + 3\cos2\delta_{3P_1} - \cos2\delta\ {3D_1} +$$
$$4\sqrt{2}\sin(\delta_{3S_1} + \delta_{3D_1})\ \underline{\sin2\ \varepsilon_1} + \cos2\delta_{3P_2} - 5\cos2\delta_{1D_2} + 5\cos2\delta_{3D_2} - \cos2\delta_{3F_2} +$$
$$4\sqrt{6}\ \sin(\delta_{3P_2} + \delta_{3F_2})\ \sin2\varepsilon_2. \tag{4}$$

Here, except for $^3F_2$, phase shifts with total angular momentum J≥3 were neglected, since their influence is negligible in the energy region of interest. Using Arndt's SP89 phase-shift solution, one obtains for x at 10 MeV:

$$x = -0.41283 + 5.5566\ \sin2\varepsilon_1 \tag{5}$$

Table 3 summarizes the factor x for 3 different values of $\varepsilon_1$. A $1°$ change of $\varepsilon_1$ corresponds to a 90% change in x, i.e., to 60 mb in $\Delta\sigma_L$. The sensitivity of $\Delta\sigma_T$ to $\varepsilon_1$ is half as large.

Fig. 21 presents potential model predictions and Arndt's SP89 phase shift result for $\Delta\sigma_T$ (left side) and $\Delta\sigma_L$ (right side). Sizeable differences exist between the different predictions, especially in the 8 to 18 MeV energy range.

The experimental determination of $\Delta\sigma_L$ ($\Delta\sigma_T$) requires the measurement of the difference in the attenuation of a longitudinally (transversely) polarized neutron beam through a longitudinally (transversely) polarized proton target when one or the other spin is reversed. The spin-dependent total cross-section differences can be obtained from the asymmetries $\varepsilon_L$ and $\varepsilon_T$ measured with parallel and antiparallel neutron spin orientation[31]:

$$\varepsilon_L = -\frac{1}{2}\,\omega P_p\,P_n\,\Delta\sigma_L \quad \text{and}$$
$$\varepsilon_T = -\frac{1}{2}\,\omega P_p\,P_n\,\Delta\sigma_T, \tag{6}$$

where $\omega$ refers to the number of polarized protons per $cm^2$ in the target and $P_p$ and $P_n$ to the proton target and neutron beam polarization, respectively.

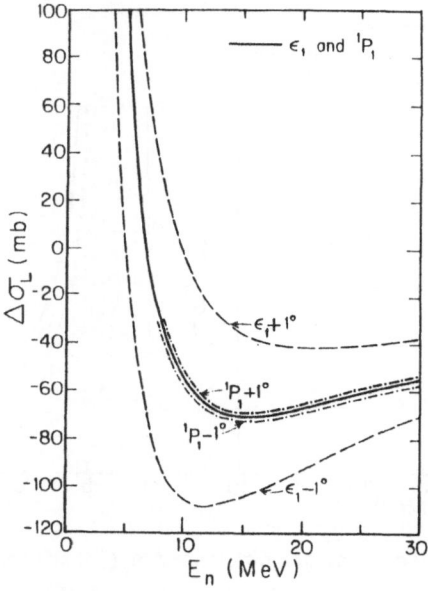

Fig. 20 Sensitivity of longitudinal total cross-section difference $\Delta\sigma_L$ to $\pm 1°$ changes of $\varepsilon_1$ and of the $^1P_1$ phase-shift.

Table 3.  Sensitivity of factor x in eqs. 3 and 4 to a $\pm 1°$ change of $\varepsilon_1$ at $E_n = 10$ MeV

| $\varepsilon_1$ (deg) | x |
|---|---|
| 2.03 | -0.019 |
| 1.03 | -0.213 |
| 0.03 | -0.407 |

As is clearly displayed in Fig. 21, $\Delta\sigma_L$ and $\Delta\sigma_T$ cross through zero at low energies. Fig. 22 gives an expanded view of the zero-crossing regions of $\Delta\sigma_T$ (left side) and $\Delta\sigma_L$ (right side) calculated from different potential models and Arndt's phase-shift solution SP89. As we pointed out in Ref. 31, the different zero-crossing energies are only partially related to the different $\varepsilon_1$ values. The zero-crossing energies are very sensitive to the angular momentum $\ell = 0$ $^1S_0$ and $^3S_1$ NN phase shifts. Since the Paris and Nijmegen NN potential models are fitted to the pp scattering length $a_{pp}$ and the Bonn potentials to the np scattering length $a_{np}$, their associated values for $^1S_0$ are slightly different at low energies.

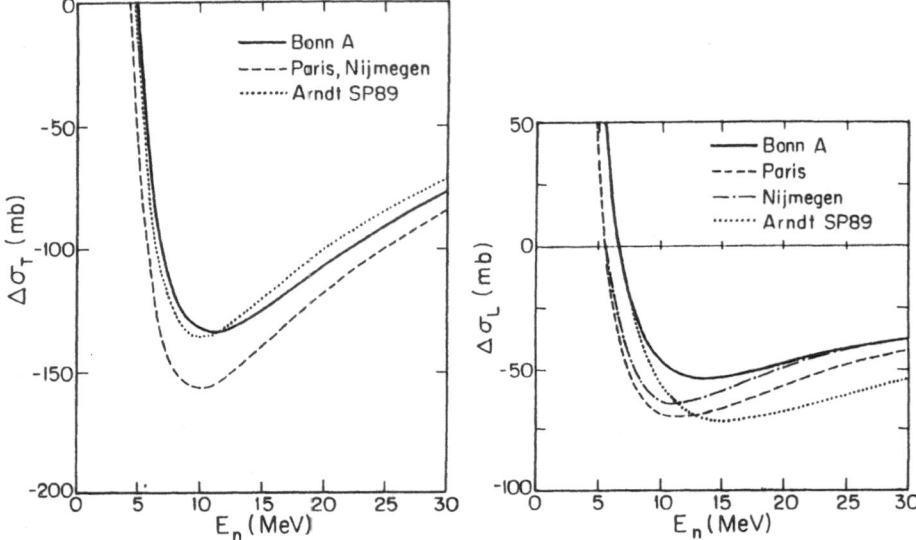

Fig. 21 Longitudinal total cross-section difference $\Delta\sigma_L$ (left side) and transverse total cross-section difference $\Delta\sigma_T$ (right side) obtained from NN potential models and Arndt's phase-shift solution SP89.

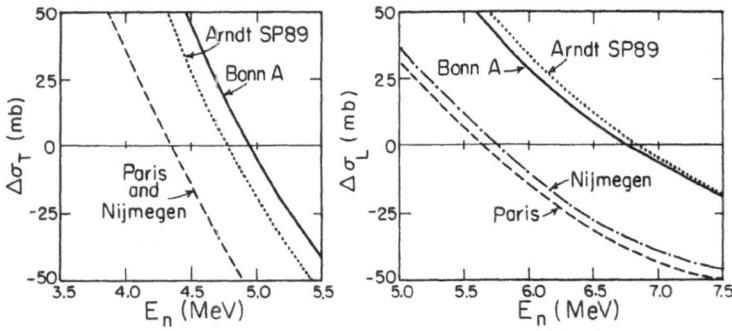

Fig. 22 Predicted zero-crossings of $\Delta\sigma_T$ (left side) and $\Delta\sigma_L$ (right side).

Although the experimental determination of the zero-crossing energies of $\Delta\sigma_T$ and $\Delta\sigma_L$ provides only limited information about $\varepsilon_1$, it is of special interest. First, as can be seen from eqs. 6, the determination of the zero-crossing energies does not require the accurate knowledge of $\omega$, $P_p$ and $P_n$. Therefore, the determination of the zero-crossing energies can be very accurate and it is limited only by the statistical uncertainties in the asymmetries $\varepsilon_L$ and $\varepsilon_T$. The second point requires a more thorough discussion. As we already mentioned above, $\Delta\sigma_L$ and $\Delta\sigma_T$ depend, like $A_{yy}$, $A_{zz}$ and $K_y^{y'}$, on the $^1S_0$ phase shift. More generally, both $\Delta\sigma_L$ and $\Delta\sigma_T$ can be expressed in terms of the spin-dependent central part $\sigma_{central}$ and the tensor part $\sigma_{tensor}$

$$\Delta\sigma_{L,T} = \sigma_{central} + \sigma_{tensor}, \qquad (7)$$

where $\sigma_{central}$ depends on the singlet phase-shift parameters $^1S_0$, $^1P_1$, $^1D_2$, $^1F_3$ etc. In order to remove the unwanted sensitivity to the singlet phase-shift parameters, one can build the difference between $\Delta\sigma_L$ and $\Delta\sigma_T$:

$$\begin{aligned}
\Delta\sigma_L - \Delta\sigma_T = &-\cos 2\delta 3_{P_0} + \frac{3}{2}\cos 2\delta 3_{P_1} - \frac{1}{2}\cos 2\delta 3_{P_2} - \frac{3}{2}\cos 2\delta 3_{D_1} \\
&+ \frac{5}{2}\cos 2\delta 3_{D_2} - 2\cos 2\delta 3_{F_2} + 3\sqrt{2}\sin(\delta 3_{S_1} + \delta 3_{D_1})\underline{\sin 2\varepsilon_1} + \\
&3\sqrt{6}\sin(\delta 3_{P_2} + \delta 3_{F_2})\sin 2\varepsilon_2
\end{aligned} \qquad (8)$$

As can be seen, $\Delta\sigma_L - \Delta\sigma_T$ depends only on the triplet phase shifts and on the mixing parameters $\varepsilon_1$ and $\varepsilon_2$. In the energy range of interest, $\varepsilon_2$ is very small and its influence is negligible. As was already shown for the individual terms $\Delta\sigma_L$ and $\Delta\sigma_T$, the difference $\Delta\sigma_L - \Delta\sigma_T$ is also extremely sensitive[31] to $\varepsilon_1$. Fig. 23 displays $\Delta\sigma_L - \Delta\sigma_T$ in the energy range of interest. Clearly, as can be seen also from eq. 8, the difference increases with increasing $\varepsilon_1$. Even at energies as low as 5 MeV, where all realistic NN potential models have practically identical $\varepsilon_1$ values, $\Delta\sigma_L - \Delta\sigma_T$ differs from one model to the other. Therefore, $\Delta\sigma_L - \Delta\sigma_T$ measurements are extremely useful to find out whether or not the modern, meson-exchange based NN potential models predict the correct $\varepsilon_1$ values at low energies.

Of course, an even more ambitious task would be to distinguish between different potential model predictions for $\Delta\sigma_L - \Delta\sigma_T$. Fig. 23 shows that the largest difference between the potential model predictions is only about 10 mb. Therefore, experimental uncertainties in $\Delta\sigma_L - \Delta\sigma_T$ must be very small. In the two special cases where one of the two individual total cross-section differences (either $\Delta\sigma_L$ or $\Delta\sigma_T$) crosses through zero, the difference $\Delta\sigma_L - \Delta\sigma_T$ can be measured very accurately. Therefore, the only energies at which $\varepsilon_1$ can be determined to sufficient accuracy are those where $\Delta\sigma_T$ and $\Delta\sigma_L$ cross through zero. Such zero-crossing experiments are underway at TUNL. (For the discussion of $(\Delta\sigma_T/\Delta\sigma_L)$ experiments see Ref. 31).

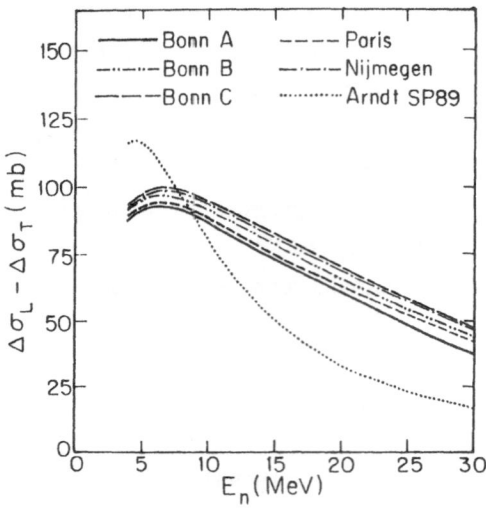

Fig. 23
Difference $\Delta\sigma_L$ - $\Delta\sigma_T$ obtained from NN potential models and Arndt's phase-shift solution SP89.

SUMMARY AND CONCLUSION

Recent computational advances in 3N bound state and 3N scattering calculations and the availability of high accuracy 3N scattering data at low energies led to a renewed interest in an accurate determination of the free NN interaction.

Phase-shift analyses of NN scattering data indicate a surprisingly large charge dependence of the $\pi$NN coupling. This CIB causes a sizeable difference between the pp and np $^3P_{0,1,2}$ NN interactions. Experimental data for the 3N observable that is most sensitive to the $^3P_{0,1,2}$ NN interactions, the nd analyzing power $A_y(\theta)$, are in clear disagreement with all meson-exchange based NN potential models. However, the large CIB in the $^3P_0$ NN interaction is supported by a recent combined analysis of 2N and 3N scattering data. The origin of this large CIB in the $^3P_0$ NN interactions is unknown.

Charge-symmetry breaking (CSB) studies in $^3P$ states are inconclusive. Due to the lack of a suitable free neutron target, CSB cannot be investigated in the 2N systems. Since the accurate inclusion of the Coulomb interaction in 3N continuum Faddeev calculations is presently not possible, the experimentally established difference between the nd and pd $A_y(\theta)$ data cannot be interpreted as a clear manifestation of CSB in the $^3P_0$ NN interaction.

Since it was shown that the Bonn A potential with its weak tensor force (low D-state probability) almost correctly predicts the triton binding energy, experimental NN studies have concentrated on the accurate determination of the $^3S_1$ - $^3D_1$ mixing parameter $\varepsilon_1$. The new generation of 2N experiments indicate, that NN potential models, the Bonn A potential included, are inconsistent with experimental results. However, some caution is necessary with respect to the uncertainty associated with the "experimental" $\varepsilon_1$ values. It should be pointed out that a dedicated 3N experiment, i.e., $K_y^{y'}(\theta)$ in pd scattering, is in close agreement with the Bonn A prediction. An accurate determination of $\varepsilon_1$ in 2N scattering seems possible only at energies below about 20 MeV where only a few phase-shift parameters contribute to the np observable used to determine $\varepsilon_1$. Here, the polarized neutron-polarized proton total cross-section differences $\Delta\sigma_L$ and $\Delta\sigma_T$ are of special interest. Experimental results for these observables are expected to be available in the near future.

## ACKNOWLEDGMENTS

We gratefully acknowledge the opportunity to use the code SAID of R. Arndt. This work was supported in part by the U.S. Department of Energy, Office of High Energy and Nuclear Physics, under Contract No. DE-AC05-76ER01067.

## REFERENCES

1. I. Slaus, Y. Akaishi, and H. Tanaka, Phys. Rep. 173:257 (1989).
2. R. A. Arndt, interactive computer code SAID (private communication).
3. P. Langacker and D. A. Sparrow, Phys. Rev. C25:1194 (1982).
4. J. R. Bergervoet, P. C. van Campen, T. A. Rijken, and J. J. de Swart, Phys. Rev. Lett. 59:2255 (1987).
5. J. R. Bergervoet, P. C. van Campen, R. A. Klomp, J. -L. de Kok, T. A. Rijken, V. G. J. Stoks, and J. J. de Swart, Phys. Rev. C41:1435 (1990).
6. R. Koch and E. Pietarinen, Nucl. Phys. A336:331 (1980).
7. V. G. J. Stoks, P. C. van Campen, T. A. Rijken, and J. J. de Swart, Phys. Rev. Lett. 61:1702 (1988).
8. R. Machleidt, I. Slaus, W. Tornow, W. Glöckle, and H. Witala, submitted to Few-Body Systems (Progress Report) and references therein.
9. R. A. Arndt, Z. Li, L. D. Roper, and R. L. Workman, Phys. Rev. Lett. 65:157 (1990).
10. R. Machleidt and F. Sammarruca, Phys. Rev. Lett. 66:564 (1991).
11. G. Höhler and E. Pietarinen, Nucl. Phys. B95:210 (1975).
12. W. Tornow and R. L. Walter, Few-Body Systems 8:11 (1990).
13. D. Holslin, J. McAninch, P. A. Quin, and W. Haeberli, Phys. Rev. Lett. 61:1561 (1988).
14. W. Tornow, C. R. Howell, M. Alohali, Z. P. Chen. P. D. Felsher, J. M. Hanly, R. L. Walter, G. Weisel, G. Mertens, I. Slaus, H. Witala, and W. Glöckle, Phys. Lett. B257:273 (1991).
15. M. Lacombe, B. Loiseau, J. M. Richard, R. Vinh Mau, J. Coté, P. Pires, and R. de Tourreil, Rev. Rev. C21:861 (1980).
16. R. Machleidt, K. Holinde, and C. Elster, Phys. Rep. 149:1 (1987).
17. H. Witala, W. Glöckle, and T. Cornelius, Few-Body Systems, Suppl. 2:555 (1987).

18. W. Glöckle, "Lecture Notes in Physics," Springer, Berlin (1987), Vol 273, p.3 ; H. Witala, T. Cornelius, and W. Glöckle, Few-Body Systems 3:123 (1988).

19. T. Cornelius, W. Glöckle, J. Haidenbauer, Y. Koike, W. Plessas, and H. Witala, Phys. Rev. C41:2538 (1990); J. L. Friar, B. F. Gibson, G. Berthold, W. Glöckle, T. Cornelius, H. Witala, J. Haidenbauer, Y. Koike, G. L. Payne, J. A. Tjon, and W. M. Kloet, Phys. Rev. C42:1838 (1990).

20. A. C. Berick, R. A. Riddle, and C. M. York, Phys. Rev. 174:1105 (1968).

21. H. Witala, W. Glöckle, and Th. Cornelius, Nucl. Phys. A491:159 (1989).

22. R. Machleidt, Adv. Nucl. Phys. 19:189 (1989).

23. W. Grüebler, V. König, P. A. Schmelzbach, F. Sperisen, B. Jenny, R. E. White, F. Seiler, and H. W. Roser, Nucl. Phys. A398:445 (1983).

24. W. Glöckle, H. Witala, and T. Cornelius, Nucl. Phys. A508:115c (1990).

25. H. Witala and W. Glöckle, Nucl. Phys. A, in press.

26. W. Tornow, C. R. Howell, R. C. Byrd, R. S. Pedroni, and R. L. Walter, Phys. Rev. Lett. 49:312 (1982).

27. J. L. Friar, B. F. Gibson, and G. L. Payne, Phys. Rev. C35:1502 (1987); Y. Wu, S. Ishikawa, and T. Sasakawa, Phys. Rev. Lett. 64:1875 (1990), and references therein; R. A. Brandenburg, G. S. Chulick, Y. E. Kim, D. J. Klepacki, R. Machleidt, A. Picklesimer, and R. M. Thaler, Phys. Rev. C37:781 (1988).

28. E. M. Henley and G. A. Miller, "Mesons in Nuclei", M. Rho and D.H. Wilkinson, ed., North Holland, Amsterdam (1979),Vol. 1, p. 405.

29. W. Tornow, C. R. Howell, and R. L. Walter, submitted to Phys. Lett. B.

30. G. H. Berthold, A. Stadler, and H. Zankel, Phys. Rev. Lett. 61:1077 (1988); Phys. Rev. C41:1365 (1990).

31. W. Tornow, O. K. Baker, C. R. Gould, D. G. Hasse, N. R. Roberson, and W. S. Wilburn, Topical Conf. on "Physics with Polarized Beams on Polarized Targets, eds. J. Sowinski and S. E. Vigdor, World Scientific Publishing Co. (1990), p.75.

32. M. Hammans, C. Brogli-Gysin, S. Burzynski, J. Campbell, P. Haffter, R. Henneck, W. Lorenzon, M. A. Pickar, I. Sick, J. A. Konter, S. Mango, and B. van den Brandt, Phys. Rev. Lett. 66:2293 (1991).

33. J. Binstock and R. Bryan, Phys. Rev. D9:2528 (1974).

34. P. Doll, V. Eberhard, G. Fink, R. W. Finlay, T. D. Ford, W. Heeringa, H. O. Klages, H. Krupp, and C. Wölfe, Contributed Papers, 12[th.] Int. Conf. on Few Body Problems in Physics, Vancouver, 1989, TRI-89-2, p.C16.

35. G. Fink, P. Doll, T. D. Ford, R. Garrett, W. Heeringa, K. Hofmann, H. O. Klages, and H. Krupp, Nucl. Phys. A518:561 (1990).

36. R. Machleidt, private communication.

37. M. Ockenfels, F. Meyer, T. Köble, W. von Witsch, J. Weltz, K. Wingender, and G. Wollmann, Nucl. Phys. A526:109 (1991).

38. M. Ockenfels, T. Köble, M. Schwindt, J. Weltz, V. Winkler, and W. von Witsch, Verhandl. DPG (V1) 26:576 (1991).

39. M. Schöberl, H. Kuiper, R. Schmelzer, G. Mertens, and W. Tornow, Nucl. Phys. A489:284 (1988).

40. M. M. Nagels, T. A. Rijken, and J. J. de Swart, Phys. Rev. D17:728 (1978).

41. R. de Tourreil, R. Rouben, and D. W. L. Sprung, Nucl. Phys. A242:445 (1975).

42. R. B. Wiringa, R. A. Smith, and T. L. Ainsworth, Phys. Rev. C29:1207 (1984).

43. R. V. Reid, Ann. Phys. (N.Y.) 50:411 (1968).

44. R. Machleidt, Contributed Papers, 12th. Int. Conf. on Few Body Problem in Physics, Vancouver 1989, TRI-89-2, p. G51.

45. M. Clajus, P. M. Egun, W. Grüebler, P. Hautle, I. Slaus, B. Vuaridel, F. Sperisen, W. Kretschmer, A. Rauscher, W. Schuster, R. Weidmann, M. Haller, M. Bruno, F. Cannata, M. D'Agostino, H. Witala, Th. Cornelius, W. Glöckle, and P. A. Schmelzbach, Phys. Lett. B245:333 (1990).

46. Nuclei, Nucleons, Quarks, Nuclear Science in the 1990's, U. S. Department of Energy, Office of Energy Research, Division of Nuclear Physics, Dec. 1989.

47. G. E. Brown and A. D. Jackson, in "The Nucleon-Nucleon Interaction", North-Holland Publ. Co., 1976, p. 74.

# DELTA AND ROPER RESONANCE EXCITATION IN HADRONIC

# INTERACTIONS WITH PARTICLES OF ISOSCALAR STRUCTURE[*]

H. Peter Morsch

Laboratoire National Saturne
F-91191 Gif-sur-Yvette Cedex, France
and
Institut für Kernphysik, Forschungszentrum Jülich
D-5170 Jülich, Germany

## 1. INTRODUCTION

In the near future several medium energy hadron and electron accelerators will be available. This will allow a detailed study of the structure of baryons. The detailed experimental investigation of the baryon properties presents a large challenge in nuclear physics for the next decade since this is related to the understanding of the structure of QCD in the non-perturbative regime. A study of baryons with both electromagnetic and hadronic probes is absolutely necessary to obtain complementary information on the different structures. Of large importance is the study of baryon resonances which determine the dynamical properties of baryons. So far, only the lowest baryon resonance, the $\Delta(1232$ MeV) has been studied in detail. This resonance is directly related to the $\pi$-degree of freedom and the gross properties of the experimental data can be understood in meson exchange models. The study of the behavior of the $\Delta$ resonance in nuclei presents a very interesting problem which is so far not well understood[1].

Another important class of baryon excitations which samples very different properties of baryons are radial excitations. These are excitations of baryon resonances with the same quantum numbers as the nucleon ground state j=1/2, t=1/2. As for these excitations the spin-isospin structure of the nucleon is not changed the radial profile has to be modified. The radial properties present one of the most basic degrees of freedom of the system. For the ground state they can be investigated by studying charge densities. The different charge densities for protons and neutrons can e.g. be understood in a model[2] which contains a baryon number density with a radius much smaller than the nucleon radius, surrounded by a meson cloud. Dynamical properties of the baryon density can be studied by radial excitations which give information on the compressibility. Radial modes with orbital motion give insight into polarizability and deformabilities. The lowest N* resonance, N(1440 MeV) is a first candidate for a radial excitation.

The energy of the radial excitations is critically dependent on the basic parameters of the different baryon models, e.g. it depends on the confining potential in a quark model or the bag size in bag models. For the chiral bag model it is interesting to see wether the lowest compressional mode presents a vibration of the bag, of the meson cloud, or shows a strong coupling between both. In the constituent quark model the radial mode is due to a 2 hω quark excitation from the 1s to the 2s shell which corresponds to about 900 MeV. In contrast, such a radial mode lies at rather low energies in Skyrmion models[3].

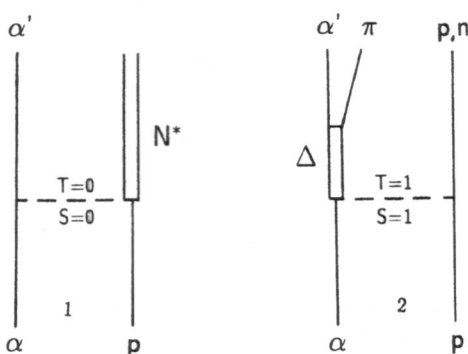

Figure 1. Graphs for target and projectile excitation which contribute to inelastic $\alpha + p \rightarrow \alpha' + X$ scattering.
1. Inelastic excitation of the target with a dominant T=0, S=0 transition in the forward scattering amplitude.
2. Projectile excitation with subsequent decay back into the $\alpha$ ground state by $\pi$-emission. This is dominated by a T=1, S=1 transition giving rise to $\Delta$ excitation.

The study of these particular excitations appears to be difficult since proton-nucleon excitations are dominated by spin-isospin modes. Further, because of their particular structure these modes are rather weakly excited in electromagnetic interactions. Therefore, in order to study these specific excitations it is important to look for selective probes which may enhance their cross sections. A favorable reaction appears to be the inelastic scattering by $\alpha$-particles because in the forward scattering isoscalar non-spin-flip (T=S=0) excitations should be dominant due to the structure of the $\alpha$-particle. This would lead to the excitation of the radial and orbital modes discussed above. Also deuteron-proton scattering appears interesting since it should also enhance isoscalar excitations.

At Saturne energies, there are complications in the investigation of these hadronic reactions due to the fact that both the target as well as the projectile may be excited during the scattering process. For the $\alpha + p \rightarrow \alpha' + X$ reaction this is seen in fig.1 which shows different graphs for target and projectile excitation. Whereas for the target the $\Delta$ excitation should be small there are no selection rules which inhibit $\Delta$ excitation in the projectile. In section 2 the problem of the $\Delta$ resonance excitation in the projectile is discussed which is related to the properties of Deltas in complex systems. This is a significant "background" contribution for the study of the Roper resonance excitation which is discussed in section 3.

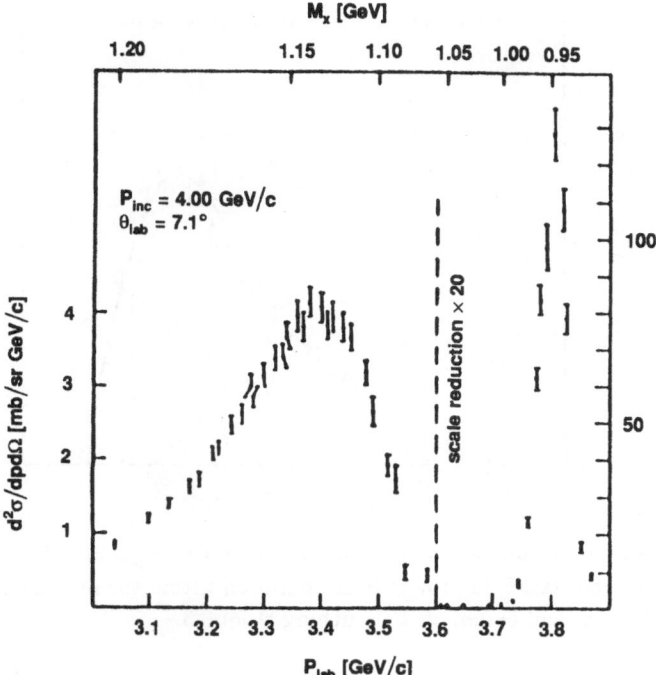

Figure 2. Momentum spectrum of inelastic scattered $\alpha$-particles on hydrogen measured at an $\alpha$-momentum of 4 GeV/c. The data are from ref.5.

## 2. DELTA EXCITATION OF THE PROJECTILE IN $\alpha$-p AND d-p SYSTEMS

Experiments on these scattering systems have been performed at Saturne[4,5]. In the missing mass spectra $\alpha + p \rightarrow \alpha' + X$ at 4 GeV/c a bump is observed above the $\pi$-threshold (see fig.2). This structure has been discussed in the past as coherent $\pi$-production and may be interpreteted as subsequent excitation and decay of the Delta resonance in the projectile. For the $\Delta$ excitation of the projectile we expect large forward angle cross sections comparable to charge exchange reactions. If further the ground state decay branch $B_o : \alpha_\Delta \rightarrow \alpha_{g.s.} + \pi$ is sufficiently large (10-20%) then this contribution of projectile excitation should be observed in the inclusive spectra. The appearance of this contribution in the missing mass spectra at a location close to the pion threshold (fig.2) is explained by a Lorentz boost in the projectile excitation due to the large velocity of the projectile. Calculations of $\Delta$ excitation in the projectile have been performed within the meson exchange model. A parametrization by Dimitriev et al.[6] has been used which describes the absolute cross sections and spectral shape for the elementary $p\,p \rightarrow n\,\Delta^{++}$ system. Using the impulse approximation this parametrization has been also successfully applied to $\Delta$ production in the ($^3$He,t) charge exchange reaction. In this formulation the matrix element is calculated for the elementary system and the effect of the complex projectile is taken into account by the $^3$He form factor. In analogy to this approach a calculation of the projectile excitation in the $\alpha - p$ system has been performed. The matrix elements for the elementary $\Delta$ production processes with protons and neutrons were introduced into the Monte Carlo code 'FOWL' and folded with the $\alpha$-particle form factor describing elastic

491

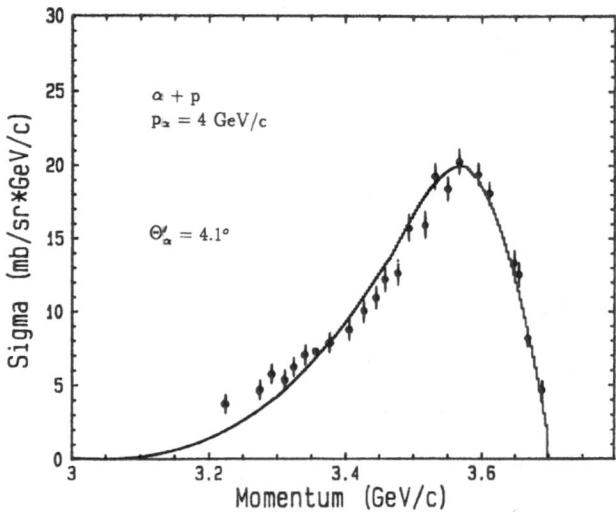

Figure 3. Similar spectrum as in fig.2 measured at an angle of 4.1°. The solid line corresponds to a calculation of projectile $\Delta$ excitation within the meson exchange model discussed in the text and assuming a branching $B_o$ of 0.3.

scattering. The ground state branching $B_o$ is obtained by the normalization of these calculations to the experimental data. The comparison with the small angle data of ref.5 is given in fig.3. Using a value of $B_o$ of 0.3 a good description of the experimental data is obtained. In these calculations non-resonant contributions in the $\Delta$ resonance region have not been considered. Dependent on these contributions which should be added, our results are consistent with the ground state branching $B_o$ of about 0.2 discussed above.

The value for $B_o$ of 0.3 extracted from the data may be compared to $\pi + {}^4$He scattering since the graph which presents $\Delta$ excitation of the ${}^4$He-projectile with subsequent decay into ${}^4$He and $\pi$ is similar to that governing elastic $\pi + {}^4$He scattering, apart from off-shell effects. The determination of $B_o$ by the ratio of the elastic to the total cross sections[7] $\sigma_{el}/\sigma_{tot}$ yields $B_o \sim 0.33$ in good agreement with the value discussed above.

For the interpretation of our data discussed below it is important to know that the spectral shape of the projectile excitation is quite independent of the $\Delta$ resonance parameters and the detailed assumptions on the background contribution. Even assuming, e.g. a width of the $\Delta$ resonance in the projectile twice as large as in the free case would hardly change the shape in the spectral distribution.

3. STUDY OF THE ROPER RESONANCE EXCITATION IN $\alpha$-p SCATTERING

To investigate the region of the Roper resonance $P_{11}(1440\ \mathrm{MeV})$, which is a candidate for the lowest radial excitation of the nucleon, we studied $\alpha$-p scattering at a beam momentum of 7 GeV/c which is close to the maximum for Saturne. Scattered $\alpha$-particles were measured in the SPES IV magnetic spectrometer. Missing energy $\Omega$ spectra ($\Omega = E_i - E_f$) are given in fig.4 measured at very small scattering angles of 0.8 and 2.0 degree. In these spectra in addition to N* excitations we expect the same features observed at low

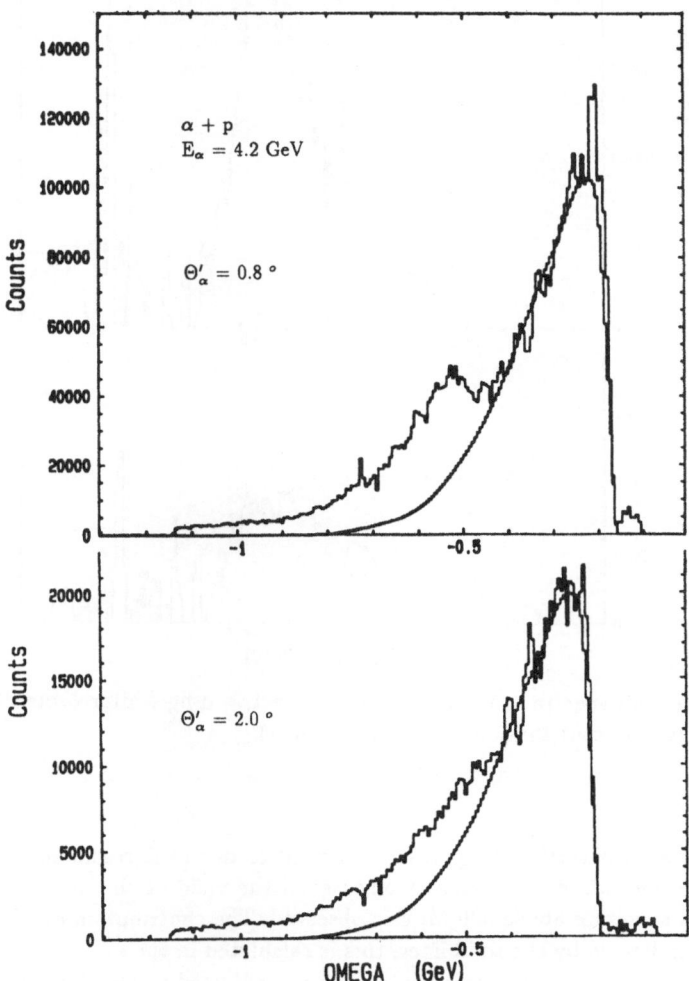

Figure 4. Missing energy $\Omega$ spectra of inelastic $\alpha$-p scattering at $E_{\alpha}$=4.2 GeV measured at 0.8° (upper part) and 2.0° (lower part). The solid lines correspond to the spectral shape calculated for projectile excitation.

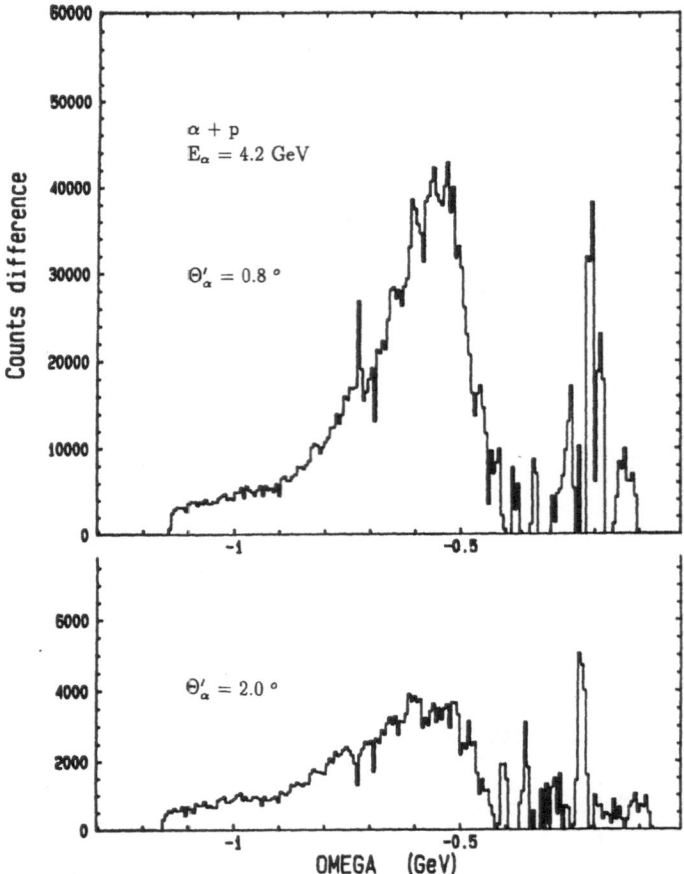

Figure 5. Difference spectra of the experimental spectra in fig.4 after subtraction of the projectile excitation contribution (solid lines in fig.4).

energies (discussed in section 2), a strong contribution due to $\Delta$-resonance excitation in the projectile. Indeed, we observe a strong rise of the yield at the pion-threshold and a pronounced structure above 500 MeV is observed. The contribution due to projectile excitation is indicated by the solid lines, this is calculated in the same way as discussed in section 2. The amplitude of this contribution is adjusted to the data. In the region above the $\pi$-threshold the shape of the spectrum is quite well reproduced by the projectile contribution. However, at larger values of $\Omega$ there is a significant excess yield indicating a strong excitation of the Roper resonance region. It is interesting to note that at the smallest angle measured this excitation is sufficiantly large to produce a bump in the inclusive spectrum. In order to see the details of this structure, the difference spectra in which the projectile excitation contribution is subtracted from the measured spectra is shown in fig.5. This shows a pronounced bump in the Roper resonance region which falls off rapidly towards larger values of $\Omega$. Also the Roper excitation decreases rapidly with increasing angle. Above 0.9 GeV the yield is rather flat. and shows a much smoother angular distribution. This is the region of the $S_{11}$ and $D_{13}$ resonances which can be excited in inelastic $\alpha$-scattering by L=1 transfer. Preliminary extraction of the differential cross sections for $P_{11}$ excitation gives values of several mb/sr. Estimates of the differential cross sections have been made using different approaches. The angular distribution observed experimentally is in good agreement with the behaviour of the calculated cross sections for monopole excitation. The large cross sections extracted can only be understood by assuming that this excitation exhausts a large fraction of the monopole sum rule strength. Utilizing sum rules the compressibility can be calculated from the central energy of the monopole strength.

In the experimental spectra the monopole strength is significantly shifted to lower values of $\Omega$ due to the momentum transfer dependance of the $\alpha$-particle form factor. To obtain the monopole strength function, the formfactor dependance has to be unfolded. Preliminary results for this procedure give the monopole strength close to the nominal energy of the Roper resonance. This yields a value for the nucleon compressibility of 1450 MeV if a radius of the nucleon of 0.85 fm from electron scattering is used. Solutions using a smaller radius, e.g. that of a quark bag in a chiral bag model in the order of 0.5 fm can be ruled out because this would lead to very small $\alpha$-scattering cross sections which are inconsistent with our data.

In summary, strong excitation of baryon resonances has been observed in inclusive $\alpha$-proton scattering. In the spectra one observes a large contribution which can be explained by $\Delta$ excitation in the projectile. This is a rather interesting two-step process which gives information on the ground state branching $B_o$ and the thermalization of the $\Delta$-resonance energy. In the target sector the Roper resonance region is strongly excited, which supports a picture of the Roper resonance being the compressional mode of the nucleon. More exclusive experiments are planned which should allow a strong reduction of the contribution due to projectile excitation. Further, completely exclusive experiments are considered to obtain detailed information on the multipole strength distributions and the decay properties of N* resonances.

References

1. C. Ellegaard et al., Phys. Rev. Lett. 50 (1983) 1745;
   D. Contardo et al., Phys. Rev. Lett. 168B (1989) 331;
   P. A. M. Guichon and J. Delorme, Proceedings 5. Journees d'Etudes Saturne, Piriac (1989).

2. U. G. Meissner, N. Kaiser and W. Weise, Nucl. Phys. A 466 (1988) 427.

3. C. Hajduk and B. Schwesinger, Phys. Lett. 140B (1984) 172;
   A. Hayashi and G. Holzwarth, Phys. Lett. 140B (1984) 175.

4. R. Baldini Celio et al., Nucl. Phys. A 379 (1982) 477.

5. F. L. Fabbri et al., Nucl. Phys. A 338 (1980) 429.

6. V. Dimitriev, O. Sushkov and C. Gaarde, Nucl. Phys. A 459 (1986) 503.

7. F. Binon et al., Phys. Rev. Lett. 35 (1975) 145.

---

* Saturne experiment at $\alpha$-momentum of 7 GeV/c performed in collaboration with:

M. Boivin, F. Plouin, Y. Yonnet, LNS Saclay.
R. Frascaria, R. Siebert, E. Warde, IPN Orsay.
B. Saghai, DPhN Saclay.
W. Jacobs, Indiana University Cyclotron Facility, USA.
P. E. Tegner, University of Stockholm, Sweden.
P. Zupranski, Institute for Nuclear Studies, Warsaw, Poland.

# THE QUASIFREE RESPONSE FOR (d,2p) ON NUCLEI

Thomas Sams [1]

*Laboratoire National Saturne, F-91191 Gif-sur-Yvette, France* and
*Niels Bohr Institute, DK-2100 Copenhagen, Denmark*

Experimental data on the nuclear spin-isospin response in the quasifree region measured with the charge-exchange reaction $(\vec{d}, 2p\,[^1S_0])$ at $1.6\,\mathrm{GeV}$ bombarding energy is presented. Some results with this reaction have previously been reported [1, 2]. The purpose of the study is the determination of the nuclear response to a pionic probe, i.e. the response to a spin-longitudinal probe.

Only hadronic probes bring in the spin-longitudinal response, but in general it is accompanied by spin-transverse contributions. The separation into spin-longitudinal and spin-transverse cross section is necessary and obtained by the use of the tensor polarized deuteron beam. This provides information of the same rank as could be obtained in an $(\vec{n}, \vec{p})$ experiment. On the nuclear targets $^{12}$C and $^{40}$Ca we find an enhancement of the spin-transverse cross section relative to the spin-longitudinal cross section as compared to the deuteron target. In a simple calculation we show that this is largely due to distortion effects. Further, the distortion dramatically reduces the capacity of the probe to distinguish spin-longitudinal and spin-transverse response.

The isospin channel of the nucleon-nucleon interaction is highly spin dependent. When the momentum transfer is larger than $m_\pi$ the spin-longitudinal and the spin-transverse channel even have opposite sign. Similar contrast between the longitudinal and transverse spin channel is expected for the residual interaction in an excited nucleus. This could generate differences in the nuclear response in the two channels. But no unambiguous signal of coherence due to the attractive residual interaction in the spin-longitudinal channel has been observed experimentally.

The experimental study of the quasi-elastic excitation of nuclei with the $(^3\mathrm{He}, t)$ reaction results in a dispersion relation deviating substantially from the free

$$\omega_{\mathrm{lab}} \;=\; \frac{-t}{2m_{\mathrm{N}}} \stackrel{\mathrm{N.R.}}{=} \frac{q^2}{2m_{\mathrm{N}}} \tag{1}$$

---

[1] The experimental data presented here were obtained in collaboration with D. Bachelier, M. Bedjidian, J. L. Boyard, A. Brockstedt, D. Contardo, P. Ekström, C. Ellegaard, C. Gaarde, C. D. Goodman, T. Hennino, J. C. Jourdain, J. S. Larsen, M. Österlund, P. Radvanyi, B. Ramstein, M. Roy-Stephan, and P. Zupranski. The results on effects of distortion have been obtained in collaboration with V. F. Dmitriev, C. Gaarde, M. Ichimura, and K. Kawahigashi.

At large momentum transfers, of order 2 - $3p_F$, the quasi-elastic peak is shifted downwards to about half of this value, whereas it qualitatively agrees with the value given by (1) at momentum transfers similar to the Fermi momentum. Also with the $(d, 2p)$ probe this behavior is seen. One major reason for studying the spin-isospin excitation of nuclei in selected spin channels is to determine whether the altered "dispersion" relation reflects a polarization of the nuclear medium. Notably, the $(p, p')$ and $(e, e')$ reactions follow the free dispersion relation, though for the latter shifted by a constant [3]. The $(p, n)$ reaction seems to fall close to $(e, e')$ [4].

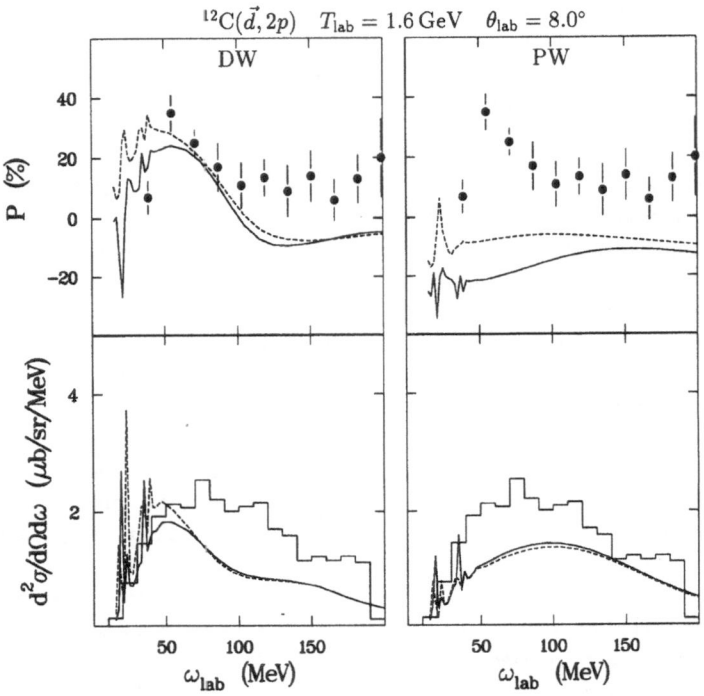

Figure 1. Cross section and tensor polarization response as measured in $^{12}C(\vec{d}, 2p\,[^1S_0])$ at laboratory bombarding energy 1600 MeV and 8.0° scattering angle. The momentum transfer is around $2.1\,\mathrm{fm}^{-1}$. The curves on the right plot are PW CRPA (full) PW C0TH (dash), the calculated PW cross sections have been divided by 4. The curves on the left plot are DW CRPA (full) DW C0TH (dash).

## Experiment

The $(\vec{d}, 2p)$ experiments have been performed using the tensor polarized deuteron beam at Laboratoire National Saturne. Quantized along the symmetry axis of the synchrotron, the tensor polarization of the beam was $\rho_{20} = 0.60 \pm 0.01$ to be compared to a maximal theoretical value of $1/\sqrt{2}$. The cuts from the $1.7° \times 2.6°$ collimator and from the spectrometer restricts the relative motion of the two protons to the $^1S_0$ state with high efficiency [1].

In "popular" terms the tensor polarization response we have measured is

$$P = \frac{\rho_{20}^{\text{beam}}}{\rho_{20}^{\text{max}}} \frac{\sigma(\uparrow) + \sigma(\downarrow) - 2\sigma(0)}{\sigma(\uparrow) + \sigma(\downarrow) + \sigma(0)} \qquad (2)$$

to be read as: spin-up + spin-down ÷ 2 times spin-0 divided by the sum. The spin is here quantized along the axis of the synchrotron. In terms of tensor analyzing powers in a co-ordinate system defined by the reaction, the polarization response may be expressed

$$P = \rho_{20}^{\text{beam}} \left( \frac{1}{2} T_{20}^{\text{M}} + \sqrt{\frac{3}{2}} \langle \cos 2\varphi \rangle T_{22}^{\text{M}} \right) \qquad (3)$$

The tensor analyzing powers $T_{\lambda\mu}^{\text{M}}$ refer to the Madison frame [5], i. e. $Z$-axis along the beam and $Y$-axis along the normal $\hat{n}$ to the reaction plane $\boldsymbol{p}_1 \times \boldsymbol{p}_3$. $\varphi$ is the angle between the normal to the reaction plane and the direction of the beam polarization, making the expectation value of $\cos 2\varphi$ essentially a function of the scattering angle. When $\cos 2\varphi = 1$ the polarization response $P$ is proportional to the cartesian analyzing power $A_{YY}$, when $\cos 2\varphi = -1$ it is proportional to the cartesian analyzing power $A_{XX}$.

As yet only experiments with the symmetry axis of the beam polarization along the symmetry axis of the synchrotron have been performed. This means that $\langle \cos 2\varphi \rangle \simeq 1$ at finite scattering angles. Experiments in which the deuteron spin is precessed in a superconducting soleonoid will make it possible to measure $T_{20}^{\text{M}}$ and $T_{22}^{\text{M}}$ separately, thus providing further information on the spin structure of the reactions. In particular the combination corresponding to $\cos 2\varphi = -1$ is of interest, since, in the plane wave description, it gives the best separation between the spin-transverse and spin-longitudinal reaction mechanism. This is due to the fact that, in quasi-elastic kinematics, the momentum transfer is essentially parallel to the $X$-axis of the Madison frame, making $A_{XX}$ the quantity that separates spin transfer according to whether there is spin flip (transverse) or not (longitudinal) along this axis. However, due to the smallness of one of the spin-transverse amplitudes, the quantity already measured gives a rather good separation [6]. We arrive at a simple picture: Spin-longitudinal has $P$ large negative and spin-transverse large positive $P$.

## Model

In the following a simple prescription to calculate the effect of distortion in the eikonal limit for a composite projectile is outlined. We take into account the absorption of the projectile nucleons individually. This is found to have a strong effect on the spin observables. The calculation is presented in more detail in [6]. It does not allow for inelastic rescattering of the nucleons in the probe, which may be the reason for the failure of the description at large energy transfers. This limitation is shared with DWBA calculations of the $(p, n)$ reaction on nuclei in the quasi-free region, and the calculations have in common that they underestimate the cross section at large energy transfers [7].

In the case of scattering of a nucleon off a nucleus the product of the incoming and outgoing distorted waves would be

$$\chi_f^\dagger(\boldsymbol{s}_1)\chi_i(\boldsymbol{s}_1) = e^{i\,\boldsymbol{q}_{\text{ex}} \cdot \boldsymbol{s}_1}\, D_1(\boldsymbol{b}_1) \qquad (4)$$

$e^{i\,\boldsymbol{q}_{\text{ex}} \cdot \boldsymbol{s}_1} = e^{i(\boldsymbol{k}_i - \boldsymbol{k}_f)\cdot \boldsymbol{s}_1}$ is the product of incoming and outgoing plane waves and the distortion function $D_1$ describes the deviation from plane wave approximation. For $(d, 2p)$ the product of the incoming and outgoing waves is written

$$\chi_f^\dagger(\boldsymbol{s}_1, \boldsymbol{s}_2)\chi_i(\boldsymbol{s}_1, \boldsymbol{s}_2) = e^{i\,\boldsymbol{q}_{\text{ex}} \cdot (\boldsymbol{s}_1 + \boldsymbol{s}_2)/2}\, D_1(\boldsymbol{b}_1) D_2(\boldsymbol{b}_2) \psi_{2p}^\dagger(\boldsymbol{s}_2 - \boldsymbol{s}_1) \psi_d(\boldsymbol{s}_2 - \boldsymbol{s}_1) \qquad (5)$$

where the intrinsic wavefunctions in entrance and exit channels are introduced. The distortion is allowed to act on the projectile particles separately through the distortion functions $D_1$ and $D_2$. It is assumed that the effect of distortion on the projectile particles can be factorized, an approximation which has previously been discussed in [8].

In the eikonal limit of Glauber theory the distortion function for the spectator is $D_2(b_2) = \exp(-\bar{\gamma}\,\tilde{\rho}(b_2))$. The nuclear thickness function $\tilde{\rho}(b) = \int \rho((b^2 + z^2)^{1/2})\,dz$ is taken from electron scattering data [9]. For the particle undergoing charge exchange the thickness is multiplied by a factor $(A-1)/A$ to take into account that the particle undergoing charge exchange is rescattered once less than the spectator [8]. The distortion parameter $\bar{\gamma}$ is essentially taken from the optical theorem. At $T_{\text{lab}} = 800\,\text{MeV}$ per nucleon the value $\bar{\gamma} = (2.1 + \text{i}\,0.3)\,\text{fm}^2$ is used (from [10]). This corresponds to $\sigma_{\text{NN}}^{\text{tot}} = 42\,\text{mb}$.

The two-dimensional distortion functions $D_1$ and $D_2$ are transformed into momentum representation where the distorting momentum transfer $\boldsymbol{\delta} = \boldsymbol{q}_{\text{ex}} - \boldsymbol{q}$ is perpendicular to the probe momentum

$$\widehat{D{-}1}(\boldsymbol{\delta}) = \int d^2b\, e^{\text{i}\boldsymbol{\delta}\cdot\boldsymbol{b}}\,(D(b){-}1) = 2\pi \int J_0(\delta\,b)\,(D(b){-}1)\,b\,db \qquad (6)$$

A convergent integral was obtained by subtracting the plane wave contribution.

The scattering matrix in momentum representation then becomes

$$\mathcal{M} = \text{i} \sum_{\mu_1 \sigma_2 \mu_2} \int \frac{d^2\delta}{(2\pi)^2}\,\langle\,\mu_1\,|\,f(\boldsymbol{q}_{\text{ex}},\boldsymbol{\delta})\,|\,M\,\rangle\; t_{1\mu_1\,\sigma_2\mu_2}(\boldsymbol{q}_{\text{ex}}{-}\boldsymbol{\delta})\,\langle\,n\,|\,\hat{j}^\dagger_{1-1\;\sigma_2\mu_2}(\boldsymbol{q}_{\text{ex}}{-}\boldsymbol{\delta})\,|\,\hat{0}\,\rangle \quad (7)$$

This is an integral over distorting momentum transfer of the product of the probe weight function $f$, the elementary charge exchange amplitude $t(\boldsymbol{q})$, and the target form factor $\text{i}\,\langle\,n\,|\,\hat{j}^\dagger(\boldsymbol{q})\,|\,\hat{0}\,\rangle$. Kinematic and spin-statistics factors were omitted, see [6] for details.

The probe weight function $f(\boldsymbol{q}_{\text{ex}},\boldsymbol{\delta})$ introduced in (7) is

$$\langle\,\mu_1\,|\,f(\boldsymbol{q}_{\text{ex}},\boldsymbol{\delta})\,|\,M\,\rangle = \int \frac{d^2\delta_2}{(2\pi)^2}\,\hat{D}_1(\boldsymbol{\delta}{-}\boldsymbol{\delta}_2)\,\hat{D}_2(\boldsymbol{\delta}_2)\,\langle\,\mu_1\,|\,S(\boldsymbol{q}_{\text{ex}}{-}2\boldsymbol{\delta}_2)\,|\,M\,\rangle \qquad (8)$$

$$= (2\pi)^2\delta^{(2)}(\boldsymbol{\delta})\langle\,\mu_1\,|\,S(\boldsymbol{q}_{\text{ex}})\,|\,M\,\rangle \qquad (9)$$

$$+\; \widehat{D_1{-}1}(\boldsymbol{\delta})\langle\,\mu_1\,|\,S(\boldsymbol{q}_{\text{ex}})\,|\,M\,\rangle$$

$$+\; \widehat{D_2{-}1}(\boldsymbol{\delta})\langle\,\mu_1\,|\,S(\boldsymbol{q}_{\text{ex}}{-}2\boldsymbol{\delta})\,|\,M\,\rangle$$

$$+\; \int \frac{d^2\delta_2}{(2\pi)^2}\,\widehat{D_1{-}1}(|\boldsymbol{\delta}{-}\boldsymbol{\delta}_2|)\,\widehat{D_2{-}1}(\boldsymbol{\delta}_2)\langle\,\mu_1\,|\,S(\boldsymbol{q}_{\text{ex}}{-}2\boldsymbol{\delta}_2)\,|\,M\,\rangle$$

First line of (9) is the plane wave contribution. Second line represents rescattering of the reactor nucleon, a term which is symmetric around the probe momentum. These two contributions are common for a pointlike and a composite probe. The third and fourth line are specific for a composite probe. Third line is a term where the spectator is rescattered. The fourth line represents rescattering of both projectile nucleons. The plane wave form factor for the transition from the deuteron to the di-proton $^1S_0$ state, $\langle\,\mu_1\,|\,S(\boldsymbol{q})\,|\,M\,\rangle = \langle\,2p\,[^1S_0]\,|\,\sigma^\dagger_{1\mu_1}\,e^{\text{i}\,\boldsymbol{q}\cdot\boldsymbol{s}/2}\,|\,d\,1M\,\rangle$, is in all cases evaluated at the difference between the momentum transfer on the reactor and the spectator.

The spin dependence of the driving force is described by resolving the interaction into its spherical tensor components. The rank $\sigma = 0, 1$ and the component $\mu = -\sigma, .., \sigma$ of the interaction is specified at both "vertices"

$$\langle\,pn\,|\,t^0(\boldsymbol{q}) + t^1(\boldsymbol{q})\,\boldsymbol{\tau}^1\cdot\boldsymbol{\tau}^2\,|\,np\,\rangle = 2\,t^1(\boldsymbol{q}) = 2\sum_{\substack{\sigma_1\mu_1 \\ \sigma_2\mu_2}} t_{\sigma_1\mu_1\,\sigma_2\mu_2}(\boldsymbol{q})\,\sigma^\dagger_{\sigma_1\mu_1}\,\sigma^\dagger_{\sigma_2\mu_2} \qquad (10)$$

The spin dependence of the interaction is in the operators $\sigma_{\sigma\mu}$ acting in the two vertices of the process. Only five amplitudes are independent for each isospin channel [12]. In the present calculation the phenomenological (on-shell) amplitudes of D. V. Bugg have been used [10].

The nuclear spin-isospin response function, which contains the physics we address experimentally, was calculated within the continuum RPA framework as described in [13]. It is defined as

$$R_{\tau\upsilon\ \sigma\mu}^{\tau'\upsilon'\ \sigma'\mu'}(\boldsymbol{q},\boldsymbol{q}',\omega) = \sum_{n\neq\hat{0}} \langle \hat{0} | \hat{j}_{\tau\upsilon\sigma\mu}(\boldsymbol{q}) | n \rangle\, \delta(\omega - (E_n - E_0))\, \langle n | \hat{j}_{\tau'\upsilon'\ \sigma'\mu'}^{\dagger}(\boldsymbol{q}') | \hat{0} \rangle \quad (11)$$

with the spin-isospin current given by

$$\hat{j}_{\tau\upsilon\sigma\mu}(\boldsymbol{q}) = i\sum_{j=1}^{A} \tau_{\tau\upsilon}^{j}\, \sigma_{\sigma\mu}^{j}\, e^{-i\,\boldsymbol{q}\cdot\boldsymbol{r}_j} \quad (12)$$

Here $\sigma_{00} = 1$, $\sigma_{10} = \sigma_z$, $\sigma_{1\pm1} = \mp(\sigma_x \pm i\,\sigma_y)/\sqrt{2}$, and similar for the isospin operator.

The nuclear states $|n\rangle$ entering in (11) are the eigenstates of the Hamiltonian

$$H = H_{\mathrm{sp}} + V \quad (13)$$

including the residual interaction $V$. If the particle-hole interaction $V$ is turned off, the response is referred to as uncorrelated, i. e. the pure shell model response. The effective particle hole interaction was parametrized as

$$V(\boldsymbol{q},\omega) = \left(\frac{f_\pi}{m_\pi}\right)^2 \left\{ \left( g' - \frac{q^2}{m_\pi^2 - t}\Gamma_\pi^2(t) \right)(\boldsymbol{\sigma}^1\cdot\hat{\boldsymbol{q}})(\boldsymbol{\sigma}^2\cdot\hat{\boldsymbol{q}}) \right. \quad (14)$$

$$\left. + \left( g' - C_\rho\frac{q^2}{m_\rho^2 - t}\Gamma_\rho^2(t) \right)(\boldsymbol{\sigma}^1\times\hat{\boldsymbol{q}})\cdot(\boldsymbol{\sigma}^2\times\hat{\boldsymbol{q}}) \right\} \boldsymbol{\tau}^1\cdot\boldsymbol{\tau}^2$$

with $m_\pi = 139\,\mathrm{MeV}$, $m_\rho = 770\,\mathrm{MeV}$, $C_\rho = 2.18$ and $f_\pi = 1.008$. Cut-off masses in the monopole form factors $\Gamma_\pi$ and $\Gamma_\rho$ were 1300 MeV and 2000 MeV respectively [14, 13]. Coupling to the $\Delta$ with coupling constant $f_\pi^* = 2.0f_\pi$ was included in the calculation. The parameter $g'$, which determines the interaction at zero momentum transfer, was in calculations shown in this manuscript set to 0.6.

The plane wave calculation shown in figure 1 demonstrates how the correlations lead to a softening and enhancement of the longitudinal response versus a hardening and quenching of the transverse response, i. e. a preference for longitudinal response in the lower end of the quasi-elastic peak: The polarization response $P$ predicted in PWIA becomes an increasing function of the energy transfer over the quasi-free region.

Even in the plane wave calculation, where the effect is strongest, there is very little effect on the position of the quasielastic peak from correlations: Correlations as the key to understanding the downward shift of the quasi-elastic peak at large momentum transfers seem ruled out.

## Conclusion

Whenever the response on nuclei deviates from the free response, it is observed to be relatively more spin transverse. Estimates of the effect of distortion have been made in the eikonal limit, but allowing for only one inelastic scattering. They are found to be large, bringing the calculated tensor analyzing power closer to the observed.

At energy transfers comparable to the Fermi energy only about half of the observed cross section is reproduced by the calculation. We are therefore reluctant to draw firm conclusions from the measured polarization response in this region. In regions where the cross section is reproduced, i. e. at small energy transfers, also the observed analyzing power is in qualitative agreement with the calculation. Distortion reduces the predicted effect of a medium polarization leaving less signal of correlations, if any, at the level of the measured quantity.

## References

[1] C. Ellegaard, C. Gaarde, T. S. Jørgensen, J. S. Larsen, C. Goodman, I. Bergqvist, A. Brockstedt, P. Ekström, M. Bedjidian, D. Contardo, J. Y. Grossiord, A. Guichard, D. Bachelier, J. L. Boyard, T. Hennino, J. C. Jourdain, M. Roy-Stephan, P. Radvanyi, and J. Tinsley, Phys. Rev. Lett. **59** (1987) 974, *The $(\vec{d}, {}^2\text{He})$ reaction at intermediate energies*.

[2] C. Ellegaard, C. Gaarde, T. S. Jørgensen, J. S. Larsen, B. Million, C. Goodman, A. Brockstedt, P. Ekström, M. Österlund, M. Bedjidian, D. Contardo, D. Bachelier, J. L. Boyard, T. Hennino, J. C. Jourdain, M. Roy-Stephan, P. Radvanyi, and P. Zupranski, Phys. Lett. **B231** (1989) 365, *Spin structure of the $\Delta$ excitation*.

[3] C. Gaarde, Nucl. Phys. **A478** (1988) 475c, *$\Delta$'s in nuclei*.

[4] J. McClelland and T. Taddeucci, private communication, 1991.

[5] *Madison Convention*, in *Proceedings of the Third International Polarization Symposium*, edited by H. H. Barschall and W. Haberli, University of Wisconsin Press, 1971.

[6] T. Sams, Ph. D. thesis, Niels Bohr Institutet, 1990, *The nuclear spin isospin response. The $(\vec{d}, 2p)$ reaction at intermediate energies*.

[7] M. Ichimura, private communication, 1991, *DWBA calculation with continuum RPA response for ${}^{40}\text{Ca}(p, n)$ in quasi-elastic region*, see also [13].

[8] V. F. Dmitriev, Phys. Lett. **B226** (1989) 219, *Do the medium effects exist for the $({}^3\text{He}, t)$ reaction at intermediate energies?*

[9] C. W. Jager, H. de Vries, and C. de Vries, Atom. Data and Nucl. Data Tables **14** (1974) 479, *Nuclear charge- and magnetization-density-distribution parameters from elastic electron scattering*.

[10] D. V. Bugg, private communication (1988), *Empirical nucleon-nucleon amplitudes at lab. bombarding energies 25 - 800 MeV*, see also [11].

[11] R. Dubois, D. Axen, R. Keeler, M. Comyn, G. A. Ludgate, J. R. Richardson, N. M. Stewart, A. S. Clough, D. V. Bugg, and J. A. Edgington, Nucl. Phys. **A377** (1982) 554, *Nucleon-nucleon phase shifts from 142 to 800 MeV*.

[12] A. K. Kerman, M. McManus, and R. M. Thaler, Ann. Phys. **8** (1959) 551, *The scattering of fast nucleons from nuclei*.

[13] M. Ichimura, K. Kawahigashi, T. S. Jørgensen, and C. Gaarde, Phys. Rev. **C39** (1989) 1446, *Excitation of spin-isospin modes in the quasifree scattering region*.

[14] W. M. Alberico, M. Ericson, and A. Molinari, Nucl. Phys. **A379** (1982) 429, *Quenching and hardening in the transverse quasi-elastic peak*.

# CONFERENCE SUMMARY

Edward F. Redish

Department of Physics
University of Maryland
College Park, MD 20742-4111

## 1. INTRODUCTION AND OVERVIEW

This series of Telluride conferences provides a good view of how nuclear physics has changed over the past decade. The trend has been away from building purely phenomenological models and towards seeking a better understanding of nuclear dynamics from a sub-nucleonic viewpoint. In this meeting, many of the talks addressed sub-nucleonic issues and their relations with nuclear phenomenologies.

In order to give us a framework for understanding the broad issues being addressed here, let me being with a brief description of "the nuclear paradigm" -- my view of what it is we are trying to do in nuclear physics.

Nuclei are complex many-body quantum systems whose size is on the order of their wavelength. The best understood and most successful example of small many-body quantum systems is atomic and molecular physics. Every physics grad student is familiar with the structure shown in Fig. 1.

The Coulomb potential is based on observations with individual charged particles. It is then put into a many-body Schrödinger equation. The equation is approximated by various well-developed and tested approximation schemes, and used to calculate and understand the factors that dominate many-body properties and dynamics. Many of the issues in atomic and chemical physics today are how to simplify, understand, and describe what is happening in various complex situations. When calculations can be carried to convergence, theory and experiment usually agree to a high accuracy. Models play an important role in helping understand what is going on, but their role is not fundamental.

In nuclear physics, for many years we have built our expectations on what I call "atomic physics envy". Our paradigm was modelled after the successful one in atomic and molecular physics. Since our situation is more complex -- nucleons are not as simple as electrons -- we had to add some elements: some input from particle theory and a three body force. This makes our paradigm look like Fig. 2. Since many of the elements in this paradigm have remained uncertain, the role of semi-phenomenological models has been very important in organizing our way of thinking about nuclei.

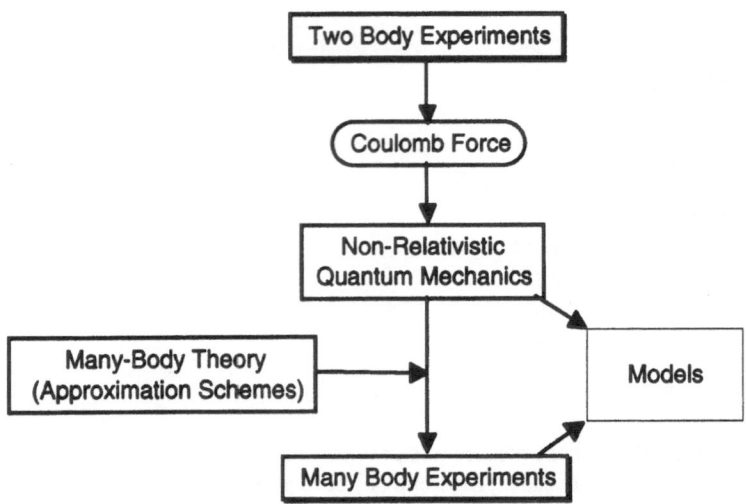

Fig. 1. The paradigm for non-relativistic atomic and molecular physics.

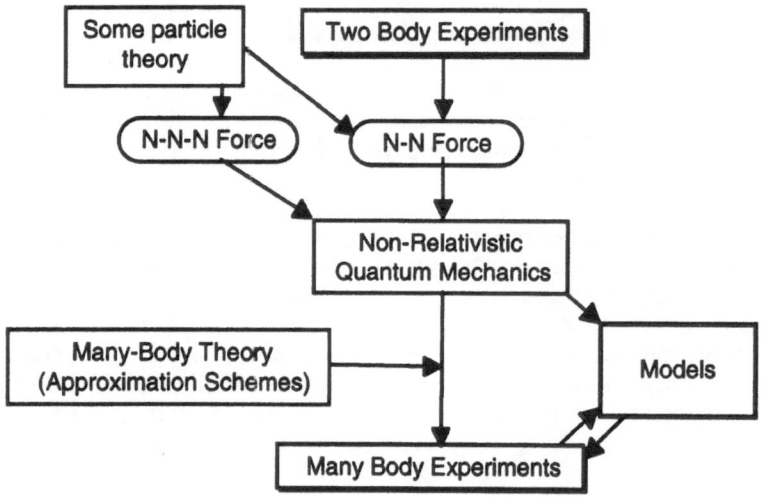

Fig. 2. The traditional paradigm for nuclear physics.

The simplified paradigm in Fig. 1 above doesn't accurately reflect the view of a sophisticated professional in the field. Rather, they know that the origin of Coulomb's law is now well understood and arises from the quantum field theory of electrons and charges: QED. Of course the nuclei of atoms are not point charges, but they may be treated as such in most circumstances, thanks to the factor of a million in the ratio of atomic to nuclear sizes. A better paradigm for atomic and molecular physics is shown in Fig. 3. By "Exotic effects" I mean those processes which depend directly on effects beyond those contained in the traditional non-relativistic model, such as two-photon processes, vacuum polarization, etc.

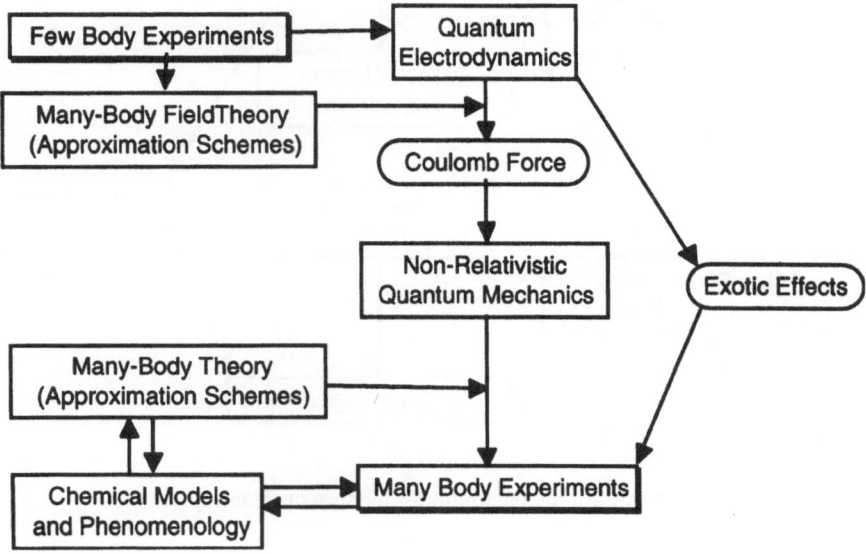

Fig. 3. A modern paradigm for atomic and molecular physics.

We now know enough about sub-nuclear physics to be able to conjecture that an analogous paradigm might serve for nuclear physics. My draft of what I would like it to look like is shown in Figure 4. We currently believe that Quantum Chromodynamics (QCD) is an appropriate starting point as a fundamental theory of strong interactions. Unfortunately, our evidence for believing this is mostly indirect -- perturbative results at large momentum transfers, inherited symmetries, etc. No simple solutions exist that demonstrate convincingly that QCD provides the correct basis for the broad range of complicated phenomena that occupy the boundary between the perturbative QCD of quarks and gluons and low energy hadron physics.

A plausible conjecture is: the understanding of hadron systems that we have built up in the past four decades can serve as an organizing phenomenology for making the link between the structure of hadrons and their behavior in nuclear systems at low energy. This means choosing effective degrees of freedom which are nucleons and mesons, using these when the hadrons are far apart (small energy and momentum transfers), and making the transition to quark and gluonic degrees of freedom at short distances (large energy and momentum transfers).

At present, it is not at all clear whether this structure is the appropriate one, though there are some strong hints that it may serve.

## 1.1. Issues in Nuclear Physics

If we accept the idea that QCD (and the standard model) is the correct theory of hadronic systems, we still have to address the question: *What are the appropriate degrees of freedom to use in describing nuclei and their interactions?*

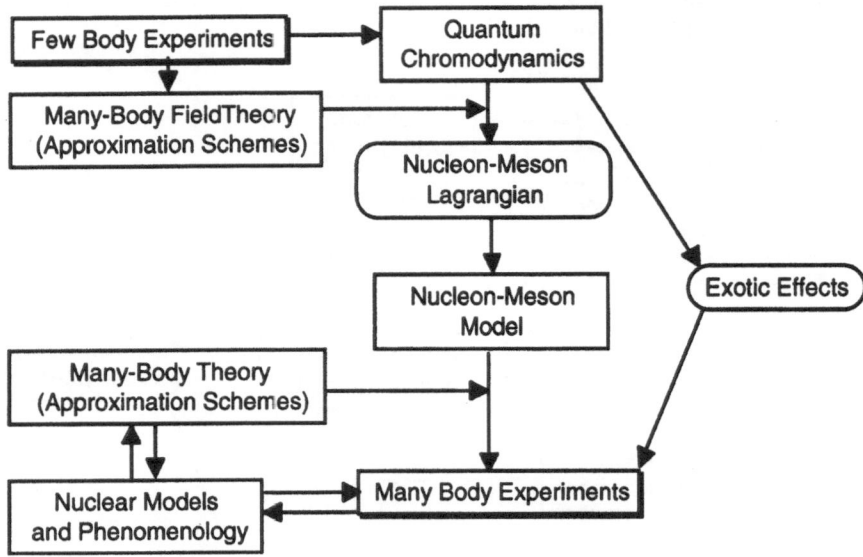

Fig. 4. A possible paradigm for modern nuclear physics.

Clearly the answer to this question depends on circumstances. In the same way that we can't expect to describe electron-atom scattering at momentum transfers of $10^6$ $nm^{-1}$ without including some information about the structure of nuclei, we can't expect to describe central collisions at RHIC in terms of hadrons alone. But the question still remains. If we want to describe "typical low energy nuclear physics", what degrees of freedom do we need? It is not appropriate to solve for the low lying states of $^{40}$Ca by treating quarks and gluons, any more than it is appropriate to solve for the chemical bonding of large atoms using all the electrons. We need to select appropriate collective degrees of freedom.

## 1.2. Degrees of Freedom

The difficulty with selecting appropriate degrees of freedom is that in principle, any problem can be crunched into any other. Suppose the operator P projects onto the effective degrees of freedom we've decided to use and Q = 1-P. Schematically, we can "Feshbach" our equations as is done in non-relativistic quantum mechanics. We project our Schrödinger equation on the P space to get an effective Schrödinger equation:

$$H \Psi = E \Psi$$

$$H (P+Q) \Psi = E (P+Q) \Psi$$

$$H_{eff} \, (P \, \Psi) = (H_{PP} + H_{PQ}(E - H_{QQ})^{-1}H_{QP} \, (P \, \Psi) = E \, (P \, \Psi)$$

Not only H, but all operators become modified to be effective operators:

$$<\Psi| \, \Theta \, |\Psi> = <P \, \Psi| \, \Theta_{eff} \, |P \, \Psi>.$$

If this were all there were to it, life would be easy. Unfortunately, we have to live with the consequences of Weinberg's third law of theoretical physics:[1]

> *You can use any degrees of freedom you want, but*
> *if you use the wrong ones you will be very unhappy.*

The difficulty is that the effective operators in a theory with the "wrong" degrees of freedom will be very peculiar and difficult to deal with -- highly non-local, rapidly energy dependent, and possibly of ranges characteristic of the whole system rather than of the parts.[2]

A number of different models of how to organize our fundamental thinking about nuclei are currently under consideration:

- The traditional model;

- Some relativistic models;

- A number of quark-parton models.

Each of these models has some successes. Each has some problems.

*The traditional model*

This model assumes that the dominant degrees of freedom will be *non-relativistic nucleons only*, so I refer to this as the NRNO model. Nucleons are described by two component spinor wave functions that satisfy a non-relativistic Schrödinger equation. The interaction is via an energy-independent short-ranged two-body (and perhaps a three-body) potential. The longest ranged part of the potential arises from one pion exchange, and there is a strong short-ranged repulsion. The parameters of the potential are fit to two-body data below the energy where pion production becomes significant (about 300 MeV). The theory is consistent and can be extended to other operators, such as electromagnetic currents.

The state of the art in this model is as follows. Many different "realistic" potential models have been developed. They are similar on-shell (but not identical). They also tend to have similar off-shell amplitudes, at least in the ranges -200 MeV < E < 300 MeV and k < 600 MeV/c. Some three-body potentials have been developed, but they are undoubtedly more uncertain than the two-body ones at this stage, and consistency between two- and three-body potentials is a problem. Many-body theories are very well developed.

*The relativistic models*

Although relativistic nuclear models have developed in parallel with the non-relativistic ones throughout the history of nuclear physics, it was only in the decade of the '80s that we saw enough significant advances to make this model competitive with the NRNO model. The most important result is that relativistic models seem to do much better in handling the spin degrees of freedom in nuclei than do non-relativistic models.

The critical element in relativistic models is the observation that the nuclear potential is a balance of nearly cancelling attractive and repulsive forces and that these have different Lorentz character. The mid-range attraction usually attributed to a "$\sigma$ meson" arises from the exchange of two correlated mesons (with some nucleon excitation in intermediate states) and is a Lorentz scalar. The short-range repulsion is attributed to $\rho$ and $\omega$ exchange and is a Lorentz vector. The scale associated with these is large ($\cong 300$ MeV). They tend to cancel for central forces but add for spin-orbit. The spin-orbit forces for nucleons are relatively much bigger than for electrons. Relativistic treatments must be performed despite the fact that in nuclei the kinetic energy is small ($KE/mc^2 \cong 0.05$), since the individual parts of the potential energy are not ($PE/mc^2 \cong 0.3$).

Two classes of relativistic models dominate the field at present: Dirac effective Lagrangian[3] (DEL) models and Quantum Hadrodynamics[4] (QHD).

The DEL models are generally not renormalizable and require form factors. The negative energy components are treated dynamically and are different from the free results. Note that non-relativistic models do not ignore the small components. Rather, they treat them kinematically instead of dynamically, choosing them to have the same ratio to the large components as for the free particle. The DEL models provide a good description of the spin dependence in intermediate energy nucleon-nucleus scattering and of the magnetic moments of nuclei.

The QHD model takes a renormalizable Lagrangian for point nucleons and mesons but starts with shifted meson operators (mean fields) and treats these mean fields dynamically. This model provides a good treatment of nuclear spin-orbit forces and the saturation of nuclear matter. In both models a full many-body theory is only beginning to be built.

*Quark-parton models*

A large number of quark-parton models have been built. Most have little if any deep connection with QCD. A list of models include: non-relativistic models with heavy ($m \approx 1/3\ m_p$) quarks bound in a confining potential; relativistic (nearly massless) quarks confined to move freely in a bubble in the vacuum; relativistic heavy quarks interacting with mean meson and gluon fields; non-relativistic and relativistic models with heavy or light quarks connected by string. Most have addressed the structure of single hadrons and have not yet been able to give any convincing new insights into how nucleons interact inside a nucleus, though the attempt is beginning to be made. Most suggestions for modifications of nuclear properties coming from quark models can also be obtained within meson-nucleon models given the uncertainties that still exist. As yet, there is no "smoking gun".

One of the things that we learn from this meeting is that we are beginning to get detailed information on the structure of the nucleon. Although it is not yet clear that the information is consistent, there are some very interesting and compelling predictions and some surprises in the structure functions.

## 1.3. Some Fundamental Questions

The various successes of and interactions between these different models lead to a number of important questions. I will view the contents of the conference in the reflection of how they shed light on these questions.

Very high energy scattering results show convincingly that quarks and gluons exist in hadrons. As nuclear physicists we are led to ask:

*Question 1: How do the properties and characteristics of quarks and gluons affect the structure and properties of nucleons and their excited states?*

The quark model is best understood at very short distances. Since we know rather well that

- At large separations (r > 2 fm) the nucleon force is well described by OPEP

- At short distances (r < 0.2 fm) the nucleon looks like free quarks (asymptotic freedom)

We are therefore led to ask where we have to make the transition from one type of model to the other.

*Question 2: At what separation of two hadrons must one abandon a model based on hadrons and hadron exchange and use degrees of freedom associated with quarks?*

A general feature of the quark models built so far is that the extent of the quark distribution in a hadron is about 0.5-1.0 fm. Since the average separation of nucleons in nuclear matter is less than 2 fm, we are led to rephrase our question in a stronger way (see Fig. 2 in Hwang and Speth):

*Question 2': How can we possibly get away with using meson exchange models of the nucleon force at an average separation of nucleons in nuclei?*

Mesons and nucleons are of course quantum systems, so what matters is not how close their quark distributions get, but how much they have to overlap before a significant excitation of their internal degrees of freedom is obtained. Still, the picture is disturbing.

We also know that the momentum distribution of quarks in a nucleon is modified when that nucleon is inside a nucleus. This leads to the third question:

*Question 3: Does the change in the quark distribution of a nucleon imbedded in a nucleus come from changes in the properties of the nucleon, of the correlations (e.g., meson exchange), from exotic effects such as percolation, or from fundamental changes inside a nucleus from hadronic to some other?*

Finally, we should not simply abandon all we have learned about nuclei over the past decades, even though we are beginning to learn new things about the nucleon's structure. This leads to a group of questions we summarize as a fourth question:

*Question 4: What constraints does our knowledge of nuclear properties put on the underlying models and how we should use them?*

Specific sub-questions in this category include:

*Question 4a: Do we need to include the small components of nucleon wave functions dynamically in our treatment of nuclei (NRNO vs. DEL / QHD)?*

*Question 4b: Do we need to include mesonic mean fields or quark / gluon condensates as dynamic degrees of freedom in our treatment of nuclei?*

*Question 4c: Does our knowledge of nuclear properties and responses limit the ways we can include sub-nucleonic degrees of freedom in a nuclear model (e.g., nucleon polarization, presence of condensates, quark percolation, presence of six- and nine-quark bags, etc.)?*

I'll review the content of the conference in the light of these questions. I'll begin with the smallest structure considered, the nucleon, and work my way up in size, through few-hadron systems to real nuclei.

## 2. THE NUCLEON

The quark-parton model of sub-nucleonic structure essentially says that nucleons are made up of valence quarks, sea quarks (quark-anti-quark pairs), and gluons. Some of these may have to be represented in hadrons by collective degrees of freedom or condensates. Most of the discussions of sub-nucleon structures at this meeting focussed on the issue of understanding the implications of the experiments that attempt to measure the distributions of quarks inside the nucleon. These distributions are represented by structure functions. The integrals of various structure functions can be combined to yield sum rules which test the validity of the underlying model.

### 2.1. Structure Functions and Sum Rules

In their presentations, Tim Londergan and Bob Jaffe discussed various aspects of nucleon structure functions and sum rules. Since the electroweak interaction only couples to quarks and not to gluons, leptonic scattering probes the quark distributions. Londergan gave a clear and compelling review showing many interesting and tantalizing results. Jaffe introduced some new structure functions.

What is happening in the world of nucleon structure functions is very much what we had hope would happen in nuclear physics nearly 40 years ago. We hoped that we could learn about the distributions of nucleons in nuclei by going to scattering at high enough energies that some kind of Born or Impulse approximation would work. This is what happens in atomic and molecular physics where the Born series becomes valid at high energies and the long range Coulomb force won't support large momentum transfers. What we learned is that the nucleon is too soft for this to work. The nucleus is essentially black until the nucleon itself begins to be transparent. There's not enough room between the nucleon scale and the quark scale to let us probe nuclear degrees of freedom perturbatively. If we look at quark behavior, we are rescued by the fact of asymptotic freedom -- quarks interact weakly at high relative momenta. There does seem to be enough room between the nucleon and quark substructure scales to let us treat quarks as elementary constituents reached perturbatively in high energy processes.[5]

Londergan and Jaffe discussed a variety of structure functions and sum rules. Where one has a conserved quantity that is being constructed (such as baryon number) or where one can match to a quantity which can be expanded in radiative corrections calculated from high energy perturbative QCD (the twist expansion), the results are excellent. Note that the twist expansion is "inside-out" from the familiar multiple scattering series. That has as its leading term the process where only a single nucleon is struck. The leading term in the twist expansion has all the valence partons sharing the (large) momentum transfer.

Londergan also reported perturbative QCD calculations of the $Q^2$ dependence of various form factors. The agreement with experiment (Figs. 17 and 18 in Londergan and Kumano) is spectacularly good. These results give us confidence that the parton picture of nucleon structure is a good one. But when it comes to rules that depend on a specific quark-parton model of low energy properties, such as the Gottfried sum rule, the results are not as good. Some of the implications extracted from combining the various rules seem counter-intuitive, such as the results that the strange quarks in the sea are significantly polarized and that the total spin carried by the quarks in a nucleon is small. Londergan and Kumano discuss a variety of possible explanations.

Jaffe stressed that we are still in an early stage of the analysis of structure functions and the error bars are still large. Confirming and improved experiments are needed. He also introduced two new chirally-odd spin structure functions that could lead to sensitive measures of the relativistic character of quarks in a nucleon. They can't be measured in deep inelastic lepton scattering because of conservation of chirality, but they might be observable in a Drell-Yan (lepton-anti-lepton production) process.

2.2. Gluons

Electroweak interactions only see the quarks. In his talk, Frank Close focussed on lower energy issues and on ways of determining the gluon content of hadrons. He pointed out that some of the successes of simple quark models do not necessarily exclude the presence of real gluons. For example, the "very-naive non-relativistic quark model" correctly predicts the ratio of proton and neutron magnetic moments: $\mu_p / \mu_n = -3/2$. However, all that's really required to get this is that the spin polarizations of the up and down quarks are related by $\Delta u = -4 \Delta d$. This is given by many models, including some with explicit gluon content. For another example, adding a gluon to three quarks leads to the same SU(6) representation for the first cluster of excited states as radial / orbital excitations of a quark. The models are not indistinguishable, however. Electroproduction of nucleonic resonances at CEBAF would do the job.

Close also made an important point about the relation between one gluon exchange, the spectrum of the nucleon, and the isospin splitting of the quark structure functions. At low energies, one gluon exchange (OGE) produces a spin dependence that splits the N and $\Delta$ states. The spin dependence, combined with the Pauli principle, leads the two same-flavor quarks (the up quarks in the proton) to get a larger fraction of the energy than the singleton quark. This results in a charge dependence in the structure functions. The OGE strength obtained from fitting low energy spectra can be used in a bag model to calculate $u(x) - d(x)$ measured in deep inelastic lepton scattering and gives a good result. This leads to the nice result that the same mechanism is responsible for N-Δ splitting and the isospin asymmetry of the structure functions.

Two additional talks dealt with nucleon structure. Louis described a proposed LAMPF experiment that would measure the strange form factor in v-p scattering. This may be able to tell us something about the strange quark contribution to the nucleon spin.

Peter Morsch described an interesting experiment. Instead of using nucleons as projectiles to study the excitation of nuclei, a nucleus projectile was used at Saturne to study the excitation of the proton. A 4.2 GeV $^4$He nucleus was used as a T=0 S=0 projectile to excite the proton. This let them select only isoscalar monopole excitations. The Roper resonance was clearly visible, and, if interpreted as a breathing mode, allowed them to evaluate the energy weighted sum rule which relates the radius of the nucleon with its compressibility. Taking the radius of the quark distribution in the nucleon to be 0.85 fm leads to a compressibility of 1450 MeV. A significantly smaller radius, such as 0.5 fm, seems to be ruled out by the absolute magnitude of the cross section. This result is important in considering the answer to Question 2.

The other results discussed here deal with my Question 1. The conclusion from the presentations is that the quark picture is well established, but the kind of model that should be used at low energies is still undetermined.

## 3. FEW-HADRON SYSTEMS

Few-hadron systems play a special role in contemporary nuclear physics and have a special importance in the boundary between hadronic and sub-hadronic models. They are complex enough to be sensitive to the details of the model used to describe them, but they are often simple enough, given modern computational tools, to permit a complete (or nearly complete) construction of the states within the model. The uncertainties due to uncalculated many-body effects, so daunting in the study of larger nuclei, can be substantially reduced.

Many few-hadron systems were discussed. Speth reported on calculations of meson-meson systems. Bradamante, Lamanna, and Eisenstein discussed experimental results on nucleon-antinucleon processes. Thé and Potterveld presented new results on the deuteron as probed by electron scattering. Tornow reported on nucleon-nucleon and nucleon-deuteron scattering experiments, and McKeown and Häusser reported on ways to probe the wave function of $^3$He.

### 3.1. Meson-Meson

Josef Speth reported on calculations he and his collaborators have performed on systems with baryon number zero. The approach is analogous to the one-boson-exchange model that has been extensively developed for nucleon-nucleon interactions. The pion and the mesonic resonances are taken as elements of an effective Lagrangian field theory and their properties are calculated by summing classes of perturbation graphs to all orders by solving dynamic equations. These calculations address Question 2. Mesonic models are taken seriously and pushed downward to distances that seriously trouble quark modelers.

Speth stresses that the size of an electromagnetic form factor should not be mistaken for the size of the quark distribution (range of the strong form factor). The largest scales in meson and nucleon electromagnetic form factors can be explained in a straightforward way in terms of the leading mesonic processes, such as vector dominance. He reported on a dynamic calculation of $\pi$–$\pi$ scattering based on a combination of $\rho$ exchange and the presence of a bare s-channel $\rho$ resonance. Fitting the $\pi$–$\pi$ phase shifts produces a T matrix which can then be used to give an excellent fit to the pion electromagnetic form factor. The strong form factor needed in this calculation is extremely hard -- about 4 GeV (corresponding to a size of about 0.05 fm)!

This work is in the spirit of a number of other calculations that take vector meson exchange seriously and show that it can correlate some detailed properties of hadronic systems very well. This is a surprising result, given the current characteristics of quark models. Even a hard-nosed meson pusher like me finds it difficult to accept the idea that we won't need quark degrees of freedom until hadron separations of 0.1 fm! What this calculation seems to be telling us, is that we're going to have to be careful: many properties of hadronic systems can be accurately described in complementary ways. The success of one model in fitting data cannot be taken as an argument for selecting that model over another model. We have to know what *both* models predict. I'll return to this theme a number of times in this talk.

3.2. Nucleon-Antinucleon

At the last Telluride meeting, we were promised $N\overline{N}$ spin data from LEAR. In their talks, Bradamante and Lamanna each presented LEAR data on $p\overline{p} \rightarrow p\overline{p}$ and $p\overline{p} \rightarrow n\overline{n}$ (preliminary) at momenta ranging from 60 MeV/c to 1900 MeV/c. These experiments are interesting as possible probes of Question 2. The exchange of an $\omega$ between two nucleons produces a repulsion, but G-parity tells us it should produce an attraction in the nucleon-antinucleon system. If a signal of this flip could be seen it would give us some information on how far down we can use the meson picture.

Unfortunately, preliminary analyses seem to show that annihilation processes are sufficiently long ranged to cover up the $\omega$-exchange part pretty well. This is what we would expect if the mesonic degrees of freedom were no longer valid and we had to go to a quark-based model. Unfortunately, it also doesn't rule out a meson theoretic model in principle. We need a better understanding of what kind of annihilation strengths and ranges are implied by taking the meson model seriously.

Bob Eisenstein presented a very lovely experiment and analysis of $p\overline{p} \rightarrow \Lambda\overline{\Lambda}$ just above threshold. Thanks to the weak spin-polarized decay of the $\Lambda$, he and his collaborators were able to get detailed information on the spin amplitudes. Since the $\Lambda\overline{\Lambda}$ channel was just above threshold, only a few partial waves are required, even though the $p\overline{p}$ system has several hundred MeV of kinetic energy. Using an effective-range approach, they demonstrate the need for both S and P wave contributions, for P wave splitting, and that the hyperon pairs are always produced in the triplet state. The excellent detail provided may yield information on gluonic vs. mesonic degrees of freedom since something like vector meson exchange may be required to fit the P wave splitting.

### 3.3. Nucleon-Nucleon

The two-nucleon system was discussed in talks by Thé, Potterveld, and Tornow. Thé reported the long-awaited measurement of the tensor polarization, $t_{20}$, in e-D scattering at Bates and Potterveld gave a preliminary report on an analogous experiment at Novosibirsk.

Plans for both deuteron experiments were discussed at the last Telluride conference and both are experimental tours-de-force. The Bates experiment involves a double scattering, analyzing the polarization of the recoiling deuteron, while the Novosibirsk experiment uses a polarized gas target.

The Bates results (Thé et al., Fig. 4) give $t_{20}$ in the momentum range 3-5 fm$^{-1}$. These results have important implications. First, the prediction of perturbative QCD for the asymptotic behavior of $t_{20}$ is not even qualitatively correct. This squelches speculations based on incomplete data that perturbative QCD might be valid at a very low momenta (of the order of $\Lambda_{QCD} \sim 200$ MeV/c). Second, it permits the detailed separation of the deuteron's monopole and quadrupole form factors and displays the first zero in the monopole form factor (at $q \approx 4.4$ fm$^{-1}$). These form factors (and $t_{20}$ itself) provide detailed tests of meson exchange and quark-based models of the deuteron. The predictions of mesonic models are qualitatively good, but most differ from the data in some details. It seems necessary to include a realistic potential, relativistic kinematics, and some explicit meson exchange currents in a consistent way.

Note that the $t_{20}$ experiment, even though it is occurring at high momenta, says little about the need for small components in a relativistic treatment of the deuteron. Small components play almost no role in the deuteron for two reasons. The deuteron is a T=0 state so it gets no one-step contribution to its wave function from virtual emission of one pion, and it is a very low density system so we don't expect any relativistic M* effects.

Tornow discussed two important issues in the study of the low-energy nucleon-nucleon scattering observables: the tensor coupling strength, in particular the S-D mixing parameter $\epsilon_1$, and charge independence breaking (CIB).

The strength and character of the tensor force is very important. Since the tensor force is quite strong, it plays a critical role in nuclear binding and spin correlations. The primary constraint on the two-nucleon tensor force comes from the quadrupole moment and asymptotic D/S ratio of the deuteron. (The $\epsilon_1$ coupling in elastic scattering is poorly determined because of the need for double polarization experiments and competition with other phase shifts.) The tensor/central force ratio at short and medium range is therefore not well determined. The tensor force is also an important factor in obtaining the answer to Question 2. The contribution of one $\pi$ exchange to the tensor force dominates at long range, but is singular at the origin and must be cut off at short distances. The exchange of the $\rho$, a heavy vector meson, is important at mid-range. Any suppression of the $\rho$ and any mechanism to cut off $\pi$ exchange at short distance due to a switch-over to quark degrees of freedom should be sensitively displayed in the tensor force.

Tornow reports on some new experiments to measure two-nucleon scattering observables sensitive to $\epsilon_1$ at Basel, Bonn, and Karlsruhe. The results seem to start from zero more slowly and rise to higher values than modern meson-exchange potentials predict (Tornow, Fig. 14). Pinning these values down more sharply is of great importance. Tornow also discusses an ongoing experiment at TUNL designed to get a better handle on $\varepsilon_1$.

The question of CIB in NN scattering has recently been raised by the Nijmegen group, who find the $\pi^0$NN coupling constant to be different from the $\pi^{\pm}$NN one. It is well known that the pp and np $^1S_0$ scattering lengths differ by about 30% ($a_{pp} = -17.3$ fm, $a_{np} = -23.7$ fm), but the pion coupling constants respond mostly to the NN P waves. Tornow reports on an important three-body study that casts some light on these issues.

The theory of the three-nucleon system has made significant strides in the past decade. Indeed, it is now the only place in nuclear physics where a completely converged calculation can be performed. In particular, the results of no-free-parameter n-d scattering calculations below 50 MeV are spectacularly good (see, for example, Tornow, Fig. 5). The only discrepancy is in the value of $A_y$ in the dip of the cross section, where the peak in $A_y$ is too small to fit the data. Witała and Glöckle show that introducing some CIB in the NN P waves can account for both the n-d and p-d data. The CIB effects being found seem to be much larger than those predicted by meson exchange theory, which involve significant contributions due to vector mesons.

Terrien reported on preliminary results from Saturne on a comprehensive full-phase-space survey of the np $\rightarrow$ pp$\pi^-$ reaction using polarized neutrons at 784 MeV. Extensive cross-section and asymmetry data will be taken in the energy range from 400-1150 MeV. This reaction is of particular interest since the contributions of $\Delta$ resonances should be suppressed by considerations of spin and isospin.

## 3.4. Trinucleons

The success of the Faddeev theory in obtaining converged calculations given an NRNO potential model has also had implications for testing the bound states. One of the surprises of the past decade has been the observation that in the NRNO model (even with three-body forces) the observables of the three-nucleon bound state are uniquely correlated.[6] (See Fig. 5.) The NRNO still has enough ambiguity to yield a fairly wide range of binding energies for the triton. The relevant uncertainties are the on-shell CIB and the strength of the tensor force[7] discussed by Tornow, and the short range cut off of the three-nucleon force. The strong correlation means that, despite the uncertainties, one can pick a wave function which will fit essentially all the low-energy bound state observables, by picking a combination of two- and three-body forces that fit the triton binding energy.[8] This should yield a good NRNO three-body wave function for use in testing effects beyond the NRNO model.

McKeown discussed plans to use polarized $^3$He as a stand-in for a neutron target to measure the electromagnetic structure of the neutron by electron scattering at Bates. Crudely, one might expect the two protons in $^3$He to be paired to spin 0, so all the spin response should be carried by the neutron. There are higher order components to the wave function (for example, pp relative $^3$P waves) but these should be reliably estimated by current Faddeev calculations. He reported on inclusive asymmetry measurements of e-$^3$He scattering at Bates that permitted a measurement of $G_E^n$. An asymmetry coincidence measurement could display the presence of a $\Delta$ in the $^3$He wave function. McKeown also discussed proposals to measure electron scattering from p, D, and $^3$He at

Hermes at high $Q^2$ to get better measurements of $g_1(x)$. These could be important for some of the quark-parton sum rules.

Häusser discussed the use of $^3$He as a neutron target in hadron scattering experiments. They have developed a new high-pressure polarized $^3$He cell which should reach polarizations of about 75% at up to 9 atm. He reported on measurements of spin asymmetries in elastic and quasi-elastic knockout by protons on $^3$He at 290 MeV. These show a sensitivity to the mixed symmetry (S') state of the trinucleon. The asymmetries in quasi-elastic knockout is reasonably well described by a standard PWIA, but a theoretical optical model is inadequate to describe the elastic scattering. Some data on the asymmetry in the elastic scattering of 100 MeV pions was also reported.

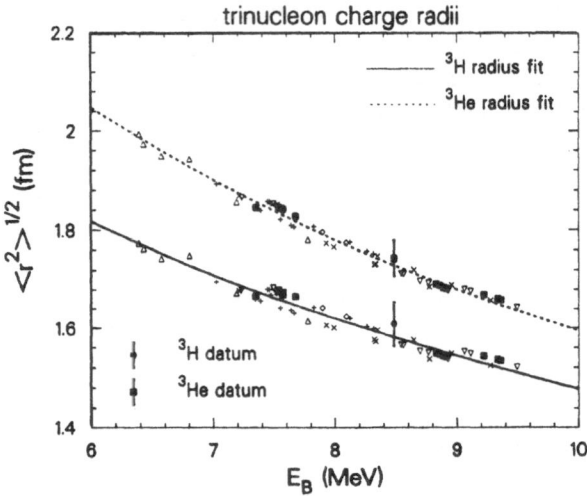

Fig. 5. The correlation of the $^3$H and $^3$He point charge radii plotted vs. the triton binding energy for various model Hamiltonians in the NRNO model. From ref. 6.

## 4. NUCLEON-NUCLEUS INTERACTIONS

If we want to understand the way nucleons behave inside a nucleus, we somehow have to probe all spin and isospin responses. Using a fast nucleon provides coupling to all those degrees of freedom, but is sufficiently complicated that it takes some work to disentangle what's going on. Elastic scattering and inelastic scattering to discrete states use the multiple scattering theory and the fact of selecting a unique discrete final state to permit what seems to be a reasonably stable theoretical description of the reaction process. Elastic scattering, especially the spin dependence, seems to be a good place to look at my Question 4a: the need for dynamic small components. Inelastic scattering to discrete states seems to be a good place to look for medium modifications to the effective interaction, Question 4c: limits on how we can test for sub-nucleonic changes to the system. It also may be a good probe of Question 2: the hadron / parton changeover distance.

Inelastic nucleon scattering to the nuclear continuum has the problem of mixing the nuclear response with the reaction mechanism. The comparison of a variety of observations with a variety of different probes have led to an understanding of the nuclear response at low energy in terms of giant resonances. The attempt to simplify the reaction mechanism by going to higher energies is producing some interesting results, but the situation has not yet cleared sufficiently to allow these processes to illuminate the fundamental issues. The hope of a few years ago that the Gamow-Teller strength distribution would tell us about nucleon vs. quark coupling through the presence of the excitation of $\Delta$'s seems to have been a will-o'-the-wisp. Even the interpretation of low-energy strength via (p,n) reactions was challenged at this meeting.

### 4.1. Elastic Nucleon Scattering

Elastic proton scattering at intermediate energies was a big story at the last two Telluride conferences. The spin response data played a substantial role in jump starting the Dirac/relativistic model of nuclei. The talks by Brieva and Ray, and recent developments in the relativistic model by Wallace and others have shown the story to be a bit more complicated than we thought.

In elastic scattering, the cross sections for spin-up and for spin-down protons are both sharply diffractive. The relative shift in these patterns as a result of the spin-orbit force in the surface turns the spin response in the dips of the cross section into a strong magnifying glass for many small effects -- relativity, off-shell behavior, Coulomb, Wigner rotations, etc. These all interact with each other in non-linear ways, even though they are each small. The bad news is that we have learned that a complete calculation must be done before we can really draw a strong conclusion. The good news is that we are almost there.

Brieva (with Arellano and Love) reported on a non-relativistic full-folding calculation of the p-A optical potential in the 200-500 MeV range. This correctly takes into account the non-locality in the effective NN force, in contrast to a factorized or even a factorized direct+exchange calculations.[9] Different realistic energy-independent potentials are used and all give essentially the same results. This is not surprising.[10] The results are much better when compared to data than those of cruder non-relativistic calculations. They are about as good as the present relativistic calculations, but those are not as sophisticated, so the comparison is still between apples and oranges. Some people found it surprising that replacing an effective T-matrix interaction by a Brueckner (local-density nuclear-matter) g matrix made little difference, even at 200 MeV.

One problem with standard non-relativistic calculations is that they can't be used reliably above about 400 MeV since they don't include $\pi$ production. Lanny Ray discussed a method to improve this. He has extended the Feshbach-Lomon boundary condition model to include nucleonic excited states via coupled channels. He adds some phenomenological short range interaction and is able to fit two-body data quantitatively between 0 and 1000 MeV. (Other similar models have been developed.[11]) He calculates a nuclear-matter g matrix and adds the shift to an on-shell factorized tρ potential (local+exchange) in order to guarantee both a good on-shell fit and the correct zero-density limit. He finds a strong density dependence, especially for the real central and

517

imaginary spin-orbit potentials, in contrast to the Arellano, Brieva, Love work. However, his fits are very good, comparable to the current relativistic state of the art.

Horowitz summarized the relativistic approach. The results for fairly simple optical model calculations are excellent. Non-locality effects have not, however, been considered in as great detail as they have in the non-relativistic calculation of Arellano, Brieva, and Love which currently sets the state of the art.

The situation here falls into the category discussed earlier: fits to the data are not enough. We need to know what each theory predicts before we can judge between them. And it is essential to not make too much of calculations that agree with data to a high accuracy when we have neglected effects that contribute at that level of accuracy. We must conclude that, at this stage, elastic scattering does not yet provide compelling evidence for small components of the wave function that differ significantly from those of the free nucleon.

4.2. Inelastic Scattering to Discrete States

Wissink and Stephenson each reported on the measurement of spin transfer observables in inelastic proton scattering at IUCF. The motivation for this work is based on the observation of Moss and Bleszynski that the matrix of spin transfer observables could be written (in PWIA) as a linear combination of the squares of the fundamental NN effective interaction multiplied by the longitudinal or transverse spin response function of the nuclear transition. They suggested that certain combinations of observables should be particularly sensitive to a single term in the NN effective interaction. This could shine a spotlight on those terms responsible for discrepancies.

The full matrix of eight spin-transfer observables were measured in inelastic proton scattering at IUCF in the neighborhood of 200 MeV for a number of interesting states. DWIA calculations show that the specific sensitivity of the proposed combinations is not lost because of distortions.

Stephenson analyzed the data for excitation of a stretch state -- the 4⁻ in $^{16}O$ at an excitation energy of 19 MeV. The standard Love-Franey interaction in DWIA does not do particularly well, a somewhat troubling result, given the simplicity of the state's structure. Stephenson used the Moss-Bleszynski separation technique to focus on the effective tensor interaction as the culprit. He added a term that corresponds to increasing the coupling of the ρ-meson in nuclear matter, a suggestion made by a number of authors in response to certain features present in QCD.[12] A change of less than 10% in the coupling constant was enough to bring the data into line. The kind of modification that results in the effective interaction is shown in Stephenson and Tostevin's Fig. 2. Although this is a fairly simple and preliminary analysis, it indicates the power of the probe and suggests that a sensitive probe of the NN effective interaction has been added to our toolkit.

In an important footnote comment, Jim Carr reminded us that in many energy regimes, including the 100-200 MeV range for the above experiments, there are very few NN spin-transfer measurements. More reliable nucleon-nucleon scattering amplitudes will be needed if we are to make full use of this tool.

Hintz reported an analysis of inelastic electron and proton scattering to near-stretch (12⁻ and 14⁻) states in $^{208}Pb$. In this work, only cross sections were studied, but an RPA calculation was performed with modified effective interaction. He concludes that the tensor force needs to be reduced and the spin-orbit force enhanced somewhat in order to fit the 12⁻ states.

### 4.3. Excitation of the Continuum

Inelastic scattering that excites the nucleus into its continuum has been studied for many years and for a variety of purposes. One reason is to study the response to the nucleus to the input of momentum, energy, angular momentum, spin, and isospin. Understanding this response requires that we both understand the collective response of the nucleus to an excitation and the reaction mechanism that leads produces it. The use of different projectiles and the study of their spin response allows us to begin to disentangle the various nuclear responses.

The excitation of the nuclear continuum by protons and the (p,n) and (n,p) charge exchange reactions were discussed. The fundamental issues that arise include the role of relativity and the validity of hadron-based models. The latter play a particularly important role at high energy loss, where the $\Delta$ can be excited and real $\pi$ production occurs.

*(p,p')*

The spin dependent inelastic scattering of protons has been measured in order to try to extract the spin response of the nucleus at low momentum transfer (q < 1 fm⁻¹) and energy transfers up to about 40 MeV. This response tests our understanding of detailed collective effects in nuclei.

Todd Baker reviewed the experimental results. At the last Telluride meeting, Kevin Jones reported the result that the spin response of $^{40}$Ca at q ≈ 0.5 fm⁻¹ and ω ≈ 40 MeV is dominated by $\Delta$S=1. Since then, longitudinal and transverse responses have been studied for $^{12}$C and $^{208}$Pb at Saturne, LAMPF, and TRIUMF. Similar results were found. A comparison of ($\Delta$S=1)/[($\Delta$S=1)+($\Delta$S=0)] for $^{40}$Ca at 319 and 800 MeV show qualitatively similar results, suggesting that issues of reaction mechanism may not overly obscure the nuclear information.

Wambach presented a rather microscopic analysis of the response. At the last meeting, Smith presented a calculation of the response in a crude "slab model" and mentioned a more sophisticated possibility. This time, we were treated to the results of the better model. The model is a DWIA reaction mechanism using a Love-Franey effective interaction and an approximate second RPA (2RPA) for the final excited states. A true 2RPA would diagonalize on a basis of 1p-1h and 2p-2h states. This is too large for present computational tools, so the effect if the 2p-2h states is taken to introduce a spreading width. These are obtained from fits to a low energy optical model analysis. The result is a smearing of the 1RPA result and fits the data reasonably well.

One interesting point Wambach raised involves the modification of the residual interaction in the medium. He applied the $\pi\pi$ scattering model of the $\rho$ discussed by Speth earlier and included the modification of the $\pi$ propagator due to the medium. The result was that the $\rho$ spectrum was shifted in a manner whose effects would be similar to those suggested by QCD arguments mentioned above. This again points out the difficulty of interpreting such an effect as a "smoking gun" for quark effects. In a many-body system there may be more than one way to modify a meson.

There is some possibility that these experiments will yield constraints on our fundamental models of nuclei, but that seems, unfortunately, rather far off. The studies are still in a relatively early stage and questions about reaction mechanisms still remain. On the theoretical side, a converged calculation with realistic forces is not in sight, so many uncertainties can be expected for some time to come.

The spin response for inclusive (p,p') in the quasifree scattering region was discussed by Horowitz in his review and may have something important to say about the relativistic vs. non-relativistic model. (Recall that the more precise question is whether the small components of the nucleon wave function need to be treated dynamically or whether they always can be taken to be the same as for the free nucleon.) Since quasifree scattering is momentum matched and occurs quickly, the nuclear response and the reaction mechanism tend to be simpler than at lower momentum transfer. Horowitz reported that a simple relativistic PWIA correctly predicts the quenching of the polarization, P, in quasifree (inclusive) scattering from a number of nuclei. This is a fairly direct response to modifications of the small component. Other amplitudes are not as simple (and not as well predicted).

For inclusive scattering, it may be very important to include a detailed treatment of the nuclear response. An exclusive (p,2p) experiment would have a cleaner response since a discrete nuclear final state is explicitly specified. In addition, more complete reaction mechanism (distorted wave) calculations are needed. This is another place where better nucleon-nucleon amplitudes would be of considerable value.

## (p,n')

The (p,n') reaction was covered by two speakers, Goodman and Taddeucci. Goodman reviewed the state of the Gamow-Teller (GT) strength problem, and Taddeucci presented new data on polarization transfer in the quasifree region.

The Gamow-Teller strength function, $B_i(GT) = |<\Psi_i|\sigma \cdot \tau|\Psi_0>|^2$, is believed to be measured fairly directly in $\beta$-decay and in the (p,n) reaction at $0^o$. Experimentally, the sum of the strengths found has consistently been below the values expected from the theoretical sum rule, sometimes by as much as a factor of two. This could be very interesting. The Gamow-Teller operator puts both spin and isospin into the nucleus. If it's absorbed by a nucleon and the strength is spread out and shifted to high excitation energy by the residual interaction between nucleons, then the residual interaction is different than we might have expected. On the other hand, the spin and isospin might be absorbed by a quark, flipping a nucleon into a $\Delta$. Then the strength would be at very high energies and the nucleus would be softer with respect to quark degrees of freedom than many of us expect. This issue was discussed vigorously at previous meetings and Goodman reviews these discussions in some detail. Not too much has happened since the last conference (only 2 of Goodman's 28 references have been written since then) and we seem to be waiting for some new ideas. The issue is not entirely dormant, however. See the discussion of Garcia's challenging $\beta$-decay results below.

In the only real "shell-model" paper of the conference, Jim Carr presented a set of calculations of the fragmentation of the M6 response in the s-d shell. This is an important issue both for calibrating reaction mechanisms and beginning to understand the quenching of sum rules. Starting from a $0^+$ ground state, there is only one way a one-step $1\hbar\omega$ transition can produce a $6^-$ state in the an s-d shell nucleus: by producing a

$f_{7/2}d_{5/2}^{-1}$ particle-hole pair. This simplifies the reaction mechanism. The 6-, T=1 strengths can be accurately measured in (e,e'), and those strengths used to calibrate nucleon scattering models which can then extract T=0 strengths.

The difficulty is that usually only a small fraction of the simple stretch configuration is found in the largest or "stretched" state. The strength is expected to be reduced by ground state correlations and spread out over many states by the residual interaction. It's very hard to calculate these effects in a shell model, since a diagonalization -- even in a severely truncated basis -- may involve tens of thousands of many-body states. Carr has investigated a new approach called the collective-vector method. This uses the collective vector, here the M6 operator acting on the ground state, as the starting vector for a Lanczos iteration. This method converges rapidly and gives excellent results for the convergence of the moments. He finds much improvement in the comparison of theoretical and experimental strengths. For the model he uses (i.e., for the choice of subspace and residual interaction), most of the "missing" strength is shifted within a few MeV of the dominant state.

Terry Taddeucci presented new polarization transfer data for (p,n) to the continuum on $^2$H, $^{12}$C, and $^{40}$Ca at 500 and 800 MeV using the newly upgraded NTOF facility at LAMPF. The spin-transfer coefficients should be sensitive to the presence of non-central interactions or the transfer of non-zero orbital angular momentum.

The inclusive (p,n) energy spectra at various angles show the quasifree peak shifted by about 25 MeV to energies higher than the free values, in contrast to the (p,p) spectra which are unshifted. These are shown in Fig. 6 below. Note that the (p,n) and (e,e') results are shifted equally, while the dispersion relation for the quasi-particle excited in the ($^3$He,t) reaction seems to be different from the others.

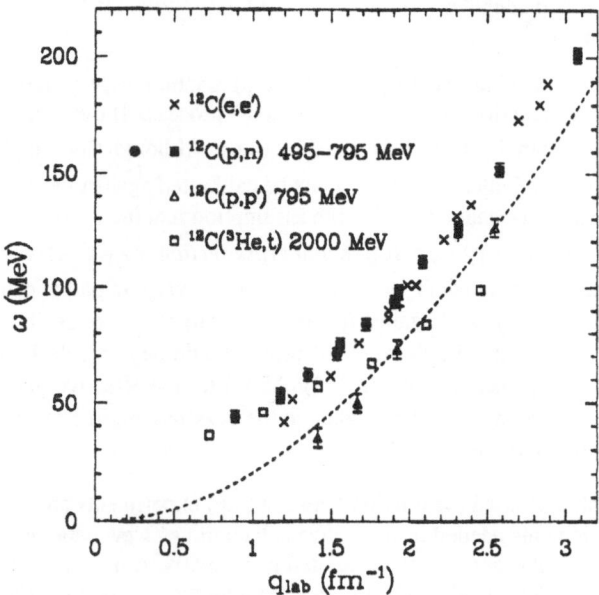

Fig. 6. Position of quasifree peak in $^{12}$C excited by a variety of projectiles.[13]

Taddeucci suggests that this may indicate the presence of important isospin dependent collective nuclear effects in the nuclear response function. We'll return to this issue in our discussion of $\Delta$ production below.

The spin transfer coefficients for quasifree (p,p') show a strong quenching of the polarization compared to free scattering. Horowitz has explained this as a relativistic small component effect arising from the near cancellation of the scalar and vector potentials. This doesn't happen in (p,n), and, as expected, the spin-transfer observables agree reasonably well with the free in the quasifree regime. Compare Taddeucci's Figs. 7 and 8 with Horowitz's Fig. 6 to see how the simple relativistic PWIA does.

The ratio of longitudinal to transverse spin flip for C and Ca at 500 MeV seems to be essentially the same as the ratio for deuterons (assumed to correspond to free scattering) for energy transfers between 20 and 100 MeV. This is somewhat surprising and does not seem to be consistent with the inclusion of nuclear collectivity.

Taddeucci noted that the free scattering observables needed to analyze this reaction at 500 MeV have been floating around significantly. Some observables in the VPI data base have changes by almost a factor of two since 1986!

Khanna, He and Umezawa discussed some speculations of a possible relation of the idea of spontaneous symmetry breaking and nuclear spin-isospin excitations.

*(n,p)*

New experiments on the (n,p) reaction were described by Jack Rapaport. The experiments were done with the new LAMPF white neutron facility. This facility produces a neutron flux with energies distributed over a wide range -- from 50 to 300 MeV. It's strikingly different from other neutron experiments in that all data in the energy range are taken at the same time. Rapaport's Fig. 2 shows the kind of data we can expect. The energy dependence of the relative strength of the giant resonances are dramatically displayed.

Since most nuclei are neutron rich, the (n,p) reaction displays a different set of states than the (p,n) reaction. It's a "stretch isospin" process. If the target has isospin $T_0$, only states with isospin $T_0+1$ can be excited by (n,p). Although this simplifies the spectrum, the Gamow-Teller excitations can't be calibrated against the $\Delta T=0$ Fermi states as in (p,n). The primary analysis tool is the assumption that the cross section can be factorized into a slowly varying part (the *unit cross section*, $\hat{\sigma}_{GT}(E,A)$) depending on the energy and nucleus, times the GT strength. The slowly varying part is calibrated using the few available $\beta$-decays and against (p,n) for T=0 targets. The results are fairly consistent. At energies about 200 MeV, the probe should be particularly selective for $\Delta S=1$, $\Delta T=1$ strength, since the ratio $(\Delta S=1)/(\Delta S=0)$ for the effective interaction strength is large. Data have also been taken in the quasifree regime on $^{12}C$ and are consistent with the (p,n) data.

Ling et al. discussed the relation between (n,p) experiments and the electron capture matrix elements needed in supernova calculations. New data on $^{64}Ni(n,p)$ were presented and the unit cross section calculated using a DWIA model. The measured unit cross sections agree reasonably well (to 20-30%) with the calculation. Sams reported on preliminary results for a measurement of the tensor polarization in the charge exchange reaction $(d,(2p)_{q\approx0})$ in the quasifree scattering region from C and Ca at Saturne. He analyzed the results using an eikonal-DWIA and obtained reasonable results for small energy loss.

*Delta Production*

A number of speakers addressed the issue of what the quasifree $\Delta$ peak looks like when it's produced by inclusive charge exchange scattering inside a nucleus by various projectiles. The peak has been studied with the (p,p'), (p,n), (d,2p) and ($^3$He,t) reactions on a variety of nuclei. A substantial shift of the peak, similar to that show in Fig. 6 is found in the charge exchange processes. In contrast, nuclear excitation of the $\Delta$ with (p,p'), (e,e') or $\gamma$-absorption doesn't show the shift. This suggests that the effect is coming from the longitudinal rather than the transverse spin-isospin response.

Theoretical attempts to explain the inclusive spectra were presented by de Pace, Udagawa, and Gareev. DePace presented a Glauber calculation using the RPA for the target. The ($^3$He,t) data were qualitatively reproduced, but the (p,p') and (p,n) were not. Udagawa et al. showed an inclusive-DWIA calculation which did reasonably well. Gareev et al. described a Glauber-based analysis which related free cross sections and an "effective number of nucleons involved" to the in-medium cross sections. Their analysis suggests that the (p,n) and ($^3$He,t) cross sections from a nucleus are proportional to those from a free nucleon and that a form factor is responsible for the flattening of the peak shift in ($^3$He,t) compared to (p,n).

As in the quasifree nucleon knockout experiments, looking at an exclusive process should help clarify what's going on. This is confirmed in new experimental results from Saturne, which were shown by Ramstein et al. and Gaarde. In her presentation, Ramstein presented the results of an exclusive experiment on $\Delta$ production in ($^3$He,t) using the $4\pi$ Diogene detector. When they trigger on final states that decay by production of $1\pi + 1p$, the resulting $\Delta$ peak for production from $^{12}$C has the same position and width as the free production peak. The data indicate that the shift in the inclusive spectrum is largely due to events that decay into two protons. This should be very helpful and allow us to focus our explanations of the inclusive shift more sharply on rescattering and FSI effects. Some questions remain. Why is the ($^3$He,t 2p) spectrum shifted so strongly? I would expect the ($^3$He,t) process to be more strongly surface localized than (p,p') and therefore to have less rescattering. What's happening in the (p,n) reaction to shift the spectrum? Lots more exclusive data would be very welcome.

## 5. WEAK INTERACTIONS

Nuclear weak interactions have been used as detailed probes both of the fundamental theory of nuclear systems and of nuclear structure information ever since the '50s. At this meeting, two papers provided very striking results on these two issues. Wilkinson discussed how to use modern electroweak interaction theory to get out information about the fundamental constituents of nuclei, and Garcia et al. reported on a study of the Gamow-Teller strength using the $\beta$-decay of exotic nuclei.

Wilkinson presented a detailed calculation as to what information could be extracted from the set of "simplest" nuclear $\beta$-decays: allowed T=1, $J^\pi = 0^+ \rightarrow 0^+$ Fermi transitions. Since these transitions carry neither spin nor orbital angular momentum, the initial and final nuclear wave functions are almost the same. Only the charge of one of the nucleons in changed. There are just 8 of these transitions, but their ft values are accurately known to almost 0.01%.

There are many small corrections to the simple model, including the effect of the charge change (Coulomb corrections), recoil, modification of the positron wave function due to the nuclear size and screening by the atomic electrons, electroweak radiative corrections, the QED running coupling constant, and the nuclear mismatch. Wilkinson is able to calculate the effect of all of these corrections, except the last, to an accuracy equal to or better than the experimental uncertainty in the ft value. A critical element in his analysis was the use of modern forms of the electroweak radiative corrections.

The nuclear mismatch is the effect on the nuclear wave function of changing the charge. This is a real many-body nuclear effect and is hard to calculate. Wilkinson divides the correction into a smoothly varying and a fluctuating part, extracts the fluctuating part from calculations and fits the smooth part with two parameters. The result is an accurate extraction of the vector coupling constant, $G_V^2 = (1.2934 \pm 0.0011) \times 10^{-10}$ GeV$^{-4}$. This is immensely interesting and important because it can be used to confirm various aspects of the standard model. For example, combining $G_V$ with the muon decay constant, $G_\mu$, gives the up-down coupling element of the Cabbibo-Kobayashi-Maskawa (CKM) matrix. The other elements of the top row have been measured. Adding their squares should give 1 by unitarity. Wilkinson's result is 0.999 ± 0.001, in excellent agreement.

One can also turn this around. One of the radiative corrections for $G_V$ depends on the number of fundamental constituents in the nucleon, $N_q$. If we assume CKM unitarity and extract $G_V$ from it, with all the other corrections are taken as before, $N_q$ can be obtained from the value for $G_V$ to be 3.0 ± 0.6![14]

One of the most surprising results of the conference was the measurement by Garcia et al. of the $\beta^+$-decay of $^{37}$Ca. The motivation for their work was to measure the efficiency of the absorption of neutrinos on $^{37}$Cl in order to calibrate Ray Davis's solar neutrino detector. The nuclei $^{37}$Ca and $^{37}$Cl are isobaric analog pairs. Thanks to a large Q value, the $^{37}$Ca $\beta$-decay has an unusually large energy window. The corrections due to breaking of isospin symmetry are expected to be small. The experiment is difficult since $^{37}$Ca has a half-life of only 175 ms. Furthermore, a $\gamma$-coincidence must be performed to distinguish final nuclear states. The experiment was performed at CERN.

The results on the calibrations were reasonable (and make the solar neutrino deficit slightly worse). But the most interesting result was the distribution of Gamow-Teller strength. The extracted strengths did not agree with the extractions of similar strengths from the (p,n) reaction on the analog nucleus. Significant strength was found at excitation energies where the (p,n) reaction finds almost nothing. Enough additional strength was found to completely eliminate the "Gamow-Teller deficit". If this bombshell can be confirmed, a large number of trusted results may be out the window. We would have to rethink the relation between inelastic scattering and the fundamental nuclear response function and our understanding of reaction mechanisms in hadron induced reactions.

# 6. CONCLUSION

To summarize, on a broad scale, the big story at this meeting was the shift in interest and stress from nuclear many-body issues to issues more fundamental to establishing a working paradigm for nuclear physics. In the past decade, nuclear physics has moved from[15] "normal science" to "revolutionary science". A variety of competing paradigms have developed. The traditional model is still alive and well; relativistic models have made substantial strides; models are developing with dynamic mesons; and multiple sub-nucleonic models are beginning to address relevant data. Although these models are occasionally contradictory, in many cases they are being used where they are strong to provide tools to shed light on each other.

The paradigm in Fig. 4 still contains many ambiguities. What do we mean by a "meson model"? This could refer to the construction of effective non-relativistic two and three-body potentials and the use of the NRNO. Or it could refer to a QHD approach using a relativistic field theory with dynamic meson fields. A more radical view has the meson Lagrangian and model "withering away" -- only being useful for very long range effects (r > 3 fm) -- with a QCD-based model being required for most of nuclear physics.

In evaluating the implications of a calculation or an experiment for a paradigm, one must remember: One cannot prove the correctness of a model by comparing an approximate calculation to data. What is critical in the relative test of two models is the comparison of what each model predicts at comparable levels of approximation. Two examples are: (1) Is the mass/coupling of the $\rho$ to the nucleon quenched/enhanced in nuclear matter? (2) Does the spin response in the cross-section dips for elastic nucleon-nucleus scattering at intermediate energy require the dynamic inclusion of the small components of nucleon wave functions?

The traditional NRNO model has had some spectacular successes with the development of converged calculations in the few-body problem and we saw some evidence that these results are being used as levers to obtain information from beyond the model. The relativistic model has reached a higher level of maturity, but the "crunch" point demonstrating the need for dynamic treatment of small components has not yet been reached.

Mesonic degrees of freedom have been applied to an increasing number of calculations with occasionally rather precise successes. Some of the presentations at this meeting suggest that vector meson exchanges occur at distances where many quark modelers expect hadronic descriptions to be substantially suppressed.

An important recurring theme was the need for additional NN scattering data, especially ones where two spins are measured. We have reached a new plateau in the power and accuracy of our measurements of spin observables beyond the analyzing power in nuclear scattering, and as a result, we can no longer be satisfied with the fairly crude NN amplitudes currently available. This was independently mentioned by different speakers for the energy regimes 0-50, 50-300, and 500-800 MeV.

The meeting proves that spin and isospin continue to give useful handles on nuclear properties, even as the focus of nuclear physics begins to shift substantially. The organizers of this meeting have done an excellent job of bringing together the diverse threads of the field. Nuclear physics is alive and well in Telluride!

## ACKNOWLEDGEMENT

I would like to thank Manoj Banerjee, Tom Cohen, and John Tjon for useful discussions. This work was supported in part by U.S.D.O.E. grant #DE-FG05-87ER-40322.

## REFERENCES

[1] Steven Weinberg, "Why the renormalization group is a good thing", in *Asymptotic Realms of Physics*, A. Guth, K. Huang, and R. Jaffe, eds. (MIT Press, 1983) 1.

[2] To understand this comment, consider what would happen if we tried to describe a metal in terms of effective interactions between atoms without breaking them up into ions and shared electrons. The effective force between two atoms has a range equal to the size of the entire block of metal!

[3] S. J. Wallace, Ann. Rev. Nucl. and Part. Sci. **37** (1987) 267.

[4] B. Serot and J. D. Walecka, Adv. in Nucl. Phys., **16** (1986) 1.

[5] Note that the structure functions of quarks in a nucleon and their integrals are *static* properties, i.e., properties of the QCD ground state with the quantum numbers of the nucleon, despite the fact that they are extracted using high energy scattering. This is like measuring the wave function of an electron in helium using high energy (e,2e).

[6] C.-R. Chen, G. Payne, J. Friar, and B. Gibson, *Phys. Rev.* **C33** (1986) 1740.

[7] R. Brandenburg, *et al.*, Phys. Rev. **C37** (1988) 1245.

[8] It also implies that if one uses a calculation that doesn't give the right binding energy, the other observables, such as radius and Coulomb energy shift, will also be given incorrectly. It's therefore not enough to select a "converged" model calculation. The model has to be adequate; and adequacy is essentially defined by one number: the triton binding energy.

[9] The off-shell amplitude in a direct+exchange effective interaction has qualitatively the wrong behavior for the imaginary part of the amplitude. See M. M. MacFarlane and E. F. Redish, Phys. Rev. **C37** (1988) 2245.

[10] E. F. Redish and K. Stricker-Bauer, Phys. Rev. **C36** (1987) 513.

[11] W. M. Kloet and R. R. Silbar, Nucl. Phys. **A338** (1980) 281; T. - S. Lee, Phys. Rev. Lett. **50** (1983) 1571; E. van Faassen and J. A. Tjon, Phys. Rev. **C30** (1984) 285; P. U. Sauer, Prog. Part. Nucl. Phys. **16** (1986) 35; P. Gonzalez and E.L. Lomon, Phys. Rev. **D34** (1986) 1351; Ch. Elster et al., Phys. Rev. **38** (1988) 1828; F. Sammarucca and T. Mizutani, Phys. Rev. **41** (1990) 2286; Ch. Elster, Nucl.Phys. **A508** (1990) 197c; W. M. Kloet and E. L. Lomon, Phys. Rev. **C43** (1991) 1575.

[12] J. Noble, Phys. Rev. Lett. **46** (1981) 412; Phys. Lett. **B178** (1986) 285; L. S. Celenza, A. Rosenthal, and C. M. Shakin, Phys. Rev. Lett. **53** (1984) 892; Phys. Rev. **C 31** (1985) 232; P. J. Mulders, Phys. Rev. Lett. **54** (1985) 2560, Nucl. Phys. **A249** (1986) 525; G. Brown and M. Rho, Phys. Lett., **B237** (1990) 3.

[13] T. Taddeucci and J. McClelland, private communication.

[14] Wilkinson's paper is very understated, but I think this result deserves an exclamation point.

[15] T. S. Kuhn, *The Structure of Scientific Revolutions, 2nd Ed.* (Univ. of Chicago Press, 1970).

## PARTICIPANTS

Eric Adelberger
University of Washington
Nuclear Physics Laboratory, GL-10
Seattle, WA 98195

W.P. Alford
University of Western Ontario
Physics Department
London, Ontario
CANADA, N6A 3K7

Hugo Arellano
University of Georgia
Department of Physics
Athens, GA 30602

David Aschman
University of Cape Town
Physics Department
Rondebosch
SOUTH AFRICA, 7700

Andrew Bacher
Indiana University Cyclotron Facility
2401 Milo B. Sampson Lane
Bloomington, IN 47405

Helmut Baer
Los Alamos National Laboratory
MP-4/H846
Los Alamos, NM 87544

F. Todd Baker
University of Georgia
Department of Physics
Physics Building
Athens, GA 30602

David Beatty
Rutgers University
Serin Physics Laboratory, Box 849
Piscataway, NJ 08854

Angela Betker
Texas A&M University
Cyclotron Institute
College Station, TX 77843

Sonya Bowyer
Indiana University Cyclotron Facility
2401 Milo B. Sampson Lane
Bloomington, IN 47405

Theodore Bowyer
Indiana University Cyclotron Facility
2401 Milo B. Sampson Lane
Bloomington, IN 47405

Franco Bradamante
Dipartimento di Fisica dell'Università
Via A. Valerio 2
34127 Trieste, ITALY

Francisco A. Brieva
University of Chile
Physics Department
Casilla 487-3  Santiago, CHILE

Daniel Carman
Indiana University Cyclotron Facility
2401 Milo B. Sampson Lane
Bloomington, IN 47405

James Carr
Florida State University
B-186
SCRI-400SCL
Tallahassee, FL 32306-4052

Frank Close
ORNL-RAL
Rutherford Laboratory
Chilton, Didcot  0X11 0QX
ENGLAND

Sidney A. Coon
New Mexico State University
Physics Department
Las Cruces, NM 88003

John Dawson
University of New Hampshire
Department of Physics
Durham, NH 03824

Arturo DePace
Instituto Nazionale Fisica Nucleare
Via P. Giuria 1
I-10125 Torino, ITALY

Geoffrey Edwards
Rutgers University
MS H841, LAMPF
Los Alamos National Laboratory
Los Alamos, NM 87545

Robert Eisenstein
University of Illinois
Nuclear Physics Laboratory
23 Stadium Dr.
Champaign, IL 61820

Thomas C. Ferrée
University of Colorado
Nuclear Physics Laboratory
Boulder, CO 80309-0446

Roger Finlay
Ohio University
Department of Physics
Athens, OH 45701-2979

Carl Gaarde
Niels Bohr Institute
Blegdamsvej 17
2100 Copenhagen O
DENMARK

Carl Gagliardi
Texas A&M University
Cyclotron Institute
College Station, TX 77843

Alejandro Garcia
University of Washington
Nuclear Physics Laboratory, GL-10
Seattle, WA 98195

Michel Garçon
CEN-Saclay
DPHN/SEPN
91191 Gif-Sur-Yvette Cedex
FRANCE

Fangill A. Gareev
Joint Institute for Nuclear Research
Dubna
USSR

Charles Goodman
Indiana University Cyclotron Facility
2401 Milo B. Sampson Lane
Bloomington, IN 47405

Andrew Green
Rutgers University
MS 4841, LAMPF
Los Alamos National Laboratory
Los Alamos, NM 87545

Otto Häusser
TRIUMF
4004 Wesbrook Mall
Vancouver, B.C. CANADA, V6T 2A3

Ernest Henley
University of Washington
Department of Physics, FM-15
Seattle, WA 98195

Norton Hintz
University of Minnesota
School of Physics
116 Church Street SE
Minneapolis, MN 55455

Charles Horowitz
Indiana University Cyclotron Facility
2401 Milo B. Sampson Lane
Bloomington, IN 47405

Robert Jaffe
Massachusetts Institute of Technology
6-306
Cambridge, MA 02168

Faqir Khanna
Department of Physics
University of Alberta
Edmonton, Alberta
CANADA, T6G 2J1

Mariana Kirchbach-Arenhovel
Technische Hochschule Darmstadt
Institut für Kernphysik
Schlossgartenstrube 9
Darmstadt D-6100, GERMANY

Klaus Koch
Los Alamos National Laboratory
LANL, MS H841 MP-10
Los Alamos, NM 87545

Massimo Lamanna
Dipartimento di Fisica dell'Universtà
Via A. Valerio 2
34127 Trieste, ITALY

James Langenbrunner
University of Minnesota
LAMPF, MS H841
Los Alamos, NM 87545

Ping Li
Indiana University Cyclotron Facility
2401 Milo B. Sampson Lane
Bloomington, IN 47405

Richard Lindgren
University of Virginia
Department of Physics
Charlottesville, VA 22901

Alan Ling
Los Alamos National Laboratory
MS-H803
Los Alamos, NM 87545

Jerry Lisantti
Indiana University Cyclotron Facility
2401 Milo B. Sampson Lane
Bloomington, IN 47405

Paul Lisowski
Los Alamos National Laboratory
P-17 H803 LANL
Los Alamos, NM 87545

Tim Londergan
Indiana University Cyclotron Facility
2401 Milo B. Sampson Lane
Bloomington, IN 47405

William Louis
Los Alamos National Laboratory
MS H846
Los Alamos, NM 87545

W. Gary Love
University of Georgia
Dept. of Physics & Astronomy
Athens, GA 30602

Oren V. Maxwell
Florida International University
Department of Physics
University Park
Miami, FL 33199

John McClelland
Los Alamos National Laboratory
LANL, MP-10, MS H841
Los Alamos, NM 87545

Robert McKeown
California Institute of Technology
Kellogg Radiation Laboratory
Pasadena, CA 91125

Alfredo Molinari
Italian General Consulate
100 Boylston Street, Suite 900
Boston, MA 02116

C. Fred Moore
University of Texas at Austin
Physics Department
RLM 5.208
Austin, TX 78712-1081

H. Peter Morsch
Laboratoire National Saturne
F-91191 Gif Sur Yvette Cedex
FRANCE

John O'Donnell
Los Alamos National Laboratory
MP-10, MS H841
Los Alamos, NM 87544

Michael Österlund
University of Lund
Department of Physics
Sölvegatan 14
223 62 Lund, SWEDEN

Brent Park
Ohio University
Department of Physics
Athens, OH 45701-2979

Stephen Pate
Indiana University Cyclotron Facility
2401 Milo B. Sampson Lane
Bloomington, IN 47405

Jorge Piekarewicz
Florida State University
452 SCL
Tallahassee, FL 32306

David Potterveld
Argonne National Laboratory
PHY 203
Argonne, IL 60439

David Prout
University of Colorado
Nuclear Physics Laboratory
Boulder, CO 80309-0446

Beatrice Ramstein
Institut Physique Nucléaire
91406 Orsay Cedex
FRANCE

Jack Rapaport
Ohio University
Department of Physics
Athens, OH 45701-2979

Lanny Ray
University of Texas at Austin
Department of Physics
RLM 5.208
Austin, TX 78712

Edward F. Redish
University of Maryland
Department of Physics
College Park, MD 20742

Philip Roos
University of Maryland
Department of Physics
College Park, MD 20742

Lawrence Rybarcyk
Los Alamos National Laboratory
LANL, MS H841  MP-10
Los Alamos, NM 87545

Hide Sakai
University of Tokyo
Department of Physics
Hongo 7-3-1, Bunkyo, Tokyo
JAPAN,  113

Thomas Sams
Laboratoire National Saturne
F91191 Gif-Sur-Yvette Cedex
FRANCE

Peter Schwandt
Indiana University Cyclotron Facility
2401 Milo B. Sampson Lane
Bloomington, IN 47405

Anil Sethi
University of Minnesota
School of Physics
Minneapolis, MN 55455

Danny Sorenson
Los Alamos National Laboratory
P17, MS-H803
Los Alamos, NM 87545

James Sowinski
Indiana University Cyclotron Facility
2401 Milo B. Sampson Lane
Bloomington, IN 47405

Joseph Speth
Institut for Kernphysik
Postfach 1913
D-5170 Jülich 1
GERMANY

Edward J. Stephenson
Indiana University Cyclotron Facility
2401 Milo B. Sampson Lane
Bloomington, IN 47405

Terry Taddeucci
Los Alamos National Laboratory
LANL, MP-10, MS H841
Los Alamos, NM 87545

Yves Terrien
CEN-Saclay
DPHN/SEPN
91191 Gif-Sur-Yvette Cedex
FRANCE

Irwan The
Massachusetts Institute of Technology
Room 26-447
77 Massachusetts Avenue
Boston, MA 02139

Werner Tornow
Duke University
Department of Physics
Durham, NC 27713

Takeshi Udagawa
University of Texas at Austin
Department of Physics
Austin, TX 78712

John Ullman
Los Alamos National Laboratory
P-17 H803 LANL
Los Alamos, NM 87545

I. J. Van Heerden
University of the Western Cape
Private Bag X17, Bellville
Cape Province 7535
SOUTH AFRICA

Barbara Von Przewoski
Indiana University Cyclotron Facility
2401 Milo B. Sampson Lane
Bloomington, IN 47405

Jochen Wambach
Forschungszentrum Jülich
Institute für Kernphysik
5170 Jülich
GERMANY

Steven Wells
Indiana University Cyclotron Facility
2401 Milo B. Sampson Lane
Bloomington, IN 47405

Sir Dennis Wilkinson
Gayles Orchard
Friston
East Bourne
BN20 0BA
England

Scott Wissink
Indiana University Cyclotron Facility
2401 Milo B. Sampson Lane
Bloomington, IN 47405

# INDEX

Absolute calibration standards, 259–260
Analyzing power, 71, 115, 120, 124–126, 164, 166, 216, 223–227, 239, 242ff, 259–260, 271–273, 275, 297, 300, 385–389, 405–406, 427, 463ff
Annihilation, 116, 118
Antineutron, 121ff, 155ff
Antiparticles (general), 417
Antiproton, 115ff, 155ff
annihilation, 115, 134
Associated production of strangeness, 132ff
Axial vector coupling constant, 18, 60, 169, 181, 189–190, 361, 365, 375, 376

Background in experimental data, 71, 172, 173, 348, 350, 348–350, 427
Bag models, 77, 78, 490
Beta decay, 172, 173, 177, 182, 185, 188ff, 207ff, 353ff, 361, 372, 375, 376, 398

Cabibbo-Kobayashi-Maskawa matrix, 372, 374–376
Cerenkov counters, 342–344
Charge exchange reactions, 63, 69, 111ff, 115, 121, 124, 155ff, 161ff, 167ff, 207ff, 215ff, 297, 353, 355, 393ff, 427, 433ff, 451, 453ff, 497, 520–523
Charge independence breaking, 461–462
Charge symmetry breaking, 468–472
Chiral symmetry, 181ff, 184, 345
odd vs. even, 146
Color transparency, 83
Computational techniques, 335ff
Configuration mixing, 333ff, 355
Conserved vector currents (CVC), 191, 361, 365
Coupled-channel models, 93, 107
relativistic, 248
Currents (neutral and charged), 345, 347, 373

Deep inelastic scattering, 1ff, 33ff, 59–60, 346, 347
kinematics, 2
and meson exchange, 33
Delta (1232) resonance, 69ff, 79, 82, 111, 161, 170, 172, 180, 197, 303, 306, 445, 489, 491
in ground state, 59
Delta-hole interaction, 69, 195–196, 295, 303, 307, 326, 445
Density dependence (of interactions), 139ff, 327, 422
Detectors, *see* Facilities
Dirac equation applications, 184ff, 231, 281–282, 364, 415ff, 508 (*see also* Relativity)
Dispersion relations, 325
Distribution functions, *see* Structure functions
Drell-Yan reaction, 15–16, 321
polarized, 145, 147

Effective interaction, 262–265, 281ff, 287ff, 355, 437–440 (*see also* Medium modifications)
Effective mass, 281, 284–285, 287ff, 416
Elastic scattering (nucleon-nucleus), 219ff, 231ff, 259–260, 388–389, 517–518
Electron capture, 207–208
Electron scattering, 2–3, 69, 85ff, 99ff, 338, 424, 445
inclusive, 53ff, 445
Electroweak processes, 3, 346, 365

Facilities
Bates, 56–58, 88–89
CERN (BCDMS), 5
CERN (EMC), 5ff, 20–22, 60, 76, 80, 321, 347
CERN (NMC), 12, 13
CERN (Isolde-3), 356
DESY (HERMES), 61, 62
IUCF, 256–258
LAMPF (LSND), 341–344
LAMPF (NTOF), 302, 394–398

QCD, 2, 10, 24–28, 33, 59, 145ff, 365, 505–506
QHD, 419, 508
Quarks, 1ff, 33ff, 145ff, 177-178, 341, 373, 489–490, 508, 510
Quasi-elastic scattering
  hadronic, 63ff, 275–276, 296, 301, 311ff, 321ff, 381, 385ff, 403ff, 424ff, 443, 456, 497ff, 519
  leptonic, *see* electron scattering

Random phase approximation (RPA), 63–65, 270, 288, 290, 317, 319, 321ff, 411, 419–420, 427, 442
Radiative corrections, 365–368
Relativity, 198, 231, 242, 415ff, 507–508 (*see also* Dirac)
Residual interaction, 326ff
Resonating group method, 93
Roper resonance, 77, 489, 492ff

SAID (phase shift solutions), 293, 386, 462, 479–482
Scaling, 3
Scattering length, 138, 461, 468
Self energy, 324
Skyrme model, 23, 94
Slater determinants, 335
Solar neutrino problem, 353, 358
Spin-dipole excitations, 270, 299, 433, 440ff
Spin observables, 16ff, 115ff, 155ff, 165, 223-227, 239–242, 245–248, 259ff, 281ff, 294, 295, 300, 305, 311, 329ff, 385–389, 393ff, 415ff, 472–478 (*see also* Analyzing power)
  robustness of, 315–318
Spin of the proton, 23, 60, 75, 145, 341
Spin-orbit interaction, 126, 237, 239ff, 267, 287, 290, 292, 317, 415
  distortions from, 270
Spin response (longitudinal and trans-verse), 4, 28, 64, 69, 111, 267, 270, 295ff, 311ff, 321, 323, 332, 393–394, 403ff, 424, 433, 497ff
Spin transfer, 155ff, 223–227, 239, 253ff, 270, 271, 273, 281ff, 311ff 321, 322ff, 329, 385–389, 393ff, 426, 428

Spreading width, 325–326, 442
Standard model, 346, 361, 373
Storage cells, 102–103
"Stretched" states
  excitation of, 262–267, 271, 273, 275, 281ff, 288–292, 521
  fragmentation of, 333ff, 520
  structure of, 262, 273, 288–292, 333ff
Structure functions (*see also* Form factors)
  deuteron, 86–87, 99–100, 514
  nucleon, 3ff, 35–37, 59–60, 145, 510
    high-twist, 145ff
  QCD predictions for, 24–28
Sum rules
  Coulomb, 424
  Gamow-Teller, 170–172, 182, 191, 195–196
  higher twist, 145ff
  quark/parton, 8ff, 510
    Adler, 10
    Gottfried, 11–15, 44–48
    Gross-Llewellyn Smith, 8–10
  spin, 16ff
    Bjorken, 18, 22, 23, 60, 80
    Drell-Hearn-Gerasimov, 80, 82
    Ellis-Jaffe, 19–21, 347
Supernova, 207, 212

Tensor interaction, 170, 267, 272, 275, 282–285, 290, 399–403, 472ff, 514
Tensor polarization (deuteron), 85ff, 99ff, 314–315, 498–499
Three-body forces, 466 (*see also* Off-shell effects)
Threshold reactions, 134ff
t-matrix, 219, 221, 233–235, 245, 262, 282–285, 288, 399, 400, 447
Transition density, 333–334
Twist, 146ff (*see also* QCD)

Vector Coupling Constant, 181, 208, 361ff, 370

Z graph, 417